The Routledge Handbook of Urban Ecology

The birds, animals, insects, trees and plants encountered by the majority of the world's people are those that survive in, adapt to, or are introduced to, urban areas. Some of these organisms give great pleasure; others invade, colonise and occupy neglected and hidden areas such as derelict land and sewers. Urban areas have a high biodiversity and nature within cities provides many ecosystem services including cooling the urban area, reducing urban flood risk, filtering pollutants, supplying food, and providing accessible recreation. Yet, protecting urban nature faces competition from other urban land uses.

The Routledge Handbook of Urban Ecology analyses this biodiversity and complexity and provides the science to guide policy and management to make cities more attractive, more enjoyable, and better for our own health and that of the planet. This Handbook contains 50 interdisciplinary contributions from leading academics and practitioners from across the world to provide an in-depth coverage of the main elements of practical urban ecology. It is divided into six parts, dealing with the philosophies, concepts and history of urban ecology; followed by consideration of the biophysical character of the urban environment and the diverse habitats found within it. It then examines human relationships with urban nature, the health, economic and environmental benefits of urban ecology before discussing the methods used in urban ecology and ways of putting the science into practice.

The Handbook offers a state-of-the-art guide to the science, practice and value of urban ecology. The engaging contributions provide students and practitioners with the wealth of interdisciplinary information needed to manage the biota and green landscapes in urban areas.

Ian Douglas is Emeritus Professor at the School of Environment and Development, University of Manchester, UK. **David Goode** is a Visiting Professor in Geography at the Environment Institute, University College London, UK. **Michael C. Houck** is Executive Director of Urban Greenspaces Institute, Portland, Oregon, USA. **Rusong Wang** is Deputy Director at State Key Lab of Urban and Regional Ecology, Research Center for Eco-Environmental Sciences, Chinese Academy of Sciences, Beijing, China.

The Routledge Handbook of Urban Ecology

Edited by Ian Douglas, David Goode, Mike Houck and Rusong Wang

The UK Man and the Biosphere Committee Urban Forum

LONDON AND NEW YORK

First published 2011
by Routledge
2 Park Square, Milton Park, Abingdon, Oxon OX14 4RN

Simultaneously published in the USA and Canada
by Routledge
270 Madison Avenue, New York, NY 10016

Routledge is an imprint of the Taylor & Francis Group, an informa business

Typeset in Bembo by
Wearset Ltd, Boldon, Tyne and Wear
Printed and bound in Great Britain by
CPI Anthony Rowe, Chippenham, Wiltshire

British Library Cataloguing in Publication Data
A catalogue record for this book is available from the British Library

Library of Congress Cataloguing-in-Publication Data
Handbook of urban ecology / edited by Ian Douglas ... [et al.].
p. cm.
Includes bibliographical references and index.
1. Urban ecology (Sociology)–Handbooks, manuals, etc. 2. Urban ecology (Biology)–Handbooks, manuals, etc. 3. Urban geography–Handbooks, manuals, etc. I. Douglas, Ian, 1936–
HT241.H35 2011
307.76–dc22 2010025749

ISBN: 978-0-415-49813-5 (hbk)
ISBN: 978-0-203-83926-3 (ebk)

Contents

List of illustrations *x*

Contributors *xv*

Acknowledgements *xix*

Prologue *xx*
Ian Douglas

PART 1
Context, history and philosophies **1**

Introduction 3
Ian Douglas

1 Urban ecology: definitions and goals 7
N. E. McIntyre

2 The analysis of cities as ecosystems 17
Ian Douglas

3 Urban ecology and industrial ecology 26
Xuemei Bai and Heinz Schandl

4 Urban areas in the context of human ecology 38
Roderick J. Lawrence

5 In livable cities is preservation of the wild: the politics of providing for nature in cities 48
Michael C. Houck

6 The human relationship with nature: rights of animals and plants in the urban context 63
Jason Byrne

7 Urban natural histories to urban ecologies: the growth of the study of urban nature 74
Ian Douglas and David Goode

8 Planning for nature in towns and cities: a historical perspective 84
David Goode

9 How much is urban nature worth? And for whom? Thoughts from ecological economics 93
Anna Chiesura and Joan Martínez-Alier

PART 2
The urban ecological environment **97**

Introduction 99
Ian Douglas

10 Climate of cities 103
C.S.B. Grimmond

11 Urban heat islands 120
T.R. Oke

12 Urban effects on precipitation and associated convective processes 132
J.M. Shepherd, J.A. Stallins, M.L. Jin and T.L. Mote

13 Urban hydrology 148
Ian Douglas

14 Urban geomorphology 159
Ian Douglas

15 Urban soils 164
Peter J. Marcotullio

16 The process of natural succession in urban areas 187
Wayne C. Zipperer

17 Recombinant ecology of urban areas: characterisation, context and creativity 198
Colin D Meurk

18 Creative conservation 221
Grant Luscombe and Richard Scott

PART 3
The nature of urban habitats 233

Introduction 235
Ian Douglas

19 Walls and paved surfaces: urban complexes with limited water and nutrients 239
C. Philip Wheater

20 Urban cliffs 252
Jeremy Lundholm

21 Suburban mosaic of houses, roads, gardens and mature trees 264
Ian Douglas

22 Urban wildlife corridors: conduits for movement or linear habitat? 274
Ian Douglas and Jonathan P. Sadler

23 Landscaped parks and open spaces 289
M. Hermy

24 Grassland on reclaimed soil, with streets, car parks and buildings but few or no mature trees 301
Tony Kendle

25 Urban contaminated land 311
Michael O. Rivett, Jonathan P. Sadler and Bob C. Barnes

26 Urban woodlands as distinctive and threatened nature-in-city patches 323
C.Y. Jim

27 Wetlands in urban environments 338
Joan G. Ehrenfeld, Monica Palta and Emilie Stander

28 Urban animal ecology 352
Peter J. Jarvis

29 Feral animals in the urban environment 361
Peter J. Jarvis

PART 4
Ecosystem services and urban ecology **371**

Introduction 373
Ian Douglas

30 Intrinsic and aesthetic values of urban nature: a journalist's view from London 377
David Nicholson-Lord

31 Intrinsic and aesthetic values of urban nature: a psychological perspective 385
Rachel Kaplan

32 Urban nature and human physical health 394
Jenna H. Tilt

33 Urban nature: human psychological and community health 408
Rod Matsuoka and William Sullivan

34 Street trees and the urban environment 424
Gerald F.M. Dawe

35 Urban gardens and biodiversity 450
Kevin J. Gaston and Sian Gaston

PART 5
Methodologies **459**

Introduction 461
Ian Douglas

36 Urban habitat analysis 465
Ian Douglas

37 Urban habitat type mapping 478
Peter J. Jarvis

38 Invasive species and their response to climate change 488
Ian Douglas

39 Urban biogeochemical flux analysis 503
Nancy B. Grimm, Rebecca L. Hale, Elizabeth M. Cook and David M. Iwaniec

40 Urban metabolism analysis 521
Shu-Li Huang and Chun-Lin Lee

PART 6
Applications and policy implications **529**

Introduction 531
Ian Douglas

41 Delivering urban greenspace for people and wildlife 537
John Box

42 Urban areas and the biosphere reserve concept 549
Pete Frost and Glen Hyman

43 Urban ecology and sustainable urban drainage 561
Peter Worrall and Sarah Little

44 Green roofs, urban vegetation and urban runoff 572
Joachim T. Tourbier

45 The role of green infrastructure in adapting cities to climate change 583
Ian Douglas

46 Creative use of therapeutic green spaces 589
Ambra Burls

47 Peri-urban ecology: green infrastructure in the twenty-first century metro-scape 599
Joe Ravetz

48 Biodiversity as a statutory component of urban planning 621
David Goode

49 Making urban ecology a key element in urban development and planning 630
John Stuart-Murray

50 Towards Ecopolis: new technologies, new philosophies and new developments 636
Rusong Wang, Paul Downton and Ian Douglas

Conclusion 652
Ian Douglas

Index 654

Illustrations

Figures

3.1	The urban social-ecological system	30
3.2	Integrated urban metabolism and urban system performance indicators	33
4.1	The holistic framework of a human ecology perspective showing the interrelations between genetic biospace, ecospace, cultural space and artefacts	40
5.1	Aerial photograph of Oaks Bottom in Oregon	51
5.2	Aerial photograph of Salmon Creek Greenway in Vancouver, Washington	51
5.3	Cedar Mill Creek Watershed in 1990	52
5.4	Cedar Mill Creek Watershed in 2002	53
5.5	Remaining natural areas in Portland, Oregon, from aerial photography	54
5.6	Map of riparian corridors indicating fish and wildlife habitats	55
5.7	Composite map of the Portland-Vancouver region showing land cover types	58
6.1	Opossum (*Didelphis marsupialis*) traversing a backyard fence in San Gabriel, California	65
6.2	Water dragon (*Physignathus lesueurii*) sunbathing on a Gold Coast metropolitan beach, Australia	66
10.1	Schematic figure to show the link between the energy and water exchanges at the urban surface	106
10.2	Schematic representation, with photographs, of the three horizontal-vertical scales commonly used in studies of urban atmospheric processes	107
10.3	Radiation and energy balance fluxes for Marseille, Miami and Ouagadougou	110
10.4	Changing Bowen ratio ($\beta = Q_H/Q_E$) as a function of plan area fraction of active vegetation cover	112
10.5	The packing of the buildings ('roughness elements') on the surface results in a change in wind regime	114
11.1	A classification of the main scales found in urban climates	121
11.2	Simplified schematic depiction of UHI_{UCL} spatial and temporal features and the main controls on its form and magnitude	124
11.3	Schematic of thermal features of the UHI_{UBL} in 'ideal' weather conditions for a large city	130
12.1	Satellite view of earth at night	133
12.2	Schematic of the downwind urban rainfall anomaly	134
12.3	Rainfall and lightning flash anomalies for the warm season in Atlanta	137
12.4	Coupled atmosphere-land model framework	138
12.5	Difference in divergence/convergence for an URBAN and NOURBAN simulation	139

12.6 Aerosol number distributions next to a source (freeway), for average urban, for urban influenced background, and for background conditions 140
13.1 Decline in ground-water levels in the sandstone aquifer, Chicago and Milwaukee areas, 1864–1980 154
16.1 Vegetation dynamics is illustrated as a hierarchy ranging from the most general phenomenon of community change to detailed interactions within a mechanism or causation 188
17.1 Degraded remnant association – a *Quercus-Ulmus-Tilia/Lonicera/Dryopteris-Aegopodium* woodland in Moscow Botanic Gardens 201
17.2 Spontaneous primary recombinant association on pavement edge (micro-wasteland) in Sheffield, UK 202
17.3 Spontaneous primary wall association in St Austell, Cornwall, UK 202
17.4 Complex recombinant association in Christchurch, NZ 203
17.5 Spontaneous primary recombinant association – a NZ cabbage tree seedling (*Cordyline australis*) self-established on top of a stone wall in St Austell, Cornwall, UK 204
17.6 Deliberative roadside shrubbery with street tree standards on an earth embankment forming a sound buffer for a residential subdivision in Christchurch, NZ 205
17.7 Deliberative hedge planting of *Muehlenbeckia complexa* (Polygonac.) and adjacent oak street trees in Melbourne, Australia 206
17.8 Deliberative novel designs on the Fresno University Campus, California, USA 206
17.9 Complex recombinant association in Christchurch, NZ 207
17.10 A deliberative planting for ecological restoration in Auckland, NZ adjacent to a small detention pond 208
17.11 Three barriers to achieving ecological restoration (deliberative indigenous associations) 209
18.1 Pickerings Pasture, Halton 223
18.2 Cornfield annuals in Kirkby 228
18.3 New meadows: making townships handsome again 229
18.4 The National Wildflower Centre, Knowsley 230
19.1 Percentage of lichen species found on different habitats and substrates (London, UK) 241
19.2 Silver birch growing from a domestic chimney 241
19.3 Plants concentrated in a sheltered basement frontage 243
19.4 Diverse vegetation growing at the base of a boundary wall 247
19.5 Using vegetation to create green roofs and walls 249
20.1 Natural rock outcrop showing cliff face and talus habitat 254
20.2 Natural high biodiversity limestone pavement 255
20.3a Extensive (shallow substrate) green roof: urban analog of rock pavement or drought tolerant grassland over bedrock 260
20.3b Green wall incorporating vascular plants and bryophytes 260
20.3c Urban cliff landscape incorporating green facades, green roofs and other cliff and talus analogs 260
22.1 Abandoned railway line at Stamford Brook, Trafford, Greater Manchester, UK, forming a potential wildlife corridor 275
22.2 Sustainable urban drainage landscaping at Stamford Brook, Trafford, Greater Manchester, UK, forming a potential wildlife corridor 275

22.3	Woodland planted by the Red Rose Forest at the peri-urban Dainewell Woods, Greater Manchester, UK. The trackway is the Trans Pennine Trail from Liverpool to Hull, used by horse riders, cyclists and walkers	276
22.4	Diagram showing the patterns of movement associated with the six key functions of wildlife corridors	278
22.5	The Bridgewater Canal at Brooklands, Trafford, Greater Manchester, UK, with broad vegetated areas on both banks offering a diversity of cover for wildlife	279
22.6	The Bridgewater Canal at Sale, Trafford, just one kilometre north of Brooklands where both banks of the canal have little vegetation, disrupting the connectivity for terrestrial fauna	280
22.7	The abundance of terrestrial (a) invertebrates and (b) beetles (Coleoptera) in contrasting plots of Balsam rich and Balsam poor areas on the Rivers Cole, Rea and Tame, Birmingham, UK	284
23.1	The ECOPOLIS concept of urban parks	290
23.2	Diversity relationships in 15 (sub)urban parks in Flanders (Belgium)	296
25.1	Diverse examples of contaminated land	312
25.2	Schematic representation of the Environment Agency's Ecological Risk Assessment framework	315
26.1	The Bois de Boulogne offers a sprawling urban natural area in the western fringe of Paris	324
26.2	The small remnant urban woodland inside Holland Park is encapsulated in the core of London	325
26.3	Some parts of the huge Phoenix Park in Dublin are managed as urban woodlands	325
26.4	The Yoyogi Park is a 70 ha urban woodland planted in the early twentieth century around the Meiji Jingu in the heart of Tokyo	326
26.5	Classification of urban synanthropic plants based on the native-alien dichotomy and human mediation	331
27.1	Anthropogenic factors modifying wetland ecology in urban regions	339
27.2	Examples of disturbed hydrologic patterns in urban forested wetlands	343
27.3	Examples of urban wetland systems (a) Hackensack Meadowlands, NJ; (b) XiXi National Wetland Park, Hangzhou, China	347
32.1	Individual and factor preference for (a) high, (b) medium and (c) low vegetation scenes	399
33.1	The human benefits, involving both individual and community benefits, of urban nature contact	410
34.1	Cartoon of the contrasts between 'Old' and 'New' street tree problems	433
39.1	Simple conceptual framework for biogeochemical fluxes in urban ecosystems	504
39.2	Diagram showing the complex and multi-faceted flows into and out of an urban ecosystem that comprise "urban metabolism."	513
40.1	The complete metabolism of a city	522
40.2	LCA framework	524
40.3	Energy circuit symbols	525
40.4	Emergy synthesis table	526
41.1	Landscape structure planting around Telford Central station provides a sense of arrival in a green town as well as screening and noise reduction	538

41.2 A wooden dragon constructed by children at Plants Brook LNR in Birmingham shows that all forms of wildlife can be appreciated even in a high quality designated site 539
41.3 Regularly mown grassland in the Town Park in Telford in 1990 545
41.4 The same area in 2005 showing scrub habitats created by natural regeneration after mowing ceased 545
42.1 Biosphere Reserve Zonation 550
42.2 Local adaptation of BR zones to manage the urban and natural reality of Florianopolis 553
42.3 An urban forest: the São Paulo metropolis and Cantareira State Park, a BR core zone 554
42.4 Geographic proximity: Melbourne and the MPWPBR 556
43.1 Depth frequency duration analysis for SUDS wetlands 562
43.2 Structurally diverse SUDS wetland 564
43.3 Reedbed SUDS at Potteric Carr 566
43.4 Sedum green roof at a Sheffield school 568
44.1 Typical hydrograph changes due to increases in impervious surface areas 573
44.2 Cross-section through an extensive roof cover 575
44.3 An egg-carton like synthetic material may be used as a drain layer 576
44.4 The Post Giro Facility in Munich, Germany 579
44.5 The University of Warsaw library has become a landmark of the city 580
47.1 World urbanization prospects 601
47.2 Peri-urban land use relationships 604
47.3 Environmental drivers in peri-urban land use 605
47.4 Globalization-localization dynamics in the peri-urban: example of Manchester city-region 612
47.5 'Spatial ecology' of land-use relationships in the peri-urban 615
47.6 Sustainability policy agendas for the peri-urban 617

Tables

1.1 Quantitative definitions of "urban" by various entities 8
10.1 Controls on urban climate – general and urban specific effects with examples of studies of urban influences 104–105
10.2 Parameters commonly used to characterise cities 108
11.1 List of potential causes of the canopy layer urban heat island (UHI_{UCL}) 123
11.2 List of causes of boundary layer urban heat island (UHI_{UBL}) 129
13.1 Partial urban water balances for selected cities 149
13.2 Annual water balances for selected urban areas 149
13.3 Relationship between imperviousness and land use in a western Washington suburb 153
14.1 Geomorphological problems for urban development 160
14.2 Sequence of fluvial geomorphic response to land use change: Sungai Anak Ayer Batu, Kuala Lumpur 161
15.1 General physical differences between urban and more natural soils 167
15.2 Potential sources of heavy metals 171
15.3 Average heavy metal concentrations in urban soils from different cities in the world 172

15.4 Comparison of total soil PAH concentration from different cities 175
15.5 Soil organisms 176
17.1 Key attributes of four broad types of recombinant vegetation with increasing levels of intervention 200
19.1 Characteristics of major microhabitats on walls 246
19.2 The influence of various factors on different zones of buildings, walls and paving on vegetation 248
20.1 Brief summary of similar features in natural and urban rock outcrop habitats 256
21.1 Comparative data on urban land cover for cities in Europe and the USA 265
22.1 Corridor design table 285
23.1 List of habitat units distinguished in (sub)urban parks in Flanders 294–295
23.2 Park area and biodiversity for 15 sub(urban) parks in Flanders 297
25.1 Example bioassays 318
27.1 Wetland area within and outside of large American cities 340–341
31.1 Preference matrix 388
32.1 Overall stated reasons for preferences of the three vegetation scenes 400
33.1 Criteria for studies being included in this review concerning the benefits of urban nature 409
33.2 Real world settings in which the studies were conducted, and the corresponding benefits 411–412
33.3 Researchers citing attention restoration, psycho-evolutionary, or both theories as underlying mechanisms to explain their findings 418
34.1 How street-tree problems are changing 431
34.2 Examples of 'Old' problems of street trees – ecocentric – tree-oriented 432
34.3 Examples of 'New' street tree – anthropocentric – problems 433
34.4 Normal power relationships in local government, and their consequences for preservation or diminution of street tree stock 436
36.1 Urban habitats described by various authors 467
36.2 Types of urban and peri-urban land cover and ecological conditions 468–469
36.3 The ten most important cover types in the City of Troy, NY, USA 470
36.4 Subdivisions of the major classes of the J category of the EUNIS habitat classification 473
36.5 The EUNIS habitats occupied by alien bryophytes in the UK 474
36.6 Association of invasive plant species with particular EUNIS urban habitats 475
38.1 The most invasive alien insect species in urban areas in Europe 491
38.2 The three most invasive alien fungi species in urban areas in Europe 492
38.3 The most successful neophytes in Berlin and in Brussels and Chicago 493
41.1 Provision of Local Nature Reserves in a selection of urban local authorities in England in 1993 and 2006 542–543
47.1 Structural dynamics and conflicts in the peri-urban 616

Box

25.1 Ecological Risk Assessment (ERA) of a gasworks site 319

Contributors

Xuemei Bai, Senior Science Leader, Senior Principal Research Scientist, CSIRO Sustainable Ecosystems, GPO Box 284, Canberra, ACT 2601, Australia

Bob C. Barnes, Environment Agency, Science Dept., Olton Court, 10 Warwick Road, Solihull, West Midlands, B92 7HX

John Box, Atkins Limited, Cornerstone House, Stafford Park 13, Telford, TF3 3AZ, UK

Ambra Burls, Ecosystem Health Subject Expert and Lecturer – North Wales UK

Jason Byrne, Griffith School of Environment, Griffith University, Gold Coast Campus, Parklands Drive, Southport, QLD, 4222, Australia

Anna Chiesura, Agronomist – Environmental consultant at ISPRA (Istituto Superiore per la Ricerca e la Protezione Ambientale), Dipartimento Stato dell'Ambiente e Metrologia Ambientale, Servizio Valutazioni Ambientali, Via Curtatone 3, 00185 Rome, Italy

Elizabeth M. Cook, School of Life Sciences, State University, Tempe, AZ 85287, USA

Gerald F.M. Dawe, Consultant Ecologist, Hereford, UK

Ian Douglas, School of Environment and Development, University of Manchester, Manchester M13 9PL, UK

Paul Downton, Ecopolis Architects, 109 Grote Street, Adelaide, SA 5000, Tandanya Bioregion, Australia

Joan G. Ehrenfeld, Department of Ecology, Evolution, and Natural Resources, SEBS, Rutgers University, 14 College Farm Road, New Brunswick, NJ 08901, USA

Pete Frost, Countryside Council for Wales, Maes y Ffynnon, Penrhosgarnedd, Bangor, Gwynedd, LL57 2DW, UK

Kevin J. Gaston, Department of Animal and Plant Sciences, University of Sheffield, Sheffield S10 2TN, UK

Sian Gaston, Tapton School, Darwin Lane, Sheffield S10 5RG, UK

David Goode, Environment Institute, University College London

Nancy B. Grimm, School of Life Sciences, State University, Tempe, AZ 85287 USA

C.S.B. Grimmond, Environmental Monitoring and Modelling Group, Geography, King's College London, The Strand, London, WC2R 2LS, UK

Rebecca L. Hale, School of Life Sciences, State University, Tempe, AZ 85287, USA

Martin Hermy, Department of Earth and Environmental Sciences, University of Leuven, Celestijnenlaan 200 E, B-3001 Heverlee, Belgium

Shu-Li Huang, Graduate Institute of Urban Planning. National Taipei University, Taipei, Taiwan 237

Michael C. Houck, Executive Director, Urban Greenspaces Institute, PO Box 6903, Portland, Oregon 97228–6903, USA

Glen Hyman, Center for Sociology of Organizations, Institut d'études politiques de Paris (Sciences-Po), 19, rue Amélie, 75007 Paris, France

David M. Iwaniec, School of Sustainability, Arizona State University, Tempe, AZ 85287 USA

Peter J. Jarvis, Geography, Earth and Environmental Sciences, University of Birmingham, Edgbaston, Birmingham, B15 2TT, UK

C.Y. Jim, Department of Geography, University of Hong Kong, Pokfulam Road, Hong Kong

M.L. Jin, Department of Meteorology and Climate Science, San Jose State University, San José, CA 95192, USA

Rachel Kaplan, School of Natural Resources and Environment, University of Michigan, 440 Church St., Ann Arbor, MI 48109–1041, USA

Tony Kendle, The Eden Project, Bodelva, Cornwall, PL24 2SG, UK

Roderick J. Lawrence, Human Ecology Group, Institute of Environmental Sciences, Faculty of Social and Economic Sciences, University of Geneva, 7 route de Drize, 1227 Carouge (GE), Switzerland

Chun-Lin Lee, Department of Landscape Architecture, Chinese Culture University, Taipei, Taiwan 11114

Sarah Little, Penny Anderson Associates Limited (Consultant Ecologists), Park Lea, 60 Park Road, Buxton, Derbyshire, SK17 6SN, UK

Jeremy Lundholm, Biology and Environmental Studies, Saint Mary's University, 923 Robie Street, Halifax, Nova Scotia B3H 3C3, Canada

Grant Luscombe, Landlife, National Wildflower Centre, Court Hey Park, Liverpool, L16 3NA, UK

Peter J. Marcotullio, Geography, Urban Affairs and Planning, 1046 Hunter North, Hunter College, City University of New York, 695 Park Ave, New York, NY 10065, USA

Joan Martínez-Alier, Department of Economics and Economic History, Autonomous University of Barcelona, Spain

Rod Matsuoka, Landscape Horticultural Group, Department of Horticulture, National Taiwan University, 138 Keelung Road, Section 4, Taipei, Taiwan, R.O.C.

Nancy E. McIntyre, Department of Biological Sciences, Texas Tech University, Box 43131, Lubbock, TX 79409–3131, USA

Colin D. Meurk, Landcare Research NZ Ltd, PO Box 40, Lincoln 7640, Canterbury, New Zealand

T.L. Mote, Department of Geography, University of Georgia, Athens, GA 30602–2502, USA

David Nicholson-Lord, Environmental author and journalist, formerly with *The Times*, *The Independent* and *The Independent on Sunday*

T.R. Oke, Emeritus Professor, Department of Geography, The University of British Columbia, Vancouver, B.C., Canada V6T 1Z2

Monica Palta, Department of Ecology, Evolution, and Natural Resources, SEBS, Rutgers University, 14 College Farm Road, New Brunswick, NJ 08901, USA

Joe Ravetz, Centre for Urban and Regional Ecology (CURE), School of Environment and Development, University of Manchester, Manchester, M13 9PL

Michael O. Rivett, School of Geography, Earth and Environmental Sciences, University of Birmingham, Birmingham, B15 2TT, UK

Jonathan P. Sadler, School of Geography, Earth and Environmental Sciences, University of Birmingham, Birmingham, B15 2TT, UK

Heinz Schandl, Senior Science Leader, Social and Economic Sciences Program, CSIRO Sustainable Ecosystems, Gungahlin Homestead, GPO Box 284, Canberra ACT 2601, Australia

Richard Scott, Landlife, National Wildflower Centre, Court Hey Park, Liverpool L16 3NA, UK

J.M. Shepherd, Department of Geography, University of Georgia, Athens GA 30602–2502, USA

J.A. Stallins, Department of Geography, Florida State University, Tallahassee, FL 32306–2190 USA

Emilie Stander, US Environmental Protection Agency, Urban Watershed Management Branch, 2890 Woodbridge Ave, MS-104, Edison, NJ 08837, USA

John Stuart-Murray, Landscape Architecture, Edinburgh College of Art, Lauriston Place, Edinburgh, EH3 9DF, Scotland, UK

William Sullivan, Department of Landscape Architecture, University of Illinois at Urbana-Champaign, 611 E. Taft Drive, 101 Buell Hall, Champaign, IL 61820, USA

Jenna H. Tilt, Department of Geosciences, Oregon State University, 104 Wilkinson Hall, Corvallis, Oregon 97331, USA

Joachim T. Tourbier, Landscape Construction, Institute of Landscape Architecture, Faculty of Architecture, Dresden University of Technology, Dresden, Germany

Rusong Wang, State Key Lab of Urban and Regional Ecology, Research Center for Eco-Environmental Sciences, Chinese Academy of Sciences, P.O. Box 2871, Beijing 100085, China.

C. Philip Wheater, School of Science and The Environment, Manchester Metropolitan University, Chester Street, Manchester, M1 5GD, UK

Peter Worrall, Penny Anderson Associates Limited (Consultant Ecologists), Park Lea, 60 Park Road, Buxton, Derbyshire, SK17 6SN, UK

Wayne C. Zipperer, USDA Forest Service, P.O. Box 110806, Gainesville, FL 32611–0806 USA

Acknowledgements

The editors thank the UK Man and Biosphere (MAB) Urban Forum for suggesting in 2007 that there should be a reference work for practitioners and students engaged in urban ecology. Members of the Urban Forum provided valuable suggestions during the planning of the book and several members have contributed chapters to the volume.

The UK MAB Urban Forum is a network of managers, planners and researchers involved with the environment and nature conservation in urban areas. Its distinctive contribution to this goal is integrated thinking. It seeks to raise awareness; stimulate research; influence policy; improve the design and management of urban systems and push urban nature conservation up the social and political agenda.

The Urban Forum reports to the UK UNESCO MAB Committee and is thus part of UNESCO's network of organizations concerned with the interactions between people and nature. The pioneering work of the UNESCO MAB Programme 11 has been an inspiration for much of the effort reported in this volume.

The editors thank all the authors for their co-operation and tolerance during the editorial process and are especially grateful to those who gave advice on the structure of the volume.

Particular thanks are due to the editorial and production staff of Routledge for all their assistance during the preparation of the book.

Prologue

Ian Douglas

More than half the world's people now living in cities have a reduced everyday contact with nature. In its place, they are surrounded by novel ecological situations in the interiors of buildings and shelters and in the spaces between those buildings. In many ways, urbanization has created a myriad of microhabitats, each of which supports its own food chain for all life forms from microbes to mammals. These have adapted to changing conditions of energy, water and nutrient availability as a function of the urban biogeochemistry created by human consumption and construction, waste disposal and demolition. Just as human activities in the urban environment change, so does nature. Adaptation by the various species varies, and is altered by deliberate and inadvertent introductions. Since 1960 living things and urban habitats have begun to show signs of responding to climate change, from the earlier flowering of garden plants to new insect pests. Yet our understanding of urban ecology now provides an opportunity to use nature in the city to assist our own adaptation to climate change, to help us live healthier lives, to bring our children closer to understanding nature, and to improve the appearance and aesthetic appeal of our cities. In the belief that all professions concerned with the design, operation, management and enjoyment of cities and towns need to understand, appreciate and use urban nature in their work, this handbook sets out what we know now, what we still need to know, and how we can use that knowledge of urban ecology to build and maintain a better, more sustainable urban future.

People enjoy urban nature in many ways, from the passive enjoyment of parks and gardens to active involvement in wildlife conservation and the creative conservation of wildflowers. Yet nature poses many problems to people, from the predations of urban foxes to the bacteria that attack food and the vectors that bring diseases. To understand and manage the complexity of nature in cities requires knowledge of the dynamics of both ecosystems and social systems. This book aims to outline that knowledge by discussing the origins of people's views of urban nature, the basic science that underlies the working of urban ecosystems, the different types of habitat found in urban areas, the ecosystem services provided by urban nature, the methods of analysing the dynamics of urban ecosystems, and the applications and policy implications of urban ecology.

Urban ecosystems range from the totally people-made, artificially watered and maintained, to the virtually undisturbed remnants of natural vegetation surrounded by the built environment. From the earliest cities, gardens were built by the powerful and anyone with access to suitable land grew at least some of their own food. In the first decade of the twenty-first century, urban people have become aware of the need for greater self-sufficiency, increased sustainability of lifestyles and more local food production. These three concepts are practical everyday realities for many of the urban poor in Africa, Asia and Latin America, but for most urban dwellers in Australasia, Europe and North America they require changes in ways of living and new thinking.

Nevertheless, many people are putting forward new ideas and are making practical examples of ways of creating new opportunities for food production, for creating novel gardens and for using vegetation to make cities more liveable and to mitigate the impacts of climate change. Managing urban ecosystems in this way brings multiple benefits, from the practical control of storm runoff to the aesthetic enjoyment of pleasing landscapes. Achieving those benefits requires understanding of the natural processes involved, from earth surface process to ecological succession, and of all the complex political, economic, social and cultural factors involved in managing cities in all their political and cultural diversity.

This book brings together the perspectives of managers, practitioners, lobbyists, journalists and scientists from many different disciplines. It thus presents a wide range of approaches to nature in the city and to the analysis and application of urban ecology to give the reader contrasting and complementary views of people with differing responsibilities for, and attitudes to, the urban environment. It is divided in to six parts, each of which has its own introduction. The history and philosophy of urban ecology and its connections to other disciplines are set out in Part 1. The second part examines the physical and biological scientific context of urban ecology, setting the scene and providing the background for the study of the biodiversity of ecological habitats in urban areas in Part 3. Part 4 begins the consideration of why urban nature is important to human society in terms of ecosystem services, particularly in relation to human well-being. Ways of examining urban ecosystems are discussed in Part 5, while the final section, largely written by people actually engaged in planning and delivering urban greenspace and policies for urban nature, considers the application and policy implications of urban nature.

Clarification of terms used in discussing urban ecology

Nowadays people frequently discuss nature in the city in terms of the 'wild', or what might be regarded as the unmanaged growth and movements of organisms in urban areas. This notion of 'naturalness' in urban ecology is set against the image of landscape design and maintenance or gardening as a correct or 'proper' way of using plants to beautify and gain the benefits of trees, flowers, shrubs and grass. Nevertheless, both the managed and the unmanaged vegetated urban sites provide ecosystem services and contribute to the natural capital of cities. All the diverse unbuilt areas provide some kind of habitat. Plants and animals also invade derelict buildings, colonize walls, exploit cracks in pavements and accumulate in unmanaged drains.

The term 'greenspace' is widely used in this book in the manner in which it is defined by Greenspace Scotland: 'greenspace is any vegetated land or water within or adjoining an urban area' (Greenspace Scotland 2009). This includes:

- green corridors like paths, disused railway lines, rivers and canals;
- woods, grassed areas, parks, gardens, playing fields, children's play areas, cemeteries and allotments;
- countryside immediately adjoining a town which people can access from their homes;
- derelict, vacant and contaminated land which has the potential to be transformed.

Urban greenspace is now recognized as having the potential to provide a green infrastructure for urban areas. Natural England defines green infrastructure as 'a strategically planned and delivered network of high quality green spaces and other environmental features'. It provides a multifunctional resource capable of delivering a wide range of environmental and quality of life benefits for local communities. Green infrastructure includes parks, open spaces, playing fields, woodlands, allotments and private gardens. Green infrastructure has become a key planning concept, guiding the provision of vegetated areas in new developments and their retrofitting into existing urban areas.

As dynamic ecosystems, urban ecosystems behave and interact similarly to natural ecosystems. Unlike natural ecosystems however, urban ecosystems are a hybrid of natural and man-made elements whose interactions are affected not only by the natural environment, but also by human culture, personal behaviour, politics, economics and social organization. To tackle this, Parts 1 and 6 of this book pay particular attention to the roles of planning, economics and political decision making in the safeguarding, promotion and management of urban greenspace.

Urban natural capital

Natural capital highlights the value and utility of natural assets, such as forests, mountains, lakes, farms and urban parks, which provide key benefits for our economic prosperity and quality of life. Natural capital, along with built, human and social capital, is an important component of the wealth of a nation. Ecosystem services are essential to human well-being and sustainable development. The value of ecosystem services in both monetary and non-monetary terms must be recognized in decision making. Like other forms of capital, natural assets require careful stewardship and investment for their value to grow and pay dividends over the long term.

Urban natural capital – everything from wild areas and water resources to soccer fields and community gardens – pays psychological, physical and financial dividends that greatly improve the lives of urban residents and help sustain the long-term economic prosperity of our cities (Wilkie and Roach 2004). People benefit from urban natural capital in many ways including better health, greater social cohesion, richer urban cultures, cleaner air and water, more recreation opportunities, and improved urban aesthetics. Urban natural capital helps to attract tourism revenue and skilled labour, and to raise property values. Urban natural capital need not be that natural at all. Community and rooftop gardens, golf courses, cemeteries, landscaped boulevards and street trees may not come to mind when one first thinks of urban natural capital (Wilkie and Roach 2004).

Urban natural capital has to be taken into account in all urban planning, development and conservation exercises. One way of doing this is to develop urban natural capital indicators alongside economic, political and social indicators (Olewiler 2006). The state of our environment is a form of capital; it can yield essential goods and services over time, but only if we understand what is happening to these goods and services as a result of our activities, and if we sustain natural capital as part of society's total capital. Establishing those natural capital services implies examining how urban greenspaces, trees and wildlife together deliver ecosystem services.

Ecosystem services

All components of the urban green environment, from extensive woodlands, wetlands, floodplains and large parks to small green spaces, such as urban gardens, generate key urban ecosystem services (Colding *et al.* 2006). More attention is paid to the larger open spaces, particularly those in public ownership than to the many small, mainly private, parcels of land that provide great heterogeneity and diversity of habitat.

Understanding the ecological processes on which the complex interrelations of ecosystems, and the ecosystem services they provide, depend, can help us be aware of how nature will respond when we try to mitigate trade-offs and enhance synergies (Bennett *et al.* 2009). Urbanization can reduce ecosystem functions, such as water infiltration and soil productivity, and therefore conserving ant communities and the ecosystem services they provide could be an important target in land-use planning and conservation efforts. When land development and disturbance affects 30 to 40 per cent of a site, ant diversity may begin to decline, revealing that ants and the ecosystem services they provide are vulnerable to degradation in response to urbanization (Sanford *et al.* 2008).

Urban landscapes are dynamic, constantly evolving, not merely through demolition and construction, but by planting and maintenance of greenspaces. Many urban gardens and parks have favoured introduced species. Non-native plants popular in suburban landscapes have not been considered a threat to biodiversity because most of them are ornamental and lack invasive traits. Regardless of their dispersal abilities, non-native ornamentals, favoured by landscapers and homeowners, now dominate the first trophic level in millions of hectares of North America (Burghardt *et al.* 2008). How the large-scale replacement of native vegetation with non-native plants in managed ecosystems affects members of higher trophic levels has yet to be determined. In south-eastern Pennsylvania, USA, urban gardens with North American native plants support significantly more caterpillars and caterpillar species and significantly greater bird abundance, diversity, species richness, biomass and breeding pairs of native species (Burghardt *et al.* 2008). Bird species of regional conservation concern were eight times more abundant and significantly more diverse in such sites. Native landscaping positively influenced the avian and lepidopteran carrying capacity of suburbia and provided a mechanism for reducing biodiversity losses in human-dominated landscapes.

Diversity and change

Now that the majority of people live in urban environments, urbanization is arguably the most severe and irreversible driver of ecosystem change on the planet. Urban change modifies floras in four main ways, by changing habitat availability, the spatial arrangement of habitats, the pool of plant species, and evolutionary selection pressures on populations persisting in the urban environment (Williams *et al.* 2009). Four filters govern the pool of species: habitat transformation, habitat fragmentation, urban environmental conditions and human preference. These four filters reflect the land cover change, the subdivision of land parcels, the modifications to urban climate, hydrology and soil, and the choices people make about what to plant and what to remove on and in vegetated parts of towns and cities.

Urbanization leads to an increase in avian biomass but a reduction in richness, indicating that avian communities change with urban development (Scott 1993). Locally the house sparrow has disappeared from places where it once was common, like in the city centres of Amsterdam and The Hague. In the UK, a massive decrease in the house sparrow population (Crick *et al.* 2002; Hole *et al.* 2002; Raven *et al.* 2003) has led to almost complete extinction in some urban centres like London, where there was a 71 per cent decline from 1994 to 2002 (Raven *et al.* 2003). In Delhi, India, the high-density urban area is high on total bird density, but low on species diversity, as against agricultural and forested areas where species diversity is relatively high (Khera *et al.* 2009). The house sparrow is relatively low in numbers in the densely built-up areas of Delhi. While changing building styles may restrict nesting opportunities and changes in food availability can affect numbers, a highly probable reason for the lower density and restricted distribution of the house sparrow may be competition from the co-occurring common bird species like common myna, rock pigeon and house crow (Khera *et al.* 2009). This example neatly illustrates the interworking of natural and societal processes in terms of urban biodiversity.

Change is now occurring at many scales from genetic modification of plants to global climate change. Patterns of movement of invasive species and the cultivation of garden plants are being modified. Further modifications to urban greenspace will occur as people adapt to climate change, both through planning to manage higher temperatures and more extreme rainfall events. Green infrastructure is now being promoted as a key tool in urban adaptation to climate change, but the climate change benefits should always be seen as one of the multiple functions of urban greenspace. In Parts 4 and 6 of this book, the multiple benefits are addressed in terms of human health, therapy and human well-being, sustainable drainage and climate change adaptation.

The editors and authors hope that this book will provide a valuable reference source for anyone engaged in promoting and caring for urban nature and urban greenspace.They also hope that it will help students of the planning, geography, ecology and sociology of urban areas to understand the interlinkages between their own fields and others in the urban environment.

References

Bennett, E.M., Peterson, G.D. and Gordon, L.J. (2009) 'Understanding relationships among multiple ecosystem services', *Ecology Letters*, 12(12): 1394–404.

Burghardt, K.I., Tallamy, D.W. and Shriver, W.G. (2008) 'Impact of native plants on bird and butterfly biodiversity in suburban landscapes', *Conservation Biology*, 23(1): 219–24.

Colding, J., Lundberg, J. and Folke, C. (2006) 'Incorporating green area user groups in urban ecosystem management', *Ambio*, 35(5): 237–44.

Crick, H.Q., Robinson, R.A., Appleton, G.F., Clark, N.A. and Rickard, A.D. (2002) *Investigation into the causes of the decline of starlings and House Sparrows in Great Britain*, Research Report No. 290, London: Department for Environment, Food and Rural Affairs (DEFRA) and London: British Trust for Ornithology.

Greenspace Scotland (2009) *State of Scotland's Greenspace 2009*, Stirling: Greenspace Scotland.

Hole D.G., Whittingham, M.J., Bradbury, R.B., Anderson, G.Q.A., Patricia L.M., Lee P.L.M., Wilson, J.D. and Krebs, J.R. (2002) 'Agriculture: widespread local House-Sparrow extinctions', *Nature*, 418: 931–2.

Khera, N., Arkaja Das, A., Srivasatava, S. and Jain, S. (2009) 'Habitat-wise distribution of the House Sparrow (Passer domesticus) in Delhi, India', *Urban Ecosystems* DOI 10.1007/s11252–009–0109–8.

Olewiler, N. (2006) 'Environmental sustainability for urban areas: the role of natural capital indicators', *Cities*, 23(3): 184–95.

Raven, M.J., Noble, D.G. and Baillie, S.R. (2003) 'The breeding bird survey 2002', BTO research report 334, London: British Trust for Ornithology.

Sanford, M.P., Manley, P.N. and Murphy, D.D. (2008) 'Effects of urban development on ant communities: implications for ecosystem services and management', *Conservation Biology*, 23(1): 131–41.

Scott, T.A. (1993) 'Initial effects of housing construction on woodland birds along the wildland urban interface', in J.E. Keeley (ed.) *Interface Between Ecology and Land Development in California*, Los Angeles, CA: Southern California Academy of Sciences, pp. 181–7.

Wilkie, K. and Roach, R. (2004) *Green among the Concrete: the Benefits of Urban Natural Capital*, Calgary: Canada West Foundation.

Williams, N.S.G., Schwartz, M.W., Vesk, P.A., McCarthy, M.A., Hahs, A.K., Clemants, S.E., Corlett, R.T., Duncan, R.P., Norton, B.A., Thompson, K. and McDonnell, M.J. (2009) 'A conceptual framework for predicting the effects of urban environments on floras', *Journal of Ecology*, 97(1): 4–9.

Part 1
Context, history and philosophies

Introduction

Ian Douglas

This part considers the nature of urban ecology, how it relates to industrial ecology and human ecology, the politics of providing nature in cities, the human relationship with nature, the history of concern and planning for urban nature and the valuation of urban nature.

The goal of urban ecology is to understand the full complexity of the relationship between the biological community and the urban environment due to the interaction between human culture and the natural environment. The words 'urban' and 'ecology' are used in different ways by different disciplines. Any working definition has to encompass the full breadth of the people: environment relationship in the urban context. Nancy McIntyre shows how urban ecology is starting to move away from basic documentation of patterns in urban ecosystems, towards a more mature examination of how multiple components – physical, socioeconomic, and biotic – interact to form urban ecosystems that can be planned and designed for conservation, aesthetic, and other purposes.

Urban ecosystems may be viewed in three ways:

- As the built up areas that are the habitat of urban people, their pets, their garden plants, the adapted animals and organisms (birds, moulds, etc.) and the pests (rats, weeds, parasites, etc.) that depend upon. These areas depend on outside (external) support in the form of energy, water and materials inputs for their survival.
- The immediate urban life-support system of the urban area and its surroundings (the peri-urban area), providing such ecological services as water supplies, sources of aggregates, areas for landfill, recreation zones, watershed protection, greenhouse gas uptake and biodiversity.
- The areas affected by urbanization as a driving force by providing life support services to urban areas, including supplies of food, energy, water and materials and also those areas affected by the emissions and waste flow from urban areas. For any individual city these may have a global outreach, with energy (coal, natural gas or oil) and food (exotic fruits, fish, meat, grain, soya, etc.) drawn from distant countries or seas. As urban populations and purchasing power grows this global outreach and its impacts on other ecosystems expand.

Ian Douglas reports important work that sees the human ecosystem as a coherent system of biophysical and social factors that can achieve adaptation and sustainability over time and argues that human ecosystems could be a starting point for the management of urban and other ecosystems.

Industrial ecology has been called 'the science of sustainability' in that knowledge important to understanding and designing a more effective path towards sustainable development can be gained through the study of the mechanisms of material and energy interchange within industrial systems using the framework of natural ecosystems. It uses material flow analysis, life-cycle assessment, substance flow analysis and energy flow models derived from metabolic aspects of the analogy to natural systems. It has parallels and overlaps with the ecological analysis of urban areas, especially with urban metabolism and urban consumption models. It also related to studies of biogeochemical flows in urban areas, linking the release of chemicals to the environment to urban economic activities. Xuemei Bai and Heinz Schandl report that, together with non-equilibrium thermodynamics and the application of metabolism analysis in both urban ecology and industrial ecology, complexity theory has the potential to integrate anthropogenic and natural activities into one framework. They also argue that a future research programme drawing on both urban ecology and industrial ecology would be well placed to deal with major urban sustainability issues.

Human ecology deals with adaptive mechanisms of human populations surviving in local ecosystems. It is often focused on how living conditions and the environment affect human health, but may be more broadly interpreted as the analysis of how people adapt to different environments. Because urban areas are the places where nature has been most changed and where multiple changing stressors affect individuals and communities, human ecology becomes closely concerned with living conditions and with practical solutions to issues that degrade the quality of urban life. Human ecology emphasizes the contribution of the individual, the sharing of skills and experiences, and the dignity and insight of social and cultural and religious experiences. From this standpoint, human ecology works to create sustainable, lasting improvements in people's lives by fostering projects that engage and enhance the skills of local communities, involve all sectors of society, improve livelihoods and maintain environmental benefits. Roderick Lawrence shows that when dealing with complex subjects, such as urban areas, it is necessary to shift from monodisciplinary to interdisciplinary and transdisciplinary concepts and methods. Such moves underlie the pioneering work in both human ecology and urban ecology, such as the Hong Kong human ecology project and UNESCO MAB Programme 11.

Natural areas, nature reserves, parks and other urban greenspaces have always been controversial. Huge debates occurred over the establishment of the great parks in cities like New York, London, Berlin and Paris. Today debate ranges about the misuse of parks and natural areas and over the priorities given to recreation, safety, organized games, dog walking, use of bicycles, places for skateboarding, playgrounds for the under-fives, areas for encouraging biodiversity, bird sanctuaries, wetlands and the development of park nurseries, income generating activities, cafés, amusements and special events. Social, economic, political and psychological processes influence the creation, design, management and use of parks and natural areas in cities, and the ways in which people act with regard to other organisms in urban ecosystems. The interplay between national and local government, industry, business interests, community groups and individual citizens means that providing for open spaces and natural areas in cities will never meet everyone's aspirations. One person's golf course is another person's loss of freedom to roam and enjoy wild nature. Mike Houck shows how in the USA attitudes have changed since the early 1980s, when urban wildlife was viewed as an oxymoron, to having elected officials embracing the need to integrate nature and green infrastructure into the urban fabric. The election of politicians who campaigned on a green agenda is largely attributable to grassroots citizen-based work and a growing desire by urban residents for access to nature, parks, and trails in their own neighbourhoods and close to work.

The way we think about ourselves in relation to other creatures and organisms may or may not lead into greater concerns about environmental protection, and greater action in our lives to shape cities, with better results for biodiversity. The dominant religion, science and mainstream philosophical traditions of modern western culture all work to smother ways of thinking that would include values directly in nature. Yet, according to Aristotle, all living things have souls, understood as the functioning of their innate characters or forms. Plants have powers of nutrition and reproduction and thus have vegetative souls; animals have these powers with the additional capacities of locomotion and sensation, giving them sensitive souls. Human beings have all the previous functions plus the conscious abilities to calculate and reflect, producing *rational souls*. For many thinkers today there is increasing sensitivity towards the rights and sensations of plants and animals. There is an emerging worldview recognizing subjective valuings beyond the human that expresses itself by thinking through and proclaiming human duties, owed directly to entities in nature, as well as to fellow human centres of value. Jason Byrne argues that genuinely recognizing and coexisting with urban nature is challenging. Pet euthanasia, wildlife extermination, pest eradication, pollution and ecosystem appropriation will need to give way to new practices that include plants and animals within the circle of moral considerations – an ethics of caring and respect based on kinship but also difference.

The natural history of urban areas and their fringes developed from the late nineteenth century interests in the natural world and concern about the preservation of natural open spaces on the fringes of cities. Natural history societies were established in Liverpool, Manchester and Bristol during the 1860s. Such natural history societies have been extremely important in documenting and recording the flora and fauna of their local areas. Richard Jeffries published *Nature Near London* in 1883. The Hampstead Heath Society published a natural history of the heath in 1910. R.S.R. Fitter's *London's Natural History* appeared in 1945. Such works were the precursors of the development of urban ecology. Ian Douglas and David Goode trace this history in the UK and in other countries showing urban ecology has become a world-wide concern with bright future prospects.

In Britain, Section 21 of the National Parks and Access to the Countryside Act, 1949, gave local authorities powers to declare and manage Local Nature Reserves (LNR) on land in their area of jurisdiction and owned or leased by them or subject to appropriate management arrangements with the owners and occupiers of the land concerned. It also encourages Nature Conservation Strategies (NCS), the first of which – the West Midlands NCS – was produced in the early 1980s. This legislation enabled areas within cities to be protected and made accessible to the public. Many urban local nature reserves were on derelict, brownfield sites. For example, Saltwells LNR was designated as the first Local Nature Reserve in the West Midlands metropolitan county in 1981; now covering over 100 hectares, it forms one of the largest urban nature reserves in the country. The process has also promoted Green Networks and Green Infrastructure which integrate greenspace firmly into development planning for new residential areas and new towns. David Goode traces how ecological concerns and natural greenspace values have gradually been incorporated into planning guidance and practice, emphasizing the important roles played by key organizations such as the urban ecology school in Berlin led by Herbert Sukopp.

The way in which the economic benefits of urban greenspace have been calculated in the past might be criticized as the 'sums have been done wrongly'. Nevertheless, such economic assessments have become powerful drivers for deleterious change in cities, badly affecting biodiversity. Adequate economic examination of the 'externalities' through environmental economics, will show how economic valuation should have led to better, greener cities. Anna Chiesura and Joan Martínez-Alier argue that if urban ecology is to contribute to urban sustainability, distributional and ethical issues cannot be ignored. They point out that ecological economics moves beyond the obsession of 'taking nature into account' in money terms, and attempts to avoid clash among incommensurable values by accommodating in the discourse concern for value pluralism, distributional and social equity issues.

Together these contributions show that urban ecology has wide connections, requires collaboration between varied professions and works for both society and nature to improve human and ecological well-being.

1

Urban ecology

Definitions and goals

N.E. McIntyre

"Urban" is in the eye of the beholder

The place where I live – Lubbock, Texas – is home to just over 200,000 people. I moved there from one of the largest cities in the United States: Phoenix, Arizona, home to nearly 3 million people. Coming from Phoenix, Lubbock did not appear very urban to me – there were few high-rise buildings, only a small selection of restaurants and shops, a commuter rush hour that lasted only 15 minutes, hardly any high-density housing, and little crime, graffiti, or noise. Instead, there were patches of wildflowers within undeveloped lots inside the city limits and a multitude of stars visible in a night sky untroubled by light pollution. I settled myself in for a more bucolic lifestyle than that to which I had become accustomed in Phoenix. One of my first students, however, came from Loving County, Texas – only about 270 kilometers away, this is the least-populated county in the U.S., discounting the vast wilderness of Alaska: there were just 67 people living in this 1753 km^2 county in the year 2000. To that student, Lubbock was an imposing city, a true urban environment.

The term "urban" is one of those readily comprehensible words that, upon reflection, lack precision (like "disturbance" and "heterogeneity"). Understanding urbanization's influence on the world will be hindered without an acknowledgment that the same term can mean different things to different people. This point was first brought home to me when I started working on an urban ecology project involving dozens of people from a wide variety of fields – there were biologists, sociologists, landscape architects, land-use planners, economists, and many others. Our meetings were clumsy affairs at first, characterized by a vague sense of déjà vu – hadn't we already discussed this? Isn't that group of project workers reinventing the wheel? We soon realized that many of our frustrations stemmed from a lack of consistent and coherent communication, particularly between the social and natural sciences, about what was truly urban (as opposed to the broader concept of being human-dominated but not urban per se) and thus within our purview. This led two colleagues and me to write a paper discussing this problem and suggesting several variables beyond population size that should be included in any study on urban ecology (McIntyre *et al.* 2000). In this chapter, I shall expand upon that initial "urban definition" paper by exploring traits that are characteristic of urban areas and approaches that have been taken in studying cities as unique ecosystems. Because urban areas are synthetic ecosystems that encompass natural and anthropogenic components, any working definition of what it means to be urban must encompass the full breadth of the people-environment relationship in order to guide action such as land-use planning, development, and future research.

What do we mean by "urban"?

In its usual bloated fashion, the Oxford English Dictionary (2007 online edition) defines urban as "pertaining to or characteristic of, occurring or taking place in, a city or town." Following this custom, throughout this chapter I use the terms "city" or "town" as synonyms for an urban area. However, this does not resolve the question as to what is urban and therefore a subject for urban ecology. Many authors in the urban ecology literature do not specify what they consider to be urban (e.g. Berry 1990; Alberti *et al.* 2003). As with obscenity, it is usually assumed that something that is urban can be recognized when one sees it. However, defining with precision what is urban is important in examining the effects of urbanization. The fundamental properties of urban areas – the presence of lots of people, anthropogenic forms of land use, and altered forms of land cover – have ramifications on biota, ecosystem processes, social patterns (e.g. sense of community), and economic and demographic characteristics. And these factors may be affected to differing degrees depending on the degree and type of urbanization. Therefore, a quantified description is necessary.

For most entities, there are two primary components to a definition of what is urban: population size and population density. However, different entities use different limits to define what is urban, in part because of different goals (Table 1.1). For example, the United Nations defines an urban area as one with >20,000 people, whereas various national census agencies around the world may include areas with as few as 1000 people.

Even with separated components of population size and density, look at the difference between Australia and Japan, for example: a place that would be considered urban in Australia might not be included as such in Japan. Another approach is to consider any named, incorporated area with its own governance as urban—even when these areas have fewer than 100 residents. These differences make it difficult to compare statistics such as percent land urbanized or urban population size among nations and ultimately mean that something merely described as urban is insufficient: a quantitative description must also be included.

Table 1.1 Quantitative definitions of "urban" by various entities

Entity	*Defining characteristics*
Australian Bureau of Statistics[1]	≥1000 people, with a density of ≥200 people/km^2
Statistics Bureau of Japan[2]	>5000 people, with a density of >4000 people/km^2
Statistics Canada[3]	≥1000 people, with a density of ≥400 people/km^2
United States Census Bureau[4]	≥1000 people/mi^2 (385/km^2) immediately surrounded by areas with ≥500 people/mi^2 (193/km^2)

Notes

1 www.abs.gov.au/AUSSTATS/abs@.nsf/lookup/6DB91BD08C425487CA256F190012EEF4?opendocument.

2 www.stat.go.jp/english/index/official/202.htm.

3 www12.statcan.ca/english/census06/reference/dictionary/geo049.cfm.

4 www.census.gov/geo/www/ua/ua_2k.html.

Variations on a theme: suburban, exurban, rural, wildland, megalopolis, mega-city, and megapolitan

Even this prescription is difficult to fill because there is much jargon tossed about in urban ecology, including the terms suburban, exurban, rural, wildland, megalopolis, mega-city, and megapolitan. In general, a suburban area is one that is on the fringe of an urban area and has lower population density and physical infrastructure. Even newer, smaller, and farther out are the exurbs. The suburbs and exurbs are generally included as part of the urban area, as a symptom of sprawl (rapid and scattered development, especially with leap-frogging; Ewing 1994). Sprawl typically occurs because the cost of spreading out is cheaper than building up or infilling existing space, which entails renovation costs, zoning limits to size and/or design, and association with existing "old" or "outmoded" surroundings. Beyond the suburbs and exurbs are rural areas, typically defined by the national census agencies listed above as any area other than urban. Such a binary definition fails to distinguish between areas with and without human presence, however, so it may be helpful to use Bourne and Simmons' (1982) definition of rural (≤1 person/ha), as distinguished from wildlands with no permanent human presence. At the other end of the spectrum are very large urban areas, often comprised of two or more formerly separate cities that have grown and merged to form metropolitan areas or megalopolises. And there are terms for all the hair-splitting in between – the U.S. Census Bureau, for example, uses the term "urbanized area" to denote an area with ≥50,000 people; areas with fewer than 50,000 people (but still urban) are called "urban clusters."

What criteria separate urban from non-urban areas?

Because of this hodgepodge of inconsistency, two colleagues and I tried to identify which demographic and physical variables must be considered and measured in order to accurately compare cities (McIntyre *et al.* 2000). What we found was that this list was far from concise. For example, even the two key demographic variables of population size and density fail to capture aspects of growth rate, ethnicity, and socioeconomic structure, all of which may play a role in the ecology of organisms inhabiting an urban environment (see e.g. Kinzig *et al.* 2005). A description of the physical environment can become a veritable thicket of traits to measure, such as historic, current, and projected future land-cover and land-use types, extents (percentage coverage), and degree of patchiness (number, sizes, and configuration of patches), to give just one (pruned) example. And should we report average or median values? Even when traits can be agreed upon, then we are back at the problem of different entities using different thresholds as to what is urban.

As an alternative, Odum (1997) proposed energy use of >100,000 kcal $m^2 y^{-1}$ as a universal metric of what is urban. However, measuring energy use is much easier said than done. A surrogate may be found in ecological footprint analysis. A city's footprint is the land area needed to sustain current levels of resource consumption; this footprint may be >100 times larger than the spatial extent of the city itself (Folke *et al.* 1996; Rees and Wackernagel 1996). The footprint of Phoenix, for example, is the largest in the United States and is several orders of magnitude larger than the size of the city's physical boundaries (Luck *et al.* 2001). But even footprint analysis fails to satisfy because, like most indices, it distills a complex form into a relatively simple number. So what is to be done?

For a start, we can acknowledge the fact that "urban" is an imprecise term and then describe each setting as quantitatively as possible. A quantitative description of what is urban will necessarily incorporate techniques from many disciplines to encompass physical, biological, and social facets.

Methodologies from another multidisciplinary field may be able to guide us in this. Landscape ecology has been touted as having a natural affinity with urban ecology, since the urban system can be viewed as simply another type of landscape (Musacchio and Wu 2004, Wu 2008). As landscape ecology integrates "pattern, process, and design" (Nassauer 2007), it must quantify the composition and configuration of a site. For a review of the kinds of landscape ecology metrics that may be applicable to urban ecology, see Alberti (2008).

Why is it important to define what is urban?

It would be all too easy to feel frustrated and overwhelmed at this point. However, urbanization is going to continue whether we want it to or not, so we must get a handle on its effects if we are to understand and plan. Although cities and towns only comprise about 2 percent of the Earth's land surface (Wu 2008), they contain an ever-increasing proportion of the world's population. In the year 1700, the largest city in the world was Constantinople at 700,000 residents. By 1800, one city had more than one million inhabitants (Peking). By 1900, 16 cities exceeded one million residents. By 2000, over 300 cities in the world had more than a million inhabitants, with several cities merging into even larger metropolitan areas (statistics from Berry 1990, Pickett *et al.* 2001). These multi-city megalopolises are often named for their largest city (e.g. "Phoenix" for a conglomeration of >20 autonomous cities, "Tokyo" for the area including both Tokyo and Yokohama and every settlement in between).

Growth in the urban population comes from two sources: human population growth and also from coalescing human distributions. Urban areas have something akin to a gravitational pull, drawing population away from small towns in many regions. Consequently, urban population growth has far exceeded rural population growth in the past century (Wu 2008). This urban growth has influenced climate, biodiversity, ecosystem functions, and many other factors—hence the development of urban ecology (Collins *et al.* 2000).

Effects of urbanization on climate, biodiversity, and ecosystem functions

Urban areas generate their own microclimate by altering wind and water currents, a smog-generated greenhouse effect, and use of building materials that have high insulative properties and albedo (Arnfield 2003). Particularly well-known is the urban "heat island" effect (Landsberg 1981; Brazel *et al.* 2000; Kuttler 2008). This effect is especially felt at night, when the surrounding areas cool down while the city reluctantly releases its built-up heat. In Phoenix, for example, the mean daily air temperature within the city is 3.1°C warmer than the surrounding undeveloped area, and the mean nighttime temperature is 5°C warmer (Baker *et al.* 2002). The heat island effect induces stresses on plants, animals, and humans, with resultant effects on productivity, activity patterns, floral phenology, biotic distributions, and human quality of life (Baker *et al.* 2002; Gehrt and Chelsvig 2003; Parris and Hazell 2005; Neil and Wu 2006). For example, an increase in temperature in an already hot climate can boost "human misery hours" (Baker *et al.* 2002). This and other climate-related effects will become of even greater concern as global warming occurs (Alcoforado and Andrade 2008) (see also Chapters 10 and 11, below).

Biodiversity is affected by urbanization, with various effects depending on taxonomic group (Douglas 1983; Gilbert 1989; Nilon and Pais 1997). For example, whereas soil fungi and microbes may decrease in diversity and abundance with increased urbanization (Pouyat *et al.* 1994), bird species richness may remain relatively constant but with a decrease in evenness (Emlen 1974; Reale and Blair 2005). Plant species richness may increase, but this effect is often due to the presence of exotic species (Zipperer *et al.* 1997). Even within a taxonomic group, different species that may show differing tolerances or affinities for urban development (Blair 1996, 2001; McIntyre 2000).

Cities around the world often contain the same species – dandelions, pigeons, cockroaches, and rats are well-known examples of species that are able to thrive in urban environments. This effect has been called biotic homogenization (Blair 2001; Lockwood and McKinney 2001; McKinney 2006). Certain species are able to live in cities because their resource needs are being met. Indeed, for some species, cities create an "oasis effect" (Bock *et al.* 2008) whereby limiting resources such as water, food, or breeding sites are more consistent or abundant in urban areas than elsewhere as a direct consequence of human activities such as watering lawns and gardens, putting out bird feeders and houses, or planting trees and shrubs suitable for nesting or roosting (Shochat *et al.* 2004, 2006; Cook and Faeth 2006; Bock *et al.* 2008) (see also Chapters 22 and 36, below).

Most research on how urbanization affects organisms has been conducted on birds (see e.g. Beissinger and Osborne 1982; Blair 1996, 2001; Marzluff *et al.* 2001), plants (e.g. Dorney *et al.* 1984; Hope *et al.* 2003; Martin *et al.* 2004; Grove *et al.* 2006), and terrestrial arthropods (e.g. Blair and Launer 1997; McIntyre 2000; McIntyre *et al.* 2001; Niemelä *et al.* 2002; Ishitani *et al.* 2003; Alaruikka *et al.* 2003; Shochat *et al.* 2004). Much less research has been done on reptiles and amphibians (but see e.g. Germaine and Wakeling 2001; Parris 2006; Pillsbury and Miller 2008), mammals (e.g. Nilon and VanDruff 1987; Baker *et al.* 2003), fish (e.g. Wolter 2008), and aquatic invertebrates (e.g. Paul and Meyer 2001).

Because organisms are affected by urbanization, so are the ecosystem functions that they perform. For example, because transportation and industry produce pollutants such as heavy metals that may affect soil chemistry, soil organisms may be reduced in number and type (Pouyat and McDonnell 1991; Pouyat *et al.* 1995). As a result, decomposition rates are typically lower in urban than non-urban sites (Pouyat *et al.* 1997). Similarly, although urban plantings may provide floral resources to support a variety of pollinators (McIntyre and Hostetler 2001), pollination efficiency is often lower in urban than non-urban sites due to pollinator limitation from lack of nesting sites (Liu and Kopter 2003).

Various approaches to try to encapsulate what is urban

Effects may depend on the degree of urbanization, a city's age, growth pattern, and other factors that are still as yet seldom examined in much detail, meaning that the ecological, economic, and social costs of urban development are still poorly understood. There are numerous heuristic models of how urban ecosystems are structured and how they function (see e.g. Pickett *et al.* 1997, 2005; Alberti 2008). One of the most commonly used is the gradient approach (McDonnell and Pickett 1990). As one moves away from a city's centre, differences are encountered in the type, size, shape, and spacing of various forms of land use; metrics from landscape ecology are often used to describe this gradient (e.g. in terms of patch size, shape, and variety; Luck and Wu 2002). This approach can be used to determine degrees of urbanization (Luck and Wu 2002), which can provide a quantitative cutoff for what is considered urban. The gradient is usually assumed to be concentric, which would be inappropriate for megalopolises and for cities that have grown in an anisotropic fashion (e.g. due to boundaries such as coastlines). Biota are assumed to respond in a gradient fashion, although sometimes rapid transitions and threshold responses are observed. Unlike elevation or other natural gradients, the urban gradient is established very quickly relative to evolutionary processes. With other gradients, we may observe coevolved pattern-process relationships that have become established after many generations of adaptation. With urban gradients, especially for recent urbanization, we may be witnessing relationships that are still in the process of crystallizing, with extinction debts still being paid. Cross-city comparisons of cities that differ greatly in age should keep this in mind, since we do not know how quickly it takes for the effects of urbanization to manifest themselves.

Another commonly used approach is to compare different forms of urban land-use types (e.g. residential, industrial, etc.) or urban land use to natural forms of land cover (see e.g. Cicero 1989). Examining a single area over time (urban succession or patch dynamics) is more difficult (Berling-Wolff and Wu 2004) but should be facilitated by the establishment of long-term ecological research projects designed to monitor cities as dynamic ecosystems (see e.g. http://caplter.asu.edu, www.beslter.org; Grimm *et al.* 2000).

There are some commonalities to these approaches: all include an aspect of land use; social factors related to income, demography, or governance; and/or physical factors related to impervious surfaces, climate, and the flows of materials, energy, and information. Urban ecology is starting to move away from foundational work documenting patterns in urban ecosystems, towards a more mature examination of how multiple components – physical, socioeconomic, and biotic – interact to form urban ecosystems that can be planned and designed for conservation, aesthetic, and other purposes.

Goals and future research agendas

Urbanization is both a pattern and a process. Perhaps this dualism is the ultimate source of confusion. But despite imprecision in the term "urban," we know that relative to non-urban areas, urban ecosystems are characterized by high human population density (with associated built structures and services), an altered climate (usually being warmer, especially at night), anthropogenic impervious surfaces, a high concentration of chemicals of anthropogenic origin (e.g. heavy metals, atmospheric gases and particulates), altered productivity regimes (with dampened fluctuations), and a large ecological footprint. These traits are associated with a changed biota (usually containing more exotic species), altered soil biogeochemistry and local hydrology, and altered rates of ecosystem functions. Yet what we do not know outweighs what we do. Future research agendas for urban ecology must embrace the unique and synthetic nature of urban ecosystems by integrating the human and natural components, and by merging science with design and policy. This will be facilitated by conducting more explicitly comparative work from various locales, on less-well-represented taxa (chiefly non-avian vertebrates and non-arthropod invertebrates).

Urban ecology is not just for ecologists – its findings will be used by planners, architects, and others in the design of cities of tomorrow. With their enormous footprints, cities are currently not sustainable (Rees and Wackernagel 1996; McGranahan and Satterthwaite 2003). This represents an enormous challenge as well as an enormous opportunity (Loucks 1994; Wu 2008).

Cities are already providing insight into how urban planning can affect human quality of life (Pacione 2003), how socioeconomic variables are just as important as biological ones, if not more so, in terms of urban biodiversity (Kinzig *et al.* 2005), and how to live in a warmer world, a pressing issue as the effects of global climate change materialize (Grimm *et al.* 2008).

As stated by geographer Chauncy Harris in 1956: "We stand today in the midst of a gigantic and pervasive revolution, the urbanization of the world." Caught up in the maelstrom of revolution, defining what is "urban" is surprisingly difficult. As the human population continues to grow, cities will continue to grow, and what is considered urban (and suburban, and so forth) will continue to change. Cities need and deserve study – they are ecosystems of our own making and part of an inadvertent global experiment on design and sustainability.

Acknowledgments

My thinking about urban ecosystems has been shaped largely by my interactions with fellow members of the Central Arizona-Phoenix Long-Term Ecological Research Project, particularly Amy Nelson Gonzalez, Nancy Grimm, Diane Hope, Kim Knowles-Yánez, and Chuck Redman.

References

Alaruikka, D., Kotze, D.J., Matveinen, K., and Niemelä, J. (2003) Carabid beetle and spider assemblages along a forested urban-rural gradient in southern Finland, *Journal of Insect Conservation*, 6(4): 195–206.

Alberti, M. (2008) *Advances in Urban Ecology: Integrating Humans and Ecological Processes in Urban Ecosystems*, New York, NY: Springer.

Alberti, M., Marzluff, J., Shulenberger, E., Bradley, G., Ryan, C., and ZumBrunnen, C. (2003) Integrating humans into ecology: Opportunities and challenges for studying urban ecosystems, *BioScience*, 53(12): 1169–79.

Alcoforado, M.J. and Andrade, H. (2008) Global warming and the urban heat island, in J.M. Marzluff, E. Shulenberger, W. Endlicher, M. Alberti, G. Bradley, C. Ryan, C. ZumBrunnen, and U. Simon (eds.), *Urban Ecology: An International Perspective on the Interaction Between Humans and Nature*, New York, NY: Springer, pp. 249–62.

Arnfield, A.J. (2003) Two decades of urban climate research: a review of turbulence, exchanges of energy and water, and the urban heat island, *International Journal of Climatology*, 23(1): 1–26.

Baker, L.A., Brazel, A.J., Selover, N., Martin, C., McIntyre, N., Steiner, F.B., Nelson, A., and Musacchio, L. (2002) Urbanization and warming of Phoeniz (Arizona, USA): Impacts, feedbacks and mitigation, *Urban Ecosystems*, 6(3): 183–203.

Baker, P.J., Ansell, R.J., Dodds, P.A.A., Webber, C.E., and Harris, S. (2003) Factors affecting the distribution of small mammals in an urban area, *Mammal Review*, 33(1): 95–100.

Beissinger, S.R. and Osborn, D.R. (1982) Effects of urbanization on avian community organization, *Condor*, 84: 75–83.

Berling-Wolff, S. and Wu, J. (2004) Modeling urban landscape dynamics: A case study in Phoenix, USA, *Urban Ecosystems*, 7(3): 215–40.

Berry, B.J.L. (1990) Urbanization, in B.L. Turner, W.C. Clark, R.W. Kates, J.F. Richards, J.T. Mathews, and W.B. Meyer (eds.) *The Earth as Transformed by Human Action*, Cambridge, UK: Cambridge University Press, pp. 103–19.

Blair, R.B. (1996) Land use and avian species diversity along an urban gradient, *Ecological Applications*, 6(2): 506–19.

Blair, R.B. (2001) Birds and butterflies along urban gradients in two ecoregions of the United States: Is urbanization creating a homogeneous fauna? in J.L. Lockwood and M.L. McKinney (eds.), *Biotic Homogenization: The Loss of Diversity through Invasion and Extinction*, New York, NY: Kluwer, pp. 33–56.

Blair, R.B. and A.E. Launer (1997) Butterfly diversity and human land use: Species assemblages along an urban gradient, *Biological Conservation*, 80(1): 113–25.

Bock, C.E., Jones, Z.F., and Bock, J.H. (2008) The oasis effect: Response of birds to exurban development in a southwestern savanna, *Ecological Applications*, 18(5): 1093–106.

Bourne, L.S. and Simmons, J.W. (1982) Defining the area of interest: Definition of the city, metropolitan areas and extended urban regions, in L.S. Bourne (ed.), *Internal Structure of the City*, New York, NY: Oxford University Press, pp. 57–72.

Brazel, A., Selovar, N., Vose, R., and Heisler, G. (2000) The tale of two climates—Baltimore and Phoenix urban LTER sites, *Climate Research*, 15: 123–35.

Cicero, C. (1989) Avian community structure in a large urban park: Controls of local richness and diversity, *Landscape and Urban Planning*, 17(3): 221–40.

Collins, J.P., Kinzig, A., Grimm, N.B., Fagan, W.F., Hope, D., Wu, J., and Borer, E.T. (2000) A new urban ecology, *American Scientist*, 88(5): 416–25.

Cook, W.M. and Faeth, S.H. (2006) Irrigation and land use drive ground arthropod community patterns in an urban desert, *Environmental Entomology*, 35(6): 1532–40.

Dorney, J.R., Guntenspergen, G.R., Keough, J.R., and Stearns, F. (1984) Composition and structure of an urban woody plant community, *Urban Ecology*, 8(1–2): 69–90.

Douglas, I. (1983) *The Urban Environment*, Baltimore, MD: Edward Arnold.

Emlen, J.T. (1974) An urban bird community in Tucson, Arizona: Derivation, structure, regulation, *Condor*, 76(2): 184–97.

Ewing, R.H. (1994) Characteristics, causes and effects of sprawl: A literature review, *Environmental and Urban Studies*, 21(2): 1–15.

Folke, C., Holling, C.S., and Perrings, C. (1996) Biological diversity, ecosystems and the human scale, *Ecological Applications*, 6(4): 1018–124.

Gehrt, S.D. and Chelsvig, J.E. (2003) Bat activity in an urban landscape: Patterns at the landscape and microhabitat scale, *Ecological Applications*, 13(4): 939–50.

Germaine, S.S. and Wakeling, B.F. (2001) Lizard species distributors and habitat occupation along an urban gradient in Tucson, Arizona, USA, *Biological Conservation*. 97(2): 229–37.

Gilbert, O.L. (1989) *The Ecology of Urban Habitats*, New York, NY: Chapman & Hall.

Grimm, N.B., Grove, J.M., Pickett, S.T.A., and Redman, C.L. (2000) Integrated approaches to long-term studies of urban ecological systems, *BioScience*, 50(7): 571–84.

Grimm, N.B., Faeth, S.H., Golubiewski, N.E., Redman, C.L., Wu, J., Bai, X., and Briggs, J.M. (2008) Global change and the ecology of cities, *Science*, 319(5864): 756–60.

Grove, J.M., Troy, A.R., O'Neil-Dunne, J.P.M., Burch, W.R., Cadenasso, M.L., and Pickett, S.T.A. (2006) Characterization of households and its implications for the vegetation of urban ecosystems, *Ecosystems*, 9(4): 578–97.

Harris, C.D. (1956) The pressure of residential-industrial land use, in W.L. Thomas, C.O. Sauer, M. Bates, and L. Mumford (eds.), *Man's Role in Changing the Face of the Earth*, Chicago, IL: University of Chicago Press, pp. 881–95.

Hope, D., Gries, C., Zhu, W., Fagan, W.F., Redman, C.L., Grimm, N.B., Nelson, A.L., Martin, C., and Kinzig, A. (2003) Socioeconomics drive urban plant diversity, *Proceedings of the National Academy of Sciences of the United States of America* 100(15): 8788–92.

Ishitani, M., Kotze, D.J., and Niemelä, J. (2003) Changes in carabid beetle assemblages across an urban-rural gradient in Japan, *Ecography*, 26(4): 481–9.

Kinzig, A.P., Warren, P.S C., Martin, C.D., Hope, D., and Katti, M. (2005) The effects of human socioeconomic status and cultural characteristics on urban patterns of biodiversity, *Ecology and Society*, 10(1): Article 23, online, available at: www.ecologyandsociety.org/vol.10/iss1/art23/.

Kuttler, W. (2008) The urban climate – basic and applied aspects, in J.M. Marzluff, E. Shulenberger, W. Endlicher, M. Alberti, G. Bradley, C. Ryan, C. ZumBrunnen, and U. Simon (eds.), *Urban Ecology: An International Perspective on the Interaction Between Humans and Nature*, New York, NY: Springer, pp. 233–48.

Landsberg, H.E. (1981) *The Urban Climate*, New York, NY: Academic Press.

Liu, H. and Koptur, S. (2003) Breeding system and pollination of a narrowly endemic herb of the Lower Florida Keys: impacts of the urban-wildland interface, *American Journal of Botany*, 90(8): 1180–7.

Lockwood, J.L. and McKinney, M.L. (eds.) (2001) *Biotic Homogenization*, New York, NY: Springer.

Loucks, O.L. (1994) Sustainability in urban ecosystems: Beyond an object of study, in R.H. Platt, R.A. Rowntree, and P.C. Muick (eds.), *The Ecological City: Preserving and Restoring Urban Diversity*, Amherst, MA: University of Massachusetts Press, pp. 49–65.

Luck, M. and Wu, J. (2002) A gradient analysis of urban landscape pattern: a case study from the Phoenix metropolitan region, Arizona, USA, *Landscape Ecology*, 17(4): 327–39.

Luck, M.A., Jenerette, G.D., Wu, J., and Grimm, N.B. (2001) The urban funnel model and the spatially heterogeneous ecological footprint, *Ecosystems*, 4(8): 782–96.

Martin, C., Warren, P.S., and Kinzig, A.P. (2004) Socioeconomic characteristics are useful predictors of landscape vegetation in small urban parks and surrounding neighbourhoods, *Landscape and Urban Planning*, 69(4): 355–68.

Marzluff, J.M., Bowman, R., and Donnelly, R. (eds.) (2001) *Avian Ecology and Conservation in an Urbanizing World*, Norwell, MA: Kluwer.

McDonnell, M.J. and Pickett, S.T.A. (1990) Ecosystem structure and function along urban-rural gradients: an unexploited opportunity for ecology, *Ecology*, 71(4): 1231–7.

McGranahan, G. and Satterthwaite D. (2003) Urban centers: An assessment of sustainability, *Annual Review of Environment and Resources*, 28: 243–74.

McIntyre, N.E. (2000) The ecology of urban arthropods: a review and a call to action, *Annals of the Entomological Society of America*, 93(4): 825–35.

McIntyre, N.E. and Hostetler, M.E. (2001) Effects of urban land use on pollinator (Hymenoptera: Apoidea) communities in a desert metropolis, *Basic and Applied Ecology*, 2(3): 209–17.

McIntyre, N.E., Knowles-Yanez, K., and Hope, D. (2000) Urban ecology as an interdisciplinary field: differences in the use of "urban" between the social and natural sciences, *Urban Ecosystems*, 4(1): 5–24.

McIntyre, N.E., Rango, J., Fagan, W.F., and Faeth, S.H. (2001) Ground arthropod community structure in a heterogeneous urban environment, *Landscape and Urban Planning*, 52(4): 257–74.

McKinney, M.L. (2006) Urbanization as a major cause of biotic homogenization, *Biological Conservation*, 127(3): 247–60.

Musacchio, L.R. and Wu, J. (2004) Collaborative landscape-scale ecological research: Emerging trends in urban and regional ecology, *Urban Ecosystems*, 7(3): 175–8.

Nassauer, J. (2007) *Keynote address, World Congress for Landscape Ecology*, Wageningen, The Netherlands, July 9, 2007.

Neil, K. and Wu, J. (2006) Effects of urbanization on plant flowering phenology: A review, *Urban Ecosystems*, 9(3): 243–57.

Niemelä, J., Kotze, J., Venn, S., Penev, L., Stoyanov, I., Spence, J., Hartley, D., and Montes de Oca, H. (2002) Carabid beetle assemblages (Coleoptera, Carabidae) across urban-rural gradients: an international comparison, *Landscape Ecology*, 17(5): 387–401.

Nilon, C.H. and Pais, R.C. (1997) Terrestrial vertebrates in urban ecosystems: Developing hypotheses for the Gwynns Falls Watershed in Baltimore, Maryland, *Urban Ecosystems*, 1(4): 247–57.

Nilon, C.H. and VanDruff, L.W. (1987) Analysis of small mammal community data and applications to management of urban greenscapes, *Proceedings of the National Symposium on Urban Wildlife*, 2, pp. 53–9.

Odum, E.P. (1997) *Ecology: A Bridge between Science and Society*, Sunderland, MA: Sinauer.

Pacione, M. (2003) Urban environmental quality and human wellbeing—a social geographical perspective, *Landscape and Urban Planning*, 65(1–2): 19–30.

Parris, K.M. (2006) Urban amphibian assemblages as metacommunities, *Journal of Animal Ecology*, 75(3): 757–64.

Parris, K.M. and Hazell, D.L. (2005) Biotic effects of climate change in urban environments: The case of the grey-headed flying-fox (*Pteropus poliocephalus*) in Melbourne, Australia, *Biological Conservation*, 124(2): 267–76.

Paul, M.J. and Meyer, J.L. (2001) Streams in the urban landscape, *Annual Review of Ecology and Systematics*, 32: 333–65.

Pickett, S.T.A., Cadenasso, M., and Grove, J.M. (2005) Biocomplexity in coupled human-natural systems: A multi-dimensional framework, *Ecosystems*, 8(3): 1–8.

Pickett, S.T.A., Burch, W.R., Dalton, S., Foresman, T., Grove, J.M., and Rowntree, R. (1997) A conceptual framework for the study of human ecosystems in urban areas, *Urban Ecosystems*, 1(4): 185–99.

Pickett, S.T.A., Cadenasso, M.L., Grove, J.M., Nilon, C.H., Pouyat, R.V., Zipperer, W.C., and Costanza, R. (2001) Urban ecological systems: Linking terrestrial ecological, physical, and socioeconomic components of metropolitan areas, *Annual Review of Ecology and Systematics*, 32: 127–57.

Pillsbury, F.C. and Miller, J.R. (2008) Habitat and landscape characteristics underlying anuran community structure along an urban-rural gradient, *Ecological Applications*, 18(5): 1107–18.

Pouyat, R.V. and McDonnell, M.J. (1991) Heavy metal accumulations in forest soils along an urban-rural gradient in Southeastern New York, USA, *Water, Air and Soil Pollution*, 57–58(1): 797–807.

Pouyat, R.V., McDonnell, M.J., and Pickett, S.T.A. (1995) Soil characteristics of oak stands along an urban-rural land use gradient, *Journal of Environmental Quality*, 24(3): 516–26.

Pouyat, R.V., McDonnell, M.J., and Pickett, S.T.A. (1997) Litter decomposition and nitrogen mineralization in oak stands along an urban-rural land use gradient, *Urban Ecosystems*, 1(2): 117–31.

Pouyat, R.V., Parmelee, R.W., and Carreiro M.M. (1994) Environmental effects of forest soil-invertebrate and fungal densities in oak stands along an urban-rural land use gradient, *Pedobiologia*, 38: 385–99.

Reale, J.A. and Blair, R.B. (2005) Nesting success and life-history attributes of bird communities along an urbanization gradient, *Urban Habitats*, 3: 1–24.

Rees, W. and Wackernagel, M. (1996), Urban ecological footprints: Why cities cannot be sustainable and why they are a key to sustainability, *Environmental Impact Assessment Review*, 16: 223–48.

Shochat, E., Stefanov, W.L., Whitehouse, M.E.A., and Faeth, S.H. (2004) Urbanization and spider diversity: Influences of human modification of habitat structure and productivity, *Ecological Applications*, 14(1): 268–80.

Shochat, E., Warren, P.S., Faeth, S.H., McIntyre, N.E., and Hope, D. (2006) From patterns to emerging processes in mechanistic urban ecology, *Trends in Ecology and Evolution*, 21(4): 186–91.

Wolter, C. (2008) Towards a mechanistic understanding of urbanization's impacts on fish, in J.M. Marzluff, E. Shulenberger, W. Endlicher, M. Alberti, G. Bradley, C. Ryan, C. ZumBrunnen, and U. Simon (eds.)

Urban Ecology: An International Perspective on the Interaction Between Humans and Nature, New York, NY: Springer, pp. 425–36.

Wu, J. (2008) Making the case for landscape ecology: An effective approach to urban sustainability, *Landscape Journal*, 27(1): 41–50.

Zipperer, W.C., Foresman, T.W., Sisinni, S.M., and Pouyat, R.V. (1997) Urban tree cover: An ecological perspective, *Urban Ecosystems*, 1(4): 229–46.

2

The analysis of cities as ecosystems

Ian Douglas

Cities as systems

The city as a system has been a central metaphor for urban management since the 1950s (Marcotullio *et al.* 2004). The systems approach is essentially a formalized method of determining the role of components within the overall operation of a system (Exline *et al.* 1982). A system can be defined as 'a structured set of objects (components), or a structured set of attributes, or a structured set of objects and attributes combined together' (Dury 1981: 4). The components of a system are physical objects, the attributes non-physical but rationally definable characteristics. In cities we are not dealing with isolated systems cut off from external influences, but with open systems involving the transfer of both energy and matter with their surroundings. All biological systems have this characteristic. They require energy to operate and nutrients in order to exist. They have outputs of waste products and materials that are carried away (Douglas 1983) or which accumulate with, or at the edge of, cities in dumps or landfills that sometimes create problems for future generations.

Cities may be seen as various types of systems. As economic systems, on the one hand, cities can be interpreted in terms of flows of money, goods, services and materials. As ecosystems, on the other hand, cities can be viewed in terms of flows of energy, water and chemical elements, or alternatively as a habitat for organisms, including human beings (Douglas 1983). The urban systems approach to urban planning focuses on the articulation of various components of a city and the flows and processes between them (Marcotullio *et al.* 2004). The systems approach leads to more dynamic, adaptive thinking about the future of cities and to more awareness of possible consequences, especially for the environment and human and ecosystem health, of planning decisions and the character of specific urban developments.

Ecological studies of terrestrial urban systems have involved a variety ways of examining cities in contrasting ways, some examining ecology within cities, others the overall ecology of cities. They may take land use planning versus biological perspectives, or disciplinary versus interdisciplinary, or biogeochemical compared to organismal. Increasingly the characteristic spatial diversity and patchwork nature of land cover in cities has been seen as particularly important in the study of urban systems. This characteristic heterogeneity of the urban surface and diversity of soils and substrates, including both natural, usually displaced, materials and people-made substances, including wastes, creates a host of habitats and ecological niches which contribute to the high level of urban biodiversity.

However, since about 1990, there has been a much more conscious effort to understand the true complexity of urban systems through the integration of the organic, biogeochemical and energetic approaches to the city, emphasizing throughout the need for understanding the social dimensions of urban ecology and integrating humans into the study of urban ecosystems (Grimm *et al.* 2000). The work of the Long Term Ecological Research (LTER) projects in Baltimore and Phoenix in the USA (Redman 1999; Redman *et al.* 2004) has used the watershed approach to analyse urban systems, because it embodies both the integrated dynamics of the overall topographic unit, and encompasses the heterogeneity, both natural and resulting from human action, within it. Like any watershed or river basin that has been modified by human activity, an urban area is part of the coupled human and natural system (CHANS) (Liu *et al.* 2007). In CHANS, people and nature interact reciprocally across diverse organizational levels. They form complex webs of interaction that are embedded in each other. Thus by 2010 we have full recognition of the global environmental significance of urban ecology in its widest sense. Urban ecosystems can be defined at different scales, but they depend upon and impact upon the remainder of the world. From the litter in the world's oceans from cruise ships carrying urbanites on holiday and tankers bringing oil and gas to heat and power buildings and movement in cities, to the atmospheric pollutant fallout embedded in the world's icecaps, the evidence of the effects of urban life is scattered around the globe.

Cities as economic systems

As economic systems, cities can be interpreted in terms of flows of money, goods and services that are regulated by national laws and international agreements, but which are also often greatly influenced by the ideas and desires of entrepreneurs and investment institutions, including the insurance companies that hold huge financial resources in terms of pension investments. Trade has been the essence of great cities for centuries. Ancient Rome required resources, such as olive oil, from North Africa to survive (Lane Fox 2006). In the Middle Ages, goods travelled great distances to places like Venice, Venetian merchants sending wax, pepper, sandalwood and ginger to the city from Syria, the Indies, Malabar and Timor from 1300 onwards (Ackroyd 2009). This exchange led to biological changes. The Roman introduced the domestic cat to Gaul (now France) and took cherry and walnut trees, celery and carrots to Britain (Lane Fox 2006). Along with these deliberate movements of plants and livestock came the pests and invasive species, the inadvertent introductions that still accompany world trade. Economic activity leads to ecological change in many ways, both obvious and less readily apparent. Thus the workings of cities as economic systems are highly relevant to their operations as ecosystems.

Cities in North America and Europe changed substantially after 1960 when the old heavy industries, such as iron and steel making, motor vehicle manufacture and ship building, declined. While the most successful became post-industrial, with an ever-increasing service sector economy, superimposed upon a diminished base of manufacture, shipping and skilled trades (Savitch and Kantor 2002). Less successful cities shrank: Detroit's population fell from 1,670,144 in 1960 to 951,270 in 2000. Many cities affected this way managed to secure some post-industrial activity (small downtowns, tourism, stadiums and exhibition centres). Regeneration of declining industrial areas has become a whole employment area of its own. Meanwhile the new-age boomtowns thrived on a combination of office employment, services, electronics, and light industry all located close to universities, research centres, and low-density development, as in California's 'Silicon Valley' and around Cambridge, England.

This change in the relative status and social structure of cities has taken them all into a competitive scramble to secure their economic well-being. The decline of old industries and the emergence of new investment patterns, often including subsidies and subventions from government or transnational bodies, such as the European Union, draws citizens and politicians in to a search for a niche for their communities in the new economic order. In the process, cities may be engulfed in internal conflicts over means and ends, a belief that if a community does not grow it will surely die, and a rush to develop and grow faster (Savitch and Kantor 2002).

Meanwhile, in other parts of the world, huge urban expansion, with large scale rural to urban migration has happened in countries such as China, India and Brazil. In many cases this has been accompanied by increases in pollution and environmental degradation, but China too is seeing its own set of 'rust-belt' cities emerge as coal mines become exhausted and as old heavy industries are closed down in favour of modern, more efficient plants elsewhere. This constant urban dynamic reflects the changing relative importance of consumption, commerce and production in each particular urban centre.

As commercial systems, cities reflect their classic origins as centres of trade and exchange. Land is a key commodity in such cities, that closest to the main centre of activity being the most valuable. Cities develop patterns of land use, and thus building densities, rates of energy use, pollutant emissions and open space that reflect their commercial functions. Commonly a central business district with banking, insurance and main offices of corporations adjacent to major retails shopping areas is found at the city centre and has high land values. Away from this central area, land values tend to decrease. In these terms, the city can be divided into a set of zones and sectors, essentially based on land values or rent, rentals gradually decreasing away from the highest rent areas. However, government intervention, such as the siting of social housing or the allocation of land for industry or utilities, can so alter any trend towards concentric or zonal growth as to make these theories irrelevant in many cases. Provision of parks and open spaces, such as public gardens and squares, goes against the land value and rent trend, but could be said to be an example of the value that society places on urban greenspace.

Commercial and industrial structural changes have led to a reversal of the traditional pattern of higher land values towards the city centre, with much inner city land becoming vacant and rental values declining. This led to the so-called 'doughnut city' phenomenon, in which the city has a ring of high value activities around a peripheral freeway and a relatively empty core. The vacant greenspaces provide temporary habitats for invasive flora and fauna, but become less and less attractive as places to live and work unless initiatives and planning decisions are taken to counteract the trend. In Melbourne, Australia, in 1998 the metropolitan planning agency published a report *From Doughnut City to Café Society* (Victorian Department of Infrastructure 1998) that documented the trend of younger people to leave the monotonous suburbs to live close to the city centre and enjoy great diversity and entertainment that it offers (Sandercock 2003). Similar events have occurred elsewhere, especially in major European cities such as Manchester and Leeds (Douglas 2007). However, the 'doughnut' effect persists in many others, for example Detroit and Syracuse, USA (Orfield 1996).

In addition to being a commercial system, the city is also a production system where materials and labour are combined to produce goods which earn wealth. As a production system, the city puts pressure on the surrounding countryside. Demands for food and recreational space will come from city residents who are essentially the labour force in the production system, while demands for raw materials and space for industrial production come indirectly from consumers in the city and in markets further afield via the firms within the city (Hodder and Lee 1974).

The production function of the city involves the concentration in the space occupied by the city of both raw and processed materials gained from outside that space. Equally the processing of materials and the consumption of goods involves the disposal of wastes and by-products from the city. Many of those wastes are dumped in landfills and land raise, or mine and industrial spoil heaps which have become legacies from the past and will continue to pose land management problems for the future, despite the much greater efforts to reduce, reuse and recycle industrial materials in modern society. These waste residues in turn affect the character of parcels of land in the city, being both temporary wildlife habitats and sites for land restoration in the future, providing always that the communities involved have the funds to carry out the necessary remediation works on brownfield and superfund sites.

In terms of consumption, cities in North America have seen their dominant commercial cores challenged by the spread of superstores and out-of-town shopping centres and the freedom of office location made possible by information technology and revolutions in telecommunications. These have disrupted the workplace-residence links once expressed by rail and tramway routes radiating from city centres. Concentric rings of freeways now link suburbs and commercial and industrial complexes permitting a complex web of journeys throughout the built-up area. Elements of this pattern are now found in Europe and have emerged in many Asian cities, including Beijing and Shanghai. Significant portions of urban greenspace are now associated with this landscape of consumerism, including tree-planting along freeways and ring roads, patches of vegetation in business parks, and, where planning authorities are alert to the value of urban vegetation, in shopping complexes.

Thus in terms of these economic functions, the city is constantly introducing new organic and inorganic components of the landscape altering, modifying and creating new habitats, new urban land surface biogeochemistries and living spaces for all kinds of organisms. This leads to greater heterogeneity in the urban space, greater opportunities for specialized niche habitats and for unusual or novel combinations of plants and animals (see Chapter 17 below).

Cities as urban ecosystems

The multi-faceted notion of an ecosystem can be applied in many contexts. The key idea is the functional linking of organisms and a specific physical environment. Ecosystems can be described at many scales; both a rotting log on a forest floor or the entire Amazon rain forest can be delimited as ecosystems (Pickett *et al.* 1997).

The complex interactions between urban patterns, derived mainly from the city's economic and cultural functions, and ecological processes occur across multiple scales. Understanding of the ways species populations and community characteristics change in response to urban development requires knowledge about drivers and effects of ecosystem structure and functions in urban landscape.

With the continuing growth of urban population around the world, the nature of urban ecosystems plays a larger and larger role in shaping people's views about natural ecosystems. From an ecological perspective, urban ecosystems are highly dynamic and thus able to give valuable insights on how to manage biodiversity in other ecosystems. Biodiversity concerns related to urban ecosystems can be divided into three major groups: (1) those related to the impact of the city itself on adjacent ecosystems; (2) those dealing with how to maximize biodiversity within the urban ecosystem; and (3) those related to the management of undesirable species within the ecosystem (Savard *et al.* 2000). Although species abundance and diversity are often related to the quality of urban life, excessive numbers of some species may become intolerable (Clergeau *et al.* 1996).

Urban ecosystems may be viewed at many scales. Four of the more generally used are:

- In terms of patches within the complex urban mosaic of habitats, for example garden lawns, urban stream channels, or fragments of contaminated derelict land. Examples of these are described in detail in Chapters 20 to 28. However, patches vary greatly in size from flower pots and window boxes to green roofs and to the great urban parks, such as Central Park in New York and the Bois de Boulogne in Paris.
- The built-up areas are the habitat of urban people, their pets, their garden plants, the adapted animals and organisms (birds, moulds, etc.) and the pests (rats, weeds, parasites, etc.). These areas depend on outside (external) support through inputs of energy, water and materials inputs for their survival.
- The immediate urban life-support system of the urban area and its surroundings (the peri-urban area), providing such ecological services as water supplies, sources of aggregates, areas for landfill, recreation zones, watershed protection, greenhouse gas uptake and biodiversity.
- The areas affected by urbanization as a driving force by providing life support services to urban areas, including supplies of food, energy, water and materials and also those areas affected by the emissions and waste flow from urban areas. For any individual city these may have a global outreach, with energy (coal, natural gas or oil) and food (exotic fruits, fish, meat, grain, soya, etc.) drawn from distant countries or seas. As urban populations and purchasing power grow, this outreach and the impacts on other ecosystems expand.

Each of these scales of urban ecosystem can be described quantitatively by models that typically specify the amounts living tissue present, dead organic matter stored in the soil and above ground, nutrients in the soil and biomass, and toxins in organisms and the environment are present. Such models, when applied to urban areas, begin to look at both natural, but modified, flows and deliberate human flows of energy, water and materials occur for urban areas as a whole, or parts of urban complexes.

Some of the earliest ecosystem modelling was based on materials flows, developing a mass-balance of the city in which all raw material inputs end up as goods and wastes. Wastes may be disaggregated into process wastes and residual wastes from production and consumption. The UNESCO/MAB programme to consider cities as ecological systems prompted an outstanding project in Hong Kong (Newcombe 1975; Boyden *et al.* 1981). The work involved establishing the input-output characteristics, distribution and use of each item to be examined. Energy, for example, was studied in terms of its use in various sectors of the economy, socio-economic patterns of consumption, and spatial and temporal variations in use. Computer modelling was used to determine the interaction, dispersal and concentration of materials in the atmosphere and marine environments with respect to pollution problems, taking as a base the information gathered in the patterns of flow and distribution of relevant materials.

The experiences that came from this work, embracing over 100 studies in all regions of the world, covering a wide range of bioclimatic, biogeographic, social, cultural, economic political and development situations, helped to establish the bases for an ecological paradigm of urban/peri-urban/industrial systems (Celicia 1996). The assessments of flows of energy and materials were complemented by studies of urban nature in cities like Berlin, Rome and Vienna. People's responses to the urban environment were examined in detail in psychosocial studies in Rome (e.g. Bonaiuto 1999).

The urban metabolism approach was popularized by William Rees' development of the ecological footprint concept (Wackernagel and Rees 1996) which showed, for example, that London's total footprint extends to 125-times its surface areas of 159,000 ha, or approximately the whole productive area of Britain (Girardet 1996). The details of the footprint concept may be criticized, but it powerfully shows how most large cities are dependent on ecosystems elsewhere. Any country or city importing food can be said to be consuming the 'embodied' water involved in the growth and preparation of the food, as well as all the energy consumed in its cultivation, transportation and delivery. These ideas demonstrate how cities are critically dependent on CHANS elsewhere and are thus vulnerable to natural and human factors affecting the stability of those ecosystems and the production systems they support.

An important dimension of this approach to urban ecosystems has been the analysis of urban energy flows. Energy can be used as a common numerator to evaluate the work of nature, either with or without including humans. It provides a sophisticated and elegant way of analysing interacting ecological and economic systems. This type of analysis views urbanization in terms of ecological energetics, following Odum and Peterson (1972) who related the complexity of cities to ecological principles and energy flows. To assess the services nature provides to urban people, financial cost accounting should be complemented by ecological energetic analysis (Odum 1971).

This form of analysis has been extended to the relationship between spatial organization and energy hierarchy of city-regions and its implications for urban planning (Huang 1998a, 1998b; Odum and Odum 2001). Progress towards an integrated theory of the interdependence of urban development and energetic flow has now begun to be made (Huang and Chen 2005). Analogies can be made between the stages of urban development and the development stages of ecosystems. In the early stages of development, urban systems exhibit early successional characteristics of rapid growth and inefficient use of resources. As they mature, urban systems generate more structure; higher empower density, greater entropy production and larger information flows (see Chapters 3 and 41). Analysis of the metabolism of Taipei, Taiwan, has clearly demonstrated this trend (Huang and Chen 2005).

A somewhat different approach, but also aimed at modelling urban ecosystem dynamics in a manner that is useful to planners has been developed in Washington State, USA. Most operational urban models are still somewhat crude in the way they represent ecological processes. However, the available environmental models used to evaluate the ecological impact of an urban region generally do not represent human systems well. In a multidisciplinary project at the University of Washington examining the environmental and human systems in the Puget Sound an urban ecological model (UEM) addresses the human dimension of the Puget Sound regional integrated simulation model (PRISM). UEM simulates environmental pressures arising from human activities under alternative demographic, economic, policy and environmental scenarios. Specifically UEM sets out to: quantify the major sources of human-induced environmental stresses (such as land-cover changes and nutrient discharges); determine the spatial and temporal variability of human stressors in relation to changes in the biophysical structure; relate the biophysical impacts of these stressors to the variability and spatial heterogeneity in land uses, human activities, and management practices; and predict the changes in stressors in relation to changes in human factors (Alberti 2008).

Cities as integrated biosocial systems

A newer set of biosocial approaches to urban ecosystems has evolved since 1970 and has been elaborated since 1995. This work sees the human ecosystem as a coherent system of biophysical and social factors that can achieve adaptation and sustainability over time and argues that human ecosystems could be a starting point for the management of urban and other ecosystems (Pickett *et al.* 1997). This human ecosystem framework incorporates three types of critical resources:

natural resources (such as energy, water and materials), socio-economic resources (including labour and capital) and cultural resources (for example, myths and beliefs). Rather than being a 'self-regulating' system, the flows of the resources within the human ecosystem are regulated by an unpredictable social system. Those working on the ground to protect the urban 'green' and 'wild' and thereby to contribute to making cities more sustainable and pleasant to live in, know that both regulatory and non-regulatory policies evolve over time in response to 'environmental champions' leveraging changes within institutions. In places such as Portland, Oregon, USA, such action has had a massive impact on the urban ecosystem. Ultimately, the social interactions matter most. All that models can do is to predict 'what if' scenarios, such as what happens if planning regulations are tightened or the price of oil reaches a certain level. However, to make such predictions valuable, sound knowledge of all the components of the human ecosystem is required.

The social science understanding of social structure and the social allocation of natural and institutional resources can be readily incorporated into ecosystem models of large urban areas. The deep awareness of the spatial dimensions of social differentiation has similarities to concepts and data on patch dynamics in ecology. These aspects of our knowledge of cities can be brought together in the human ecosystem framework to advance our understanding of urban ecosystems. The framework recognizes the openness of ecosystems, and does not assume equilibrium or self-regulation of urban ecosystems. It recognizes spatial heterogeneity in both the natural and social components of urban ecosystems. This heterogeneity can be dynamic, with different components changing at different rates, such the onset of rapid change in urban wetlands during a flood and slower change when land is cleared for housing development upstream. The social and natural heterogeneity may also be visualized hierarchically. As mentioned above, this can be tackled by using a watershed approach that provides a special and functional framework for assessing the interaction of the social and ecological components of the integrated system.

An urban ecosystem is therefore a separate kind of biosocial system that shares certain theoretical similarities with other types of human ecological systems and CHANS, but also exhibits specific, unique properties. For example, although we may use universal ecosystem theories to study a tropical rain forest or taiga forest system, our interest is to understand how these ecosystems are unique variations of common ecological themes. Similarly, this same integrated approach to urban ecosystems avoids the notion of cities as the zone of greatest 'human impact' espoused by many environmental scientists on the one hand and the 'human centred approach' that many social scientists favour on the other. The urban ecosystem has to be studied in its own right, not as an aberration or an evolutionary end point of nature (Grove and Burch 1997).

Further thought on such a system leads to the conclusion that all urban plans and designs, whether intentionally ecological or not, can be evaluated for their contribution to, or deduction from, ecological functioning in urban areas. The clarion call from Ian McHarg in the 1970s to 'design with nature' (McHarg 1971) has been recognized in terms of the need to quantify all the environmental services, and variables affected by any given ecological design should be quantified (Pickett and Cadenasso 2008). There is an urgent need to understand ecologically the individual and incremental impacts of urban planning. Such a move would be an important contribution to improvement in the quality of urban life. It could lead to a reduction in the urban sprawl that consumes so much land and so many habitats for wildlife. Engaging built environment professionals in the adaptive ecological design cycle would benefit not only residents and so the global environment, but also ecological and built environment professionals and scientists alike.

Further reading

Alberti, M. (2008) *Advances in Urban Ecology: Integrating Humans and Ecological Processes in Urban Ecosystems*, New York, NY: Springer.

Douglas, I. (1983) *The Urban Environment*, London: Arnold.

Odum, H.T. (1971) *Environment, Power, and Society*, New York, NY: Wiley.

Odum, H.T. and Peterson, L.L. (1972) 'Relationship of energy and complexity in planning', *Architecture and Design*, 43: 624–9.

Pickett, S.T.A and Cadenasso, M. L. (2008) 'Linking ecological and built components of urban mosaics: an open cycle of ecological design', *Journal of Ecology*, 96: 8–12.

Pickett, S.T.A., Burch, W.R. Jr., Dalton, S.E., Foresman, T.W., Grove, J.M. and Rowntree, R. (1997) 'A conceptual framework for the study of human ecosystems in urban areas', *Urban Ecosystems*, 1(4): 185–99.

References

Ackroyd, P. (2009) *Venice: Pure City*, London: Chatto & Windus.

Alberti, M. (2008) 'Modeling the Urban Ecosystem: A Conceptual Framework', in J.M. Marzluff, E. Shulenberger, W. Endlicher, M. Alberti, G. Bradley, C. Ryan, U. Simon, and C. ZumBrunnen (eds), *Urban Ecology: An International Perspective on the Interaction Between Humans and Nature*, New York, NY: Springer, pp. 623–46.

Bonaiuto, M., Aiello, A., Perugini, M., Bonnes, M. and Ercolani, A.P. (1999) 'Multi-dimensional perception of residential environment quality and neighbourhood attachment in the urban environment', *Journal of Environmental Psychology*, 19(4): 331–52.

Boyden, S., Millar, S., Newcombe, K. and O'Neill, B. (1981) *The Ecology of a City and its People: The Case of Hong Kong*, Canberra: Australian National University Press.

Celecia, J. (1996) 'Towards an urban ecology', *Nature and Resources*, 32(2): 3–6.

Clergeau, P., Esterlingot, D., Chaperon, J. and Lerat, C. (1996) 'Difficultés de cohabitation entre l'homme et l'animal: le cas des concentrations d'oiseaux en site urbain', *Natures-Sciences-Sociétés*, 4(2): 102–15.

Douglas, I. (1983) *The Urban Environment*, London: Arnold.

Douglas, I. (2007) 'Reconcentration', in I. Douglas, R. Huggett and C.R. Perkins (eds), *Companion Encyclopaedia of Geography: From Local to Global*, London: Routledge, pp. 483–95.

Dury, G. (1981) *An Introduction to Environmental Systems*, London: Heinemann.

Exline, C.H., Peters. G.L. and Larkin, R.P. (1982) *The City, Patterns and Processes in the Urban Ecosystem*, Boulder, CO: Westview Press.

Girardet, H. (1996) 'The metabolism of cities', *Nature and Resources*, 32(2): 6–7.

Grimm, N.B., Redman, C.L., Grove, J.M. and Pickett, S.T.A. (2000) 'Integrated Approaches to Long-Term Studies of Urban Ecological Systems', *BioScience*, 50(7): 571–84.

Grove, J.M. and Burch Jr., W.R. (1997) 'A social ecology approach and applications of urban ecosystem and landscape analyses: a case study of Baltimore, Maryland', *Urban Ecosystems*, 1(4): 259–75.

Hodder, B.W. and Lee, R. (1974) *Economic Geography*, London: Methuen.

Huang, S.-L. (1998a) 'Urban ecosystems, energetic hierarchies, and ecological economics of Taipei metropolis', *Journal of Environmental Management*, 52(1): 39–51.

Huang, S.-L. (1998b) 'Ecological energetics, hierarchy, and urban form: A system modeling approach to the evolution of urban zonation', *Environmental Planning B: Planning and Design*, 25(3): 391–410.

Huang, S.-L. and Chen, C.-W. (2005) 'Theory of urban energetics and mechanisms of urban development', *Ecological Modelling*, 189(1–2): 49–71.

Lane Fox, R. (2006) *The Classical World: An Epic History of Greece and Rome*, London: Penguin Books.

Liu, J., Dietz, T., Carpenter, S.R., Folke, C., Alberti, M., Redman, C.L., Schneider, S.H., Ostrom, E., Pell, A.N., Lubchenco, J., Taylor, W.L., Ouyang, Z., Deadman, P., Kratz, T. and Provencher, W. (2007) 'Coupled Human and Natural Systems'. *Ambio*, 36(8): 639–49.

Marcotullio, P.J., Boyle, G., Ishii, S., Karn, S.K., Suzuki, K., Yusuf, M.A. and Zandaryaa, S. (2004) 'Defining an ecosystem approach to urban management and policy development', *Proceedings of International Ecopolis Forum: Adaptive Ecopolis Development:* Ningbo, China, pp. 28–51.

McHarg, I. (1971) *Design with Nature*, New York: Doubleday & Company.

Newcombe, K. (1975) 'Energy use in Hong Kong, part I: an overview', *Urban Ecology*, 1(1): 88–113.

Odum, H.T. and Odum, E.C. (2001) *A Prosperous Way Down*, Boulder, CO: University Press of Colorado.

Odum, H.T. and Peterson, L.L. (1972) 'Relationship of energy and complexity in planning', *Architecture and Design*, 43: 624–9.

Orfield, M. (1996) *Metropolitics: A Regional Agenda for Community and Stability*, Washington, DC: Brookings Institution Press.

Pickett, S.T.A., Burch Jr., W.R. Dalton, S.E., Foresman, T.W., Grove, J.M. and Rowntree, R. (1997) 'A conceptual framework for the study of human ecosystems in urban areas', *Urban Ecosystems*, 1(4): 185–99.

Pickett, S.T.A. and Cadenasso, M. L. (2008) 'Linking ecological and built components of urban mosaics: an open cycle of ecological design', *Journal of Ecology*, 96: 8–12.

Redman. C.L. (1999) 'Human dimensions of ecosystem studies', *Ecosystems*, 2(4): 296–8.

Redman. C.L., Grove, J.M. and Kuby, L.H. (2004) 'Integrating Social Science into the Long-Term Ecological Research (LTER) Network: Social Dimensions of Ecological Change and Ecological Dimensions of Social Change', *Ecosystems*, 7(2): 161–71.

Sandercock, L. (2003) 'The sustainable city and the role of the city-building professions', in L.F. Girard, B. Forte, M. Cerreta, P. De Toro and F. Forte (eds), *The Human Sustainable City: Challenges and Perspectives from the Habitat Agenda*, Aldershot: Ashgate, pp. 375–85.

Savard, J.-P.L., Clergeau, P. and Mennechez, G. (2000) 'Biodiversity concepts and urban ecosystems', *Landscape and Urban Planning*, 48(3): 131–42.

Savitch, H.V. and Kantor, P. (2002) *Cities in the International Marketplace: The Political Economy of Urban Development in North America and Western Europe*, Princeton NJ: Princeton University Press.

Victorian Department of Infrastructure (1998) *From Doughnut City to Café Society,* Melbourne: State Government of Victoria.

Wackernagel, M. and Rees, W. (1996) *Our Ecological Footprint: Reducing Human Impact on the Earth* (The New Catalyst Bioregional Series, 9), Gabriola Island, BC: New Society Publishers.

3

Urban ecology and industrial ecology

Xuemei Bai and Heinz Schandl

Introduction

Industrial ecology and urban ecology are distinct disciplines with different professional communities, but they share many common concepts and approaches. This chapter explores the ways industrial and urban systems and the science relating to them interact, through reviewing the history and main areas of research within each discipline, common and distinctive approaches, cross contributions, and emerging frontiers. Material and energy flow analysis, which is a common approach in both industrial and urban ecology, is used as an example to articulate linkages between the disciplines. While they have different starting points and remain as separate disciplines, there is also a merging and coevolving trend in research scope and methodology in urban ecology and industrial ecology, with industrial ecology extending its scope to include literature dealing with social and management dimensions, and an increasing volume of literature addressing production and consumption subsystems within cities appearing in urban ecology.

Industrial and urban ecosystem

There are strong spatial and functional linkages between cities and industries. Many industries are located in or nearby cities, taking advantage of the concentration of labour and infrastructure provided by urban agglomerations. Conversely cities rely on industrial activities, whether manufacturing or services, to maintain their economic vitality and competitiveness. These spatial and functional linkages often result in strong environmental linkages too, with the environmental performance of industries within cities affecting the environmental quality of cities and the livelihoods and well-being of urban dwellers, and cities sometimes making strategic decisions around the development or relocation of particular polluting industries (Bai 2002). The nature and complexity of the relationship between the industrial system and the urban system can vary due to functional differences between cities, their stages of economic development, external factors such as globalization, and the institutional and governance settings within which they operate.

Both urban and industrial ecology have drawn parallels between cities and ecosystems, and industries and ecosystems. While urban ecology has used the ecological metaphor to reflect the complexity that occurs within an urban system, industrial ecology has mainly tried to model industrial processes as an analogy for biological processes. An industrial system located in a city can be considered as a nested subsystem within an urban ecosystem (Ma and Wang 1984), with complex interactions between the city and the subsystem often regulated at a higher level by the urban ecosystem (Kaye *et al.* 2006; Bai 2007). From an ecosystem point of view, both the industrial system and the urban are viewed as open systems that import low entropy energy and raw materials from their surroundings and export high entropy energy, products and waste beyond their system boundaries to sustain their function. These metabolic interactions provide one foundation for the methodological similarities among the two disciplines of industrial ecology and urban ecology.

A brief history of industrial ecology and urban ecology

Industrial ecology aims to model industrial systems to operate like biological systems. In ecological systems, the wastes produced by one species tend to be a resource for another species. By the same token, industrial ecology designs clusters where the output that one industry considers to be waste can be used as a valuable input, i.e. raw material for another industry, thus reducing the use of raw materials, pollution and saving on waste treatment (Frosch and Gallopoulos 1989; Levine 2003).

Industrial ecology is a relatively new discipline. Seminal research such as Ayres and Kneese (1969), the *Limits to Growth* report by the Club of Rome (Meadows *et al.* 1972), as well as Commoner (1971), set the scene for a new interdisciplinary research field linking natural scientists, physicists and engineers with planners and social scientists to look at the interface between industrial and environmental systems, the material and energy flows that enable production and consumption activities and the potential for improving the effectiveness and efficiency of resource use by mimicking ecological processes, such as closing the loop by reuse and recycling.

While precursors existed in the late 1960s and early 1970s, a new research community was formed in the late 1990s under the banner of industrial ecology.

Urban ecology, in contrast, has evolved over a much longer time frame. Some of the earliest research efforts that can be classified as urban ecology were made by the Chicago School of Sociology, which applied ecological concepts to study the patterns and dynamics of cities. For example, Park *et al.* (1925) borrowed the concept of succession from plant ecology for describing land use change in cities. Odum (1975; Odum and Odum 1980) described cities as open systems, which establish a metabolic interaction with their environment. Wolman (1965) conducted the first metabolism analysis for an imaginary city of one million, estimating the total input and output of various energy, materials and waste streams. The first metabolism analyses in real world cities were conducted under the UNESCO Man and the Biosphere (MAB) Program during the 1970s and 1980s (Boyden *et al.* 1981).

Many urban ecologists have focused on studying the natural components of cities, i.e. the flora and fauna within cities (Sukopp and Werner 1982; Sukopp 1998), which is referred to as "ecology in cities". While this remains an important part of urban ecology, especially in the application of urban and landscape planning and design, recent studies have focused increasingly on the "ecology of cities" (Grimm *et al.* 2000; Grimm *et al.* 2008). This more fundamental approach treats cities as distinctive ecological systems and studies their patterns, processes, functions and dynamics. In this research tradition, cities are conceptualized as coupled social-ecological systems which are complex, adaptive, dynamic and constantly changing, with their overall performance and trajectories shaped by both internal and external factors (Ma and Wang 1984; Alberti *et al.* 2003; Bai 2003; Alberti 2008; Ernstson *et al.* 2008).

Current scope of interests

According to Graedel and Allenby (2002), research questions in industrial ecology usually fall under four major domains.

The first research domain looks at industrial systems and their potential for improvements in resource use and emissions intensity when modelled after ecological systems. Research in this area has looked at the potential for linking industrial sectors to allow for synergies and to establish cycles for materials that are used in industrial processes. The aim of this research has been to moderate resource use and environmental impacts through technological improvements and product design.

The second domain has looked at resource use in socio-economic systems at various functional and geographic scales to empirically assess the magnitude of resource use and emissions, in order to improve the management of society-environment interactions through informed environmental management at national, sectoral and corporate levels.

The third research domain has investigated the future of technology-environment relations by employing scenario analysis techniques for the development of future technology and its relationship to the environment, to better understand how changes in environmental systems affect technological systems.

The fourth research domain has looked at how to operationally define and address sustainability, and how to measure it.

Research has been guided by a number of quite influential theoretical concepts or notions including industrial metabolism, industrial symbiosis and life cycle approaches, which have become associated with neighbouring research fields such as ecological economics. Most of the research has had a strong empirical basis and evolved around a family of methodological tools including material and energy flow analysis, substance flow analysis, life cycle assessment, ecological footprint assessment, input-output analysis and design for the environment. Much of the research has focused on industry or policy impact with regard to redesigning industrial processes, industrial design and improving firm based standards, or through informing policy planning and programmes, and evaluating past policies.

Interest in urban research in the industrial ecology research community has grown since 2000, examining regional and global impacts of cities through industrial ecology tools, sustainable production and consumption, and research on urban metabolism and how urban form, density, transportation and design choices affect these flows.

It is difficult to produce a handful of key questions in urban ecology, given its diversity and interdisciplinary nature. However, three main domains of interest can be identified from the scholarly literature:

Ecology in cities

This research domain deals with the natural component of cities, studying ecological structure and the functioning of habitats or organisms within cities. This research domain has the longest history in urban ecology, and remains an important part of urban studies. In addition to traditional vegetation, flora and fauna studies in cities (Zipperer *et al.* 1997; Sukopp 1998), there has been increasing interest in looking at the urban physical environment as affected by human activities (Douglas 1983), e.g. urban heat islands and anthropogenic energy use (Oke 1995), and urban hydrological modification (Paul and Meyer 2001); as well as the ecological functions and services provided by the natural component of cities, e.g. urban green space and its role in mitigating the heat island effect (Wong and Yu 2005), and the role of nature in human health and wellbeing (Tzoulas *et al.* 2007).

Human dimensions in cities

While traditionally ecologists have viewed human influence in an ecosystem as a "disturbance", it is increasingly recognized that the structure and function of the Earth's ecosystems cannot be understood without accounting for the dominant influence of human activities in ecosystems (Vitousek *et al.* 1997; Redman 1999). This is especially valid in an urban setting, where social activities are dominant, and drive ecosystem function and change. Grimm *et al.* (2000) suggest a conceptual framework to address three fundamental drivers underlying human action: flows of information and knowledge; incorporation of culturally based attitudes, values and perceptions; and creation and maintenance of institutions and organizations. Based on this framework, core topics for investigation include: demographic patterns, economic systems, power hierarchies, land use and land management, and environmental design (Redman 1999). Modern studies linking consumption behaviour to social psychology may also fall into this category (Jackson 2005).

Ecology of cities

This third strand of research views cities as coupled social-ecological systems, which include humans living in cities and urbanized landscapes (Alberti 2008). This approach examines entire cities or metropolitan areas from an ecological perspective, including biogeochemical cycles, material and energy flows, ecosystem patterns and processes, which collectively are referred to as the ecology of cities (Grimm *et al.* 2000). This research approach recognizes close linkages and multiple interactions among different components of the system, which in turn generates opportunities for realizing co-benefits, but also creates the risk of unintended consequences and trade-offs (Bai *et al.* forthcoming).

Urban metabolism as linking and/or common approach

One of the most obvious common concepts shared between the industrial ecology and urban ecology communities is the notion of "metabolism". Ayres and Simonis introduced the notion of industrial metabolism in their 1994 book which has since been used to guide research within the industrial ecology research community and has led to applications at various functional and geographic scales, including national economies, economic sectors, businesses, products, regions and cities. The vibrant research activity and the close links which could be established to policy and multilateral institutions such as, for instance, the Organization of Economic Cooperation and Development (OECD) or the United Nations Environment Program (UNEP) enabled standardization and harmonization of methodologies most successfully at a national scale, but with implications for research at other levels such as cities. In 2007, the OECD in collaboration with the European Statistical Office (EUROSTAT) released a methods guidebook for national material flow accounts. This guidebook has set a standard for application of metabolism research and accounting for material and energy flows at urban scales (EUROSTAT 2007).

Fischer-Kowalski (1998) further elaborated the metabolism concept, focusing on the interface between society and nature in terms of the exchange of materials and energy. According to Fischer-Kowalski's seminal overview of the origins of the social metabolism concept, the theory looks back on a long history in a number of scientific disciplines such as sociology, human geography, cultural and ecological anthropology and economics starting around 1860 (Fischer-Kowalski 1998). The metabolism concept describes labour as a process between humans (society) and nature in which society organizes and satisfies its material demands. It is within the labour process, and through the application of technologies, that socio-economic systems organize essential supplies of materials and energy for production and reproduction of human societies (Schandl and Schulz 2002). The metabolism approach views social systems as analogous to ecosystems with regard to their fundamental requirement to organize and maintain a throughput of materials, water and energy which are extracted from natural sources, transformed in economic activities, used to build up stocks of buildings, vehicles and capital and consumer goods and ultimately disposed of into the environment in the form of waste and emissions via different paths (including water, air and land).

The social metabolism approach is based on a broader understanding of structurally coupled social-ecological systems (Boyden 1987; Fischer-Kowalski and Weisz 1999). As Figure 3.1 shows, the built urban environment and people living in cities are at the interface of the urban cultural system and the urban natural environment. Urban stocks are established and maintained by a continuous throughput of materials and energy, which are channelled through socio-technical systems organizing the provision of housing, mobility, food, energy and water.

The way in which the society-nature interface is organized depends on urban culture and institutions, and the way in which the urban-political economy operates. The notion that culture is a system of communication which guides the design and investment decisions that play out in the built urban system and deliver a very specific urban life world is highly important in the approach suggested by Fischer-Kowalski and Weisz (1999) that we draw upon here.

Overall, the industrial ecology strand of metabolism analysis focuses mostly on production systems, and is often applied in the analysis of national scale, major economic sectors.

The application of the metabolism approach to cities also has historical precedents. Wolman was the first to introduce the notion of urban metabolism in his classic 1965 study (Wolman 1965). He calculated energy, water and materials throughput for a hypothetical city of one million inhabitants using national United States production and consumption data. The first application of urban metabolism to an actual city was done for the Belgian capital, Brussels, by Duvigneaud and Denaeyer-De Smet (1977). Their analysis of energy and water flows as well as air pollution found that the city of Brussels was depending on its hinterland for the supply of energy and water and did experience comparatively high rates of pollution. The city of Hong Kong has also been a focus of urban metabolism research because the city boundaries match with the state boundaries and hence economic statistics have been readily available. This has enabled a series of urban metabolism studies including Newcombe (1977) and Warren-Rhodes and Koenig (2001), and especially the work of Boyden *et al.* (1981) who aimed to integrate indicators for relative quality of life in the city with the urban metabolic analysis.

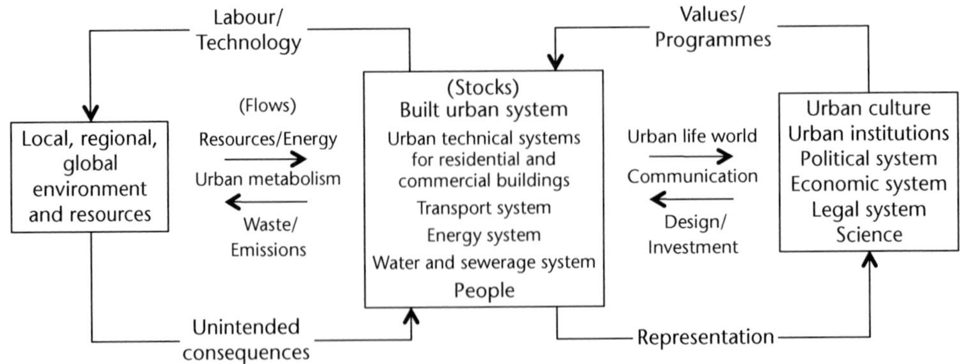

Figure 3.1 The urban social-ecological system.

Since the late 1990s, there has been increased effort in urban metabolism research, resulting in studies for various cities in the United States (Tarr 1996; Grove and Burch 1997; Kennedy 2002; Tarr 2002; Jenerette *et al.* 2006), Europe (Hendriks *et al.* 2000; Pauleit and Duhme 2000; Ravetz 2000; Barles 2009; Bramley *et al.* 2009; Niza *et al.* 2009) and urban metabolism studies for Santiago de Chile (Wackernagel 1998), Sydney (Newman 1999) and Taipei (Huang and Hsu 2003).

There is another body of literature examining the environmental impacts of urban metabolic flows beyond direct material and energy requirements. Ecological footprint analysis starts with resource flow accounting but reinterprets resource use in terms of the land that was required to produce the respective resources in the first place (Wackernagel and Rees 1995). Studies on the urban ecological footprint include Collins *et al.* (2006), Folke *et al.* (1997), Luck *et al.* (2001), White (2001), and Jenerette *et al.* (2006). In general their major findings coincide, indicating that in terms of the primary resources cities need to service their urban metabolism, urban areas show a significant and increasing ecological footprint through their interaction with regional and global resource flows. Examining the carbon footprint of cities, Kennedy *et al.* (2009) conclude that because of small physical territories and related high-density and consumption levels, emissions outside the administered territory, i.e. embedded or up-stream emissions tend to dominate the urban carbon footprint.

All of these research efforts are surrounding a major research topic, namely the regional and global impact of cities, via their dependence on resource flows from elsewhere and their increasing emissions, driven by the globalization of economic activity, the production and consumption of goods and services, and employment. Cities play a major role in global metabolism as centres of consumption. Whilst cities already appear as significant resource users and greenhouse gas emitters on a per-capita basis when direct flows are taken as the basis of accounting, their impacts need to be adjusted upwards when research considers the raw material equivalent of cities' resource consumption. Recently the research methodology for embedded, up-stream flows has been advanced considerably by Munoz *et al.* (2009) who used an input-output economics approach to account for the raw material equivalents of trade in Brazil, Chile, Columbia, Ecuador and Mexico, showing the considerable waste and emission stream that remains in countries with large extractive or industrial activities. Lenzen and Peters (2009) using a similar methodological approach looked at the resource mobilization of urban consumption in Australia from a spatially explicit point of view. While the issue of embedded flows has increasingly become an important research topic in metabolism research most of the methodological development has focused on the scale of the national economy. Urban metabolism research has been complicated by a lack of data for urban system boundaries resulting in a severe limitation for case studies. This has hampered progress in methodological harmonization among urban metabolism studies and between urban and national metabolism studies.

A very important research question is which factors influence a city's metabolism? Five major factors can be indentified.

1 Natural factors, including a city's geographic location and the associated spatial and climatic features, or the natural resource endowment and geomorphology of the city. For example, colder cities tend to have more energy use for heating.
2 Functional role of the city, e.g. the type and intensity of major economic activities. Industrial cities tend to have larger metabolic flows than financial or political centres.
3 Income level also affects the type and amount of urban metabolism. With economic development, cities attract more consumer goods and discharge more greenhouse gases (Bai and Imura 2000; Warren-Rhodes and Koenig 2001; Bai 2003).

4 Urban policy and management practices have the potential to make a difference. Cities have also been identified as important sinks for strategic materials such as copper and iron, and the industrial ecology literature has looked at the potential for urban mining in comparison to consuming natural reserves of these resources (van Beers and Graedel 2007; Halada *et al.* 2009). There is potential for significant recycling and reuse of these materials within cities, which would reduce the overall inflow and outflow of resources. Some recent studies have started to explore spatially explicit representations of urban resource stocks and flows, in order to predict future construction waste flows and their potential use in urban infrastructure renewal (Tanikawa and Hashimoto 2009).
5 Urban planning and design choices. While, traditionally, urban metabolism studies have focused on total or per capita amounts, the importance of spatial distribution of building and infrastructure stocks is increasingly recognized, and spatial issues are known to affect transportation choice, building energy efficiency, etc. (Doherty *et al.* 2009; Newman and Kenworthy 2006).

Common frontiers of industrial ecology and urban ecology

Urban Ecology and Industrial Ecology have had many points of contact and cross influences over time, allowing for fruitful dialogue between the two neighbouring interdisciplinary research fields. In the light of growing concern about global environmental change and urban living, building more coherent linkages would seem appropriate. These linkages would best be built around the increasing recognition of the social and governance dimension both in research communities, and around the integration of complex system thinking into research (Korhonen *et al.* 2004). This would enable a stronger conceptual position and provide guidance for empirical studies aiming to inform urban policies that deal with the challenges of globalization and global environmental change.

In the current situation of continuing population growth, rapid urbanization and industrialization and the convergence of major resource use and global environmental issues (Raupach *et al.* 2007; Weisz and Schandl 2008; Schandl and West, forthcoming) there is an increasing research and policy focus on the transformability of urban structures to use less natural resources and energy and produce less waste and emissions. This poses a serious public policy issue and has led to the development of urban sustainability strategies in many cities around the world. Despite being well-intentioned, many strategies lack a solid baseline assessment of historical and current trends for resource use and impacts, and lack a comprehensive and conceptually sound basis. Policymakers may misunderstand the complexity of urban systems and thus arrive at policy conclusions which are simplistic and sometimes even counterproductive for achieving urban sustainability.

Industrial ecology started as the study of material and energy flows underpinning and resulting from socio-economic activities by looking at different scales (national, sectoral, businesses and firms, products) and cross-scale interactions. Based on the insights from metabolic analysis, the research has looked at ways and methodologies to close cycles in order to minimize the environmental impact of these activities. There has been a strong focus on design, engineering and technological innovation within industrial systems and a lesser focus on the analysis of institutional arrangements and related public policy questions.

The research field has, however, moved significantly towards integrating social sciences into the interdisciplinary endeavours to analyse and improve resource use and environmental impact of production and consumption across scales. Increasingly, the scholarly literature from neighbouring research fields such as economic sociology, institutional analysis and institutional economics, as well as science and technology studies, has referred to industrial ecology research to reflect the social embeddedness of industrial ecology (Boons and Howard-Grenville 2009); see also Thomas *et al.* (2003) for a focus on policy.

Despite this growing interest in the social domain of industrial ecology, the urban metabolism concept already addresses the fact that society-nature interactions and related material and energy flows in a modern industrial society are organized and managed at the social level, well above the individual. Industrial societies have established structural coupling between society and resources via socio-technological systems for major areas of provision including water and sewerage, energy, housing, mobility and food. Each such system consists of institutional and infrastructural elements and mobilizes a bulk resource. These interactions tend to be complex or at least complicated, and show inertia towards change because of the stability of institutional arrangements and the long-term use of large infrastructure such as urban transport infrastructure, commercial and residential buildings, and water and sewerage infrastructure. Major improvements in socio-technological systems may occur through the improvement of existing systems and efficiency gains but often more fundamental change is required, such as has been envisaged by the systems innovation literature (Elzen and Wieczorek 2005; Geels 2005).

The urban ecology research community has also increasingly recognized the important roles of the social behaviour of citizens, and urban policy making and management decisions, in shaping and regulating linkages between cities and the regional and global environment (Bai *et al.* forthcoming). There have been attempts to extend the traditional urban metabolism approach beyond a resource flow perspective (Newman 1999; Newton and Bai 2008). Figure 3.2 builds on these attempts, and shows the input of natural, capital and information resources into a city, and the output of products, knowledge and services, waste and other emissions. Urban system structures and functions, which include the natural and built environments within cities, distribution of goods and provision of services, urban design and governance, and urban lifestyles and consumption activities, mobilize and shape the input and output. The performance of an

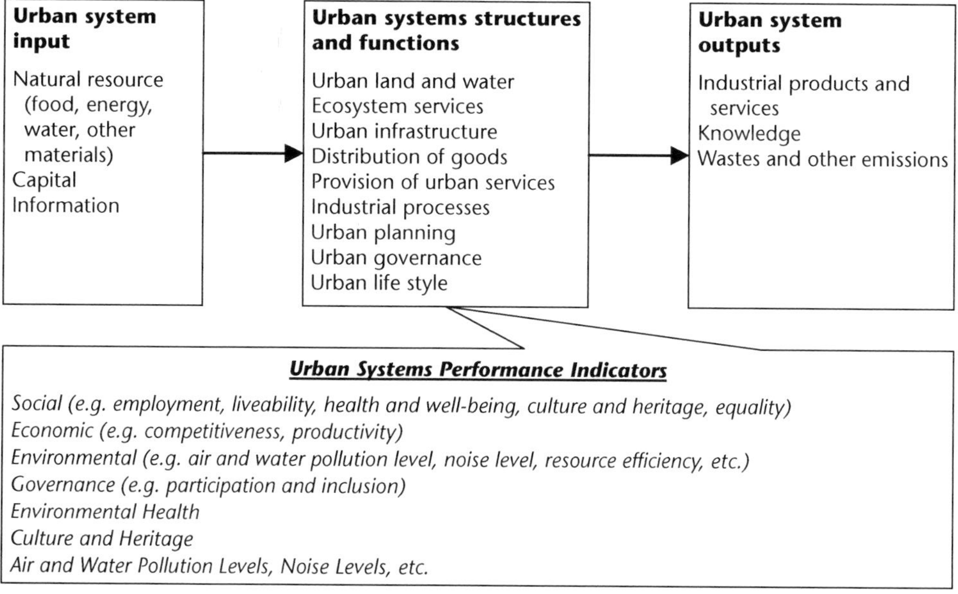

Figure 3.2 Integrated urban metabolism and urban system performance indicators.

urban system can be measured by a set of indicators that includes social, economic, environmental and governance aspects, which can then be used to inform and guide policy.

The integration of social, economic and governance dimensions into urban and industrial ecology expands the boundary of the disciplines as well as adding complexity in terms of modelling and analysing them. This leads to another common frontier in both urban and industrial ecology, which is the increasing adoption of a complex systems view. Jonathan Spiegelman (2003) argues that together with non-equilibrium thermodynamics and the application of metabolism analysis in both urban ecology and industrial ecology, complexity theory has the potential to integrate anthropogenic and natural activities into one framework. Complex system science aims to describe and understand systems that are characterized by nonlinear behaviour, feedbacks, self-organization, irreducibility and emergent properties (Anderson 1972). Related theories such as systems dynamics, cellular automata, agent-based modelling, and network analysis have been applied to urban problems such as population, land use change and transportation modelling (Baynes 2009). Other recent applications have included a study of social interactions in cities and their implications, which concluded that these interactions follow a sublinear scale and are strong drivers of innovation, economic productivity, and the spread of disease (Bettencourt *et al.* 2007).

A future research programme drawing from the rich scholarly knowledge within both urban ecology and industrial ecology would be well placed to deal with major urban sustainability issues. Such research could investigate how linear flows of resources might be changed to a circular mode, where resources are recycled and reused within and among subsystems of cities. Research would look at the role of economic growth and industrial activities for urban metabolism, the increasingly important interface between urban governance and management and industrial activities; and how to model and analyse the complex interactions and feedbacks in an urban system. It would also explore how the resilience of cities, in terms of sustaining functions, can be maintained and enhanced to weather a changing climate and to deal with major social issues by understanding cities as complex adaptive systems (Pickett *et al.* 2004). There is a need for a dynamic perspective on urban metabolism and urban development, the modelling of urban metabolism in a spatially explicit way, and for a much better understanding of the interrelationship of urban forms and structures with urban metabolism. This needs to be accompanied by a sound conceptual and empirical analysis of the role of cities in national and global resource flows and emissions, and would involve the attribution of up-stream flows of resources and emissions to urban centres, as well as providing better understanding of the role of cities in decision making and the framing of a dominant production and consumption culture. Finally, an improved understanding, informed by social science, of the challenges involved in transforming urban structure and form by urban development strategies and planning policies will be essential to drive sustainable urban development.

References

Alberti, M. (2008), *Advances in Urban Ecology: Integrating Humans and Ecological Processes in Urban Ecosystems*, New York, Springer.

Alberti, M., J.M. Marzluff, E. Shulenberger, G. Bradley, C. Ryan and C. ZumBrunnen (2003), "Integrating humans into ecology: opportunities and challenges for studying urban ecosystems", *BioScience*, 53(12): 1169–1179.

Anderson, P.W. (1972), "More is different", *Science*, 177(4047): 393–396.

Ayres, R.U. and A.V. Kneese (1969), "Production, consumption, and externalities", *American Economic Review* 59(3): 282–297.

Ayres, R.U. and U.E. Simonis (1994), *Industrial Metabolism: Restructuring for Sustainable Development*, Tokyo, UN University Press.

Bai, X.M. (2002), "Industrial relocation in Asia: a sound environmental management strategy?" *Environment*, 44(5): 8–21.

Bai, X.M. (2003), "The process and mechanism of urban environmental change: an evolutionary view", *International Journal of Environment and Pollution*, 19(5): 528–541.

Bai, X.M. (2007), "Integrating global concerns into urban management: the scale and the readiness arguments", *Journal of Industrial Ecology*, 11(2): 15–29.

Bai, X.M. and H. Imura (2000), "A comparative study of urban environment in East Asia: stage model of urban environmental evolution", *International Review for Global Environmental Strategies*, 1(1): 135–158.

Bai, X.M., R.R.J. McAllister, M. Beaty and B. Taylor (forthcoming), "Urban policy and governance in a global environment," *Current Opinion in Environmental Sustainability*.

Barles, S. (2009), "Urban metabolism of Paris and its region", *Journal of Industrial Ecology*, 13(6): 898–913.

Baynes, T.M. (2009), "Complexity in urban development and management", *Journal of Industrial Ecology*, 13(2): 214–227.

Bettencourt, L.M.A., J. Lobo, D. Helbing, C. Kuhnert and G.B. West (2007), "Growth, innovation, scaling, and the pace of life in cities", *Proceedings of the National Academy of Sciences of the United States of America*, 104(17): 7301–7306.

Boons, F. and J. Howard-Grenville (2009), *The Social Embeddedness of Industrial Ecology*, Cheltenham, Edward Elgar.

Boyden, S. (1987), *Western Civilization in Biological Perspective*, New York, Oxford University Press.

Boyden, S., S. Millar, K. Newcombe and B.J. O'Neill (1981), *The Ecology of a City and its People: the Case of Hong Kong*, Canberra, ANU Press.

Bramley, G., N. Dempsey, S. Power, C. Brown and D. Watkins (2009), "Social sustainability and urban form: evidence from five British cities", *Environment and Planning A*, 41(9): 2125–2142.

Collins, A., A. Flynn, T. Weidmann and J. Barret (2006), "The environmental impacts of consumption at a subnational level: the ecological footprint of Cardiff", *Journal of Industrial Ecology*, 10(3): 9–24.

Commoner, B. (1971), *The Closing Circle: Nature, Man, and Technology*, New York, Knopf.

Doherty, M., H. Nakanishi, X.M. Bai and J. Meyers (2009), "Relationships between form, morphology, density and energy in urban environments", *Global Energy Assessment Working Paper*.

Douglas, I. (1983), *The Urban Environment*, Baltimore, MD, Edward Arnold.

Duvigneaud, P. and S. Denaeyer-De Smet (1977), "L'Ecosystème Urbs: L'Ecosystème urbain Bruxellois", in P. Duvigneaud and P. Kestemont (eds), *Productivité Biologique en Belgique*, Paris, Editions Duculot, pp. 581–597.

Elzen, B. and A. Wieczorek (2005), "Transitions towards sustainability through system innovation", *Technological Forecasting and Social Change*, 72(6): 651–661.

Ernstson, H., S. Sörlin and T. Elmqvist (2008), "Social movements and ecosystem services: the role of social network structure in protecting and managing urban green areas in Stockholm", *Ecology and Society*, 13(2): Article 39.

EUROSTAT (2007), *Economy-Wide Material Flow Accounting, A Compilation Guide*, Luxembourg, European Statistical Office.

Fischer-Kowalski, M. (1998), "Society's metabolism: the intellectual history of materials flow analysis. Part I, 1860–1970", *Journal of Industrial Ecology*, 2(1): 61–78.

Fischer-Kowalski, M. and H. Weisz (1999), "Society as a hybrid between material and symbolic realms: toward a theoretical framework of society-nature interrelation", *Advances in Human Ecology*, 8: 215–251.

Folke, C., A. Jansson, J. Larsson and R. Costanza (1997), "Ecosystem appropriation by cities," *Ambio*, 26(3): 167–172.

Frosch, R.A. and N.E. Gallopoulos (1989), "Strategies for manufacturing", *Scientific American*, 261(3): 144–152.

Geels, F.W. (2005), "Processes and patterns in transitions and system innovations: refining the co-evolutionary multi-level perspective", *Technological Forecasting and Social Change*, 72(6): 681–696.

Graedel, T.E. and B.R. Allenby (2002), *Industrial Ecology*, second edition, Upper Saddle River, NJ, Prentice Hall.

Grimm, N.B., J.M. Grove, S.T.A. Pickett and C.L. Redman (2000), "Integrated approaches to long-term studies of urban ecological systems", *Bioscience*, 50(7): 571–584.

Grimm, N.B., S.H. Faeth, N.E. Golubiewski, C.L. Redman, J.G. Wu, X.M. Bai and J.M. Briggs (2008), "Global change and the ecology of cities", *Science*, 319(5864): 756–760.

Grove, J.M. and W.R. Burch (1997), "A social ecology approach and application of urban ecosystem and landscape analysis: a case study of Baltimore, Maryland", *Urban Ecosystems*, 1(4): 259–275.

Halada, K., K. Ijima, M. Shimada and N. Katagiri (2009), "A possibility of urban mining in Japan", *Journal of the Japan Institute of Metals*, 73(3): 151–160.

Hendriks, C., R. Obernosterer, D. Mueller, S. Kytzia, P. Baccini and P.H. Brunner (2000), "Material flow analysis: a tool to support environmental policy decision making: case study of the city of Vienna and the Swiss lowlands", *Local Environment*, 5(3): 311–328.

Huang, S.L. and W.L. Hsu (2003), "Materials flow analysis and energy evaluation of Taipei's urban construction", *Landscape and Urban Planning*, 63(2): 61–74.

Jackson, T. (2005), "Live better by consuming less? Is there a 'double dividend' in sustainable consumption?" *Journal of Industrial Ecology*, 9(1–2): 19–36.

Jenerette, G.D., W.A. Marussich and J.P. Newell (2006), "Linking ecological footprints with ecosystem valuation in the provisioning of urban freshwater", *Ecological Economics*, 59(1): 38–47.

Kaye, J.P., P.M. Groffman, N.B. Grimm, L.A. Baker and R.V. Pouyat (2006), "A distinct urban biogeochemistry?" *Trends in Ecology & Evolution*, 21(4): 192–199.

Kennedy, C., J. Steinberger, B. Gasson, Y. Hansen, T. Hillmann, M. Havranek, D. Pataki, A. Phdungslip, A. Ramaswami and G.V. Mendez (2009), "Greenhouse gas emissions from global cities", *Environmental Science & Technology*, 43(19): 7297–7302.

Kennedy, C.A. (2002), "A comparison of the sustainability of public and private transportation systems: study of the Greater Toronto Area", *Transportation*, 29(4): 459–493.

Korhonen, J., F. von Malmborg, P.A. Strachan and J.R. Ehrenfeld (2004), "Management and policy aspects of industrial ecology: An emerging research agenda", *Business Strategy and the Environment*, 13: 289–305.

Lenzen, M. and G.M. Peters (2009), "How city dwellers affect their resource hinterland", *Journal of Industrial Ecology*, 14(1): 73–90.

Levine, S.H. (2003), "Comparing products and production in ecological and industrial systems", *Journal of Industrial Ecology*, 7(2): 33–42.

Luck, M.A., G.D. Jenerette, J. Wu and N.B. Grimm (2001), "The urban funnel model and the spatially heterogeneous ecological footprint", *Ecosystems*, 4(8): 782–796.

Ma, S.J. and R.S. Wang (1984), "Social–economic–natural complex ecosystem", *Acta Ecol Sin* 4(1): 1–9 (in Chinese).

Meadows, D.H., D.L. Meadows, J. Randers and W.W. Behrens III (1972), *The Limits to Growth*, Washington, D.C., Potomac Associates, New American Library.

Munoz, P., S. Giljum and J. Roca (2009), "The raw material equivalents of international trade", *Journal of Industrial Ecology*, 13(6): 881–897.

Newcombe, K. (1977), "Nutrient flow in a major urban settlement: Hong Kong", *Human Ecology*, 5(3): 179–208.

Newman, P. and J. Kenworthy (2006), "Urban design to reduce automobile dependence", *Opolis*, 2(1): 35–52.

Newman, P.W.G. (1999), "Sustainability and cities: extending the metabolism model", *Landscape and Urban Planning*, 44(4): 219–226.

Newton, P. and X. Bai (2008), "Transitioning to sustainable urban development", in P. Newton (ed.), *Transitions: Pathways Towards Sustainable Urban Development in Australia*, Canberra, CSIRO Press, pp. 3–19.

Niza, S., L. Rosado and P. Ferrão (2009), "Urban metabolism: ethodological advances in urban material flow accounting nased on the Lisbon case study", *Journal of Industrial Ecology*, 13(3): 384–405.

Odum, E.P. (1975), *Ecology: The Link between the Natural and Social Sciences*, New York, Holt, Rinehart and Winston.

Odum, H.T. and E.C. Odum (1980), *Energy Basis for Man and Nature*, New York, McGraw-Hill.

Oke, T.R. (1995), "The heat islandof the urban boundary layer: characteristics, causes and effects", in J.E. Cermak, A.G. Davenport, E.J. Plate and D.X. Viegas (eds), *Wind Climate in Cities*, Netherlands, Kluwer Academic Publisher, pp. 81–107.

Park, R.E., E.W. Burges and R.D. McKenzie (eds) (1925), *The City*, Chicago, IL, University of Chicago Press.

Paul, M.J. and J.L. Meyer (2001), "Streams in the urban landscape", *Annual Review of Ecology and Systematics*, 32: 333–365.

Pauleit, S. and F. Duhme (2000), "Assessing the environmental performance of land cover types for urban planning", *Landscape and Urban Planning*, 52(1): 1–20.

Pickett, S.T.A., M.L. Cadenasso and J.M. Grove (2004), "Resilient cities: meaning, models, and metaphor for integrating the ecological, socio-economic, and planning realms", *Landscape and Urban Planning*, 69(4): 369–384.

Raupach, M.R., G. Marland, P. Ciais, C. Le Quere, J.G. Canadell, G. Klepper and C.B. Field (2007), "Global and regional drivers of accelerating CO2 emissions", *Proceedings of the National Academy of Sciences of the United States of America* 104(24): 10288–10293.

Ravetz, J. (2000), "Integrated assessment for sustainability appraisal in cities and regions", *Environmental Impact Assessment Review*, 20(1): 31–64.

Redman, C.L. (1999), "Human dimensions of ecosystem studies", *Ecosystems* 2(4): 296–298.

Schandl, H. and N. Schulz (2002), "Changes in the United Kingdom's natural relations in terms of society's metabolism and land-use from 1850 to the present day", *Ecological Economics*, 41(2): 203–221.

Schandl, H. and J. West (2010), "Resource use and resource efficiency in the Asia-Pacific region", *Global Environmental Change*, 20(4): 636–647.

Spiegelman, J. (2003), "Beyond the food web connections to a deeper industrial ecology", *Journal of Industrial Ecology*, 7(1): 17–23.

Sukopp, H. (1998), Urban ecology-scientific and practical aspects, in J. Breuste, H. Feldman and O. Uhlmann (eds), *Urban Ecology*, Berlin, Springer, pp. 3–16.

Sukopp, H. and P. Werner (1982), *Nature in Cities*, Strasbourg, Council of Europe.

Tanikawa, H. and S. Hashimoto (2009), "Urban stock over time: spatial material stock analysis using 4d-GIS", *Building Research and Information*, 37(5–6): 483–502.

Tarr, J.A. (1996), *The Search for the Ultimate Sink: Urban Pollution in Historical Perspective*, Akron, OH, The University of Akron Press.

Tarr, J.A. (2002), "The metabolism of the industrial city: the case of Pittsburgh", *Journal of Urban History*, 28(5): 511–545.

Thomas, V., T. Theis, R. Lifset, D. Grasso, B. Kim, C. Koshland and R. Pfahl (2003), "Industrial ecology: policy potential and research needs", *Environmental Engineering Science*, 20(1): 1–9.

Tzoulas, K., K. Korpela, S. Venn, V. Yli-Pelkonen, A. Kazmierczak, J. Niemelä and P. James (2007), "Promoting ecosystem and human health in urban areas using Green Infrastructure: a literature review", *Landscape and Urban Planning*, 81(3): 167–178.

van Beers, D. and T.E. Graedel (2007), "Spatial characterisation of multi-level in-use copper and zinc stocks in Australia", *Journal of Cleaner Production*, 15(8–9): 849–861.

Vitousek, P.M., H.A. Mooney, J. Lubchenco and J.M. Melillo (1997), "Human domination of Earth's ecosystems", *Science*, 277(5325): 494–499.

Wackernagel, M. (1998). "The ecological footprint of Santiago de Chile", *Local Environment*, 3(1): 7–25.

Wackernagel, M. and W. Rees (1995), *Our Ecological Footprint: Reducing Human Impact on the Earth*, Gabriola Island, BC, New Society Publishers.

Warren-Rhodes, K. and A. Koenig (2001), "Escalating trends in the urban metabolism of Hong Kong: 1971–1997", *Ambio*, 30(7): 429–438.

Weisz, H. and H. Schandl (2008), "Materials use across world regions: inevitable pasts and possible futures", *Journal of Industrial Ecology*, 12(5/6): 629–636.

White, R.R. (2001), *Building the Ecological City*, Cambridge, Woodhead Publishing.

Wolman, A. (1965), "The metabolism of cities", *Scientific American*, 213(3): 178–193.

Wong, N.H. and C. Yu (2005), "Study of green areas and urban heat island in a tropical city", *Habitat International*, 29(3): 547–558.

Zipperer, W.C., T.W. Foresman, S.M. Sisinni and R.V. Pouyat (1997), "Urban tree cover: an ecological perspective", *Urban Ecosystems* 1(4): 229–247.

4

Urban areas in the context of human ecology

Roderick J. Lawrence

Introduction

Cities and towns can be interpreted as human-made ecosystems that result from the interrelations between ecological, economic, material, political and social factors. Cities such as Bangkok, Cape Town or London are interesting examples of the intersection between these sets of factors, because they illustrate their diversity as well as change over long periods. Given the changing nature of these factors it is appropriate to discuss ways and means of sustaining urban areas as human ecosystems that are meant to ensure an acceptable quality of life. This approach underlines the fundamental principle that all human societies regulate their relation to the biosphere and the local environment by using a range of codes, practices and principles based on scientific knowledge, professional know-how and tacit knowledge. Societies use legislation, surveillance, monetary incentives and taxes, as well as behavioural rules and socially agreed conventions in order to ensure their sustenance (Lawrence 1996).

Human ecology refers to the study of the relations, especially the reciprocal relations between people, their habitat and the environment beyond their immediate surroundings. Human groups and societies establish and maintain viable relationships with their habitat through collective mechanisms that stem from their *anthropos* and generate a system of relations and networks rather than independent action. Hence, the methodological framework of human ecology studies is the analysis of people in their habitual living conditions using a systemic framework that explicitly examines the reciprocal relations between individuals, groups, the components of their habitat and larger environmental conditions (Lawrence 2001). This framework is particularly appropriate for the study of urban areas because an increasing number of people live in urban ecosystems.

Human ecology is explicitly anthropocentric. The *anthropos* refers to all the material and non-material dimensions of urban areas. These dimensions include genetic patrimony, especially the capacity of the human brain to interpret and transform land and other natural resources into urban areas; demographic characteristics such as the size and composition of human populations in cities and towns; the social organisation of human groups in urban neighbourhoods (including kinship relations and household structure); institutions including associations, rules and customs that regulate individual and collective behaviours; the local economy including all consumption and production processes; and, last but not least, the beliefs, knowledge, religion and values which collectively constitute a world-view of individuals, groups and societies (Lawrence 2001). World-views influence the way people perceive and interpret their surroundings.

What is human ecology?

The author of this chapter has noted that human ecology transgresses traditional disciplinary boundaries by explicitly applying a broad conceptual and methodological framework that integrates contributions from the natural and the social sciences (Lawrence 2001). This broader interdisciplinary framework is represented in Figure 4.1. This figure is not meant to be a detailed model of people-environment relations that can provide a complete understanding of a complex and vast subject. Instead it represents an enlarged interpretation because it includes concepts and principles from both the natural and the human sciences. Second, it underlines the systemic interrelations between sets of biotic, abiotic and anthropogenic factors that are combined together in any human ecosystem. Hence it concentrates not only on specific components but also considers the whole system as the unit of study for people-environment relations. This conceptual model can be applied at different geographical scales including urban areas. It is a synchronic representation of a human ecosystem that is open and linked to others. The model is meant to be reapplied at different times to explicitly address short- and long-term historical perspectives. This dual temporal perspective can identify change to any of the specific components as well as the interrelations between them.

Figure 4.1 presents a conceptual reference model which can be used by researchers in diverse disciplines and policymakers in different sectors. It illustrates the principle that disciplinary knowledge and specialisation have hindered the development of a broader understanding of the contextual conditions of human habitats and larger ecological and environmental subjects. For example, at the beginning of the twenty-first century, it is unfortunate that the division still exists in both academic research and teaching between human and physical geography; likewise between physical/human and cultural/social anthropology.

This chapter considers core principles of human ecology that can be applied to analyse urban areas. These principles include cultivation and adaptation which are meant to provide and sustain an acceptable quality of life in a changing world. Given that urban development and building construction have accelerated since the mid-nineteenth century in all continents of the world, both positive and negative impacts have occurred at both local and global levels. The UNESCO-MAB programme included an innovative series of projects to identify these impacts in specific localities. The pioneer case study of Hong Kong is briefly presented in this chapter for this reason. In addition, it highlights the added value of applying core principles of human ecology using an interdisciplinary research agenda.

Applying core principles of human ecology

One basic principle of biological life is that all living organisms (irrespective of their species) impact on their surroundings (Odum 1994). The livelihood of organisms is dependent on the interrelations between them and their surroundings, the volume and quality of the available local resources, the discharge of waste products, and the creation of new resources. By their existence, all living organisms change the conditions upon which they depend for their subsistence. In addition, all organisms are components of ecological systems and, therefore, they explicitly influence the living conditions of other species (Marten 2001). For example, when human beings construct cities and towns they change their habitat as well as the habitats of animals, insects and plants to such an extent that they can jeopardise their existence over the long term.

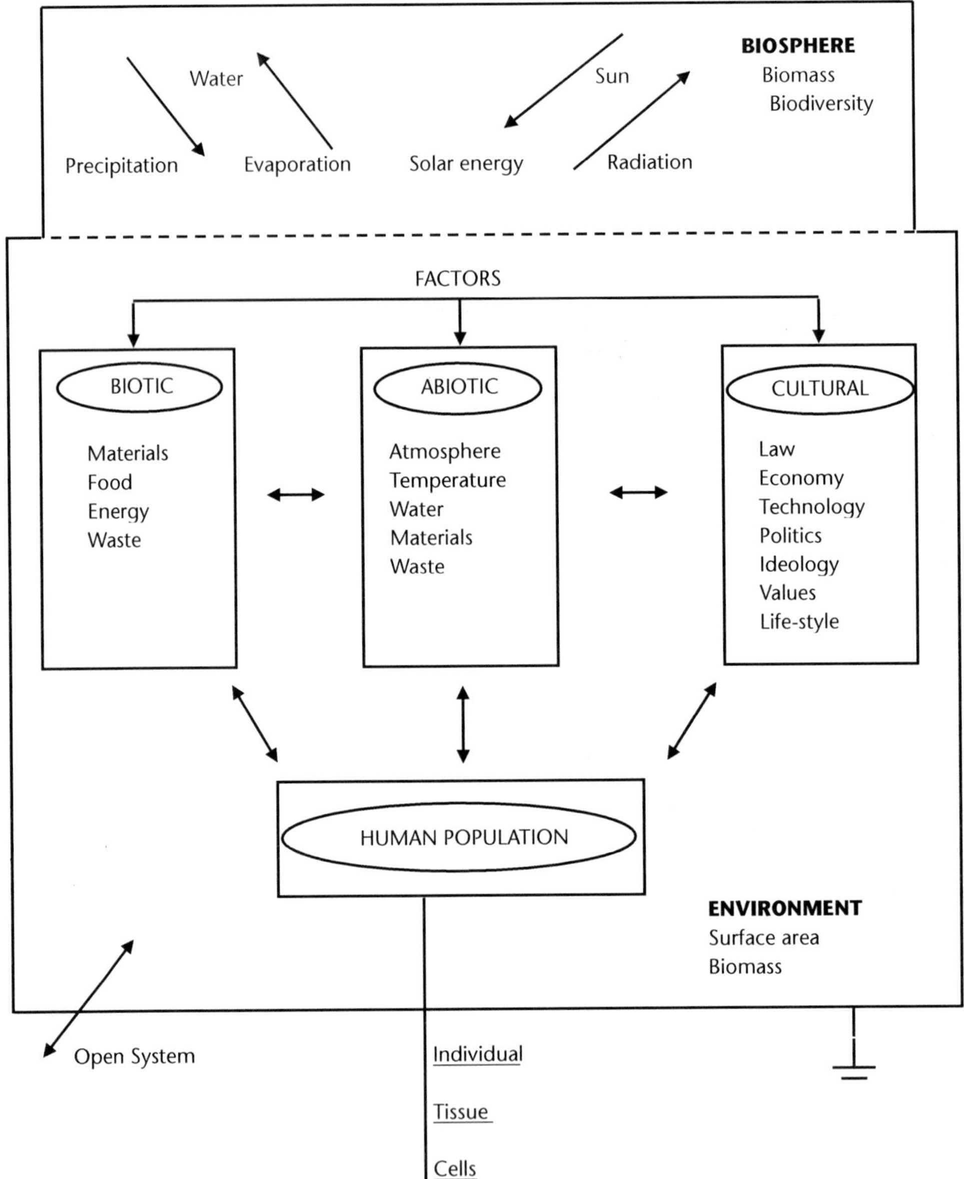

Figure 4.1 The holistic framework of a human ecology perspective showing the interrelations between genetic biospace, ecospace, cultural space and artefacts.

There are certain conditions and limits overriding the sustenance of human groups and societies that are defined by some fundamental principles that should be an integrated part of human ecology and incorporated into sustainability research. First, the biosphere and the Earth are finite (Boyden 1992). Both natural and human ecosystems at all scales of the planet and its atmosphere are circumscribed by certain immutable limits, such as the surface of land, its biomass and biodiversity, the water cycle, biochemical cycles and thermodynamic principles about the production and transformation of energy, including the accumulation and radiation of heat from the Earth.

Second, human ecosystems are *not* closed, finite systems because they are open to external influences of an ecological kind (like solar energy and earthquakes), as well as influences of a biological kind and an anthropological kind, such as disease and warfare (McMichael 1993). This means that research on urban ecosystems that deals with internal conditions and processes should also consider those external factors that impact upon them in order to conform to Figure 4.1. Unfortunately, recent contributions on this subject include misconceptions about the autonomy of cities and the ability of modern technology to overcome ecological constraints.

Third, humans must create and transform energy by using materials, energy and acquired knowledge to ensure their livelihood (Boyden 1987). The increasing disparity between ecological and biological processes and products on the one hand, and the products and processes of human societies, on the other hand, is largely attributed to the rapid growth of urban populations, the creation of many synthetic products that cannot be recycled into natural processes, plus increases in energy consumption based on the use of non-renewable and renewable resources at a greater rate than their replacement. At the global level, the negative consequences of these trends include the depletion of the ozone layer, a reduction in biodiversity, the accumulation of wastes, the enhancement of the 'green house effect' and the incidence of environmental catastrophes including floods, landslides and famine.

Fourth, human beings can be distinguished from other biological organisms by the kinds of regulators they commonly use to define, modify and control their living conditions (Laughlin and Brady 1978). Humans have several mechanisms that enable them to adjust to specific environmental conditions. These mechanisms include thermo-regulation and circadian rhythms, which are used to ensure and maintain vital needs, such as nutrition. This fundamental need is not only guaranteed by biological and physiological mechanisms, because cultural rules and practices that vary between ethnic groups, across cultures and within societies are also used.

Understanding cultivation

Cultivation is a multi-dimensional process in which implicit cognitive structures, individual and group practices, social rules and conventions, institutional structures, and human consciousness are purposely interrelated. The intentional use of space, time and resources at any geographical scale implies that a part of the world is appropriated psychologically and physically (Lawrence and Low 1990). The term appropriation has etymological roots in the Latin word *appropriare*, which means 'to make one's own'. Cultivation processes may be conducted to express and communicate adherence to cultural traditions, or new social trends, or to express individualism rather than collectivism. Rituals, roles and a wide range of conventional practices are commonly used during the construction of cities in order to express and transmit cultural codes and social identities (Wheatley 1971). The transformation of landscapes into urban ecosystems implies that geographical space, resources and time are cultivated by people.

Cultivation implies that researchers and practitioners should identify and understand the active, perhaps mobile interrelations between individuals and their habitat. This concept can also account for the cognitive and symbolic interrelations between individuals, groups and their past and present (Lawrence and Low 1990). Cultivation also stresses the importance of intentionality within the ongoing practices of domesticity, especially the way that individual, social, and cultural identities are expressed and communicated.

Acculturation involves the interaction between individuals and groups from different parent cultures. This collective process leads to the foundation of hybrid culture traits. Acculturation processes have been well illustrated in Australia during the first half of the nineteenth century and especially since the 1950s. One of the first outcomes of acculturation processes in the nineteenth century was the construction of vernacular houses with verandas and detached kitchens by migrants from England, Germany and Scotland.

Understanding adaptation

Adaptation is a set of interrelated processes that sustain human ecosystems in the context of a continual change (Moran 1982). Evolutionary adaptation refers to processes of natural selection. It is only applicable to populations and it is trans-generational. Innate adaptation refers to physiological and behavioural changes that occur in individuals that are genetically determined and do not depend on learning. Cultural adaptation refers to adaptation by cultural processes that are not innate, such as legal measures, or changes in lifestyle and therefore it includes institutional adaptations. The outcome of adaptation depends on a complex set of biological, ecological, cultural, societal and individual human mechanisms.

Behaviours, concepts, ideas and goals are used implicitly and explicitly in all kinds of human activities and relations concerning individuals, groups, societies and their habitat (Lawrence 2001). These components of human culture not only influence the relations between people and natural ecosystems, but also the interrelations between different human groups. Some human ecologists refer to the 'social environment' in order to distinguish the latter from the 'natural environment' (Marten 2001). In order to understand the complexity of urban areas both the natural and social dimensions of these human ecosystems need to be considered simultaneously. For example, social rules prohibiting the long-term use of land or other natural resources used for building construction and urban development are meant to sustain these natural resources over the long term. In addition, these rules should be fair and equitable so that no one benefits at the expense of others.

The following principles are essential components of human ecology and its application for the analysis of urban areas. First, the interrelations between humans and their urban environment are manifested through a wide range of physiological, psychological and cultural processes (Lawrence and Low 1990). All these kinds of processes include not only sensations and perceptions (which animals also share), but also beliefs, doctrines, ideas and values, which are unique characteristics of the human intellect. Therefore, human ecology is also a subjective ecology and it should include human perception and cognition of the environment, such as values and uses attributed to all natural resources before they are modified to become components of urban areas (Lawrence 2001). In principle, the interrelations between people and the environment are not just spatial nor static but subjective and, therefore, variable over relatively short and long periods of time.

Second, the urban environment can be contrasted with the habitat of other biological organisms by the instrumental functions and by the symbolic values that have been attributed to it (Lawrence 2001). Human processes and products transform the constituents of the environment to meet prescribed aspirations, goals and needs. In addition, human activities can provoke unintended consequences on abiotic and biological constituents of ecosystems and in return, have an impact on human health and well-being.

Health and quality of urban life

Urban health is a vast and complex subject that transgresses disciplinary, geographical and professional boundaries (Lawrence 2000). In order to understand the multidimensional nature of health and quality of life, numerous sets of risk factors and the interrelations between them should be considered across a range of geographical scales from the local to the international level as well as in terms of the time period of human exposure to them. These risk factors include four interrelated sets of hazards:

1 Environmental risk factors including ambient air quality, ambient noise levels, soil and water contamination, and solid waste disposal.
2 Economic risk factors comprising the lack of affordable housing, food and water for poor households, permanent unemployment and inequalities of access to diverse kinds of resources and services including affordable primary health care.
3 Technological risk factors including traffic accidents, industrial and chemical disasters on industrial sites and contamination from mass produced foods and synthetic products.
4 Social and individual risk factors including criminality, violence and social exclusion, as well as the lack of education and training especially for immigrants, women and children.

These four main sets of risk factors are variable within and between buildings and urban neighbourhoods as well as over short and relatively long periods of time. Their interrelated nature is complex: for example, the internal conditions of a room in a housing unit may be related to the physical condition of the residential building and its site and the urban neighbourhood in which they are located. In turn, the neighbourhood, the city, the region and the country are all directly or indirectly interrelated. In essence, local and international levels are linked across cultural, geographical, political and temporal boundaries.

The exposure of different groups (e.g. children, ethnic groups, the elderly and the unemployed) to these four sets of risk factors needs to be understood (Galea and Vlahov 2005). It cannot be assumed that an individual, household or a particular social group exposed to one or more of the above sets of hazards will necessarily suffer negative impacts to their health. In principle, biologically inherent mechanisms are mediated by the social and environmental circumstances of residential neighbourhoods (Hartig and Lawrence 2003). Hence someone exposed to a hazard can have a different risk of health damage compared with another person also exposed to these hazards. Therefore, it is necessary to interpret the health of urban populations in terms of both individual and social differences by explicitly accounting for age, gender, socio-economic class, occupational status, and the geographical and temporal distribution of the population in terms of the four sets of risk factors presented above.

The human ecology research agenda considers four main sets of interrelated factors: the individual, who has a specific genetic code with a susceptibility and immunity to illness and disease, as well as lifestyle traits; the agent or vector of illness and disease, including not only biogeophysical components of the environment but also the social and psychological dimensions of human settings; the physical and social environment of the individual which affects the susceptibility of the host, the virulence of biophysical agents and the exposure, quantity and nature of the contact between host and vector; the available resources used by the individuals and households including housing, nutrition, money, information, and access to health and medical services which ought to be affordable for all groups of the population. This broad perspective implies that an analysis of the interrelations between multiple components of any human ecosystem is necessary. Systemic interpretations of human illness, health and local environments have a long history. They can be traced back as least as far as the Hippocratic treatise *On Airs, Waters, and Places* published initially about 2,600 years ago (Hippocrates 1849).

The distinction between biomedical models and ecological interpretations of health is fundamental. The germ theory, for example, is an incomplete explanation of human illness and disease because it ignores the contribution of numerous physical and social dimensions of the environment that can impact on health. Ecological interpretations maintain that the presence of a germ is a necessary but not a sufficient condition for an individual to become ill. They accept that some individuals become more susceptible to certain illnesses because of their differential exposure to numerous environmental, economic and social factors that can promote or be harmful to health and well-being. This interpretation does not ignore the influence of genetics, individual behaviour or primary health care. However, it maintains that, alone, these do not address possible relations between social problems and illness (e.g. inequalities) or positive social dimensions and health promotion (e.g. public education). The distinction between potential and actual health status can be the foundation for a new interpretation of health which includes the way environmental, economic, social, and technological risk factors transgress traditional disciplinary boundaries while remaining locality and temporally specific. Cues for this interpretation were already provided by human ecologists in the 1920s.

The early studies in human ecology at the University of Chicago plotted the geographical distribution of some characteristics of the resident population of Chicago including their ethnic origin, socio-economic status, birth and mortality rates, delinquency, mental and other illnesses. These cartographic studies enabled the authors to overlay the maps of the spatial distribution of these characteristics in order to identify those that occurred in the same urban area. This approach established many correlations (Park *et al.* 1925). For example, the distribution of cases of tuberculosis corresponded with the highest incidence of delinquency. This finding led the authors to suggest that cities comprise 'natural' areas that are defined geographical, economic, social and cultural dimensions. The term 'natural' was used because these areas were not planned but were constituted as the city developed. Each 'natural' area is characterised not only by its geographical location but also the market price of land and property, and sets of customs, norms and non-monetary values that are part of the lifestyle of the residents.

The first generation of studies in human ecology from the 1920s marked a growing interest in collaboration between researchers from disciplines who worked to deal with complex multidimensional questions including those of urban health. This kind of interdisciplinary collaboration has been extremely fruitful since the 1970s.

Ecological impacts of architecture, urban ecosystems

During the 1960s and 1970s, the growing concern about the negative ecological impacts of human activities led some architects and town planners to consider how urban development impacts on the natural environment at both the local and global levels. One leader in this movement was the late Constantinos Doxiadis (1913–1975). He was educated as an architect and town planner. Throughout his career as a civil servant, a private practitioner and a consultant, he dealt with the challenge of urban development, especially its ecological impacts at the global level.

Doxiadis (1977) made an important contribution. He first used the word ekistics in 1942 during a lecture at Athens Technical University to refer to 'the science of human settlements which are the territorial arrangements made by *anthropos* (Man) for his own sake' (Doxiadis 1977: xv). He devoted his professional career to developing a model to show how principles of ecology and ekistics can be analysed systematically in order to mitigate the negative impacts of human settlements. Unlike other disciplines which focus only on one component of human habitats – such as human society (sociology), or buildings and infrastructure (architecture) – ekistics incorporates the knowledge of several disciplines including anthropology, economics, geography and other disciplines in both the natural and social sciences. Doxiadis proposed a systemic framework – the *anthropocosmos* model – that combined ecology with ekistics. This model includes a taxonomy of all known kinds of human settlements, the five main ekistics elements of settlements, the ekistics human populations scale, units of time, and five sets of driving forces that underlie the construction of cities and towns. These driving forces are economic/ financial; social/group; political; technical; and cultural. The five characteristic elements of human settlements are Nature; individual human beings; human society; human-made structures; and infrastructure networks including roads, railways, pipelines and communication channels.

In order to analyse systematically the interrelations between all the components of the model at the global level, Doxiadis proposed a global ecological balance (GEB) which is his interpretation of the capacity of the Earth and biosphere to accept human population growth and associated activities without compromising the future of both human life and nature. Although some mathematical calculations (such as the maximum world population of 22 billion people) are flawed, Doxiadis made an important contribution which has largely been rejected or ignored. In fact, his work was a precursor for sustainable development, especially recent contributions on carrying capacity and ecological footprints (Wackernagel and Rees 1996) as well as the impacts of innovative technologies.

Monitoring evo-deviation and techno-addiction

Boyden (1992) has studied the impacts of the activities of human societies on biophysical components of the biosphere and on humans themselves. He and his colleagues are also interested in the adaptation processes used in response to changes in human ecosystems. The term evo-deviation is used to refer to a general biological principle that describes life conditions of human and other species that are different from those in the natural habitat of the species. When these differences become large sudden, perhaps irreversible, behavioural and physiological maladjustments may occur (Laughlin and Brady 1978). Some physiological maladjustments during the last 10,000 years include diseases such as scurvy, typhoid, cholera, smallpox and influenza.

Another approach used by Boyden and his colleagues analyses the advent and acceptance of new technologies including those that are not essential for basic human needs and sustenance. These inventions become an integral part of human societies by a process termed techno-addiction. The spatial and social organisation of societies can become dependent on them as shown by the reliance on combustion engines using fossil fuels. For example, this has led to transportation policies in many countries based heavily on road traffic using private cars in preference to public transport by rail. Some consequences of this kind of techno-addiction include the extensive use of arable land for urban and regional development, the incidence of air pollution and noise, as well as non-active lifestyles. An in-depth case study of Hong Kong illustrates the principles of evo-deviation and techno-addiction. This case study was completed as part of the UNESCO Man and the Biosphere Program (MAB Program) dealing with ecological studies of human settlements (Boyden *et al.* 1981).

The case study of Hong Kong by Boyden and his colleagues identifies and measures the changing ecological characteristics of the human settlement and its hinterlands prior to the formal founding of the British Colony in 1842 until the 1970s. This study notes that with the advent of colonialism traditional rules and customs were challenged by foreign ones leading to the introduction of novel forms of land tenure, new resources including materials, machinery and techniques, and the proliferation of cash and market economies. The specialisation of tasks in Hong Kong increased. The economic consequences of these kinds of changes are important, because the total cost of energy consumption, materials and labour increased.

This case study examines patterns of energy production and consumption flows, the production and disposal of wastes including pollutants, the densification and dispersal of the built environment, and impacts on the health and well-being of the population. The case study also includes a comparison of traditional Chinese and contemporary imported methods of food production, processing and packaging. The outcomes of the change from one set of process to the other include impacts on land uses, agricultural production methods, the import and export of diverse kinds of materials, and the diet and health of the population. The authors note that, at the beginning of the twentieth century, inadequate nutrition was common in Hong Kong owing to a diet almost exclusively limited to polished rice. By the late twentieth century, however, the local environment provided a range of food including fresh fruit, fish, meat and vegetables. Although nutritional diseases like beri-beri have been eradicated, in recent decades there have been changes in the diet of the population especially the increasing consumption of refined carbohydrates (e.g. white flour and refined sugar) which imply reduced dietary fibre. This change can lead to increased dental caries, over-nutrition and obesity, as well as an increase in diabetes. The authors note that the death rate from diabetes was 4.3 per 100,000 in 1974, which is a triple increase since 1961 and a four-fold increase since 1949.

Boyden and his colleagues raise many questions about the costs and benefits of industrialised food production, the toxicity of fertilisers, the ecology of soil and water catchment areas, and deforestation. In essence, this kind of case study can identify both the intended and unintended consequences of human products and processes over the long term. It also shows that it is important to distinguish between the tacit know-how of the local population prior to colonisation (in which knowledge and practice are indistinguishable) and the explicit know-how of the colonial administration (in which theory and practice are interrelated but clearly distinguishable). The gradual suppression of implicit instruments of regulation by a growing number of explicit ones has occurred with urbanisation in Hong Kong and elsewhere. This example illustrates that there is a need to reconsider both implicit and explicit regulatory means and measures, which are fundamental constituents of people-environment relations.

Interdisciplinary and transdisciplinary contributions

The relationship between researchers in different disciplines, especially in the human/social and the basic/natural sciences, is often considered to be a source of conflict. Nonetheless, this need not be the case as shown already more than 20 years ago by the contribution of Boyden and his colleagues in their applied human ecology research about Hong Kong (Boyden *et al.* 1981). Innovative contributions of this kind can lead to the development of new terminology, innovative concepts and new knowledge. This is an important challenge for those who wish to apply human ecology in research on urban areas.

When dealing with complex subjects, such as urban areas, it is necessary to shift from monodisciplinary to interdisciplinary and transdisciplinary concepts and methods (Lawrence and Despres 2004). In order to be effective, this shift should be founded on a clarification of definitions, goals and methods. In this chapter disciplinarity refers to the specialisation of academic disciplines especially since the nineteenth century (Lawrence 2001). Multidisciplinary refers to an additive research agenda in which each researcher contributes but remains within his discipline and applies its concepts and methods without necessarily sharing the same goal. Interdisciplinary studies are those in which concerted action and co-ordination are accepted by researchers in different disciplines as a means to achieve a shared goal that usually is a common subject of study. In contrast, transdisciplinarity refers to a research agenda that incorporates a combination of concepts and knowledge not only used by academics and researchers but also other actors in civic society, including representatives of the private sector, public administrators and the public. These contributions enable the cross-fertilisation of knowledge and experiences from diverse groups of people that can promote an enlarged vision of a subject, as well as new explanatory theories. Rather than being an end in itself, this kind of research is a way of achieving innovative goals, enriched understanding and a synergy of new methods.

Multidisciplinarity, interdisciplinarity and transdisciplinarity are complementary rather than being mutually exclusive (Lawrence and Despres 2004). Without specialised disciplinary studies there would be no in-depth knowledge and data. Transdisciplinary research and practice require a common conceptual framework and analytical methods based on shared terminology, mental images and common goals. Once these have been formulated then the next requirement is to develop a research agenda based conceptually and pragmatically on diverse sources of data and information that can be organised in ways to help understand, interpret and act in urban areas. This chapter has presented the core principles of human ecology that have been tested and should be reapplied to define that research agenda.

NB: This chapter has links with other chapters in Part 1, notably Chapter 1, 2 and 3.

References

Boyden, S. (1987) *Western Civilisation in Biological Perspective: Patterns in Bio-history*, Oxford: Oxford University Press.

Boyden, S. (1992) *Biohistory: The Interplay between Human Society and the Biosphere: Past and Present*, Paris: UNESCO.

Boyden, S., Millar, S., Newcombe, K. and O'Neill, B. (1981) *The Ecology of a City and its People: The Case of Hong Kong*, Canberra: Australian National University Press.

Doxiadis, C. (1977) *Ecology and Ekistics*, London: Elek Books.

Hartig, T. and Lawrence, R. (eds) (2003) 'The residential context of health', *Journal of Social Issues*, 59(3): 455–676 (special issue).

Hippocrates (1849) 'On air, waters and places', in F. Adams (trans. and ed.), *The Genuine Works of Hippocrates*, London: The Sydenham Society

Galea, S. and Vlahov, D. (eds) (2005) *Handbook of Urban Health: Populations, Methods and Practice*, New York NY: Springer.

Laughlin, C. and Brady, I. (eds) (1978) *Extinction and Survival in Human Populations*, New York: Columbia University Press.

Lawrence, D. and Low, S. (1990) 'The built environment and spatial form', *Annual Review of Anthropology*, 19: 453–505.

Lawrence, R. (1996) 'Urban environment, health and the economy: cues for conceptual clarification and more effective policy implementation', in C. Price and A. Tsouros (eds), *Our Cities, Our Future: Policies and Action Plans for Health and Sustainable Development*, Copenhagen: World Health Organization, OECD, pp. 38–64.

Lawrence, R. (ed.) (2000) 'Urban health', *Reviews on Environmental Health*, 15: 1–271 (special issue).

Lawrence, R. (2001) 'Human Ecology', in M.K. Tolba (ed.), *Our Fragile World: Challenges and Opportunities for Sustainable Development Volume 1*, Oxford: Eolss Publishers, pp. 675–93.

Lawrence, R. and Despres, C. (eds) (2004) 'Futures of transdisciplinarity', *Futures*, 36: 97–526 (special issue).

Marten, G. (2001) *Human Ecology: Basic concepts for sustainable development*, London: Earthscan.

McMichael, A. (1993) *Planetary Overload: Global environmental change and the health of the human species*, Cambridge: Cambridge University Press.

Moran, E. (1982) *Human Adaptability: An Introduction to Ecological Anthropology*, Boulder CO: Westview Press.

Odum, H. (1994) *Ecological and General Systems: An Introduction to Systems Ecology*, Niwot CO: University Press of Colorado.

Park, A., Burgess, E. and McKenzie, R. (1925) *The City: Suggestions for the Study of Human Nature in the Urban Environment*, Chicago IL: University of Chicago Press.

Wackernagel, M. and W. Rees (1996) *Our Ecological Footprint: Reducing Human Impact on Earth*, Gabriola Island: New Society Publishers.

Wheatley, P. (1971) *The Pivot of the Four Quarters*, Edinburgh: Edinburgh University Press.

5

In livable cities is preservation of the wild

The politics of providing for nature in cities

Michael C. Houck

> The belief that the city is an entity apart from nature and even antithetical to it has dominated the way in which the city is perceived and continues to affect how it is built. The city must be recognized as part of nature and designed accordingly.
>
> (Spirn 1984: 5)

Launched in 1999 by the Wildlife Conservation Society based at the Bronx Zoo, the *Mannahatta Project*, is an effort to catalog the flora and fauna and ecosystems that once dominated the world's fifth most populated metropolis. The 2000 wintertime appearance of a beaver in New York's Bronx River, the first in two centuries, stimulated a renewed interest in reconnecting the region's 16 million residents with nature (Miller 2009: 126). Increasingly, cities throughout the United States have realized how essential nature is in creating livable cities. Integrating nature into cities, a relatively new phenomenon in the United States, has been common in many European cities for decades.

David Nicholson-Lord, in his book *Green Urbanism, Learning from European Cities*, makes a persuasive case that Europe went through an attraction to and revulsion by the city that predated a similar phenomenon in the US. He argued that, while appreciating the city's function of promoting social cohesion, economic opportunities and "beating back the wilderness," as cities grew denser and quality of life decreased they began to lose their allure (Nicholson-Lord 1987).

By the late 1880s the Garden City movement, led by Ebenezer Howard, envisioned cities with populations limited to 30,000 residents. They would be surrounded by an agricultural greenbelt, the precursor of the United Kingdom's greenbelt program, which was ultimately applied to 15 urban areas totaling 1.8 million ha or 15 percent of England's total area (Beatley 2000: 38). By the late 1800s, public parks, and gardens in Britain were touted as a mechanism to keep the population content and "improve morals and reduce disease and crime." (Nicholson-Lord 1987: 28). At the time Frederick Law Olmsted, Sr. was visiting Liverpool's Birkenhead Park, gaining the inspiration for his own work in public parks throughout the US. While there was a convergence of thought regarding the need for ample parks and open space on both sides of the Atlantic, urban planning and design could not have been more divergent.

European models

Much of the impetus for building more sustainable cities emanated from the European Union whose 1990 *Green Paper on the Urban Environment* called for "more integrated, holistic approaches to planning, and the need to view cities as a necessary part of the solution to global environmental problems." This spawned a more ecologically focused approach via The EU Expert Group's *European Sustainable Communities* report that declared "The city must be viewed as a complex, interconnected and dynamic system. Cities are both a threat to the natural environment and an important resource in their own right" (Beatley 2000: 16).

European cities are well known for excellence in urban design, compact urban form, mixed used development and recycling of derelict land. In one study the average density of six European cities was 281 residents per ha, contrasted with 86 per ha in six US jurisdictions of similar size. The same cities also contrasted per capita annual Vehicle Miles Traveled (VMT) with Europeans averaging 26,200 km and US cities averaging 66,300 km (Beatley 2000: 30). By almost every measure of sustainability Europeans have long been ahead of US cities. Beatley does point out, however, that even Europeans have not been universally diligent regarding access to nature and ecosystem protection (Beatley 2000: 407).

Amsterdam's 1000 ha Amsterdamse Bos Park, represents one of those exceptions where access to nature has enjoyed coequal status with other planning policies. The Bos park, about the same size as Paris' Bois de Boulogne, creates a large "green wedge" between highly developed urban fingers. It and other green wedges converge on the "Green Heart" (*Groene Hart*), a group of small towns that have incorporated significantly more agricultural land and open spaces into their development than the surrounding cities of Amsterdam, Utrecht, Rotterdam and The Hague (Beatley 2000: 35).

In Germany, the eco-city of Freiburg also utilizes a "green wedge" strategy to integrate natural areas and open spaces into the city. Development has occurred on only 32 percent of the land within the city, which sits at the edge of the Black Forest. Helsinki, Finland's 10 km long Central Park extends from the city center north into an old growth forest. Helsinki has myriad nature oriented programs including nature preserves, important bird, amphibian, reptile, and bat areas, areas of floral interest, and protected habitats (City of Helsinki, Finland website 2009).

Regional planning

In Copenhagen, Denmark's regional "finger plan" was developed during World War II to both contain urban sprawl and develop a transportation network. The *Finger Plan*, modeled after Britain's Garden Cities movement, called for green wedges that alternate between built up areas that form the "fingers" (Matthiessen 2008).

Copenhagen has also established a goal that 90 percent of its residents will be a 15 minute walk to a park, a beach or an ocean pool by 2015. Their vision plan asserts that, "a sustainable city is also a city in harmony with nature. The city's trees, parks, and natural areas contain an environment for a rich plant and animal life, thereby contributing to the city's biological diversity" (City of Copenhagen website 2007: 15). Copenhagen is also developing a network of "green bicycle routes" and greenways to create safe and pleasant bicycle network that will cover more than 60 miles when completed. The mobility and connectivity elements of this network are being used as a template for the Portland, Oregon regional plans for an interconnected regional trail network to provide increased access to nature and enhanced transportation mobility and connectivity (Metro website 2008).

Portland, Oregon a case study

Beatley (2000: 224) cites Portland, Oregon, as one example of progressive regional, bioregional, and metropolitan-scale greenspace planning in the US. Portland is also known for its land use planning and sustainable practices. Indeed, the city has more LEED (Leadership in Environmental Design) buildings than any other city. While the nation had increased greenhouse gases by 13 percent, Portland's fell by 12 percent between 1990 and 2001. During a comparable period public transit ridership grew by 75 percent and bicycle commuting by 500 percent. Between 1990 and 2000 the Portland region's population grew by 31 percent but consumed only 4 percent more land to accommodate that growth. By contrast the Chicago region grew by 4 percent yet consumed 36 percent more land (Chicago Wilderness 1999a: 21).

However, Portland's urban nature agenda lagged behind other sustainability initiatives. Competing policies pitted otherwise progressive planning objectives against natural resource protection. Urban planners maintained, inappropriately, that protecting too much greenspace would deplete the buildable lands inventory inside the Urban Growth Boundary. Many politicians also made this argument. As a result, until recently, the Portland metropolitan region failed to adequately protect natural resources within the region's Urban Growth Boundary (Houck and Labbe 2006: 40; Wiley 2001).

Henry David Thoreau's aphorism, "In wildness is the preservation of the world" has driven the nation's conservation agenda to focus almost exclusively on protection of wilderness and pristine habitats, at the expense of urban nature protection. A twenty-first century corollary to Thoreau's call to protect nature should be "in livable cities is preservation of the wild." It will only be by protecting and restoring a vibrant urban green infrastructure of healthy watersheds, fish and wildlife habitat, parks, and recreational trails where the vast majority of our population lives – in our cities – that rural wetlands will be protected.

The regional landscape view

> Connectivity is needed both within a particular network and across many networks of human, built, and natural systems in a region. Some structures and patterns would be more appropriately understood at a regional and metropolitan scale; others, at the city or neighborhood scale; and still others at the site scale.
>
> *(Gerling and Kellett 2006)*

Portland, Oregon and Vancouver, Washington sit at the confluence of the Columbia and Willamette rivers. The Columbia River extending more than 1,900 km from its Canadian headwaters, drains a region the size of France. The 306 km-long Willamette River drains Oregon's Willamette Valley, Cascade Mountains and Coastal Range.

In the 13,200 km^2 bi-state region urban and rural landscapes unite the region and provide a shared sense of place. While proximity to spectacularly beautiful and wildlife-rich hinterlands adds to the region's mystique and quality of life, the more proximate landscapes, those within our immediate radius of reach and treasured most by the region's residents are the streetscapes, neighborhood parks, and urban natural areas (see Figures 5.1 and 5.2 for examples).

Early urban nature planning

Comprehensive efforts to describe and protect Portland's special landscapes and assure access to nature date back to 1903 when John Charles Olmsted, landscape architect and step-son of

Figure 5.1 Aerial photograph of Oaks Bottom in Oregon.

Figure 5.2 Aerial photograph of Salmon Creek Greenway in Vancouver, Washington.

Fredrick Law Olmsted, Sr., observed that Portland was "most fortunate, in comparison with the majority of American cities, in possessing such varied and wonderfully strong and interesting landscape features" (Olmsted and Olmsted 1903: 34). Olmsted proposed that Portland should create a "system of public squares, neighborhood parks, playgrounds, scenic reservations, rural or suburban parks, and boulevards and parkways" built around features that are today among Portland's most treasured natural landscapes (Olmsted and Olmsted 1903: 36–68).

Park and landscape planning at the regional scale began in 1971 when the Columbia Region Association of Governments (CRAG) laid out a bi-state *Urban-Wide Park and Open Space System* based on the premise that "open spaces are needed ... for immediate enjoyment and use within the urban complex" (CRAG 1971: 3–4).

The CRAG report, unfortunately, gathered dust for the next 17 years because it had neither the political nor public support necessary for its implementation. By the mid-1980s public alarm at the loss of local greenspaces (compare Figures 5.3 and 5.4) led to the proliferation of grassroots citizen organizations that grew from a mere handful in the 1980s to a robust network of more than 100 grassroots organizations (Audubon Society of Portland website 2009). Grassroots advocacy resulted in the current effort to create a regional Portland-Vancouver parks and greenspaces system (Howe 1992: 2).

Metro, CRAG's successor and the only directly elected regional government in the United States, initiated a bi-state inventory of natural areas in 1989 (personal communication). Under the direction of Portland State University Geography Department's Dr. Joe Poracsky, color infrared

Figure 5.3 Cedar Mill Creek Watershed in 1990.

Figure 5.4 Cedar Mill Creek Watershed in 2002.

photography of the four county region was used to create a map that for the first time depicted the remaining natural areas (see Figure 5.5). Three years later Metro adopted the *Metropolitan Greenspaces Master Plan* calling for "a cooperative regional system of natural areas, open space, trails, and greenways for wildlife and people" in the four-county metropolitan area that would protect and manage significant natural areas through a partnership with governments, nonprofit organizations, land trusts, businesses and citizens. The system would also preserve the diversity of plant and animal life in the urban environment using watersheds as the basis for ecological planning and restore green and open spaces in neighborhoods where natural areas had all but been eliminated (Metro1992).

In 1995 several Metro Councilors threw their political weight behind the passage of a region-wide bond which produced $135.6 million for urban natural area acquisition. By June, 2002, Metro had acquired over 3321 ha of land far exceeding the original target of 2439 ha. In the fall of 2006 Metro passed a second bond that provides an additional $227.4 million for acquisition of another 2025 ha.

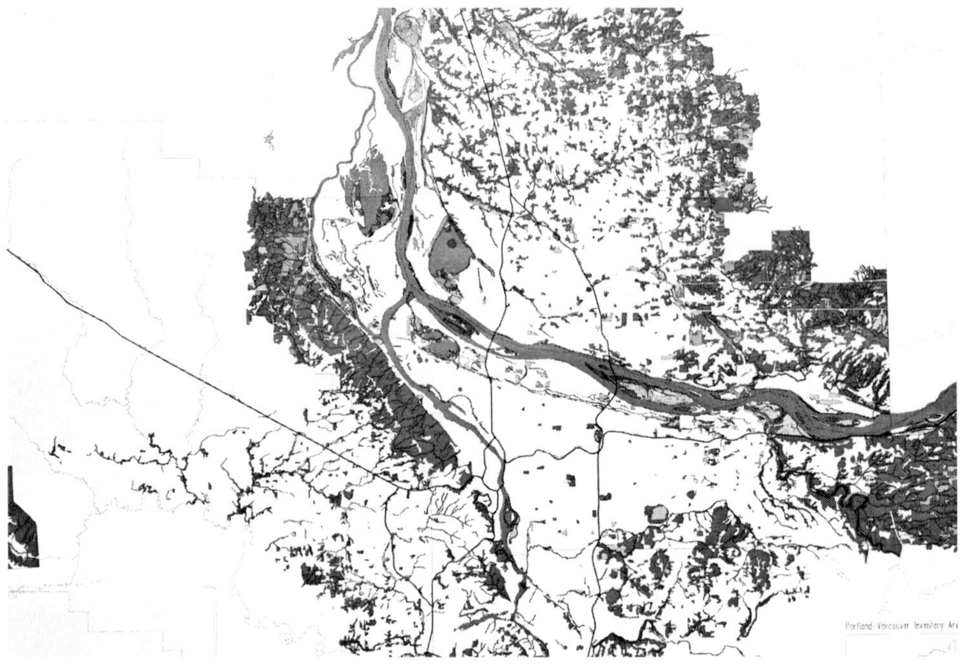

Figure 5.5 Remaining natural areas in Portland, Oregon, from aerial photography.

Role of land use planning in urban nature preservation

In addition to this non-regulatory park and greenspaces effort, land use regulations have been adopted to protect water quality and fish and wildlife habitat and to reduce natural hazards as part of the region's growth management strategy. In August, 2005 Metro Council established a regional *Nature in Neighborhoods* habitat protection and restoration program covering 32,619 ha inside and just beyond the Urban Growth Boundary (UGB) (see Figure 5.6).

This program also includes performance measures such as

> preserving and improving streamside, wetland, and floodplain habitat and connectivity, increasing riparian forest canopy by 10 percent; limiting floodplain development to 10 percent; and preserving 90 percent of forested wildlife habitat within 300 feet [91.5 m] of streams by the year 2015.
>
> *(Metro 2005: 44–46)*

Learning from others

London, England

In 1990, Dr. David Goode, at the time Director of the London Ecology Unit, was invited by citizen park and natural area advocates to speak to the City Club of Portland, the city's pre-eminent civic organization. Goode, an internationally recognized expert on urban nature schemes, shared his experiences with the London Ecology Unit on nature conservation efforts (Goode 1990).

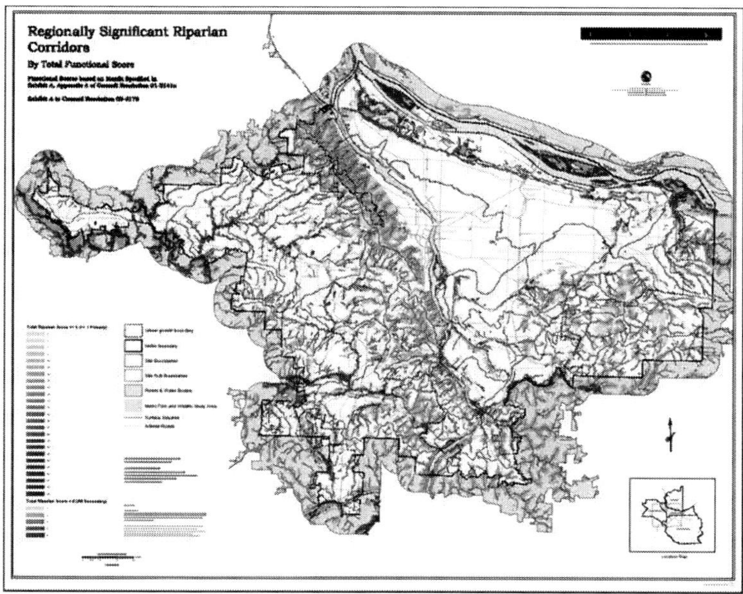

Figure 5.6 Map of riparian corridors indicating fish and wildlife habitats.

The Unit put forth a comprehensive rationale for integrating nature into the city and providing access to nature for multiple benefits including emotional, intellectual, social and physical (London Ecology Unit 1990: 2–3). This work presaged the recent *No Child Left Inside* movement in the United States, which calls for an environmental literacy plan for pre-kindergarten through grade 12 to ensure that elementary and secondary school students are environmentally literate (Government Track website 2009).

A second visit by Goode focused on restoration of what most would consider a hopelessly degraded inner city waste site at what became known as Camley Street Natural Park. Camley Street is a two-acre greenspace situated on the Regent's Canal in a heavily industrialized area of central London next to the Eurostar station. This green gem in the midst of a bleak brownfield site had been a tipping site for many years. The garbage was removed and a beautiful, small wetland with boardwalk and nature center created, which is now managed by the London Wildlife Trust (London Wildlife Trust website 2009).

Camley Street was an inspiring demonstration of how a seemingly irredeemably blighted urban site can be rehabilitated as a "naturalistic" nature site. It also represented an instance where an otherwise nature- and park-impoverished section of London, one with no greenspaces for the nearby low income housing residents, could provide a bit of green for children who had no other avenue for connecting to nature. The Camley Street model greatly accelerated the Portland region's commitment to addressing environmental and social equity and inner city park and greenspace needs.

East Bay, San Francisco

The East Bay Regional Park District (EBRPD) in the San Francisco Bay area of California also provided Portland's greenspace advocates with an on-the-ground demonstration of urban nature preservation in an intensely developed metropolitan region. The District, created in 1934, had passed a $225 million bond measure in 1988 with which they added 8910 ha to their existing 16,200 ha natural areas system.

In 1990 the Portland region's elected officials, citizen advocates, and park managers attended two tours of EBRPD. District staff met with elected officials and took two extended on-the-ground tours which gave politicians, professional park staff, and citizen advocates specific examples that they would soon emulate in their own park bureaus and districts.

East Bay continues to lead the US in urban nature conservation. In 2008 the District passed another bond for $500 million that will allow it to add thousands more hectares to its current 40,000 hectares of urban wilds that are distributed over 4519 km^2 in one of the most heavily populated regions in the United States.

Assessing biodiversity in the Portland-Vancouver metropolitan region

In 1997 the Oregon Natural Heritage Database Project undertook an analysis of biodiversity in the three states of Oregon, Washington and Idaho. The Oregon Gap Analysis Program was a scientifically based program to identify how well native animal species and habitats are represented in the modern landscape. Species and habitats not represented are referred to as conservation "gaps."

The GAP analysis measured biodiversity across the three-state region with mapping units of 101.25 ha spread over a study area of 253,000 km^2 in Oregon. The analysis identified the entire Portland metropolitan area as having low species diversity, not surprising given the city is a mosaic of built and natural environment containing few natural areas exceeding GAP's minimum mapping size (Oregon Natural Heritage Center 1999).

To remedy this problem the US Fish and Wildlife Service contracted with the GAP research team in 2003 to remap the 1036 km^2 within the Portland region's Urban Growth Boundary with a corresponding smaller mapping unit. This refined analysis revealed that there was, in fact, tremendous biodiversity within the urban area, even in the heart of downtown Portland.

This region is home to 289 avian species, 17 amphibians, 16 reptiles, 64 mammals, and 39 fish (Metro 2002; Portland Bureau of Environmental Services 2008; Portland Planning Bureau 2008). Cleary, urban nature was anything but an oxymoron in Portland.

Roughly 5 percent of all peregrine falcons in Oregon nest on the bridges of downtown Portland and the state's most productive nest is on a downtown bridge (Sallinger 2009). Ospreys, which until 15 years ago were rare, now nest throughout downtown Portland along the Willamette River. Bald eagles, also rare until quite recently, are a routine sighting from downtown skyscrapers. At more than 25,000 birds, the world's largest Vaux's swift roost is in a local school chimney near downtown Portland. River otter, deer, beaver, bobcat, fox, coyotes, black bear, and even cougar are all among the region's diverse wildlife.

Grey to green, the role of green infrastructure in naturalizing the city

Green infrastructure is defined by the Green Infrastructure Network as

> strategically planned and managed networks of natural lands, working landscapes and other open spaces that conserve ecosystem values and functions and provide associated benefits to human populations. Green infrastructure networks work together as a whole to sustain ecological values and functions.
>
> *(The Conservation Fund and Sprawl Watch Network website 2002)*

Through its *Grey to Green, Going for Clean Rivers* program Portland acknowledges the city has far to go to adequately protect and restore its watersheds and improve fish and wildlife habitat. The city has committed $55 million to create 17.4 ha of ecoroofs, plant 33,000 yard trees and 50,000 street trees; build 920 Green Street facilities; purchase and protect 170 ha of high priority fish and wildlife habitat; control the spread of invasive plants; and restore native vegetation.

Creating an ecologically sustainable metropolitan region means ecological processes must be considered from a "nested" perspective, telescoping up and down the scale, integrating the built and natural environment, from large regional landscapes to watersheds and sub-watersheds, down to individual neighborhoods and streetscapes. Portland's Watershed 2005 Plan (City of Portland 2005), seeks to "incorporate stormwater into urban development as a resource that adds water quality benefits and improves livability, rather than considering it a waste that is costly to manage and dispose of."

The plan "is built on the principle that urban areas do not have to cause damage to watershed health" and that "a healthy urban watershed has hydrologic, habitat, and water quality conditions suitable to protect human health, maintain viable ecological functions and processes, and support self-sustaining populations of native fish and wildlife species" (City of Portland 2005: 38).

A new look at regional growth management and landscape ecology

> I have found that people who feel very strongly about their own landscape are more often than not the same people who are pushing for better comprehensive planning. But it is the landscape that commands their emotions. Planning that becomes too abstract or scornful of this aspect will miss a vital motivating factor. The landscape element of any long-range regional plan, more than any other element can enlist a personal involvement. People are stirred by what they can see.
>
> *(Whyte 2002: 327)*

Integrating urban and rural landscapes

In June, 2006 a regional mapping charette resulted in an ecologically based map delineating landscapes that ecologists and park planners hope will:

- Preserve significant natural areas for wildlife habitat and public use.
- Enhance the region's air and water quality.
- Connect the region's communities with trails and greenways.
- Provide sense of place and community throughout the bi-state metropolitan region.
- Support an ecologically sustainable metropolitan area.

Information from this charette was integrated with other natural resource data to create a composite map, covering 9375 km^2. The final composite map (see Figure 5.7) included the highest priority conservation areas from state fish and wildlife, The Nature Conservancy, local park providers, and Oregon State Willamette River Basin Planning Atlas (Hulse *et al.* 2002).

This Natural Resources Lands Inventory (NRLI) will be used to make decisions regarding future Urban Growth Boundary expansion through the regional Rural and Urban Reserves planning process (Metro website 2009). Urban reserves will be designated on lands currently outside the urban growth boundary that are suitable for accommodating urban development over the next 40 to 50 years. Rural reserves will be designated on high value working farms,

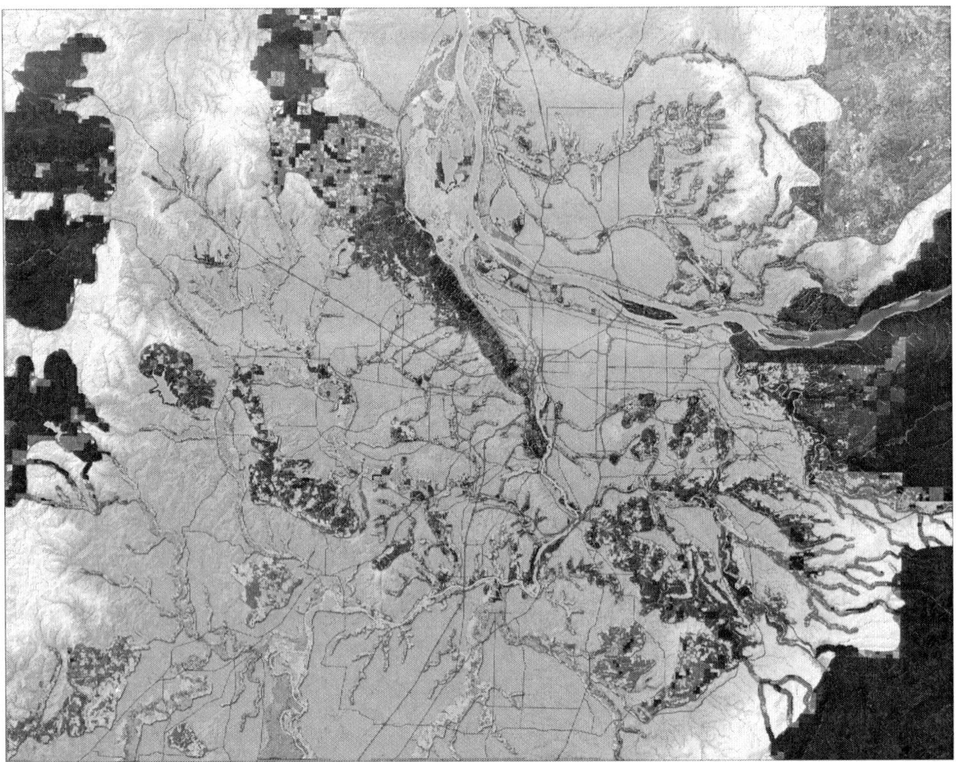

Figure 5.7 Composite map of the Portland-Vancouver region showing land cover types.

forests, and natural features like rivers, wetlands, buttes and floodplains. These areas are to be protected from urbanization for at least the next 40 to 50 years.

Addressing equity and social and environmental justice

The conservation movement has long been criticized for not integrating issues of equity into the region's ecologically based initiatives. In 1994 several NGOs created the Coalition for a Livable Future whose mission is to "protect, restore, and maintain healthy, equitable, and sustainable communities, both human and natural, for the benefit of present and future residents of the greater metropolitan region." The Coalition's *Regional Equity Atlas, Metropolitan Portland's Geography of Opportunity* (Coalition for a Livable Future website 2009), in addition to mapping accessibility to transit, affordable housing, schools, and jobs also provides spatial analysis of access to parks and natural areas throughout the four-county region. This information has had a significant impact on park design and acquisition, with an eye to providing historically underserved communities better access to nature in the heart of the metropolitan region (Houck and Labbe 2007: 44–46).

The Intertwine Alliance

Over the years the single most important ingredient for a successful campaign to bring nature to the city has been establishing political will among elected officials. While much has been accomplished over the past 30 years in the Portland metropolitan region, these gains have come for the most part in spite of, or often in opposition to, the political establishment.

There has, however been a significant shift from the early 1980s when urban wildlife was viewed as an oxymoron, to elected officials embracing the need to integrate nature and green infrastructure into the urban fabric. This is due in large measure to the election of politicians who have recently campaigned on a green agenda. This in turn is largely attributable to grassroots citizen-based work and a growing desire by urban residents for access to nature, parks, and trails in their own neighborhoods and close to work.

In June, 2007 Metro Council President David Bragdon, one of the chief architects of the successful 2006 regional bond measure, dedicated his final term of office to creating "the world's greatest regional system of parks, trails, and natural areas" (Metro 2007). A new, broader coalition of NGOs, including members of the Coalition for a Livable Future, government agencies, and representatives of the health and business sectors has emerged as *The Intertwine Alliance*. The *Alliance* is modeled on the highly successful Chicago Wilderness and is collaborating with Chicago, Houston Wilderness, and the Cleveland, Ohio region's Lake Erie Allegheney Partnership for Biodiversity (LEAP) to share information and to become national models for protecting urban nature and biodiversity.

Chicago Wilderness

One of the preeminent models for urban nature conservation in the United States is Chicago Wilderness, broad based coalition that champions the cause of biodiversity in the Chicago Metropolitan region. Their federal fish and wildlife and forestry partners have ensured that Chicago Wilderness' focus would be on urban biodiversity. Their effort is based on an Atlas of Biodiversity (Chicago Wilderness 1999a) and a Biodiversity Recovery Plan (Chicago Wilderness 1999b). Another hallmark of the Chicago Wilderness program is active engagement with the business community. A Corporate Council helps integrate the region's economic goals with issues related to ecosystem health, and sustainable development.

Houston Wilderness

While one might not immediately link efforts to protect urban nature with Houston, Texas, the Houston Wilderness initiative in southeast Texas seeks to connect the region's six million residents to their natural landscapes. Houston Wilderness consists of 63 organizations that promote protection of biodiversity in the 49,210 km^2, 24 county, Gulf Coast region (Houston Wilderness website 2009).

As with Chicago Wilderness, much of their work is informed by the *Houston Atlas of Biodiversity* which highlights the diversity, cultural importance, and global environmental values found in the Houston region. Taking the lead from the national *No Child Left Inside* movement, the *Children In Nature* program provides local children and families with nature experiences close to home.

Lake Erie Allegheny Partnership for Biodiversity (LEAP)

The Lake Erie Allegheny Partnership for Biodiversity (LEAP) is a consortium of organizations with diverse missions that was initiated in March of 2004. They too have a vast geographic scope that stretches from southern Canada to the Allegheny Mountains and the glaciated region of northeastern Ohio, northwestern Pennsylvania, and western New York. LEAP's membership includes government agencies and multiple conservation organizations involved with conservation of biodiversity, environmental education, and public outreach. Their effort mirrors Chicago, Houston and Portland in focusing on the enhancement of biodiversity and ecosystems (LEAP website 2009).

New York City

New York City's focus on protecting and restoring urban natural areas began long before the *Mannahatta Project*. Of the city's 11,745 park land hectares, more than 4000 ha are wetlands, forests and other natural landscape attractions – small bits of nature that cumulatively are as significant as the more well known Central and Prospect Parks.

In 1984, Parks Commissioner Henry J. Stern founded the Natural Resources Group (NRG) and established as its mission, "To conserve New York City's natural resources for the benefit of ecosystem and public health through acquisition, management, restoration, and advocacy using a scientifically supported and sustainable research" (City of New York Natural Resources Group 1987: 10). The NRG's team of biologists, natural resource managers, mapping scientists and restoration ecologists, develop and implement management programs for protection, acquisition, and restoration of the City's natural resources. Their work has long been recognized for their pioneering research in urban ecology restoration and management.

Conclusions

Attitudes toward the role of nature in the city have changed dramatically, particularly in the United States. Efforts like the *Mannahatta Project*, *The Intertwine*, Chicago Wilderness, Houston Wilderness and Cleveland's LEAP initiative are expanding throughout the United States. The US owes much of the inspiration for the renaissance in urban nature planning to Europe. Recent trips to Amsterdam and Copenhagen have stirred planners, developers and citizen activists in Portland to emulate their European counterpart's cycling, transit, urban design, and urban greenspace policies. Urban nature is no longer considered an oxymoron. It is clear to urban planners, architects and landscape architects, and elected officials that protection and restoration of nature in cities is the key to making cities livable, loveable and ecologically sustainable.

Further reading

Houck, M.C. and Cody, M.J. (2000) *Wild in the City, A Guide to Portland's Natural Areas*, Portland, OR: Oregon Historical Society Press. (A description of 100 of the Portland-Vancouver region's natural areas and historical background to parks, trails, and greenspace planning in the metropolitan region.)

Little, C. (1995) *Greenways for America*, Baltimore, MD: The Johns Hopkins University Press. (A comprehensive history of parks and greenspace planning in Europe and the United States. An inspirational book on greenway and ecosystem planning.)

Louv, R. (2006) *Last Child in the Woods, Saving Our Children from Nature-Deficit Disorder*, Chapel Hill, NC: Algonquin Books. (Louv launched the *No Child Left Inside* movement in the United States, and effort to ensure school children have access to nature near their homes and schools.)

Mabey, R. (1973) *The Unofficial Countryside*, London: Collins. (A natural history of the city and suburbs, where town meets country. Mabey explores bomb sites, docks, garbage sites, and factory walls where he finds a wealth of nature in the city.)

Platt, R.H. (2006) *The Humane Metropolis, People and Nature in the 21st Century*, Amherst, MA: University of Massachusetts Press. (Explores ways to create a more humane metropolis through essays and case studies and builds on the legacy of William H. Whyte to whom the book is dedicated.)

Pyle, R.M. (1993) *The Thunder Tree, Lessons From An Urban Wildland*, New York, NY: Houghton Mifflin

Company. (Everyone should have a ditch, asserts Pyle, who poses the provocative question, "why should a child care about the extinction of the Condor if they haven't seen a wren?" His chapter "The Extinction of Experience" makes the case for the current *No Child Left Inside* movement in the US.)

Schuyler, D. (1986) *The New Urban Landscape, The Redefinition of City Form in Nineteenth-Century America*, Baltimore, MD: The Johns Hopkins University Press. (Schuyler traces the origins of some of America's most important public landscapes and urban parks.)

References

Audubon Society of Portland (2009) *Urban Natural Resources Directory*, online, available at: www.audubonportland.org/issues/metro/urban [accessed September 7, 2009].

Beatley, T. (2000) *Green Urbanism, Learning from European Cities*, Washington DC: Island Press.

Chicago Wilderness (1999a) *An Atlas of Biodiversity, Protected Land in the Chicago Region*, Chicago IL: Chicago Region Biodiversity Council, Chicago Wilderness.

Chicago Wilderness (1999b) *Biodiversity Recovery Plan*, Chicago IL: Chicago Region Biodiversity Council, Chicago Wilderness, pp. 21–22.

Coalition for a Livable Future (2009) *Equity Atlas*, online available at: www.equityatlas.org [accessed September 7, 2009].

City of Copenhagen (2007) *Eco-Metropolis, Our Vision for Copenhagen, 2015*, online, available at: www.sfu.ca/city/PDFs/Eco_Metropolis_gb_2009_web.pdf [accessed September 7, 2009].

City of Helsinki Website (2009) *Central Park, Close to the Forest, Nature in the City. The Planning of Central Park*, online available at: www.hel2.fi/keskuspuisto/eng/5history/ [accessed September 7, 2009].

City of New York Natural Resources Group (1987) *Country In the City*, New York NY: New York City Parks and Recreation, Natural Resources Group.

Conservation Fund and Sprawl Watch Clearinghouse (2002) *Green Infrastructure: Smart Conservation for the 21st Century*, online, available at: www.sprawlwatch.org/greeninfrastructure.pdf [accessed September 7, 2009).

CRAG (Columbia Region Association of Governments) (1971) *A Proposed Urban-Wide Park and Open Space System*, Portland, OR: CRAG, pp. 3–4.

Gerling, C.L. and Kellett, R. (2006), *Skinny Streets & Green Neighborhoods, Design for Environment and Community*, Washington DC: Island Press.

Goode, D. (1990) *Wild in London*, London: Michael Joseph.

Government Track (2009) *Senate Bill 266, U. S. Senate, No Child Left Inside Act of 2009*, online, available at: www.govtrack.us/congress/bill.xpd?bill=s111-866 [accessed September 7, 2009].

Houck, M.C. and Labbe, J. (2007) *Ecological Landscape: Connecting Neighborhood to City and City to Region*, Portland OR: Metropolitan Briefing Book, Institute for Metropolitan Studies, Portland State University.

Houston Wilderness (2009) *Goals and Objectives*, online, available at: www.houstonwilderness.org/default.asp?Mode=DirectoryDisplay&id=99 [accessed September 7, 2009].

Howe, D.A. (1992) *The Environment as Infrastructure: Metropolitan Portland's Greenspaces Program*, Portland, OR: Portland State University.

Hulse, D., Gregory, S., and Baker, J. (2002) *Oregon State Willamette River Basin Planning Atlas, Trajectories of Environmental and Ecological Change*, Corvallis, OR: OSU Press.

LEAP (Lake Erie Allegheny Partnership for Biodiversity) (2009) *Regional Biodiversity Plan*, online, available at: www.leapbio.org/bioplan.php [accessed September 7, 2009].

London Ecology Unit (1990) *Nature Areas for City People, A Guidebook to the Successful Establishment of Community Wildlife Sites, Ecology Handbook 14,* London: London Ecology Unit.

London Wildlife Trust Nature Reserves (2009) *Camley Street Natural Park*, online, available at: www.wildlondon.org.uk/Naturereserves/CamleyStreetNaturalPark/tabid/124/Default.aspx [accessed September 7, 2009].

Matthiessen C.W. (2008) *Infrastructure: Factsheet Denmark*, Copenhagen: Ministry of Foreign Affairs, online, available at: www.netpublikationer.dk/UM/8583/pdf/Infrastructure.pdf [accessed September 16, 2010].

Metro Regional Government Portland (1992) *Greenspaces Master Plan*, Portland OR: Metro Parks and Greenspaces Department.

Metro Regional Government Portland (2002) *Metro's Riparian Corridor and Wildlife Habitat Inventories August 8, 2002, Appendix* 1, Portland OR: Metro Regional Government Portland.

Metro Regional Government Portland (2005) *EXHIBIT 2, Ordinance No. 05.1077B Urban Growth Management Functional Plan, Title 13, "Nature in Neighborhoods,"* Portland OR: Metro Regional Government Portland.

Metro Regional Government Portland (2007) *A Bold New Goal: Connecting Green, Leaders for a Regional Wide Park System, A Call to Action*, Portland OR: Metro Regional Government Portland.

Metro Regional Government Portland (2008) *The Case for an Integrated Mobility Strategy, November, 2008*, online, available at: http://library.oregonmetro.gov/files/08478_brc_final_report_pks_11-12.pdf.

Metro Regional Government Portland (2009) *Urban and Rural Reserves Work Program Overview (July 10, 2009)*, online, available at: www.oregonmetro.gov/index.cfm/go/by.web/id=26257 [accessed September 7, 2009].

Miller, P. (2009) "Before New York, Rediscovering the Wilderness," *National Geographic Magazine*, 126(September): 122–137.

Nicholson-Lord, D. (1987) *The Greening of Cities, Learning from European Cities*, London: Routledge & Kegan Paul.

Olmsted, J.C. and Olmsted Jr., F.L. (1903) *Outlining a System of Parkways, Boulevards and Parks for the City of Portland*, Portland OR: Report of the Park Board.

Oregon Natural Heritage Center (1999) *Oregon GAP Analysis Project, A Geographic Approach to Planning for Biological Diversity*, Corvallis OR: Oregon Natural Heritage Center, Oregon State University.

Portland Bureau of Environmental Service (2008) *Vertebrate Wildlife Habitat Inventory*, Portland OR: Portland Bureau of Environmental Services.

Portland Planning Bureau (2005) *Actions for Watershed Health, 2005 Portland Watershed Management Plan*, Portland OR: City of Portland Planning Bureau.

Portland Planning Bureau (2008) *Willamette River Natural Resources Inventory: Riparian Corridors and Wildlife Habitat, Proposed Draft Report*, Portland OR: City of Portland Planning Bureau.

Sallinger, B. (2008) *Peregrine Falcons in Portland, Oregon* (personal communication, February 2008).

Spirn, A.W. (1984) *The Granite Garden, Urban Nature and Human Design*, New York: Basic Books.

Whyte, W.H. (2002) *The Last Landscape*, Philadelphia: University of Pennsylvania Press.

Wiley, P. (2001) *No Place for Nature, The Limits of Oregon's Land Use Program*, in Protecting Fish and Wildlife Habitat in the Willamette Valley, Washington, DC: Defenders of Wildlife.

6

The human relationship with nature

Rights of animals and plants in the urban context

Jason Byrne

> What have they done to the earth? What have they done to our fair sister? Ravaged and plundered and ripped her and bit her, stuck her with knives in the side of the dawn, and tied her with fences, and dragged her down.
>
> *(Morrison 1967)*

Introduction

Most city dwellers tend to go about their daily lives rarely thinking about the impacts their actions might have upon the 'natural' world. Maybe this is because in cities there are fewer opportunities for encounters with wild animals and plants, or perhaps it is because we seldom associate nature with cities (Douglas 1981; Miller 2005). Our interactions with urban nature are typically limited to:

i exchanges with pets and pests;
ii chance encounters with the few hardy native plants and animals able to co-exist with us; and
iii gardening, recreating, or watching nature documentaries on television.

Yet increasing numbers of scholars have begun to suggest that our interactions with nature are formative in how we see the world, how we treat each other and how we relate to the environment that surrounds us (Kibert 1999; Jackson 2003; Kellert 2004; Kahn Jr. 2005; Miller 2005; Heynen *et al.* 2006). If this is true, urban dwellers' diminished interactions with nature and depauperate understanding of the natural world is alarming. For instance Miller (2005: 430) has recently reported that: "adolescents in south ... Los Angeles [were] more likely to identify correctly an automatic weapon by its report than they [were] a bird by its call". With half of the world's population now living in urban areas this disconnection between urbanites and nature will probably have profound consequences for global ecosystems (Newman *et al.*

2009). We must begin to rethink human relations with nature lest we cause irreparable harm to ourselves, to the biogeochemical systems that sustain us, and to other species with whom we share planet Earth.

Ethics, rights and values are appealing concepts that hold hope for solutions to our current global environmental crisis – and more specifically to the disquieting loss of species from our cities. But looking to "nature's rights" for salvation could invite trouble. Not only do many urbanites have a deeply ambivalent relationship with the natural environment, the whole idea of finding redemption through rights is a veritable "Pandora's Box". This essay is a foray into the tricky realm of the "moral considerability" of nature (Goodpaster 1978), and human obligations to animals and plants. It is beyond the scope of the essay to trace the various Western philosophers and philosophies that have informed debates about whether nature has rights; others have already done that (see for example Dobson 1995; Coates 1998; Low and Gleeson 1998; Torrance 1998; Varner 1998; Cafaro 2001; Palmer 2001; Hay 2002; Kahn Jr. 2005). Seeking guidance from the Greco-Roman foundations of Western thought is equally problematic. As we will see, the Greeks were also deeply divided about their relationship with nature. While Aristotle believed that animals had souls, Plato regarded them as the lowest form of life (Glacken 1967; Plumwood 1993; Dobson 1995; Coates 1998; Torrance 1998; Hay 2002). Instead, this essay considers more recent contributions to the natural rights debate from philosphers like Peter Singer, Arne Naess, Luc Ferrie, Donna Haraway and Val Plumwood, lawyers like Christopher Stone, ecologists like Edward Wilson and Tim Flannery, conservationists like Aldo Leopold and Rachael Carson, environmental historians like William Cronon, and new animal geographers like Jennifer Wolch.

We begin by considering the realities of daily human interactions with nature in urban environments, looking at some of the problematic relationships between people and nature in the city, relationships that date back to at least the Neolithic revolution when humans began to profoundly reconfigure their interactions with nature (Plumwood 1993; Kellert 1997; Coates 1998). As nature became commoditized – capable of exchange and ownership, humans began to see themselves as outside of nature, or at the very least as elevated above plants and animals (Flannery 1994; Leakey and Lewin 1996). Next we take a philosophical turn, concisely examining how the Greeks got us into this dilemma, before exploring what we mean by "rights", "values" and "ethics". We then probe the implications these ideas have for the moral considerability of nature. The essay concludes by considering the rights of non-endemic species like weeds and feral animals, and highlights some problems associated with trying to use the "rights" concept to determine what belongs where, what should be protected, and what is "out of place" and – by some accounts – should be exterminated. These forays are necessarily cursory, and open more questions than they answer, but we must take care before sentencing other life-forms to death in our cities. If nothing else, pondering the complex and oftentimes contradictory ideas of rights and values might help us make more informed decisions.

Urban animals and animal urbanism

Animals have been crucial to the development of human civilizations, playing major roles in transportation, warfare, fashion, religion, entertainment, communication, companionship and sustenance. For example, the bodies of animals have yielded: fat for soaps, perfumes and cosmetics and flesh, bone, sinews and feathers for food, medical and religious purposes. We have used skins for clothing, book binding, bags, shoes, drums and furniture; and sinews, bone, teeth, feathers and wool for tools, pens, jewellery, musical instruments, blankets and paintbrushes (Wolch *et al.* 2003). Animal muscle power has tilled fields, drawn carriages, and hauled timber and stone. In many ways, our cities are founded on animals (McShane and Tarr 2007; Shepard 1997).

Although it has been fashionable throughout the ages to claim that cities are inherently "unnatural", denying the presence of the myriad species that share urban spaces with us is misleading. Our cities are not "dead zones"; nature clearly permeates our urban environments, and animals still inhabit most cities in surprisingly large numbers (Douglas 1981; Platt *et al.* 1994; Davis 2003; Heynen *et al.* 2006; Wolch 2007). Opportunistic species in particular seem to flourish in cities. Our houses, backyards, parks and landfills create many opportunities for a wide variety of plants and animals. Old trees, garden sheds, roof cavities, abandoned vehicles and decrepit factories provide spaces for hibernation, denning, nesting and foraging. Roof gutters offer a source of drinking water, garden ponds provide habitat for aquatic species, and vacant lots and abandoned car bodies provide shelter and habitat for terrestrial ones (Byrne and Wolch 2009). Urban environments may actually provide better prospects for the flourishing of some species than many "wildland" areas (Hoffman and Gottschang 1977; Rebele 1994; Riley *et al.* 1998; Schaefer 2003; Gehrt 2004; Mannan and Boal 2004).

In Los Angeles for example, feral parrots screech across suburban skies, in the Hollywood Hills coyotes prey on pets; mountain lions stalk Orange County trail users, and opossums and skunks raise their young in San Gabriel backyards (see Figure 6.1). In downtown New York, falcons dine on pigeons; Londoners share their city with sparrows, foxes, deer and the occasional badger. A multitude of birds and animals still flourish in Australian cities. It is not uncommon to see magpies or crows in inner-city Sydney; kangaroos frequenting suburban golf courses in Perth, Canberra and Brisbane, pythons, possums and fruit bats in Gold Coast yards, with water

Figure 6.1 Opossum (*Didelphis marsupialis*) traversing a backyard fence in San Gabriel, California.

dragons patrolling the city's beaches and bull-sharks menacing its canal estates (see Figure 6.2). But human-animal interactions in cities are characterized by both affection and antimony. Many opportunistic species face massive eradication efforts (e.g. seagulls in landfills, Canada Geese near airports and White Ibis in parks) (Belant 1997; Gosser *et al.* 1997; Martin *et al.* 2007). The ability of urban wildlife to coexist with humans depends upon the time, place and scale of human-animal interactions. Issues of seasonality – such as breeding cycles, and the duration, intensity and predictability of interactions are important, as are the types of animals involved, their overall health, and their body size, behavioural adaptability, social group size, age and sex (Seymour *et al.* 2006; Byrne and Wolch 2009).

Unfortunately cities can have severe impacts on urban wildlife, typically through exploitation, disturbance, habitat modification and pollution. Exploitation results in the death of animals as a direct result of human interaction, including hunting, trapping, fishing or collection. Disturbance may be either unintentional (e.g. accidentally scaring a nesting bird) or intentional (e.g. frightening a deer to get a good photograph). Habitat modification typically results from vegetation clearing or damage, the introduction of invasive plant species or the release of diseases, predators or competitors. Pollution may occur in a variety of forms including noise pollution, light pollution, visual intrusion, and air, water and soil contamination (through activities such as applying pesticides, dumping trash, or contaminants from storm-water runoff) (Rich and Longcore 2006; Seymour *et al.* 2006). Other negative impacts include electrocution from overhead powerlines, poisoning from insecticides, avicides, or rodenticides, and collision with vehicles or with glass

Figure 6.2 Water dragon (*Physignathus lesueurii*) sunbathing on a Gold Coast metropolitan beach, Australia.

windows. While some animals consequently modify their behaviour in cities in response to these problems, for example coyotes and bobcats become more nocturnal in the presence of humans, others are driven away entirely (Byrne and Wolch 2009).

But these factors are not the only determinants of animal-flourishing in cities. Ideologies surrounding the control of nature underpin many of our interactions with urban plants and animals, and are central to how we perceive and behave towards them. In her examination of the multiple meanings of domestication, Kay Anderson (1997) has argued that all animal practices are connected to power and identity. Zoos for instance are not just stationary animal exhibits; they are central to the formation of human and even national identity – representing colonial conquest over exotic species and places. And certain animal practices conceal undercurrents of power; there are strong connections between race, gender and representations of "animality" (Anderson 1997).

For example, observing that Nazi Germany was the first country to develop nature conservation legislation, French philosopher Luc Ferry (1995) has shown how Nazi environmentalism was born not of enlightened ethics but rather paranoia over foreign incursion and a xenophobic drive to protect the "purity" of the Aryan nation. Animal geographer Chris Philo (1998) has similarly shown how in the nineteenth century, particular urban places such as slaughterhouses, animal markets and proximate working class residents became coded as "impure", "unhygienic", "promiscuous" or even "wild and savage" based upon the putative habits of the animals that inhabited these spaces. More recently, Jennifer Wolch – also an animal geographer – and her colleagues have noted that in the United States, some Asians and Latinos have been maligned for animal practices that transgress white cultural norms; practices such as dog-eating, some types of hunting, and some religious activities which stigmatise these groups as "other", "beastly" and even "inhuman" (Wolch *et al.* 2003). So what are the origins of these ideas of nature?

Origins of Western domination of nature

A longstanding intellectual and moral schism between humans and "nature" has underpinned much Western thought (Collingwood 1960). We inherited this dualistic thinking from philosophers like Plato, Kant and Descartes (Plumwood 1993), thinkers who believed that humans were "rational" creatures whereas animals and plants were riven by base instincts. While there is insufficient space to visit all these thinkers in any detail here, two warrant closer attention – Plato and Descartes.

According to the late Val Plumwood (1993) – a philosopher and ecofeminist – the ontological separation of human from nature can be traced back to Greek philosophy and to Plato specifically. Plumwood has unravelled the evolution of dualistic thinking in Western culture to show how Plato (and later Descartes) endowed Western thought with a series of dualisms including, among many others, the separation of mind from body, male from female, master from slave, rationality from emotion, universal from particular, and culture from nature (Plumwood 1993: 43). A hierarchical reasoning underpins these dualisms, positing one as superior to the other, and naturalizing multiple oppressions such as sexism, racism and speciesism (Singer 2002).

Dualistic thinking is inherently premised upon what Plumwood calls "backgrounding" and "denial"; the dominant and oppressive ignores its dependence upon the subordinated, and backgrounds it to privilege the "master view". For example, Plato saw the rightful place of humans as being with "the divine", yet militarism, misogyny and elitism underpinned much of his thought. Plato regarded women as primitive, chaotic, emotional, incompetent, animal-like and gripped by base appetites (Plumwood 1993: 77). Strikingly, Plato also thought that animals descended from humans in a bizarre evolutionary inversion: animals' lack of reason deformed their bodies and drew them close to the earth, away from the divine above (one does wonder though how he resolved the transgressions of birds). Indeed, the deprecation of nature, evident in much of Plato's writing, stems from his ideal(ist) ontology which valued death over life, denied dependency on the natural world, and promoted hyperseparation.

René Descartes, a seventeenth century French philosopher, mathematician and pre-Enlightenment thinker, has also been strongly influential in shaping the dualistic thinking undergirding modern human-animal interrelations. Like Plato, Descartes saw reason, the opposite of nature, as separating humans from animals. But for Descartes, the "basis of [the] mind [shifted] from rationality to consciousness" (Plumwood 1993: 112) and being "unconscious", nature was thus rendered mindless. Denying animals as "mindful" beings, Descartes recast them as organic machines, to be controlled and used. As machines, animals could feel no pain, removing the "possibility for mutual recognition and exchange" (Plumwood 1993: 117). For Descartes, since animals could possess no "true" sensation, even their "aliveness" came into question. His mechanistic conception of nature paved the way for contemporary understandings of humans as being outside nature, and contemporary practices that result in the thoughtless destruction of countless animal and plant lives.

But Marxist geographer David Harvey (2000) in his essay on architects, bees and "species being", has cautioned us to the perils of dualistic thinking. He says that we have much in common with organisms like beavers, termites and even cyano-bacteria, organisms that also modify their environments for their own benefit. And a bevy of scientists have recently corroborated these ideas, adding to overwhelming evidence that humans are less unique than we once thought, with startling discoveries that challenge the indelible marks of humanity (e.g. emotion, planning, self-awareness). Our supposedly "human" qualities may not be so exclusive. We share many of our capabilities with other animals. Cuttlefish can conceal their private "conversations" from conspecifics (Palmer *et al.* 2006); crows are able to solve difficult spatial problems (Emery and Clayton 2004); some primates appear capable of premeditated actions (e.g. chimpanzees that cache stones to throw at zoo visitors) (Osvath 2009); and elephants may be self-aware (Plotnik *et al.* 2006).

Whether these examples offer glimpses into the "souls" of other organisms – as Aristotle would have it (Collingwood 1960) – or simply reflect genetic adaptations and predispositions is open to debate. What these radical findings challenge though is entrenched and archaic notions that non-human organisms are incapable of feeling pain, having emotions or planning for the future. Based on findings like these, some scientists including biologist Edward O. Wilson and ecologist Stephen Kellert have argued that humans have a kinship with non-human species – a "biophilia", and as fellow animals and "ecological citizens" we are morally obliged to care for other species (Wilson 1992; Kellert 1997, 2004). Arguably what makes us different is this capacity for caring – our ability to ponder our impacts, and to contemplate the "rightness", "goodness" or "appropriateness" of our actions. This is why some people believe that the idea of "animal rights" is the solution to our environmental problems.

The rights of urban animals and the boundaries of moral considerability

Clearly, our ideas of "nature" inform how we interact with nature (Talbot and Kaplan 1984; Kellert 1997; Kaplan and Kaplan 2008; Matsuoka and Kaplan 2008). Most ordinary people – conditioned by philosophers like Bentham, Kant, and Descartes see no problem with using "natural resources" to benefit humanity, even if this means harming other species (Dobson 1995; Low and Gleeson 1998; Palmer 2001; Hay 2002). The problem seems to be inconsistencies in what we see as our obligations to nature, and incommensurable differences in the ethics and values we use to guide how we treat other organisms. For instance, we may protest the bashing of fur seals or harpooning of whales – donating money to organizations like Sea Shepherd to intervene on our behalf, but then conveniently overlook the destruction of tropical rainforests when we purchase teak furniture for our lounge-rooms or use palm sugar in our kitchens (Low and Gleeson 1998). We often struggle when we try to determine how best to consider the "rights" of other species inhabiting our urban environments. Possibly this may have something to do with the whole concept of "rights".

Does nature have rights?

Does nature have rights? Are plants and animals worthy of moral considerability? What are the moral obligations of humans to nature? These questions have ignited fierce and impassioned debates among philosophers, ecologists, clergy, property developers, town planners, conservationists, aboriginal peoples and many others. The notion of the "rights" of nature and of ecological ethics and values fits within a frame known as ecological justice (Baxter 2005). Ecological justice is a philosophical and moral position that addresses how humans relate with non-human species and the natural world. Sometimes called justice to nature, it seeks to delineate our moral obligations to other species. Justice here refers to the idea of fair treatment, of equity or even equality. Many proponents of ecological justice assert that nature has intrinsic value – it is valuable in and of itself, outside of any benefits to humans or human measures of worth. Many ecological justice advocates also acknowledge the interconnections and mutual interdependence of all species. They follow in footsteps of Aldo Leopold (1989: 225) who stated that: "a thing is right when it tends to preserve the integrity, stability, and beauty of the biotic community. It is wrong when it tends otherwise." Some commentators like lawyer Christopher Stone have even sought to expand this domain of "moral considerability" beyond humans to encompass animals, plants and even inanimate objects like rocks, rivers and oceans (Stone 1972), with mixed results – a point we return to shortly.

There are several foundations of ecological justice. Religious grounds posit humans as custodians of the natural world; they are typically founded on humans' moral responsibility to other species, which stem from supernatural entities (God, Buddha, Allah, Dreamtime beings, etc.) (Low 1999; Palmer 2001; Baxter 2005). Instrumental grounds – founded on Jeremy Bentham's idea of utility – that is the greatest good for the greatest number – acknowledge that current and future generations of humans are reliant upon the natural world for their needs (e.g. food, medicine and clothing) and nature should be protected to prevent the mental, physical or emotional suffering of other humans. Finally, rights-based notions of ecological justice are founded on the idea that an individual is deserving of moral and legal protection.

Christopher Stone (1972: 451) suggests that several criteria must be satisfied for rights to exist. First, a public body must be able to hold an individual, corporation or other entity accountable for their actions. Second, there must be "procedural safeguards" to ensure that if an entity is wronged, it has recourse to punitive action. Third, for rights to exist an entity must also be able to initiate legal actions, courts must recognize that the entity is capable of being harmed/injured, and "relief" or remedial actions must benefit the entity.

But a problem with rights thinking is that extending rights to animals is inherently anthropocentric and represents an egoistic extension of the human self (Plumwood 1993: 179), tantamount to erasing the "otherness" of animals. Similar problems occur with most deep ecology thinking, which results in what Plumwood (1993: 160) refers to as a blurring of the boundaries between self and other: "[a] difference-denying assimilation" of nature and "devouring [of] the other" (Plumwood 1993: 192). Another issue is that proponents of rights typically apply them to the individual (Regan 1999), whereas many environmental problems occur at the level of ecosystems. What this means is that animal rights activists consider that a feral fox may be worthy of being protected from poisoning due to the pain it would suffer, but a vulnerable ecosystem will have no rights-based protection.

Worse still, if we extend rights to nature, we can get into some pretty ridiculous quandaries. For instance, what right does a lion have to eat a zebra? (Regan 1999). What right do we have to grow vegetables, undertake experiments or take vaccines? Should bacteria be able to sue us for taking antibiotics, or mold for cleaning our bathrooms? And what rights might the atmosphere have against pollution, even from volcanoes? Perhaps the best approach is to do away with the concept of rights altogether and to look for better alternatives. But what are these alternatives? Goodpaster (1978: 316) offers us a clue; he suggests that we must move beyond notions of whether a being is capable of experiencing suffering to considering whether a being has a drive, intentionality or other purpose to live. As if anticipating some of the problems described here, Goodpaster argues that although there are: "limits to the operational character of respect for living things ... the regulative character of ... moral consideration ... [for] all living things asks ... for sensitivity and awareness, not for suicide" (Goodpaster 1978: 324).

Conclusion: are animals just "strange people"?

Drawing on some of the scientific evidence discussed earlier in this essay, the new animal geographers tell us that animals have subjectivity and agency (Philo and Wilbert 2000). They are capable of complex thoughts and emotions, can resist human interventions, will follow their own agendas, and have their own will to flourish. Animals and even plants are not automatons driven by instinct or biochemical processes alone – they actively interact with their environment in complex ways. To paraphrase Wolch *et al.* (2003), people, animals and plants are enmeshed in intricate webs of relations upon which their mutual wellbeing depends.

To arbitrarily discriminate against animals and plants then, on the basis of supposed "inherently innate characteristics" like sentience, cognition, emotion or genes can take us into very troubling moral and legal terrain. Inter-species organ transplants, zoonotic diseases, genetically modified crops, human-animal tissue cultures, and other transgressions of species boundaries have blurred the distinction between human, animal and plant – we are increasingly living in a hybridized or "cyborg" world where machines act intelligently, animals speak in sign language, and genes can be patented by multi-national corporations (Haraway 1997). A rights-based approach to human-nature interrelations creates all sorts of irreconcilable dilemmas. Is a person with pig organs still human? Is someone whose ear has been grown on the back of a mouse still a person (Haraway 1997)? Why does a severely mentally disabled child have moral status when a more intelligent octopus does not (Singer 1999, 2002)? And could self-aware computers be murdered? What about the moral considerability of species we designate as feral or weeds, simply because they are "out of place"? Whatever the ecological harm these species might cause, do we really have the "right" to exterminate them? Could alternatives such as immuno-contraception or predator-aversive conditioning work better than mass-poisoning, shooting, trapping and infection with viral-control agents – and the pain and suffering they cause (Gustavson *et al.* 1974; Miller *et al.* 1998)? As Wolch *et al.* (2003: 192) argue: "[s]ince humans cannot be disentangled from non-humans, non-humans, [are] best seen as 'strange persons' to be treated ... in the same way as human groups".

What we are left with then is the need for a "situated" or "relational" ethics, an ethics that recognizes the interconnections between human, animal and plant, an ethics that is context-dependent (Warren 1999). Such an ethics will enable us to break out of the constraints of dualistic thinking, and recover notions of both continuity and difference with the natural world, without falling into the trap of assimilation. We need an ethics premised upon the idea of mutualism, a virtue-based ethics of care (Michel 1998), where respect, sympathy, concern, gratitude, kinship/friendship and love guide our actions towards nature. An integral part of such an ethics is recognizing nature's telos or intentionality through the drive of other species for growth and flourishing.

But genuinely recognising and coexisting with urban nature – what Wolch (1996) has termed "zoöpolis" – is challenging. Pet euthanasia, wildlife extermination, pest eradication, pollution and ecosystem appropriation will need to give way to new practices that include plants and animals within the circle of moral considerability – an ethics of caring and respect based on kinship but also difference. Urban forests, wildlife corridors, adaptive re-use of buildings, green-roofs, ecological restoration, permaculture and other practices central to sustainable cities will take us some of the way towards zoöpolis. But we will also need to change land use regulations, landscaping practices, building design, transportation systems, food production and distribution, medical technologies, cosmetic manufacturing and energy generation to actively accommodate the needs of non-human species who share our cities (Wolch 2007). Perhaps most of all, we will need to engender a relational understanding of the natural world where humans and human practices are reconfigured as part of nature – not apart from nature.

References

Anderson, K. (1997) "A walk on the wild side: a critical geography of domestication", *Progress in Human Geography*, 21(4): 463–85.

Baxter, B. (2005) *A Theory of Ecological Justice*, London: Routledge.

Belant, J.L. (1997) "Gulls in urban environments: landscape-level management to reduce conflict", *Landscape and Urban Planning*, 38(3–4): 245–58.

Byrne, J. and Wolch, J. (2009) "Urban habitats/nature", in N. Thrift and R. Kitchin (eds), *International Encyclopedia of Urban Geography Vol. 12*, Oxford: Elsevier, pp. 46–50.

Cafaro, P. (2001) "Thoreau, Leopold, and Carson: toward an environmental virtue ethics", *Environmental Ethics*: 22(spring): 3–17.

Coates, P. (1998) *Nature: Western Attitudes Since Ancient Times*, Berkeley: University of California Press.

Collingwood, R.G. (1960) *The Idea of Nature*, New York: Galaxy Books.

Davis, M. (2003) *Dead Cities: And Other Tales*, New York: The New Press.

Dobson, A. (1995) *Green Political Thought*, London: Routledge.

Douglas, I. (1981) "The city as an ecosystem", *Progress in Physical Geography*, 5(3): 315–67.

Emery, N.J. and Clayton, N.S. (2004) "The mentality of crows: Convergent evolution of intelligence in corvids and apes", *Science*, 306(5703): 1903–7.

Ferry, L. (1995) *The New Ecological Order*, Chicago: University of Chicago Press.

Flannery, T. (1994) *The Future Eaters: An Ecological History of the Australasian Lands and People*, Kew, Victoria: Reed Books.

Gehrt, S.D. (2004) "Ecology and management of striped skunks, raccoons and coyotes in urban landscapes", in N. Fascione, A. Delach and M.E. Smith (eds) *People and Predators: From Conflict to Coexistence*, Washington, D.C.: Island Press, pp. 81–104.

Glacken, C.J. (1967) *Traces on the Rhodian Shore: Nature and Culture in Western Thought from Ancient Times to the end of the Eighteenth Century*, Berkeley, CA: University of California Press.

Goodpaster, K. (1978) "On being morally considerable", *The Journal of Philosophy* 75(6): 308–25.

Gosser, A.L., Conover, M.R. and Messmer, T.A. (1997) *Managing Problems Caused by Urban Canada Geese*, Logan, UT: Berryman Institute, Utah State University.

Gustavson, C.R., Garcia, J., Hankins, W.G. and Rusiniak, K.W. (1974) "Coyote predation control by aversive conditioning", *Science* 184(4136): 581–3.

Haraway, D.J. (1997) *Modest_Witness@Second_Millennium. FemaleMan©_Meets_OncoMouse™*, New York: Routledge.

Harvey, D. (2000) *Spaces of Hope*, Berkeley: University of California Press.

Hay, P. (2002) *Main Currents in Western Environmental Thought*, Sydney: University of New South Wales Press.

Heynen, N., Kaika, M. and Swyngedouw, E. (eds) (2006) *In the Nature of Cities: Urban Political Ecology and the Politics of Urban Metabolism*, London: Routledge.

Hoffman, C.O. and Gottschang, J.L. (1977) "Numbers, distribution, and movements of a raccoon population in a suburban residential community", *Journal of Mammalogy*, 58(4): 623–36.

Jackson, L.E. (2003) "The relationship of urban design to human health and condition", *Landscape and Urban Planning*, 64(4): 191–200.

Kahn Jr., P.H. (2005) "Encountering the other", *Children, Youth and Environments*, 15(2): 392–97.

Kaplan, R. and Kaplan, S. (2008) "Bringing out the best in people: a psychological perspective", *Conservation Biology*, 22(4): 826–29.

Kellert, S. (1997) *Kinship to Mastery: Biophilia in Human Evolution and Development*, Washington, D.C.: Island Press.

Kellert, S. (2004) "Ordinary nature: the value of exploring and restoring nature in everyday life", in W.W. Shaw, L.K. Harris and L. Vandruff (eds) *4th International Urban Wildlife Symposium*, Tucson, AZ: The University of Arizona, pp. 9–19.

Kibert, C.J. (ed.) (1999) *Reshaping the Built Environment: Ecology, Ethics and Economics*, Washington, D.C.: Island Press.

Leakey, R. and Lewin, R. (1996) *The Sixth Extinction: Patterns of Life and the Future of Humankind*, New York: Anchor Books.

Leopold, A., with Schwartz, C.W. and Finch, R. (1989) *A Sand County Almanac, and Sketches Here and There*, New York: Oxford University Press.

Low, N. (ed.) (1999) *Global Ethics and Environment*, London: Routledge.

Low, N. and Gleeson, B. (1998) *Justice, Society and Nature: An Exploration of Political Ecology*, London: Routledge.

Mannan, R.W. and Boal, C.W. (2004) "Birds of prey in urban landscapes", in N. Fascione, A. Delach and M.E. Smith (eds) *People and Predators: from Conflict to Coexistence*, Washington, D.C.: Island Press, pp. 105–17.

Martin, J.M., French, K. and Major, R.E. (2007) "The pest status of Australian white ibis (Threskiornis molucca) in urban situations and the effectiveness of egg-oil in reproductive control", *Wildlife Research*, 34(4): 319–24.

Matsuoka, R.H. and Kaplan, R. (2008) "People needs in the urban landscape: Analysis of Landscape and Urban Planning contributions", *Landscape and Urban Planning*, 84(1): 7–19.

McShane, C. and Tarr, J.A. (2007) *The Horse in the City: Living Machines in the Nineteenth Century*, Baltimore, MD: Johns Hopkins University Press.

Michel, S.M. (1998) "Golden Eagles and the environmental politics of care", in J. Wolch and J. Emel (eds), *Animal Geographies: Place, Politics and Identity in the Nature-Culture Borderlands*, London: Verso, pp. 162–87.

Miller, J.R. (2005) "Biodiversity conservation and the extinction of experience", *Trends in Ecology and Evolution*, 20(8): 430–4.

Miller, L.A., Johns, B.E. and Elias, D.J. (1998) "Immunocontraception as a wildlife management tool: some perspectives", *Wildlife Society Bulletin*, 26(2): 237–43.

Morrison, J. (1967) "When the music's over", *Strange Days*, Los Angeles, CA: Elektra Records.

Newman, P., Beatley, T. and Boyer, H. (2009) *Resilient Cities: Responding to Peak Oil and Climate Change*, Washington, D.C.: Island Press.

Osvath, M. (2009) "Spontaneous planning for future stone throwing by a male chimpanzee", *Current Biology*, 19(5): R190–1.

Palmer, J.A. (ed.) (2001) *Fifty Key Thinkers on the Environment*, London: Routledge.

Palmer, M.E., Calvé, M.R. and Adamo, S.A. (2006) "Response of female cuttlefish Sepia officinalis (Cephalopoda) to mirrors and conspecifics: evidence for signaling in female cuttlefish", *Animal Cognition*, 9(2): 151–5.

Philo, C. (1998) "Animals, geography and the city: Notes on inclusions and exclusions", in J. Wolch and J. Emel (eds.), *Animal Geographies: Place, Politics and Identity in the Nature-Culture Borderlands*, London: Verso, pp. 51–71

Philo, C. and Wilbert, C. (eds) (2000) *Animal Spaces, Beastly Places: New Geographies of Human-Animal Relations*, London: Routledge.

Platt, R.H., Rowntree, R.A. and Muick, P.C. (eds) (1994) *The Ecological City: Preserving and Restoring Urban Biodiversity*, Amherst, MA: The University of Massachusetts Press.

Plotnik, J.M., de Waal, F.B.M. and Reiss, D. (2006) "Self-recognition in an Asian elephant", *Proceedings of the National Academy of Sciences of the United States of America:* 103(45): 17053–7.

Plumwood, V. (1993) *Feminism and the Mastery of Nature*, London: Routledge.

Rebele, F. (1994) "Urban ecology and special features of urban ecosystems", *Global Ecology and Biogeography Letters*, 4(6): 173–87.

Regan, T. (1999) "Mapping human rights", in N. Low (ed.), *Global Ethics and Environment*, London: Routledge, pp. 158–174.

Rich, C. and Longcore, T. (2006) *Ecological Consequences of Artificial Night Lighting*, Washington, D.C.: Island Press.

Riley, S.P.D., Hadidian, J. and Manski, D.A. (1998) "Population density, survival, and rabies in raccoons in an urban national park", *Canadian Journal of Zoology*, 76(6): 1153–64.

Schaefer, V. (2003) "Green links and urban biodiversity – an experiment in connectivity", in S. Brace (ed.), *The Georgia Basin/Puget Sound Research Conference*, Vancouver, BC: Puget Sound Action Team, pp. 1–9.

Seymour, M., Byrne, J., Martino, D. and Wolch, J. (2006) *Green Visions Plan for 21st Century Southern California: A Guide for Habitat Conservation, Watershed Health, and Recreational Open Space. 9. Recreationist-Wildlife Interactions in Urban Parks*, Los Angeles, CA: University of Southern California, GIS Research Laboratory and Center for Sustainable Cities.

Shepard, P. (1997) *The Others: How Animals made us Human*, Washington, D.C.: Island Press.

Singer, P. (1999) "Ethics across the species boundary", in N. Low (ed.), *Global Ethics and Environment*, London: Routledge, pp. 146–157.

Singer, P. (2002) *Animal Liberation: A New Ethics for Our Treatment of Animals*, New York: Harper Collins.

Stone, C. (1972) "Should trees have standing? – Towards legal rights for natural objects", *Southern California Law Review*, 45(2): 450–501.

Talbot, J.F. and Kaplan, R. (1984) "Needs and fears: The response to trees and nature in the inner city", *Journal of Arboriculture* 10(8): 222–8.

Torrance, R.M. (ed.) (1998) *Encompassing Nature, A Sourcebook: Nature and Culture from Ancient Times to the Modern World*, Washington, D.C.: Counterpoint.

Varner, G. (1998) *In Nature's Interests?: Interests, Animal Rights and Environmental Ethics*, Oxford: Oxford University Press.

Warren, K. (1999) "Care-sensitive ethics and situated universalism", in N. Low (ed.), *Global Ethics and Environment*, London: Routledge.

Wilson, E.O. (1992) *The Diversity of Life*, New York: W.W. Norton and Company.

Wolch, J. (1996) 'Zoopolis', *Capitalism Nature Socialism*, 7(2): 21–47.

Wolch, J. (2007) "Green urban worlds", *Annals of the Association of American Geographers*, 97(2): 373–84.

Wolch, J., Emel, J. and Wilbert, C. (2003) "Reanimating cultural geography", in K. Anderson, M. Domosh, S. Pile and N. Thrift (eds), *Handbook of Cultural Geography*, London: Sage, pp. 184–206.

7

Urban natural histories to urban ecologies

The growth of the study of urban nature

Ian Douglas and David Goode

The wild, that which is free from human control, has always had a fascination for people. It sets up ambivalent emotions, of beauty and attraction on the one hand, and of fear and even loathing on the other. Contrasted with the wild, is the image of the well-ordered place, where flowers blossom, trees bear fruit and the birds sing. Some claim the latter is the image of Eden, a beautiful calm garden set against the barren landscape of the desert, or wilderness (Mitchell 2001). The cities of Mesopotamia, developed in such a setting and supported by a highly regulated irrigation agriculture had parks and gardens, some with freely ranging wild animals, for their wealthier citizens. To such people, the idea of valuing wild nature in the city, as opposed to the selected flora and fauna of the gardens parks, would have been strange and uncivilized. In contrast, from the Tang dynasty onwards scholars, poets and officials in China created town gardens containing all the elements of nature, bringing a sense of harmony with the natural world to the places where they lived and worked.

From ancient times people have been intrigued by the way that nature is always present in the city and many have commented on the arrival of new species and their tendency to spread, whether people want them or not. Weeds were well known in Biblical times.

By the Middle Ages, most attention to plants was for their medicinal value. Plants were listed in 'Herbals' so that they could be recognized by those seeking to make potions and cures. Local plants descriptions were developed as part of the medical curriculum, particularly for those cities where students could be taken out into the field to learn to recognize the commoner plants believed to have healing properties (Allen 2010). From these herbals, local descriptions of plants gradually came to place more emphasis on other uses, particularly those relevant to agriculture and horticulture. For example in his flora of the London region, Curtis (1777) wrote about the uses and sowing of various meadow grasses, providing far more information than most of the general illustrated floras of that time.

By the eighteenth century, following the advent of more scientific exploration, such as the voyage of Sir John Banks with Captain Cook to Australia, the recording and classifying of fauna and flora began to be more widespread and more organized. Not only did new works on European cities and the environs appear (e.g. on London (Curtis 1777), Paris (Bulliard 1780)) but also the new colonial cities and their surroundings were investigated. One of the first accounts of the flora of New York City and its environs was published by Cadwallader Colden with his *Plantae Coldenhamiae* (1743, 1751).

By the mid-nineteenth century, investigations of fauna and flora in and around major cities became more rigorous and more concerned with the conditions under which organisms survived. Young scientists were encouraged by enthusiastic publishers. A new flora of Middlesex, which then included London (Trimen and Thiselton-Dyer 1868), reflected this emerging trend towards a more ecological botany.

By the last two decades of the nineteenth century London had developed its railway-linked suburban growth and the redevelopment of inner London had important consequences for wildlife. In Kensington Gardens, the great rookery based on a grove of 700 large trees became deserted in 1880 because most of the trees were cut down. The naturalist W.H. Hudson was furious: 'I can now feel nothing but horror at the thought of the unspeakable barbarity the park authorities were guilty of in destroying this noble grove' (Hudson 1898). However, the new transport infrastructure was providing opportunities for some plants and animals. In the 1890s Jefferies described how the plants of the countryside had been displaced by the suburban spread of London and had taken refuge on railway embankments or in cuttings (Jefferies 1883).

Hudson's books, not only on London, but particularly his *Nature in Downland* (Hudson 1900), helped to spread a changing attitude to nature at a time when the countryside was becoming more accessible by public transport and when more people began to live in suburban houses with gardens (Allen 2010).

The formal study of the flora of New York City began with John Torrey and his catalogue of plants found in the vicinity of New York (Torrey 1819). Beginning in the 1880s with the creation of the New York (Bronx) Botanical Garden, a significant effort was made under the aegis of the Torrey Botanical Club to collect plant species from the metropolitan area (Rusby 1906). Historically many plant species throughout New York City, were found or collected from locales that were not parkland at the time; these areas have since been developed (see Sefferien 1932; Kieran 1959). In the Bronx and Manhattan, most of the parks with the largest natural areas were established in the nineteenth century, while in Brooklyn, Queens and Staten Island, most parks were established in the twentieth century (DeCandido *et al.* 2004).

Around the same time, the vision of naturalistic parks as a key element in city planning had taken root in the USA, following Frank Law Olmsted's creation of Central Park in New York. His stepson, John Charles Olmsted, went to Seattle to develop a plan for parks for the city wanting to carry forward the idea of parks as designed for the contemplation of natural beauty. On the other hand, the municipal engineering tradition was concerned with efficiency and systems building (Klingle 2007). This could be seen as the same contrast as pervades today in people's attitudes to vegetated open space with groups of trees. Some are positive about such scenery, for others the shade and obscurity of the dense vegetation induces a sense of fear. Several parks were created, but some were dissected by new boulevards built to carry the newly arrived motor cars. J.C. Olmsted did not always preserve the natural landscape, in one instance working with the city engineer to culvert a stream so that he could develop more magnificent lawns for his parks. The parks were the products of landscape design, of people's conceptions of natural beauty, but nevertheless they provided opportunities for wildlife. By the 1920s there was a public outcry about illegal shooting of birds and hunting of mammals in Seattle's parks (Klingle 2007). This was an urban ecology of conflict, between social groups who saw the parks in different ways.

In the UK there has been a long tradition of botanical exploration, initially for the production of herbals, and later through development of the natural sciences, particularly during Victorian times. Hampstead Heath illustrates these changes rather well. John Gerard, the Elizabethan herbalist whose *Great Herball* was published in 1597, made frequent forays into the countryside surrounding London, including Hampstead Heath. He provides the earliest description of the heath together with locations of many notable plants. Close on his heels came Thomas Johnson, a leading member of the Society of Apothecaries, whose description of a visit to the Heath by a party of botanists in 1629 (during which 72 plant species were recorded) is the first published account of a botanical excursion in the London area (Fitter 1945). Plant and insect collecting by Victorian naturalists brought a spate of new records, which became more organized locally with formation of the Hampstead Scientific Society in 1899. In 1913 this new body published a detailed account of the geology and natural history of the heath which, in addition to describing the flora and fauna included considerable discussion of trees in streets around the heath, including the presence of alien species. Popular volumes on Wimbledon Common and Battersea Park in London also appeared at that time, the latter emphasizing nature study (Johnson 1910, 1912).

Meanwhile there had been an enormous growth of interest in natural history in British towns and cities. Natural history societies were established in Liverpool, Manchester and Bristol during the 1860s and their field excursions attracted hundreds of participants. Manchester had 550 people attending one such gathering. Many of these societies have been instrumental in organizing the systematic recording of Britain's natural history, and because they were based on cities there was a wealth of information about nature within urban areas. Most of these records were collected by amateur naturalists specializing in particular groups of organisms, and indeed most natural history societies are still organized this way today. It was not until the second half of the twentieth century that ecological studies became established which cut across taxonomic boundaries.

However, the taxonomic approach has produced an important legacy of studies relating to individual cities, especially those resulting in distribution maps and atlases. Examples include the butterflies, birds and plants of London (Plant 1987; Hewlett 2002; Burton 1983), the bird atlas of Berlin (OAB 1984) and the Red Data Book for Moscow (Anon. 2001). The natural history societies have been extremely important in documenting and recording the flora and fauna of their local areas, and some of their studies, like those on the status of foxes and badgers in London were of considerable significance (Teagle 1967, 1969).

With the strong influence of taxonomy and its inevitable compartmentalization it is no surprise that early investigations concentrated on specific habitats to describe nature in the city, rather than the nature of a city as a whole (Sukopp 2008). The plants on the ruins of castles and city walls attracted a succession of naturalists and were described in many cities, such as Rome (Panarolis 1643; Sebastiani 1815), Paris (Vaillant 1727; Bulliard 1780). This fascination with walls has continued, from Rome (Anzalone 1951) to Varanasi, India (Varshney 1971) and their ecology was explored in some detail by Darlington (1981).

The term urban ecology was introduced by the Chicago school of social ecology within sociology (Park *et al.* 1925). It looked at the human side of characterizing cities, but was highly relevant to the debate over parks in cities like Seattle. The degree to which a park was vandalized, suffered illegal felling of trees or shooting of wildlife was greatly influenced by the human social context in which it was set. Knowledge of urban social differentiation and of the human and environmental characteristics of cities would help park managers to understand why things happen and to work out the best ways in which to involve communities in having a say, and indeed becoming stakeholders, in park and natural area management and policies.

By the 1930s there was concern about the loss of natural vegetation and the apparently relentless spread of suburbia. Leading authorities wrote about the impact of human activity on vegetation (Salisbury 1933, 1938). Public concern though was less about what was happening in cities but on gaining access to the countryside for recreation. In the UK, rambling and access to the hills and coastline were major preoccupations. Part of this outdoor experience was stimulated by a growth of popular interest in natural history. The extraordinary growth of interest in bird-watching which we see today had its early roots in books by Nicholson (1926a, 1926b) and others including Jefferies (1883) and Hudson (1898).

Things changed with the onset of the Second World War. Bombing brought an opportunity for interesting alien plants to appear in city centres. The bombed sites became important in the development of studies on urban flora. Less than three years after the Blitz, Salisbury (1943) described the plants that colonized the ruined houses in London. Across many cities the changes that followed war damage gave rise to studies of the flora and fauna of derelict sites and ruins (Erkamo 1943; Balke 1944; Lousley 1944; Burges and Andrews 1947; Scholz 1960; Pfeiffer 1957). The warmer and drier conditions among the urban dereliction allowed many previously rare plants to become permanent members of the urban flora in war-damaged European cities.

The 1939–45 War had many legacies. It prompted new visions of cities, new town and national park movements, and greater attention to a good standard of living for all, typified by ideas such as the welfare state. For some the end of the World War meant continued disturbance. In China, this culminated with the Revolution of 1949. After the revolution large efforts were made to improve the environment, especially through tree planting. At the time of the Revolution, Guangzhou had only 5 ha of parks: by 1979 it had 161 ha of parks and gardens. The number of trees in Nanjing increased from 2,000 to 20,000 in the same 30 years (Hoa 1981). Historically, Beijing had many gardens and tree-lined streets, but since 1949, the city has grown enormously. Nevertheless, the total area of public greenspace increased by 250 per cent between 1949 and 1986 (Office of the Capital Afforestation Commission 1986).

The understanding of urban flora and fauna was greatly helped by the appearance of *London's Natural History* (Fitter 1945). This brought together the evolution of the city and the development of the flora and fauna in different urban habitats. It remains a good introduction to the city's urban ecology. In 1950, the natural regeneration of Birmingham's derelict industrial land gained a place in the Handbook produced for the annual meeting of the British Association (Rees and Skelding 1950). Gradually other studies of individual cities appeared, each showing the diversity of habitats and wildlife in cities such as Paris (Jovet 1954), New York (Kieran 1959), Vienna (Kühnelt 1955; Schweiger 1962), Saarbrücken (Müller 1972), Brussels (Duvigneaud 1974), Berlin (Kunick 1974) and Birmingham (Teagle 1978).

By the 1970s systems thinking had permeated much of science. At the same time a number of incidents such as the Donora, Pennsylvania smog of 1948 and the London smog of December 1952 aroused interest in severe environmental impacts in urban areas. Academic symposia on human impacts on the environment began to appear (Thomas 1956). Rachel Carson's *Silent Spring* (1962) brought popular attention to environmental damage. Governments began to take action, culminating in the US National Environmental Policy Act of 1969. New policies emerged and the 1970s became a decade of environmental activism.

In this context people began to think of cities as integrated systems. An ecological mass balance of Sydney, Australia, was published in 1972 (Ecological Society of Australia 1972). This project is concerned with systematic and integrated studies on the flows of matter and of energy that constitute the functionality or the functional order of urban centres considered as ecosystems. It is part of the new integrated ecology that places energy in the dynamics of an urban system considered in all its complexity (Giacomini 1978). A pilot project was carried out in Hong Kong (Boyden *et al.* 1981), the first results appearing in Duvigneaud (1974) who applied the methods of forest and lake analysis used in the International Biological Program (IBP) to analyse a big city. Taking Brussels as an example, flows of matter and energy were calculated, treating the city like a black-box. The inner habitats of the city were subdivided in a similar manner to those of Berlin

(Kunick 1974): densely built up zone, partly built up central areas, inner suburbs, outer suburbs (see also Chapter 39, below).

Urban ecology became established as a separate subdiscipline in the early 1970s with systematic studies of climate, soil, water and organisms. The journal *Urban Ecology* started publication in 1975 (later to be merged with *Landscape Planning*). It is striking that by the time the first European symposium on Urban Ecology was held in Berlin in 1980 there was a wealth of detailed work on a wide range of newly developing topics in the field of urban ecology (Bornkamm *et al.* 1982). Since then a surge in the scientific ecological study of cities occurred with many programmes focusing on the mapping of biotopes (Schulte *et al.* 1993). Hejný (1971) distinguished 68 habitat types in Prague. 223 German urban areas (all cities and many medium-sized towns) and 2,000 villages and small towns were biotope mapped (Schulte and Sukopp 2000). Vegetation mapping of urban areas such as Sydney, Australia (Benson and Howell 1994), Kuala Lumpur, Malaysia and Singapore was carried out at this time.

In Germany, Sukopp led the development of a school of urban ecology in Berlin (Sukopp and Werner 1982) and began to encourage the application of urban ecology in city planning (Sukopp *et al.* 1995). The leadership saw the second European Ecological Symposium being held in Berlin in 1980 and being devoted to urban ecology (Bornkamm *et al.* 1982).

In the UK, by the late 1970s, urban wildlife projects were underway in several cities (Goode 1989). The William Curtis Ecological Park, providing a range of habitats for study by London schoolchildren, was opened near Tower Bridge in 1978. Its success inspired similar projects elsewhere in the UK. The park was always intended to be temporary and in 1985 the land was returned to its owners. But by this time, the Ecological Parks Trust was managing seven nature parks in London, several of which are still there today. By that time too there were many similar projects elsewhere in the UK.

In 1985 the Ecological Parks Trust became the Trust for Urban Ecology, building on experience gained in running such parks to advise more broadly on urban ecology.

The UK Nature Conservancy Council actively promoted urban nature conservation at this time, producing a series of publications, the first of which, *The Endless Village* (Teagle 1978) was an outstanding study of wildlife in the West Midlands Conurbation. George Barker, the urban officer of the Nature Conservancy, established the UK MAB Urban Forum in 1987 to promote urban nature and urban ecology with a mission to:

- Raise awareness
- Stimulate research
- Influence policy
- Improve the design and management of urban systems
- Push urban nature conservation up the social and political agenda

By the 1980s concern for wildlife provision in new residential developments became prominent (Goldstein *et al.* 1983; Goode and Smart 1986). Naturalistic planting and new wildlife habitat creation became part of the planning of new towns in the UK, for example in Warrington (Scott *et al.* 1986). However, some felt there was a danger that such schemes may produce little more than a green veneer (Goode and Smart 1986). Attention needed to be paid to the autecology of a range of species in order to accommodate a greater variety of species within an urban setting. Furthermore, it was recognized that the right habitat alone is insufficient: consideration must also be given to the management of people (Goldsmith and Warren 1993).

Comparison of the new with the old permits evaluation of urban impacts on the native flora and helps to identify where management of the urban wild has been successful. A new wave of books about urban wildlife has emerged including those of London (Goode 1986), Glasgow (Dickson 1991), Cardiff (Gillham 1992, 1998, 2002, 2006), Belfast (Scott 2004) in the UK and Portland, OR (Houck and Cody 2000, 2011). One of the most comprehensive studies was in London where the London Ecology Unit produced a series of Ecology Handbooks describing the natural habitats of each London Borough in considerable detail (see also Chapter 8). The last of these (Yarham and Game 2000) contains a full list of the 31 published handbooks.

In the UK government papers promoted greening the city (Department of Environment 1966) and more recently provide planning guidance on accessible open space including natural areas (Department of Communities and Local Government 2006). In 2007, the UK Royal Commission on Environmental Pollution recommended that:

> The Department for Communities and Local Government and its devolved equivalents amend their planning policy statements and guidance to reflect a broader definition of the natural environment in urban areas and to recognize and protect the role that urban ecosystems can play in improving towns and cities. Planning policy and guidance should describe the range of functions and benefits associated with the natural environment of urban areas, promote the use of green infrastructure and provide a menu of options for planners and developers to use, including:
>
> - creation of green networks and green infrastructure;
> - urban river restoration;
> - the use of green and built infrastructure for flood storage and redirection;
> - the use of sustainable drainage systems, including green roofs; and
> - the promotion of urban trees and woodland.
>
> *(Royal Commission on Environmental Pollution 2007: 147)*

This interest by the UK government had parallels in many other countries. In the science community, national funding bodies supported major programmes, particularly the long-term urban ecological research programs in Baltimore and Phoenix, USA (Grimm *et al.* 2000). In Britain, three of the national research councils funded programmes on aspects of the urban environment (Owens *et al.* 2006), while the European Union supported several major projects (e.g. James *et al.* 2009). Many other projects, both national and international are making major new contributions to urban ecology.

Interest in Urban Ecology is now global. China publishes several journals related to urban ecology, including *Urban Environment and Urban Ecology*. The Chinese Academy of Science has a large group specializing in urban ecology that works closely with municipalities to deliver green infrastructure and more sustainable cities (Wang and Ye 2004; Li *et al.* 2005). The first four volumes of the *Ecology of Indonesia Series* all contain chapters on urban ecology. That on Sumatra argues that Sumatran towns are not nearly as interesting ecologically as they could be because their vegetation is largely foreign and therefore supports few birds (Whitten *et al.* 2000). That on Kalimantan lists 13 types of urban habitat, from high water towers to drainage ditches, which offer varying opportunities for wildlife (Mackinnon *et al.* 1996). That on Java and Bali extols the value of the urban environment for education and has a detailed section on urban pests, such as rats, cockroaches, mosquitoes and feral dogs carrying rabies (Whitten *et al.* 1996). That on Sulawesi examines the diversity of plants in urban gardens and peculiarities of urban trees (Whitten *et al.* 1987). A major urban ecology conference in Paris (Lizet *et al.* 1997) had contributions from Brazil, Greece and Madagascar as well as European countries.

Rebele (1994) saw ecological research in the urban setting falling into two broad categories: social sciences oriented and ecology oriented. Although traditionally separate (Wittig and Sukopp 1993), these two approaches are being brought together to the benefit of all science (Blood 1994; Rees 1997). The human ecosystem model has provided a framework for urban ecological studies addressing questions of varying specificity. However, the specific ecological theories within the model can be the same for both urban and non-urban areas. (Niemalä 1999). The concept of the city as an ecosystem and studying cities as ecosystems within new paradigms of ecosystem science (Pickett *et al.* 1992; Wu and Loucks 1995; Flores *et al.* 1997) has raised the collective consciousness of ecologists about urban ecosystems and will contribute to the further development of concepts that apply to all ecosystems. The prospects for urban ecology are bright.

References

Allen, D.E. (2010) *Books and Naturalists*, London: Collins.

Anon. (2001) *Red Data Book of Moscow*, Moscow: ABF.

Anzalone, B. (1951) 'Flora e vegetazione dei muri di Roma', *Annali di Botanica*, 23(3): 393–497.

Balke, N.P.W. (1944) 'Vegetatie op het Rotterdamse puin', *Weer en Wind*, 8(2): 33–7.

Benson, D. and Howell, J. (1994) 'The natural vegetation of the Sydney 1:100,000 map sheet', *Cunninghamia*, 3(4): 677–787.

Blood, E. (1994) 'Prospects for the development of integrated regional models', in P.M. Groffman and G.E. Likens (eds), *Integrated Regional Models*, New York: Chapman and Hall, pp. 145–53.

Bornkamm, R., Lee, J.A. and Seaward, M.R.D. (eds) (1982) *Urban Ecology*, Oxford: Blackwell Scientific Publications.

Boyden, S., Millar, S., Newcombe, K. and O'Neill, B. (1981) *The Ecology of a City and its People: The Case of Hong Kong*, Canberra: Australian National University Press.

Bulliard, M. (1780) *Flora Parisiensis ou descriptions et figures des plantes qui croissent aux environs de Paris*, Paris: Didot.

Burges, R.C.L. and Andrews, C.E.A. (1947) 'Report of the bombed sites survey subcommittee', *Birmingham Natural History and Philosophical Society, Proceedings*, 18(1): 1–12.

Burton, R.M. (1983) *Flora of the London Area*, London: London Natural History Society.

Carson, R. (1962) *Silent Spring*, New York: Fawcett Crest.

Colden, C. (1749, 1751) 'Plantae Coldenhamiae in provincia noveboracensi americes sponte crescentes', *Acta Societatis Regiae Scientiarum Upsaliensis*, 1743: 81–136 (1749 ed.); 1744–50: 47–82 (1751 ed.).

Curtis, W. (1777) *Flora Londinensis:: or, Plates and descriptions of such plants as grow wild in the environs of London; with their places of growth and times of flowering; their several names according to Linnaelis and other authors; with a particular description of each plant in Latin and English*, London: printed for and sold by the author; and B. White, bookseller.

Darlington, A. (1981) *Ecology of Walls*, London: Heinemann Educational Books.

DeCandido, R., Muir, A.A. and Gargiullo, M.B. (2004) 'A first approximation of the historical and extant vascular flora of New York City: Implications for native plant species conservation', *Journal of the Torrey Botanical Society*, 131(3): 243–51.

Department of Communities and Local Government (2002) *Assessing Needs and Opportunities: A Companion Guide to PPG17*, Norwich: The Stationery Office.

Department of the Environment (1996) *Greening the City: A Guide to good practice*, Norwich: Her Majesty's Stationery Office.

Dickson, J.H. (1991) *Wild Plants of Glasgow*, Aberdeen: Aberdeen University Press.

Duvigneaud, P. (1974) 'L'ecosysteme "Urbs"', *Mémoires de la Société Royale de Botanique de Belgique*, 6: 5–35.

Ecological Society of Australia (1972) *The City as a Life System: ESA Conference Proceedings Vol. 7*, Sydney: Surrey Beatty and Sons.

Erkamo, V. (1943) 'Über die Spuren der Bolschewikenherrschaft in der Flora der Stadt Viipuri', *Annals Botanical Society of Vanamo*, 18: 1–24.

Fitter, R.S.R. (1945) *London's Natural History*, London: Collins.

Flores, A., Pickett, S.T.A., Zipperer, W.C., Pouyat, R.V. and Pirani, R. (1997) 'Application of ecological concepts to regional planning: A greenway network for the New York metropolitan region', *Landscape and Urban Planning* 39(4): 295–308.

Gerard, J. (1597) *Great Herball, or Generall Historie of Plantes*, London: John Norton.

Giacomini, V. (1978) 'Man and the biosphere: an amplified ecological vision', *Landscape Planning*, 5(2–3): 193–211.

Gillham, M.E. (1992) *The Garth Countryside, Part of Cardiff's Green Mantle*, Cardiff: M.E. Gillham.

Gillham, M.E. (1998) *Town Bred, Country Nurtured: A Naturalist Looks Back Fifty Years*, Cardiff: M.E. Gillham.

Gillham, M.E. (2002) *A Natural History of Cardiff: Exploring along the River Taff, being an Account of the Animal and Plant Life in and around our Capital City*, Caerphilly: Lazy Cat Publishing.

Gillham, M.E. (2006) *A Natural History of Cardiff 3: Exploring along the Rivers Rhymney and Roath*, Llandybie: Dinefwr.

Goldsmith, F.B. and Warren, A. (eds) (1993) *Conservation in Progress*, Chichester: John Wiley.

Goldstein, E.L., Gross, M. and DeGraaf, R.M. (1983) 'Wildlife and green space planning in medium-scale residential developments', *Urban Ecology* 7(3): 201–14.

Goode, D.A. (1986) *Wild in London*, London: Michael Joseph.

Goode, D.A. (1989) 'Urban nature conservation in Britain', *Journal of Applied Ecology*, 26(3): 859–73.

Goode, D.A. and Smart, P.J. (1986) 'Designing for wildlife', in A.D Bradshaw, D.A. Goode and E.H.P. Thorp (eds), *Ecology and Design in Landscape*, Oxford: Blackwell, pp. 219–35.

Grimm, N.B., Grove, J.M., Pickett, S.T.A. and Redman, C.L. (2000) 'Integrated approaches to long-term studies of urban ecological systems', *BioScience*, 50(7): 571–84.

Hampstead Scientific Society (1913) *Hampstead Heath: Its Geology and Natural History*, London: T Fisher Unwin.

Hejný S. (1971) 'Metodologický příspěvek k výzkumu synantropní květeny a vegetace velkoměsta (na příkladu Prahy)' Bratislava: Zborn. Pred. Zjazdu Slov. Bot. Spoloč. Tisovec, 2: 545–67.

Hewlett, J. (2002) *The Breeding Birds of the London* Area, London: London Natural History Society.

Hoa, L. (1981) *Reconstruire la Chine: trente ans d'urbanisme*, Paris: Editions du Moniteur.

Houck, M.C. and Cody, M.J. (2000) *Wild in the City: A Guide to Portland's Natural Areas*, Portland OR: Oregon Historical Society Press.

Houck, M.C. and Cody, M.J. (2011) *Wild in the City: Exploring the Intertwine*, Portland, OR: Audubon Society of Portland.

Hudson, W.H. (1898) *Birds in London*, London: Longmans, Green & Co.

Hudson, W.H. (1900) *Nature in Downland*, London/New York: Longmans Green.

James, P., Tzoulas, K., Adams, M.D., Barber, A., Box, J., Breuste, J., Elmqvist, T., Frith, M., Gordon, C., Greening, K.L., Handley, J., Haworth, S., Kazmierczak, A.E., Johnston, M., Korpela, K., Moretti, M., Niemelä, J., Pauleit, S., Roe, M.H., Sadler, J.P. and Ward Thompson, C. (2009) 'Towards an integrated understanding of green space in the European built environment', *Urban Forestry and Greening*, 8(2): 65–75.

Jefferies, R. (1883) *Nature near London*, London: Chatto and Windus.

Johnson, W. (1910) *Battersea Park as a Centre for Nature-Study*. London: T. Fisher Unwin.

Johnson, W. (1912) *Wimbledon Common: Its Geology, Antiquities and Natural History*, London: T. Fisher Unwin.

Jovet, P. (1954): 'Paris, sa flore spontanee, sa vegetation'. – In: *Notices botaniques et itineraires commentés publiés à l'occasion du VIIIe congres International de Botanique*, Paris and Nice, 21–60.

Kieran, J. (1959) *A Natural History of New York City*, New York: Houghton Mifflin Company.

Klingle, M. (2007) *Emerald City: An Environmental History of Seattle*, New Haven, CT/London: Yale University Press.

Kühnelt, W. (1955) 'Gesichtspunkte zur Beurteilung von Groÿstadtfauna (mit besonderer Berücksichtigung der Wiener Verhältnisse)', *Österreichliche Zoologische Zeitschrift*, 6: 30–54.

Kunick W. (1974) *Veränderungen von Flora und Vegetation einer Groÿstadt, dargestellt am Beispiel von Berlin (West)*, dissertation, Berlin: Technische Universität.

Li, F., Wang, R., Paulussen, J. and Liu, X. (2005) 'Comprehensive concept planning of urban greening based on ecological principles: a case study in Beijing, China', *Landscape and Urban Planning*, 72(4): 325–36.

Lizet, B., Wolf, A.-E. and Celecia, J. (1997) *Sauvages dans la ville*, Paris: JATBA, Laboratoire d'Ethnobiologie, Biogéographie, Éditions du Muséum National d'Histoire Naturelle.

Lousley, J.E. (1944) 'The pioneer flora of bombed sites in Central London', *Reports of the Botanical Exchange Club, 1941/42*, pp. 528–31.

Mackinnon, K., Hatta, G., Halim, H. and Mangalik, A. (1996) *The Ecology of Kalimantan*, Singapore: Periplus.

Mitchell, J.H. (2001) *The Wildest Place on Earth*, Washington, D.C.: Counterpoint.
Müller, P. (1972): 'Probleme des Ökosystems einer Industriestadt, dargestellt am Beispiel von Saarbrücken' in *Proc Belastung und Belastbarkeit von Ökosystemen*, pp. 123–32, Ges. f. Ökologie, Gieÿen.
Nicholson, E.M. (1926a) *Birds in England: An Account of the State of our Bird-Life and a Criticism of Bird Protection*, London: Chapman & Hall.
Nicholson, E.M. (1926b) *The Study of Birds: An Introduction to Ornithology*, London: Ernest Benn.
Niemalä, J. (1999) 'Is there a need for a theory of urban ecology?' *Urban Ecosystems*, 3(1): 57–65.
OAB (Ornithologische Arbeitsgruppe Berlin (West)) (ed.) (1984) Brutvogelatlas Berlin (West), *Ornithologischer Bericht für Berlin (West)*, 9, Sonderheft.
Office of the Capital Afforestation Commission (1986) *Beautiful Beijing*, Beijing: China Photographic Publishing House.
Owens, S., Petts, J. and Bulkeley, H. (2006) 'Boundary work: knowledge, policy, and the urban environment', *Environment and Planning C: Government and Policy*, 24(5): 633–43.
Panaroli, D. (1643) *Jatrologismi Sive Medicae Observationes Quibus Additus est in Fine Plantarum Amphitheatralium Catalogus*, Rome: Typis Dominici Marciani.
Park, R.E., Burgess, R.W. and McKenzie, R.D. (1925) *The City*, Chicago: Chicago University Press.
Pfeiffer, H. (1957): 'Pflanzliche Gesellschaftsbildung auf dem Trümmerschutt ausgebombter Städte', *Vegetatio*, 7(5–6): 301–20.
Pickett, S.T.A., Parker, V.T. and Fiedler, P. (1992). 'The new paradigm in ecology: Implications for conservation biology above the species level', in P. Fiedler and S. Jain (eds), *Conservation Biology: The Theory and Practice of Nature Conservation, Preservation, and Management*, New York: Chapman & Hall, 65–88.
Plant, C.W. (1987) *The Butterflies of the London Area*, London: London Natural History Society.
Rebele, F. (1994) 'Urban ecology and special features of urban ecosystem', *Global Ecology and Biogeography Letters*, 4(6): 173–87.
Rees, W.E. (1997) 'Urban ecosystems: the human dimension', *Urban Ecosystems*, 1(1): 63–75.
Rees, W.J. and Skelding, A.D. (1950) 'Vegetation' in M.J. Wise (ed.) *Birmingham and its Regional Setting: A Scientific Survey*, Birmingham: British Association Local Executive Committee, 65–76.
Royal Commission on Environmental Pollution (2007) *Twenty Sixth Report: The Urban Environment Cm7009*, Norwich: The Stationery Office.
Rusby, H.H. (1906) 'A historical sketch of the development of botany in New York City', *Torreya*, 6: 101–11, 133–45.
Salisbury, E.J. (1933) 'The influence of man on vegetation (Presidential Address)', *Transactions South Eastern Union of Scientific Societies*, pp. 1–17.
Salisbury, E.J. (1938) 'Plants in relation to the human environment', *Science Progress*, 33: 230–9.
Salisbury, E.J. (1943) 'The flora of bombed areas', *Nature*, 151: 462–6.
Scholz, H. (1960) 'Die Veränderungen in der Ruderalflora Berlins. Ein Beitrag zur jüngsten Florengeschichte', *Willdenowia*, 2: 379–97.
Schulte, W. and Sukopp, H. (2000): 'Stadt und Dorfbiotopkartierungen. Erfassung und Analyse ökologischer Grundlagen im besiedelten Bereich der Bundesrepublik Deutschland – ein Überblick (Stand: März 2000)', *Naturschutz und Landschaftsplanung*, 32(5): 140–7.
Schulte, W., Sukopp, H. and Werner P. (eds) (1993) 'Flächendeckende Biotopkartierung im besiedelten Bereich als Grundlage einer am Naturschutz orientierten Planung', *Natur und Landschaft*, 68(10): 491–526.
Schweiger, H. (1962) 'Die Insektenfauna des Wiener Stadtgebiets als Beispiel einer Kontinentalen Groÿ-Stadtfauna', *Verhandlungen XI. Intern.ationaler Kongress Entomologie* 3: 184–93.
Scott, D., Greenwood, R.D., Moffatt, J.D. and Tregay, R.J. (1986) 'Warrington new town: an ecological approach to landscape design and management' in A.D. Bradshaw, D.A. Goode and E.H.P. Thorp (eds), *Ecology and Design in Landscape*, Oxford: Blackwell, pp. 143–60.
Scott, R. (2004) *Wild Belfast – On Safari in the City*, Belfast: Blackstaff Press.
Sebastiani, A. (1815) *Enumeratio Plantarum Sponte Nascentium in Ruderibus Amphitheatri Flavii*, Rome: Typis Pauli Salviucci et Filii.
Sefferein, M.L. (1932) 'Wild flowers of the Spuyten-Duyvil and Riverdale sections of New York City', *Torreya*, 32: 119–27.
Sukopp, H. (2008) 'On the early history of urban ecology in Europe', in J.M. Marzluff, E. Shulenberger, W. Endlicher, M. Alberti, G. Bradley, C. Ryan, U. Simon, and C. ZumBrunnen (eds), *Urban Ecology: An International Perspective on the Interaction Between Humans and Nature*, New York: Springer, pp. 79–97.
Sukopp, H. and Werner, P. (1982) *Nature in Cities: A Report and Review of Studies and Experiments Concerning*

Ecology, Wildlife and Nature Conservation in Urban and Suburban Areas, Strasbourg: Council of Europe Nature and Environment Series 28.
Sukopp, H., Numata, M. and Huber A. (eds) (1995) *Urban Ecology as the Basis for Urban Planning*, The Hague: SPB Academic.
Teagle, W.G. (1967) 'The fox in the London suburbs', *London Naturalist*, 46: 44–68.
Teagle, W.G. (1969) 'The badger in the London area', *London Naturalist*, 48: 48–75
Teagle, W.G. (1978) *The Endless Village*, Birmingham: Nature Conservancy Council, West Midlands.
Thomas, W.L. (1956) *Man's Role in Changing the Face of the Earth*, Chicago, IL: University of Chicago Press.
Torrey, J. (1819) *Catalogue of plants growing spontaneously within 30 miles of New York City*, Albany, NY: Lyceum of Nature Museum of New York, and Webster and Skinner.
Trimen, H. and Thiselton-Dyer, W.T. (1869) *Flora of Middlesex: a topographical and historical account of the plants found in the county: with sketches of its physical geography and climate, and of the progress of Middlesex botany during the last three centuries*, London: Robert Hardwicke.
Vaillant, S. (1727) *Botanicon Parisiense*, Leiden: Boorhaave.
Varshney, C.K. (1971) 'Observations on the Varanasi wall flora', *Plant Ecology*, 22(6): 355–72.
Wang, R. and Ye, Y. (2004) 'Eco-city development in China', *Ambio*, 33(6): 341–2.
Whitten, T., Mustafa, M. and Henderson, G.S. (1987) *The Ecology of Sulawesi*, Singapore: Periplus.
Whitten, T., Soeriaatmadja, R.E. and Afiff, S.A. (1996) *The Ecology of Java and Bali*, Singapore: Periplus.
Whitten, T., Damanik, S.J., Anwar, J. and Hisyam, N. (2000) *The Ecology of Sumatra*. Singapore: Periplus.
Wittig, R. and Sukopp, H. (1993) 'Was ist Stadt Ökologie?' in H. Sukopp and R. Wittig, *Stadtökologie*, Stuttgart: Gustav Fischer Verlag, pp. 1–9.
Wu, J. and Loucks, O.L. (1995) 'From balance of nature to hierarchical patch dynamics: a paradigm shift in ecology', *Quarterly Review of Biology*, 70(4): 439–66.
Yarham, I. and Game, M. (2000) *Nature Conservation in Brent*, Ecology Handbook 31. London: London Ecology Unit.

8

Planning for nature in towns and cities

A historical perspective

David Goode

Introduction

One of the features of urban nature conservation is that many different and totally unrelated initiatives started at about the same time in places as far apart as Portland Oregon, New York, Toronto, London, Birmingham, Berlin, Tokyo and Cape Town. These initiatives had contrasting origins, but their objectives were much the same, namely to provide a place for nature in towns and cities. During the 1940s and 1950s several writers described the ecology of big cities, including *London's Natural History* by Richard Fitter and *The Natural History of New York City* by John Kieran. These influential publications described, for the first time, the vast array of species and habitats that exist within the urban environment. The scene was set for others to follow and it was not long before specialists involved in urban design were advocating the need to take a more ecological approach. Notable amongst these were Ian McHarg in *Design with Nature* (1969) and Nan Fairbrother's *New Lives, New Landscapes* (1970). Another influential book was Richard Mabey's *Unofficial Countryside* (1973) which first drew attention to the value of wild corners in British towns and cities.

It was a time when new ideas were developing. One of the first ecologists to espouse the value of urban wildlife was Ray Dasmann in a speech entitled *Wildlife and the New Conservation* given in Maryland, USA in 1966. He pointed out that generations were growing up in cities, with no roots in the land and little experience of the natural world. Dasmann felt that the wildlife profession in America was too closely identified with game animals and hunters, and urged naturalists to get out of the forests and into the cities. He argued that they should work with city and metropolitan regional planners, with landscape architects and others concerned with the urban environment, to make towns and cities into places where each person's everyday life could be enriched by contact with nature. It was all the more significant coming from one of the leading figures in world conservation (Dasmann 1966).

Not long after this, in 1968, a national conference on *Man and Nature in the City* was held in Washington, D.C., sponsored by the Bureau of Sport Fisheries and Wildlife, at which the director admitted that if the Bureau continued to focus on wide open spaces and neglected the people of the city, as it had in the past, then it would find itself in a very questionable position with society (Gottschalk 1968). Other conferences on urban wildlife followed in the USA and

Canada, and in 1973 a National Institute for Urban Wildlife was established in the USA. One of its first publications was a report on planning for wildlife in cities and suburbs (Leedy *et al.* 1978). By this time a number of cities had started to take urban wildlife seriously, including New York, one of the first to develop an urban wildlife programme with inventories being made of urban habitats (Miller 1978). By the mid-1980s a number of projects were underway in the USA which involved planning as the principal means of achieving urban nature conservation. A conference on *Wildlife Conservation and New Residential Developments* (Stenberg and Shaw 1986) included a review of such initiatives (Leedy and Adams 1986). The Director of the National Institute, Lowell Adams, must take credit for the surge of activity at this time, as he actively promoted the involvement of professional planners and landscape architects (Adams and Leedy 1987).

The Washington conference of 1968 led to a similar conference in Manchester UK nearly ten years later, organised by landscape architect Ian Laurie (1975). It is fascinating that in Britain it was members of the landscape profession who led the way in making proposals for urban nature conservation, rather than ecologists or wildlife conservationists. The meeting in Manchester was the first in the UK to bring together all the different professions involved, including urban planners, landscape designers, ecologists and even social scientists. It was a significant event which led to a spate of new initiatives across a wide range of disciplines during the 1980s.

Meanwhile, in Germany and Poland there were already well established research schools of urban ecology associated with universities in Berlin and Warsaw, investigating many different aspects in great depth. Professor Herbert Sukopp established the Institute of Ecology at the Technical University in Berlin which concentrated largely on botany and vegetation, with a strong emphasis on the phytosociology of urban areas. Dr Luniak and colleagues in Poland concentrated mainly on animal ecology. These two centres effectively led the way in urban ecological studies in Europe.

West Berlin has a special place in the history of urban ecology and conservation. During the 1970s strategic nature conservation programmes were developed in several states in the Federal Republic of Germany, and West Berlin was included as a city state. Isolated as an island in East Germany it had to provide for nature conservation alongside other land uses within the confined boundaries of the city. Comprehensive biotope mapping was carried out for the whole city, providing the ecological basis of the land use plan. The result was an extremely detailed strategic plan for conservation, adopted in 1979 (Henke and Sukopp 1986). Not only was Berlin the first city to have such a plan, but it was supported by an immense amount of ecological information. No other city has been subject to the same degree of investigation and the studies carried out over the years have provided major advances in our understanding of urban ecology (Sukopp 1990).

It is striking that by the time the first European symposium on Urban Ecology was held in Berlin in 1980 there was a wealth of detailed work on a wide range of newly developing topics in the field of urban ecology (Bornkamm *et al.* 1982). Most contributions were specific autecological or habitat related studies. Examples include urban fox populations, invertebrate diversity of urban habitats, and the botanical importance of industrial habitats. There were also attempts to look in a holistic way at the urban environment as an ecosystem. The contribution by Numata (1982) is a notable example. The organisers endeavoured to include contributions on the ecological effects of human activity in urban areas, and on the application of ecological knowledge in urban design and planning. But it is clear that at that time, with the notable exception of Berlin, planning for nature in European cities was still in its infancy.

An important initiative dealing with the ecology of cities at this time was the UNESCO Man and the Biosphere programme known as MAB 11, led by John Celecia. This was an ambitious programme which involved environmental system analysis of selected cities around the world, including Tokyo, Hong Kong, Rome, Madrid, Buenos Aires and Seoul, as well as some smaller European cities such as Delft and Valencia. The broad objective was to examine the way in which cities function, in order to find more sustainable solutions. One of the aims was to demonstrate through ecosystem modelling the dependency of large cities on their underlying ecological features, and to encourage greater interaction between environmental scientists and political decision makers. Many of these projects involved detailed analysis of urban habitats and their flora and fauna as part of the modelling process. The investigation of Tokyo, referred to above is a good example (Numata 1982). Whilst such studies did not lead directly to systems of planning for nature conservation, the methodologies provided useful tools for production of inventories and classification of biotopes (Celecia 1990).

The situation changed dramatically during the 1980s when planning for nature in cities suddenly became a mainstream activity with a wide range of initiatives in many different countries. The way that this developed in the UK illustrates very well how strategic planning can be used as a basis for conserving important sites.

Conserving nature through strategic planning in the UK

The protection of wildlife sites in urban areas through strategic planning began with the formation of new metropolitan counties in the 1970s. These covered large conurbations such as London, Birmingham, and Manchester. They were required to produce Structure Plans to provide a strategic assessment of their resources, including proposals for nature conservation. When planners in the West Midlands asked the Nature Conservancy Council (the UK government agency for nature conservation) for a list of important sites for inclusion in the Structure Plan, it was realised that practically nothing was known about sites within urban areas. So the NCC commissioned an ecological survey of the West Midlands, which resulted in publication of a seminal document on urban conservation, *The Endless Village* (Teagle 1978). This drew attention to the vast array of high quality wildlife sites surviving amidst the industrial dereliction of the Birmingham conurbation and identified important sites, like the Blackbrook Valley, for inclusion in the Structure Plan. But the document also included something more significant. It provided the blueprint for a national strategy on urban nature conservation written by George Barker, who later became the NCC's specialist advisor on this subject. All this activity resulted in another important initiative, the development of an Urban Wildlife Group in the West Midlands, which led to a large number of similar groups springing up in towns and cities across the country.

Other metropolitan counties followed Birmingham's lead and a series of nature conservation strategies was developed in the early 1980s in London (GLC 1984), Manchester (GMC 1986), Tyne and Wear (NCC 1988) and the West Midlands (1984). Many other large cities soon followed with major programmes in Bristol, Edinburgh, Leicester and Sheffield. Common strands running through these strategies were the need for protection of valuable habitats, enhancement of existing areas of open land for wildlife, and the creation of new habitats in areas deficient in nature. They all placed considerable emphasis on the local value of nature to urban residents. All recognised the need for comprehensive ecological survey and evaluation as the basis for selection of important sites. For details on the content of such strategies see Goode (1989, 1993, 1994).

The process which started with the large urban conurbations has since become standard practice in most towns and cities. Policies for the protection of urban wildlife sites are achieved largely through designation by local planning authorities of Sites of Importance for Nature Conservation (SINCs) in their strategic plans. Many of these programmes have been remarkably successful in ensuring the protection of important sites through the planning process.

Nature conservation in London

One of the most successful was the programme developed in London, which is used here as an example to illustrate the processes involved. Originally instigated by the Greater London Council in 1982 it was implemented from 1986 to 2000 by the London Ecology Unit. It was adopted by the Mayor of London in his Biodiversity Strategy for the capital (GLA 2002), and now forms part of the statutory London Plan (GLA 2004). The programme involved many different players including official agencies, especially the London boroughs, together with voluntary bodies such as the London Wildlife Trust. The process benefited from strong public support, particularly at the local level. New approaches with a strong social dimension, at first seen as a radical departure from traditional nature conservation, have been adopted as an integral part of city management.

A main objective was to ensure that nature conservation was built into strategic planning processes in London. When the work started in 1982 there was no reference to nature or wildlife in London's strategic development plan. Although provision was made for protection of open space for public enjoyment, the emphasis was largely on landscape and visual amenity rather than the natural environment or ecology. A priority task for the GLC, therefore, was to produce a set of policies on ecology and nature conservation for use by London boroughs in their local plans (GLC 1984).

One of the key policies recommended identification and protection of sites of nature conservation value. But knowledge of London's ecology at that time was patchy and incomplete and so it was necessary to undertake a comprehensive survey and evaluation of wildlife habitats throughout the capital. This was carried out in 1984/85. Priority was given to areas of open land of potential significance for nature conservation. Formal parks and cemeteries, private gardens, playing fields and open areas with little wildlife interest, such as arable land, were all excluded. An initial desk study using air photography resulted in over 1,800 'sites' being selected for survey, totalling about 20 per cent of the land area of Greater London. For each site information was collected on the types of habitat present and the dominant species, richness of plant species, presence of rare or unusual species, current land use and accessibility.

The results provided the basis for the first comprehensive strategic nature conservation plan for Greater London and provided the starting point for selection of Sites of Importance for Nature Conservation. The rationale for deciding which areas were important was published (GLC 1985) and this remained the basis of the system used by the London Ecology Unit from 1986–2000. A standardised set of criteria was used for comparing and evaluating sites. Although many of these criteria are similar to those developed by UK government agencies for selecting sites of national importance (such as species richness, size and presence of rare species), there are some essential differences. Public access and value for environmental education are examples. Details of these criteria and the way in which they have been applied are given in Goode (1999, 2005).

Although some changes have occurred in the detailed approach, the rationale remains much the same and it has been widely accepted as the basis for nature conservation planning in London. Though non-statutory, it was endorsed in 1995 by the London Planning Advisory Committee for use by London boroughs in their Unitary Development Plans. The same policy, criteria and procedures for identifying nature conservation sites were adopted by the Mayor of London in 2000, and are set out in full in his statutory Biodiversity Strategy (GLA 2002).

During the 1990s the strategy was successfully tested at numerous public inquiries. The results set important precedents for London in favour of nature conservation. Examples included disused railway land and industrial sites, as well as long established habitats such as woods, meadows and marshland. In many cases it was the value of these places to local people that won the day, rather than scientific arguments about rare habitats or species. Such cases were important in illustrating newly emerging values and helped to establish the validity of nature conservation in heavily built-up urban areas (Goode 2005).

The data have been updated periodically through more detailed surveys of each individual London borough. Over the past 25 years the database has provided a vital tool in strategic planning and for advising on the ecological implications of proposed new developments. It was also used to produce detailed nature conservation strategies for most of the London boroughs. Between 1985 and 2000 31 strategies were published. Probably the most detailed ecological database of any part of the UK, it now provides essential information for implementation of the Mayor's biodiversity strategy for the capital.

Categories of protected sites

The strategy is based on a hierarchy of sites at three levels London-wide, borough and local. Those of London-wide strategic significance are called Sites of Metropolitan Importance for Nature Conservation. They include nationally protected sites, such as National Nature Reserves and Sites of Special Scientific Interest, as well as many other important sites, which together represent the full range of habitats in London. The second category comprises sites of significance to individual London boroughs, and a third category of Local Sites are those important at neighbourhood level.

The use of these three different levels of importance is an attempt not only to protect the best sites in London but also to provide each area of London with accessible wildlife sites so that people are able to have access to nature within their local neighbourhood. Sites of London-wide importance are chosen in the context of the geographical area of Greater London. Those of borough importance are chosen from the range of sites in each individual borough. Sites of local importance are those valued by local residents, schools or community groups at the neighbourhood level.

At the London-wide level about 140 Sites of Metropolitan Importance are identified. They are distributed throughout London and vary in size from only a few hectares to over 1,000 hectares. Most (90 sites) are less than 100 ha, of which 55 are less than 50 ha. A few Sites of Metropolitan Importance have been lost to development since the list was first endorsed by the London Ecology Committee in 1988. Most of these were wasteland sites which were already scheduled for development. Additional sites have been added as individual boroughs have been surveyed in greater detail. The list of Metropolitan Sites was endorsed by the Mayor in 2002 and these sites are given statutory protection by policies in the London Plan (GLA 2004), which now provides the strategic planning framework for London.

As a result of the detailed surveys for individual boroughs the strategy identifies over 1,500 sites, all which are recognised as being of importance for biodiversity conservation in borough plans. This includes all three categories of protection i.e. metropolitan, borough and local. A significant number of sites designated through this process are also protected as Statutory Local Nature Reserves (LNRs). This is a designation made by the boroughs to give a greater degree of protection in the long term and to promote greater public access. Over 100 such sites are now designated by London boroughs as LNRs, compared with only two in 1980.

Areas of deficiency

The government agency English Nature encouraged the development of standards for accessible natural greenspace in towns and cities (Box and Harrison 1993; Harrison *et al.* 1995) which placed considerable emphasis on the values to be gained for people's health and well-being. Although open space standards were widely used by local authorities, such standards focused almost exclusively on the provision of sport and recreation facilities, to the exclusion of natural greenspace. The London strategy is, however, one example where *Areas of Deficiency* in access to nature have been mapped and are actually used to improve people's access to nature as part of daily life.

Those parts of London which do not have good access to high quality wildlife sites are identified as Areas of Deficiency, which are defined as built-up areas more than one kilometre from an accessible Metropolitan or Borough Site. Detailed maps defining such areas help boroughs to identify priority sites for provision of new habitats and aid the choice of Sites of Local Importance. Where no such sites are available, opportunities can be taken to provide them by habitat enhancement or creation, by direct acquisition of land to fulfil this function, or by negotiating access and management agreements and improving access routes in the surrounding urban area (Goode 2007). Recognition of such areas of deficiency goes some way to meeting the proposals made by Box and Harrison (1993) for provision of accessible natural greenspace. The approach adopted in London remains one of the few examples of such a scheme actually implemented as part of regional planning.

Implementing the strategy through strategic planning

The Mayor's Biodiversity Strategy published in 2002 includes specific policies and proposals to protect and enhance biodiversity through strategic planning. These policies are also contained in the statutory London Plan (GLA 2004). Protection of Sites of Importance for Nature Conservation is covered by the following policies:

1 The Mayor will identify Sites of Metropolitan Importance for Nature Conservation. Boroughs should give strong protection to these sites in their Unitary Development Plans. The Metropolitan Sites include all sites of national or international importance for biodiversity.
2 Boroughs should use the procedures adopted by the Mayor to identify and protect Sites of Borough and Local Importance for Nature Conservation. The Mayor will assist and advise them in this.

The effect of these policies is that the hierarchy of designations in London is now subject to statutory planning procedures. But the Biodiversity Strategy also states that the Mayor will measure the success of his strategy against two targets, to ensure first, that there is no net loss of Sites of Importance for Nature Conservation and, second, that the Areas of Deficiency in accessible wildlife sites are reduced. Monitoring of these targets is addressed in the *Mayor's State of the Environment Report* (GLA 2003). Details of the programme for improving people's access to nature in London, and explaining how this is implemented by the London boroughs as part of strategic planning is given in GLA (2008). For more detailed accounts of the development of this whole programme see Goode (1989, 1999, 2005).

Green infrastructure and ecosystem services

During the past ten years sustainability has become a critical issue in the management of cities, particularly with regard to the impacts of climate change.

In the past it has been convenient for planning purposes to compartmentalise types of urban greenspace according to their primary functions, such as parks and other amenity areas, nature reserves, river and canal corridors, and more extensive tracts of metropolitan open land or green belt. Other categories include derelict or contaminated land, much of which is classified as brownfield land. Each has its own set of criteria for planning purposes and there has in the past been little attempt to find a unifying approach by which benefits of greenspace could be considered in a more holistic way. However, a more unified treatment has started to emerge as part of sustainable approaches to urban planning and design.

English Nature outlined the multiple benefits of green networks in a research report (Barker 1997) which recognised that a range of functions could be accommodated, including river and wildlife corridors, together with local cycle and walking routes and extensive areas of amenity greenspace. Such networks could benefit biodiversity by connecting locally important wildlife sites and provide greater opportunities for people to have access to natural areas in towns and cities. The potential health benefits of urban greenspace were recognised as being part of an integrated package.

More recently, with the need for adaptation to climate change, it is recognised that ecosystem functions of green infrastructure provide benefits which promote more sustainable conditions in the urban environment. These include provision of sustainable drainage systems and enhanced flood alleviation, local climatic amelioration, improved air quality, encouraging conditions for urban biodiversity, provision of greenspace for pubic uses, and associated health benefits.

Conservation of biodiversity as part of a broader sustainability agenda was addressed by the TCPA (2004) in its design guide for sustainable urban areas. This recognised the functional value of natural areas and other greenspace in towns and cities in terms of local climatic regulation, flood alleviation and health. It provided numerous case studies demonstrating the value of the ecological services provided by such areas, and demonstrated how biodiversity can be built into design at every level from master-planning down to provision of green roofs. These different levels were summarised as follows:

Existing greenspace infrastructure

- Regional parks, green grids and community forests
- Greenway linkages including both woodlands, waterways and wetlands
- Parks, and natural greenspaces

Green infrastructure within the built environment

- Street trees
- Communal and neighbourhood greenspace
- Green roofs and other habitats within the built environment

New urban developments

- Newly created green infrastructure as part of new urban developments, including greenway linkages and sustainable urban drainage systems.

The first category provides an ecological framework for spatial planning, from the large scale to more local provision of greenspace. This takes advantage of the benefits of existing greenspace and also offers opportunities for its enhancement through creation of new elements at varying scales from community forests and extensive green grids, to local parks and nature areas. Examples of cities which have adopted policies for protection of the green infrastructure at a regional scale for spatial planning include Glasgow, and London (GLA 2002). In London the green infrastructure is being used as the basis for adaptation strategies to counter the effects of climate change in revisions to the London Plan.

Yet again Berlin was ahead of the game. Since 1994 the biotope-based strategy for city-wide planning has been further developed, with the primary objective of using the green infrastructure to deliver ecological services. For example different climatic zones within the city have been mapped, illustrating variations in average air temperature, humidity and soil moisture. These have been correlated with differences in the major biota, from heavily built-up areas to others dominated by natural greenspace. Five broad zones have been identified which reflect the moderating influences of different kinds of urban greenspace (TCPA 2004).

In the UK the Royal Commission on Environmental Pollution has argued that there is a very significant potential for ecosystem services to be delivered by green infrastructure, and that this deserves greater recognition through government guidance on urban development (RCEP 2007 and Goode 2006). The Commission recommended specifically that relevant government departments,

> amend their planning policy statements and guidance to reflect a broader definition of the natural environment in urban areas, and to recognise and protect the role that urban ecosystems can play in improving towns and cities. Planning policy and guidance should describe the range of functions and benefits associated with the natural environment of urban areas, promote the use of green infrastructure and provide a menu of options for planners and developers to use, including:
>
> - creation of green networks and green infrastructure
> - urban river restoration
> - use of green and built infrastructure for flood storage and redirection
> - use of sustainable drainage systems, including green roofs
> - promotion of urban trees and woodland.
>
> *(RCEP 2007: 147)*

As the need for action to counter the effects of climate change becomes more critical, the value of the ecosystem services provided by biodiversity will become ever more apparent.

References

Adams, L.W. and Leedy, D.L. (eds) (1987) *Integrating Man and Nature in the Metropolitan Environment*. Proceedings of the National Symposium on Urban Wildlife, National Institute for Urban Wildlife, Columbia, MD.

Barker, G. (1997) *A Framework for the Future: Green Networks with Multiple Uses in and around Towns and Cities*, English Nature Research Report 256, Peterborough.

Bornkamm, R., Lee, J.A. and Seaward, M.R.D. (eds) (1982) *Urban Ecology*, Blackwell, Oxford.

Box, J. and Harrison, C. (1993) 'Natural spaces in urban places', *Town and Country Planning*, 62: 231–5.

Celecia, J. (1990) 'Half the world and increasing: cities, ecology and Unesco's action', *The Statistician*, 39(2): 135–41.

Dasmann, R.F. (1966) 'Wildlife and the new conservation', *Wildlife Society News*, 105: 48–9.

Fairbrother, N. (1970) *New Lives, New Landscapes*, Architectural Press, London.

Fitter, R.S.R. (1945) *London's Natural History*, Collins, London.

GLA (2002) *Connecting with London's Nature: The Mayor's Biodiversity Strategy*, Greater London Authority, London.

GLA (2003) *Green Capital: The Mayor's State of the Environment Report for London*, Greater London Authority, London.

GLA (2004) *The London Plan: Spatial Development Strategy for Greater London*, Greater London Authority, London.

GLA (2008) *Improving Londoners' Access to Nature*, London Plan Implementation Report, Greater London Authority, London.
GLC (1984) *Ecology and Nature Conservation in London*, Ecology Handbook 1, Greater London Council, London.
GLC (1985) *Nature Conservation Guidelines for London*, Ecology Handbook 3, Greater London Council, London.
GMC (1986) *A Nature Conservation Strategy for Greater Manchester*, Greater Manchester Council, Manchester.
Goode, D.A. (1989) 'Urban nature conservation in Britain', *Journal of Applied Ecology*, 26(3): 859–73.
Goode, D.A. (1993) 'Local authorities and urban conservation', in F.B. Goldsmith and A.Warren (eds), *Conservation in Progress*, Chichester, Wiley, pp. 335–45.
Goode, D.A. (1994) 'Ecological planning', in J. Agyeman and B. Evans (eds), *Local Environmental Policies and Strategies*, Longman, Harlow, pp. 190–209.
Goode, D.A. (1999) 'Habitat survey and evaluation for nature conservation in London', *Deinsea*, 5: 27–40, Natural History Museum of Rotterdam.
Goode, D.A. (2005) 'Connecting with nature in a capital city: The London Biodiversity Strategy', in T. Trzyna (ed.), *The Urban Imperative*, California Institute of Public Affairs and IUCN, pp. 75–85.
Goode, D.A. (2006) *Green Infrastructure*, report commissioned by the Royal Commission on Environmental Pollution, online, available at: www.rcep.org.uk/reports/26-urban/documents/green-infrastructure-david-goode.pdf.
Goode, D.A. (2007) 'Nature conservation in towns and cities', in H. Clout (ed.), *Contemporary Rural Geographies*, London: Routledge, pp. 111–28.
Gottschalk, J.S. (1968) 'Opening remarks', in Symposium on *Man and Nature in the City*, Bureau of Sport Fisheries and Wildlife, Washington, D.C.
Harrison, C., Burgess, J., Millward, A. and Dawe, G. (1995) *Accessible Natural Greenspace in Towns and Cities: A Review of Appropriate Size and Distance Criteria*, English Nature Research Reports Number 153, English Nature, Peterborough.
Henke, H. and Sukopp, H. (1986) 'A natural approach in cities', in A.D. Bradshaw, D.A. Goode and E.H.P. Thorp (eds), *Ecology and Design in Landscape*, Symposium of the British Ecological Society, Blackwell, Oxford, pp. 307–24.
Kieran, J. (1959) *A Natural History of New York City*, Houghton Mifflin, Boston, MA.
Laurie, I.C. (ed.) (1975) *Nature in Cities*, Proceedings of a Symposium of the Landscape Research Group, Manchester University Press, Manchester.
Leedy, D.L. and Adams, L.W. (1986) 'Wildlife in urban and developing areas: an overview and historical perspective', in K. Stenberg and W.W. Shaw (eds), *Wildlife Conservation and New Residential Developments*, Proceedings of the National Symposium on Urban Wildlife, University of Arizona, Tucson, AZ, pp. 8–20.
Leedy, D.L., Maestro, R.M. and Franklin, T.M. (1978) *Planning for Wildlife in Cities and Suburbs*, American Society of Planning Officials, Planning Advisory Service Report 331, US Government Printing Office, Washington, D.C.
Mabey, R. (1973) *The Unofficial Countryside*, Collins, London.
McHarg, I. (1969) *Design with Nature*, Doubleday Natural History Press, Garden City, NY.
Miller, R. (1978) 'New York's urban wildlife programme', in J. Galli (ed.), *Proceedings of a Meeting on Non-Game, Urban Wildlife and Endangered Species Programmes*, The Wetlands Institute, Stone Harbor, NJ, pp. 27–30.
NCC (1988) *Tyne and Wear Nature Conservation Strategy*, Nature Conservancy Council, Peterborough.
Numata, M. (1982) 'Changes in ecosystem structure and function in Tokyo', in R. Bornkamm, J.A. Lee and M.R.D. Seaward (eds), *Urban Ecology*, Blackwell, Oxford, pp. 139–47.
RCEP (2007) *The Urban Environment*, 26th Report of the Royal commission on Environmental Pollution, Cm 7009, online, available at: www.rcep.org.uk/ reports/index.htm.
Stenberg, K. and Shaw, W.W. (1986) *Wildlife Conservation and New Residential Developments*, Proceedings of the National Symposium on Urban Wildlife, University of Arizona, Tucson, AZ.
Sukopp, H. (1990) *Stadtökologie: das Beispiel Berlin*, Reimer, Berlin.
TCPA (2004) *Biodiversity by Design: A Guide for Sustainable Communities*, Town and Country Planning Association, London.
Teagle, W.G. (1978) *The Endless Village*, Nature Conservancy Council, Shrewsbury.
West Midlands (1984) *The Nature Conservation Strategy for the County of West Midlands*, West Midlands County Council, Birmingham.

9

How much is urban nature worth? And for whom?

Thoughts from ecological economics

Anna Chiesura and Joan Martínez-Alier

Introduction

Official statistics estimate that by the year 2030, 80 per cent of the world's people will live in urban conglomerations (United Nations 2006). In Europe this percentage will be reached by 2020, reaching peaks of 90 per cent in seven countries (EEA 2006). The increasing pace of the urbanization process and its sprawling pattern, described by the European Environment Agency as 'the physical pattern of low-density expansion of large urban areas, mainly into the surrounding agricultural areas', are particularly worrying phenomena, because of the many adverse environmental effects they generate (air pollution, soil degradation, resource consumption, etc.), and the negative impacts they have on the quality of life and well-being of millions of (urban and rural) citizens. The loss of green spaces as a consequences of urbanization challenges us to consider the importance of urban nature more closely (Pelkonen-Yli and Kohl 2005). The issue is not trivial, if we consider urban nature (parks, neighbourhood green spaces, and other semi-natural and un-built urban surfaces) not just as a mere residue of the urbanization process, waiting to be 'used' for some profitable purpose, but as a vital component of the larger urban mosaic, an everyday source of valuable services to citizens, a strategic resource for sustainable urban planning.

The valuation of urban nature is even less trivial if we do it from an ecological economics point of view: the matter becomes then not a mere question of giving monetary figures to the value(s) and the benefits society gets from urban green areas, but it should also take into account who actually enjoys them, and it should look at whose values are taken into consideration. When attempting to value urban nature from an ecological economics point of view, one should be aware not only of the variety of beneficial services and values (social, economical and environmental) provided by nature in urban contexts, but also of the plurality of stakeholders involved (including citizens, planners, administrators, business, birds and plants) and – consequently – of the complexity of the legitimate perspectives that can be taken. Therefore, the central question to be asked is not only 'How much is urban nature worth?', but also 'for whom?' Thoughts from ecological economics suggest that technical and methodological argumentations about the best valuation approach to be applied cannot be separated from – or, better, it should not leave apart – ethical considerations and distributional aspects.

Ecological economics: environmental valuation through value pluralism

Building on fundamental works of various thinkers and scientists (among which the biologist and urban planner Patrick Geddes, but also the (bio)economist Nicholas Georgescu-Roegen), ecological economics (EE) – in opposition to conventional economics – sees the economy as embedded in a larger biophysical system subjected to the laws of thermodynamics (Martínez-Alier and Schlupman 1987), as well as a part of a social structure of property rights, distribution of power and income (Martínez-Alier 2001). In a little more than two decades, EE has introduced and developed new topics and methods such as

a new physical (un)sustainability indicators (Vitousek *et al.* 1986; Wackernagel and Rees 1995);
b the application of ecological notions of carrying capacity and resilience to human ecosystems;
c the valuation of environmental services in monetary terms, but also the discussion of the weak comparability of values (Martínez-Alier *et al.* 1998; O'Neil 1993) and the application of multi-criteria evaluation (Munda 1995), or integrated assessment (Faucheux and O'Connor 1998);
d risk assessment, uncertainty, complexity and "post-normal" science (Funtowicz and Ravetz 1994);
e ecological distribution conflicts, 'ecological debt' and environmental justice; and
f dematerialization and industrial ecology.

We believe that some of these lines of thought can offer useful perspectives to better understand urban nature's values. The assumption of value pluralism, for example: EE rests on a foundation of 'weak comparability of values' (O'Neil 1993).

As we all know, and as has been extensively documented in the literature, urban green areas (ranging from the smallest neighbourhood green lots up to the bigger urban parks or peri-urban forests and green belts) provide human societies with valuable products and services: beside their well-known social functions (recreation, health, etc.), they contribute to mitigating soil, water and air pollution, ameliorating urban microclimate, maintaining biodiversity and enriching the aesthetic quality of urban landscapes (Chiesura 2004). However, attempts made to obtain money values of such services have, from one side, had the merit of making the plurality of nature's functions (and values) more explicit to decision-making, but, from the other (see among others Farber *et al.* 2002) they have shown the limitations of monetary measurements when it comes to non-material benefits (spiritual and artistic inspiration, beauty, relaxation and recreation, psycho-physical health, social integration, cognitive development etc.), intrinsic and non-consumptive values of nature.

Furthermore, economic valuation tells little about its significance for marginalized groups or low-income social classes, and hardly reveals relationships and areas of conflict among different stakeholders' interests that are equally important for policy formulation. Therefore, the social and environmental benefits of urban green areas and the costs of their un-management often remain largely unaccounted for within urban planning processes, resulting in social and environmental externalities, leading – among others – to social conflicts and environmental injustice at different temporal (present-future) and spatial (north-south, city-countryside, etc.) scales.

Konijnendijk (2000) reports various examples of urban forest conflict cases occurring in some European cities: these conflicts have concerned protests by local urban inhabitants when they felt that 'their' forests were under threat, for instance by urban development (e.g. highway constructions, commercial building complexes). In Italy, for example, various citizens and local active groups are fighting through street protests and legal actions to save the natural urban areas of their neighbourhoods from the threats of continuous urbanization. The strong economic interests of state agencies, supported by local policies that often use urban planning instruments to justify land conversion and additional concreting of urban and peri-urban areas, often threaten the very survival and functionality of green spaces and parks, which are vital to local communities. In this case, the values at stake are economic, environmental and social: public interests weighted against private interests.

In a cost-benefit language, the choice of converting a public green area or an urban forest to a (public?) infrastructure of some kind (e.g. for highways, social housing or commercial centres) may be deemed sustainable because of the benefits accruing to the region (in terms of better services, reduced travelling distances, more shopping malls, income generation and job opportunities) outweigh the damages/losses suffered (e.g. loss of biodiversity, reduced quality of life and recreation-health opportunities for children and elders, the banalization of the landscape). But in ecological economics terms, the valuation process has a different normative premise (weak comparability of values, indeed) and that forces us to ask ourselves: how can we be sure that the song of a songbird is less valuable than a supermarket in the long term, or that the loss of peace and relaxation will one day generate more public health costs than the benefits gained today from job creation? Furthermore, are we sure that this choice really reflects the changing lifestyles of urban dwellers and that it represents a democratic compromise of all the values at stake among them?

Conclusions

When planning urban development, questions like who will actually enjoy it, whose needs and interests are to be met, and who will bear the costs of unsustainable urbanization are largely unasked in due time. If urban ecology is to contribute to urban sustainability, distributional and ethical issues cannot be ignored. Ecological economics moves beyond the obsession of 'taking nature into account' in money terms, and attempts to avoid clash among incommensurable values by accommodating in the discourse concern for value pluralism, distributional and social equity issues.

Bibliography

Bolund, P. and Hunhammar, S. (1999) 'Ecosystem services in urban areas', *Ecological Economics*, 29(2): 293–301.

Chiesura, A. (2004) 'The role of urban parks for the sustainable city', *Landscape and Urban Planning*, 68(1): 129–38.

Chiesura, A. and de Groot, R. (2003) 'Critical natural capital: a socio-cultural perspective', *Ecological Economics*, 44(2–3): 219–31.

EEA (European Environmental Agency) (2006) *Urban Sprawl in Europe: The Ignored Challenge*, Copenhagen: European Environmental Agency.

Farber, S.C., Costanza, R. and Wilson, M.A. (2002) 'Economic and ecological concepts for valuing ecosystems' services', *Ecological Economics*, 41(3): 375–92.

Faucheux, S. and O'Connor, M. (eds) (1998), *Valuation for Sustainable Development: Methods and Policy Indicators*, Cheltenham: Edward Elgar.

Folke, C., Jansson, A., Larsson, J., and Costanza, R. (1997) 'Ecosystem appropriation of cities', *Ambio*, 26(3): 167–72.

Funtowicz, S.O. and Ravetz, J.R. (1994) 'The worth of a songbird: ecological economics as a post-normal science', *Ecological Economics*, 10(3): 197–207.

Grimm, N.B., Grove, J.M., Pickett, S.T.A. and Redman, C.L. (2008) 'Integrated approaches to long-term studies of urban ecological systems', in J.M. Marzluff, E. Shulenberger, W. Endlicher, M. Alberti,

G. Bradley, C. Ryan, U. Simon and C. ZumBrunnen (eds), *Urban Ecology: An International Perspective on the Interaction Between Humans and Nature*, New York, NY: Springer, pp. 123–41.

Martínez-Alier, J. and Schlupmann, K. (1987) *Ecological Economics: Energy, Environment and Society*, Oxford: Blackwell.

Munda, G. (1995) *Multicriteria Evaluation in a Fuzzy Environment: Theory and Applications in Ecological Economics*, Heidelberg: Physicky Verlag.

Konijnendijk, C. (2000) 'Adapting forestry to urban demands – role of communication in urban forestry in Europe', *Landscape and Urban Planning*, 52(2–3): 89–100.

Martínez-Alier, J. (2001) 'Ecological conflicts and valuation: mangroves vs. shrimp in the late 1990s', *Environment and Planning C, Government and Policy*, 19(5): 713–28.

Martínez-Alier, J., Munda, G. and O'Neil, J. (1998) 'Weak comparability of values as a foundation for ecological economics', *Ecological Economics*, 26(3): 277–86.

O'Neil, J.F. (1993), *Ecology, Policy and Politics: Human Well-Being and the Natural World*, London: Routledge.

Pelkonen-Yli,V. and Khol, J. (2005) 'The role of local ecological knowledge in sustainable urban planning: perspectives from Finland', *Sustainability: Science, Practice, & Policy*, 1(1): 3–14, online, available at: http://ejournal.nbii.org/archives/vol1iss1/0410-007.yli-pelkonen.html.

Rees, W.E. (1992) 'Ecological footprints and appropriated carrying capacity: what urban economics leaves out', *Environment and Urbanization*, 4(2): 121–30.

United Nations (2006) *World Urbanization Prospects, The 2005 Revision: Data Tables and Highlights*, New York, NY: United Nations, Department of Economic and Social Affairs, Population Division.

Vitousek, P., Elrich, P., Elrich, A. and Matson, P. (1986) 'Human appropriation of the products of photosynthesis', *BioScience*, 34(6): 368–73.

Wackernagel, M. and Rees, W. (1995) *Our Ecological Footprint*, Gabriola Island, BC: New Society Publishers.

Part 2

The urban ecological environment

Introduction

Ian Douglas

To understand the ecological character of urban areas, a good knowledge of the biophysical character of the urban area is required, including the climate, hydrology, geomorphology, soils and special character of ecological processes in the most-modified environment. The chapters in this part set out how urban activities have modified the impacts of the natural circulation of air, water and materials, and so created the potential for a great diversity of habitats. It also examines the way ecological processes, such as succession, occur in the urban environment, and assesses the contributions that can be made to those processes by recombinant and creative ecology.

Urban development radically alters the nature of the ground surface. Buildings of all types profoundly influence the conversion of solar radiation. They also affect wind flow at ground level, reducing wind velocity, causing changes in wind direction, greater turbulence, and localized acceleration. Built-up areas also differ from the countryside in terms of their thermal regime and in levels of, and periodic changes in, relatively humidity and water vapor content. In the urban energy budget, heat from combustion processes can play a significant role, especially in winter in high latitudes. Being areas of concentrated emissions of pollutants and fine particles, cities modify the receipt of sunlight, create conditions for fog (smog) and produce condensation nuclei for rain formation. Sue Grimmond demonstrates how urban climates are due to the surface-atmosphere exchanges of energy, mass, and momentum. Understanding these exchanges, and the effects of a particular urban setting on their spatial and temporal dynamics, are key to understanding urban climates at the scale of the city, neighborhood, or individual street or property level, and to predicting and mitigating negative effects.

Urban temperatures tend to be warmer, especially at night around the city center than at the edges of the built-up area. Some of the radiation received during the day is stored in buildings and released at night. Tim Oke points out that the convoluted configuration of the urban building materials exposes a much larger surface area for exchange than a flat site and because they are often dry (due to their ability to shed not store water) the heat they absorb is used efficiently to warm the material rather than to evaporate water. In addition heat from combustion intensifies during peak traffic periods, and also comes from domestic heating on winter evenings. The heat island generates its own wind system, with wind flow converging on the city centre and then rising, assisting the development on convectional clouds under calm conditions, and subsequently flowing outwards and descending at the city edge. Parks and other greenspaces modify the intensity of the urban heat island, a phenomenon found in large cities in all latitudes, but being most marked under calm weather conditions, such as those that prevail for much of the year in the sub-tropics and mid latitudes.

Urban areas can develop their own rainfall through the formation of convectional clouds as a result of the urban heat island effect. Well-documented in Europe and North America, the effect was also reported by the world's first space-based rain radar aboard NASA's Tropical Rainfall Measuring Mission (TRMM) satellite. Mean monthly rainfall rates within 30–60 kilometers downwind of the cities were, on average, about 28 percent greater than the upwind region. In some cities, the downwind area exhibited increases as high as 51 percent. Marshall Shepherd and his co-authors show how the urban environment alters aspects of regional hydroclimates, particularly precipitation and related convective processes. Although at times the human influence has been exaggerated, studies have established the scientific basis for how human activity in urban environments shapes their pattern of convection, precipitation, and lightning. Higher urban rainfalls, which may increase further under climate change, have implications for urban drainage design, but because of the large impermeable areas, they may not have much implication for soil-plant-water relations.

Urban areas have a dual hydrologic system: the modified natural drainage system, including canals and river diversions, and the artificial water supply and waste water disposal system. Some parts of the original drainage system may be buried and some of the buried streams may become interconnected with the sewer system. This introduces complexities for runoff management, especially when large roofed and paved areas feed into combined sewers that are designed to overflow into rivers after heavy rains. For aquatic life, much depends on the management of the discharges from the artificial system. Just one overflow of an old combined sewer system can create a depletion of oxygen that lasts long enough to greatly reduce fish stocks. Ian Douglas shows that climate change is possibly already causing more urban flooding from surface water and sewer overflows which suggests a need for creative conservation, stream daylighting and sustainable urban drainage systems to create multi-functional spaces that can support wildlife, provide recreation and, for a few days every few years, provide stormwater runoff storage.

Good urban planning requires a sound understanding of the ground on which cities are built. Some are built on glacial deposits that are liable to move when excessively wet. Others overlie cavernous limestone terrain that poses foundation problems and can be subject to sinkhole collapse. Some sit upon shrink-swell clays that can dry out and cause subsidence. Almost everywhere in hilly terrain, urban construction can potentially trigger landslides, but the potential for such mass movements all too often goes unrecognized. Removal of the original forest vegetation and exposure of bare soil or weathered rock leads to erosion and sediment production. The sediment is washed into rivers and aggravates flood problems, often causing damage to water supply intakes, while its deposition can disrupt both urban and rural activities. Ian Douglas demonstrates that urban development also creates landforms, whether they are the result of deliberate encroachment by filling on to floodplains or shorelines, or just the accumulation of material on the surface by the dumping of construction debris, domestic waste and landfill operations. All such changes to landforms have to be made with an understanding of the geomorphic processes that will affect them. They all have implications for urban plant and animal life.

Urban soils are highly varied in character, not only because they reflect the original soil associations of the rural landscape, but because a whole variety of materials have been added to the soil as a result of urban change. Often land has been bulldozed and filled or excavated. Soils have been compacted, and their organic matter has been lost. Soil parent materials now include debris from construction and demolition, fragments of concrete, sand, and bricks. Peter

Marcotullio shows how soils in highly built-up areas have high nitrogen and phosphate levels, due to excreta from domestic animals, and runoff from car-washing. Many roads soils in higher latitudes are affected by de-icing salt. They can also have high lead concentrations, resulting from the former use of lead in petrol. Soil pH can be high in areas of acid rain. Many soils on old brownfield sites are contaminated with heavy metals and complex compounds derived from hydrocarbons.

Succession is a dynamic and continuous process, often occurring gradually over time. Urbanization and its associated activities have a profound impact on natural succession, with the end result that little natural succession occurs in most metropolitan areas. However, Wayne Zipperer notes that from a species performance perspective, the increase in CO_2 in urban areas resulting from fossil fuel burning, coupled with the increase in temperature from the urban heat island effect, has a significant effect on species productivity. This increase in productivity is further augmented by nitrogen deposition (wet and dry) in the urban landscape. These effects may offset the decrease in productivity likely to be caused by O_3 in urban areas. Much depends on management. For example, a widespread practice in urban forests is to clean out the understory by raking leaves, branches, seeds, and seedlings on the forest floor. Such a loss of the understory may have negative consequences for many wildlife species. Likewise, the extensive use of ornamental invasive species and 'weed-free' lawn areas has similar impacts.

Recombinant ecology is a concept that acknowledges the dynamic reconfiguration of urban ecologies through the ongoing relationships between people, plants, and animals. Recombinant design is the interweaving of ecological, urban, architectural, and social systems to produce metadisciplinary results not possible in each system's parent organization. It involves exploration of the practical implications of exotic species in urban floras and faunas in relation to natural communities. Colin Meurk uses examples from New Zealand to show that recombinant ecosystems can be a vehicle for landscape legibility (or eco-revelation) as much as pure indigenous communities may have once been. The purpose and value of recombinant ecosystems lies in their intrinsic ecological interest, but also they can be seen as a means of maintaining biodiversity through managed coexistence or 'reconciliation'.

The notions of 'creative ecology' and 'creative conservation' have encouraged reassessment of the purpose and practice of wildlife-resource management in the built environment. They apply ideas of people-made nature and ecosystem modification to create improved urban space, such as wildflower meadows. Creative conservation and urban design have been used to generate empirical recommendations for enhancing biodiversity in urban gardens. The Greenwich Peninsula Ecology Park in London is frequently cited as an example of such creative conservation. Grant Luscombe and Richard Scott use examples from their LANDLIFE projects to show that creative conservation focuses on using common core species of native origin that occur widely across the country. It is an opportunity to put back simple habitats suited to soil type, for people to enjoy: the buttercup meadow, the poppy field, and the cowslip bank. Such wildflower spectacles are supposed to be common, yet few have seen them, and fewer are there to be seen.

These chapters show how the climatic, geologic, and ecological aspects of the urban environment, as modified by human action, create opportunities for organisms to colonize, adapt and create habitats. Our work in cities can lead to new combinations of plants and animals in the very varied habitats to be discussed in Part 3.

10

Climate of cities

C.S.B. Grimmond

Introduction

Urban areas create distinct local and micro-scale climates. Commonly cited effects are presented in Table 10.1. What this summary does not make clear, however, is that urban climates vary significantly both within and between cities. Urban climates result from changes in the nature of the urban surface (the materials, its morphology, the fraction of built and vegetated cover, etc.) and the activities of the cities' inhabitants (generating heat, greenhouse gases, aerosols, etc.) as they move around, work and live in the city. Ultimately urban climates are due to the surface-atmosphere exchanges of energy, mass and momentum (represented conceptually in Figure 10.1). Understanding these exchanges, and the effects of a particular urban setting on their spatial and temporal dynamics, are key to understanding urban climates at the scale of the city, neighbourhood or individual street or property level, and to predicting and mitigating negative effects. This chapter describes these energy and mass exchanges and highlights key urban controls with data and examples of studies that document effects.

Global, regional and local effects

At the global scale, cities cover only a small fraction of the Earth, approximately 2 per cent of the land surface (Shepherd 2005). Thus, in terms of direct surface changes individual cities do not impact global weather or climate patterns. However, given the large and ever increasing fraction of the world's population living in cities (the majority of the world's population now are urban dwellers and in the developing world rates of growth of cities exceed 200,000 people per day) and the disproportionate share of resources used by these urban residents, cities and their inhabitants are key drivers of global climatic change. Cities affect greenhouse gas sources and sinks both directly and indirectly – urban areas are the major sources of anthropogenic carbon dioxide emissions from the burning of fossil fuel for heating and cooling; from industrial processes; transportation of people and goods etc., and the demand for goods and resources by city dwellers, both historically and today, are the major drivers of regional land use change such as deforestation. While the exact values are subject to debate, it is widely suggested that more than 70 per cent of anthropogenic carbon emissions can be

Table 10.1 Controls on urban climate – general and urban specific effects with examples of studies of urban influences

Variable	*General controls*	*Urban controls/effects*	*Examples of studies documenting urban effects*
Incoming solar radiation (K↓)	Latitude; synoptic conditions/cloud cover	Air quality/industrial sources influence scattering	Gomes *et al.* (2008), Robaa (2009)
Outgoing solar radiation (K↑)	Incoming solar radiation, albedo	Surface materials; surface morphology/ geometry	Aida (1982), Kanda *et al.* (2005), Fortuniak (2008)
Incoming long wave radiation (L↓)	Synoptic conditions/ cloud cover	Air quality/industrial sources affect absorption	Nunez *et al.* (2000), Jonsson *et al.* (2006)
Outgoing long wave radiation (L↑)	Surface temperatures, emissivity, sky view factor	Thermal and radiative properties of materials; surface morphology/ geometry	Kobayashi and Takamura (1994), Sugawara and Takamura (2006)
Net all-wave radiation (Q*)	Latitude; synoptic conditions/cloud cover	Materials, morphology, air quality	Oke (1988, 1997), Offerle *et al.* (2003), Harman *et al.* (2004)
Sensible heat flux (Q_H)	Temperature gradient; atmospheric stability; synoptic conditions	Building volume; built fraction	Oke (1988), Grimmond and Oke (1995,2002), Christen and Vogt (2004), Offerle *et al.* (2005a, 2006a, 2006b), Roth (2007)
Latent heat flux (Q_E)	Moisture gradient; atmospheric stability; synoptic conditions	Fraction greenspace; irrigated surfaces; piped/channelled water systems; detention ponds etc	Grimmond and Oke (1991,1995,1999a, 2002), Christen and Vogt (2004), Offerle *et al.* (2005a,2006a,b), Mitchell *et al.* (2007), Roth (2007)
Storage heat flux (ΔQ_S)	Air-ground temperature gradients; thermal properties surface materials	Materials and morphology urban surface; orientation walls; mass/volume urban surface	Grimmond and Oke (1999b), Offerle *et al.* (2005b), Roberts *et al.* (2006)
Anthropogenic heat flux (Q_F)	Latitude; continentality; regional setting; economic development status	Heating/cooling requirements; industrial activity; socio-economic conditions; population/ building density; Transportation routes and methods	Oke (1988), Grimmond (1992), Sailor and Lu (2004), Klysik (1996), Offerle *et al.* (2005b), Pigeon *et al.* (2007), Moriwaki *et al.* (2008), Flanner (2009)

Table 10.1 continued

Variable	*General controls*	*Urban controls/effects*	*Examples of studies documenting urban effects*
Air temperature	Latitude; continentality; regional setting	Materials and morphology of urban surface release of anthropogenic heat; air quality	Oke Chapter 11 this volume
Humidity	Latitude; continentality; regional setting – proximity to water bodies	Reduced vegetation; fewer moist surfaces localised releases (industrial sources) as by-product combustion; urban air temperature	Hage (1975), Holmer and Eliasson (1999), Fortuniak *et al.* (2006), Kuttler *et al.* (2007)
Wind field	Synoptic conditions	Building density; morphology of buildings and roofs affect roughness and displacement lengths; channelling through urban canyons	DePaul and Sheih (1986), Nakamura and Oke (1988), Rotach (1995), Grimmond and Oke (1999c), Roth (2000), Martilli *et al.* (2002), Eliasson *et al.* (2006), Offerle *et al.* (2007), Nelson *et al.* (2007)
Precipitation	Latitude (solid, liquid); synoptic conditions; topographic variations	Air quality/industrial-traffic sources→cloud condensation nuclei; roughness elements/ surface heating→ convection	Jauregui and Romales (1996), Lowry (1998) Bornstein and Lin (1998), Shepherd *et al.* (2002); Rosenfeld (2000); Shepherd *et al.* Chapter 12 this volume

attributed to cities (Svierejva-Hopkins *et al.* 2004 place this value in excess of 90 per cent). Thus despite their small surface area globally; the effects of cities are significant regionally and globally, as well as locally.

A city's climate, and its effects on climate, is influenced by its geographical setting. Latitude has an influence through basic solar forcing; continentality influences seasonal extremes; while the sequence of expected fronts, low pressures systems, etc., forms synoptic scale influences, affecting the range of meteorological conditions a city experiences. These all influence the design of a city (e.g. building styles) and the behaviours and activities of inhabitants (demands for heating and cooling, etc.).

At the meso-scale, the geographic setting of a city will influence regional wind systems that are forced by topography (e.g. mountain–valley) or the presence of water bodies (e.g. sea or lake – land breezes). These in turn affect such things as the redistribution of air pollutants. The presence of the city itself can, under some synoptic conditions (e.g. anticyclones), create regional

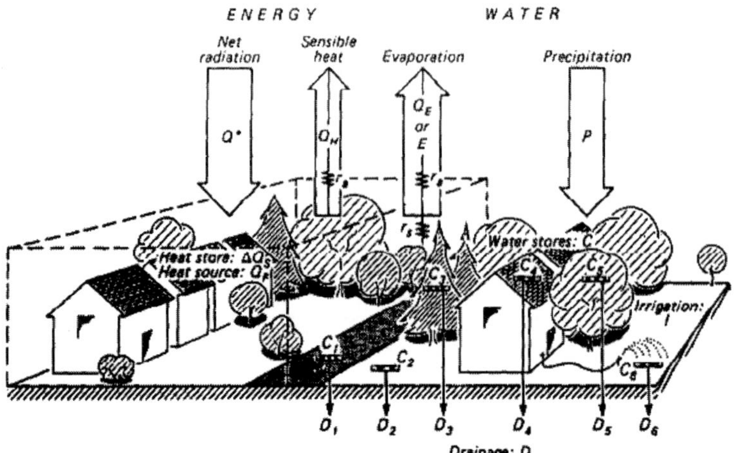

Figure 10.1 Schematic figure to show the link between the energy and water exchanges at the urban surface (*source:* Grimmond and Oke 1991).

winds, the so-called rural-to-city breezes (e.g. in Paris documented by Lemonsu and Masson 2002). Cities also influence areas downwind. They are a source of warm, polluted air and have been documented to modify precipitation patterns (Lowry 1998, Shepherd 2005).

Within a city, neighbourhoods with similar land-use and land-cover, generate distinct local scale climates. These are a function of the urban morphology, built materials, amounts of vegetation, and human activity. Repeated patterns, based on features such as the height of the buildings, width of the canyons between them, the shape of roofs etc., the areal fraction of vegetation can be clearly identified (Figure 10.2). Urban climatologists commonly characterise cities at this scale in terms of the height, width, density of buildings, the fraction of greenspace, and amounts of heat released by human activities (Table 10.2). Many numerical models simulate urban climates at this scale (e.g. the Weather Research and Forecasting Model WRF, MM5, Meso-NH) (see further description in Grimmond *et al.* 2010).

Within neighbourhoods arrays of micro-scale climates exist. At these smaller spatial scales (10^0–10^1 m) a person walking down the street can experience a range of conditions: the sunlit or shaded sides of the street; the channelling or blockage of wind; the influence of a park or shade trees on boulevards. At this scale the effects of individual properties have an influence on the urban climate.

Thus, key to an understanding of urban climates is a clear understanding of spatial scale, both horizontally and vertically (Figure 10.2). This vertical dimension also has implications for how to conduct urban measurements (the height and siting of instruments, Grimmond 2006; Oke 2006) or run numerical simulations (Masson *et al.* 2002). What is important to stress here is that fundamental in any study of urban climate is the question as to 'what is actually of interest' – the overall effect of the city, conditions in a neighbourhood, or the climate of an individual house or garden. Confusion results if this is not clear and a spatial mismatch is introduced between the entity of interest, observations collected or used, or the spatial resolution of models run to simulate or predict effects.

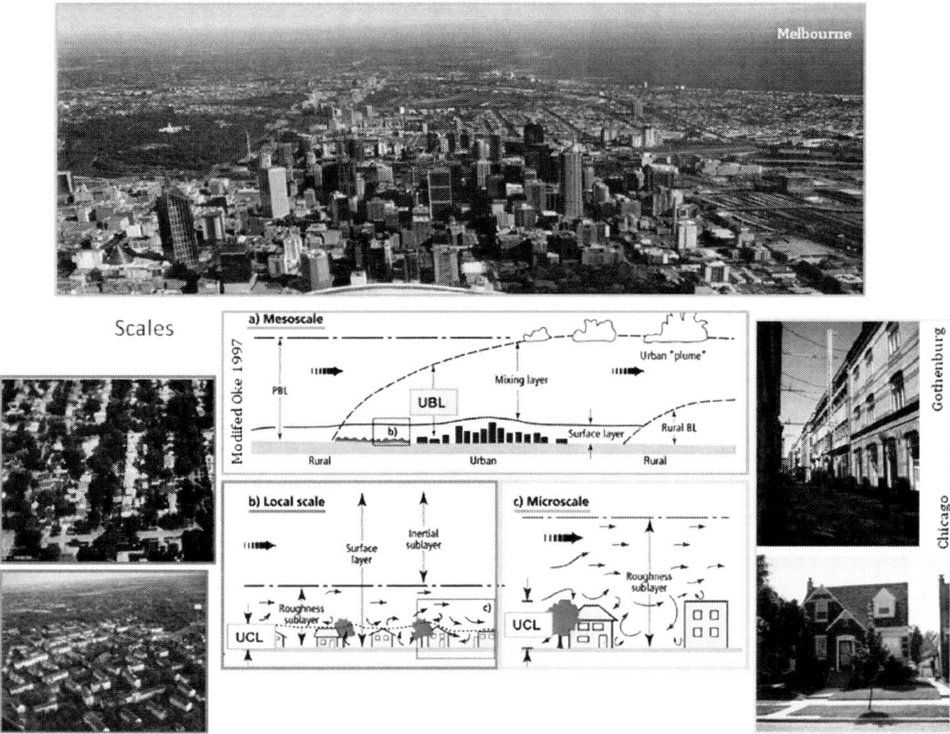

Figure 10.2 Schematic representation, with photographs, of the three horizontal-vertical scales commonly used in studies of urban atmospheric processes: (a) the planetary boundary layer (PBL) and urban boundary layer (UBL) (shown here for daytime convective conditions); (b) the urban canopy layer (UCL) with vertical dimensions related to the mean height of the roughness elements – depending on the local scale area or 'neighbourhood' the relative proportions of buildings and trees will vary; (c) the micro or individual property scale. The bold arrows in the sub-figures indicate the mean wind direction. The smaller arrows shown in (b) and (c) indicate the nature of the mean and turbulent flow (figure modified after Oke 1997, reprinted with kind permission of Kluwer Academic Publishers from Piringer *et al.* 2002, Fig. 1, p. 3).

The surface energy and water balance

As noted above the most powerful conceptual framework within which to understand urban climates is the surface energy balance. The available energy at any location, urban or rural, to evaporate water, to heat the air, or heat the ground, is dependent on the balance of radiative fluxes. This simple statement of the conservation of energy can be defined (Oke 1988):

$$Q^* = K^* + L^* = K\downarrow - K\uparrow + L\downarrow - L\uparrow \qquad \text{Units: W m}^{-2}$$

where Q^* is the net all wave radiation, K the short wave or solar fluxes, L the long wave or terrestrial fluxes, with the arrows indicating whether the flow of energy is towards ($\downarrow$) or away ($\uparrow$) from the surface. The net all-wave radiation, with the additional source of energy in cities,

Table 10.2 Parameters commonly used to characterise cities. Each subscript refers to a separate parameter. Subscript f = roof; r = road; w = wall; v = pervious; t = tree; H = building; g = grass, s = soil, m = momentum; h = heat, u = urban, c = canyon.

Category	*Parameter*	*Name*
Radiative characteristics	$\alpha_{f,r,w,v,t,g,c}$	Albedo
	$\varepsilon_{f,r,w,v,t,g,c}$	Emissivity
Roughness	$z0m,_{c,u}$; $z0h,_{c,u}$	Roughness length for momentum and heat
	$d0_{f,r,w}$	Zero-place displacement height
Thermal characteristics	$c_{Pf,r,w,v,t,g,s}$	Volumetric heat capacity
	$K_{f,r,w,v}$	Thermal conductivity
	$nl_{f,r,w,v,s}$	Number of layers
	$dl_{f,r,w,v,s}$	Layer thickness
Urban morphology	$z_{H,g,t,u}$	Mean height
	W_X	Averaged building separation/canyon width
	L_X	Average width of buildings
	SVF	Sky view factor
	L_Y	Mean block length
	az	Mean long axis azimuth of walls
	λ_F	Frontal area index
Plan area	$\lambda_{f\,v,g,t,H}$	Fraction of area
Urban land use	–	Terrain classification Ellefsen (1991), Oke (2006) or Stewart and Oke (2009)

the anthropogenic heat flux (Q_F), drives non-radiative exchanges between the surface and the atmosphere (Oke 1988):

$$Q^{\star} + Q_F = Q_H + Q_E + \Delta Q_S + \Delta Q_A \qquad \text{Units: W m}^{-2}$$

where Q_H is the turbulent sensible heat flux (heating of the air), Q_E is the turbulent latent heat flux (linked to E evapotranspiration (mm of water) through the latent heat of vaporization), ΔQ_S is the net storage heat flux (heating of the urban fabric), and ΔQ_A is the net horizontal advective heat flux (Table 10.1). For the energy balance, the sign convention is that fluxes away from the surface are positive on the right hand side. The radiative fluxes towards the surface are positive.

Net all-wave radiation

Key factors influencing each of the radiative fluxes in any city are defined in Table 10.1. These factors are separated into those common across all land covers/land uses (column 2), those influenced more specifically by urban conditions (column 3), and examples of documented urban effects (column 4).

Above a city, latitude influences the typical incoming shortwave or solar radiation (K↓) at the top of the atmosphere. The synoptic and regional settings influence the probability of cloud and the nature of air pollution within the region, thus affecting solar radiation received at the surface. Increased concentrations of aerosols alter scattering and therefore the relative amounts of direct and diffuse radiation received, as well as the overall transmissivity of the atmosphere and the shortwave flux received at the surface.

The nature of the surface, both materials and morphology, alters the albedo (α) (reflectivity) of the surface, and thus the outgoing shortwave radiation ($K\uparrow$). In a city, if surface materials are kept constant, taller buildings result in lower 'bulk' albedo if street widths are kept constant (larger height: width ratios – H:W) (Aida 1982; Kanda *et al.* 2005). Lighter materials tend to have higher albedos than darker building fabrics. For many urban settings, significant investments are being directed to development of urban building materials (paints, roofing covers, etc.) which have higher albedos to try and reduce the radiative loading of urban areas and thus mitigate urban heat islands (Akbari and Taha 1992; Synnefa *et al.* 2008). These so-called 'cool' materials no longer have to be light coloured (Levinson *et al.* 2007) and are being used for all elements in the city, roof and wall materials, and for vehicles as well.

Net longwave radiation ($L^\star$) similarly is dependent on atmospheric conditions, through influences on incoming longwave radiation ($L\downarrow$), and surface conditions on outgoing longwave radiation ($L\uparrow$). Surface materials and morphology influence the surface temperature and emissivity. The trapping of longwave radiation in areas with low sky view factors (larger H:W ratio of buildings and vegetation) results in increased $L^\star$. Developments of cool materials, noted above, take into account the spectral response of short and long wave length radiative fluxes.

While the influence on individual fluxes can be significant, and can vary largely at the micro or local scale, overall the effects tend to balance out and the net radiative flux in cities tends to be close to that documented in nearby rural settings (Oke 1997).

Figure 10.3 a–c provides a comparison of radiative fluxes at two sites (Marseille and Miami). A comparison is also shown between clear sky and all sky conditions (Figure 10.3 b, c). These data give a sense of the magnitude and diurnal course of fluxes.

Anthropogenic heat flux

The anthropogenic heat flux is an additional energy term introduced in the urban energy balance to account for energy that is used and released in cities. This flux derives largely from mobile transportation sources; fixed buildings whether industrial, commercial or residential (for lighting, heating, air conditioning, industrial manufacturing, etc.), and that energy attributable directly to human metabolism (Oke 1988; Grimmond 1992; Sailor and Lu 2004; Moriwaki *et al.* 2008). Dense urban areas with tall buildings that intensively use energy for air conditioning in the summer and/or heating in the winter can result in significant amounts of energy being released into the urban environment. In central Tokyo, for example, fluxes of over 1000 W m^{-2} have been calculated (Ichinose *et al.* 1999). In areas where the buildings are smaller and less dense, and there is minimal use of heating and cooling, values for Q_F are less than 5 W m^{-2} (Grimmond 1992). In terms of the mobile transport sources of heat key controls are the nature of the vehicle fleet (e.g. balance of trucks versus cars and the fuel used), the density of traffic (e.g. major versus minor roads), and the level of usage and commuting behaviours (e.g. related to employment hours, when school is in session, etc.). Heat released directly by metabolism is controlled by where people are likely to be (e.g. work, home) and their levels of activity (e.g. active or resting).

Turbulent sensible heat flux

The turbulent sensible heat flux is driven by the net available energy, the gradient in air temperature between the surface and the air above it, and the ability of the air to transport the energy away from the 'warm' location (towards or away from the surface). Atmospheric stability

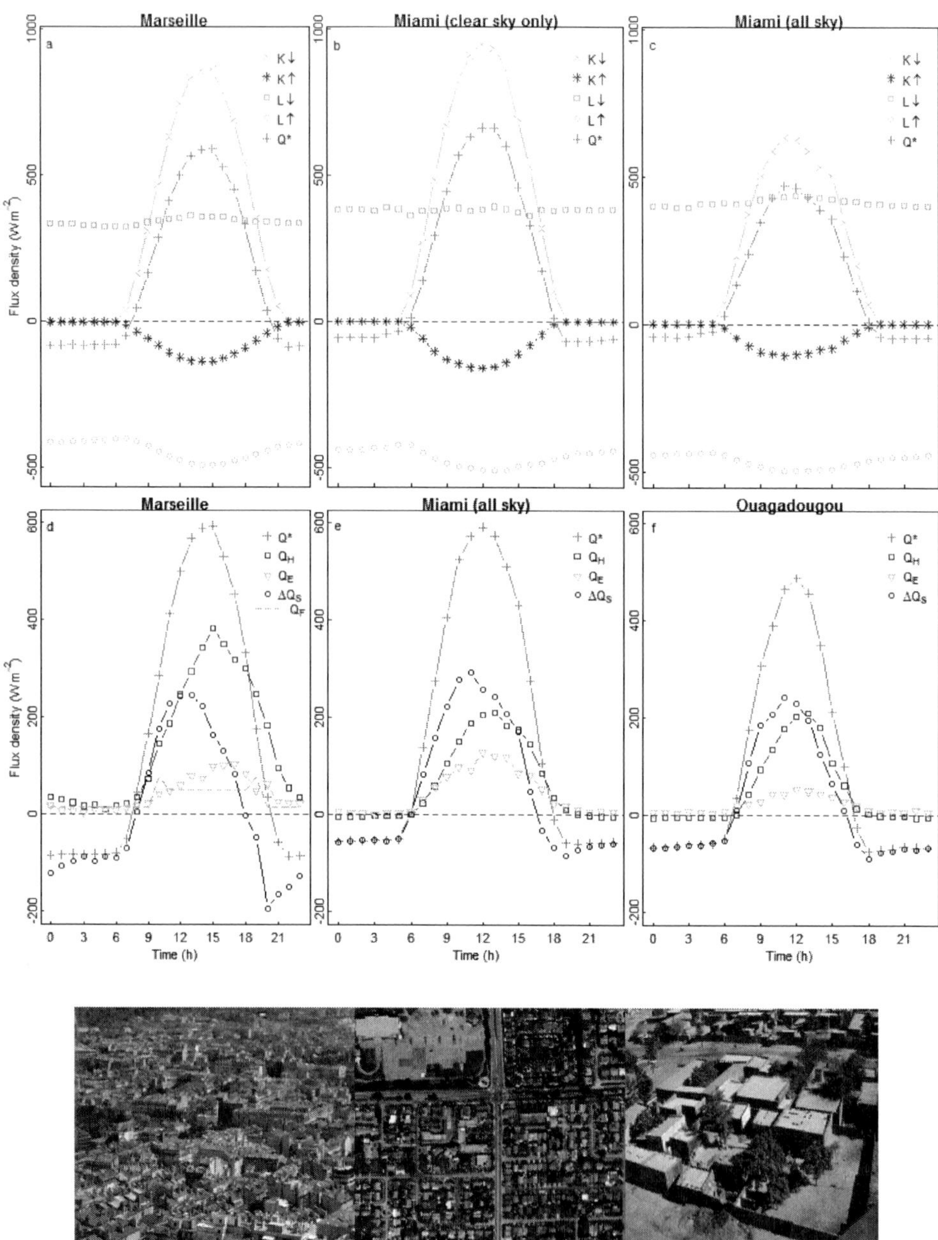

Figure 10.3 Radiation and energy balance fluxes for Marseille, Miami and Ouagadougou. The top row are observed radiation balance data for (a) all sky Marseille (July) (b) clear sky Miami and (c) all sky Miami conditions (May–June). The middle row shows energy balance data (d) Marseille (e) Miami and (f) Ouagadougou (February). Photographs on the bottom row are of the sites where data were collected (g) Marseille central business district (h) Miami residential (source Google Earth) and (i) Ougadougou residential. See text for symbol definitions. Data are from Grimmond *et al.* (2004), Newton *et al.* (2007), Grimmond and Oke (1995) and Offerle *et al.* (2005a).

is a measure of the ability of the air to mix the air or transport heat – typically classified as unstable, neutral and stable. Under unstable conditions there is a lot of vertical motion which helps to move heat away from the surface. Under such conditions, large positive Q_H fluxes are expected. Under neutral conditions there is suppression of the buoyant transport of heat and Q_H tends to be close to zero. Such conditions are common under overcast, windy conditions or in the evening. Under stable conditions there is net transport of sensible heat towards the surface.

Typically in cities in the summer time in extensively built-up areas, unstable conditions prevail during the daytime and mildly unstable or neutral conditions at night. In high latitudes in the winter, for example in Scandinavian or northern European cities, Alaska, northern Canada or in areas of very low density, negative Q_H values may be observed. Wind regimes, which are influenced by the urban surface morphology, can enhance or dampen heat (and moisture) transport away from the surface (see further discussion about wind fields below).

The relative importance of the sensible heat flux for three contrasting cities is shown in Figure 10.3. These illustrate clearly the typical diurnal course of the flux and key differences between residential areas of cities related to the nature of the building fabric (a key control on the storage heat flux) and the available moisture/fraction of greenspace (a key control on the latent heat flux).

Turbulent latent heat flux

The turbulent latent heat flux is dependent on the availability of moisture at the surface and the ability of the atmosphere to move the moisture away from the surface. For the latter, the influence of atmospheric stability and wind fields discussed above apply. Thus a critical control is the strength of the moisture gradient. Unlike the spatial pattern of surface temperatures in a city where differences always are present, albeit with varying contrasts, in urban environments it is possible to have areas/times where there is no surface moisture (e.g. a totally sealed parking lot with no vegetation a long period after rain) and areas where it is freely available (e.g. irrigated parks, detention ponds). Human activities such as street cleaning, allowing/banning/regulating garden irrigation etc. can significantly modify water availability and thus rates of Q_E. When irrigation bans are put in place to conserve water, rates of evapotranspiration drop accordingly. Bans may be complete or impose limitations on water use (e.g. day of week, time of day, type of irrigation). Typically the central business district of a city has less vegetation and residential areas greater vegetation so patterns of Q_E will reflect this (Figure 10.4). It is important to stress though, that even in the driest urban settings, water is present and urban evapotranspiration does occur (see for example the fluxes from Ouagadougou shown in Figure 10.3).

Of course, latent heat flux rates also are influenced by the frequency and intensity of precipitation events, and the methods used to detain or rapidly drain rainwater away. In many urban areas water is retained in neighbourhood detention ponds (particularly common in USA) (see Chapter 14 this volume) or recycled into local wetlands (see Chapter 27 this volume), or held at individual properties (becoming common in Australia) to irrigate vegetation or for internal water use. Under very humid conditions small moisture gradients will reduce evaporation rates. Immediately following rain, or early morning after dewfall, there can be large latent heat flux values for a short period of time (e.g. Richards and Oke 2002; Richards 2004).

As the latent heat flux is the energy equivalent to the evaporation term in the water balance (see Chapter 19 this volume), the size of the flux influences not only the partitioning of the convective energy fluxes (Bowen ratio (β) = Q_H/Q_E) (heating air: evaporating water), but also other water balance fluxes such as ground water recharge.

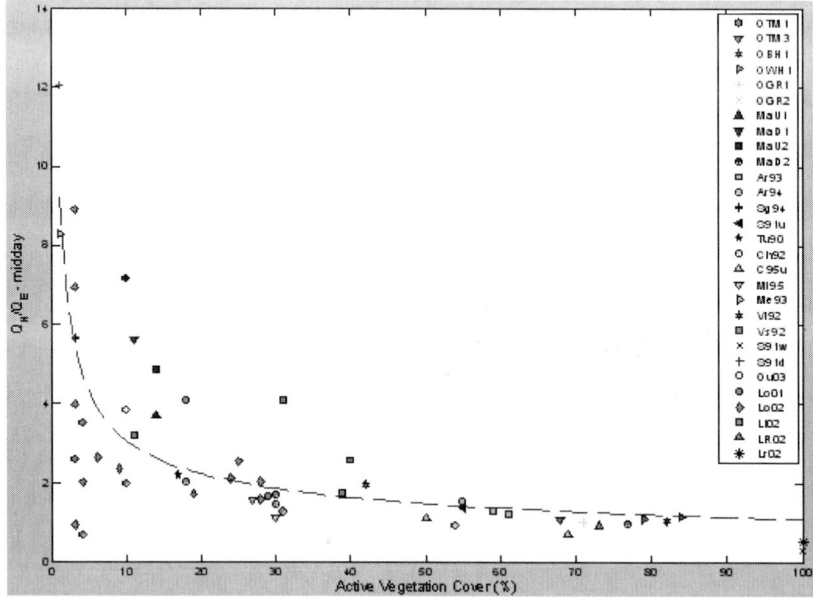

Figure 10.4 Changing Bowen ratio ($\beta = Q_H/Q_E$) as a function of plan area fraction of active vegetation cover. Larger Bowen ratios are associated with smaller amount of active vegetation in the urban area. Each dot is for an urban site. The data are for the middle of the day (10–14 h). Data are from Oklahoma city (OTM1, OTM3, OBH1, OWH1) (Grimmond unpublished); Marseille (MaU1, MaD1, MaU2, MaD2) (Grimmond *et al.* 2004); Los Angeles (Ar93, Ar94, Sg94), Sacramento (S91u), Tucson (Tu90), Chicago (Ch92, C95u), Miami (Mi95), Mexico City (Me93), Vancouver (Vl92, Vs92)(Grimmond and Oke 1995, 2002), Ouagadougou (Ou3) (Offerle *et al.* 2005a), Lodz (Lo01, Lo02, Ll02, LRO2) (Offerle *et al.* 2005b, 2006a, b). For comparison rural areas or completely vegetated areas are included near Oklahoma (OGR1, OGR2), Sacramento (S91w, S91d) (Grimmond *et al.* 1993), Lodz (Lr02). Where there are multiple points for a site each is for an individual month.

The spatial patterns of atmospheric moisture (various measures of humidity) within cities are also influenced by temperature patterns (see Chapter 12 this volume), as well as the surface moisture status and latent heat fluxes. Typically, urban areas are described as having an atmospheric urban moisture deficit compared to surrounding rural areas (see, for example, the studies cited in Kuttler *et al.* 2007). This is because of the limited vegetated surfaces, the enhanced air temperatures (which increases the amount of moisture required to saturate air), and engineered pipe networks designed to rapidly remove precipitation from urban areas. Care needs to be taken when comparing moisture metrics as there are a number of different measures (e.g. relative humidity, specific humidity, dew point temperature, absolute humidity, vapour pressure) and these may be a function of other variables (e.g. pressure, temperature) as well as actual moisture content change in the air.

Net heat storage flux

Typically the heat storage flux is considerably larger in urban areas relative to rural surroundings (e.g. Grimmond and Oke 1999b). This flux is the net uptake or release of energy (per unit area and time) by sensible heat changes in the urban canopy air layer, buildings, vegetation and the ground. Given that in cities there is a significant amount of mass to heat up and cool down, plus there are vertical faces that are being directly heated in the morning (east facing walls), middle of the day (south facing in the northern hemisphere, north facing in the southern hemisphere) and in the afternoon (west facing walls) the flux is significant. While in a rural area the soil heat flux may be about 5 per cent of the net all-wave radiation, in cities this value may be up to 40–50 per cent. Moreover, there is a distinct diurnal trend in cities. The flux is typically larger in the morning, before solar noon, as heat is moved away from the surface into the building volume. However, by mid- to late afternoon heat is being transferred back towards the surface and released into the atmosphere. This helps to maintain a positive sensible heat flux in cities in the evening and at night which results in a warmer air temperature. This large conductive heat store also helps to increase the energy available for longwave radiative exchanges (see discussion above).

Key characteristics of the urban environment that influence the size of the storage heat flux are the surface materials, morphology and thermal mass. The surface materials influence the ability of the heat to be conducted into (and out of) the urban fabric. As conduction is not as efficient as convection, typically there are steep thermal gradients relative to the surface temperature. Very high surface temperatures have been frequently observed (e.g. by thermal remote sensing), but away from the surfaces (e.g. inside building cavities or air temperatures) the range of temperatures are considerably less. Thus, just as with the radiative characteristics of built materials, the thermal characteristics (heat capacity, thickness of layers, density) of built materials provide opportunities for architects, planners and engineers, to manipulate the energy exchanges both internally and externally for a building, thus affecting urban climates at micro- to local scales. The morphology (spacing, heights and orientations) of buildings also have a key influence on initial solar gains by the surface and radiative trapping.

Net heat advection flux

Advection, the net horizontal flow of energy, results from spatial differences of surface characteristics. Contrasting surface temperatures, moisture availability or roughness will create a spatial gradient which drives horizontal transport of energy. The setting of a city, next to water, for example, will dictate the magnitude of these exchanges (into or out of) a city at the larger scale. Within the city, patchiness of urban surfaces (e.g. at the lot or neighbourhood scale) affects horizontal energy exchanges and mixing. For example, patchy vegetation may allow contrasts of air temperature and moisture to occur in close proximity, which then result in a net horizontal flux cooling warmer areas. On a hot summer's day, well irrigated grass next to a road or other paved surface will result in advection. This induces spatial variability of evapotranspiration rates in parks (Spronken-Smith *et al.* 2000). Such patchiness and advection have important implications for the stress of vegetation in urban settings.

Wind fields

Underlying an understanding of any of the convective heat exchanges already described (Q_H, Q_E) is the need to know about urban wind fields. One very distinct characteristic of cities is the roughness elements – the objects that the air has to flow over. These are both bluff bodies (e.g. buildings) and porous elements (e.g. vegetation). Clearly porous objects occur over most land surfaces, so it is the bluff bodies and their characteristics that are important in creating distinct wind environments in cities. As air flow is three dimensional (towards an object, at right angles to the object, and vertically – usually referred to as u, v, w components, respectively) it is the response of the wind in each of these directions that is important to urban wind fields at the micro-scale.

The presence of bluff bodies force air to go up, over and around the edges of the objects. This can cause an increase in wind speeds at points upwind and a cavity of low flow behind the object(s) (Figure 10.5a). With an increase in the number of objects (e.g. increasing building density), the surface reaches a peak in roughness, which is associated with the largest roughness length for momentum. After that any further increase in density is associated with effectively lifting the 'surface' higher. Under such conditions the canyon flows are little influenced by above

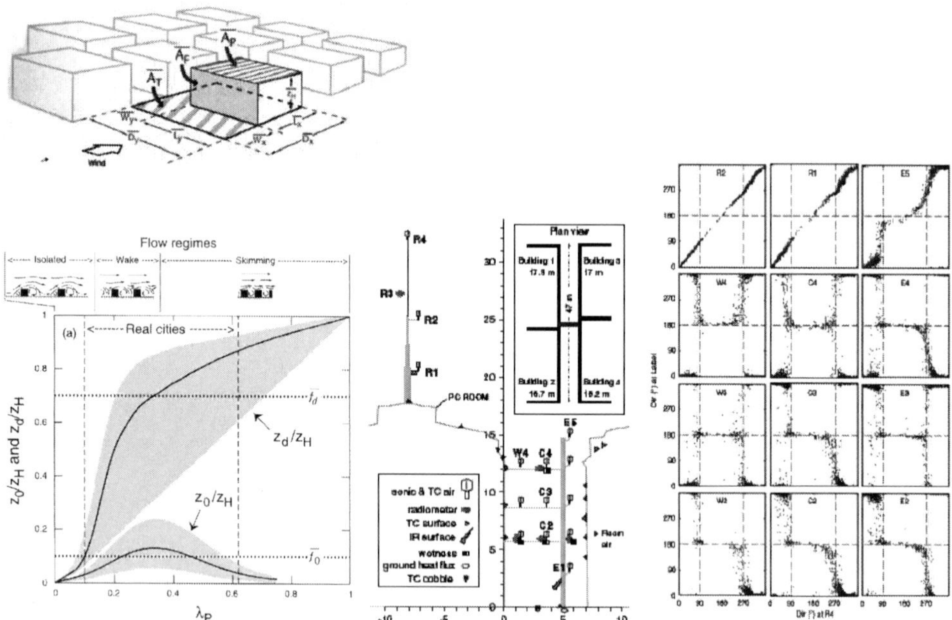

Figure 10.5 The packing of the buildings ('roughness elements') on the surface results in a change in wind regime. Upper left figure shows selected characteristics to describe the urban morphology (see also Table 10.2). Middle left figure shows that as the roughness elements become closer together the interaction of the wind with the surface changes. Lower left shows the impact of increasing density of buildings (λ_p is the fraction of the area that is buildings) results in a change of roughness length for momentum and zero plane displacement length. Middle figure is the instrument arrangement for wind observations above (R4 to R1) a narrow canyon (H:W 2.1), at mean canyon height (E5) and within the canyon (W4 to E1). The observations are located in the centre of the block (i.e. approximately 23.5 m from the intersections). Right hand figures are for the data observed in the preceding instrument setup. All have the wind direction observed at R4 on the X-axis and then each of the individual wind direction at the 12 other instrument locations on the Y-axis. As the observations become further removed from the R4 site (closer to the roof and then in the canyon) the wind direction comparison varies by a greater amount from that observed above. Note at the lowest levels (W2, C2, E2) the wind direction is almost the reverse of that observed at two times canyon height. These data are for conditions when the R4 wind speed was greater than 2 m s^{-1}. Figures/ data are from Grimmond and Oke (1999c) and Eliasson *et al.* (2006).

canyon cross flow (Figure 10.5a). The zero-plane displacement length for momentum, the mean level of momentum transfer between the flow and the roughness elements, continues to increase with increasing building height. Thus the roughness length and zero-plane displacement length are neighbourhood scale characteristics which will vary across a city. At the scale of the whole city, the city itself will act as regional scale bluff body causing flow to be deflected around it.

At the micro-scale wind fields can be complex and result in wind directions and wind speeds that are quite different from those that are above the roughness sublayer (Figure 10.5c). This can mean that a small change in measurement height results in quite different wind components (u, v, w) being observed. Because of the complexity of this, extensive efforts have directed to the modelling of airflow and dispersion at these scales to be able to predict where, for example, a pollutant released within a canyon might disperse to give different stability conditions (see, for example, the work cited in Vardoulakis *et al.* 2003; Kanda 2006; Klein *et al.* 2007; Cai *et al.* 2008). These three-dimensional models (e.g. CFD – computational fluid dynamic models; LES – large eddy simulations) can provide ways of understanding flow conditions in complex situations where it would not be possible to actually document the real world without changing the flow patterns because of interference by the instruments (Vardoulakis *et al.* 2003; Li *et al.* 2006; Klein and Clark 2007).

Air pollution

Urban areas are well known for having other mass exchanges, those involved in air pollution, which can have major impacts on human health and well-being, and also affect vegetation. Many of the sources of anthropogenic heat are also sources of air pollution. These pollutants are normally classified as primary, i.e. those that are directly emitted, or secondary, i.e. those that occur because of chemical interactions. Notable secondary pollutants are ozone and smog. Primary pollutants often of concern are sulphur oxides (SO_x), nitrogen oxides (NO_x), carbon monoxide (CO) and carbon dioxide (CO_2), and volatile organic compounds (VOCs). From a health perspective, the chemical characteristics of the substances released, along with the physical properties of the aerosols themselves, are important. Particulates are particularly relevant in considerations of respiratory problems. Increasingly data on PM10 (which refers to particles which are less than 10 μm in diameter) and PM2.5 ($<=$ 2.5 μm in diameter) are documented throughout urban environments.

The presence of the particles or aerosols in the atmosphere also influences physical atmospheric processes; e.g. radiative transfer, scattering, absorption. The presence of these particles is fundamental to precipitation processes (they act as condensation nuclei and freezing nuclei), but their chemical composition can lead to secondary pollutants through, for example, oxidation processes and change the chemical composition of precipitation (e.g. wet deposition). The exact urban impact thus depends on what type of particles are being released, at what temperature and velocity (which will influence their transport), the meteorological conditions that are occurring at the time of release, where they are released (e.g. tailpipe of a vehicle, from a tall industrial stack), and the previous chemical state of the atmosphere (which will influence the atmospheric chemical reactions that can occur).

Final comments

In the previous sections the nature of the processes that occur in any urban setting of any size and location are discussed. Note the urban heat island is discussed separately in the next chapter.

In terms of the dominant seasonal controls, the urban energy balance in tropical cities will be affected most by changes in precipitation conditions, whereas for high latitudes the most distinct differences occur because of major differences in solar radiation availability and variations in albedo (related to snow cover). Mid-latitude cities typically will have strong seasonal variations associated with changes in radiation receipt also (exacerbated by winter-time snow).

Across a city significant changes in energy partitioning will occur because of varying surface characteristics. Key is the amount of active vegetation, which typically results in a decrease in Bowen ratio (Q_H/Q_E) as greenspace, particularly irrigated greenspace, increases. Increasing the built fraction increases the probability of unstable conditions at night and positive turbulent sensible heat flux.

Acknowledgements

The author would like to acknowledge funding support provided by the UK Met Office (P003174), European Union FP7-ENV-2007–1 BRIDGE (211345) and MegaPoli (212520) projects; US National Science Foundation (ATM-0710631). Dr Catherine Souch provided useful comments on a draft of this chapter. Thomas Loridan replotted some of the figures for this chapter.

References

Aida, M. (1982) Urban albedo as a function of the urban structure – A model experiment, Part 1, *Boundary-Layer Meteorology*, 23(4): 405–13.

Akbari, H. and Taha, H. (1992) The impact of trees and white surfaces on residential heating and cooling energy use in four Canadian cities, *Energy*, 17(2): 141–9.

Bornstein, R. and Lin, Q.L. (1998) Urban heat islands and summertime convective thunderstorms in Atlanta: three case studies, *Atmospheric Environment*, 34(3): 507–16.

Cai, X.-M., Barlow, J.F. and Belcher, S.E. (2008) Dispersion and transfer of passive scalars in and above street canyons – Large-eddy simulations, *Atmospheric Environment*, 42(3): 5885–95.

Christen, A. and Vogt, R. (2004) Energy and radiation balance of a central European city, *International Journal of Climatology*, 24(11): 1395–421.

DePaul, F.T. and Sheih, C.M. (1986) Measurements of wind velocities in a street canyon, *Atmospheric Environment*, 20(3): 455–9.

Eliasson, I., Offerle, B., Grimmond, C.S.B. and Lindqvist, S. (2006) Wind fields and turbulence statistics in an urban street canyon, *Atmospheric Environment*, 40(1): 1–16, doi:10.1016/j.atmosenv.2005.03.031.

Ellefsen, R. (1991) Mapping and measuring buildings in the canopy boundary layer in ten U.S. cities, *Energy and Buildings* 16(3–4), 1025–49.

Flanner, M.G. (2009) Integrating anthropogenic heat flux with global climate models, *Geophysical Research Letters*, 36, L02801, doi:10.1029/2008GL036465.

Fortuniak, K. (2008) Numerical estimation of the effective albedo of an urban canyon, *Theoretical and Applied Climatology*, 91(1–4): 245–58.

Fortuniak, K., Kłysik, K. and Wibig, J. (2006) Urban-rural contrasts of meteorological parameters in Łódź, *Theoretical And Applied Climatology*, 84(1–3): 91–101.

Gomes, L., Mallet, M., Roger, J.C. and Dubuisson, P. (2008) Effects of the physical and optical properties of urban aerosols measured during the CAPITOUL summer campaign on the local direct radiative forcing, *Meteorology and Atmospheric Physics*, 102(3–4): 289–306.

Grimmond, C.S.B. (1992) The suburban energy balance: methodological considerations and results for a mid-latitude west coast city under winter and spring conditions, *International Journal of Climatology*, 12(5): 481–97.

Grimmond, C.S.B. (2006) Progress in measuring and observing the urban atmosphere, *Theoretical and Applied Climatology*, 84: 3–22, DOI: 10.1007/s00704–005–0140–5.

Grimmond, C.S.B. and Oke, T.R. (1991) An evaporation-interception model for urban areas, *Water Resources Research*, 27(7): 1739–55.

Grimmond, C.S.B. and Oke, T.R. (1995) Comparison of heat fluxes from summertime observations in the suburbs of four North American cities, *Journal of Applied Meteorology*, 34(4): 873–89.

Grimmond, C.S.B. and Oke, T.R. (1999a) Rates of evaporation in urban areas: impacts of urban growth on surface and ground waters, *International Association of Hydrological Sciences Publication*, 259: 235–43.

Grimmond, C.S.B. and Oke, T.R. (1999b) Heat storage in urban areas: observations and evaluation of a simple model, *Journal of Applied Meteorology*, 38(7): 922–40.

Grimmond, C.S.B. and Oke, T.R. (1999c) Aerodynamic properties of urban areas derived from analysis of surface form, *Journal of Applied Meteorology*, 38(9): 1262–92.

Grimmond, C.S.B. and Oke, T.R. (2002) Turbulent heat fluxes in urban areas: Observations and local-scale urban meteorological parameterization scheme (LUMPS), *Journal of Applied Meteorology*, 41(7): 792–810.

Grimmond, C.S.B., Oke, T.R. and Cleugh, H.A. (1993) The role of 'rural' in comparisons of observed suburban – rural flux differences. Exchange processes at the land surface for a range of space and time scales, *International Association of Hydrological Sciences Publication*, 212: 165–74.

Grimmond, C.S.B, Salmond, J.A., Oke, T.R., Offerle, B. and Lemonsu, A. (2004) Flux and turbulence measurements at a dense urban site in Marseille: Heat, Mass (water, carbon dioxide) and Momentum, *Journal of Geophysical Research*, 109: D24101, doi:10.1029/2004JD004936.

Grimmond, C.S.B., Blackett, M., Best, M., Barlow, J., Baik, J.J., Belcher, S., Bohnenstengel, S., Calmet, I., Chen, F., Dandou, A., Fortuniak, K., Gouvea, M., Hamdi, R., Hendry, M., Kondo, H., Krayenhoff, S., Lee, S.H., Loridan, T., Martilli, A., Miao, S., Oleson, K., Pigeon, G., Porson, A., Salamanca, F., Shashua-Bar, L., Steeneveld, G.J., Tombrou, M., Voogt, J. and Zhang, N. (2010) The International Urban Energy Balance Models Comparison Project: First results, *Journal of Applied Meteorology and Climatology*, 49(6): 1268–92, doi: 10.1175/2010JAMC2354.1.

Hage, K.D. (1975) Urban-rural humidity differences, *Journal of Applied Meteorology*, 14(7): 1277–83.

Harman, I.N., Best, M. and Belcher, S.E. (2004) Radiative exchange in an urban street canyon, *Boundary-Layer Meteorology*, 110(2): 301–16.

Holmer, B. and Eliasson, I. (1999) Urban-rural vapour pressure differences and their role in the development of urban heat islands, *International Journal of Climatology*, 19(9): 989–1009.

Ichinose, T., Shimodozono, K. and Hanaki, K. (1999) Impact of anthropogenic heat on urban climate in Tokyo, *Atmospheric Environment*, 33(24): 3897–909.

Jauregui, E. and Romales, E. (1996) Urban effects on convective precipitation in Mexico City, *Atmospheric Environment*, 30(20): 3383–9.

Jonsson, P., Eliasson, I., Holmer, B. and Grimmond, C.S.B. (2006) Longwave incoming radiation in the Tropics: results from field work in three African cities, *Theoretical and Applied Climatology* 85(3–4): 185–201 doi10.1007/s00704-005-0178-4.

Kanda, M. (2006) Progress in the scale modeling of urban climate: review, *Theoretical and Applied Climatology*, 84(1–3): 23–33.

Kanda, M., Kawai, T. and Nakagawa, K. (2005) A simple theoretical radiation scheme for regular building arrays, *Boundary-Layer Meteorology*, 114(1): 71–90.

Klein, P. and Clark, J.V. (2007) Flow variability in a North American downtown street canyon, *Journal of Applied Meteorology and Climatology*, 46(6): 851–77.

Klein, P., Leitl, B. and Schatzmann, M. (2007) Driving physical mechanisms of flow and dispersion in urban canopies, *International Journal of Climatology*, 27(14): 1887–907.

Klysik, K. (1996) Spatial and seasonal distribution of anthropogenic heat emissions in Łódz, Poland, *Atmospheric Environment*, 30(20): 3397–404.

Kobayashi, T. and Takamura, T. (1994) Upward longwave radiation from a non-black urban canopy, *Boundary-Layer Meteorology*, 69(1–2): 201–13.

Kuttler, W., Weber, S., Schonnefeld, J. and Hesselschwerdt, A. (2007) Urban/rural atmospheric water vapour pressure differences and urban moisture excess in Krefeld, Germany, *International Journal of Climatology*, 27(14): 2005–105.

Lemonsu, A. and Masson, V. (2002) Simulation of a summer urban breeze over Paris, *Boundary-Layer Meteorology*, 104(3): 463–90.

Levinson, R., Akbari, H. and Reilly, J.C. (2007) Cooler tile-roofed buildings with near-infrared-reflective non-white coatings, *Building and Environment*, 42(7): 2591–605.

Li, X.-X., Liu, C.H., Leung, D.Y.C. and Lam, K.M. (2006) Recent progress in CFD modelling of wind field and pollutant transport in street canyons, *Atmospheric Environment*, 40(29): 5640–58.

Lowry, W.P. (1998) Urban effects on precipitation amount, *Progress In Physical Geography*, 22(4): 477–520.

Martilli, A., Clappier, A. and Rotach, M.W. (2002) An urban surface exchange parameterisation for mesoscale models, *Boundary-Layer Meteorology*, 104(2): 261–304.

Masson, V., Grimmond, C.S.B. and Oke, T.R. (2002) Evaluation of the Town Energy Balance (TEB) scheme with direct measurements from dry districts in two cities, *Journal of Applied Meteorology*, 41(10): 1011–26.

Mitchell, V.G., Cleugh, H.A., Grimmond, C.S.B. and Xu, J. (2007) Linking urban water balance and energy balance models to analyse urban design options, *Hydrological Processes*, 22(6): 2891–900, DOI: 10.1002/hyp. 6868.

Moriwaki, R., Kanda, M., Senoo, H., Hagishima, A. and Kinouchi, T. (2008) Anthropogenic water vapor emissions in Tokyo, *Water Resources Research*, 44:W11424.

Nakamura, Y. and Oke, T.R. (1988) Wind, temperature and stability conditions in an east west oriented urban canyon, *Atmospheric Environment*, 22(12): 2691–700.

Nelson, M.A., Pardyjak, E.R., Klewicki, J.C., Pol, S.U. and Brown, M.J. (2007) Properties of the wind field within the Oklahoma City Park Avenue street canyon Part I: Mean flow and turbulence statistics, *Journal of Applied Meteorology and Climatology*, 46(12): 2038–254.

Newton, T., Oke, T.R., Grimmond, C.S.B. and Roth, M. (2007) The suburban energy balance in Miami, Florida, *Geografiska Annaler A*, 89(4): 331–47.

Nunez, M., Eliasson, I. and Lindgren, J. (2000) Spatial variations of incoming longwave radiation in Göteborg, Sweden, *Theoretical and Applied Climatology*, 67(3–4): 181–92.

Offerle, B., Grimmond, C.S.B., and Fortuniak, K. (2005b) Heat storage and anthropogenic heat flux in relation to the energy balance of a central European city center, *International Journal of Climatology*, 25(10): 1405–1419, DOI: 10.1002/joc.1198.

Offerle, B., Grimmond, C.S.B. and Oke, T.R. (2003) Parameterization of net all-wave radiation for urban areas, *Journal of Applied Meteorology*, 42(8): 1157–73.

Offerle, B., Eliasson, I., Grimmond, C.S.B. and Holmer, B. (2007) Surface heating in relation to air temperature, wind and turbulence in an urban street canyon, *Boundary-Layer Meteorology*, 122(2): 273–92, DOI 10.1007/s10546–006–9099–8.

Offerle, B., Grimmond, C.S.B., Fortuniak, K. and Pawlak, W. (2006b) Intraurban differences of surface energy fluxes in a central European city, *Journal of Applied Meteorology and Climatology*, 45(1): 125–36.

Offerle, B., Jonsson, P., Eliasson, I. and Grimmond, C.S.B. (2005a) Urban modification of the surface energy balance in the west African Sahel: Ouagadougou, Burkina Faso, *Journal of Climate*, 18(19): 3983–95.

Offerle, B., Grimmond, C.S.B., Fortuniak, K., Kłysik, K. and Oke, T.R. (2006a) Temporal variations in heat fluxes over a central European city centre, *Theoretical and Applied Climatology*, 84(1–3):103–116, DOI 10.1007/s00704–005–0148-x.

Oke, T.R. (1988) The urban energy balance, *Progress in Physical Geography*, 12(4): 471–508.

Oke, T.R. (1997) Urban climates, in *The Surface Climates of Canada*, W.B. Bailey, T.R. Oke and W.R. Rouse (eds), McGill-Queen's University Press, Montréal.

Oke, T.R. (2006) Initial guidance to obtain representative meteorological observations at urban sites, *Instruments and Observing Methods Report No* 81, World Meteorological Organization, online, available at: www.wmo.int/pages/prog/www/IMOP/publications/IOM-81/IOM-81-UrbanMetObs.pdf, [accessed 24 February 2009].

Pigeon, G., Legain, D., Durand, P., Masson, V. (2007) Anthropogenic heat release in an old European agglomeration (Toulouse, France), *International Journal of Climatology*, 27(14): 1969–81.

Piringer, M., Grimmond, C.S.B., Joffre, S.M., Mestayer, P., Middleton, D.R., Rotach, M.W., Baklanov, A., De Ridder, K., Ferreira, J., Guilloteau, E., Karppinen, A., Martilli, A., Masson, V. and Tombrou, M. (2002) Investigating the surface energy balance in urban areas – recent advances and future needs, *Water, Air, & Soil Pollution: Focus*, 2(5–6): 1–16.

Richards, K. (2004) Observation and simulation of dew in rural and urban environments, *Progress in Physical Geography*, 28(1): 76–94.

Richards, K. and Oke, T.R. (2002) Validation and results of a scale model of dew deposition in urban environments, *International Journal of Climatology*, 22(15): 1915–33.

Robaa, S.M. (2009) Urban-rural solar radiation loss in the atmosphere of Greater Cairo region, Egypt, *Energy Conversion and Management*, 50(1): 194–202.

Roberts, S., Oke, T.R., Grimmond, C.S.B. and Voogt, J. (2006) Tests of four methods to estimate urban heat storage in central Marseille, *Journal of Applied Meteorology and Climatology*, 45(12): 1766–81.

Rosenfeld, D. (2000) Suppression of rain and snow by urban and industrial air pollution, *Science*, 287(5459): 1793–6.

Rotach, M.W. (1995) Profiles of turbulence statistics in and above an urban street canyon, *Atmospheric Environment*, 29(13): 1473–86.

Roth, M. (2000) Review of atmospheric turbulence over cities, *Quarterly Journal of The Royal Meteorological Society*, 126(564): 941–90.

Roth, M. (2007) Review of urban climate research in (sub)tropical regions, *International Journal of Climatology*, 27(14): 1859–73.

Sailor, D.J. and Lu, L. (2004) A top-down methodology for developing diurnal and seasonal anthropogenic heating profiles for urban areas, *Atmospheric Environment*, 38(17): 2737–48.

Shepherd, J.M. (2005) A review of current investigations of urban-induced rainfall and recommendations for the future, *Earth Interactions*, 9(12): 1–27.

Shepherd, J.M., Pierce, H. and Negri, A.J. (2002) Rainfall modification by major urban areas: Observations from spaceborne rain radar on the TRMM satellite, *Journal of Applied Meteorology*, 41(7): 689–701.

Spronken-Smith, R.A., Oke, T.R. and Lowry, W.P. (2000) Advection and the surface energy balance across an irrigated urban park, *International Journal of Climatology*, 20(9): 1033–47.

Stewart, I. and Oke, T.R. (2009) Newly developed 'thermal climate zones' for defining and measuring urban heat island magnitiude in the canopy layer, *American Meteorological Society Annual Meeting*, J8.2A, online, available at: http://ams.confex.com/ams/pdfpapers/150476.pdf.

Sugawara, H. and Takamura, T. (2006) Longwave radiation flux from an urban canopy: Evaluation via measurements of directional radiometric temperature, *Remote Sensing of Environment*, 104(2): 226–37.

Svirejeva-Hopkins, A., Schellnhuber, H.J. and Pomaz, V.L. (2004) Urbanised territories as a specific component of the global carbon cycle, *Ecological Modelling*, 173(2–3): 295–312.

Synnefa, A., Dandou, A., Santamouris, M., Tombrou, M. and Soulakellis, N. (2008) On the use of cool materials as a heat island mitigation strategy, *Journal of Applied Meteorology and Climatology*, 47(11): 2846–56.

Vardoulakis, S., Fisher, B.E.A., Pericleous, K. and Gonzalez-Flesca, N. (2003) Modelling air quality in street canyons: a review, *Atmospheric Environment*, 37(2): 155–82.

11
Urban heat islands

T.R. Oke

The urban heat island (UHI) effect, i.e. the characteristic warmth of a settlement compared with its surroundings, is the best-known climatic response to disruptions caused by urban development. If your car has a thermometer the warmth is relatively easy to observe from the profile of temperatures as you cross a town or city, especially at night and if winds are weak and clouds are sparse.

Scales of urban climates

Everyday living demonstrates the notion of thermal microclimate. Even in a small garden or around a single house the range of microclimates is stunning. A good gardener, or the family cat, knows some spots are warmer. Beyond being more or less open to the receipt of sunshine, wind or rain each microenvironment possesses its own mix of surface properties (radiative, thermal, moisture, and aerodynamic) which govern heat and water transfer and airflow. A good gardener knows where best to place plants to match their tolerance for daytime heat or moisture and the cat knows where to bask in a sunny, sheltered corner of the patio. At night if sites do not retain the daytime heat, or they are exposed to the sky, they may become colder than average. Of course microclimates are dynamic; changing through the day and year. The plants are rooted and hence experience the full range of conditions whereas the cat and gardener can opt to move to optimize habitat. The gardener has the additional possibilities of changing clothes and controling the thermal climate of buildings and the whole garden.

This thermal complexity of a single dwelling and its grounds is a microcosm of the climates of villages, towns and cities. Such units are repeated with varying spacing, density, and height across an urbanized landscape. Houses are often organized along streets giving 'canyon-like' structures following transport routes, and there are patches of similar development that form distinct neighborhoods and land cover types. These residential, industrial, and commercial districts have their own mix of properties leading to a myriad of thermal climates at scales set by key dimensions (Figure 11.1).

The microscale is described above; where every surface and object has a unique microclimate. Surface and air temperatures vary several degrees in short distances and airflow is greatly perturbed in speed and direction by even small objects. The defining scales are the dimensions of individual

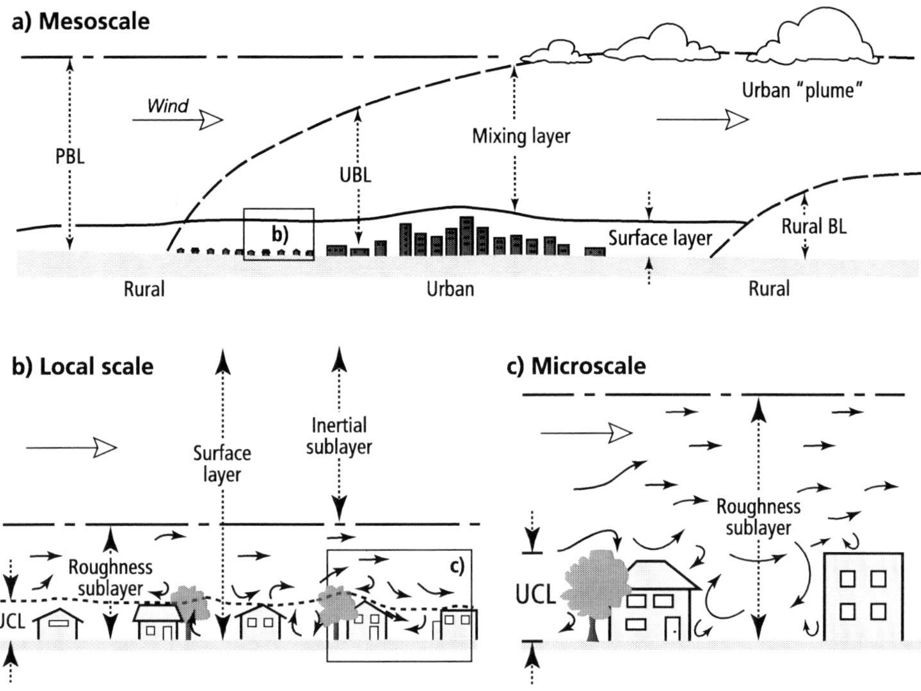

Figure 11.1 A classification of the main scales found in urban climates (*source:* modified after Oke 1997).

elements (buildings, trees, streets, courtyards, gardens, etc.), extending horizontally from less than one to hundreds of meters and vertically up to about roof-level in what is known as the urban canopy layer (UCL, Figure 11.1). Surface temperatures depend on the energy balance of the object. Air temperatures near a surface are affected but the influence decays rapidly with distance. In a few meters, and certainly by a hundred meters, the effect is lost by turbulent mixing.

The local scale is what a standard climate station is designed to monitor. In non-urban terrain it is designed to exclude microscale effects of individual surfaces by measuring over a simple extensive surface, usually grass, well away from obstacles and at about 1.5 m where the strongest effects of the underlying surface are muted. In cities such idealized sites are both difficult to find and not representative of urbanized land anyway. Instead it is recommended to expose the instruments to represent the climate of neighborhoods consisting of similar urban development (surface cover, size and spacing of buildings, human activity) known as Urban Climate Zones, UCZ (WMO 2008). One approach is to carefully select a location typical of the UCZ and expose the instruments in the UCL, in the center of a space with dimensions and surrounding objects characteristic of the zone. Whilst the spatial variability of microclimates is large, in the UCL if the surroundings are not too anomalous the averaging provided by the variation of wind direction gives reasonable estimates. Alternatively one must use many stations or move the sensor so as to adequately sample conditions in the UCL. Another way is to let the turbulent mixing action of the wind and convection in and just above the urban canopy do the blending. Observations show that readings are reasonably spatially invariant at about twice the mean height of the buildings (the extent of the roughness sublayer, RSL, Figure 11.1). The horizontal dimension of an UCZ depends on the morphology of the city, but is typically one to several kilometers.

Each UCZ generates a plume-like internal boundary layer with similar properties. These bend over downwind due to transport by the wind. In turn these blend into the mixing layer, which is a much larger 'urban plume' from the whole city. This urban boundary layer (UBL, Figure 11.1) is a *mesocale* feature that is simply the part of the planetary boundary layer affected by the urban area. In fine weather the UBL extends vertically to 1 to 2 km by day, and perhaps 100 to 300 m at night, over a large city. The depth is less in windy and/or cloudy conditions and its horizontal dimensions are set by the extent of the city (typically 10 to 50 km). The urban plume persists downwind. Thermal effects are evident for less than ten kilometers, but pollutants persist for hundreds of kilometers. Megalopolitan pollutant plumes may remain for long distances and vertical mixing creates a regional haze or brown cloud. Measuring thermal effects in the UBL requires thermometers on balloons or aircraft, or remote sensing techniques.

The methodological pitfalls involved in assessing the magnitude of urban effects on temperature (the heat island) are several (Lowry 1977). Lowry notes that ideally effects can only be assessed using measurements if pre-urban observations are available and the effects of any changes in local, regional, and global climate are accounted for and the analysis is normalized for synoptic weather. The data to achieve this hardly ever exists so a common surrogate is to define a heat island as the relative warmth of urban areas compared with their surrounding terrain. In turn this is simplified to temperature differences between 'urban' and 'rural' sites. Care must be exercised to ensure the sites are representative of those environments, the instruments are exposed correctly, that differences of topography and weather do not contaminate the results, and that the sites and instrument array used to measure the difference are at the same scale.

Scale dictates that there are not one but several heat island types. A simple classification suggests there are four main types using urban-rural temperature differences (ΔT): the UHI_{soil} in the soil beneath the city, the UHI_{surf} at the atmospheric interface with the urban elements (buildings, trees, ground), and the UHI_{UBL} and UHI_{UCL} both in the air. Each UHI requires its own observation methods, is created by its own set of heat transfer processes, is modulated by its own set of controls, and potentially affects different aspects of city infrastructure, environmental phenomena and ecological systems (Table 11.1). One should also be aware that the exact magnitude of ΔT differs depending on the operational definition and the nature of the available data. For example, spatially, is the difference between single fixed points, or between averages of multiple fixed points in each environment, or is it derived from data continuously monitored along mobile traverses, or from two-dimensional spatial images? Similarly, temporally, are the temperatures recorded continuously, or at fixed periods or only at the times of the daily maximum and minimum? In essence these possibilities define UHI subtypes based on the methods used. Care should be exercised if results are compared using UHI from different subtypes.

Surface UHI

As mentioned, every surface has a distinct thermal response governed by its surface radiative, thermal, moisture, and aerodynamic properties. The actual temperature of each surface depends on its unique energy balance due to the exchanges of radiation, sensible, and latent heat at the interface. These in turn are set by exposure to the streams of solar and infrared radiation, to and from the sun, sky, and any elements of its surroundings in its field of view (walls, trees, roofs, road, grass, and so on). It also depends on the temperature and moisture content of its underlying substrate and overlying atmospheric boundary layer to which it is coupled through the processes of conduction and convection, respectively. Convection further depends on the

Table 11.1 List of potential causes of the canopy layer urban heat island (UHI_{UCL})

Cause	*Description*
1. Surface geometry	a) Increased surface area b) Closely-spaced buildings • multiple reflection and greater solar absorption; • small sky view reduces net infrared radiation loss especially at night • wind shelter in canopy reduces heat losses by convection
2. Thermal properties	Building materials often have greater capacity to store and later release sensible heat
3. Surface state	a) Surface moisture – waterproofing by buildings and paving reduces soil moisture and surface wetness. Hence, convection favors sensible rather than latent (evaporation) heat fluxes to air b) Snow cover – much lower albedo in city (vertical surfaces snow-free; snow clearance; snow soiling) increases solar absorption
4. Anthropogenic heat	Heat released by combustion processes is much greater in city
5. Urban 'greenhouse'	Warmer, polluted and often more moist urban atmosphere emits more downward infrared radiation to surface of city

Sources: Oke (1982); Oke *et al.* (1991); Voogt (2002).

degree of exposure the surface has to the wind, the wind strength and the intensity of turbulent activity. Turbulence derives from the temperature difference between the surface and the air and the roughness of the surface in combination with the wind. The variety of properties, controls, and interactions possible in a city results in a myriad of surface temperatures since every leaf and roof tile is different when viewed in detail.

The richness and complexity of this surface temperature field is revealed by infrared imagery. Depending on the optics of the camera, and its field of view it is possible to monitor surface temperatures at all of the scales in Figure 11.1. Such thermal images potentially give a marvellous view of urban surface temperature distributions. These are significant because the interface temperature is responsible for sending a thermal wave down into the soil and up into the air that forces the soil, canopy layer, and boundary layer heat islands.

At the scale of urban neighborhoods and the whole city, the most obvious feature seen in thermal images is the strong contrast between dry surfaces and moist or vegetated ones. By day vegetation, especially if irrigated, and water bodies are almost always coolest. Conversely open, dry and dark surfaces like roofs, roads and parking lots are warmest. Differences can be large, for example, in the mid-latitudes in summer at midday roof temperatures of 50–55°C are not uncommon, at the same time exposed paved roads and parking areas may be 40–45°C, sunlit walls may be less, perhaps 30–40°C, whereas shaded surfaces and under trees it may be closer to air temperature (say 25°C) and irrigated grass and water bodies may be at 15–20°C, i.e. a range of about 35 to 40 degrees Celsius. Ecologically that represents a huge diversity of habitat possibilities and stresses. At night the spatial range of surface temperatures is less, because the differences are driven by passive infrared cooling rather than active solar heating by day. The most thermally responsive surfaces are exposed rooftops that are much cooler than the walls and canyon floor.

Figure 11.2b gives an idea of the results of these surface differences across a large city during fine weather. The difference between the warmest urban and the background rural temperatures

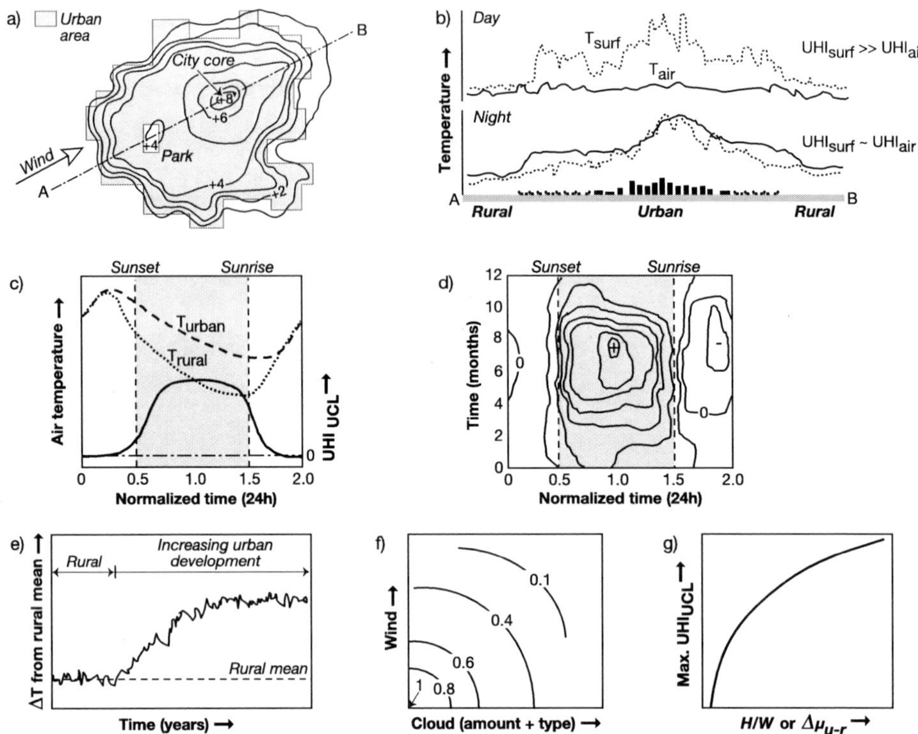

Figure 11.2 Simplified schematic depiction of UHI_{UCL} spatial and temporal features and the main controls on its form and magnitude. a) Spatial distribution on 'ideal' night; b) cross-section along line AB of both UHI_{UCL} and UHI_{surf} by day and night; c) course of urban and rural air temperatures through an 'ideal' day and the resulting UHI_{UCL}; d) hypothetical magnitude of an annual UHI_{UCL} in all weather conditions; e) trend of annual mean urban temperature excess (relative to long-term rural mean) as a station becomes developed; f) control exerted by wind and cloud, represented as UHI magnitude relative to that in 'ideal' conditions; g) generalized form of control exerted by street geometry (H/W) in city center, and also of rural soil moisture (urban-rural difference of thermal admittance, μ) in setting the maximum magnitude of the UHI in 'ideal' conditions (*sources:* Böhm, 1998; Oke, 1982, 1987, 1998; Voogt and Oke, 2003).

Note

The time scale in c and d is the fraction of the 24-hour period when the sun is above or below the horizon, i.e. daylight and night. This normalizes seasonal differences in daylength between sites.

defines the UHI_{surf}. By day this might be 5 to 10 degrees for a large city, but this depends on the nature of the rural area and its wetness. Spatial variability in the city is larger by day than at night, but the magnitude of UHI_{surf} may not be greatly different (11.2b).

Soil UHI

The UHI_{soil} is little-studied but must exist under most settlements. The surface energy balance of a city favors heat uptake relative to rural areas (See Chapter 12 this volume) hence the near surface layer of soil in a city is likely to be warmer. The resulting vertical temperature gradient creates a warm zone under the city due to heat conduction. However, it is difficult to demonstrate this given the disturbed nature of urban soils (see Chapter 15 this volume) which means that heat fluxes are horizontal as well as vertical and the soil is desiccated, due to waterproofing of the surface. This variability makes it difficult to obtain statistically significant soil temperatures or to predict them. Nevertheless the UHI_{soil} has been studied in Tokyo using multiple temperature probes and in Cologne using a network of water wells. Both show concentric isotherms decreasing outward from the city core. Even deep soil temperatures are likely to show the city as a warm node because the soil must approach equilibrium with the annual mean air temperature of the location including its UHI.

Urban soil temperatures are important ecologically because they affect root growth, the activity of soil organisms and also inorganic chemical processes. There is increasing interest in the carbon balance of cities because of their focus as a source of the majority of so-called greenhouse gases, including CO_2. Whilst emissions from combustion of fossil fuels dominate, there is interest in the role of urban vegetation in sequestering carbon and in the respiration of carbon from urban soils. Increased temperatures hasten decomposition of leaf litter and the mineralization of nitrogen (McDonnell *et al.* 1997).

Canopy layer UHI

Observation

The heat island of the urban canopy layer (UHI_{UCL}) has been studied extensively. Its nature is revealed by air temperature measurements made in standard weather screens (at about 1.5 to 2 m) above ground at fixed sites, or by mounting thermometers on vehicles and traversing them across the urban area. The exact siting of the fixed stations is critical if they are to represent the local scale (see WMO 2008). If urban-rural differences are used as a surrogate of the UHI_{UCL}, obviously the urban station should be chosen to avoid thermal anomalies, but equal or greater attention must be given to the representativeness of the rural site. The range of possible 'rural' surroundings is huge, including natural ecosystems from all biomes, managed agricultural and forest systems, and areas of mixed uses. Further, the thermal effects of relief, proximity to water bodies, seasonal wetness and snow cover adds to the complexity. However, a suitable network of fixed stations operated over extended periods can reveal the spatial and temporal features of the UHI_{UCL}. Traverses can fill in spatial details.

Genesis

The suggested possible 'causes' of an UHI_{UCL} are listed in Table 11.1. Their relative importance in a given city varies. In most cases Causes 1 and 2 are thought to be the main ones. In large cities with a temperate climate and moist or dry rural soils they are equally capable of explaining urban-rural temperature differences (Oke *et al.* 1991). Cause 4 is potentially able to produce an UHI on its own and in high latitude settlements with no solar input it does. The causes are not isolated, there is considerable interaction between them. (For heat fluxes mentioned in this section see Chapter 10 this volume.)

A good surrogate measure of urban geometry is the street height to width ratio (H/W). It is closely related to the fraction of sky able to be seen from within the canyons (the sky view factor) which is critical to radiation exchange. By day this sets both solar access into the canopy layer and the probability of multiple reflections between walls and floor which affects the albedo and absorption. The view factor also sets the infrared radiation balance because for any point in the canopy layer it defines what the point can 'see'. Typically it can see either the very cold sky or the usually much warmer buildings. At night deep street canyons with a small sky view (large H/W) hinder heat loss by radiation. On the other hand rural surfaces are generally more open to the sky and cool more easily. Further, H/W is related to the airflow regime including the extent to which above-roof flow penetrates into the UCL. Hence it is related to the shelter in the canyons which reduces convective heat loss; another cause of the UHI_{UCL}.

Cities are better heat stores than rural areas but this is not necessarily because construction materials have greater thermal conductivity and heat capacity. In fact urban values of those properties are not greatly superior to those of moist soils. What is significant is the convoluted configuration of the materials that exposes a much larger surface area for exchange than a flat site and the fact they are often dry (due to their ability to shed not store water) hence absorbed heat is efficiently used to warm the material rather than to evaporate water. Similarly, Cause 3a is a statement that the urban energy balance is biased towards sensible rather than latent heat because of the lower availability of surface moisture.

Urban activity is driven by energy use. The 'waste' by-products of this metabolism are the release of heat that leaks from space heating/cooling in buildings, chimney stacks, tailpipes, humans and animals. These give an anthropogenic source term in the energy balance which directly boosts the UHI. Depending on the city, time of year, and weather, Cause 4 can vary from insignificant to dominant. The sources most relevant to the canopy layer are those from vehicles and building walls. Emissions from chimneys and tall stacks are unlikely to be important to the UCL. In cities with cold winters this heat source can rival or exceed that from the sun. At high latitudes it may be the only source of heat in mid-winter. In hot cities waste heat from air conditioners can be a significant extra burden, especially if exhausted into the street air volume. Cause 5 exists because there is an increase in incoming infrared radiation from the polluted UBL compared to that in rural areas. This stream tempers the net infrared loss. However, in general this is not a large contributor to the UHI_{UCL}. These causes modify the daily course of air temperature at urban and rural locations (Figure 11.2c). The net result is the urban temperature wave has significantly smaller amplitude due to warmer nights.

Figure 11.2 gives a simplified view of typical UHI_{UCL} features in a large city (say one million inhabitants) located on a plain. The following account initially looks at the simplest and most striking case: when skies are clear and winds are weak or calm – the 'ideal' case. This allows maximum differentiation of microclimates and UHI features are best displayed. Later the modulating influence of less ideal (cloudy, windy, rainy) weather is added. Figure 11.2 is purposely generic, it gives idealized features and relations and should not be used to abstract exact numbers or functions.

Spatial form

The 'island' name refers to the fact the spatial field of air temperature isotherms in the canopy layer resembles the height contour map of an island (Figure 11.2a). The heat island has relatively sharp temperature gradients around the urban-rural periphery. Patterns inside the island respond to the degree of urban development (Figure 11.2a). All other things being equal greater warmth occurs in areas of taller buildings that are packed together so the soil is sealed over with built materials, there is relatively little vegetation and there is intense traffic. Hence nodes of greater development are relative hot spots, whereas open spaces like parks and lakes are relatively cool. Large vegetated parks can be a source of welcome cool and cleaner air for surrounding districts, indeed they may even spawn mini-breezes.

A temperature cross-section illustrates the island form and defines the overall magnitude of the heat island as the difference between the temperature of warmest part of the city and the general background temperature of the surrounding countryside (Figure 11.2b). Given the spatial variations of temperature in both environments it is often not easy to place a precise magnitude on the UHI_{UCL}. If there is a regional wind, advection and mixing of cooler rural air with that over the city reduces the UHI magnitude and transports the position of thermal features in the downwind direction by a few kilometers.

Temporal dynamics

Figure 11.2c shows the daily course of the urban and rural canopy layer temperature waves and the difference between them; the temporal march of UHI_{UCL} in the 'ideal' case. This clearly shows this heat island is largely a nocturnal phenomenon, i.e. the UHI_{UCL} is due more to *cooling* differences than *heating* differences between the two environments. In the afternoon urban temperatures decline gently (typical cooling rates of 0.5–1°C h^{-1}) due to the decreased rate of radiation loss in the canyons (Cause 1) plus the large store of the previous day's heat input available to be dissipated (Cause 2). At the same time in the open, rural environment cooling after sunset is strong (typically 2–3°C h^{-1}), so the 'ideal' nocturnal UHI_{UCL} grows rapidly reaching a maximum in the middle of the night period. A large city in these 'ideal' conditions could have a UHI_{UCL} of 10 to 12 degrees. Later in the night as the heat store is expended the two rates become similar. After sunrise the urban canopy responds more slowly because the local sunrise for surfaces beneath roof-level is later and because of the greater thermal inertia of the materials. Rural surfaces respond quickly, thereby eroding the urban-rural difference, so that by mid-morning both environments are at fairly similar temperatures. The daytime UHI_{UCL} is small, indeed it is not uncommon for there to be a small negative heat island (cool island) (Figures 11.2c, d). When it is windy and/or cloudy these cooling/warming rates are less and so is the absolute magnitude of UHI_{UCL}.

The dynamics of a typical annual heat island are illustrated in Figure 11.2d. Seasonal variations in weather, soil moisture and anthropogenic heat determine the magnitude and dynamics of the UHI_{UCL} in a particular city. The example in 11.2d is a hypothetical city in the northern hemisphere where heat island creation is most favored in the summer because there are fewer storms and drier rural soils. The diurnal dynamics show the largest UHI_{UCL} at night and cool islands possible by day. The actual pattern may greatly differ from this case depending on the seasonal variation in synoptic weather, the need for anthropogenic heat and the state of rural land cover and management practices.

Extending to long term climatology over many years, as a settlement grows so does the UHI_{UCL}, rapidly at first when the initial conversion of land to urban uses has a greater relative impact (Figure 11.2e). This also means that newer developments on the edge of a city produce a greater rate of change than inner city areas (Figure 11.2e).

Controls on heat island magnitude and dynamics

The magnitude and temporal march of the UHI_{UCL} are modulated by the actual weather state. The most significant controls are exerted by wind and cloud. If cloud remains fixed, increasing the wind speed decreases the UHI_{UCL} magnitude approximately exponentially. Wind speed is really a surrogate for the degree of turbulent mixing and transport. The effects of cloud depend on both the cloud cover fraction and the cloud type. Low, usually warmer, cloud (e.g. fog) is far more effective than high altitude, much colder, cloud (e.g. cirrus). Cloud is a surrogate for the strength of the surface radiation balance both due to its effects on solar transmission and infrared emission and absorption. Cloud both limits the daytime receipt of solar heat to stock the UHI heat store and at night it controls the strength of the sky heat sink. The combined effects of wind and cloud on the UHI_{UCL} are shown in Figure 11.2f. The bottom left hand corner labeled unity, is the maximum UHI magnitude at that place and time, and the isolines indicate the fraction of that due to the combined effects of wind and cloud. The wind scale increases from calm to speeds greater than $10\,m\,s^{-1}$, and the cloud scale increases from clear to overcast low cloud.

The size of the city is a control but for the canopy layer it is not the horizontal dimensions of the city that are important. Rather it is the building density, and especially the canyon aspect ratio (H/W), because it sets the sky view factor and hence the strength of nocturnal cooling (i.e. Table 11.1, Cause 1; Figure 11.2g). Actually H/W is an indirect measure of several properties (wind shelter, type of building materials, fraction of vegetation cover, and waste heat, i.e. Table 11.1, Causes 1b, 2, 3, and 4) that also boost UHI magnitude. The UHI_{UCL} is also related to the population of the settlement, which is another, but cruder, measure of city size that indirectly incorporates multiple causes.

The thermal properties of both environments are also important controls. The role of construction materials in setting the urban heat store has already been mentioned, but the thermal properties of the rural environment are just as important. First, they are part of the urban-rural difference (Table 11.1, Cause 2; Figure 11.2g) and, second, the rural value sets the base for cooling. For example, if rural soils are dry and sandy the potential for nocturnal cooling is large, perhaps as great as 20 degrees Celsius, but if the surroundings are water-logged soils the overnight cooling may be only five degrees. That immediately limits the magnitude of the UHI. Further, in general the thermal properties of cities are relatively similar, on the other hand the range of rural values is largely due to the seasonal variability of soil moisture conditions.

Naturally real cities do not neatly follow the patterns summarized in Figure 11.2. Each has its own topographic and cultural context, architectural form, building materials, and economic activity. Each has its own global and regional climatic character and is embedded in its own managed or natural rural ecosystem. The possibilities are endless.

The extra warmth of the UCL has bioclimatic and resource use implications, but whether those are deemed positive or negative depends on the general thermal climate of the city. If the city is in a hot climate, or has a hot season, the UHI_{UCL} is an unwelcome extra burden, however, if it has a cold season it could be seen as favorable. Air temperature is an important contributor to the sensation of human comfort but if the threshold for heat stress is exceeded it can lead to heat-related deaths. The value of the threshold for a given population depends on its acclimatization. Greatest concern centers on whether the nocturnal heat island disturbs sleep sufficiently to deny overnight relief.

Human discomfort due in part to high temperatures increases the desire for air conditioning, which is both expensive and a drain on energy resources. It also increases outdoor water use for irrigation which places pressures on water supplies. If the city has a cold climate or season the warmth may be favorable to human comfort by reducing wind chill on pedestrians. There may also be savings in energy use for space heating (typically by 5 to 10 percent in a large mid-latitude city). Road climate may benefit from lower frequency of ice and the reduced relative humidity gives less disruption by fog. The frost-free and growing periods are lengthened giving earlier germination and blossoming and allowing less hardy plants to thrive. Similarly, warmth may encourage over-wintering of some migratory birds and permit introduced insects and wildlife to exist.

Boundary layer UHI

The mesoscale urban heat island is connected to that in the canopy layer but is created by a different set of processes (Table 11.2). The UHI_{UBL} is the result of internal heating due to radiation

Table 11.2 List of causes of boundary layer urban heat island (UHI_{UBL})

Cause	*Description*
1. Polluted boundary layer	Aerosol and gaseous pollutants disrupt radiation transmission resulting in greater absorption and scattering of solar and greater absorption and emission of infrared radiation
2. Urban sensible heat flux	Greater turbulent flux of sensible heat from rougher and warmer city. Upward mixing of warmer canopy layer air (i.e. UHI_{UCL})
3. Anthropogenic heat	Heat injected upward into UBL from chimneys and factory stacks
4. Entrainment	Greater convection causes downward injection of warmer and drier air from above capping inversion into UBL

absorption in the polluted UBL, plus both bottom-up injection of heat from the canopy layer and anthropogenic heat from sources exiting above the canopy layer, and top-down injection of heat from the atmosphere above the UBL. This top-down process is called entrainment which acts at the top of the mixing layer (Figure 11.1). The top of the UBL is marked by a temperature inversion i.e. it is warmer aloft. Entrainment produces mixing across the inversion resulting in a net downward transport of heat. Entrainment in the daytime is driven by the strength of the surface thermal convection, which is stronger over the city. At night it can be caused by the convection created by air flowing over the rough city, plus the weak thermal convection caused by the nocturnal UHI_{UCL}.

If regional winds are near calm this heat island is shaped like a self-enclosed dome. When winds increase it tilts as the internal boundary layers from each UCZ blend to form a well mixed heat 'plume' (Figure 11.3). In fine weather the height of the daytime UBL reaches to 1 to 1.5 km above the surface. The extra strength of urban heat fluxes cause it to be about 25 percent higher than the regional boundary layer (Figure 11.3a). At night the heat island over a large city only extends up to 0.3 km or less (Figure 11.3b). The advective nature of the UHI_{UBL} is evident in the nocturnal temperature profiles. In Figure 11.3c the upwind rural profile shows a strong ground-based inversion, probably due to radiative cooling of the surface. In the city center temperatures are warmer and the profile is almost vertical, indicating a neutral boundary layer. This is due to the accumulation of heat as the air traverses the city. The difference between the two profiles gives the magnitude of the heat island with height. It is largest at the surface (i.e. the UHI_{UCL}) and decreases rapidly with height (Figure 11.3d).

The existence of the UHI_{UBL} has several impacts. For example, the extra warmth in winter may tip the balance as to whether precipitation arrives at the ground as rain- or snowfall. In summer, by enhancing photochemical reaction rates, it leads to increased ozone concentrations. The relative warmth of the city can also induce a local thermal breeze circulation not unlike a sea breeze that converges on the center from all directions. The modified vertical temperature structure shifts the stability more towards neutrality which enhances mixing and explains the fairly uniform haze over cities. The depth of the UHI_{UBL} is critical to the urban air quality because it sets the available volume into which pollutants are mixed. This smog layer is capped by an inversion giving a sharp transition to cleaner air. The smog plume may extend for many, even hundreds of kilometers, downwind from the city. Convective clouds usually develop with their bases at or near the top of the mixed layer (Figure 11.1). In favorable conditions the potential exists for these to lead to precipitation (see Chapter 12 this volume).

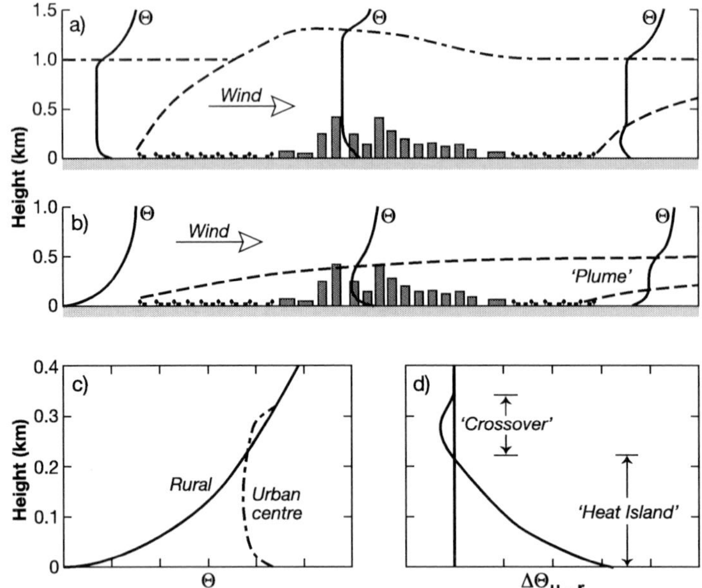

Figure 11.3 Schematic of thermal features of the UHI_{UBL} in 'ideal' weather conditions for a large city: a) Daytime, and b) nocturnal profiles of potential temperature (θ) and the height of the associated internal boundary layers (note each θ profile has its own scale), c) The rural and urban center nocturnal temperature profiles (on the same scale), whose difference defines d) the depth and magnitude of the UHI_{UBL} (*source:* Oke 1982).

Note
Potential temperature is the temperature the air would have if the effect of the pressure decrease with height is removed.

Acknowledgments

I am grateful to the Natural Science and Engineering Research Council for financial support and Dr. James Voogt for advice and review.

List of relevant chapters in the handbook

2 The analysis of cities as ecosystems
10 Climate of cities
12 Urban effects on precipitation and associated convective processes
15 Urban soils
36 Urban habitat analysis
42 Urban areas and the biosphere reserve concept
43 Urban ecology and sustainable urban drainage
46 Creative use of therapeutic greenspaces
48 Biodiversity as a statutory component of urban planning

Further reading

Chandler, T.J. (1965) *The Climate of London*, London: Hutchinson.
EPA (2008) *Reducing Urban Heat Islands: Compendium of Strategies Urban Heat Island Basics*, Washington, DC: US Environmental Protection Agency.
Landsberg, H.E. (1981) *The Urban Climate*, New York, NY: Academic Press.
Oke, T.R. (1973) 'City size and the urban heat island', *Atmospheric Environment*, 7(8): 769–79.
Oke, T.R. (1981) 'Canyon geometry and the nocturnal heat island. Comparison of scale model and field observations,' *International Journal of Climatology*, 1(3): 237–54.
Oke, T.R. (1995) 'The heat island of the urban boundary layer: characteristics, causes and effects', in J.E. Cermak, A.G. Davenport, E.J. Plate and D.X. Viegas (eds), *Wind Climate in Cities*, Dordrecht: Kluwer Academic Publishers, pp. 81–107.

References

Böhm, R. (1998) 'Urban bias in temperature series – a case study for the city of Vienna,' *Climatic Change*, 38(1): 113–28.
Lowry, W.P. (1977) 'Empirical estimation of urban effects on climate: A problem analysis', *Journal of Applied Meteorology*, 16(2): 129–35.
McDonnell, M.J., Pickett, S.T.A., Groffman, P., Bohlen, P., Pouyat, R.V., Zipperer, W.C., Parmelee, R.W., Carreiro., M.M., and Medley, K. (1997) 'Ecosystem processes along an urban-to-rural gradient,' *Urban Ecosystems*, 1(1): 21–36.
Oke, T. R. (1982) 'The energetic basis of the urban heat island,' *Quarterly Journal of the Royal Meteorological Society*, 108(455): 1–24.
Oke, T. R. (1987) *Boundary Layer Climates*, second edition, London/New York, NY: Routledge.
Oke, T. R. (1997) 'Urban environments,' in W.G. Bailey, T.R. Oke and W.R. Rouse (eds.), *Surface Climates of Canada*, Montréal: McGill-Queen's University Press, pp. 303–27.
Oke, T. R. (1998) 'An algorithmic scheme to estimate hourly heat island magnitude,' *Second Urban Environment Symposium*, 6.1, American Meteorological Society, Albuquerque, NM, pp. 80–3.
Oke, T.R., Johnson, G.T., Steyn, D.G. and Watson, I.D. (1991) 'Simulation of nocturnal surface urban heat islands under "ideal" conditions: Part 2. Diagnosis of causation,' *Boundary-Layer Meteorology*, 56(4): 339–58.
Voogt, J.A. (2002) 'Urban heat island,' in I. Douglas (ed.), *Causes and Consequences of Global Environmental Change Vol. 3* of *Encyclopedia of Global Environmental Change*, R.E. Munn (ed-in chief), Chichester: Wiley, pp. 660–6.
Voogt, J.A. and Oke, T.R. (2003) 'Thermal remote sensing of urban climates,' *Remote Sensing of Environment*, 86(3): 370–84.
WMO (2008) 'Urban observations,' Chapter 11 in *Guide to Meteorological Instruments and Methods of Observation Part II – Observing Systems*, WMO-No. 8, seventh edition, Geneva: World Meteorological Organization, II 11.1–II 11.25.

12

Urban effects on precipitation and associated convective processes

J.M. Shepherd, J.A. Stallins, M.L. Jin, and T.L. Mote

Introduction

Scholars from antiquity until today have observed and speculated on how human activity shapes the Earth system (Ruddiman 2003; Von Storch and Stehr 2006; Yow 2007). Noted scholar Paul Crutzen has stated,

> humans have become a geologic agent comparable to erosion and eruptions ... it seems appropriate to emphasize the central role of mankind in geology and ecology by proposing to use the term "anthropocene" for the current geological epoch.
>
> *(Crutzen and Stoermer 2000: 17–18)*

Though the physical footprint of urban areas are local to regional in scale, it is increasingly evident that they have impacts spanning local to global scales by altering moisture and energy exchanges at the surface, changing atmospheric composition, and perturbing components of the water cycle (Seto and Shepherd 2009). The scale of the global impact can be judged from Figure 12.1.

The most comprehensively studied effect of the built environment on climate is the urban heat island (UHI) (see Chapter 11 this volume). While the UHI is a clearly established signature of anthropogenic alteration of climate, the effects of the built, urban environment on other aspects of the Earth's climate system are less well understood (Mills 2007; Seto and Shepherd 2009; Dabberdt *et al.* 2000). Herein, we review how the urban environment is altering aspects of regional hydroclimates, particularly precipitation and related convective processes. Although at times the human influence has been exaggerated (e.g. "rain follows the plow" during the settling of the central US plains (Worster 1979)), studies have established the scientific basis for how human activity in urban environments shapes their pattern of convection, precipitation, and lightning.

Convection and precipitation are key components in the global water cycle and a proxy for changing climate. Proper assessment of the urban environment's impact on clouds, precipitation, and associated hazards (e.g. lightning, flash flooding) will be increasingly important in ongoing climate diagnostics and prediction, global water and energy cycle (GWEC) analysis and

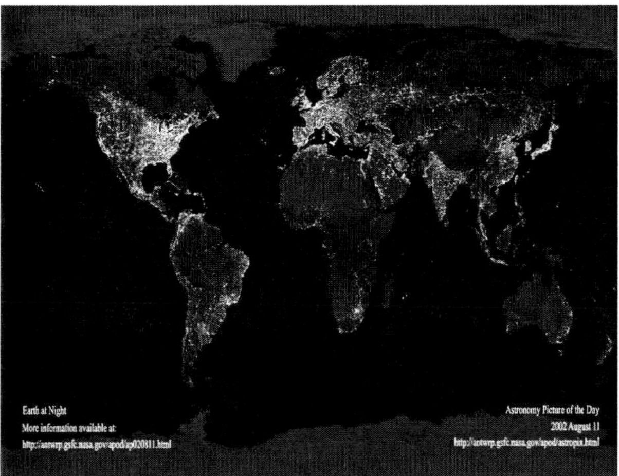

Figure 12.1 Satellite view of Earth at night.

modeling, weather forecasting, freshwater resource management, urban planning-design, and land-atmosphere-ocean interface processes. These assessments are particularly critical if current projections for global urban growth are accurate.

An historical perspective of urban effects on precipitation and convection

Understanding of the potential urban effects on precipitation or convection can be traced to the early 1900s. Horton (1921) noted a tendency for thunderstorm formation over large cities rather than the rural environment. Kratzer (1937, 1956) further corroborated the notion that urban environments could enhance precipitation. Helmut Landsberg, a pioneering urban climatologist, discussed the potential impacts of large urban areas on rainfall patterns in his classic text "The Climate of Towns" (1956). In the 1960s, Stout (1962) and Changnon (1968) began to examine the so-called "La Porte Anomaly." Hypotheses were posited that industrialized urban regions in northwest Indiana and southeast of Chicago, Illinois, received increased precipitation due to industrial aerosols. Though results concerning the La Porte Anomaly were generally inconclusive (Lowry 1998), the work stimulated an important period of activity for urban hydroclimate studies.

Observations of increased rainfall downwind of urban areas were the most common evidence of precipitation enhanced by urban effects (Landsberg 1970; Huff and Changnon 1972a, b). In North America, one of the first and most well known investigations of how urban areas modify their weather and climate was the METROpolitan Meteorological EXperiment (METROMEX). METROMEX synthesized the research on urban climates undertaken in North America, and developed an observational network to examine urban-rural differences for a suite of atmospheric properties. Observations from METROMEX showed enhanced precipitation by urban effects typically 25–75 km downwind of a city during summer months (Braham *et al.* 1975; Huff and Vogel 1978; Changnon 1979). Precipitation amounts were enhanced by urban effects between 5 and 25 percent over background values (Changnon *et al.* 1981, 1991). The size of an urban area was shown to influence the horizontal extent and magnitude of urban enhanced precipitation (Changnon 1992). However, Lowry (1998) published a review questioning whether appropriate methodologies or data were utilized during the METROMEX and post-METROMEX years.

He also called into question whether methodological deficiencies had been responsible for resolving the so-called anomalies or enhancement regions. Other investigators have subsequently presented results counter to the downwind enhancement hypothesis. Tayanc and Toros (1997) found no urban rainfall effects in their analysis of data in four large cities in Turkey. Robaa (2003) even suggested that an inverse relationship existed between the degree of urbanization and rainfall around Cairo, Egypt. Kaufmann *et al.* (2007) used a statistical model and ground-based rainfall data to establish that urban land cover growth in the Pearly River Delta region of China was associated with decreased rainfall.

Post-METROMEX studies generally supported the findings in METROMEX. Balling and Brazel (1987) reported more frequent late-afternoon storms in Phoenix during a period of rapid transformation from arid to urban land cover. Bornstein and LeRoy (1990) presented some of the first studies of a megalopolis and its impact on convective activity. They found that New York City affects summer daytime thunderstorm formation, distribution, and movement. Their radar analysis established a tendency for echo maxima on the lateral edges and downwind of the city. Jauregui and Romales (1996) observed that the daytime UHI seemed to be correlated with intensification of rain showers during the wet season (May–October) in Mexico City. They also showed that the frequency of intense rain showers has increased in recent decades in conjunction with the growth of the city. Selover (1997) found similar results for moving summer convective storms over Phoenix. Changnon and Westcott (2002) provided evidence of increasing heavy rainstorms from 1960 onwards and suggested that intensity and frequency could continue to increase in the future. Shepherd *et al.* (2002) evaluated the application of spaceborne radar-derived rainfall data to the urban precipitation problem. This capability mitigates one of the primary criticisms of urban precipitation studies (Lowry 1998) by enabling replication for multiple cities simultaneously. Shepherd *et al.* (2002) found evidence of rainfall anomalies downwind of major US cities. Figure 12.2 shows a conceptualization of the "urban rainfall effect" (URE) as presented in Simpson (2006) based on the original figure from Shepherd *et al.* (2002). Diem *et al.* (2004) and Shepherd (2004) also debated some of the initial findings and methodologies presented in the satellite-based study by Shepherd *et al.* (2002). Although the scales of observation and analysis in Shepherd *et al.* (2002) limited fine-scale generalization,

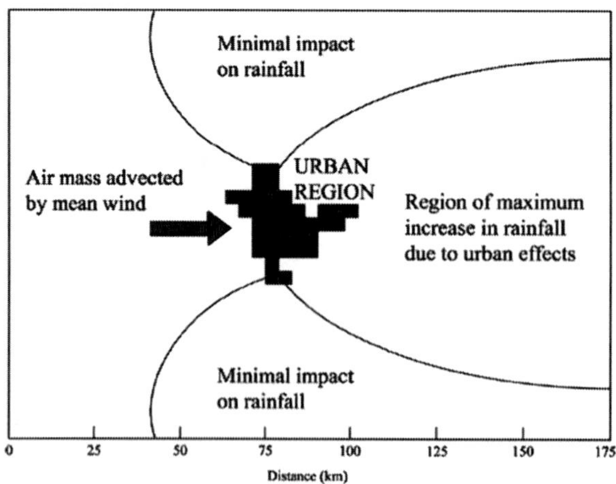

Figure 12.2 Schematic of the downwind urban rainfall anomaly (following Shepherd *et al.* (2002) as presented in Simpson (2006)).

this paper reinvigorated questions about the propensity for cities to modify the rainfall in their vicinity (Souch and Grimmond 2006).

Dixon and Mote (2003) examined urban initiation of convection around Atlanta, Georgia, and concluded that under a priori moist conditions some nocturnal convective events may be initiated by the city. Around the same time period, increased convection downwind of large urban areas such as Tokyo was attributed to enhanced surface convergence over the urban region (Fujibe 2003). Shepherd and Burian (2003) extended the satellite-based methodology to Houston, Texas, and found rainfall anomalies that were spatially consistent with lightning flash density anomalies reported by Orville *et al.* (2001).

Lowry (1998) presented several deficiencies in much of the historical literature, and Shepherd's review of urban effects on rainfall (2005) provided recommendations on how to move forward. Lowry's key recommendations argued that urban rainfall studies required:

1 Designed experiments – especially legitimate controls and, where appropriate, stratification schemes – in which explicitly stated hypotheses are tested by means of standard statistical methods.
2 Replication of the experiments in several urban areas.
3 Use of spatially small, and temporally short, experimental units reflecting the discontinuous nature of precipitating systems.
4 Disaggregation of standard climatic data to increase sample size and avoid merging effects between dissimilar synoptic weather systems.

(Lowry 1998: 515)

Importantly, post-2000 studies highlighting the influence of urbanization on rainfall throughout the world are explicitly or implicitly addressing several of the methodological deficiencies that Lowry (1998) used to question findings from the METROMEX era.

The urban area and downwind regions of Houston, Texas, had 59 percent and 30 percent respectively greater rainfall between noon and midnight in the warm season compared to the upwind region (Burian and Shepherd 2005). In the monsoon season, northeastern suburbs and exurbs of the Phoenix metropolitan area experienced statistically significant increases in mean precipitation of 12–14 percent from a pre-urban (1895–1949) to post-urban (1950–2003) period (Shepherd 2006). The so-called Lower Verde anomaly in Phoenix (Diem and Brown 2003) is probably caused by urban-topographical interactions (Shepherd 2006). A similar urban-topographical interaction may cause a nocturnal rainfall anomaly in Taipei, Taiwan (Chen *et al.* 2007). Mote *et al.* (2007), using ground-based radar, showed that the downwind suburbs of Atlanta, Georgia, receive up to 30 percent more warm-season rainfall during evening and early morning hours. Extending previous findings in Diem and Mote (2005) using a long-term climatological analysis of rain gauge data, Diem (2008) reconfirmed the existence of a northeast suburban Atlanta rainfall anomaly.

The recognition of the propensity for cities to modify precipitation stimulated investigations that integrate rainfall observations with other convective phenomena. Rose *et al.* (2008) coupled cloud-to-ground lightning data with the North American Regional Reanalysis dataset and found enhancement of lightning and precipitation around Atlanta as a function of prevailing wind. Hand and Shepherd (2009) used satellite and rain gauge estimates to reveal a statistically significant rainfall anomaly in the north-northeast (e.g. downwind based on prevailing wind analysis) sections of Oklahoma City, Oklahoma, and suburbs. More importantly, they established that merged infrared-microwave satellite estimates are relatively accurate compared to ground-based rain gauge networks for identifying spatially variant rainfall around a city. Therefore, satellite data can replicate analyses in rapidly urbanizing regions around the globe, particularly if robust ground-based data is not available. A radar-based analysis by Meng *et al.* (2007) also showed that, due to the urban effects, thunderstorms associated with a tropical cyclone strengthened when moving over Guangzhou City, China, with maximum radar echoes observed right over the urban area.

Even as methods and data availability improve, there is still uncertainty and ongoing debate regarding the role of urban environments on rainfall distribution. Trusilova *et al.* (2008) used a regional model to simulate European climate sensitivity to urban land cover. They found statistically significant increases (decreases) in winter (summer) rainfall in their urban simulations as compared to the pre-urban settlement runs. Their study used relatively coarse (~10 km) model resolution and heavily relied on cumulus-parameterized rainfall that might not explicitly resolve urban convection. Kaufmann *et al.* (2007) suggested that urbanization in the Pearl River Delta of China has reduced local precipitation due to changes in surface hydrology. Zhang *et al.* (2008) argued that replacement of natural surfaces with Beijing's urban land cover reduced ground evaporation and evapotranspiration, which was conducive to the occurrence of the rainfall. Guo *et al.* (2006) also found decreased cumulative rainfall around Beijing. The studies that found decreasing rainfall around Chinese cities did not consider the effects of pollution.

Urban pollution has been shown to affect precipitation processes. Smaller cloud droplet size distributions and delayed or suppressed rainfall have been shown to occur due to increased aerosol concentrations from anthropogenic sources over and downwind urban areas (Rosenfeld *et al.* 2008). The size of aerosol particles and their concentration determine the extent rainfall is suppressed or enhanced (van den Heever and Cotton 2007). Giant cloud condensation nuclei (CCN) act to enhance rainfall, while small-diameter pollutants act as CCN to suppress precipitation. Investigations of how polluted versus relatively unpolluted urban air modifies convective processes is in part an extension of the studies of how aerosols impact global climate, particularly in the context of how convection differs over oceans and continental regions.

Overwhelmingly, the "urban rainfall effect" has been linked to convective rainfall rather than the lighter, stratiform rainfall. Therefore, it is not surprising that researchers have expressed curiosity about whether an "urban lightning effect" exists. With the advent of the ground-based lightning detection networks, a spatially explicit means for accurately assessing cloud-to-ground lightning activity became available. Two general findings for lightning have remained consistent across many cities (Stallins and Rose 2008). First, cloud-to-ground negative polarity flash densities increase, particularly downwind of the city center. Second, the percentage of positive polarity cloud-to-ground flashes has been found to decrease in the vicinity of cities. Most of these studies observed this modified flash activity within 100 km of the city center (Stallins and Rose 2008). In São Paulo, Brazil, patterns of lightning followed the outlines of the urban area, underscoring the dual influence of atmospheric thermodynamics associated with surface cover variation on urban convection (Naccarato *et al.* 2003). Thus urban precipitation anomalies have similar areal distributions to lightning anomalies (Figure 12.3).

Several plausible hypotheses for urban effects on convection have emerged over time (Shepherd 2005):

1 enhanced convergence due to increased surface roughness in the urban environment;
2 enhanced sensible heat fluxes;
3 destabilization due to UHI perturbation of the boundary layer and resulting downstream translation of the UHI circulation or UHI-generated convective clouds;
4 enhanced aerosols in the urban environment for cloud condensation nuclei sources; or
5 bifurcation or diversion of precipitating systems by the urban canopy or related processes.

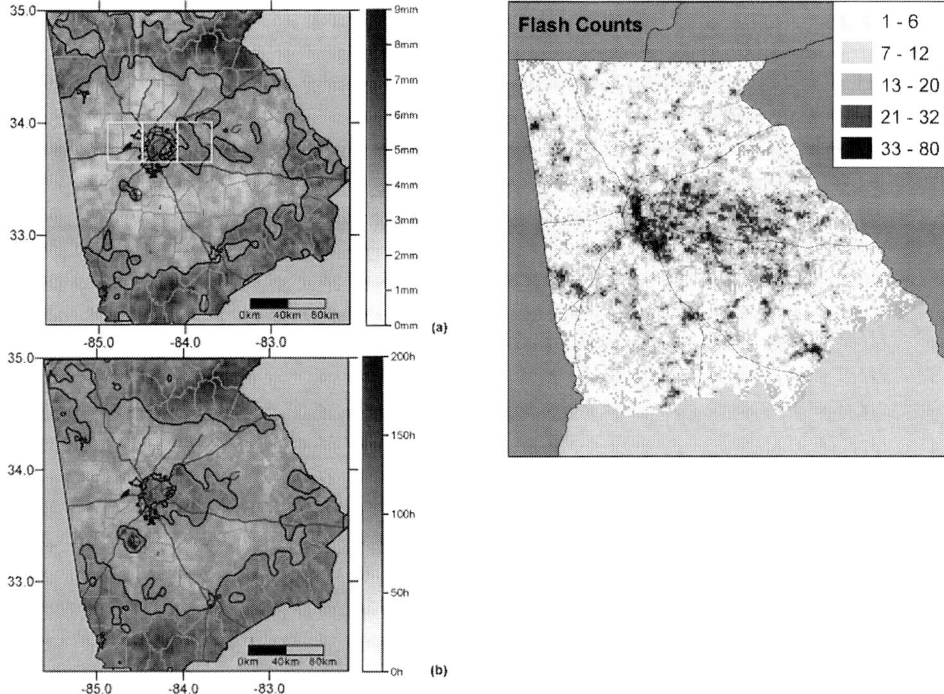

Figure 12.3 Rainfall (left) and lightning flash anomalies (right) for the warm season in Atlanta.

Because there is no universally accepted theory on which of these mechanisms, independently or synergistically, controls urban convective processes, we will hereafter refer to the "UCE" as some combination of the relevant processes. A newer generation of studies emphasize the relative importance of different mechanisms (Teller and Levin 2008).

Mechanistic linkages among urbanization, rainfall and lightning

Enhanced convergence, surface fluxes, and UHI destabilization

Contemporary work has relied upon simulations incorporating physically based mathematical solutions to land-atmosphere dynamic and thermodynamic processes (Pielke *et al.* 2007). Such physically based coupled atmosphere-land surface (CALS) models enable controlled, designed experiments. Figure 12.4 is a schematic describing numerical model simulations involving land-atmosphere interactions.

CALS model investigations began in the 1970s. Vukovich and Dunn (1978) used a three-dimensional primitive equation model to demonstrate that intensity of the urban-rural temperature gradient and boundary layer atmospheric stability play dominant roles in the development of UHI-induced meso-circulations. Huff and Vogel (1978) argued that the urban circulation is enhanced by the exchange of heat from the land surface to the atmosphere through increased sensible heat fluxes and surface roughness. Hjelmfelt (1982) simulated the UHI of St. Louis and attributed positive vertical velocities downwind of the city to surface roughness-induced (e.g. mechanical turbulence from buildings) convergence effect and the downwind translation or enhancement of the UHI circulation by the synoptic flow. Shafir and Alpert's

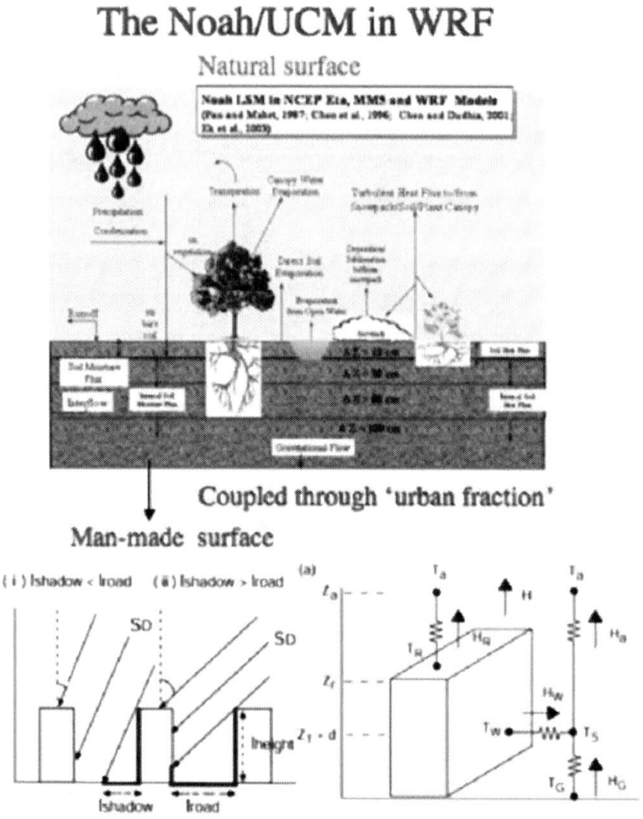

Figure 12.4 Coupled atmosphere-land model framework (*source:* UCAR).

(1990) simulations of the Jerusalem rainfall anomaly focused on elevated temperatures, humidity, and cloud condensation nuclei over the city.

Thielen *et al.* (2000) used a two-dimensional (2D) model to further corroborate earlier findings related to convergence and fluxes. They showed that surface parameters like sensible heat flux affected the development of precipitation over Paris, France. When UHIs were weak, surface sensible heat fluxes, convergence, and buoyancy variations that influence rainfall development were most effective at a distance from the central heat source. Craig and Bornstein's (2002) three-dimensional mesoscale simulations showed how the UHI induces convergence and convection around Atlanta. Rozoff *et al.* (2003) examined a 1999 storm case in St. Louis to ascertain the role of urban-generated surface convergence mechanisms on initiating deep, moist convection. They found that nonlinear interactions among the friction of urban surfaces, momentum drag, and urban heating could induce downwind convergence.

Niyogi *et al.* (2006) simulated a mesoscale precipitating storm in Oklahoma City, Oklahoma, using a land surface-urban canopy model system. Urban canopy models allow the land surface to represent the three-dimensional urban morphology and its heterogeneity rather than using simplified slab-type representation of the urban surface. Like previous studies, Niyogi *et al.* (2006) found that the urban area concentrated the precipitation in the downwind region. Ikebuchi *et al.* (2007) found that urban land cover and anthropogenic heating affected the location and intensity of a convective rainfall event in the Tokyo, Japan, metropolitan area.

Baik *et al.* (2007) and Han and Baik (2008) have provided some of the most compelling modeling work to describe the downwind enhancement observed in urban convective studies. Using a nonlinear numerical model and a two-layer linear analytical model, Baik *et al.* (2007) found that as the atmospheric boundary layer becomes less stable, a downwind updraft cell induced by the UHI intensifies. It is known that UHI magnitude is greater in the evening, so their results offered a plausible explanation for possible daytime urban-induced rainfall anomalies. They argued that as the boundary layer becomes unstable, the height of the maximum updraft velocity and the depth of the downwind updraft circulation increase. During daytime nearly neutral or weakly stable conditions, UHI circulations can be relatively strong even though the UHI magnitude itself might be relatively weak. Han and Baik's (2008) work used a more realistic three-dimensional framework. This study produced consistent results with the aforementioned study but also resolved an internal gravity wave field with an upward component downwind of the theoretical heating center (e.g. the city).

Shem and Shepherd (2009) found that combination of enhanced convergence on the perimeter of Atlanta's urban land cover footprint (Figure 12.5) and increased sensible heat flux enhanced pre-existing convection around Atlanta, Georgia. By running simulations with and without the urban land cover, they showed that the storms during their case days were not initiated by Atlanta but that subsequent cumulative rainfall might have been enhanced by the mesoscale dynamic interactions. Shepherd *et al.* (2010) found similar results for a storm in Houston. Using an urban growth model to project Houston's land cover to the year 2025, this study sought to understand how the convective activity would respond to future land cover. Future land cover was inserted in the CALS model and run under the "current" conditions. The results suggested that dynamic-thermodynamic response to the larger urban heat source significantly altered the precipitation

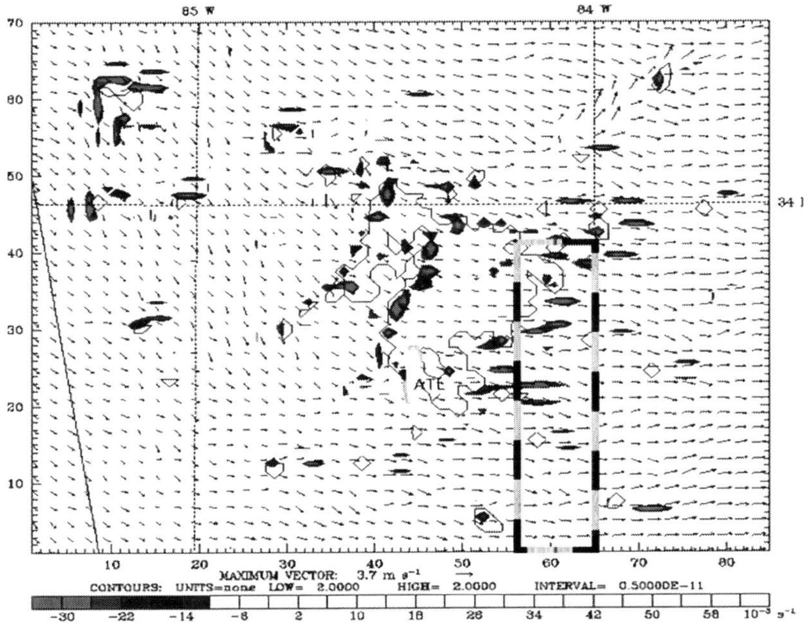

Figure 12.5 Difference in divergence/convergence for an URBAN and NOURBAN simulation. Negative values represent net convergence (following Shem and Shepherd 2009). The rectangle represents the region of enhanced rainfall for their case day.

evolution. This study further emphasized that other human-induced changes beyond greenhouse gas emissions can alter regional hydroclimates.

The aforementioned modeling studies have advanced our understanding of urban-atmosphere interactions significantly. However, they all lack adequate representation of aerosol or pollution effects within the microphysical (e.g. cloud hydrometeors, interactions, and precipitation) processes. Only a few studies like van den Heever and Cotton (2007) have considered the integrated effects of urban land cover and aerosols on precipitation. Yet, the discussion in the next section will elucidate why the next generation of CALS model studies must integrate both urban-related factors.

Aerosol or pollution effects

Over urban regions, there are various sizes of aerosols, ranging from 0.01 μm to 10 μm (Figure 12.6), and various types of aerosols, from a range of human and natural sources. The main components of urban aerosols are sulfates, nitrates, ammonium, organics, crustal rock, sea salt ions and water (Seinfeld and Pandis 1998). Aerosols interact with clouds by serving as cloud condensation nuclei (CCN). In this mechanism, aerosols modify the microphysical interaction between CCN and atomospheric water. Urban regions, in general, have high aerosol content, and these aerosols may serve as CCN to form clouds. Twomey (1977) theorized that under the condition of limited water vapor content in the air, more CCN means more particles compete

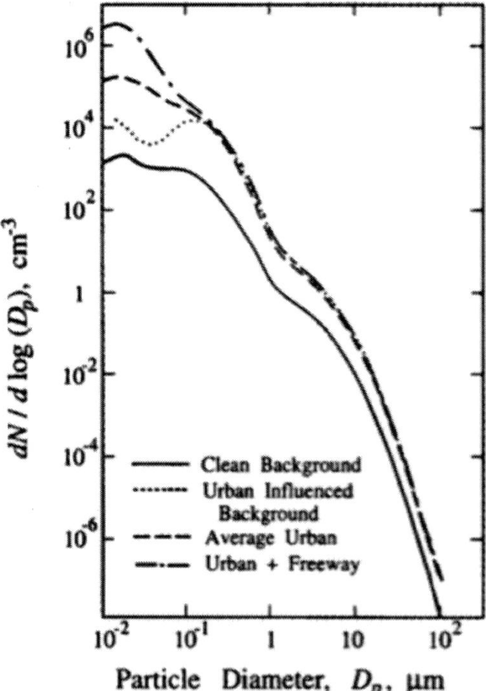

Figure 12.6 Aerosol number distributions next to a source (freeway), for average urban, for urban influenced background, and for background conditions (reproduced from *Atmospheric Physics and Chemistry*, Seinfeld and Pandis, 1998.)

for water vapor to form cloud droplets. Therefore, the size of clouds droplet would decrease. Increasing pollution generally means increasing CCN concentrations, hence increasing cloud optical thickness, finite cloud thicknesses, and increasing cloud albedo. Satellite observations validate Twomey's theory (Jin and Shepherd 2008).

Because aerosols are essential to the formation of clouds through their role as CCN, aerosols indirectly influence precipitation, and lightning. A considerable body of knowledge has accumulated on how oceanic and continental clouds differ in their aerosol properties, cloud development, and precipitation production (Rosenfeld and Lensky 1998; Rosenfeld *et al.* 2008). Large maritime aerosols enhance the collision coalescence process, promote large droplet size, and lead to early rainout. Finer and more concentrated aerosols over continents slow the collision coalescence process, leading to postponement of rainout, but deeper vertical cloud development. This maritime-continental difference in convective processes has been extended to the urban-rural gradient of air quality (Freud *et al.* 2008; Givati and Rosenfeld 2004). Increased anthropogenic aerosol content ultimately modifies convection by shaping how atmospheric water droplets coalesce and are transported by updrafts. Fine urban aerosols slow the collision and coalescence of water droplets. Because of the ensuing delay in the formation of downward-moving precipitation, stronger, deeper updrafts transport more precipitable water to the mixed-phase levels where the non-inductive charge separation process of lightning production takes place. Under this scenario, initial suppression of precipitation by urban aerosols may be followed by increased electrification and modification of rainfall rates (Rosenfeld *et al.* 2008).

The strongest arguments for the role of aerosols in lightning production come from studies finding a spatial correspondence among aerosols-generating facilities, higher flash densities, and decreased positive flashes in nonurban locations. Stallins *et al.* (2006) established a decrease in the percentage of positive flashes downwind of large stretches of a heavily traveled north-south trending interstate passing through Atlanta, Georgia. This aerosol effect may be seen in rural areas 100 km downwind from the congested traffic emerging out of the Atlanta city center. Steiger and Orville (2003) detected higher flash densities and a lowered percentage of positive flashes in the Lake Charles–Baton Rouge, Louisiana, corridor. This location has a dense infrastructure of aerosol-producing refineries and chemical production facilities, but overall low population numbers. For these nonurban locales, the spatial association among atmospheric particulate matter, increased flash densities, and decreases in the percentage of positive flashes support a mechanistic role for aerosols.

Bifurcation

The vertical structure and thermodynamic signature of urban land cover/buildings and their effects upon local circulation are hypothesized to bifurcate or redirect thunderstorms around the periphery as they move into an urban region. Ntelekos *et al.* (2007) observed that cell motion for a thunderstorm in Baltimore-DC reflected the influence of urban canopy form and the outflow boundaries of the thunderstorm as it approached the city. Bornstein and LeRoy (1990) hypothesized that moving thunderstorms bifurcated around New York City due to building-barrier effects. Air mass thunderstorms have different patterns of lightning than frontal thunderstorms in Atlanta, Georgia (Stallins and Bentley 2006). Frontally associated thunderstorms had peaks in flash density on the perimeter of the city, while high flash densities in air mass thunderstorms retracted to the urban core.

Reconciling mechanisms and methods

It is likely that the UCE is explained by a combination of factors. For example, Bell *et al.* (2008) used the stratification of observations according to proxies for aerosol loads (day of the week, traffic volume) to document midweek precipitation increases over the southeastern US. Sato and Takahaski (2000) also identified a weekly cycle in precipitation, although Schultz *et al.* (2007) and DeLisi *et al.* (2001) found no evidence for a weekly cycle. Lacke *et al.* (2009) found no statistically significant weekly cycle in aerosols and rainfall around Atlanta, but their results suggested a tendency for a midweek peak in aerosol activity. Stallins and Bentley (2006) hypothesized that aerosol mechanisms contribute to the variability in flash counts exhibited over an urban region, while the number of thunderstorms observed is a better indicator of the influence of mechanisms evolving directly from urban land surfaces. Model simulations by van den Heever and Cotton (2007) suggest how aerosols and wind convergence interact in urban areas. Aerosol size and relative abundance influences the timing of convection and its location. However, wind convergence driven by urban effects on local circulation was the central factor determining whether thunderstorms actually developed in the downwind direction. Once they developed, aerosols influenced the amount of liquid water and ice present in the atmosphere, the surface precipitation totals, the strength and timing of updrafts and downdrafts, and the longevity of updrafts. Although lightning was not incorporated into these simulations, the propensity of aerosols to modify updrafts and water transport suggests that altered cloud electrification is possible.

As a strategy to unfold mechanistic understanding, emphasis has also shifted to the coordinated study of multiple weather phenomena. These integrated approaches implicitly or explicitly recognize that clouds, rainfall, and lightning develop from related physical processes and that their coordinated expression may reveal subtleties of mechanism. For example, Atlanta has been the subject of several independent studies of urban precipitation and lightning. Nevertheless, there remains a paucity of urban studies that conjointly detail the variability in the timing of precipitation and flash production outside of the context of urban flash flood prediction. How do peaks in precipitation and lightning production coincide geographically? Are there dry urban thunderstorms that produce dense lightning with delayed precipitation? For a given level of stability, do aerosols enhance updrafts, allowing a relatively small level of instability to produce disproportionately more cloud electrification and delayed rainfall production (Rosenfeld *et al.* 2008)? Boussaton *et al.* (2007) found variability in the temporal sequencing of radar reflectivity intensity with lightning production over Paris. In some thunderstorm events, high reflectivities indicative of heavy precipitation developed after peaks in lightning production. In other storms, reflectivities peaked and then decreased before maxima in flash production. These contrasts may be illustrative of how urban aerosols modify the convective processes by influencing updrafts, precipitation timing, and transport of water to the levels where the non-inductive charge separation process of cloud electrification occurs.

Societal implications and recommendations

Given the burgeoning migration of human populations to urban habitats, many sectors of society are directly affected by changes in precipitation or lighting. Lightning is a widely acknowledged public safety risk addressed by many governmental entities. The combination of increased thunderstorms, rainfall, and heating may be linked to increased hospital admissions for respiratory distress and asthma for the region around cities (see Grundstein *et al.* 2008) or more water and vector-borne diseases such as dysentery, cholera, and malaria.

In the developed and developing world urban flooding is a reality for many, and its effects upon the public safety and health of city dwellers is considerable (Shepherd *et al.* 2010). Burian *et al.* (2004) discussed the implications of urban-induced precipitation on the design of urban drainage systems. Urban weather and climate change is an opportunity for urban planning and design. Can inadvertent weather modification ever be used strategically? For example, Chen *et al.* (2007) note how increased rainfall and thunderstorm activity in the Taipei basin in Taiwan worsens traffic hazards but eases the stress on water supply, reduces ground subsidence, and cleans up pollution. Shepherd and Mote (2009) argued that the state of Georgia could consider urban rainfall anomalies when planning future water supply reservoir locations. Such considerations also need to be borne in mind when establishing urban forests and restoring degraded areas as accessible greenspace.

In closing, urban weather and climate studies should use multiple methods, especially those that balance description and mechanism (Schröder and Seppelt 2006; Grimaldi and Engel 2007). Geographic investigations that couple nomothetic (generalizing) and ideographic (context specific) perspectives (Phillips 2001) will foster more understanding of mechanism as well as prediction. A synoptic typology for cities, deployed across a large range of latitudes and maritime versus continental locations will help disentangle the aerosol, thermodynamic, and surface roughness-based mechanisms and local mitigating influences. Improvements in observing systems (Shepherd 2005), urban canopy and aerosol representation in regional-scale models (Burian *et al.* 2004; Khain *et al.* 2004; van den Heever and Cotton 2007; Kusaka and Kimura 2004a, b), and leveraged field experimentation will also be required. Scholars must also bridge the gap between regional hydrometeorological and global hydroclimate processes. Urban footprints are converging with the increasingly high spatial resolution of regional and global climate models. The climate modeling community should begin to represent urban land surface and aerosol processes to better understand the aggregate influences of built-up land and urban aerosols on short- and long-term climate change (see Jin and Shepherd 2008; Jin *et al.* 2007; Jin and Shepherd 2005).

Acknowledgments

Dr. Shepherd acknowledges support from NASA grant NNX07AF39G, Precipitation Measurement Missions Program. Dr. Stallins acknowledges support from NSF BCS (Award Number 0241062). Dr. Jin acknowledges support from NSF ATM (Award Number 0701440). Dr. Mote and Dr. Shepherd acknowledge support from the Southern High Resolution Modeling Consortium and USDA Forest Service contract AG-4568-C-08–0063.

Bibliography

Baik, J.-J., Kim, Y.-H., Kim, J.-J., and Han, J.-Y. (2007) Effects of boundary-layer stability on urban heat island-induced circulation, *Theoretical and Applied Climatology*, 89(1–2): 73–81.

Balling, R.C. and Brazel, S.W. (1987) Diurnal variations in Arizona monsoon precipitation frequencies, *Monthly Weather Review*, 115(1): 342–6.

Bell, T.L., Rosenfeld, D., Kim, K.-M., Yoo, J.-M. Lee, M.-I., and M. Hahnenberger (2008) Midweek increase in US summer rain and storm heights suggests air pollution invigorates rainstorms, *Journal of Geophysical Research-Atmospheres* 113: D02209, doi:10.1029/2007JD008623.

Bornstein, R. and LeRoy, M. (1990) Urban barrier effects on convective and frontal thunderstorms, *Extended Abstracts, Fourth Conference on Mesoscale Processes, Boulder, CO, January 25–29, American Meteorological Society*: 120–1.

Boussaton, M.P., Soula. S., and Coquillat, S. (2007) Total lightning activity in thunderstorms over Paris, *Atmospheric Research*, 84(3): 221–32.

Braham, R.R., Morris, T.R., Dungey. M., Changnon, S.A., and Huff, F.A. (1975) Summary of project Metromex radar findings, *Bulletin of The American Meteorological Society*, 56: 155.
Burian, S.J. and Shepherd, J.M. (2005) Effect of urbanization on the diurnal rainfall pattern in Houston, *Hydrological Processes*, 19(5): 1089–103.
Burian, S., Stetson, W., Han, W.S., Ching, J.K.S., and Byun, D.W. (2004) High-resolution dataset of urban canopy parameters for Houston, Texas, *Proceedings of the American Meteorological Society 9.3, Fifth Conference on the Urban Environment, Vancouver.*
Changnon, S.A. (1968) The La Port weather anomaly – fact or fiction? *Bulletin American Meteorological Society*, 49: 4–11.
Changnon, S.A. (1979) Rainfall changes in summer caused by St. Louis, *Science*, 205(4404): 402–4.
Changnon, S.A. (1992) Inadvertent weather modification in urban areas: Lessons for global climate change, *Bulletin American Meteorological Society*, 73(5): 619–27.
Changnon, S.A. and Westcott, N.E. (2002) Heavy rainstorms in Chicago: Increasing frequency, altered impacts, and future implications, *Journal of the American Water Resources Association*, 38(5): 1467–75.
Changnon, S.A., Shealy, R.T., and Scott, R.W. (1991) Precipitation changes in fall, winter and spring caused by St. Louis, *Journal of Applied Meteorology and Climatology*, 30(1): 126–34.
Changnon, S.A., Semonin, R.G., Auer, A.H., Braham, R.R., and Hales, J. (1981) METROMEX: A review and summary, *Meteorological Monograph, 18(40), American Meteorological Society.*
Chen, T.-C., Wang, S.-Y., and Yen, M.-C. (2007) Enhancement of afternoon thunderstorm activity by urbanization in a valley: Taipei, *Journal of Applied Meteorology and Climatology*, 46(9): 1324–40.
Craig, K. and Bornstein, R. (2002) MM5 simulation of urban induced convective precipitation over Atlanta, *Preprint Volume, Proceedings of the Fourth American Meteorological Society Symposium on the Urban Environment, Norfolk, VA*: 5–6.
Dabberdt, W.F., Hales, J., Zubrick, S., Crook, A., Krajewski, W., Doran, J.C., Mueller, C., King, C. Keener, R.N., Bornstein, R., Rodenhuis, D., Kocin, P., Rossetti, M.A., Sharrocks, F., and Stanley Sr., E.M. (2000) Forecast issues in the urban zone: Report of the 10th Prospectus Development Team of the US Weather Research Program, *Bulletin of the American Meteorological Society*, 81(9): 2047–64.
DeLisi, M.P., Cope, A.M., and Franklin, J.K. (2001) Weekly precipitation cycles along the northeast corridor? *Weather and Forecasting*, 16(3): 343–53.
Diem, J.E. (2008) Detecting summer rainfall enhancement within metropolitan Atlanta, Georgia USA, *International Journal of Climatology*, 28(1): 129–33.
Diem, J.E. and Brown, D.P. (2003) Anthropogenic impacts on summer precipitation in central Arizona, USA, *Professional Geographer*, 55(3): 343–55.
Diem, J.E. and Mote, T.L. (2005) Interepochal changes in summer precipitation in the southeastern United States: Evidence of possible urban effects near Atlanta, Georgia, *Journal of Applied Meteorology and Climatology*, 44(5): 717–30.
Diem, J.E., Coleman, L.B., Digirolamo, P.A., Gowens, C.W., Hayden, N.R., Unger, E.E., Wetta, G.B., and Williams, H.A. (2004) Comments on "Rainfall Modification by Major Urban Areas: Observations from Spaceborne Rain Radar on the TRMM Satellite", *Journal of Applied Meteorology and Climatology*, 43(6): 941–50.
Dixon, P.G. and Mote, T.L. (2003) Patterns and causes of Atlanta's urban heat island-initiated precipitation, *Journal of Applied Meteorology and Climatology*, 42(9): 1273–84.
Freud, E., Strom, J., Rosenfeld, D., Tunved, P., and Swietlicki, E. (2008) Anthropogenic aerosol effects on convective cloud microphysical properties in southern Sweden, *Tellus Series B-Chemical and Physical Meteorology*, 60(2): 286–97.
Fujibe, F. (2003) Long-term surface wind changes in the Tokyo metropolitan area in the afternoon of sunny days in the warm season, *Journal of the Meteorological Society of Japan*, 81(1): 141–9.
Givati, A. and Rosenfeld, D. (2004) Quantifying precipitation suppression due to air pollution, *Journal of Applied Meteorology and Climatology*, 43(7): 1038–56.
Grimaldi, D.A. and Engel, M.S. (2007) Why descriptive science still matters, *BioScience*, 57(8): 646–47.
Grundstein, A., Sarnat, S.E., Klein, M., Shepherd, J.M., Naeher, L., Mote, T.L., and Tolbert, P. (2008) Thunderstorm-associated asthma in Atlanta, Georgia, *Thorax*, 63(7): 659–60.
Guo, X., Fu, D., and Wang, J. (2006) Mesoscale convective precipitation system modified by urbanization in Beijing City, *Atmospheric Research*, 82(1–2): 12–126.
Han, J.-Y. and Baik, J.-J. (2008) A theoretical and numerical study of urban heat island-induced circulation and convection, *Journal of the Atmospheric Sciences*, 65(6): 1859–77.
Hand, L. and Shepherd, J.M. (2009) An investigation of warm season spatial rainfall variability in Oklahoma

City: Possible linkages to urbanization and prevailing wind, *Journal of Applied Meteorology and Climatology*, 48(2): 251–69.

Hjelmfelt, M.R. (1982) Numerical-simulation of the effects of St. Louis on mesoscale boundary-layer air-flow and vertical air motion – simulations of urban vs non-urban effects, *Journal of Applied Meteorology and Climatology*, 21(9): 1239–57.

Horton, R.E. (1921) Thunderstorm breeding spots, *Monthly Weather Review*, 49(4): 193.

Huff, F. and Changnon, S.A. (1972a) Climatological assessment of urban effects on precipitation at St. Louis, *Journal of Applied Meteorology and Climatology*, 11(5): 823–42.

Huff, F. and Changnon, S.A. (1972b) Climatological assessment of urban effects on precipitation St. Louis: Part II, *Final Report. NSF Grant GA-18781. Illinois State Water Survey.*

Huff, F.A. and Vogel, J.L. (1978) Urban, topographic and diurnal effects on rainfall in St. Louis region, *Journal of Applied Meteorology and Climatology*, 17(5): 565–77.

Ikebuchi, S., Tanaka, K., Ito, Y., Moteki, Q., Souma, K., and Yorozu, K. (2007) Investigation of the effects of urban heating on the heavy rainfall event by a cloud resolving model CReSiBUC, *Annals of the Disaster Prevention Research Institute of Kyoto University*, 50C.

Jauregui, E. and Romales, E. (1996) Urban effects on convective precipitation in Mexico City, *Atmospheric Environment*, 30(20): 3383–9.

Jin, M. and Shepherd, J.M. (2005) Inclusion of urban landscape in a climate model – How can satellite data help? *Bulletin of the American Meteorological Society*, 86(5): 681–9.

Jin, M. and Shepherd, J.M. (2008) Aerosol relationships to warm season clouds and rainfall at monthly scales over east China: Urban land versus ocean, *Journal of Geophysical Research-Atmospheres* 113: D24S90, doi:10.1029/2008JD010276.

Jin, M., Shepherd, J.M., and Peters-Lidard, C. (2007) Development of a parameterization for simulating the urban temperature hazard using satellite observations in climate model, *Natural Hazards*, 43: 257–71.

Kaufmann, R.K., Seto, K.C., Schneider, A., Liu, Z., Zhou, L., and Wang, W. (2007) Climate response to rapid urban growth: Evidence of a human-induced precipitation deficit, *Journal of Climate*, 20(10): 2299–306.

Khain, A., Pokrovsky, A., Pinsky, M., Seifert, A., and Phillips, V. (2004) Simulation of effects of atmospheric aerosols on deep turbulent convective clouds using a spectral microphysics mixed-phase cumulus cloud model. Part I: Model description and possible applications, *Journal of the Atmospheric Sciences*, 61(24): 2963–82.

Kratzer, P.A. (1937) *Das Stadtklima*, Braunschweig: F. Vieweg und Sohne.

Kratzer, P.A. (1956) *Das Stadtklima*, second edition, Braunschweig: F. Vieweg und Sohne (trans. by the US Air Force, Cambridge Research Laboratories, Bedford, M).

Kusaka, H. and Kimura, F. (2004a) Thermal effects of urban canyon structure on the nocturnal heat island: Numerical experiment using a mesoscale model coupled with an urban canopy model, *Journal of Applied Meteorology and Climatology*, 43(12): 1899–910.

Kusaka, H. and Kimura, F. (2004b) Coupling a single-layer urban canopy model with a simple atmospheric model: Impact on urban heat island simulation for an idealized case, *Journal of the Meteorological Society of Japan*, 82(1): 67–80.

Lacke, M., Mote, T.L., and Shepherd, J.M. (2009) Aerosols and Associated Precipitation Patterns in Atlanta, *Atmospheric Environment*, 43(28): 4359–73.

Landsberg, H. (1956) The climate of towns, in W.L. Thomas Jr. (ed.), *Man's Role in Changing the Face of the Earth*, Chicago, IL: The University of Chicago Press.

Landsberg, H. (1970) Man-made climate changes, *Science*, 170: 1265–74.

Lowry, W.P. (1998) Urban effects on precipitation amount, *Progress In Physical Geography*, 22(4): 477–520.

Meng, W., Yen, J., and Hu, H. (2007) Urban effects and summer thunderstorms in a tropical cyclone affected situation over Guangzhou city, *Science in China Series D Earth Sciences*, 50(12): 1867–76.

Mills, G. (2007) Cities as agents of global change, *International Journal of Climatology*, 27(14): 1849–57.

Mote, T.L., Lacke, M.C., and Shepherd, J.M. (2007) Radar signatures of the urban effect on precipitation distribution: A case study for Atlanta, Georgia, *Geophysical Research Letters*, 34: L20710, doi:10.1029/2007GL031903.

Naccarato, K.P., Pinto, O., and Pinto, I. (2003) Evidence of thermal and aerosol effects on the cloud-to-ground lightning density and polarity over large urban areas of Southeastern Brazil, *Geophysical Research Letters*, 30(13): 1674–7.

Niyogi, D., Holt, T., Zhong, S., Pyle, P.C., and Basara, J. (2006) Urban and land surface effects on the 30 July 2003 mesoscale convective system event observed in the southern Great Plains, *Journal of Geophysical Research-Atmospheres* 111(D19): D19107, doi:10.1029/2005JD006746.

Ntelekos, A.A., Smith, J.A., and Krajewski, W.F. (2007) Climatological analyses of thunderstorms and flash floods in the Baltimore metropolitan region, *Journal of Hydrometeorology*, 8(1): 88–101.

Orville, R.E., Huffines, G., Nielsen-Gammon, J., Zhang, R.Y., Ely, B., Steiger, S., Phillips, S., Allen, S., and Read, W. (2001) Enhancement of cloud-to-ground lightning over Houston, Texas, *Geophysical Research Letters*, 28(13): 2597–600.

Pielke, R.A., Adegoke, J., Beltran-Przekurat, A., Hiemstra, C.A., Lin J., Nair, U.S., Niyogi, D., and Nobis, T.E. (2007) An overview of regional land-use and land-cover impacts on rainfall, *Tellus Series B-Chemical And Physical Meteorology*, 59(3): 587–601.

Phillips, J.D. (2001) Human impacts on the environment: unpredictability and the primacy of place, *Physical Geography*, 22(4): 321–32.

Robaa, S.M. (2003) Urban-suburban/rural differences over greater Cairo, Egypt, *Atmósfera* 16(3): 157–71.

Rose, L.S., Stallins, J.A., and Bentley, M.L. (2008) Concurrent cloud-to-ground lightning and precipitation enhancement in the Atlanta, Georgia (United States), urban region, *Earth Interactions*, 12(11): 1–30.

Rosenfeld, D. and Lensky, I.M. (1998) Satellite-based insights into precipitation formation processes in continental and maritime convective clouds, *Bulletin of the American Meteorological Society*, 79(11): 2457–76.

Rosenfeld, D., Lohmann, U., Raga, G.B., O'Dowd, C.D., Kulmala, M., Fuzzi, S., Reissell, A., and Andreae, M.O. (2008) Flood or drought: How do aerosols affect precipitation? *Science*, 321(5894): 1309–13.

Rozoff, C.M., Cotton, W.R., and Adegoke, J.O. (2003) Simulation of St. Louis, Missouri, land use impacts on thunderstorms, *Journal of Applied Meteorology and Climatology*, 42(6): 716–38.

Ruddiman, W.F. (2003) The anthropogenic greenhouse era began thousands of years ago, *Climatic Change*, 61(3): 261–93.

Sato, N. and Takahashi, M. (2000) A weekly cycle of summer heavy rainfall events in Tokyo, *Tenki*, 47: 643–8.

Schröder, B. and Seppelt, R. (2006) Analysis of pattern-process interactions based on landscape models – overview, general concepts, and methodological issues, *Ecological Modelling*, 199(4): 505–16.

Schultz, D.M., Mikkonen, S., Laaksonen, A., and Richman, M.B. (2007) Weekly precipitation cycles? Lack of evidence from United States surface stations, *Geophysical Research Letters*, 34(22): L22815, doi:10.1029/2007GL031889.

Seinfeld, J.H. and Pandis, S.N. (1998) *Atmospheric Chemistry and Physics: From Air Pollution to Climate Change*, New York, NY: Wiley-Interscience.

Selover, N. (1997) *Precipitation patterns around an urban desert environment topographic or urban influences?* Paper presented at Association of American Geographers Convention. Fort Worth, TX, May 1997.

Seto, K. and Shepherd, J.M. (2009) Global urban land-use trends and climate impacts, *Current Opinion in Environmental Sustainability*, 1(1): 89–95.

Shafir, H. and Alpert, P. (1990) On the urban orographic rainfall anomaly in Jerusalem – a numerical study, *Atmospheric Environment Part B-Urban Atmosphere*, 24(3): 365–75.

Shem, W. and Shepherd, J.M. (2009) On the impact of urbanization on summertime thunderstorms in Atlanta: Two numerical model case studies, *Atmospheric Research*, 92(2): 172–89.

Shepherd, J.M. (2004) Comments on "Rainfall modification by major urban areas: Observations from spaceborne rain radar on the TRMM satellite" – Reply, *Journal of Applied Meteorology and Climatology*, 43(6): 951–7.

Shepherd, J.M. (2005) A review of current investigations of urban-induced rainfall and recommendations for the future, *Earth Interactions*, 9(12): 1–27.

Shepherd, J.M. (2006) Evidence of urban-induced precipitation variability in arid climate regimes, *Journal of Arid Environments*, 67(4): 607–28.

Shepherd, J.M. and Burian, S.J. (2003) Detection of urban-induced rainfall anomalies in a major coastal city, *Earth Interactions*, 7(4): 1–14.

Shepherd, J.M. and Jin, M. (2004) Linkages between the Urban Environment and Earth's Climate System, *EOS*, 85(23): 227–8.

Shepherd, J.M. and Mote, T.L. (2009) Urban effects on rainfall variability: Potential implications for Georgia's water supply, *Proceedings of the 2009 Georgia Water Resources Conference, Athens, Georgia*, online, available at: www.gwri.gatech.edu/uploads/proceedings/2009/4.6.1_Shepherd.pdf

Shepherd, J.M., Pierce, H., and Negri, A.J. (2002) Rainfall modification by major urban areas: Observations from spaceborne rain radar on the TRMM satellite, *Journal of Applied Meteorology and Climatology*, 41(7): 689–701.

Shepherd, J.M., Carter, W.M., Manyin, M., Messen, M., and Burian, S. (2010) The impact of urbanization on current and future coastal convection: A case study for Houston, *Environment and Planning B*, 37(2): 284–304.

Simpson, M.D. (2006) *Role of Urban Land Use on Mesoscale Circulations and Precipitation*, dissertation, Raleigh, NC: North Carolina State University.

Souch, C. and Grimmond, S. (2006) Applied climatology: Urban climatology, *Progress in Physical Geography*, 30(2): 270–9.

Stallins, J.A. and Bentley, M.L. (2006) Urban lightning climatology and GIS: An analytical framework from the case study of Atlanta, Georgia, *Applied Geography*, 26(3–4): 242–59.

Stallins, J.A. and Rose, L.S. (2008) Urban lightning: Current research methods, and the geographical perspective, *Geography Compass*, 2(3): 620–39.

Stallins, J.A., Bentley, M.L., and Rose, L.S. (2006) Cloud-to-ground flash patterns for Atlanta, Georgia (USA) from 1992 to 2003, *Climate Research*, 30: 99–112.

Steiger, S.M. and Orville, R.E. (2003) Cloud-to-ground lightning enhancement over southern Louisiana, *Geophysical Research Letters*, 30(19), doi: 10.1029/ 2003GL017923.

Stout, G.E. (1962) Some observations of cloud initiation in industrial areas, in *Air Over Cities*, Technical Report A62–5. Washington, DC: US Public Health Service, pp. 147–53.

Tayanc, M. and Toros, H. (1997) Urbanization effects on regional climate change in the case of four large cities of Turkey, *Climatic Change*, 35(4): 501–24.

Teller, A. and Levin, Z. (2008) Factorial method as a tool for estimating the relative contribution to precipitation of cloud microphysical processes and environmental conditions: Method and application, *Journal of Geophysical Research-Atmospheres*, 113(D2), D02202, doi:10.1029/2007JD008960.

Thielen, J.W., Wobrock, A., Gadian, A., Mestayer, P.G., and Creutin, J.-D. (2000) The possible influence of urban surfaces on rainfall development: A sensitivity study in 2D in the meso-gamma scale, *Atmospheric Research*, 54(1): 15–39.

Trusilova, K., Jung, M., Churkina, G., Karstens, U., Heimann, M., and Claussen, M. (2008) Urbanization impacts on the climate in Europe: Numerical experiments by the PSU-NCAR Mesoscale Model (MM5), *Journal of Applied Meteorology and Climatology*, 47(5): 1442–55.

Twomey, S. (1977) On the minimum size of particle for nucleation in clouds, *Journal of the Atmospheric Sciences*, 34(11): 1832–5.

Van den Heever, S.C. and Cotton, W.R. (2007) Urban aerosol impacts on downwind convective storms, *Journal of Applied Meteorology and Climatology*, 46(6): 828–50.

Von Storch, H. and Stehr, N. (2006) Anthropogenic climate change: A reason for concern since the 18th century and earlier, *Geografiska Annaler Series A-Physical Geography*, 88(2): 107–13.

Vukovich, F.M. and Dunn, J.W. (1978) Theoretical-study of St. Louis heat island – some parameter variations, *Journal of Applied Meteorology and Climatology*, 17(11): 1585–94.

Worster, D. (1979) *Dust Bowl: The Southern Plains in the 1930s*, New York, NY: Oxford University Press.

Yow, D.M. (2007) Urban heat islands: Observations, impacts, and adaptation, *Geography Compass*, 1(6): 1227–51.

Zhang, H., Sato, N., Izumi, T., Hanaki, K., and Aramaki, T. (2008) Modified RAMS-Urban Canopy Model for Heat Island Simulation in Chongqing, China, *Journal of Applied Meteorology and Climatology*, 47(2): 509–24.

13
Urban hydrology

Ian Douglas

The urban hydrological cycle

The circulation of water in the city involves two interlinked systems: the people-modified hydrological cycle and the people-created artificial supply and waste-water disposal system. The natural circulation of water is modified by the nature of the urban surface, with large impermeable areas that encourage rapid runoff and decrease infiltration. The urban heat island and energy balance affect rain-producing mechanisms and the rate of snow-melt over and within cities. Urbanization affects stream channels and flood plains, often causing water to flow through cities at higher velocities. These hydrological effects of urbanization may be summarized under four major headings (Leopold 1968):

1 a change in total runoff
2 an alteration of peak flow characteristics
3 a decline in the quality of water
4 changes in the hydrological amenities of stream and their ecology.

To these changes to the natural hydrological system must be added the network of water collecting, treating, transmitting, regulating and distributing pipes and channels of the urban water supply system and the gutters, culverts, drains, pipes, sewers and channels of the urban waste-water disposal and stormwater drainage system. These systems abstract water from the natural hydrological cycle, deliver it to points of use, and discharge it back into other sectors of the natural hydrological cycle. The supply may be abstracted from one river basin and the same water may later, after use, be discharged into a completely different river basin, as in the case of Birmingham, England, where the water is taken from the rivers Wye and Severn which flow southwestwards to the Bristol Channel, and the waste water mainly flows into the River Tame and thus to the Trent which drains north-eastwards to the North Sea. Other supplies are drawn from groundwater aquifers, some of which are naturally draining towards rivers, such as the chalk aquifers under Paris and London. Other aquifers, such as those exploited by the Great Libyan Man-Made River to supply coastal cities as well as to irrigate crops (Gijbsers and Loucks 1999), contain fossil water that once abstracted is not likely to be replaced by infiltrating rainwater. Waste water from all ground water sources is discharged into surface watercourses, unless it is used, after treatment, to replenish shallow aquifers. Inevitably, some the groundwater used for irrigation or watering urban gardens is lost back to the atmosphere by evapotranspiration. Thus the overall urban water balance is highly complex, yet absolutely vital for the whole urban ecosystem. In some cases, the waste water is discharged into a different river basin from that from which it is derived. Chicago, Illinois, actually reversed the flow of the Chicago River in order to protect its drinking water in Lake Michigan, thus diverting the polluted Chicago River south into the Mississippi basin, so that others would have to live with the problem.

To assess these diverse sources, pathways and destinations of urban water, an overall urban water balance may be calculated. Such a balance may be expressed as:

$$P + D + A + W = E + R_s + S$$

where P is precipitation including rain, snow and hail; D is dew and hoar frost; A is the water released from anthropogenic sources, more especially combustion; W is the piped, surface and subsurface water brought into the city; E is evaporation (including transpiration); R_s is the natural and piped surface and subsurface flow out of the city; and S is the change of water storage in the city (Chandler 1976). Complete data are difficult to obtain, but some idea of the relative importance of natural local and imported piped water can be gained from looking at a few cities, ignoring the A, D and S terms (Table 13.1). More complete analyses of parts of cities or small urban catchments have been undertaken (Table 13.2).

Table 13.1 Partial urban water balances for selected cities (after Douglas 1983) (all values are $10^6 m^3$)

City	*P*	*W*	*Total input*	*E*	*S*	*Total output*
All Swedish cities	2820	945	3765	1449	2316	3765
Sydney	1200	402	1602	632	970	1602
Hong Kong	2000	68	2068	1180	888	2068
Mexico City	6704	1082	7786	5518	2268	7786
New York SMSA	10,907	1928	12,835	1650	2155 (sewers only)	data incomplete

Table 13.2 Annual water balances for selected urban areas (based on Mitchell *et al.* 2003)

	P	*W*	*E*	*Stormwater runoff*	*Waste water runoff*	*S*
Curtin, Canberra, mean	630	200	508	203	118	1
Curtin driest year	247	269	347	74	107	–23
Curtin wettest year	914	141	605	290	126	33
Subiaco-Shenton Park Perth WA	788	285 + 96[a]	766	104	154	117[b]
Sunninghill, South Africa	724	114	457	107	95	217[c]
Oakridge, Canada	1215	576	578	1210[d]		3

Notes

a Inflow of stormwater from upstream area.

b Adjusted for groundwater movement.

c Change in storage in this case is due to groundwater storage.

d The number quoted combines stormwater and wastewater runoff.

The modified natural hydrological system

Precipitation

The urban environment changes rainfall patterns. For many decades, observations of increased rainfall downwind of cities (e.g. Gates 1972) or higher thunderstorm frequencies over urban areas (e.g. Atkinson 1968, 1977) were queried (e.g. Lowry 1977; Tabony 1980). However, more detailed analyses using large satellite instrument-derived data sets indicate that urban heat island-rainfall effects are real and satellite rainfall estimates from TRMM (the Tropical Rainfall Measuring Mission satellite's precipitation radar) can detect them (Shepherd *et al.* 2002; Chapter 12 this volume).

Urban areas affect precipitation by increases in hygroscopic nuclei, in turbulence via the increased surface roughness, in convection because of increased surface temperatures, and through the addition of water vapour by combustion sources. Precipitation tends to increase on the downwind side of cities or large industrial complexes, but for a long time the effect of cities on rainfall was difficult to determine as few rural stations remained unaffected to some degree by human activity. Today, radar tracking of rainfall makes it possible to assess rainfall quantities and locations over large areas and there is much greater certainty about the urban influences.

Urban influences change as the size and nature of the urban areas is altered. Loss of the greenspace that modifies urban temperatures influences the heat island effects on precipitation. Tall structures, when sufficiently high and sufficiently massed together, alter urban wind flows. In urban areas buildings may cause surface winds to decrease under significant synoptic flow and increase under weak synoptic flow. Under nearly calm regional flow, a relative low pressure may be created over the city by the anomalous high temperatures of the Urban Heat Island (UHI) (see Chapter 11 this volume), and cooler air rushes into the urban area, which causes warm air to rise. This vertical motion can create convective thunderstorms that produce precipitation maxima over the city, and it is most pronounced at night when the UHI is strongest (Bornstein and Lin 2000; Dixon and Mote 2003). Both the temperature and wind effects have to be added to broader regional changes that may be a consequence of climate change. Thus in urban areas both local and regional, globally driven, change factors combine to make predicting future hydrologic regimes difficult.

Increased magnitude and higher intensity storm rainfalls will lead to larger volumes of stormwater runoff being carried into urban drainage systems, whether flood channels, sewers or grassed waterways. Unless the design of city drainage systems takes account of possible storm rainfall increases, the drains may not be able to cope with the flow from the worst storms (Douglas 1983). Failed stormwater drains are not infrequently found in poorly designed suburban housing estates (Quilty 1977).

Climate change is likely to increase storminess generally. Depending on the regional climate, the severity of urban storm events may be greatest either in winter or in summer. The impacts of this will be coupled to the increased imperviousness and extent of built-up areas, producing higher peak urban stormwater discharges. Mitigation measures and steps to adapt to climate change will be required, involving attention to the green infrastructure, green roofs and sustainable urban drainage systems (SUDS) (see Chapters 43 and 44 this volume). Such work provides an opportunity to gain multiple benefits from improved urban drainage networks as outlined in the excellent book by Ann Riley (1998). Improving the hydrological system can go hand-in-hand with creating new wildlife habitats and making new places for urban recreation. This suite of techniques for managing water in cities is a major contribution to the functional green infrastructure.

SUDS involve a set of management practices and structures designed to drain water in a more sustainable manner than the conventional piped subsurface conveyance systems (Butler and Davis 2004). Aiming to mimic natural drainage systems as far as possible, SUDS also provide wetland habitats that are good for biodiversity and are attractive features of the urban environment. SUDS may comprise combinations of green roofs, areas of grass that convey water (swales), wooded areas, treatment basins, including ponds and wetlands, that absorb runoff. Other components of the water detention and conveyance system may be infiltration trenches, soakaways, permeable surfaces and rainwater storage reservoirs beneath pavements, car parks and roads. SUDS can help the development of integrated urban drainage management which aims to manage the various drainage conveyance methods, from river to sewer, so as to reduce pollution and flooding (see also Chapters 43 and 44).

Despite the localized thunderstorm increase in, or immediately adjacent to, urban areas, urban pollution can depress precipitation in areas downwind, for example in California and Israel (Givati and Rosenfeld 2003). The suppression of precipitation occurs mainly in relatively shallow clouds within the cold air masses of cyclones, which ingest the pollution from the boundary layer while rising over the mountains. The decreasing trends occur at the western slopes of the hills located downwind of pollution sources, for example in the mountains to the east of San Diego, California, during the twentieth century, most of it after 1940 when San Diego started to grow, and more recently with the explosive growth of Tijuana just across the Mexican border. The suppression that occurs over the upslope side is coupled with a similar amount of enhancement on the much drier downslope side of the hills, probably because more cloud water passes over the divide. Many major hydrological recharge zones for groundwater resources coincide with the areas for which significant suppression of precipitation has been established. This has major implications for groundwater resources for the cities that cause the effects. The additional precipitation over the leeward side of the mountains is much less than that lost on the windward side, further threatening the water resources for cities in regions already experiencing severe water shortages (Givati and Rosenfeld 2003). These examples indicate the need to examine urban effects on precipitation in a regional context and to recognize how rainfall regimes and seasonal weather contrasts affect the overall nature of the urban hydrological cycle in any one place.

Water released by combustion may be significant in large urban areas. In the Chicago region, in winter when combustion for heating, industrial processes and motor vehicles is high and evapotranspiration is minimal, urban sources of water vapour become relatively more important (Ackerman 1987). While the amount of such emitted water vapour may be less than 1 per cent of that entering the region from other sources it does contribute to the precipitable water over the city.

Evaporation from cities

Evapotranspiration is a significant part of the urban hydrological cycle, varying with the tree and plant cover and the amount of irrigation of that vegetation. Trees can cover large parts of suburban areas (see Chapter 22 this volume) and in many cities there are more trees than in the surrounding rural areas. A single large tree can transpire 450 litres of water per day, consuming 1000 MJ of heat energy to drive the evaporation process, thereby lowering the city's temperature (Bolund and Hunhammar 1999). Many urban gardens and public parks and lawns are irrigated, such that in some cities evapotranspiration exceeds precipitation and is sustained by water from the piped urban supply (Grimmond and Oke 1999). An example of this is the driest year for the Curtin Catchment in Canberra, Australia shown in Table 13.2.

Across a city, evapotranspiration is highly variable, depending not only on the types and spatial extent of vegetation but how that plant cover is distributed in relation to paved surfaces that heat up rapidly (see Chapter 11 this volume).

Evaporation from paved surfaces can be a high proportion of the rainfall in small showers that wet areas like roads and parking lots with water being held in puddles in depressions and irregularities in the surface. This water evaporates rapidly as the sun heats the surface. The steam rising from such surfaces after rain is a characteristic feature of tropical and subtropical city life. This rapid post-storm evaporation helps to explain why a 95 per cent paved area in central London yielded only 50 per cent of the rainfall as runoff (Oke 1974).

As cities keep their air warm by nocturnal heat loss from buildings and paved surfaces, the process of transpiration from plants may continue during the night, often depleting the moisture content of the soil. As heat loss among streets and squares is inhibited by reflection, condensation of water vapour, dew formation, occurs rarely, so there is virtually no break in the demands made on the water regime of plants. Despite the strain on their water regime, deep-rooted trees and shrubs continue to transpire, producing a cooling effect on their urban surroundings through their use of heat energy in the conversion of water to vapour (Miess 1979).

Infiltration

Paving and roofing an area reduces the opportunities for water to infiltrate. The fundamental hydrologic effect of urban development is the loss of water storage in the soil column. Compaction or stripping of soil during development and the conversion of what was once the subsurface flow within the soil to overland flow on impermeable surfaces combine to cause this loss. The net result is that in a small catchment, runoff from a storm reaches stream channels within minutes, instead of taking hours or days to do so by subsurface flow. Thus stream flows change rapidly, with peaks twice as high as before development and discharges from smaller rain events, ten times that the same rainfall would have produced under natural conditions (Booth *et al.* 2002).

Lack of infiltrating water to replenish soil moisture may lead to a lowering of the water table and a reduction of groundwater levels beneath the city. However, the diversity of the green infrastructure matrix in the suburban mosaic (see Chapter 21 this volume) with its corridors and patches of vegetation has a major impact on infiltration. Classically, the impermeable area of a city has been defined as that which is roofed or paved. It is frequently mapped by interpreting remotely sensed imagery and the data so derived are incorporated into a geographical information system (GIS). Although nearly all the precipitation falling on to a sealed surface, less that lost by evaporation, runs off, many sealed areas are not connected to a storm water pipe drainage network. The more sealed areas are surrounded by gardens and greenspaces, the more the rainwater from sealed areas runs laterally into depressions, hollows or drains where it can infiltrate into the ground (Douglas *et al.* 2007). Thus when examining the formation of flood flows from storm runoff it is important to distinguish the proportion of the sealed area that is connected to drainage pipes. This is the distinction between the *total impervious area* (TIA) and the *effective impervious area* (EIA) (Schueler 1995).

TIA, the constructed, non-infiltrating surface, does not include nominally 'pervious' surfaces that are so compacted that they shed water as if they were concrete. On the other hand it includes those sealed surfaces that shed water into pervious areas such as lawns and the places where rain in the down pipes from roofs falls onto splash blocks that disperse runoff into gardens at each corner of the building. Clearly, if infiltration and stormwater runoff are to be calculated and modelled accurately, the EIA has to be used. However, its direct measurement is complicated, requiring measurements of both TIA and EIA which can be generalized either as a correlation

Table 13.3 Relationship between imperviousness and land use in a western Washington suburb (after Dinicola, 1989)

Land use	*TIA %*	*EIA%*
Low density residential (1 unit per 1–1.5 ha)	10	4
Medium density residential (2.5 units per ha)	20	10
'Suburban' density (10 units per ha)	35	24
High density (multi-family or 20+ units per ha)	60	48
Commercial and industrial	90	86

between the two parameters or as a 'typical' value for a given land use (Table 13.3) (Douglas *et al.* 2007).

Surface detention of water

Most cities use detention ponds to slow down and hold back storm runoff to reduce flooding. The ponds can cover as much as 4 per cent of a city's area, as those with surfaces of 0.5 to 20 ha do in the Ile-de-France, the greater Paris metropolitan area. These detention ponds are linked to green corridors of high ecological value that also serve as recreational areas (Brilly 2007). The goals of stormwater detention have become progressively more ambitious as the consequences of urban-altered flow regime have become better recognized and understood (Booth *et al.* 2002). Drainage authorities and local governments have prescribed how developers must design and install detention ponds, but even the largest detention ponds are likely to be overtopped by the rarest storms. There are also questions as to whether detention ponds reduce the impacts of urban storm runoff on aquatic life. In Delaware, USA, eight stormwater management pond facilities (BMPs) failed to prevent the almost complete loss of sensitive taxa (e.g. mayflies, stoneflies and caddisflies) after urban construction. They did not reduce the impacts of development once the catchment was more than 20 per cent impervious (Maxted and Shaver 1999).

Ponds mitigate, to some degree, the hydrologic changes resulting from urban development, but changes to simulate pre-development hydrology (through conservation design) are necessary to protect aquatic resources. There are also problems of sediment and loss of capacity through accumulation of debris. However some multi-purpose flood storage schemes are highly efficient in providing ecological and social benefits as well as coping with all but the rarest storms (those likely to occur on average less than once in 100 or 200 years).

Groundwater in the urban hydrological cycle

The redirecting of rainfall in the built environment greatly affects the recharge of groundwater aquifers, mainly by evacuating most stormwater before it can infiltrate. However, urbanization also creates some new, unintended pathways for recharge. These include leaking or overflowing guttering around roofs, leaking downpipes, water mains, sewers, septic tanks and soakaways (Bolund and Hunhammar 1999). The loss of recharge affects the discharge of water from the groundwater body to stream channels during dry periods. This groundwater outflow, or baseflow, decreases with urbanization, while direct runoff, stormwater discharge, increases. These two trends combine to make streams much more 'flashy' in character, with higher peak flows and lower dry weather flows. This effect has been well documented for the southern part of Long

Island, NY, USA, which has been affected by the steady expansion of the New York urban areas for nearly 150 years.

Under natural conditions rain that fell on Long Island would infiltrate and eventually reach the stream as groundwater. While many stormwater drains now carry urban runoff directly into streams, some on Long Island flow into specially constructed collection basins, known as recharge basins, designed to allow rapid infiltration of the runoff carried into them. Where these recharge basins exist, groundwater recharge from precipitation is about equal to recharge under natural conditions. Where the stormwater is delivered directly to streams, the baseflow is greatly reduced and the storm flow increased.

Groundwater changes in urban areas are not due to changes in infiltration alone. Abstraction by pumping lowers the water table eventually leading to the subsidence that has happened in cities ranging from Bangkok to Venice, Tokyo and Houston to Winnipeg, with many geomorphological consequences (see Chapter 14 this volume). The first documented water well was completed in 1864, when Chicago's first well for drinking water was sunk. Since then, ground water has been the sole source of drinking water for about 8.2 million people in the Great Lakes watershed. This long-term pumping has lowered groundwater levels in the sandstone aquifer underlying the Chicago metropolitan area by as much as 200 metres (Figure 13.1). Reduction of groundwater withdrawals in much of the area because of worries over aquifer depletion have produced recovery of water levels in some areas, but declines continue in others (Grannemann *et al.* 2000).

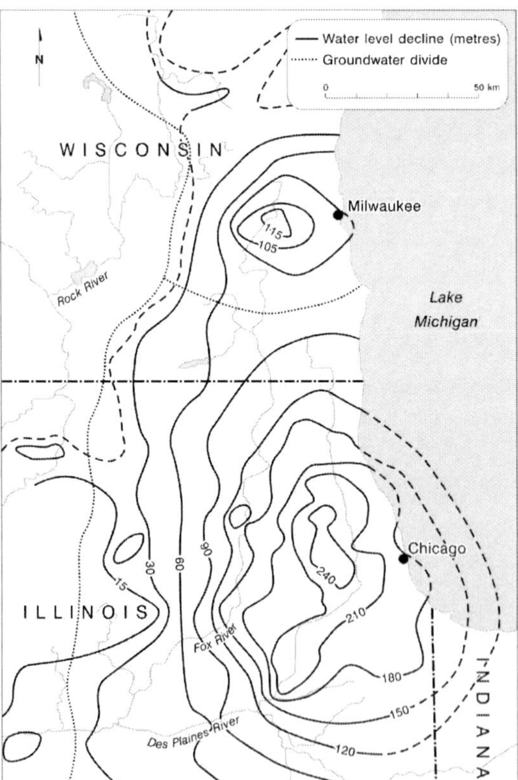

Figure 13.1 Decline in groundwater levels in the sandstone aquifer, Chicago and Milwaukee areas, 1864–1980 (Based on Alley *et al.* 1999).

However, there can be complex problems of access to water supplies. As water tables fall, older shallow wells and boreholes no longer reach the groundwater. Competition for water leads to deeper and deeper wells and inevitably the water goes to those who can afford the deepest. Sometimes this leads to international rivalry, such as the competition for the water of the Hueco Bolson aquifers beneath El Paso and Ciudad Juarez on the Rio Grande at the USA–Mexico border where huge population growth has increased the demand for water as well as increased industrial effluent into the river, straining the integrity of the aquatic system. The Rio Grande cannot be used as a recreational resource and does not support much wildlife due to its poor quality in the El Paso–Ciudad Juarez reaches (Rios-Arana *et al.* 2003).

The El Paso water company (EPWU), which has relied on the Hueco Bolson for municipal water supply, has had concerns over the ability of the Hueco to meet the demands of a growing city since the 1920s. The EPWU faces declining groundwater levels, with the water table dropping over 50 m and brackish groundwater intrusion into wells that had historically pumped fresh groundwater. Surface subsidence of 0.3 m has occurred since the 1950s (Heywood 1995). Since 1990, EPWU groundwater pumping in the Hueco has been reduced. The reduction in pumping has stabilized groundwater levels in many areas of El Paso, greatly reducing the annual groundwater storage decline. However, the one natural system is being used by two independent urban systems, one on each side of the border. Treaties for sharing the water of the Rio Grande between the two countries and arrangements for joint management were concluded in 1906 and 1944. However, differences in law and levels of development between Mexico and the United States make it difficult to develop basin-wide management strategies (Schmand 2000).

Up to 1936 groundwater fed the Rio Grande in dry periods. After then more and more water seeped through bed of the river into aquifers, until in 1968 the river channel was lined with concrete and the seepage reduced. However this led to a sharp fall in the water table in the alluvium, mainly because of the reduced inflow from the river, but also because of increased pumping (Meyer 1976). The twin cities face a complex set of issues of ground and surface water quantity and quality. The river water quality is deteriorating and aquifers are becoming contaminated with agricultural and urban runoff. Future urban supplies may depend on diverting water from agricultural to urban uses, using poor quality groundwater for some urban purposes or reusing waste water for the cities through treatment with modern microfiltration.

Urbanization greatly affects ground water quality. On Long Island, volatile organic compounds, associated particularly with hydrocarbons such as petrol (gasoline), were detected in 33 out of 60 groundwater wells in suburban areas, but not in wells from agricultural or undeveloped areas. On the other hand, carbamate insecticide residues in wells were directly related to the amount of agricultural land around the wells, reflecting that such compounds were only being used on farms. However, organochlorine insecticide residues, mainly dieldrin, chlordane and heptachlor epoxide were found in wells in both residential and agricultural areas. These insecticides have long been used by both urban gardeners and rural farmers (Eckhardt and Stackelberg 1995). Such findings indicate both the vulnerability of shallow aquifers to contamination and the importance of considering the implications of all human activities, both rural and urban, on sources of urban water supplies.

Changes to rivers

In nature river channels are dynamic, adjusting to changing flows and sediment loads. In cities, river channels are often constrained by engineering works, such as weirs, embankments, bridge spaces and sewer outlets. People have always tried to tame rivers and to prevent them from flooding valuable property. When urban populations were small, there were few major urban encroachments on to floodplains, save for river ports. However, since the 1900s, large scale urban development has often taken place on floodplains in Europe and North America. In some countries, this encroachment seems to have become more severe since 1980. In other continents, extreme poverty and lack of access to land leads many urban dwellers to build shelters on floodplains or to cultivate them for food.

Floodplains are essentially seasonally inundated wetlands, which may be suitable for playing fields, nature reserves, woodland or golf courses 90 per cent or more of the year, but which are likely to be flooded on average once a year. Within the flood plain there may be abandoned river meanders or old branches of a braided channel each of which provides a slightly different degree of habitat wetness from the other. The channel of the river will show signs of change with slumping banks on the outer bends of meanders and deposition of sandy or gravely point bars on the inside of the curve. These deposits shift during floods as banks collapse and add new sediment to the flow.

Urban development alters the rate and magnitude of these processes, especially downstream of works that constrain rivers. Many small urban streams pass under streets through circular culverts. Immediately downstream of the culvert outlet the stream scours its banks as the energy of the high speed flow through the culvert is expended on eroding the banks. Similarly, downstream of embanked urban reaches, rivers with sufficient gradient erode their banks. When additional sediment loads are added to urban streams, they adjust their channels to cope, even changing from a meandering to a braided state. Land use changes in urban areas all tend to aggravate these channel dynamics, whether they are simply paving over a front garden to provide more parking or replacing a 100-year old large family house and garden with a block of apartments and car parking for 20 households. These works release sediment to the drainage system while they are in progress and alter stormwater flows for the future.

All this has double ecological implications, for the site of the works and for ecosystems downstream. Increasingly river managers and engineers have realized that it is better to design and work with nature rather than to fight against it. Davenport *et al.* (2004) have developed an urban river survey (URS) from the River Habitat Survey (RHS) that is applied routinely to UK rivers. The URS recognizes that most urban rivers have to have some engineering works and that management decisions have to take account of 'previous channel improvements'. The URS looks at factors particularly significant in urban rivers, such as indicators of pollution, records habitat features in a more detailed manner than the RHS on rural streams and separates information on engineered reaches from that on more natural sectors of streams (Douglas *et al.* 2007). Taking such information into account enables more sustainable design and management of rivers for multipurpose river and floodplain use to occur.

Urban streams have often been buried, turned into sewers or completely culverted. Today some of the past changes to small streams are almost forgotten and their importance only emerges when combined sewers overflow during storms because they are in fact ancient water courses. These buried streams and sewers of course have an urban ecology of their own. Not only are they a critical part of the urban hydrological cycle, but they remain essential for human health and well-being.

References

Ackerman, B. (1987) 'Climatology of Chicago area: urban-rural differences in humidity', *Journal of Climate and Applied Meteorology*, 26(3): 427–30.

Alley, W.M., Reilly, T.E. and Franke, O.L. (1999) 'Sustainability of ground-water resources', *U.S. Geological Survey Circular* 1186, Denver, CO, pp. 1–7.

Atkinson, B.W. (1968) 'A preliminary examination of the possible effect of London's urban area on the distribution of hunder rainfall 1951–60', *Transactions of the Institute of British Geographers*, 44(May 1968): 97–118.

Atkinson, B.W. (1977) 'Urban effects on precipitation: an investigation of London's influence on the severe storm in August 1975', *Queen Mary College, Department of Geography Occasional Papers*, 8.

Bolund, P. and Hunhammar, S. (1999) 'Ecosystem services in urban areas', *Ecological Economics*, 29(2): 293–301.

Booth, D., Hartley, D. and Jackson, R. (2002) 'Forest cover, impervious-surface area, and the mitigation of stormwater impacts', *Journal of the American Water Resources Association*, 38(3): 835–45.

Bornstein, R. and Lin, Q. (2000) 'Urban heat islands and summertime convective thunderstorms in Atlanta: Three case studies', *Atmospheric Environment*, 34(3): 507–16.

Brilly, M. (2007) 'Local flood defense systems in Europe', in R. Ashley, S. Garvin, E. Pasche, A. Vassilopoulos and C. Zevenbergen (eds), *Advances in Urban Flood Management*, London: Taylor & Francis, pp. 321–38.

Butler, D. and Davis, J.W. (2004) *Urban Drainage*, second edition, London: E. & F.N. Spon.

Chandler, T.J. (1976) 'Urban climatology and its relevance to urban design', *World Meteorological Organization Technical Note 149*, Geneva.

Davenport, A.J., Gurnell, A.M. and Armitage, P.D. (2004) 'Habitat survey and classification of urban rivers', *River Research and Applications*, 20(6): 687–704.

Dinicola, R.S. (1989) 'Characterization and simulation of rainfall-runoff relations for headwater basins in western King and Snohomish Counties Washington state', *US Geological Survey, Water Resources Investigation Report* 89–4052.

Dixon, P.G. and Mote, T.L. (2003) 'Patterns and causes of Atlanta's urban heat island–initiated precipitation', *Journal of Applied Meteorology*, 42(9): 1273–84.

Douglas, I. (1983) *The Urban Environment*, London: Arnold.

Douglas, I., Kobold, M., Lawson, N., Pasche, E. and White, I. (2007) 'Characterisation of urban streams and urban flooding', in R. Ashley, S. Garvin, E. Pasche, A. Vassilopoulos and C. Zevenbergen(eds) *Advances in Urban Flood Management*, London: Taylor & Francis, pp. 29–58.

Eckhardt, D.A.V. and Stackelberg, P.E. (1995) 'Relation of ground-water quality to land use on Loond Island, New York', *Ground Water*, 33(6): 1019–33.

Gates, D.M. (1972) *Man and his environment: climate*, New York: Harper & Row.

Gijsbers, P.J.A. and Loucks, D.P. (1999) 'Libya's choices: desalination or the Great Man-Made River Project', *Physics and Chemistry of the Earth, Part B: Hydrology, Oceans and Atmosphere*, 24(4): 385–89.

Givati, A. and Rosenfeld, D. (2003) 'Quantifying Precipitation Suppression Due to Air Pollution', *Journal of Applied Meterology and Climatology*, 43(7): 1038–48.

Grannemann, N.G., Hunt, R.J., Nicholas, J.R., Reilly, T.E. and Winter, T.C. (2000) 'The importance of ground water in the Great Lakes region', *US Geological Survey Water-Resources Investigations Report 00–4008*, Lansing, MI.

Grimmond, C.S.B. and Oke, T.R. (1999) 'Heat storage in urban areas: local-scale observations and evaluation of a simple model', *Journal of Applied Meteorology*, 38(7): 922–40.

Heywood, C. (1995) 'Investigation of aquifer-system compaction in the Hueco basin, El Paso, Texas, USA', *International Association of Hydrological Sciences*, Publication 234: 35–45.

Leopold, L.B. (1968) 'Hydrology for urban planning – A guidebook on the hydrological effects or urban land use', *US Geological Survey Circular 554*, Washington, DC.

Lowry, W.P. (1977) 'Empirical estimation of urban effects on climate: a problem analysis', *Journal of Applied Meterology and Climatology*, 16(2): 129–35.

Maxted, J.R. and Shaver, E. (1999) 'The use of retention basins to mitigate stormwater impacts to aquatic life', *National Conference on Retrofit Opportunities for Water Resource Protection in Urban Environments, Proceedings Chicago, IL February 9–12, 1998*. Washington, DC: United States Environmental Protection Agency, EPA/625/R-99/002: 6–15.

Meyer, W.R. (1976) 'Digital model for simulated effects of ground-water pumping in the Hueco Bolson, El Paso Area, Texas, New Mexico, and Mexico', *US Geological Survey, Water Resources Investigations Report 58–75*, Austin, TX.

Miess, M. (1979) 'The climate of cities', in I.C. Laurie (ed.), *Nature in Cities,* Chichester: Wiley, pp. 91–114.

Mitchell, V.G., McMahon, T.A. and Mein, R.G. (2003) 'Components of the total water balance of an urban catchment', *Environmental Management*, 32(6): 735–46.

Oke, T.R. (1974) 'Review of urban climatology 1968–73', *World Meteorological Organization Technical Note 134*, Geneva.

Quilty, J.A. (1977) 'Erosion hits the suburbs', *Journal of the Soil Conservation of New South Wales*, 33(3): 156–64.

Riley, A.L. (1998) *Restoring Streams in Cities: A Guide for Planners, Policymakers, and Citizens*, Washington, DC: Island Press.

Rios-Arana, J.V., Walsh E.J. and Gardea-Torresdey, J.L. (2003) 'Assessment of arsenic and heavy metal concentrations in water and sediments of the Rio Grande at El Paso–Juarez metroplex region', *Environment International*, 29(7): 957–71.

Schmand, J. (2000) 'Bi-national water issues in the Rio Grande/Rio Bravo Basin', *Proceedings of the Water Environment Federation, WEFTEC 2000: Session 1 through Session 10*, pp. 756–75.

Schueler, T. (1995) 'The importance of imperviousness', *Watershed Protection Techniques*, 1(3): 100–11.

Shepherd, J.M., Pierce, H. and Ngeri, A.J. (2002) 'Rainfall modification by major urban areas: observations from spaceborne rain radar on the TRMM satellite', *Journal of Applied Meterology and Climatology*, 41(7): 689–701.

Tabony, R.C. (1980) 'Urban effects on trends of annual and seasonal rainfall in the London area', *Meteorological Magazine*, 109: 189–202.

14

Urban geomorphology

Ian Douglas

The form of the land and ocean floor affects the geologic processes, erosion and sedimentation operating upon them, and affect the ecosystems that they support. The role of human beings as agents of geologic and geomorphic change is nowhere more marked than it is in urban areas. People constantly alter the landforms in urban areas, digging foundations, filling quarries with waste, levelling playing fields, building barrier mounds along highways and constructing flood defences, harbours and wharves. The vast quantities of materials brought into cities gradually raise the level of the ground, to the extent that many old buildings now have their entrances a metre or more below the modern street level. If a development contractor drills a borehole to investigate foundation conditions, the layers of material found usually show a variety of 'made ground', rubble, fill material, and remains of human food and other consumption, overlying any natural soil or rock. Even urban gardens and parks often have such an anthropogenic stratigraphy beneath them. The processes and forms involved in urban activity as an earth surface process are the subject of urban geomorphology.

Urban geomorphology examines the geomorphic constraints on urban development and the suitability of different landforms for specific urban uses; the impacts of urban activities on earth surface processes, especially during construction; the landforms created by urbanization, including land reclamation and waste disposal; and the geomorphic consequences of the extractive industries in and around urban areas. The diversity of urban substrates is a consequence of their geomorphic history, the ways in which past environmental changes, including climatic and sea level changes have affected the form of land and the types of surface materials.

Constraints on urban development

The original founders of towns and cities carefully chose their sites for defensive, strategic, resource exploitation, navigation or cultural reasons. Great attention was given to finding sites with adequate water supplies and protection from obvious environmental hazards. However, the growth of these settlements often led to the spread of urban development on to less suitable ground and overstretched the capacity of the local environment to support the community. Many environments have particular conditions that make conditions for modifying slopes or establishing foundations difficult (Table 14.1).

Table 14.1 Geomorphological problems for urban development (based on data in Marker 1996; and Bennett and Doyle 1997)

Environment	*Chief problems*
A Climatic	
Periglacial	Permanently frozen ground and overlying active layer require special types of construction and foundations for buildings and infrastructure
Arid	Water supply problems; wind erosion; flash floods; possibility of salt weathering of building materials and foundations
Humid tropical	Rapid weathering and decomposition of building materials; deep, uneven weathering of most rocks in tectonically stable areas; frequent rain events causing rapid water erosion of exposed ground surfaces
B Topographic	
Mountainous	Risk of unstable slopes, rockfalls, debris flows and avalanches; potential for flash floods
Flood plains	Liable to periodic flooding; variable foundation conditions over former, buried river channels and alluvial deposits
Coastal plains	Storm surge and flooding risk likely to increase with rising sea levels; complex ground conditions reflecting former shorelines and old drainage channels; possible salt penetration in groundwater affecting foundations
Coasts with weak rock cliffs	Liable to rapid coastal erosion, cliff undercutting and collapse; eroded debris often deposited in ports and harbours causing dredging expenditure
Islands	Particular storm-surge, rising sea-level and salt water penetration risks on low-lying atolls and coastal plains
C Tectonic/lithological	
Active plate margins	Major risks associated with coastal urban developments, especially on Pacific rim, special foundation requirements on filled areas, lake sediments and other unconsolidated materials; major earthquake triggered landslide hazards; volcanic debris and lahar risks requiring awareness of flow pathways on lower volcanic slopes likely to have urban settlements
Shrink-swell clays	Cracking clay problems likely to be accentuated by climate change
Karst	Buried karst a major problem for foundations of tall buildings and for sinkhole development; need for knowledge of buried karst plains and effects of lowered Quaternary sea levels

Application of geomorphological mapping to the classification of the suitability of land for different types of urban development is now part of the work of geological and soil surveys in many countries. Such mapping considers the steepness of slopes, their colluvial and weathered mantles, their drainage and depth to bed rock and provides guidance on the type of development suited to different parts of the slope.

Knowledge of landform evolution is particularly important, as modern earthmoving can reactivate features inherited from past conditions, such as fossil periglacial landslides in Europe and North America. Loading of peat with urban structures formed after the retreat of ice sheets can result in significant subsidence and building damage. Karstic features formed when sea levels were lower in the Quaternary, but now buried under alluvium can pose severe problems for the foundations of high-rise buildings.

Many present day conditions constrain urban development. Widespread soils rich in montmorillonitic clays are subject to 'shrink-swell' cracking clay phenomena which require special foundations if buildings are not to become unstable. Climate change is likely to shift the areas where these problems are severe, if summers become drier. Mobile dune systems and sources of wind-blown sand pose problems for the siting of many structures. Alluvial fans may normally be inactive with the local stream confined to a narrow channel, but they may be reactivated, flooded and covered with debris if an extreme flood descends from the adjacent mountains. Building in permafrost areas has to isolate the heated structures from the frozen ground and be careful not to disturb the permafrost during the construction process.

Geomorphic impacts during urban construction

Urban construction involves removal of the natural vegetation cover and excavation of the topsoil and often much of the underlying weathered rock and bedrock layers. In new urban developments, small streams are often diverted into culverts or urban drains and minor depressions and valleys are filled in. Steep hillsides may be terraced into a series of home sites by cut and fill operations. Major rivers may be embanked and artificially straightened. In extreme cases, as in Palma de Mallorca, Spain, and Winnipeg, Canada, large new flood channels may be built around the urban centre to divert flows away from the city. The new features, replacing the original landforms are often designed to direct water away from the new developments more effectively, so producing off-site, downstream consequences.

The earthmoving operations during urban construction frequently lead to severe erosion problems and consequent channel modifications (Table 14.2). Erosion control guidelines suggest that construction should be carried out in phases to avoid disturbing too much of the land at any one time. No unnecessary clearing should be undertaken. Immediately below any cleared area, detention ponds should be constructed to retain any sediment washed off the site and to hold back stormwater runoff so that peak discharges in streams below are not increased.

Increased sediment loads and peak storm discharges lead to channel modifications (Table 14.2) with formerly meandering streams becoming braided, steeper and shallower. Sometimes

Table 14.2 Sequence of fluvial geomorphic response to land use change: Sungai Anak Ayer Batu, Kuala Lumpur

Land cover/land use	*Channel condition*
Forest	Narrow, meandering with low sediment load
Rubber plantation	Gullying during clear weeding; peak discharge increased; channel slightly widened; later stabilized; few cut-offs
Urban Construction	High sediment yield; high peak discharge; metamorphosis to wider, steeper, shallower braided channel
Channelization and stable urban built-up area	Higher peak discharge; less sediment load; channel enlargement downstream; bank erosion, minor channel incision; loss of fine bed material by scour
Post channelization siltation	Where large quantities or organic debris enter concrete channels and are deposited, vegetation can become established and build up deposits that reduce channel capacity

channel modification, often with expensive structural works, may help to control the extent of the changes. However, even these are not always successful as siltation of the channel can follow, with large accumulations of weeds and silt building up if the stream receives discharges with high nutrient loadings. Further downstream, rivers may adjust in response to upstream channelization, eroding their banks, developing new gravel bars and threatening bridge abutments and riverine structures.

Landforms of extraction

Meeting the demand for construction materials changes the land surface, by creating pits and quarries. The largest excavations take up many square kilometres of the land surface, often creating areas of brick pits that sometimes are used for waste disposal, or gravel pits that become peri-urban wetlands, often serving combined recreational and flood control purposes. Not all the filling of former opencast mines is without incident. In the past, escaping methane gas from landfills in old opencast coal pits has caused problems for the houses built upon them. In karstic terrain, the subsidence of filled material in chalk pits or in old tin mines overlying cavernous limestone has led to severe damage to houses and urban infrastructure.

Removal of mineral resources and water from beneath the ground leads to subsidence creating new surface topographies and, often, new water bodies. Groundwater pumping has put the historic world heritage buildings of Venice at risk. Built at sea level on the lagoon, Venice has subsided some 22 cm since 1900. Most of that surface lowering occurred between 1950 and 1970. High water ('aqua alta') has occurred more frequently since 1970. Whilst the people-induced subsidence is part of the problem, higher extreme sea levels related to global climate change may possibly be another factor. In the Los Angeles area, extraction of oil beneath Long Beach created severe subsidence that had to be halted by the injection of water into abandoned wells.

Landforms of deposition

Much modern urban development involves land reclamation and major landform modification. In extreme cases, huge quantities of material are moved, for example in the development of major airport sites such as Kansai, Singapore and Hong Kong. At Kansai, the fill material has caused some subsidence of the original seabed, with allowance having to be made for this in the operation and maintenance of the airport. The problems of subsidence of the second stage runway are expected to be more severe than in the first runway, with a prediction that after 50 years subsidence will have been 18 m compared to 11 m for the first stage. Detailed analyses have been made of the way landing aircraft cause small temporary depressions in the runway that in turn affect the drag on aircraft moving along the runway.

As disposal of solid waste moves from landfill to land raise, new hills appear on the edges of floodplains, above former gravel pits and quarries and on offshore islands.

In some urban areas, waste dumps are prominent features of the landscape. Although the older dumps are the result of coal, slate and china clay production, modern land raise features dominate many low relief areas. Whilst much of this waste is deposited in disused open pit mines and quarries, land raise mounds are probably the fastest growing artificial landforms in many countries today. The greatest geomorphological impact of landfill is in river valleys, sections of which are being filled, raising the height of the ground surface well above the former floodplain level. This effectively reduces the flood storage capacity of the floodplain, shifting the flood problems downstream.

Many of the old dumps are being closed or modified, from the huge dumps on the edges of cities like Istanbul and Manila to the managed disposal areas, such as Freshkills on Staten Island, New York, which has been taking nearly all the 17,000 tons of waste the city collects each day. As events at the Payatas tip in Manila have shown, some of these urban wastes mounds are unstable, prone to massive slumps and landslides. The loss of life and property that ensues is a challenge to the management of the waste disposal and the application of geomorphology to the construction of land raise mounds.

Urban regeneration itself involves creating new landforms as the old buildings are demolished and construction and demolition waste is used for on-site fill or is taken short distances to sites that have to be raised to be above known flood levels. Many historic city centres have been so rebuilt that the average level of streets is now above that of the entrances to mediaeval buildings. These changes in landform may often be individually small, but collectively they are the result of two of the main human drivers of global environmental change: increasing urban development and the mining and quarrying which supplies minerals for industry and the construction materials required to build all the infrastructure, homes, offices and factories of cities. Urban geomorphology is thus a key element in supplying the guidance needed to achieve a better quality of urban life and working towards more sustainable use of resources.

References

Bennett, M.R. and Doyle, P. (1997) *Environmental Geology: Geology and the Human Environment*, Chichester: Wiley.

Marker, B.R. (1996) 'The role of the earth sciences in addressing urban resources and constraints', in G.J.H. McCall, E.F.J. de Mulder and B.R. Marker (eds), *Urban Geoscience*, Rotterdam: Balkema, pp. 163–79.

15
Urban soils

Peter J. Marcotullio

Introduction

While urban ecological studies began during the fifteenth century, with the identification of species diversity in cities such as London and Paris, systematic studies of urban soils are much more recent, largely undertaken since 1945 (Sukopp 2008). For a long time, substrates in cities were regarded by pedologists as heterogeneous and too young to develop into soil systems and therefore were not subjects of scientific interest, Spirn (1984) regarding them as one of the most misunderstood and least studied aspects of the urban environment.

Some of the first systematic studies of soils in large cities attempted to identify the unique conditions of urban soils (Craul 1985; Craul and Klein 1980; Patterson 1976). Many focused on trace element concentrations and fluxes in urban street dust and urban soils (Purves 1966, 1972; Purves and Mackenzie 1969). Emphasis was often given to the engineering aspect of soils and the accompanying challenges for urban development (Gray 1972).

Urban soil studies multiplied with the growing interest in urban ecology. Gilbert (1989) devoted an entire chapter to soils in his comprehensive study of the impact of cities on various environmental resources. By the 1990s, books devoted to urban soils appeared (see for example, Brown *et al.* 2000; Bullock and Gregory 1991; Craul 1992). Recommendations for mapping soils in cities were published and by 1998 an International Working Group on "Soils of urban, industrial, traffic and mining areas" was founded (Sukopp 2008).

Since 2000 ecosystem studies have paid more attention to ecosystem services and the relationship between these services and human well-being (Millennium Ecosystem Assessment 2005; Daily 1997). Interest in urban soils has increased with the notion that these systems provide a number of different and important ecological services, even in cities. Palm *et al.* (2007) argue that soils underpin the production of food, feed, fiber and fuels and play a central role in determining the quality of our environment (including provisioning, regulatory, cultural, and supporting ecosystem services). In urban landscapes, soil services include regulating the retention and supply of nutrients and adsorption and storage of water, and provisioning growth media and substrate for soil fauna and flora. Urban soils are the foundation for ecological processes such as biogeochemical cycling, distribution of plant communities and provide purification services by intercepting contaminants such as pesticides and other toxics generated through human activities (Effland and Pouyat 1997).

This chapter summarizes selected key literature on urban soils. The first section describes the changing perspectives taken by soil scientists in the study of soils in urban areas. This is followed by sections that briefly identify and describe the general character of urban soils including selected physical and chemical characteristics of urban soils, pollutants in urban soils and urban soil biota. There then follows a review of studies on urban soil organic carbon densities and fluxes, while the final section identifies some new areas for research.

Changing perspectives on urban soils

Perspectives on the soils and of urban soils has changed dramatically over the past 50 years (Fenton and Collins 2000; Palm *et al.* 2007; Effland and Pouyat 1997). In the urban setting, researchers originally focused on descriptions of urban substrate. These studies focused on gaining knowledge of the geologic characteristics and engineering parameters of the material. Concern over the loss of agricultural land to urban settlements (Bogue 1956; Raup 1975), sedimentation and erosion that accompanied infrastructure development (Douglas 1974, 1968, 1978), soil contamination (Klein 1972; Purves 1972) and the influence of soils on urban functions (Gray 1972) motivated much of this early work.

The notion of urban soils developed with interest in the integration of economic and ecological considerations in urban planning (McHarg 1992). At this time, however, interest in the intrinsic value of soils remained largely focused on the value of soils for limited human purposes, including landscaping, gardening, and urban wildlife management and forestry (Craul 1985, 1992; Gill and Bonnett 1973). These studies noted the difference between natural and urban soils and assumed that all urban soils were degraded.

As soil scientists reflected on the conditions of soils in cities, they began to include them in soil surveys. Typically urban soils were categorized differently from "natural" soils (Short *et al.* 1986a). Urban soils were characterized as including thick human-made A horizon, a disturbed subsurface layer with modified, mixed, truncated or buried structure, new soil parent materials, non-agronomic functions and profound modification through human activities such as addition of organic materials or wastes (Craul 1992; Gilbert 1989; Hollis 1991; Hiller 2000; Blume 1989; Driessen *et al.* 2001). As such urban soils were called "deposit," "made," "disturbed" or "anthropogenic" soils. New names for an urban soil order include "Urbic," "Anthrosols" and "Anthroposols."

With increasing attention on the ecology of urban areas (see, for example, Douglas 1981; Hough 1984; Spirn 1984), a number of new perspectives drove urban soils research. Studies of urban soils increasingly focused on fundamental ecological characteristics. Examining urban soils provided unique opportunities to examine how these systems perform in highly stressed environments (White and McDonnell 1988; McDonnell and Pickett 1990, 1991). As ecological studies of soils included human influences detailed studies have noted that all urban soils are not disturbed, but that there was great intra-urban soil diversity (Pouyat *et al.* 2007). As cities have been seen as a collection of "patches" or mosaics of ecological communities, and not only as homogenous communities that are different from their surroundings (Pickett *et al.* 2008), soil scientists have focused on the distribution of disturbed patches with distance from the urban center (McDonnell *et al.* 1997) and how soils within cities vary (Pouyat *et al.* 2007). This variation is not only due to natural pedogensis, but also to indirect human influences (Effland and Pouyat 1997) and management and land use differences (Pouyat *et al.* 2002). As such, new soil series names have been introduced to describe different soils within some cities. In New York City, for example, a new soil survey identified a number of different names for soils formed on human constructed landforms with names such as Shea, Central Park and Big Apple (New York City Soil Survey Staff 2005).

Researchers examining socio-economic factors in the study of influences on soil dynamics have generated a number of new theories. These include the importance of the cumulative effects of urban management that create conditions across the world forcing convergence of conditions (Pouyat *et al.* 2009). This notion suggests that urban soil conditions, in different climatic regions, will converge due to strong management influence. In a related manner, research also suggests that this convergence creates homogenization (Niemelä *et al.* 2000). That is, as hypothesized for biodiversity (McKinney 2006) soil homogenization theory suggests that given the strength of similar driving forces and increasing global connectedness, including the transmission of technologies, management practices and invasive species, soil conditions within cities are becoming more alike to each other than to their respective native counterparts.

Finally, urban soil research has responded to the concerns over global environmental change. Not only can changes in soils have cumulative effects on the global environment, but they also have a role in systemic effects through their importance to global biogeochemical cycles, such as the carbon cycle (Grimm *et al.* 2008; Pataki *et al.* 2006) including, for instance, the potential role of urban soils to sequester carbon (Pouyat *et al.* 2006).

Selected physical and chemical characteristics of urban soils

Notwithstanding variation, the physical characteristics of soils in urban areas are uniquely different from those found in rural and natural areas. It is not soil texture and mineralology that change with indirect, management and land use change influences, although soil texture can be altered by erosion, but rather it is the secondary soil properties, such as bulk density, nutrient ions, and pH that are highly modified by human activities (Palm *et al.* 2007). Wheater (1999) and Craul (1992) summarized differences in the physical properties of urban and non-urban soils. They suggested that urban soils were characterized by reduced compaction, surface crusting, restricted aeration and drainage and temperature regimes. Table 15.1 summarizes each of these effects. This section briefly overviews each topic, discusses the causes of the problem and in some cases provides examples. To this list, this review adds modified pH levels.

Compaction

Compaction is the process of reduction of the specific volume (or porosity) of a soil. Soil compaction inhibits drainage, aeration and root growth and thus has received attention in the agricultural literature (Mullins 1991). Compacted soil often behaves like impervious surfaces, concrete or asphalt. Good agricultural soil has about 50 percent pore space. When a soil is compacted so that it has less than 25 percent pore space it becomes a poor medium for supporting plant growth (Arnold 1993).

Bulk density is a measure used to determine the degree of compaction and indicates how closely the soil particles are packed together. Bulk density is often expressed as porosity (the volume of voids per unit volume of soil) in terms of mass per unit volume; grams per cubic centimeter ($g cm^{-3}$ or $Mg m^{-3}$). Well-aggregated soils, rich in organic matter, have bulk density values less than 1.0 and highly compacted soil values exceed 2.0. Many arable soils have values up to 1.6. Urban soils that have been thoroughly cultivated such as those of allotments and flower beds have bulk densities within the range of 1.0–1.6 (Patterson 1976). The "ideal" soil for plant growth ranges from 1.45 $Mg m^{-3}$ for clays to 1.85 $Mg m^{-3}$ (Brady and Weil 2002).

In urban areas, however, typical conditions tend to destroy structure and increase bulk density. Craul (1992) identified six conditions within urban areas that promote compaction including:

Table 15.1 General physical differences between urban and more natural soils

Characteristics of urban soils in comparison with natural soils	*Causes*	*Resulting problems*
Harsh boundaries between soil layers	Artificial origins produce layering of different materials	Lack of continuity for tooting plants and burrowing soil animals
Compaction	Tramping and pressure from vehicles, etc.	Reduced water passage and lack of air spaces. Plants produce shallow roots
Low water drainage	Diversion of run-off to drains, and interruption of natural flow through soils	Reduced water availability for plants
Crusting and water repellency	Compaction, chemical dispersion and creation of waxy soil surface	Barriers to gaseous and water exchange between soil and atmosphere
High pH	Effects of de-icing salts and water running over calcareous building materials (e.g. concrete)	Problematic if highly alkaline, because some nutrients (e.g. phosphates) are immobilized
High soil temperatures and moisture regimes	Higher ambient air temperatures and little buffering effect of vegetation	Reduce moisture in upper layers for plants growth

Source: Wheater 1999, table 1.2; Craul 1992.

1 the partial destruction of their structure and horizon arrangement within the profile, enhancing compaction;
2 low organic matter, which is an aggregating, structure-forming agent;
3 limited soil organism populations which promote soil structure and increase soil porosity;
4 elevated urban temperatures reduce the frequency of complete freeze-thaw cycles, thus further preventing soil structure formation;
5 various physical activities over a range of moisture conditions, including trampling that destroy vegetative cover and leaving the soil surface bare and unprotected from further compacting forces; and
6 different wetting and drying cycles that further exacerbate conditions (i.e. when wet surface conditions are exposed to traffic by foot or machine).

Measurements of bulk densities in urban soils have demonstrated compaction as compared to rural areas. Short *et al.* (1986) measured bulk densities of 1.25–1.85 $Mg\,m^{-3}$ (mean 1.61 $Mg\,m^{-3}$) of the surface horizon and bulk densities of 1.4–2.3 $Mg\,m^{-3}$ (mean 1.74 $Mg\,m^{-3}$) at 0.3 m depth for open parkland in the Mall of Washington DC. Craul (1985) found values of 1.52–1.96 $Mg\,m^{-3}$ for subsoils in Central Park, New York City. Patterson (1976) found average values in Washington DC ranging from 1.74 to 2.18 $Mg\,m^{-3}$. Hiller (2000) found that soils in abandoned shunting yards in the Ruhr area, Germany, had bulk densities ranging from 1.0 to 2.1 $Mg\,m^{-3}$ depending upon the site and depth of soil where the measurements were taken.

In tropical areas, soil bulk densities range from 1.1–1.4 $Mg\,m^{-3}$ (Jim 1998b). In tropical developing cities, however, bulk densities are higher. Jim (1998b), for example, found similar ranges, 1.6–1.8 $Mg\,m^{-3}$ for parks in Hong Kong. Jim (1998a) also found an average bulk density of 1.66 $Mg\,m^{-3}$ for 50 street side soil samples in Hong Kong. A recent study in Ibadan, Nigeria, suggests that bulk density of urban soils range from 1.05–2.18 $Mg\,m^{-3}$, with a mean of 1.62 $Mg\,m^{-3}$. While these values are lower than those found in developed countries they are still statistically different (and higher than) those values for soil bulk densities found outside Ibadan in agricultural land and from those found in suburban zones (Gbadegesin and Olabode 2000).

Poaching, an extreme form of compaction, is often seen in cities around narrow pathways and entrances wherever people or machinery are forced to tread repeatedly or track the soil at a water content close to field capacity (Mullins 1991).

Surface crusting and water repellency

Soil surface crusting, or the development of a thin hard crust on bare soil, is caused by several factors, the most important being compaction of the surface layer by foot and light wheel traffic and the associated lack of vegetative groundcover. The absence of the groundcover eliminates the cushioning effect of the plant shoots, the binding and lightening effect of the root system, and the contribution of organic matter to the soil surface. The force of raindrop splash disintegrates soil aggregates and the dispersed particles of very fine sand, silt and clay fill the adjacent pores (Hillel 1980). A second factor cause of crusting is chemical dispersion enhanced by the low electrolyte concentration of rainwater. The result is the formation of one or two micro-layers of horizontally oriented particles in the upper 2 cm of soil that have a thin skin seal. Compaction and infill of fine particles reduces water infiltration and gaseous exchange of oxygen and carbon dioxide between the atmosphere and the soil. A third possible cause is the atmospheric deposition of petroleum-base aerosols and particulates on the soil surface, which form waxy and oily substances that are water repellant. The resultant hydrophobic nature of such crust can be observed at the initiation of a light rain. During treatment, samples of soil from New York City forests did not absorb water during contact with it for up to two hours (White and McDonnell 1988).

Restricted aeration and drainage

Urban soils are relatively confined, as land in cities is covered with nonporous material. Where they do emerge they are subject to crusting and compaction. Moreover, the urban soil body is often spatially interrupted by walls, sidewalks, curbs, pipe-shafts, streets paths, repeated construction disturbances, foundations, or sharp change in grade. Urban soils lack the horizontal continuity between bodies present in the natural soil mantle. Together these conditions restrict gaseous diffusion and water movement in the soil profile (Craul 1992).

The results of these restrictions are twofold. First, urban settings experience a shift from the natural local hydrological flows. Covered areas produce intensive runoff (Douglas 1983; Arnold and Gibbons 1996). Bare soils, as a result of crusting and compaction, do not absorb water as quickly and they also dry more slowly. Covered soil when moist will remain so for prolonged periods of time. Lateral movements of water may be slow from effects of compaction and physical barriers. As such aeration and drainage of urban soils is disrupted.

Given these characteristics, urban soil conditions therefore do not present optimal conditions for plant growth. For example, tree roots often cannot penetrate very deep in these types of soils and instead grow alongside or are exposed to the ground surface. The results are that trees fall easily in windy conditions. There are many examples of this in tropical cities, such as Singapore. Special trees, such as the London Plane tree thrive, because of their ability to cope with restriction on root space, but most street trees suffer from structural or physiological problems (Jim 1989).

Soil temperature

The temperature of cities is different from that of rural areas as expressed by the urban heat island model (Oke 1973). While the incoming radiation may be less in the city than in rural areas, because of increased haze and cloudiness, the amount of heat absorbed and reradiated by building and street surfaces is greater than vegetation, raising both daytime and nighttime urban air temperatures (Landsberg 1981). The city heat island is typically conterminous with the built up area of an urban center although wind direction can change temperatures of other areas temporarily. Urban temperature excess results in a bubble of maximum temperature difference at the surface trailing off to no difference somewhere between 300 and 500 meters above the city (Bornstein 1968).

Temperatures of urban soils can be 2° to 3°C higher during the growing season than in suburban and rural soils, which could account for differences in litter decomposition rates found in these sites (Pouyat *et al.* 1997). Hence higher temperature differences play a significant role in biota activity and biogeochemical cycling in urban soils.

Moreover, high temperatures affect soil-weathering processes (Hillel 1980). Soil surfaces form gradients where surface heat builds. Soil temperature is highly influenced by moisture content because water affects the heat capacity of the soil. The greatest temperature extremes occur in the surface horizons. High temperatures dry the soil surface through evaporation, creating moisture gradients that cause water to flow from lower to upper horizons (Craul 1992). Hence water movement within heat-affected soils is increased, arguably impacting translocation processes.

Modified pH

Soil chemistry includes macro and micro nutrient cycling and the interactions of soil systems with the hydrologic cycle, soil organisms, soil physical properties, human influences and climate, among other factors. This review focuses only on changes in pH in urban soils.

Urban soils tend to have soil reaction (pH) values somewhat different from their natural counterparts. In most cases, in urban areas pH values are higher, more alkaline, than found in rural areas (Craul and Klein 1980). There are several possible explanations for the elevated pH values of urban soils. First is the application of calcium or sodium chlorides as sidewalk and street de-icing salts in northern cities, as in Syracuse (Craul 1992). Second, in many urban areas, urban soils are irrigated with calcium-enriched water. The third explanation is atmospheric pollution. Ash residue from burning of fossil fuels and organic matter often contains calcium, which falls to the ground. This effect is further evidenced in the decrease in pH values with increasing depth. The fourth explanation relates to the release of calcium from weathering of construction rubble comprised of bricks, cement, plaster, etc. or the washing of calcium from building facades by polluted acid precipitation. Gilbert (1989) has stressed the chemical properties of rubble soils as the source of high pH in the UK. Brick rubble contains calcium, important as a nutrient and controller of pH, which maintains soil pH in the range 6.5–8.0. Finally, liming of soil, particularly in parks and gardens, to correct suspected soil deficiencies raises pH.

Examples of high pH values in developed world cities are found in the soils in the UK, locally roadside and path side soils have pH around 9.0 as a result of sodium and calcium chloride applied as de-icing salt (Gilbert 1989). Sukopp *et al.* (1979) report high values for pH city streets in Berlin (8.0), which decrease as one moves to city forests. Mean pH values for five different land uses (agriculture, ornamental gardens, parks, riverbanks and roadsides) in Seville was approximately 7.21 (Ruiz-Cortés *et al.* 2005). Soils in Palermo, Italy exhibited an alkaline range of pH (7.2–8.3) (Manta *et al.* 2002). Soil samples in the western district of Naples, Italy had average pH levels of 8.01, while those in the eastern district averaged 7.4 and the central district samples averaged 6.8 (Imperato *et al.* 2003).

At the same time, not all soils in cities are alkaline. Some urban soils have been found to be acidic. A study of the railroad shunting areas in the Ruhr region, Germany, suggested that pH levels were lower than those of agricultural land (Hiller 2000). Soils in the Mall in Washington DC, which include fill materials contaminated with building rubble are acidic, with pH of the horizons varying from 6.4 in the surface to 6.7 in the fifth horizon (Short *et al.* 1986b). In Central Park, NY, soil pH levels were identified as lower (4.38) than suburban (4.57) and rural (4.61) areas (Pouyat *et al.* 1997). Garden soils in Salamanca, Spain also demonstrated low pH values, between 3.4 and 7.6 (mean 6.7) (Sanchéz-Camazano *et al.* 1994).

Within developing world cities, Jim (1998b) found alkaline humid-tropical soil in Hong Kong with pH exceeding 8.0 in some cases. In Caracas, Venezuela, soils along roadsides polluted had pH levels of between 7.5 and 7.8 (Garcia-Miragaya *et al.* 1981). Krishna and Govil (2005) found neutral to alkaline pH levels (pH 7.5–8.5) in Thane-Belapur, an industrial development area of Mumbai. In satellite cities of Seoul, Chon *et al.* (1998) found pH levels of 5.2, 5.7, 6.0 and 7.6 for soils in forest, paddy, dry-field and roadsides, respectively. They suggested the relatively high pH values in roadside soils were due to materials associated with the road surface.

Just as there are examples of acidic soils in the developed world, acid soil pH levels have also been found in developing world cities. For example, Gbadegesin and Olabode (2000) found soil pH in Ibadan to be between 5.1 and 7.6 with a mean of 6.62. These values however, were not significantly different from pH values in soils in suburban and agricultural areas. Soil studies in Bangkok also found low pH levels ranging from 3.6 to 7.4 with a mean of 6.7 (Wilcke *et al.* 1998).

Pollutants in urban soils

This review focuses only on pollution concentrations of heavy metals, de-icing salts and polycyclic aromatic hydrocarbons (PAHs). Each of the following subsections introduces these topics generally, provides selected examples of differences between concentration levels in urban and rural settings when possible and also describes intra-urban variability.

Heavy metal soil contamination

Heavy metals are a subset of persistent toxics that retain their environmental impact for a relatively long period of time after release into the environment. Heavy metals found in soils include Cd, Co, Cr, Cu, Fe, Hg, Mn, Mo, Pb, and Zn (Brady and Weil 2002). These elements do not decay, but change to different forms that may increase or decrease their toxicity. Additionally, they can disperse, accumulate, or undergo other physical transformations leading to changes in the likelihood of human exposure thus altering associated risks (Davies and Mazurek 1999).

Trace metal distribution in soils is well documented for many industrialized countries such as Japan, Germany and the USA and there are also examples from the developing world. Both natural and human activities introduce trace metals into the environment (Table 15.2).

Purves (1966, 1972) and Purves and Mackenzie (1969) were among the first to examine metals in urban and industrial soils, noting elevated levels of potentially toxic trace elements such as Cu,

Table 15.2 Potential sources of heavy metals

Element	*Source*
Arsenic	Pesticides, fertilizers, plant desiccants, animal feed additives, copper smelting, sewage sludge, coal combustion, incineration and incineration ash, detergents, petroleum combustion, treated wood, mine tailings, parent rock material
Cadmium	Phosphate fertilizers, farmyard manure, industrial processes (electroplating, non-ferrous metal, iron and steel production), fossil fuel combustion, incineration, sewage sludge, lead and zinc smelting, mine tailings, pigments for plastics and paint residues, plastic stabilizers, batteries, parent rock material
Chromium	Fertilizers, metallurgic industries, electric arc furnaces, ferrochrome production, refractory brick production, iron and steel production, cement, sewage sludge, incineration and incineration ash, chrome-plated products, pigments, leather tanning, parent rock material
Nickel	Fertilizers, fuel and residual oil combustion, alloy manufacture, nickel mining and smelting, sewage sludge, incineration and incineration ash, electroplating, batteries, parent rock material
Copper	Fertilizers, fungicides, farmyard manures, sewage sludge, industrial processes, copper dust, incineration ash, mine tailings, parent rock material
Lead	Mining, smelting activities, farmyard manures, sewage sludge, fossil fuel combustion, pesticides, batteries, paint pigment, solder in water-pipes, steel mill residues
Manganese	Fertilizers, parent rock material
Mercury	Fertilzers, pesticides, lime, manures, sewage sludge, catalysts for synthetic polymers, metallurgy, thermometers, coal combusion, parent rock material
Zinc	Fertilizers, pesticides, coal and fossil fuel combustion, non-ferrous metal smelting, galvanized iron and steel, alloys, brass, rubber manufacture, oil tires, sewage sludge, batteries, brass, rubber production, parent rock material

Source: Alloway 1995; Brady and Weil 2002.

Pb, Zn and B. Working in Edinburgh area, they found that the average urban soil contained more than twice as much available B, fives times as much Cu, 17 times as much Pb and 18 times as much Zn as those collected from adjacent rural areas. Work in this area was further stimulated by concerns over high levels of heavy metals found in urban homes. For example, during the 1980s, in the United Kingdom, a national survey of house dusts in 53 representative villages, towns and city boroughs revealed that in 10 percent of the 5,228 homes tested, the lead concentration was in excess of 2,000 $\mu g g^{-1}$ (Davies *et al.* 1987; Thornton 1991). Since then, numerous studies have been undertaken to assess anthropogenic sources of heavy metals in soils and house and street dusts. Table 15.3 presents a summary of several of the results of selected studies around the world.

Given that these studies were performed in different parts of the respective cities, direct comparisons between cities must be made with caution. At the same time, it is interesting to note that levels found in cities of low-income and emerging and transition economies, in some cases, compare with those of the now developed world. We might expect these metals to accumulate over time and therefore be found in higher concentrations in the now developed world cities. Yet, some soils in developing world cities demonstrate high levels of heavy metal contamination. Two examples are important to note. First, trace metal distribution of soils in the Danang-Hoian

Table 15.3 Average heavy metal concentrations (mg/kg) in urban soils from different cities in the world

City	*Country*	*PB (mg/kg)*	*Zn (mg/kg)*	*Cu (mg/kg)*	*Cd (mg/kg)*	*Cr (mg/kg)*	*Co (mg/kg)*	*Ni (mg/kg)*
Developed								
Brussels	Belgium	113	34	19	0.7	–	–	–
Naples	Italy	262	251	74	–	11.0	–	–
Aviles	Spain	188	476	104	2.6	27.1	14.3	19.4
Coruna	Spain	309	206	60	0.3	39.0	11.0	28.0
Madrid	Spain	161	210	72	–	74.7	6.4	14.1
Mieres	Spain	92	233	57	1.0	20.4	15.8	33.8
Salamanca	Spain	53	–	–	0.5	–	–	–
Seville	Spain	156	120	55	2.9	38.4	–	21.2
Stockholm	Sweden	104	157	47	0.4	27.0	–	9.0
Aberdeen (parkland)	UK	94	58	27	–	23.9	6.4	14.9
Birmingham	UK	180	205	143	0.7	–	–	–
Glasgow	UK	216	207	97	0.5	–	–	–
Pittsburg	USA	398	–	–	1.2	–	–	–
Average		*179*	*196*	*69*	*1*	*33*	*11*	*20*
Developing and Transition								
Gaborone	Botswana	112	248	36	1.6	72.0	17.0	48.0
Hong Kong	China	90	59	16	0.9	–	–	–
Hong Kong	China	93	168	25	2.2	–	–	–
Hong Kong (roadside)	China	117	129	–	1.3	–	–	12.5
Xuzhou	China	43	144	38	0.5	78.4	11.7	34.3
Zagreb	Croatia	26	78	21	0.7	–	–	49.5
Cuenca	Ecuador	293	509	–	0.4	–	–	–
Thane-Belapur, Mumbai	India	–	191	105	521.3	–	68.7	183.6
Amman	Jordan	1,042	408	242	0.9	11.9	–	17.5
Ibadan	Nigeria	5	37	1	–	–	–	–
Manila	Philippines	214	440	99	0.6	114.0	–	–
Warsaw	Poland	53	140	25	1.0	12.9	–	12.0
Koyang	South Korea	88	238	59	2.1	47.0	23.9	47.0
Uijeongbu	South Korea	65	204	41	1.4	27.0	10.4	23.0
Bangkok	Thailand	48	118	42	0.3	26.4	–	24.8
Danang-Hoian	Vietnam	4	142	76	0.8	92.2	34.0	22.6
Average		*153,203*	*59*	*36*	*54*	*28*	*43*	–

Sources: Cal-Prieto *et al.* 2001; Carey *et al.* 1972; Chen *et al.* 1997; Chon *et al.* 1998; Davies and Houghton 1984; de Miguel *et al.* 1997; Gbadegesin and Olabode 2000; Hewitt and Candy 1990; Imperato *et al.* 2003; Jim 1998b; Jiries 2003; Krishna and Govil 2005; Li *et al.* 2001; Linde *et al.* 2001; Loredo *et al.* 2003; Peterson *et al.* 1996; Pichtel *et al.* 1997; Pizl and Josens 1995; Ruiz-Cortes *et al.* 2005; Sanchez-Camazano et al. 1994; Thuy et al. 2000; Wang et al. 2005; Wilcke et al. 1998; Zhai *et al* 2003.

area (Vietnam) was studied by Thuy *et al.* (2000), who compared industrial and urban, rural and cropland areas. The researchers demonstrated extremely high levels of Pb (up to 742 $\mu g\, g^{-1}$) are observed in industrial soils, which show an enrichment factor of 114 compared to rural soils. Cd shows only a relative local enrichment with a maximum level of 4.6 $\mu g\, g^{-1}$ in urban soils. Industrial soils have the highest values of Pb, Zn, Ni and Cu. Other metals Zn, Zr, Cu and Ni have similar distribution patterns. The highest concentrations of Zn (717 $\mu g\, g^{-1}$) and Ni (240 $\mu g\, g^{-1}$) were found in industrial soils.

A second recent study in India investigated heavy metal pollution of soils near the Thane-Belapur industrial belt of Mumbai. Mumbai is a heavily populated industrial island city on the west coast of India, which is known as the commercial capital of the country. Fundamental to rapid urbanization are the numerous industries at Thane-Belapur industrial area. It houses 400 industries within 20 km area with all types of process industries, including chemical, pharmaceutical, textile, steel, paper, plastic, and fertilizers (Krishna and Govil 2005). Soil samples analyses indicate enrichment of the soils with Cu, Cr, Co, Ni and Zn. The concentration ranges were: Cu 3.10–271.2 $mg\, kg^{-1}$ (average 104.6 $mg\, kg^{-1}$), Cr 177.9–1,039 $mg\, kg^{-1}$ (average 521.3 $mg\, kg^{-1}$), Co 44.8–101.6 $mg\, kg^{-1}$ (average 68.7 $mg\, kg^{-1}$), Ni 64.4–537.8 $mg\, kg^{-1}$ (average 183.6 $mg\, kg^{-1}$) and Zn 96.6–763.2 $mg\, kg^{-1}$ (average 191.3 $mg\, kg^{-1}$).

Both these studies indicate the high levels of heavy metals found in cities of the developing world and in particular around industrial areas. Concentrations of heavy metals in urban topsoil very considerably across and within cities depending, among other things, on the density of industrial activities in the area and technologies employed, as well as on local weather conditions and wind patterns (Ordóñez *et al.* 2003).

De-icing salts

De-icing salts are deposited in northern urban areas to raise the melting point of ice and snow. These compounds commonly are sodium chloride or calcium chloride applied unmixed or mixed with sand, fine gravel or coarse tailings. The amounts applied have increased over time. For example, Gilbert (1989) noted an increase from 132,000 metric tons in 1956 to 2.3 million metric tons in 1985, used in roads in the UK. Throughout the US, rock salt for use on highways has risen from 149,000 metric tons in 1940 to 18 million metric tons (Jackson and Jobbagy 2005).

Craul (1992) notes that de-icing salts potentially affect the soil environment in several ways including:

1 the sodium replaces other cations on the soil-exchange sites creating a potential nutrient imbalance;
2 the absorbed sodium tends to disperse soil colloids, filling small pores, which reduces water infiltration and gaseous diffusion;
3 dissolved salts decrease the osmotic potential of the soil water so that plants wilt at a higher moisture content than in "unsalted" soil;
4 being highly soluble, the sodium chloride is easily carried into natural water bodies, contaminating the water;
5 salts themselves corrode concrete, various metals, asphalt, and other unprotected surfaces;
6 sodium chlorides stimulate the release of heavy metals in sediments; and
7 salt spray does the most harm to plants by coating the twigs, buds, and leaves of plants.

Evidence for the environmental effects of de-icing have been reviewed by Ramakrishan and Viraraghavan (2005). They conclude that impacts are highly localized. At the same time, however, others have noted the high accumulation of salts damage ecosystem structure and function and may have long term consequences. A recent study suggests that the high usage of salts in the Northeast USA, for example, has long term consequences on water quality, including drinking water supply (Kaushal *et al.* 2005). The use of salts on roads has led some to observe a change in the ecology of roadsides to include more maritime species of plants (Gilbert 1989).

Much, however, remains unknown about the actual effect of road salts on soils. Studies to date note that salt effects are attributed to a road de-icing during the winter and these effects diminish with the distance from a road and are seasonal. For example, the chemical effects of road de-icing on soil were studied in five small, forested catchments in southeast Sweden. The use of de-icing salt on roads applied during the winter season 1998/99 had a profound effect on the soil chemistry. The salt applications caused increased salinity in direct proportion to the accumulated amounts (Lofgren 2001). In the Mohawk River Basin of New York, USA, concentrations of Na^+ and Cl^- have increased by 130 and 243 percent, respectively, from 1951 to 1998, while other constituents have decreased or remained constant and the use of de-icing salt on roads within the watershed appears to be the primary mechanism responsible for these increases (Godwin *et al.* 2003).

A recent study focused on the patterns and distribution of de-icing salts in Poughkeepsie, a moderately dense urban settlement in New York State, USA, supports the notions that salt cations accumulate highest near impervious surfaces and decline with distance and that levels are greater in shallow than deep samples (Cunningham *et al.* 2008). The research suggests that the highest concentrations are adjacent to roadways or downhill from parking lots, although some sites adjacent to paved surfaces had low cation concentrations. Mg^{2+} levels were found to be higher in near surface samples compared to deeper samples, while Na^+ levels were not significantly different. Importantly, detectable levels of salt cations were found at the end of the summer months after application.

Polycyclic aromatic hydrocarbons

Polycyclic aromatic hydrocarbons (PAHs) naturally occur in fossil fuels (oil and coal) and in fossil deposits (tar) and therefore are released as by-products of burning these materials. The US Environmental Protection Agency has classified seven PAH compounds as probable human carcinogens: benz[a]anthracene, benzo[a]pyrene, benzo[b]fluoranthene, benzo[k]fluoranthene, chrysene, dibenz[a,h]anthracene, and indeno[1,2,3-cd]pyrene. The focus of PAHs environmental studies research has largely been restricted to identifying atmosphere concentrations (Zheng and Fang 2000) and river sediment concentrations (van Metre *et al.* 2000; Jiries *et al.* 2000).

Given the sources of PAHs, several studies have been conducted in cities. These studies examine the level of different PAH compounds in street dust and soils (Table 15.4). A common finding is the association between motor vehicles use and PAH concentration and distribution. For example, the PAH content of surface soils in the vicinity of heavy vehicular traffic near the English Midlands motorway interchange M6-A38(M) indicated that the top (0–4 cm) of soil at a distance of 1 m from the hard shoulder of the M6 contain 20,000 μg/kg of PAH comprising pyrene, fluoranthene, chrysene, benz(a)anthracene, benzo(a)pyrene, benzo(e)pyrene and coronene. Typical concentrations of PAH in surface soils (0–4 cm) at a distance of 600 m from the interchange are in the range 4,000–8,000 μg/kg, more than doubling the PAH surface soil concentrations found on the edge of the conurbation (2,300 μg/kg) (Butler *et al.* 1984).

Direct urban-rural comparisons suggest that concentrations are higher in urban areas. For example, vegetation samples from urban, suburban, and rural locations indicate that the total PAH burdens in the rural vegetation samples were, on average, ten times lower than the urban

Table 15.4 Comparison of total soil PAH concentrations (ug/kg) from different cities

City	*PAH number*	*Soil depth (cm)*	*PAH concentration*
Nagoya, Sapporo, and Tomakomai, Japan	9–10	0–3	300–2,100
Tokyo, Japan	9	street dust	1–3
Glowna railway station, Poland	16	0–20	910–2,240
Amman, Jordan	16	street dust	23
New Orleans, USA	16	2.5	650–40,700
Niteroi, Brazil	21	street dust	570–1,250
Brisbane roadway, Australia	14	0–5	3,350
Welsh soils	16	0–23	110–54,500

Sources: Jiries 2003; Jones *et al.* 1989; Malawska and Wilkomirski 2001; Mielke *et al.* 2001; Netto *et al.* 2006; Spitzer and Kuwatsuka 1993; Takada *et al.* 1990; Yang *et al.* 1991.

samples (Wagrowski and Hites 1997). Moreover, PAH concentration varies with temperature and therefore can fluctuate seasonally (Netto *et al.* 2006). Research into PAHs concentration and distributions in cities suggest that these compounds are part of the soil mixture of accumulated substances and the effect of current levels of concentration in urban soils, vegetation and the associated risk to humans has yet to be determined accurately (Mielke *et al.* 2001).

Biological characteristics of urban soils

Soil organisms are vital to a range of soil ecosystem services including nutrient cycling, soil organic matter formation and decomposition, soil carbon sequestration, and greenhouse gas emissions. Soil biota modifies soil physical structure and hydraulic properties and therefore determines root growth and plant nutrient uptake. In addition, many invertebrates including pollinators, pest predators and their enemies spend part of their life cycle in the soil. The living components of soil are fundamental to the development and maintenance of the soil ecosystem (Brady and Weil 2002).

The physical and chemical characteristics of urban soils described above, however, have altered the composition of organisms found in urban soil ecosystems. For example, as Sukopp (2004) explains, the removal of plants and reduced primary production alters population dynamics and species composition. These changes have impacts on energy and matter fluxes, which are vital to soil communities. Soil excavation and sealing, changes in bulk density, pH and the concentration of pollutants have variable, but poorly understood effects on biodiversity. Moreover, the introduction of invasive species has helped to alter species assemblages and soil dynamics (Steinberg *et al.* 1997).

Given the changes to the environment that accompany urban growth, soil scientists suggest that there is a reduction in both species numbers and biomass of soil organisms (Harris 1991; Santas 1986). Moreover, as is hypothesized for biodiversity as a whole, urban areas tend to favor specific types of soil organisms (McKinney 2002). That is, as with animals that do well living off human resource and energy waste, and plants that do well in warmer, more alkaline soils and can withstand soil contamination (Gilbert 1989), the composition of urban soil organisms may be both unique and homogenized within urban areas (Houge 1987). This homogenization, however, does not completely reduce biodiversity and indeed there is surprisingly rich biodiversity found in some urban soils (Luniak 2008).

For general descriptive purposes, soil biota may be classified into three groups, on the basis of size as defined by Harris (1991) and Brady and Weil (2002) including: 1) the macrobiota (such as moles, prairie dogs, earthworms and the larger arthropods and enchytraeids; (2) the mesobiota, including smaller arthropods, and enchytraeids and springtails, and (3) microbiota, including nematodes, bacteria, fungi, algae, protozoa and viruses. An estimation of soil organism biodiversity within these categories has been provided by Brussard (1997) (Table 15.5). This review will describe research directed at all these different sized organisms in urban soils.

Macrobiota

Earthworms are the primary animal biomass and the most important macro-animal in soils (Marino *et al.* 1992). Because of their feeding process, excretions and secretions they modify the mineral components and organic matter of soil. Given their intimate contact with soils, they are affected by the concentration of pollutants (Rozen and Mazur 1997). Standen *et al.* (1982) report on earthworm populations in areas restored from opencast mines and spoil from deep mines. They note the lumbricid populations were small compared to permanent pasture areas nearby, even in areas 11 years after restoration. A review of studies suggests that soil disturbance leads to decreases in community size and structure (Harris 1991). Pizl and Josens (1995) recorded earthworm (*Lumbricus terrestris*) biomass declines along a gradient of increasing urbanization in Belgium.

In an urban-rural transect study in New York City however, average earthworm density was significantly higher in the urban land use types than in rural areas, while in suburban land use types earthworm populations were intermediate in number (Steinberg *et al.* 1997). For example,

Table 15.5 Soil organisms

Biota by size category	*Number of species described*
Microorganisms	
Bacteria and archaea[1]	3,200
Fungi	18–35,000
Amfungi	200
Ectomycorrhizal fungi	10,000
Protozoa	1,500
Ciliates	400
Nematodes	5,000
Mesofauna	
Mites	30,000
Collembola	6,500
Enchytaeids	>600
Root herbivorous insects	~40,000
Termites	2,000
Ants	8,800
Macrofauna	
Earthworms	3,627

Source: modified from Brussaard (1997).

Note

1 Total number, soil dwelling fraction is unknown.

urban forests had 25.1 worms per m^2 and 2.16 g of worms per m^2 whereas the rural forests had only 2.1 worms per m^2 and 0.05 g of worms per m^2. The reduced leaf litter depth, mass, and density of the forest floor (0_2 horizon) (Kostel-Hughes *et al.* 1998), and increased organic matter levels at depth (Pouyat *et al.* 1997) provide further support for increased earthworm activity in the urban forests along the study transect (McDonnell *et al.* 1997). Earthworm biomass was probably higher in New York City due to repeated introduction of earthworms from Europe and Asia. Indeed, earthworms in the northern parts of the USA were exterminated during the Quaternary climate (Gates 1976).

Mesobiota

A significant, but poorly known aspect of the urban fauna includes invertebrates of the soil and litter. In a review of 79 English language, peer-reviewed journal articles on the ecology of arthropods from 1933 to 1999, representing 51 cities from 14 countries, McIntyre (2000) found that there is a lack of basic understanding of the mechanisms accounting for distributional and abundance patterns of urban arthropods.

In general, urbanization is considered one of the primary causes for declines in arthropod populations. Pyle *et al.* (1981) suggest that pollution, drainage and diversion of watercourses, and fragmentation and destruction of habitat that accompanies city building are the main factors for insect declines. Pouyat *et al.* (1994) found that soil micro-invertebrates, of which they included the taxonomic groups Mesostigmata, Orbatida, Collembola, and other micro-insects, abundances were higher in the rural forests than in either the urban or suburban sites during the fall season, but exhibited no significant differences during the spring in the New York City area. They also found that their abundance was inversely correlated with soil heavy metal concentrations.

On the other hand, however, urbanization not only destroys and modifies habitats, it also creates habitat for invertebrates (Frankie and Ehler 1978). The result can be a surprisingly high level of soil invertebrate biodiversity. For example, one study from a Warsaw park noted that a 1 m^2 of soil holds an average of several thousand (in some cases even 40,000) individual invertebrates. Among the most numerous animal groups recorded were soil mites (Acari) (~20,000), enchytreids (Enchytreidae) (5–25,000), springtails (Colembola) (2–7,500), ants (Formicoidea) (~500), earthworms (Lumbricidae) (60–160), click beatles (Elateridae) (40–100), and many (over one hundred) insect larvae, particularly aphids (Aphidodea), beetles (Staphylinidae), flies (Dpitera), spiders (Arenei) and snails (Gastropoda) (Luniak 2008). This research suggests that the abundance of invertebrate communities in urban areas could, in many cases, exceed that in rural habitats.

Research on specific invertebrates has uncovered more complex urban-community structure relationships. A study of communities of carabid beetles in residual forest patches along urban-suburban-rural gradients in three cities (Helsinki, Finland; Sofia, Bulgaria; and Edmonton, Canada) demonstrated that only in the Finnish case did carabids show a marked division of community structure along the gradient (Niemelä *et al.* 2002). In Bulgaria and Canada, carabids did not separate into distinct urban, suburban, and rural communities. These results provide some support for the predictions that species richness will decrease, that opportunistic species will gain dominance, and that small-sized species will become more numerous under disturbance such as that provided by urbanization. The rather weak and varied response of carabids to this disturbance suggests that local factors and their interaction are of primary importance for community composition. Occurrence of reasonably similar carabid communities across the gradient at each of the three levels of urbanization suggest that habitat changes commonly associated with urbanization have not affected the ecological integrity of carabid assemblages in residual urban forest patches.

In a study of ground arthropod communities in sites representing the four most abundant forms of urban land use (residential, industrial, agricultural, and desert remnant) in a rapidly growing metropolitan area (Phoenix, AZ) researchers found that three taxa (springtails, ants, and mites) were extremely widespread and abundant among land-use types. Excluding these taxa, however, revealed that trophic dynamics varied with land use. Specifically, predators, herbivores, and detritivores were most abundant in agricultural sites, while omnivores were equally abundant in all forms of land use but that community composition differed, with certain taxa being uniquely associated with each form of land use. The researchers concluded that the arthropod community structure is affected by habitat structure and land use (McIntyre *et al.* 2001).

Another study, also in Phoenix, noted that the more mesic environments residential yards were characterized by higher abundances but lower spider diversity (dominanted by wolf spiders, *Lycosidae*, and sheet-web weavers, *Linyphiidae*) than the natural xeric surrounding environments (Shochat *et al.* 2004). Arthropods may be distributed patchily within a city, reflecting heterogeneity of urban land-use types (Kozlov 1996). Importantly, as pointed out by McIntrye (2000), reports that observe declines in invertebrate taxa do not always differentiate between numeric and proportionate declines. In a numeric decline, a taxon decreases in absolute number. In a proportionate decline, a taxon decreases in its importance to an overall assessment of diversity. An absolute decline might mean impacts to entire assemblages of invertebrates with proportional declines signaling specific impacts on the taxa.

Microbiota

Total microbial biomass (bacteria and fungi) typically composes <3 percent of the organic carbon in soils (Schlesinger 1997). Soil decomposer communities are often more species rich than the plants and vertebrate animals with which they interact and the diversity of these microorganisms is assumed to be extraordinarily high but is largely unidentified (Prosser 2002). The number of bacterial species is on the order of hundreds to thousands in 1 g of soil; total species number is estimated to be two to three million (Dejonghe *et al.* 2001). Species diversity of soil fungi is probably only slightly less than that of bacteria (Bridge and Spooner 2001; Hawksworth 2001). Some likely reasons for the enormous diversity of soil microorganism is their high fecundity combined with very short generation times and rapid growth. These factors promote a fast speciation in response to relatively small environmental changes.

Soils conditions in urban areas create unique environments for microbials. While microorganism activity may be enhanced by human intervention (De Kimpe and Morel 2000), studies suggest that microbiota activity decreases with density of settlements. For example, microbial activity of soils in northwest Germany declines with increasing urbanization (Beyer *et al.* 1996). Carreiro *et al.* (1999) demonstrated that oak litter from the urban forests decayed 25 percent slower and supported 50 percent less cumulative microbial biomass than did rural litter during laboratory bioassay tests.

Some reasons for this include the level of toxic, sub lethal, or stress effects of the urban environment on microbial activity (Yuangen *et al.* 2006). Contaminants can significantly affect the quality of organic matter and subsequent soil processes (Pouyat *et al.* 1997). Air pollution can also affect the abundance and activities of decomposers. Pollutants cause both short-term and long-term reductions in the abundance of soil organisms, rates of litter decay and oil nutrient return. In some cases, pollutant deposition has resulted in increasing the depth of the litter layer in urban forests by reducing decay rates (Carreiro 2005).

At the same time, higher temperatures in dense settlements, given that moisture is not limiting, can increase the metabolic rate of microbes and invertebrates, thus increasing decomposition

(Craul 1992). Pouyat *et al.* (1997) found that along an urban-rural gradient in New York City, decomposition of litter occurred almost twice as fast in the urban forest than it did in rural forests. At the same time, forest-floor litter mass by weight was almost three times greater in the rural stand than in the urban stands, with suburban stands as intermediate (Kostel-Huges *et al.* 1998). While there are a number of other likely contributors to these differences heat island is an important one (Carreiro 2005). A number of complex factors (including both human and natural influences) within urban areas contribute to microbiota abundance and activity.

Urban soil carbon

Interest in soil organic matter and soil carbon has been given extra emphasis over the past few decades for two different reasons. First, soil organic matter is an integrator of many soil properties. Soil organic matter levels demonstrate a balance between organic inputs to the soil and the decomposition of these inputs by soil biota. Soil organic matter provides carbon and energy for soil organisms and therefore supports the biological functions of soil. As such the level and quality of soil ecosystem services are ultimately determined by soil organic matter (Palm *et al.* 2007). Soil organic matter and soil organic carbon are directly related, but as organic matter decays relative to temperature and precipitation, among other variables, there is a wide variation in soil organic carbon among life zones (Post *et al.* 1982).

At the global scale, it is now a universally accepted proposition that humans are affecting the carbon cycle in ways that will, if unaltered, dramatically alter conditions on Earth. On a global scale soil carbon pools are roughly three times larger than the carbon stored in all land plants (Schlesinger and Andrews 2000) and twice as much stored in plants and the atmosphere combined (Brady and Weil 2002).

As of 2007, most humans live in urban settlements and hence so the majority of human activities are either organized, coordinated or occur in these areas (UNFPA 2007). As such, urbanization is considered a significance influence on global environmental change (Grimm *et al.* 2008). Moreover, the decrease in urban densities experienced around the world suggests that urban land use is expanding more rapidly than population. Not enough is known of the net effect of urban activities on the carbon cycle for proper assessments. For example, as Pataki *et al.* (2006) point out, estimates for the net effect of urbanization on net primary productivity in the USA were negative (~40 Mt C yr^{-1} of carbon or the equivalent to annual fossil fuel emission of a large North American city) in the mid-1990s due to attributed losses of productive agricultural land (Imhoff *et al.* 2004), but this value did not include increases in organic carbon due to constructed urban, suburban, and exurban ecosystems that contain managed vegetation and soils. These increases can be significant. For example, Nowak and Crane (2002) estimated that in the coterminous USA, urban organic carbon pools in trees are on the order of approximately 700 Mt with a gross sequestration rate of approximately 22.8 Mt of carbon per year. The carbon pools in urban trees amounts to about 5.5 months of total USA emissions. That said, Nowak *et al.* (2002) suggest that if fossil fuels are used to maintain urban vegetation, the net effect of urban trees will eventually become negative (i.e. net emissions of carbon). The dynamics of this trade-off vary with tree species, disposal techniques, and maintenance intensity, although the net effect can be altered if trees are planted in energy conserving locations that offset maintenance intensity.

Urbanization also potentially modifies soil carbon pools and fluxes through changes in soil chemistry, temperature regimes, soil community composition, and nitrogen and carbon fluxes. Given the expanding size of cities, the potential impact of urban land use change on soil carbon is an increasing concern of soil scientists (Pouyat *et al.* 2002, 2009, 2006). In general, soil scientists believe that urban land conversions result in poor conditions for plant growth, and therefore reduce soil organic matter inputs. Furthermore, large areas of urban land are not typically abandoned for long periods, making recovery of soil carbon slow. As such, there has been a consensus that urbanization reduces soil carbon for areas within the city for long periods of time. For example, organic carbon storage in Chuncheon, Korea soils averaged 31.6 t/ha for natural lands and 24.8 t/ha for urban lands (Jo 2002).

Recent research, however, suggest that dense settlement patterns and urban activities may increase *or* decrease soil organic carbon pools (Pouyat *et al.* 2002). In general, soil organic matter and soil organic carbon pools in cities appear to rebound after an initial loss of carbon from disturbances during conversion to urban land. The increases varied, however, with the largest observed in the most highly managed soils. For example, Qian and Follet (2002) found an increase in soil organic matter in golf course fairways after turfgrass planting. After one year they found a 1.76 percent increase, after 20 years, they found a 3.8 percent increase and after 31 years they found a 4.2 percent increase. Others have found that soil organic carbon in urban soils could increase to more than that found in surrounding native grasslands (Golubiewski 2006). Pouyat *et al.* (2006) found that cities in the Northeast USA, located on native soils with high carbon concentrations (Boston and Syracuse), lost soil organic carbon pools following urbanization. By contrast, cities located in warmer and/or drier climates, such as Chicago and Oakland, showed increases in soil organic carbon pools compared to native soil levels. Moreover, studies of variability of soil organic carbon pools within cities suggest that land use and management are significant influences (Pouyat *et al.* 2002). In New York City, for example, the highest soil organic carbon density levels occurred on a golf course (28.5×10^{-9} Mt m^{-2}), while the lowest density occurred in an old dredge site (2.9×10^{-9} Mt m^{-2}), with nearly a ten-fold difference. Pouyat *et al.* (2006) used these and other measurements to estimate that the total carbon storage of urban soils in the coterminous US was approximately 1.9×10^{3} Mt and the average soil organic carbon density was approximately 7.7×10^{-9} Mt m^{-2}.

Takahashi *et al.* (2008) in a study of 19 parks in Tokyo suggest the soil organic carbon varied among four different types of management regimes. The lowest amount was found in bare land (approximately 40.1 Mg C ha^{-1}) and the highest in coppice land (approximately 120.9 Mg C ha^{-1}). This research concluded that urban soils are potentially carbon sinks.

In a recent comparison of soil organic carbon stocks in residential turf grass and native soils, soil scientists have shown that the management of turfgrass is having a homogenizing affect on soil organic carbon densities in cities of different climates (Pouyat *et al.* 2009). In a comparison of soil organic carbon in Denver, CO and Baltimore, MD, despite different levels in the native soils, different parent material, topography, and climate, turfgrass densities of soil organic carbon were similar. This is important because the total estimated areal amount of turf grass for the coterminous US is approximately 163,800 km^2, which exceeds by three times the area of irrigated corn (Milesi *et al.* 2005).

Conclusions

This brief review of urban soils presented selected research on different topics of contemporary interest. It described a selection of the rich variety of research projects that have been undertaken on urban soils over the past few decades. Much work on urban soils has been undertaken, but much more needs to be done. Most researchers remark of the lack of basic understanding of fundamental processes in soils within cities.

Certainly, new research tools and new areas of theoretical and practical importance are stimulating work and the field is growing rapidly. Pavao-Zukerman and Byrne (2009) have recently pointed out that urban soils studies are growing due to increased attention to underground processes as important for ecosystem functioning and the provision of ecosystem services. Moreover, integrated scaled models of human-environment interactions are providing new insights into the importance of soils for not only local populations, but also for global systems. At the same time, concerns over the regional and global climate and the need to reduce the impacts of urbanization are also helping to provide new research in this area. For example, local urban agriculture is of increasing interest and the work of soil scientists can play a major role in identifying the promises and limits of this new trend. Furthermore, remediation of soils in cities is another area that will demand further research. As urban sustainability and re-use of land is increasingly emphasized, our ability to understand how soils can play a role in sustainable urban landscapes will increase in importance. Urban soils research is moving forward to address both questions of ecological importance as well as practical planning.

References

Arnold, C.L. and C.J. Gibbons (1996), Impervious surface coverage, the emergence of a key environmental indicator, *Journal of the American Planning Association*, 62(2): 243–258.

Arnold, H.F. (1993), *Trees in Urban Design*, New York: Van Nostrand Reinhold.

Beyer, L., E. Cordsen, H.-P. Blume, U. Schleuss, B. Vogt, and Q. Wu (1996), Soil organic matter composition in urbic anthrosols in the city of Kiel, NW-Germany, as revealed by wet chemistry and CPMAS 13C-NMR spectroscopy of whole soil samples, *Soil Technology*, 9(3): 121–132.

Blume, H.-P. (1989), Classification of soils in urban agglomerations, *Catenai*, 16(3): 269–275.

Bogue, D.J. (1956), *Metropolitan Growth and Conversion of Land to Non-Agricultural Uses*, Oxford, OH, published jointly by Scripps Foundation for Research in Population Problems, Miami University; and Population Research and Training Center, University of Chicago.

Bornstein, R.D. (1968), Observations of the urban heat island effect in New York City, *Journal of Applied Meteorology and Climatology*, 7(4): 576–582.

Brady, N.C. and R.R. Weil (2002), *The Nature and Properties of Soils*, 13th edition, Upper Saddle River, NJ: Prentice Hall.

Bridge, P. and B. Spooner (2001), Soil fungi: diversity and detection, *Plant and Soil*, 232(102): 147–154.

Brown, R.B., J.H. Huddleston, and J.L. Anderson (eds.) (2000), *Managing Soils in an Urban Environment*, Madison, WI: American Society of Agronomy, Inc, Crop Science Society of America, Inc, Soil Science Society of America, Inc.

Brussaard, L. (1997), Biodiversity and ecosystem functioning in soil, *Ambio*, 26(8): 563–570.

Bullock, P. and P.J. Gregory (eds.) (1991), *Soils in the Urban Environment*, Oxford: Blackwell Scientific Publications.

Butler, J.D., V. Butterworth, S.C. Kellow, and H.G. Robinson (1984), Some observations on the polycyclic aromatic hydrocarbon (PAH) content of surface soils in urban areas, *The Science of the Total Environment*, 33(1–4): 75–85.

Cal-Prieto, M.J., A. Carlosena, J.M. Andrade, M.L. Martinez, S. Muniategui, P. Lopez-Mahia, and D. Prada (2001), Antimony as a tracer of the anthropogenic influence on soils and estuarine sediments, *Water, Air, and Soil Pollution*, 29(1–4): 333–348.

Carey, A.E., J.A. Gowen, T.J. Forehand, H. Tai, and G.B. Wiersma (1980), Heavy metal concentrations in soils of five United States cities, 1972 Urban Soils Monitoring Program, *Pesticides Monitoring Journal*, 13(4): 150–154.

Carreiro, M.M. (2005), Effects of urbanization on decomposer communities and soil processes in forest remnants, in E.A. Johnson and M.W. Klemens (eds.), *Nature in Fragments, the Legacy of Sprawl*, New York: Columbia University Press, pp. 125–143.

Carreiro, M.M., K. Howe, D.F. Parkhurst, and R.V. Pouyat (1999), Variation in quality and decomposability of red oak leaf litter along an urban-rural gradient, *Biology and Fertility of Soils*, 30(3): 258–268.

Chen, T.B., J.W.C. Wong, H.Y. Zhou, and M.H. Wong (1997), Assessment of trace metal distribution and contamination in surface soils of Hong Kong, *Environmental Pollution*, 96(1): 61–68.

Chon, H.-T., J.-S. Ahn, and M.C. Jung (1998), Seasonal variations and chemical forms of heavy metals in soils and dusts from the satellite cities of Seoul, Korea, *Environmental Geochemistry and Health*, 20(2): 77–86.

Craul, P.J. (1985), A description of urban soils and their desired characteristics, *Journal of Arboriculture*, 11(11): 330–339.

Craul, P.J. (1992), *Urban Soil in Landscape Design*, New York: John Wiley.

Craul, P.J. and C.J. Klein (1980), Characterization of streetside soils of Syracuse, New York, *Metropolitan Tree Improvement Alliance (METRIA) Proceedings*, 3: 88–101.

Cunningham, M.A., E. Snyder, D. Yonkin, M. Ross, and T. Elsen (2008), Accumulation of deicing salts in soils in an urban environment, *Urban Ecosystems*, 11(1): 17–31.

Daily, G.C. (1997), *Nature's Services: Societal Dependence on Natural Ecosystems*, Washington, DC: Island Press.

Davies, B.E. and N.J. Houghton (1984), Distance-decline patterns in heavy metal contamination of soils and plants in Birmingham, England, *Urban Ecology*, 8(4): 285–294.

Davies, D.J.A., J.M. Watt, and I. Thornton (1987), Lead levels in Birmingham dusts and soils, *The Science of the Total Environment* 67(2–3): 177–185.

Davies, J.C., and J. Mazurek (1999), *Pollution Control in the United States, Evaluating the System*, Washington DC: Resources for the Future.

Dejonghe, W., N. Boon, D. Seghers, E.M. Top, and W. Verstraete (2001), Bioaugmentation of soils by increasing microbial richness: missing links, *Environmental Microbiology*, 3(10): 649–457.

De Kimpe, C. and J.-L. Morel (2000), Urban soil management: a growing concern, *Soil Science*, 165(1): 31–40.

de Miguel, E., J.F. Llamas, E. Chacon, T. Berg, S. Larssen, O. Royset, and M. Vadset (1997), Origin and patterns of distribution of trace elements in street dust: Unleaded petrol and urban lead, *Atmospheric Environment*, 31(17): 2733–2740.

Douglas, I. (1968), Erosion in the Sungei Gombak catchment, Selangor, *The Journal of Tropical Geography*, 26: 1–16.

Douglas, I. (1974), The impact of urbanization on river systems, *Proceedings of the International Geographical Union Regional Conference and Eight New Zealand Geography Conference*, pp. 307–317.

Douglas, I. (1978), The impact of urbanization on fluvial geomorphology in the humid tropics, *International Journal of Tropical Ecology and Geography*, 2: 229–242.

Douglas, I. (1981), The city as an ecosystem, *Progress in Physical Geography*, 5(3): 315–367.

Douglas, I. (1983), *The Urban Environment*, London: Edward Arnold.

Driessen, P., J. Deckers, and O. Spaargaren (2001), Lecture notes on soils of the world, *World Soil Resources Reports – 94*, Rome: FAO.

Effland, W.R., and R.V. Pouyat (1997), The genesis, classification, and mapping of soils in urban areas, *Urban Ecosystems*, 1(4): 217–228.

Fenton, T.E. and M.E. Collins (2000), The soil resource and its inventory, in R.B. Brown, J.H. Huddleston, and J.L. Anderson (eds.), *Managing Soils in an Urban Environment*, Madison, WI: American Society of Agronomy, Inc, Crop Science Society of America, Inc, Soil Science Society of America, Inc., pp. 1–32.

Frankie, G.W., and L.E. Ehler (1978), Ecology of insects in urban environments, *Annual Review of Entomology*, 23: 367–387.

Garcia-Miragaya, J., S. Castro, and J. Paolini (1981), Lead and zinc levels and chemical fractionation in roadside soils of Caracas, Venezuela, *Water, Air and Soil Pollution*, 15(3): 285–297.

Gates, G.E. (1976), More on earthworm distribution in North America, *Proceedings of the Biological Society of Washington*, 89(40): 467–476.

Gbadegesin, A. and M.A. Olabode (2000), Soil properties in the metropolitan region of Ibadan, Nigeria: implications for the management of the urban environment of developing countries, *The Environmentalist*, 20(3): 205–214.

Gibson, M.J. and F.G. Farmer (1986), Multi-step sequential chemical extraction of heavy metals from urban soils, *Environmental Pollution B*, 11(2): 117–135.

Gilbert, O.L. (1989), *The Ecology of Urban Habitats*, London: Chapman and Hall.

Gill, D. and P. Bonnett (1973), *Nature in the Urban Landscape: A Study of City Ecosystems*, Baltimore, MD: York Press.

Godwin, K.S., S.D. Hafner, and M.F. Buff (2003), Long-term trends in sodium and chloride in the Mohawk River, New York: the effect of fifty years of road-salt application, *Environmental Pollution*, 124(2): 273–281.

Golubiewski, N.E. (2006), Urbanization increases grassland carbon pools: Effects of landscaping in Colorado's Front Range, *Ecological Applications*, 16(2): 555–571.

Gray, D.H. (1972), Soil and the city, in M.G. Marcus and T.R. Detwyler (eds.), *Urbanization and Environment*, Belmont, CA: Wadsworth Publishing Company, pp. 135–168.

Grimm, N., S.H. Faeth, N.E. Golubiewski, C.L. Redman, J. Wu, X. Bai, and J.M. Briggs (2008), Global change and the ecology of cities, *Science*, 319(5864): 756–760.

Harris, J.A. (1991), The biology of soils in urban areas, in P. Bullock and P. J. Gregory (eds.), *Soils in the Urban Environment*, Oxford: Blackwell Scientific Publications, pp. 139–152.

Hawksworth, D.L. (2001), The magnitude of fungal diversity: the 1.5 million species estimate revisited, *Mycological Research*, 105(12): 1422–2432.

Hewitt, C.N. and G.B.B. Candy (1990), Soil and street dust heavy metal concentrations in and around Cuenca, Ecuador, *Environmental Pollution*, 63(2): 129–136.

Hillel, D. (1980), *Fundamentals of Soil Physics*, New York: John Wiley.

Hiller, D.A. (2000), Properties of urbic anthrosols from an abandoned shunting yard in the Ruhr area, Germany, *Catena*, 39(4): 245–266.

Hollis, J.M. (1991), The classification of soils in urban areas, in P. Bullock and P.J. Gregory (eds.), *Soils in the Urban Environment*, Oxford: Blackwell Scientific Publications, pp. 5–27.

Houge, C.L. (1987), Cultural entomology, *Annual Review of Entomology*, 32: 181–199.

Hough, M. (1984), *City Form and Natural Process*, New York: Van Nostrand Rienhold Company, Ltd.

Imhoff, M.L., L. Bounoua, R. DeFries, W.T. Lawrence, D. Stutzer, C.J. Tucker, and T. Ricketts (2004), The consequences of urban land transformation on net primary productivity in the United States, *Remote Sensing of Environment*, 89(4): 434–443.

Imperato, M., P. Adamo, D. Naimo, M. Arienzo, D. Stanzione, and P. Violante (2003), Spatial distribution of heavy metals in urban soils of Naples city (Italy), *Environmental Pollution*, 124(2): 247–256.

Jackson, R.B. and E.G. Jobbagy (2005), From icy roads to salty streams, *Proceedings of the National Academy of Sciences of the United States of America*, 102(41): 14487–14488.

Jim, C.Y. (1989), Tree-canopy characteristics and urban development in Hong Kong, *Geographical Review*, 79(2): 210–225.

Jim, C.Y. (1998a), Physical and chemical properties of a Hong Kong roadside soil in relation to urban tree growth, *Urban Ecosystems*, 2(2–3): 171–181.

Jim, C.Y. (1998b), Soil characteristics and management in an urban park in Hong Kong, *Environmental Management*, 22(5): 683–695.

Jiries, A. (2003), Vehicular contamination of dust in Amman, Jordon, *The Environmentalist*, 23(3): 205–210.

Jiries, A., H. Hussian, and J. Lintelmann (2000), Determination of polycyclic aromatic hydrocarbons in wastewater, sediments, sludge and plants in Karak Province, Jordan, *Water, Air, and Soil Pollution*, 121(1–4): 217–228.

Jo, H.-K. (2002), Impacts of urban greenspace on offsetting carbon emissions in middle Korea, *Journal of Environmental Management*, 64(2): 115–126.

Jones, K.C., J.A. Stratford, K.S. Waterhouse, and N.B. Vogt (1989), Organic contaminants in Welsh soils: polynuclear aromatic hydrocarbons, *Environmental Science and Technology*, 23(5): 540–550.

Kaushal, S.R., P.M. Groffman, G.E. Likens, K.T. Belt, W.P. Stack, V.R. Kelly, L.E. Band, and G.T. Fisher (2005), Increased salinization of fresh water in the Northeastern United States, *Proceedings of the National Academy of Sciences of the United States of America*, 102(38): 13517–13520.

Klein, D.H. (1972), Mercury and other metals in urban soils, *Environmental Science and Technology*, 6(6): 560–562.

Kostel-Huges, F., T.P. Young, and M.M. Carreiro (1998), Forest leaf litter quanity and seedling occurrence along an urban-rural gradient, *Urban Ecosystems*, 2(4): 263–278.

Kozlov, M.V. (1996), Patterns of forest insect distribution within a large city: microlepidoptera in St. Petersburg, Russia, *Journal of Biogeography*, 23(1): 95–103.

Krishna, A.K. and P.K. Govil (2005), Heavy metal distribution and contamination in soils of Thane-Belapur industrial development area, Mumbai, Western India, *Environmental Geology*, 47(8): 1054–1061.

Landsberg, H. (1981), *The Urban Climate*, New York: International Geophysics Series 28.

Li, X., C.-S. Poon, and P.S. Liu (2001), Heavy metal contamination of urban soils and street dusts in Hong Kong, *Applied Geochemistry*, 16(11–12): 1361–1368.

Linde, M., H. Bengtsson, and I. Oborn (2001), Concentrations and pools of heavy metals in urban soils in Stockholm, Sweden, *Water, Air, and Soil Pollution: Focus*, 1(3–4): 83–101.

Lofgren, S. (2001), The chemical effects of deicing salt on soil and stream water of five catchments in Southeast Sweden, *Water, Air, and Soil Pollution*, 130(1–4): 863–868.

Loredo, J., A. Ordonez, S. Charlesworth, and E. de Miguel (2003), Influence of industry on the geochemical urban environment of Mieres (Spain) and associated health risk, *Environmental Geochemistry and Health*, 25(3): 307–323.

Luniak, M. (2008), Fauna of the big city – Estimating species richness and abundance in Warsaw, Poland, in J.M. Marzluff, E. Shulenberger, W. Endlicher, M. Alberti, G. Bradley, C. Ryan, U. Simon, and

C. ZumBrunnen (eds.), *Urban Ecology, An International Perspective on the Interaction between Humans and Nature*, New York: Springer, pp. 349–354.

Malawska, M. and B. Wilkomirski (2001), An analysis of soil and plant (*Taraxacum officinale*) contamination with heavy metals and polycyclic aromatic hydrocarbons (PAHs) in the area of the railway junction Iława Głowna, Poland, *Water, Air, and Soil Pollution*, 127(1–4): 339–349.

Manta, D.S., M. Angelone, A. Bellanca, R. Neri, and M. Sprovieri (2002), Heavy metals in urban soils: a case study from the city of Palermo (Sicily), Italy, *The Science of the Total Environment*, 300(1–3): 229–243.

Marino, F., A. Ligero, and D.J.D. Cosin (1992), Heavy metals and earthworms on the border of a road next to Santiago (Galicia, Northwest of Spain): Initial results, *Soil Biology and Biochemistry*, 24(12): 1705–1709.

McDonnell, M.J. and S.T.A. Pickett (1990), Ecosystem structure and function along urban-rural gradients: an unexploited opportunity for ecology, *Ecology*, 71(4): 1232–1237.

McDonnell, M.J. and S.T.A. Pickett (1991), Comparative analysis of ecosystems along gradients of urbanization: Opportunities and limitations, in J.J. Cole, G.M. Lovett, and S.E.G. Findlay (eds.), *Comparative Analyses of Ecosystems, Patterns, Mechanisms and Theories*, New York: Springer, pp. 351–255.

McDonnell, M.J., S.T.A. Pickett, P. Groffman, P. Bohlen, R.V. Pouyat, W.C. Zipperer, R.W. Parmelee, M.M. Carreiro, and K. Medley (1997), Ecosystem processes along an urban-to-rural gradient, *Urban Ecosystems*, 1(1): 21–36.

McHarg, I.L. (1992), *Design with Nature* (25th Anniversary Edition, originally published in 1969), New York: John Wiley and Sons, Inc.

McIntyre, N. (2000), Ecology of urban arthropods: a review and call to action, *Ecology and Population Biology*, 93(4): 821–835.

McIntyre, N.E., J. Rango, W.F. Fagan, and S.H. Faeth (2001), Ground arthropod community structure in a heterogeneous urban environment, *Landscape and Urban Planning*, 52(4): 257–274.

McKinney, M.L. (2002), Urbanization, biodiversity, and conservation, *BioScience*, 52(10): 883–890.

McKinney, M.L. (2006), Urbanization as a major cause of biotic homogenization, *Biological Conservation*, 127(3): 247–260.

Mielke, H.W., G. Wang, C.R. Gonzales, B. Le, V.N. Quach, and P.W. Mielke (2001), PAH and metal mixtures in New Orleans soils and sediments, *The Science of the Total Environment*, 281(1–3): 217–227.

Milesi, C., S.W. Running, C.D. Elvidge, J.B. Dietz, B.T. Tuttle, and R.R. Nemani (2005), Mapping and modeling the biogeochemical cycling of turf gasses in the United States, *Environmental Management*, 36(3): 426–438.

Millennium Ecosystem Assessment (2005), *Ecosystems and Human Well-Being, Synthesis*, Washington, DC: Island Press.

Mullins, C.E. (1991), Physical properties of soils in urban areas, in P. Bullock and P.J. Gregory (eds.), *Soils in the Urban Environment*, Oxford: Blackwell Scientific Publications, pp. 87–118.

Netto, A.D. P., T.M. Krauss, I.F. Cunha, and E.P. Rego (2006), PAHs in SD: Polycyclic aromatic hydrocarbons levels in street dusts in the central area of Niteroi City, RJ, Brazil, *Water, Air, and Soil Pollution*, 176(1–4): 57–67.

New York City Soil Survey Staff (2005), *New York City Reconnaissance Soil Survey*, Staten Island, NY: United States Department of Agriculture, Natural Resources Conservation Service.

Niemelä, J., D.J. Kotze, S. Venn, L. Penev, I. Stoyanov, J. Spence, D. Hartley, and E.M. de Oca (2002), Carabid beetle assemblages (Coleoptera, Carabidae) across urban-rural gradients: an international comparison, *Landscape Ecology*, 17(5): 387–401.

Niemelä, J., J. Kotze, A. Ashworth, P. Brandmayr, K. Desender, T. New, K. Penev, M. Samways, and J. Spence (2000), The search for common anthropogenic impacts on biodiversity: a global network, *Journal of Insect Conservation*, 4(1): 3–9.

Nowak, D.J. and D.E. Crane (2002), Carbon storage and sequestration by urban trees in the USA, *Environmental Pollution*, 116(3): 381–389.

Nowak, D.J., J.C. Stevens, S.M. Sisinni, and C.J. Luley (2002), Effects of urban tree management and species selection on atmospheric carbon dioxide, *Journal of Arboriculture*, 28(3): 113–122.

Oke, T.R. (1973), City size and urban heat island, *Atmospheric Environment*, 7(8): 69–779.

Ordóñez, A., J. Loredo, E.D. Miguel, and S. Charlesworth (2003), Distribution of heavy metals in the street dusts and soils of an industrial city in northern Spain, *Archives of Environmental Contamination and Toxicology*, 44(2): 160–170.

Palm, C., P. Sanchez, S. Ahamed, and A. Awiti (2007), Soils: A comptemporary perspective, *Annual Review of Environment and Resources*, 32: 99–129.

Pataki, D.E., R.J. Alig, A.S. Fung, N.E. Golubiewski, C.A. Kennedy, E.G. McPherson, D.J. Nowak, R.V. Pouyat, and P.R. Lankao (2006), Urban ecosystems and the North American carbon cycle, *Global Change Biology*, 12(11): 2092–2102.

Paterson, E., M. Sanka, and L. Clark (1996), Urban soils as pollutant sinks: A case study from Aberdeen, Scotland, *Applied Geochemistry*, 11(1): 129–131.

Patterson, J.C. (1976), *Soil compaction and its effects upon urban vegetation*, paper read at Better Trees for Metropolitan Landscapes Symposium, Proceedings, USDA Forest Service General Technical Report NE-22.

Pavao-Zukerman, M.A., and L.B. Byrne (2009), Scratching the surface and digging deeper: Exploring ecological theories in urban soils, *Urban Ecosystems*, 12(1): 9–20.

Pichtel, J., H.T. Sawyerr, and K. Czarnowska (1997), Spatial and temporal distribution of metals in soils in Warsaw, Poland, *Environmental Pollution*, 98 (2): 169–174.

Pickett, S.T.A., M.L. Cadenasso, J.M. Grove, P.M. Groffman, L.E. Band, C.G. Boone, W.R. Burch Jr, C.S.B. Grimmond, J. Hom, J.C. Jenkins, N.L. Law, C.H. Nilon, R.V. Pouyat, K. Szlavecz, P.S. Warren, and M.A. Wilson (2008), Beyond urban legends: An emerging framework of urban ecology as illustrated by the Baltimore Ecosystem Study, *BioScience*, 58(2): 139–150.

Pizl, V. and G. Josens (1995), Earthworm communities along a gradient of urbanization, *Environmental Pollution*, 90(1): 7–14.

Post, W.M., W.R. Emanuel, P.J. Zinke, and A.G. Stangenberger (1982), Soil carbon pools and world life zones, *Nature*, 298(5870): 156–159.

Pouyat, R.V., M.J. McDonnell, and S.T.A. Pickett (1997), Litter decomposition and nitrogen mineralization in oak stands along an urban-rural land use gradient, *Urban Ecosystems*, 1(2): 117–131.

Pouyat, R.V., R.W. Parmelee, and M.M. Carreiro (1994), Environmental effects of forest soil-invertebrate and fungal densities in oak stand along an urban-rural land use gradient, *Pedobiologia*, 38: 385–399.

Pouyat, R.V., I.D. Yesilonis, and N.E. Golubiewski (2009), A comparison of soil organic carbon stocks between residential turf grass and native soil, *Urban Ecosystems*, 12(1): 45–62.

Pouyat, R.V., I.D. Yesilonis, and D.J. Nowak (2006), Carbon storage by urban soils in the United States, *Journal of Environmental Quality*, 35(4): 1566–1575.

Pouyat, R.V., P. Groffman, I. Yesilonis, and L. Hernandez (2002), Soil carbon pools and fluxes in urban ecosystems, *Environmental Pollution*, 116(Suppl. 1): S107–S118.

Pouyat, R.V., I.D. Yesilonis, J. Russell-Anelli, and N.K. Neerchal (2007), Soil chemical and physical properties that differentiate urban land-use and cover types, *Soil Science Society of America Journal*, 71(3): 1010–1019.

Prosser, J.I. (2002), Molecular and functional diversity in soil micro-organisms, *Plant and Soil*, 2442(1–2): 9–17.

Purves, D. (1966), Contamination of urban garden soils with copper and boron, *Nature*, 210 (5040): 1077–1078.

Purves, D. (1972), Consequences of trace-element contamination of soils, *Environmental Pollution*, 3(1): 17–24.

Purves, D. and E.J. Mackenzie (1969), Trace-element contamination of parklands in urban areas, *Journal of Soil Science*, 20(2): 288–296.

Pyle, R., M. Bentzien, and P. Opler (1981), Insect conservation, *Annual Review of Entomology*, 26: 233–258.

Qian, Y. and R.F. Follett (2002), Assessing soil carbon sequestration in turfgrass systems using long-term soil testing data, *Agronomy Journal*, 94(4): 930–935.

Ramakrishna, D.M. and T. Viraraghavan (2005), Environmental impact of chemical deicers – A review, *Water, Air, and Soil Pollution*, 166(1–4): 49–63.

Raup, P.M. (1975), Urban threats to rural lands: background and beginnings, *Journal of the American Institute of Planners*, 41(6): 371–378.

Romic, M. and D. Romic (2003), Heavy metals distribution in agricultural topsoils in urban area, *Environmental Geology*, 43(7): 795–805.

Rozen, A. and L. Mazur (1997), Influence of different levels of traffic pollution on haemoglobin content in the earthworm "Lumbricus terrestris," *Soil Biology and Biochemistry*, 29(3–4): 709–711.

Ruiz-Cortés, E., R. Reinoso, E. Dias-Barrientos, and L. Madrid (2005), Concentrations of potentially toxic metals in urban soils of Seville: relationship with different land uses, *Environmental Geochemistry and Health*, 27(5–6): 465–474.

Sánchez-Camazano, M., M.J. Sánchez-Martín, and L.F. Lorenzo (1994), Lead and cadmium in soils and vegetables from urban gardens of Salamanca (Spain), *The Science of the Total Environment*, 146–147: 163–168.

Santas, P. (1986), Soil communities along a gradient of urbanization, *Revue D'Ecologie et de Biologie du Sol*, 23(4): 367–380.

Schlesinger, W.H. (1997), *Biogeochemistry, An Analysis of Global Change*, San Diego, CA: Academic Press.

Schlesinger, W.H., and J.A. Andrews (2000), Soil respiration and the global carbon cycle, *Biogeochemistry*, 48(1): 7–20.

Shochat, E., W.L. Stefanov, M.E.A. Whitehouse, and S.H. Faeth (2004), Urbanization and spider diversity: Influences of human modification of habitat structure and productivity, *Ecological Applications*, 14(1): 268–280.

Short, J.R., D.S. Fanning, J.E. Foss, and J.C. Patterson (1986a), Soils in the Mall in Washington DC: II. Genesis, classification and mapping, *Soil Science Society of America Journal*, 50(3): 705–710.

Short, J.R., D.S. Fanning, M.S. McIntosh, J.E. Foss, and J.C. Patterson (1986b), Soils in the Mall in Washington DC: I. Statistical summary of physical properties, *Soil Science Society of America Journal*, 50(3): 699–705.

Spirn, A.W. (1984), *The Granite Garden*, New York: Basic Books.

Spitzer, T. and S. Kuwatsuka (1993), Residue levels of polynuclear aromatic comounds in urban surface soil from Japan, *Journal of Chromatography*, 643(1–2): 305–309.

Standen, V., G.B. Stead, and A. Dunning (1982), Lumbricid populations in opencast reclamation sites and colliery spoil heaps in county Durham, UK, *Pedobiologia*, 24(1): 57–64.

Steinberg, D.A., R.V. Pouyat, R.W. Parmelee, and P.M. Groffman (1997), Earthworms abundance and nitrogen mineralization rates along an urban-rural land use gradient, *Soil Biology and Biochemistry*, 29 (3–4): 427–430.

Sukopp, H. (2004), Human-caused impact on preserved vegetation, *Landscape and Urban Planning*, 68(4): 347–355.

Sukopp, H. (2008), On the early history of urban ecology in Europe, in J.M. Marzluff, E. Shulenberger, W. Endlicher, M. Alberti, G. Bradley, C. Ryan, U. Simon and C. ZumBrunnen (eds.), *Urban Ecology: An International Perspective on Interaction Between Humans and Nature*, New York: Springer, pp. 79–97.

Sukopp, H., H.-P. Blume, and W. Kunick (1979), The soil, flora, and vegetation of Berlin's waste lands, in I.C. Laurie (ed.), *Nature in Cities, The Natural Environment in the Design and Development of Urban Green Space*, Chichester: John Wiley & Sons, pp. 115–132.

Takada, H., T. Onda, and N. Ogura (1990), Determination of polycyclic aromatic hydrocarbons in urban street dusts and their source materials by capillary gas chromatography, *Environmental Science and Technology*, 24(8): 1179–1186.

Takahashi, T., Y. Amano, K. Kuchimura, and T. Kobayashi (2008), Carbon content of soil in urban parks in Tokyo, Japan, *Landscape and Ecological Engineering*, 4(2): 139–142.

Thornton, I. (1991), Metal contamination of soils in urban areas, in P. Bullock and P.J. Gregory (eds.), *Soils in Urban Areas*, Oxford: Blackwell Scientific Publishers, pp. 47–75.

Thuy, H.T.T., H.J. Tobschall, and P.V. An (2000), Distribution of heavy metals in urban soils: A case study of Danang-Hoian Area (Vietnam), *Environmental Geology*, 39(6): 603–610.

UNFPA (2007), *State of the World Population: Unleashing the Potential of Urban Growth*, New York: United Nations Population Fund.

van Metre, P.C., B.J. Mahler, and E.T. Furlong (2000), Urban sprawl leaves its PAH signature, *Environmental Science and Technology*, 34(19): 4064–4070.

Wagrowski, D.M. and R.A. Hites (1997), Polycyclic aromatic hydrocarbon accumulation in urban, suburban and rural vegetation, *Environmental Science and Technology*, 31(1): 279–282.

Wang, X.S., Y. Qin, and S.X. Sang (2005), Accumulation and sources of heavy metals in urban topsoils: a case study from the city of Xuzhou, China, *Environmental Geology*, 48(1): 101–107.

Wheater, C.P. (ed.) (1999), *Urban Habitats*, London: Routledge.

White, C.S. and M.J. McDonnell (1988), Nitrogen cycling processes and soil characteristics in an urban versus rural forest, *Biogeochemistry*, 5(2): 243–262.

Wilcke, W., S. Muller, N. Kanchanakool, and W. Zech (1998), Urban soil contamination in Bangkok: heavy metal and aluminum partitioning in topsoils, *Geoderma*, 86(3–4): 211–228.

Yang, S.Y.N., D.W. Connell, D.W. Hawker, and S.I. Kayal (1991), Polycyclic aromatic hydrocarbons in air, soil and vegetation in the vicinity of an urban roadway, *The Science of the Total Environment*, 102: 229–240.

Yuangen, Y., C.D. Campbell, L. Clark, C.M. Cameron, and E. Paterson (2006), Microbial indicators of heavy metal contamination in urban and rural soils, *Chemosphere*, 63(11): 1942–1952.

Zhai, M., H.A.B. Kampunzu, M.P. Modisi, and O. Totolo (2003) Distribution of heavy metals in Gaborone urban soils (Botswana) and its relationship to soil pollution and bedrock composition, *Environmental Geology*, 45(2): 171–180.

Zheng, M. and M. Fang (2000), Particle-associated polycyclic aromatic hydrocarbons in the atmosphere of Hong Kong, *Water, Air, and Soil Pollution*, 117(1–4): 175–189.

16

The process of natural succession in urban areas

Wayne C. Zipperer

Introduction

Succession has been a fundamental concept in ecology. Its classical definition is the orderly change in vegetation at a site that is predictable and directional towards a climax state or end point (Clements 1916). A general assumption of succession is that early seres are governed by allogenic processes, environmental processes external to the site, and early seres facilitate later successional stages. As the community matures, autogenic processes, biotic interactions, become important in facilitating later-successional assemblages and the movement of the community towards an end point (Connell and Slatyer 1977). Field studies often showed trajectories were not always predictable, end points were not always achieved, and allogenic processes played an important role in community dynamics throughout successional development. To shift the focus of successional studies away from descriptive to mechanisms or interactions that contribute to successional change, Connell and Slatyer (1977) proposed three distinct mechanisms—facilitation, tolerance, and inhibition—at the community level. This shift enabled experimental approaches to studying succession, but it failed to capture the complexity of vegetation dynamics (Pickett *et al.* 1987a).

Clements (1916) lists five basic causes of succession: (1) a disturbance opening a site; (2) migration of propagules to the site; (3) establishment of species; (4) biotic interactions; and (5) modification of the site by organisms. Pickett *et al.* (1987a, 1987b) synthesized this list into a single-organized framework that captures vegetation dynamics at multiple scales. At the highest level of hierarchical organization, three general and universal conditions exist: (1) site availability; (2) differential species availability; and (3) differential species performance (Figure 16.1) (Pickett *et al.* 1987b). At the next lower or intermediate level, an individual mechanism or causation is defined for each condition as, for example, the availability of species depends on dispersal and propagule pools. This level in turn can be examined further for a particular process.

Most successional studies have been conducted in non-urban landscapes. Although these studies can provide a framework to study succession in an urban landscape, they do not capture the uniqueness of the urban matrix (see Rebele 1994). In this chapter, I will use the hierarchy of causes of vegetation dynamics—site availability, differential species availability, and differential species performance—to examine factors and mechanisms influencing succession in urban landscapes and overview some examples of successional studies in cities.

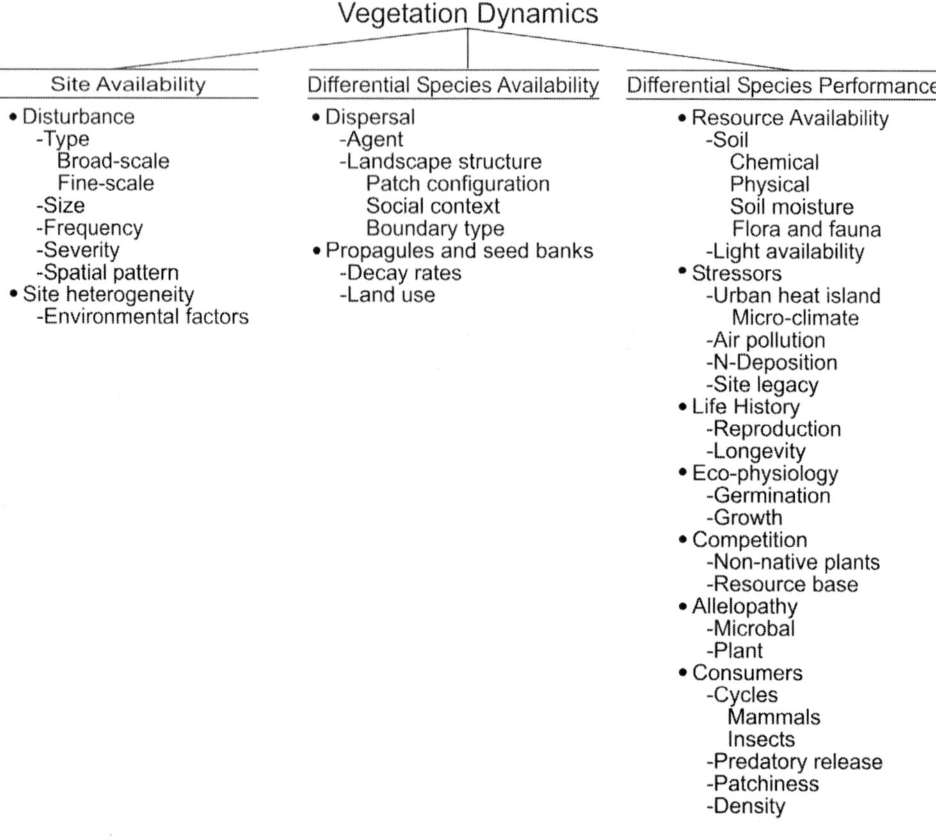

Figure 16.1 Vegetation dynamics is illustrated as a hierarchy ranging from the most general phenomenon of community change to detailed interactions within a mechanism or causation. Adapted from Pickett *et al.* (1987b) and Pickett and Cadenasso (2005).

Site availability

Site availability includes not only sites created by disturbances (broad scale) (Pickett and White 1985), but also the death of an individual and existing site conditions (fine scale) (Brand and Parker 1995). A disturbance is defined as "any discrete event in time that disrupts [landscape], ecosystem, community, or population structure and changes resources, substrate availability, or physical environment" (Pickett and White 1985). Natural disturbances that affect urban landscapes include wind and ice storms, fire, and flooding as well as herbivory (e.g. insect outbreaks), pathogens, and stress-induced mortality. A disturbance can be classified by its seasonality, distribution, frequency, magnitude, and severity (Pickett and White 1985) and, collectively, all of the disturbances affecting a landscape are called the disturbance regime.

The primary site disturbance in the urban landscape is land clearing for development. Site clearing can range from an area less than 0.25 hectare for a single structure, thus possibly leaving the adjacent vegetation relatively intact, to clearing the entire area for an entire subdivision (e.g. >50 ha). Site clearing can include not only removing existing vegetation cover but also removing soil and possibly altering drainage patterns (Effland and Pouyat 1997). Such extensive clearing can create a new substrate as A and B horizons of the soil profile are removed. Successional processes on these sites would be defined as primary succession rather than secondary succession since this new substrate has not previously supported vegetation, does not contain a seed bank, and possibly has not accumulated organic matter. In general, primary succession is associated with glaciated and volcanic sites rather than urban sites. In the urban landscape, primary succession is associated with stone and brick walls, demolition sites, spoil heaps from industrial waste, and abandoned roads and sidewalks (Rebele 1992, 1994). Contrary to vegetation dynamics in adjacent rural areas where secondary succession dominates, both primary and secondary successional processes are important ecological processes in urban landscapes.

The disturbance regime changes as a landscape is urbanized. Some disturbances such as fires and flooding are suppressed because of their devastating effects on property and the potential for loss of life. For instance, fire suppression reduces the frequency and alters the intensity of fire events. A subsequent outcome is a shift away from fire-dependent species to more mesic species and new plant communities. In the Chicago Metropolitan area, fire suppression has changed the oak savannah from a community dominated by *Quercus* spp. and grasses to one containing *Acer saccharum* and exotic species (Kline 1997).

Similarly, humans introduce new disturbances. Sharpe *et al.* (1986) observed an increase in frequency of fire because of arson, trampling, dumping of yard and garden waste, and vandalism in urban woodlands in the Milwaukee metropolitan area. A study of reforested vacant lots in Syracuse, New York revealed that 80 percent of sites sampled were disturbed intensively and extensively by human activities such as vandalism, fire, trampling, and dumping (Zipperer 2002). Matlack (1993) called this human effect a sociological-edge effect because most human activities occurred within 82 m of the edge.

Even though these novel disturbances create new sites for colonization, they affect the germination and growth of species. For instance, trampling is a novel, small-scale, frequent disturbance which many native species have not evolved mechanisms to cope with. Without natural barriers such as high stones and down-woody debris to created micro-safe sites, trees did not regenerate in remnant forest in Helsinki, Finland because of trampling (Lehvävirta *et al.* 2004).

Management activities can also be considered a type of disturbance. The type, frequency, and intensity of management will significantly influence vegetation dynamics (Zipperer *et al.* 1997). For instance, if a vacant lot is regularly managed by mowing, its structure and composition would differ from a vacant lot that is left unmanaged (*sensu* Godefroid and Koedam 2007). Overall, site management activities, such as mowing, clearing, and weeding, inhibit or alter successional processes.

Brand and Parker (1995) recognized that seasonality, environmental heterogeneity within the site, and environmental factors of the site can influence germination and growth. Environmental factors particularly important to urban landscapes include urban heat island, atmospheric pollution (e.g. nitrogen deposition and ozone), altered soil properties (physical and chemical), soil moisture, and light availability (Rebele 1994; Pickett *et al.* 2001) (see species performance section for more detail).

Another aspect of site condition is its landscape structure and social context. Landscape structure refers to site location in an urban landscape with respect to adjacent land uses as well as site configuration (e.g. shape, size, and orientation). Both attributes have been shown to influence site composition (see Godefroid and Koedam 2003). Social context refers to the social attributes of the site and adjacent sites. A study of plant diversity in Phoenix, Arizona, a city in the Sonoran desert of the United States, showed that in addition to elevation, former land use, home owner income and housing age were significant site attributes influencing vegetation dynamics (Hope *et al.* 2006).

Differential species availability

Differential species availability depends on the species' ability either to survive the disturbance, vegetative or in the seed bank, or disperse to the site (Pickett and Cadenasso 2005). In an urban landscape, species pool includes native and non-native species (Kühn *et al.* 2004). Rather than overview the quite extensive plant dispersal and seed bank literature (e.g. Leck *et al.* 1989; Poschlod *et al.* 2005), I will focus on three human-mediated dispersal (HMD) mechanisms—(1) vehicle, (2) horticultural stock, and (3) footwear—that augment natural dispersal in urban landscapes (Wichmann *et al.* 2009).

Vehicles

The importance of vehicles as dispersal vectors has been recognized for some time (e.g. Ridley 1930), but it has been only recently that they have been intensively studied. In Sheffield, England, Hodkinson and Thompson (1997) collected mud in the fall and summer from wheel-wells of parked vehicles. They observed that small seeds were commonly found in the mud and their densities varied seasonally. The most frequent occurring species included *Plantago major* (29.2 percent), *Poa annua* (16.5 percent), *Poa trivialis* (10.5 percent), *Urtica dioica* (6.4 percent), and *Matricaria discoidea* (5.6 percent). To quantify deposition and the movement of seeds into and out of a city, von der Lippe and Kowarik (2007, 2008) initiated a series of studies using seed traps in traffic tunnels leading into and out of Berlin. The studies showed that egress traffic was transporting, to a greater extent, more non-native species into the peri-urban than ingress traffic was transporting native species into the city. Consequently, sites within the urban core would be colonized locally by non-native species, whereas peri-urban sites could be colonized by native and non-native species occurring locally as well as those non-native species being transported by cars.

Horticultural stock

There are two aspects of horticultural stock that influence species availability—(1) the actual species being planted, and (2) contamination of the horticultural stock with non-native seeds or plants. Although a number of invasive, non-native species were introduced because of agricultural or forestry practices during the nineteenth and early twentieth centuries, horticultural or ornamental materials are becoming the primary sources for invasive species in the urban landscape (Reichard and White 2001). These introductions often expand the native range of the species and facilitate expansion of the species as new sites become locally available (Kowarik 2003). Without these introductions and secondary expansions, many of the invasive species would not have expanded their natural range as quickly as observed (Pyšek and Hulme 2005).

The contamination of plotted soil can be thought of as a subset of a much broader category of topsoil, which may contain seeds that have been removed from one site and transported to another (Hodkinson and Thompson 1997). By planting contaminated plots or augmenting existing soil with imported topsoil, new species can be added to a site where they otherwise may not have occurred.

Footwear

Although footwear can be a dispersal agent for any habitat visited by humans, I have included it here because of the density of footwear in urban areas and the movement of people from private to public lands and from urban to rural areas. Footwear serves as a secondary dispersal mechanism. Seeds, which are picked up by footwear, have the potential to be carried farther than if dispersed by natural means only. Wichmann *et al.* (2009) observed for two species of *Brassica* that the maximum distance of dispersal by wind was 250 m, whereas dispersal distances by footwear exceeded 5 km.

Urban landscape

In the previous section, social context and landscape position were identified as important site factors. Similarly, the urban landscape, itself, affects species availability. For instance, the size of the city will determine the number of available native and non-native species. Larger cities will have more non-native species, whereas smaller communities and villages have a higher proportion of native species (Pyšek 1998). Similarly, Williams *et al.* (2009) report that the urban landscape serves as a filtering mechanism through habitat transformation, fragmentation, urban environmental conditions, and human preferences. These filters work synergistically to influence floristic composition. Habitat transformation and fragmentation remove or alter existing habitat patches, thus eliminating species or drastically reducing their density. The urban environment, as previously discussed and will be discussed in greater detail in the section, creates a unique environment to which many species have not adapted, thus they are not able to survive or compete. Human preference is a socio-economic filter that captures human preferences for specific phenotypic characteristics, thus many new species (i.e. non-native species) are introduced into the urban landscape that otherwise would not have occurred (Williams *et al.* 2009).

Differential species performance

Once a species has colonized a site, its performance depends on its interactions with abiotic and biotic conditions (Pickett and Cadenasso 2005) (Figure 16.1). Unfortunately, auteological studies of native and non-native species in urban landscapes are lacking. Rather than focusing on how an individual species may perform within the urban matrix, I will overview those factors—soils, physical stressors, and consumers—that have unique attributes in the urban landscape and significantly affect species performance.

Soils

Soil factors influencing species performance include higher soil temperature due to the urban heat island effect; lower soil moisture due to soil hydrophobicity; higher concentrations of heavy metals due to emissions; and greater nitrogen and calcium deposition also due to emissions (White and McDonnell 1988; McDonnell *et al.* 1997; Lovett *et al.* 2000). In addition, the biotic environment also has been altered. When compared to rural woodlands, urban woodlands had lower densities of soil micro-invertebrates and fungi (Pouyat *et al.* 1994) but a higher density of earthworms (Steinberg *et al.* 1997) (see also Chapter 15 this volume). Nitrogen studies of rural and urban woodland soils have indicated that soils in urban woodlands have a substantial amount of extractable soil nitrate and nitrify rapidly (Pouyat *et al.* 1996; Carreiro *et al.* 1999). Vallet *et al.* (2008) observed also that soil pH and nitrogen content were important indicators of species occurrence in remnant forest patches along urban-to-rural gradients in Angers and Rennes, France.

When overlaying the disturbance regime of a site, anthropogenic inputs and materials, environmental factors, and atmospheric deposition, a mosaic of soil patches occur across the urban landscape that range from highly disturbed in industrial wastelands to relatively undisturbed conditions in remnant forest patches. A comparison of soil attributes across this matrix reveals significant differences for bulk density, pH, and concentrations of potassium, phosphorus, sodium and aluminum (Pouyat *et al.* 2007). Bulk density was the discerning variable between forest and other land uses, whereas soil pH differentiated residential land use from other non-forested land uses.

Physical stressors

The primary physical stressors in the urban landscape are the urban-heat island, elevated carbon dioxide (CO_2), and atmospheric pollution. It is important to note that these stressors do not act independently of each other but rather synergistically. Likewise, their effects can vary across urban landscape because of the spatial heterogeneity within that landscape.

The urban heat island is an important anthropogenic climate modification because of its effect on biological processes. It is the temperature differential between the rural and urban landscape (often the urban core) at dusk. This differential can be as great as 12°C depending on the size of the city and its morphology (Oke 1973 and Chapter 11 this volume). In general, larger cities have a greater temperature differential from adjacent rural landscape than smaller cities and towns. When one considers the potential additive effect of the projected increases in mean global temperatures by 1.7 to 4.9°C from global climate change with the urban-heat island effect, the additional heat loading on species will significantly affect ecosystem structure and function (Carreiro and Tripler 2005).

In addition to temperature, urban landscapes have elevated CO_2 concentrations because of fossil fuel consumption, a phenomenon called urban CO_2 dome (Idso *et al.* 2002). In Phoenix, Arizona, Idso *et al.* (2002) observed that the urban core and surrounding suburban landscape had a 67 percent and a 33 percent increase, respectively, in CO_2 concentration over the value (370 ppmV) observed for the rural landscape. From a species performance perspective, the increase in CO_2 coupled with the increase in temperature from the urban heat island effect has a significant effect on species productivity. Ziska *et al.* (2004) observed an increase in plant height, biomass and seedling number for *Chenopodium album*, a native forb species in the United States, in urban sites when compared to rural sites. Similar enhanced-growth patterns have been observed for non-native forb species, and this urban effect may be one of the possible factors leading to their invasiveness (Ziska 2003).

This increase in productivity is further augmented by nitrogen deposition, wet and dry in the urban landscape. Lovett *et al.* (2000) observed that N inputs for urban-remnant forests in New York City were 17 times greater than found in rural remnant forests. Analysis of throughfall over the course of the growing season for remnant forest in Louisville, Kentucky showed that N deposition was 33 percent greater than rural remnant forest (Carreiro and Tripler 2005). Nitrophilous species have been identified as a primary functional group occupying early successional sites (Godefroid 2001).

Urban areas also have high concentrations of many gaseous, particulate, and photochemical pollutants including NO_x, HNO_3, SO_x, H_2SO_4, O_3 and volatile organic compounds (Lovett *et al.* 2000; Gregg *et al.* 2003). O_3 may be detrimental to plant growth (Krupa *et al.* 1998); however, the effect of O_3 on growth may be offset by warmer temperatures from the urban heat island effect and higher concentrations of nitrogen and CO_2 (Ziska 2003). The synergetic effect of these physical stressors on a species performance needs to be a focus of future research if we are to gain a better understanding of vegetation dynamics for different site conditions.

Herbivory

Published studies examining the effect of wildlife populations on regeneration on urban sites are limited. Parker and Nilon (2008) have observed squirrel densities in urban areas exceeding 33 individuals per hectare. At these high densities, intensive seed predation by squirrels can eliminate opportunities for regeneration. In fact, high squirrel densities have been attributed to the poor regeneration of oaks in urban woodlands in New York City (Sisinni and Emmerich 1995).

Likewise, the number of species and density of insectivores may be reduced in urban landscapes because of habitat fragmentation and loss. For instance, Christie and Hochuli (2005) observed a decline in insectivorous birds and an increase in insect herbivory in small remnant forest patches when compared to edge sites and large forest remnant patches. Similarly, a species, whose vigor is already reduced because of the urban environment, may be more susceptible to insect attacks. In Gainesville, Florida, southern pine beetle (*Dendroctonus frontalis*), a native insect, attacked loblolly pines weakened by an extended drought (www.interfacesouth.org/products/pdf/case_studies1.pdf). The outbreak moved from the city into the surrounding areas. The mechanisms driving herbivory (e.g. predator release) in the urban landscape are complex and the diversity of responses by plant species to the synergistic effect of the urban environment and herbivory needs to be studied in more detail (Christie and Hochuli 2005).

Plant traits

In general, the urban environment tends to select for specific plant traits. For waste lands or vacant lots, these traits include a tolerance for nitrogenous conditions; soils that are dry, rocky, alkaline, and fertile; shade intolerance, and small seeds (Pyšek *et al.* 1995, 2004; Godefroid and Koedam 2007; Thompson and McCarthy 2008). Although these traits may provide insights into what species may be present, they do not provide insights into how a species will respond to the urban environment. More long-term monitoring of plant communities across a range of habitats need to be conducted for urban landscapes to successfully identify mechanisms and species responses that influence vegetation dynamics.

Succession in the urban landscape

A limited number of successional studies have focused on the urban landscape and these studies have principally examined the vacant/wasteland lots within European cities (see Rebele 1992; Pyšek *et al.* 2004; Rebele 2008). In general, successional pathways on vacant lots followed those observed for abandoned fallow lands in rural landscapes. Prach and Pyšek (2001) observed after 12 years native woody species dominating ruderal-urban sites in the Czech Republic. On nutrient poor sites, *Populus tremula* dominated, whereas on moderately rich nutrients sites *Sambucus nigra* and *Salix caprea* dominated. Similarly, site conditions rather than seasonality were observed to be important conditions for different successional pathways (Rebele 2008). In Berlin-Dahlem, Germany, Rebele (1992) tested three substrates—commercial topsoil, ruderal (landfill) soil, and sand—to assess vegetation dynamics over a five year period. On the topsoil, annuals gave way to biennials and perennials. On the ruderal soil, *Solidago canadensis* dominated the site after the first year. On sandy soils, *Betula pendula* and *Populus nigra* colonized the plot after five years. By comparison, sites that were continuously disturbed tended to remain being dominated by ruderal species (Pyšek *et al.* 2004).

Successional studies in rural landscapes, however, may not always be applicable to vegetation dynamics in urban landscapes. For instance, a comparative study of native grasslands in southeastern

Australia and South Africa showed that vegetation dynamics in remnant patches of grasslands along roads and railways in rural landscapes differed from remnant grasslands in urban landscapes (Cilliers *et al.* 2008). In rural landscapes, non-native species occurrence was principally an edge effect, whereas in the urban landscape occurrence resulted from both edge effect and gap-phase dynamics in the interior portion of the grasslands. Predicting vegetation dynamics based on the rural grassland remnants would not have identified the patterns of occurrence in the interior of grassland in urban landscapes.

For remnant forest patches, vegetation dynamics have been documented by comparing current vegetation with historical records rather than following permanent plots through time (e.g. Rudnicky and McDonnell 1989; Godefroid 2001). Rudnicky and McDonnell (1989) reported for a virgin forest stand in the Bronx Botanical Garden, New York, that composition shifted from slow growing *Quercus* spp. to fast growing *Betula* and *Prunus* spp. and structure shifted from larger diameter to smaller diameter trees. These shifts were attributed to increases in seed predation, loss of regeneration from trampling by visitors, and changes in the structure of the forest floor.

Conclusion

Vegetation dynamics in urban landscapes differ from those in rural landscapes. Both primary and secondary successional processes are important processes in urban landscapes as opposed to adjacent rural landscapes. Likewise, the availability of species shift from rural to urban. In urban landscapes non-native species play a significant role in vegetation dynamics especially in vacant/wasteland sites and there is a decline in native species richness (Kühn *et al.* 2004; McKinney 2006). Finally, the urban environment and land transformation create unique environmental conditions that species must adapt to in order to survive and mature.

The loss of native species and the increase in non-native species as a landscape becomes more urbanized have lead to the concept of biotic homogenization—the same "urban-adaptable" species becoming increasingly widespread and locally abundant in urban areas (McKinney 2006). Likewise, a comparison of species similarity across plant communities in urban and rural landscapes showed that urban communities were more similar in composition than the rural communities (McKinney 2006). Even though species composition is becoming more homogenous, the combination of native and non-native species creates novel communities that have not been documented before (Hobbs *et al.* 2006). These emergent communities create new niches and the opportunity for evolutionary development. How these combinations influence vegetation dynamics as well as ecosystem structure and function have yet to be determined.

References

Brand, T. and Parker, V.T. (1995) Scale and general laws of vegetation dynamics, *Oikos*, 73(3): 375–80.

Carreiro, M.M., Howe, K., Parkhurst, D.F., and R.V. Pouyat (1999) Variation in quality and decomposability of red oak leaf litter along an urban–rural gradient, *Biology and Fertility of Soils*, 30(3): 258–68.

Carreiro, M.M. and Tripler, C.E. (2005) Forest remnants along urban–rural gradients: Examining their potential for global climate research, *Ecosystems*, 8(5): 568–82.

Christie, F.J. and Hochuli, D.F. (2005) Elevated levels of herbivory in urban landscapes: Are declines in tree health more than an edge effect? *Ecology and Society*, 10(2): 1–10.

Cilliers, S.S., Williams, N.S.G., and Barnard, F.J. (2008) Patterns of exotic plant invasions in fragmented urban rural grasslands across continents, *Landscape Ecology*, 23(10): 1243–56.

Clements, F.E. (1916) *Plant Succession: An Analysis of the Development of Vegetation*, Washington, D.C.: Carnegie Institute, Publication 242.

Connell, J.H. and Slatyer, R.O. (1977) Mechanism of succession in natural communities and their role in community stability and organization, *American Naturalist*, 111(982): 1119–44.

Effland, W.R. and Pouyat, R.V. (1997) The genesis, classification, and mapping of soils in urban areas, *Urban Ecosystems*, 1(4): 217–28.

Godefroid, S. (2001) Temporal analysis of the Brussels flora as indicator for changing environmental quality, *Landscape and Urban Planning*, 52(4): 203–24.

Godefroid, S. and Koedam, N. (2003) How important are large vs. small forest remnants for the conservation of the woodland flora in an urban context? *Global Ecology and Biogeography*, 12(4): 287–98.

Godefroid, S. and Koedam, N. (2007) Urban plant species patterns are highly driven by density and function of built-up areas, *Landscape Ecology*, 22(8): 1227–39.

Gregg, J.W., Jones, C.G., and Dawson, T.E. (2003) Urbanization effects on tree growth in the vicinity of New York City, *Nature*, 424: 183–7.

Hobbs, R.J., Arico, S., Aronson, J., Baron, J.S., Bridgewater, P., Cramer, V.A., Epstein, P.R., Ewel, J.J., Klink, C.A., Lugo, A.E., Norton, D., Ojima, D., Richardson, D.M., Sanderson, E.W., Valladares, F., Vila, M., Regino, Z., and Zobel, M. (2006) Novel ecosystems: theoretical and management aspects of the new ecological world order, *Global Ecology and Biogeography*, 15: 1–7.

Hodkinson, D.J. and Thompson, K. (1997) Plant dispersal: the role of man, *Journal of Applied Ecology*, 34(6): 1484–1496.

Hope, D., Gries, C., Casagrande, D., Redman, C.L., Grimm, N.B., and Martin, C. (2006) Drivers of spatial variation in plant diversity across the central Arizona-Phoenix ecosystem, *Society and Natural Resources*, 19(2): 101–16.

Idso, S.B., Idso, C.D., and Balling Jr., R.C. (2002) Seasonal and diurnal variations of near-surface atmospheric CO_2 concentration within a residential sector of the urban CO_2 dome of Phoenix, AZ USA, *Atmospheric Environment*, 36(10): 1655–60.

Kline, V.M. (1997) Orchards of oak and a sea of grass, in S. Packard and C.F. Mutel (eds.), *The Tallgrass Restoration Handbook*, Washington, D.C.: Island Press, pp. 3–20.

Kowarik, I. (2003) Human agency in biological invasions: secondary releases foster naturalisation and population expansion of alien plant species, *Biological Invasion*, 5(4): 293–312.

Krupa, S.V., Tonneijck, A.E.G., and Maning, W.J. (1998) Ozone, in R.B. Flagler (ed.), *Recognition of Air Pollution Injury to Vegetation*, Pittsburgh, PA: Air and Waste Management Association, pp. 2.1–2.28.

Kühn, I., Brandl, R., and Klotz, S. (2004) The flora of German cities is naturally species rich, *Evolutionary Ecology Research*, 6: 749–64.

Leck, M.A., Parker, V.T., and Simpson, R.L. (eds.) (1989) *Ecology of Seed Banks*, San Diego, CA: Academic Press.

Lehvävirta, S., Rita, H., and Koivula, M. (2004) Barriers against wear affect the spatial distribution of tree saplings in urban woodlands, *Urban Forestry and Urban Greening*, 3(1): 3–17.

Lovett, G.M., Traynor, M.M., Pouyat, R.V., Zhu, W., and Baxter, J.W. (2000) Atmospheric deposition to oak forests along an urban–rural gradient, *Environmental Science and Technology*, 34(20): 4294–300.

Matlack, G.R. (1993) Sociological edge effects: spatial distribution of human impact in suburban forest fragments, *Environmental Management*, 17(6): 829–35.

McDonnell, M.J., Pickett, S.T.A., Groffman, P., Bohlen, P., Pouyat, R.V., Zipperer, W.C., Parmelee, R.W., and Medley, K. (1997) Ecosystem processes along an urban-to-rural gradient, *Urban Ecosystems*, 1(1): 21–36.

McKinney, M.L. (2006) Urbanization as a major cause of biotic homogenization, *Biological Conservation*, 127(3): 247–60.

Oke, T.R. (1973) City size and the urban heat island, *Atmospheric Environment*, 7(8): 769–79.

Parker, T.S. and C.H. Nilon (2008) Gray squirrel density, habitat suitability, and behavior in urban parks, *Urban Ecosystems*, 11(3): 243–55.

Pickett, S.T.A. and Cadenasso, M.L. (2005) Vegetation dynamics, in E. van der Maarel (ed.), *Vegetation Ecology*, Malden: Blackwell Science, pp. 172–98.

Pickett, S.T.A. and White, P.S. (eds.) (1985) *The Ecology of Natural Disturbance and Patch Dynamics*, New York: Academic Press.

Pickett, S.T.A., Collins, S.C., and Armesto J.J. (1987a) A hierarchical consideration of causes and mechanisms of succession, *Vegetatio*, 69(1–3): 109–14.

Pickett, S.T.A., Collins, S.C., and Armesto J.J. (1987b) Models, mechanisms and pathways of succession, *Botanical Review*, 53(3): 335–71.

Pickett, S.T.A., Cadenasso, M.L., Grove, J.M., Nilon, C.H., Pouyat, P.V., Zipperer W.C., and Costanza, R. (2001) Urban ecological systems: linking terrestrial ecological, physical, and socioeconomic components of metropolitan areas, *Annual Review of Ecology and Systematics*, 32: 127–57.

Poschlod, P., Tackenberg O., and Bonn, S. (2005) Plant dispersal potential and its relationship to species frequency and co-existence, in E. van der Maarel (ed.), *Vegetation Ecology*, Malden: Blackwell Science, pp. 147–71.

Pouyat, P.V., McDonnell, M.J., and Pickett, S.T.A. (1996) Litter and nitrogen dynamics in oak stands along an urban–rural gradient, *Urban Ecosystems*, 1(2): 117–31.

Pouyat, R.V., Parmelee, R.W., and Carreiro, M.M. (1994) Environmental effects of forest soil-invertebrate and fungal densities in oak stands along an urban–rural land use gradient, *Pedobiologia*, 38: 385–99.

Pouyat, R.V., Yesilonis, I.D., Russell-Anelli, J., and Neerchal, N.K. (2007) Soil chemical and physical properties that differentiate urban land-use and cover types, *Journal of the Soil Science Society of America*, 71(3): 1010–19.

Prach, K. and P. Pyšek (2001) Using spontaneous succession for restoration of human-disturbed habitats: Experience from central Europe, *Ecological Engineering*, 17(1): 55–62.

Pyšek, P. (1998) Alien and native species in central European urban floras: a quantitative comparison, *Journal of Biogeography*, 25(1): 155–63.

Pyšek, P. and Hulme, P.E. (2005) Spatio-temporal dynamics of plant invasions: Linking pattern to process, *Ecoscience*, 12(3): 302–15.

Pyšek, P., Prach, K., and Šmilauer, P. (1995) Relating invasion success to plant traits: An analysis of the Czech alien flora, in P. Pyšek, K. Prach, M. Rejmánek and M. Wade (eds.), *Plant Invasions-General Aspects and Special Problems*, Amsterdam: Academic Publishing, pp. 39–60.

Pyšek, P., Chocholoušková, Z., Pyšek, A., Jorošík, V., Chytrý, M., and Tichý, L. (2004) Trends in species diversity and composition of urban vegetation over three decades, *Journal of Vegetation Science*, 15(6): 781–8.

Rebele, F. (1992) Colonization and early succession on anthropogenic soils, *Journal of Vegetation Science*, 3(2): 201–8.

Rebele, F. (1994) Urban ecology and special features of the urban ecosystem, *Global Ecology and Biogeography Letters*, 4(6): 173–87.

Rebele, F. (2008) Vegetation development on deposit soils starting at different seasons, *Plant Ecology*, 195(1): 1–12.

Reichard, S.H. and White, P.S. (2001) Horticulture as a pathway of invasive plant introductions in the United States, *BioScience*, 51(2): 103–13.

Ridley, H.N. (1930) *The Dispersal of Plants Throughout the World*, Ashford: Reeve & Co.

Rudnicky, J.L. and McDonnell, M.J. (1989) Forty-eight years of canopy change in a hardwood-hemlock forest in New York City, *Bulletin of the Torrey Botanical Club*, 116(1): 52–64.

Sharpe, D.M., Stearns, F., Leitner, L.A., and J.R. Dorney, L.A. (1986) Fate of natural vegetation during urban development of rural landscapes in southeastern Wisconsin, *Urban Ecology*, 9(3–4): 267–87.

Sisinni, S.M. and Emmerich, A. (1995) Methodologies, results and applications of natural resource assessment in New York City, *Natural Areas Journal*, 15: 175–88.

Steinberg, D.A., Pouyat, R.V., Parmelee, R.W., and Groffman, P.M. (1997) Earthworm abundance and nitrogen mineralization rates along an urban-rural land use gradient, *Soil Biology and Biochemistry*, 29(3–4): 427–30.

Thompson, K. and M.A. McCarthy (2008) Traits of British alien and native urban plants, *Journal of Ecology*, 96(5): 853–9.

Vallet, J., Daniel, H., Beaujouan, V., and Rozé, F. (2008) Plant species response to urbanization: comparison of isolated woodland patches in two cities of North-Western France, *Landscape Ecology*, 23(10): 1205–17.

von der Lippe, M. and Kowarik, I. (2007) Long-distance dispersal of plants by vehicles as a driver of plant invasions, *Conservation Biology*, 21(4): 986–96.

von der Lippe, M. and Kowarik, I. (2008) Do cities export diversity? Traffic as dispersal vector across urban–rural gradients, *Diversity and Distribution*, 14(1): 18–25.

White, C.S. and McDonnell, M.J. (1988) Nitrogen cycling processes and soil characteristics in an urban versus rural forest, *Biogeochemistry*, 5(2): 243–62.

Wichmann, M.C., Alexander, M.J., Soons, M.B., Galsworthy, S., Dunne, L., Gould, R., Fairfax, C., Niggemann, M., Hails, R.S., and Bullock, J.M. (2009) Human-mediated dispersal of seeds over long distance, *Proceedings of the Royal Society B*, 276(1656): 523–32.

Williams, N.S.G., Schwartz, M.W., Vesk, P.A., McCarthy, M.A., Hahns, A.K., Clemants, S.E., Corlett, R.T., Duncan, R.P., Norton, B.A., Thompson, K., and McDonnell, M.J. (2009) A conceptual framework for predicting the effect of urban environments on floras, *Journal of Ecology*, 97(1): 4–9.

Zipperer, W.C. (2002) Species composition and structure of regenerated and remnant forest patches within an urban landscape, *Urban Ecosystems*, 6(4): 271–90.

Zipperer, W.C., Foresman, T.W., Sisinni, S.M., and Pouyat, R.V. (1997) Urban tree cover: an ecological perspective, *Urban Ecosystems*, 1(4): 229–46.

Ziska, L.H. (2003) Evaluation of the growth responses of six invasive species to past, present, and future atmospheric carbon dioxide, *Journal of Experimental Biology*, 54(381): 395–404.

Ziska, L.H., J.A. Bounce, R.V., and Goins, E.W. (2004) Characterization of an urban-rural CO_2/temperature gradient and associated changes in initial plant productivity during secondary succession, *Community Ecology*, 139(3): 454–58.

17

Recombinant ecology of urban areas

Characterisation, context and creativity

Colin D. Meurk

Definitions

Recombinant ecosystems comprise novel plant and animal associations that have been induced or created by people deliberately, inadvertently or indirectly. They are generally made up of various mixes of indigenous and exotic species, but they may also involve associations of indigenous species alone, never before seen in nature, for example plant signatures (Robinson 1993), native landscape garden designs and pictorial meadows (Dunnett and Hitchmough 2004), indigenous feature species introduced to areas beyond their natural range, or back-filled 'gaps' created by local extinctions. Hobbs *et al.* (2006, 2009) propose that 'novel ecosystems' ('synthetic vegetation' in Bridgewater 1990; 'no-analog communities' of Williams and Jackson 2007; 'hybrid ecosystems' in Mulcock and Trigger 2008) result from 'human action, introduction of species, and environmental change'. Disturbance is normal in urban environments, and urban survivors (Wittig 2004; Van der Veken *et al.* 2004) are pre-adapted to high levels of change. As time goes on, species are added, deleted and re-sorted and vegetation converges. But active manipulation and moulding of landscapes is a major influence in the evolution of recombinant ecosystems. In the most extreme case, gardens of only alien plants or horticultural varieties (Acar *et al.* 2007) would qualify, although surely some indigenous microbe lurks in every urban or rural habitat.

Milton (2003) refers to 'emergent ecosystems' as those where new or different, maybe synergistic functional relationships occur. In New Zealand (NZ) exotic animals compete with or eat honey-eating bellbirds, yet the latter prosper from new sources of nectar supplied by plantings of Australian *Eucalyptus* and proteas. On the other hand, loss of large seed dispersers may diminish regeneration of keystone plant species (Norton 2009). With climate change there are emerging consequences of diseases, pests and weeds extending poleward; and increasing drought stress in urban environments may make species more susceptible to disease and deterioration. For instance, English oak and other introduced trees in Christchurch, NZ are succumbing to 'old age' at 100 to 150 years old, long before their natural span. Ironically, an emerging fashion is to replace English with American red oaks, somewhat undermining the city's claim to Englishness.

In many ways, NZ is an unhappy experiment that demonstrates an almost infinite array of permutations and combinations of species mixes from all over the temperate and subtropical worlds. To many this is desirable enrichment, but it is devastating to the highly endemic biota (Wilson 2004; Meurk and Swaffield 2000). The NZ flora comprises about 2500 indigenous vascular plant species, over 80 per cent of which are regarded as endemic. However, since European colonisation from the 1840s onwards, over 25 000 plants have been introduced into the country. Most thrive only in sheltered cultivation, as house plants, or in the 'winterless' north, but are often biding their time before transitioning to the wild, perhaps on the back of warming climate (Chapin and Starfield 1997). About one tenth have already become naturalised and each year four more enter the wild (Esler and Astridge 1987).

So, what is new? Species have been coming and going throughout geological history (McGlone 2006), adjusting and forming novel associations continually. This accelerated during periods of landscape disruption due to tectonics or glaciations, with human effects in NZ being felt from about a millennium ago. It is thus a moving feast and every point in space and time is novel or emergent in some sense.

The issue now is biodiversity conservation (Zerbe *et al.* 2004; Erfurt Declaration 2008) and the simple reality that anthropogenic, novel ecosystems generally involve displacement of local species or genes by widespread, generalist species (Van der Veken *et al.* 2004). The resulting globalisation and reduction in biodiversity and natural patterns, especially where the indigenous biota is highly endemic, is calamitous. In NZ, unlike continental countries with highly competitive floras, the introductions do not stay put. Nevertheless, there is valid interest in the theoretical ecology of colonisation, assembly and the novel pact invaders make with indigenous species, providing they do survive and coexist. The niche windows of native species in novel ecosystems are informative and must be better understood for their continued coexistence.

A framework

Various kinds of recombinant ecosystems exist and some may be more or less common according to the region, environment and history or proximity to introduced species, and human intervention or management. Typologies of these ecosystems may be usefully based on position in stress-disturbance space (Grime 1979), their origins, genesis or relative proportions of native and exotic species and their stability or successional condition and therefore the amount of management effort required to maintain them in a quasi-steady state. The Grime model is valuable in terms of prediction and management of conditions that suit or maintain particular species combinations (Meurk 2004). Table 17.1 is a proposed framework for describing recombinant ecosystems.

The case studies presented here apply to temperate urban environments where vegetative homogeneity is the prevailing urban condition. History of recombinant vegetation is determined by presence of primary, spontaneous habitat with relatively undisturbed soils or secondary or planted (deliberative) vegetation generally on altered soils. Urban native vegetation may include remnant primary forests (Florgard 2009; Molloy 1995) or, more often in NZ over the past two decades, planted or restored habitats with usually low species diversity, biased towards structural plants, and a significant proportion of residual exotic species. Novel communities, unknown in the wild, but ostensibly composed of indigenous species may frequently exist on artificial substrates. For example, the arrival of Polynesians in NZ less than a thousand years ago was not associated with invasive, alien plant species because the few food plants and other stowaways introduced were of tropical origin and thus unable to thrive in temperate climes. However, fire removed forest from the eastern seaboard and was replaced by indigenous tussocks, forbs, ferns, shrubs and wetland sedges, rushes and NZ flax (McGlone *et al.* 2005; McGlone 2009)

Table 17.1 Key attributes of four broad types of recombinant vegetation with increasing levels of intervention. The basis of the typology is initial condition (history and relative indigenous exotic species composition), the imposed conditions or elements (bullets), and the projected condition or status as determined by successional processes and/or management (indigenous and exotic species may increase (+), decrease (–) or be in steady state (0)). The real world is infinitely complex with any number of these state and transition combinations being possible and superimposed on one another.

Initial condition • *Imposed*	*Indigenous species*	*Exotic species*	*Successional trend*
Remnant	Dominant	Subordinate	+, 0, –
• Degraded	Co-dominant	Co-dominant	+, 0, –
• Additive	Subordinate	Dominant	+, 0, –
Spontaneous	Dominant	Subordinate	+, 0, –
• Primary	Co-dominant	Co-dominant	+, 0, –
• Secondary	Subordinate	Dominant	+, 0, –
Deliberative	Dominant	Subordinate	+, 0, –
• Novel design	Co-dominant	Co-dominant	+, 0, –
• Restoration	Subordinate	Dominant	+, 0, –
Complex	Dominant	Subordinate	+, 0, –
• Sown/random	Co-dominant	Co-dominant	+, 0, –
• Multiple 1^0			
• Multiple 2^0	Subordinate	Dominant	+, 0, –
• Managed			

often on landforms and soils that had never experienced such 'combinations'. Nevertheless, imported dogs and rats decimated the avifauna which in turn led to, still disputed, effects on the vegetation (McGlone and Clarkson 1993, Lee *et al.* 2009). These modified native plant-animal associations qualify as recombinant just as do European hedgerows or lawns. The floodgates were opened when European settlement began around 200 years ago. A few lamented the inevitability of complete replacement of the 'aboriginal' plants by the 'superior' European species (Buller 1888; Guthrie-Smith 1921; Pawson and Brooking 2002), although the earlier responses were sometimes to collect as many specimens of the doomed species as possible for museums. From this situation has developed the vast array of emergent eco-systems (*sensu* Milton 2003).

'Remnants' include primary and secondary vegetation, *semi-natural/pristine*, *degraded* (elements lost, disturbed), *recovered* (secondary succession) or *additive* (new species invaded or planted). But the diagnostic condition is uncultivated or non-artificial substrate/soil/landform combinations and some genetic connection between species present and those in the primordial vegetation. In the degraded case the loss or extinction of some indigenous elements is then a negative form of novel system.

Case studies: The New York urban-rural gradients of McDonnell and Pickett (1990) (some are secondary); Riccarton Bush of podocarps (Molloy 1995) and Travis wetland in Christchurch, Claudelands Bush in Hamilton, NZ; Epping Forest of oak, beech and hornbeam, London; oak forest in Moscow (Figure 17.1); oak savannah in Chicago; Melbourne grasslands (Williams *et al.* 2006), and salt marsh in New York.

Figure 17.1 Degraded remnant association – a *Quercus-Ulmus-Tilia/Lonicera/Dryopteris-Aegopodium* woodland in Moscow Botanic Gardens. Degradation is subtle due to aerosol deposition of toxins and acid rain and changes in dynamics from wildlife to human use.

'Spontaneous recombinants' arise from natural colonisation of disturbed ground (primary) or vegetation (secondary) and ongoing succession of both native and exotic species (whatever propagules are at hand). Generally we are talking here about artificial surfaces but these may be surrogates of naturally disturbed ecosystems as well as creating totally novel situations. Invasion is a spontaneous additive condition. Sometimes these new arrivals, that may also be planted, form new basins of stability and inhibit or deflect natural successional trajectories. These progressive or stalled successions may be variously manipulated past temporary impediments.

Case studies: Wastelands, walls and pavements typically support spontaneous associations of *Poa annua*, *Plantago*, *Polycarpon*, *Sagina*, *Sedum*, *Anagallis*, *Cymballaria*, *Conyza*, *Sonchus*, *Epilobium*, *Polygonum*, *Malva*, *Senecio*, *Vulpia*, *Centranthus*, *Corydalis*, bromes and ferns (European origin) (Figures 17.2 and 17.3), *Portulaca oleraceum* and *Pseudognaphalium* (cosmopolitan), *Cotula australis* (southern), and *Lachnagrostis* (NZ) (de Neef *et al.* 2008). *Buddleja* (China) frequently grows on stone-lined railway embankments in England.

Other more sinister examples of spontaneous spread of alien species are: *Ailanthus* and garlic onion in north-east USA urban woodlands, *Acmena* (monkey apple), kahlil ginger, and woolly nightshade (*Solanum mauritianum*) in northern NZ, and *Tradescantia*, *Galeobdolon* (aluminium plant), *Hedera* (ivy), *Lythrum* (purple loose strife), and *Selaginella* throughout NZ (Figure 17.4). These plants are variously from Asia, South America and South Africa.

Figure 17.2 Spontaneous primary recombinant association on pavement edge (micro-wasteland) in Sheffield, UK, with *Lactuca*, *Conyza**, *Plantago*, *Taraxacum*, *Polygonum*, *Epilobium*, *Papaver*, *Medicago*, *Stellaria*, *Vulpia*, *Lolium*, the ubiquitous *Poa annua* and moss (species with * are regarded as introduced).

Figure 17.3 Spontaneous primary wall association in St Austell, Cornwall, UK with *Asplenium trichomanes*, *Erigeron**, *Polypodium* and *Cymbalaria* (species with * are regarded as introduced).

Figure 17.4 Complex recombinant association in Christchurch, NZ derived from originally planted floodplain willow*, underplanted indigenous species including pittosporums (own family), *Plagianthus* (Malvac.), *Coprosma* (Rubiac.), *Myrsine* (own family), *Griselinia* (own family) and *Dacrycarpus* (Podocarpac.), and spontaneous ground and shrub layers of *Hedera**, *Iris**, *Sambucus**. The smothering, alien ground cover inhibits further regeneration creating a basin of (temporary) stability (species with * are regarded as introduced).

Some NZ species have become invasive in urban environments elsewhere: An NZ cabbage tree has established itself on a Cornish roof (Figure 17.5). *Muehlenbeckia complexa* vine is becoming a nuisance in some San Franciscan parks; *Coprosma repens* and *Metrosideros* likewise in south-east Australian and South African cities.

'Deliberative recombinants' include oversown mixes on cultivated ground, designed, landscaped and planted gardens, 'plant signatures', and restoration projects. The implication is that there is a plan or vision according to which maintenance or manicuring takes place to retain something akin to the initial planted condition: the 'designer communities' of McMahon and Holl (2002).

Restorative activities may involve removal of invasive elements or pests (from remnants or secondary vegetation); they may start from scratch (planting into bare ground, herbicided or mown grass); or may be manipulation of successional processes by under- or inter-planting into pre-existing, predominantly exotic vegetation (grass, scrub, woodland). Restoration sites are generally maintained only until the indigenous elements are self-sustaining; the goal is for nature to take over, but periodic pest control may be an ongoing requirement.

Restoration ecology has grown rapidly as an applied science and profession. As compensatory ecology it is marketed as a means of offsetting and mitigation of environmental impacts. However, restoration should be treated cautiously in its ability to replicate the complexities of a nature we can only partially comprehend (Turner *et al.* 2001). It has also been termed

Figure 17.5 Spontaneous primary recombinant association – a NZ cabbage tree seedling (*Cordyline australis*)* self-established on top of a stone wall in St Austell, Cornwall, UK (species with * are regarded as introduced).

'reconciliation ecology' (Rosenzweig 2003; Miller 2006) implying some new accommodation between conservation imperatives (indigenous species) and practical or productive necessities (exotic species). Rebuilding or reinserting indigenous components into totally alien cultural landscapes is a worthy goal, but not as a justification for further removal of scarce, primary habitats.

Case studies – deliberative associations

Parklands and street trees

Plane trees, oaks, horse chestnut, birch, ash, poplars, willows, lindens, rowan and other noble European deciduous trees are the most commonly used throughout Eurasia (Saebo *et al.* 2005), but have been exported to the colonies everywhere in the temperate world – a prime contributor to globalisation both floristically and visually (Stewart *et al.* 2009b). Wellingtonias and magnolias (from North America) are ubiquitous park trees. But every city has some indigenous street species, podocarps in South Africa and NZ; *Plagianthus*, *Sophora* and *Alectryon* in NZ cities, *Eucalyptus* and *Callistemon* in Australia; maytens, *Ceiba speciosa* and *Tipuana tipu* in South America. These imported and indigenous elements dominate the novel urban vegetation of the temperate world, often with ground covers of standard European mixes of grasses and herbaceous 'weeds' (Figure 17.6).

Figure 17.6 Deliberative roadside shrubbery with street tree standards on an earth embankment forming a sound buffer for a residential subdivision in Christchurch, NZ. Dominant plants are *Cordyline, Pittosporum, Coprosma, Muehlenbeckia* (Polygonac.), *Anemanthele* (Poac.) and *Quercus** (species with * are regarded as introduced).

Shrubberies and hedges

Box, privet, holly, escallonia, camellia, prunus and hydrangea (and even the NZ *Hebe*) are some of the globalised elements (Meurk *et al.* 2009), although not all are originally from Eurasia. Local contributors to 'plant signatures' are *Coprosma, Hebe, Corokia, Pittosporum* and *Olearia* (NZ), *Grevillea* and *Coleonema* (Australia), *Leucodendron, Portulacaria* and pelargoniums (South Africa), *Camellia, Rhododendron, Prunus* (Asia), ericas (Europe), *Ceanothus* and dwarf conifers (North America), and a NZ vine (*Muehlenbeckia complexa*) is used as a hedge in Melbourne (Figure 17.7) but has become a weed in San Francisco. All these genera come from species-rich natural shrublands such as coastal or mountain heaths, garrigue, fynbos, chaparral and grey scrub.

Herbaceous borders

Again, the diverse origins and movements of plants around the world and the novel associations that are consequently formulated are demonstrated by the following examples: *Agapanthus*, restiads and daisies (Africa), *Arthropodium, Phormium, Myosotidium* (NZ), *Gunnera, Alstroemeria* (South America), *Bergenia, Corydalis* (Eurasia), *Rumex, Acanthus, Alcea* (Europe), *Solidago, Lupinus, Penstemon* and many daisies (North America) (Figure 17.8). These large showy forbs, often

Figure 17.7 Deliberative hedge planting of *Muehlenbeckia complexa** (Polygonac.) and adjacent oak* street trees in Melbourne, Australia (species with * are regarded as introduced).

Figure 17.8 Deliberative novel designs – a planted herbaceous border with some weeds (grasses and euphorbia) dominated by marigolds*, *Veronica**, *Penstemon*, *Agapanthus** and roses*. Beyond is planted parkland with the Asian golden rain tree (*Koelreuteria paniculata**) and sown lawn dominated by introduced grasses*. This is on the Fresno University Campus, California, USA (species with * are regarded as introduced).

cultivars, occurred naturally in fertile, moist sites partially shaded, coastally exposed or riparian, and are seral though often on sites protected from browsing.

Lawns

Festuca rubra, *Lolium perenne*, *Poa annua*, *Trifolium*, *Plantago*, *Prunella*, *Bellis* and *Taraxacum* are the European, now globalised species (Stewart *et al.* 2009a). Localised indigenous examples include *Leptinella* (NZ), *Hydrocotyle* (Pacific) (Figure 17.9), *Nierembergia* and *Soliva* (South America), *Modiola* (North America), *Pennisetum*, *Sporobolus*, *Cynodon* and *Arctotheca* (Africa).

Restoration projects

These planted systems broadly follow some natural blueprint or model, perhaps an idealised version. There is an almost infinite array of types around the world, more or less faithfully copying the model. However, generally the initial mixes are of a limited array of hardy, fast-growing native species and exotic remnants, grasses and forbs (Figure 17.10). The theory is that over time more indigenous species will spontaneously colonise, and that unwanted species will be eradicated or managed.

Complex – the real world is made up of continua and combinations or complexes (with multiple origins). The above categories describe nodes within this complex. All systems are changing as

Figure 17.9 Complex recombinant association in Christchurch, NZ – a lawn of semi-wild grasses with *Lolium**, *Agrostis**, *Anthoxanthum**, *Trifolium** and indigenous *Hydrocotyle* (Apiac.) (species with * are regarded as introduced).

Figure 17.10 A deliberative planting for ecological restoration in Auckland, NZ adjacent to a small detention pond. Planted indigenous species are *Leptospermum* (Myrtac.), *Cordyline*, pittosporums, *Coprosma*, *Myrsine*, *Podocarpus*, *Melicytus* (Violac.) and *Hoheria* (Malvac.); invasive exotics include bamboo* and *Ligustrum* spp*, whereas many exotic species are still being suppressed (*Pennisetum** and other grasses, *Allium**, *Trifolium**, other forbs*) (species with * are regarded as introduced).

a result of natural dynamics, invasions (Song *et al.* 2005), local extinctions (Van der Veken *et al.* 2004) and levels of control.

Case study: Ernle Reserve, Christchurch, NZ is a deciduous woodland of European and some American trees, planted on a floodplain of a small river in the early 1900s. Some ground covers of ivy, periwinkle, galeobdolon, violet and celandine were also established and spread which essentially prevented tree regeneration (Figure 17.4). These have been joined by tradescantia, and some trees have multiplied – shade-tolerant sycamore, yew, plums and holly. Indigenous species that have also established include NZ cabbage tree (*Cordyline australis*), *Coprosma robusta* (Rubiac.), *Pittosporum* spp. (Pittosporac.), a non-local species of *Hoheria* (Malvac.), *Melicytus* (Violac.), and *Pseudopanax* spp. (Araliac.) (Stewart *et al.* 2004). Some podocarps were also planted a century ago and these are seeding. In the last 20 years a full array of local forest emergent and subcanopy trees, shrubs, lianes and ferns have been under-planted and many are now well established. There has been removal or control of many of the pest plants. This is an evolving ecosystem and a combination of manipulation and succession will eventually lead to a forest dominated by noble indigenous trees, provided the more traditionalist residents do not actively oppose the prospect of cherished English trees eventually being displaced by natives! But it will take another century and perhaps, if things are allowed to take their course, by then no one will remember the origins or the debate. Rather, a more mature culture may appreciate the new configurations that more clearly represent New Zealand rather than some other part of the world.

Socio-political context: fashion rules the roost!

Urban is synonymous with people. One cannot exist without the other and any theory or practice of applied urban ecology must inevitably accommodate social dynamics. Cultural/anthropogenic landscapes are what we expect and are comfortable with and so recombinant ecosystems have become the norm. Indeed there is entrenched resistance to change of landscape form. This suggests that, in the restoration schema proposed by Jackson and Hobbs (2009), there is a third (social) threshold that has to be overcome to achieve restoration goals, in addition to their abiotic and biotic hurdles (Figure 17.11). Transition and evolution of nature to culture is accompanied by control, tidiness, predictability and security. Nassauer (1997) coined the phrase 'cues for care' which refers to tidy versions or surrogates of nature. Some typical urban equivalents (Meurk and Swaffield 2007) are:

- lawns in place of sheep meadows and avian turfs;
- hedges for coastal shrubberies;

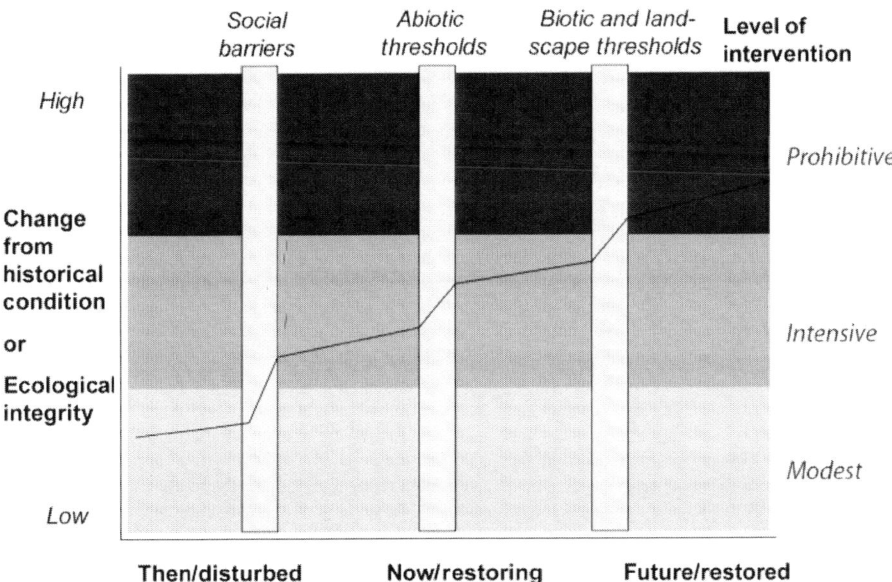

Figure 17.11 Three barriers to achieving ecological restoration (deliberative indigenous associations), restoration trajectory in relation to these major thresholds and the relative amount of resources needed to overcome these barriers – modified from Jackson and Hobbs (2009). The effort required to surpass each threshold, and the resulting convergence towards a restoration model, is represented by the vertical axis and time by the horizontal axis. In urban contexts the social resistance will often be the most difficult to breach as the abiotic and biotic barriers are largely overcome by financial and labour resources which are abundant in cities.

- parks for woodlands or savannahs;
- herbaceous borders for coastal, wet or mesotrophic meadows;
- bog gardens for heathlands;
- periodically herbicided wastelands, pavement cracks, paths and walls for riverbeds, sand dunes, limestone, basalt or granite pavements, cliffs, ledges, and canyons; and
- manicured riverbanks and ponds for riparian zones, wetlands and floodplains.

Increasingly these variants and landscape interpretations of nature are all we have of the natural world (cf. Rackham 1986). These traditionally have been shunned in the New World colonies as unworthy of conservation due to their artificial and modified state compared to the seemingly limitless pristine or wild condition of their frontier lands at the time of European contact. 'Pristine' has been the conservation benchmark. However, indigenous peoples had already had a major impact in the New World over periods of up to 50 000 years, through fire, extinction of megafauna, cultivation, and introduction of organisms and diseases (Flannery 1994), often out of balance with the original species and ecosystems. Now, with a second wave of European intensification, the frontiers have retreated to the remote distance.

What we design, build and plant is now governed by the powerful cultural influences that surround us – media, gardening magazines, advertising of products, fashion, social pressure to conform (Grove *et al.* 2006; Hostetler and Noiseux 2010), availability of plant material, and knowledge of how to manage indigenous species in the New World when most experience comes from traditions involving species imported from the motherland. And colonial cultures remain attached to the mother's apron strings for a long time (Meurk and Swaffield 2007; Swaffield *et al.* 2009). Change in general is challenging, if not threatening, to most people, but absence of nature, and particularly of indigenous nature, from the experience of society leads to 'extinction of experience' (Miller 2005) and of motivation to identify with and protect that particular type of nature (Meurk and Swaffield 2000). In the absence of frequent stimulus from, say, an indigenous thematic, the prevailing or default (exotic or virtual) style will become the familiar one, dominate the landscape of the mind, and characterise sense of place (cf. Louv 2005). During cultural development, before identity is grounded, there may be pendulum swings between competing paradigms. There has been a reaction against the dominance of exotic flora in places and a movement by progressives to liberate the land for the indigenous people and plants – a nostalgic revival of the Arcadian dream perhaps. But the traditionalist backlash is also inevitable – demanding the emergent, recombinant culture or synthesis be rolled back to more familiar 'purist', usually imported, territory. Sadly, each cycle is generally destructive of the primordial condition and operates like a ratchet – each new benchmark is beneath the last.

To take this issue further, we have to defend the view that there is a value (intrinsic) in nature beyond the utilitarian. Mulcock and Trigger (2008) posit that 'natural' is subjective and culturally bound whereas 'native' is an objective truth notwithstanding that some argue that time eventually converts all new arrivals to native status. And they ask, 'Is it justifiable to dismiss the validity or value of hybrid landscapes or hybrid communities, for example, on the basis that they are "unnatural"?' In idiosyncratic NZ, nativeness is relative (from deeply endemic organisms such as the NZ tuatara (Wilson 2004) to those merely indigenous). But it is nearly always clear which species have been introduced. On the other hand, in the UK, sycamore is believed introduced by the Romans (neophyte) (Rackham 1986), but many other species have less certain provenance. There is virtually no pristine environment in the world, especially in the predominantly biological recipient nations; all are hybrid. So these are not and should not be dismissed. From intrinsic, moral value, and legal points of view, what is unacceptable is the inexorable displacement of native species by exotics.

There is no choice – the world, and especially NZ, has and will have recombinant communities. But will they become mere homogeneous reflections of globalisation, or can they be the new face of biodiversity and preserve or regain local points of difference? The position here is that recombinant ecosystems can be a vehicle for landscape legibility (or 'eco-revelation') as much as pure indigenous communities may have once been. The purpose and value of recombinant ecosystems lies in their intrinsic ecological interest, but also they can be seen as a means of maintaining biodiversity through managed coexistence or 'reconciliation' (Rosenzweig 2003; Miller 2006). This does not preclude protecting 'natural' or 'pristine' environments or landscapes wherever possible. These would be retained through an explicit gradient management approach (Meurk and Greenep 2003) that preserves the semi-natural (as exemplified by conventional conservation practice in the New World) along with highly managed biotopes or recombinant associations that support at least some biodiversity (indigenous species). Every permutation and combination of species on the planet is information that helps to define niches, the fundamental units of ecology. A variety of mixtures is conducive to accommodating a wide range of cultural and social, as well as natural, values.

This may require a change of paradigm in the New World – from a colonial mindset that has hitherto separated nature from culture. Ironically, the classical countryside of Europe is the model, where thousands of years of accommodation between nature and culture was achieved, notwithstanding more recent regressions driven by an agro-industrial growth ethos of short term economies of scale and single value approaches to land use. Unfortunately the New World colonists took with them, not surprisingly, their own nature rather than the more generic concept of living with (indigenous) nature. Perhaps they took for granted that all their imports would stay where they were put and would work in the same way they do back home where anthropogenic combinations were almost entirely composed of indigenous species, as was also the case in Aboriginal Australia and the Americas, and Polynesian NZ. I foresee that nature conservation will, even in the New Worlds not so recently removed from vast pristine wilderness, follow the model of European cultural landscapes; not an exact replica of Europe but regional representations of the structure of English countryside and urban vegetation with increasing indigenous composition. Here will be found 'natural' patterns and ecological meaning, as much as in the remote alpine wilderness (see Swaffield *et al.* 2009; Walker *et al.* 2004).

Example: ANZAC Drive is an expressway in eastern Christchurch. ANZAC commemorates the fallen in wars of the past century and represents a revered association between Australia and New Zealand, and so it is reasonable to symbolise this relationship with a grove or two of Australian trees alongside NZ trees. But in fact, the taller Australian gum trees form continuous lines along the avenue while the indigenous mixed bushy and shrubby species occupy a subordinate role along the outer edge of the corridor adjacent to suburban fences. This suggests a culture still lacking confidence and knowledge of its own unique attributes, landscape and nationhood. There are ANZAC avenues in Australia, without a NZ tree in sight. Australian culture is extrovert by comparison, but they also have the advantage of an ancient continental endemic flora, full of the colour craved by suburbia, and fauna which has adapted to the harsh conditions of that land (dry and infertile) and diversified. It can, to some extent, repel boarders. The NZ biota has only recently been exposed to much of its contemporary disturbed and alpine environments and has a small gene pool. Inevitably the world's vast temperate flora, once introduced, provides many challenges to indigenous niches – both ecologically and culturally.

Practical pathways to recombinant futures

We do not have the luxury of hoping the recombinant 'problem' will go away, and every country and city will have to manage it and make choices. There needs to be purposes, goals, planning policies and principles, strategies, frameworks, methods, menus, designs (Ignatieva *et al.* 2008), management (Seastadt *et al.* 2008), criteria for standards, measurable outcomes and above all information – to assist involvement of the wider community. Whereas, this chapter is about vegetation; its structure and composition does not exist in isolation from other physical, fungal and faunal elements of the ecosystem. The absence of former animal components or the presence of new introductions has a profound influence on vegetation – through browsing, grazing, defoliation, trampling, fertilisation, or maybe removing competition – all varieties of 'disturbance'.

Managing the matrix – fauna, fire and reconciliation

I have presented a framework for defining and describing different types of recombinant ecosystem and trends (Table 17.1). Another useful framework for understanding and especially managing urban vegetation is the Grime matrix of stress and disturbance (Meurk 2004). Vegetation structure and composition is potentiated, disrupted, inhibited, or retarded by combinations of stresses and disturbances, natural or imposed, and the competitive interactions that follow. This is mediated through individual plant strategies. Planners and managers can apply knowledge of stress-disturbance effects on plants and competitive interactions to manipulate the survival and coexistence of chosen species. It may sound artificial, but fertilising or mowing are imperceptibly different, in terms of effects, from an ash shower in Andean forests or a herd of grazing antelope in the savannah.

Stresses

Climate can generally be ameliorated (in cities). We can provide shelter in greenhouses, release or remove nutrients by topsoil stripping, and make up water deficits by irrigation. Increasing stress can be achieved by shading, using coarse substrates on steep sunny slopes and even acidifying soils, applying toxic materials and salt, and water-logging soils (anaerobiosis). Urban substrates, rather than being novel (in terms of chemistry, moisture, texture and toxins) are generally surrogates for natural environments.

Disturbances

It is more straightforward to withhold or intensify disturbances. Examples of imposing disturbances include: gardening, weeding, mowing, burning, selective herbicides, controlled grazing (variously with cattle, sheep, rodents or birds), flooding, periodic stripping or renewal of substrates, and trampling. Most of these are forms of selective defoliation of primarily herbaceous vegetation which control or reduce competing exotic grasses and prevent litter build up, which otherwise diminishes local species diversity. Small scale, prescribed burns are even practised in front yard 'prairies' in Wisconsin and vandal-lit fires in the urban Matawai Park near Christchurch (Stewart and Wood 1995) may have rejuvenated the planted short (*Poa cita*) tussock community.

Convergence and stable states

Applying these various principles of restoration ecology, along with social considerations, will guide the most efficient way of managing 'natural' environments in the human context. Despite this apparent artificiality, there will always be plenty of room for 'urban wild' or the 'suburban safari' (Holmes 2005) where nature takes over and complements or thwarts human aspirations and endeavours to control.

Deliberative systems will incorporate an increasingly spontaneous character and recruit reproductive species from the surrounding landscape (English hedgerows are aged according to number of woody species in a 30 m stretch by Hooper's Rule (Pollard *et al.* 1974)). Left to their own devices, in time all these recombinant and remnant systems will converge along Clementsian trajectories towards some mature or quasi-stable state (Gilbert (1991) described this process for wastelands) (Lehvärvirta and Hannu 2002). Some planted or pre-established elements may retain dominance indefinitely and succession may be diverted or arrested compared to successional trajectories without planting. Thus new species combinations, that depend on particular initial conditions of vegetative cover, lead to novel successional trajectories (transitions) and basins of stability or attractor states (Briske *et al.* 2005; Standish *et al.* 2008). Various successional pathways of shade-tolerant NZ species through alien *Ulex* shrublands are well documented (Druce 1957; Wilson 1994; Sullivan *et al.* 2007), but Lee *et al.* (1986) showed in low rainfall, with browsing mammals and limited seed sources or with periodic fire, *Ulex* recycles indefinitely, or deep litter build-up arrests succession. Similarly, Lugo and Helmer (2004) report emergent forests in Puerto Rico, whereas Mascaro *et al.* (2008) found limited native plant regeneration in novel, exotic-dominated forests of Hawai'i. Meurk and Hall (2006) on the other hand modelled NZ forest dynamics showing that montane southern beech forests might be overcome by exotic trees, but lowland exotic forests would converge toward natural podocarp forests over a few centuries. Suding *et al.* (2004) describe alternative states in restoration; and planted *Kunzea* (Myrtac.) woodlands in NZ are colonised by exotic mycorrhizae with little prospect of transition to endemic forms (J. Cooper pers. comm.); Crowl *et al.* 2008 also describe the irreversible consequences of disease and other microbial invasives on natural ecosystem dynamics.

Criteria for ecologically acceptable exotic plant species

It is conceded that future regional ecologies of the world will comprise mixes of indigenous and introduced species (Hulme *et al.* 2009). Homogenisation and globalisation (Meurk 2007; Meyerson and Mooney 2007) are rampant across the world with enhanced human transport and uniformity of demand for cash crops and industrial scale agriculture and forestry. The homogenisation of culture (largely driven from the world's economic giants) is taking a heavy toll on regional and local diversity as globalised imagery becomes the mainstream identity. But, each nation has responsibility, under the Convention on Biological Diversity (CBD), to preserve their own gene pools. In continents there are typically large pools, niches are narrow, and locally adapted species make it more difficult for foreign species to infiltrate. Isolated micro-continents, like NZ, New Caledonia and Hawaii have many biological dilemmas, ecologically and socially, in contrast to similar sized islands that are largely biological extensions of their nearest continental neighbour (Britain, Sri Lanka, Japan).

If we accept the long term presence of exotic species, we still need criteria for those that are acceptable or those that pose significant biosecurity risks. Knowledge-based action is especially germane to cities because escapees from urban environments are major contributors to biosecurity problems in the hinterlands. I propose three categories (and two special cases) in NZ that may be generally applied elsewhere.

- *Beneficial and able to coexist without risk* – these in NZ comprise particularly plants that pose no current or likely future biosecurity threat usually because they are sterile, have low fecundity in the host country, poor dispersability (say, heavy woody seeds) or have virulent biological control agents or require subsidised resources to survive – and are important crop, food, pasture or forestry plants, or provide additional food resources to birds and other native wildlife at times and places that are not covered by indigenous species; or maybe they grow as planted and nurtured trees but do not reproduce due to temperature limitations. Examples are various Australian eucalyptus, protea and citrus species that in NZ commonly provide nectar resources to honey-eating birds across a broad seasonal range and into drier and colder landscapes in which most local nectar producers are unthrifty.

- *Benign and able to coexist without risk* – they have all the reproductive and dispersal limitations of the previous category, but provide no ecosystem services, other than perhaps aesthetic value or structure. Most exotic species in the UK would fall into this category; e.g. NZ *Cordyline* in Cornwall (Figure 17.5) and Athens.
- *Honorary natives* – what do we make of species or genera that were present in geological history (such as *Rhododendron* in England; *Eucalyptus*, proteas, *Casuarina*, wattles, danthonia, etc. in NZ) but subsequently became extirpated through climate change? Are these almost honorary native species? Is it merely reconnecting former functional relationships – is there a genetic memory of those relationships?
- *Pathological and invasive species that displace (at risk) biodiversity* – these plants require assiduous containment and where possible eradication. This is where human ignorance is, however, a major barrier to implementing appropriate measures. While millions of dollars may be spent on eliminating, say, the painted apple moth in Auckland, NZ, this investment can be sabotaged by undetected reintroduction of the organism into the country or region. Pest fish, possums and wallabies, and Russel lupins, tradescantia and sycamore are routinely spread by hunters, fishers and gardeners into places where they have not been seen before.
- *Non-provenance* – in NZ there are also northern indigenous spp. that are spreading and hybridising with southern species – *Hoheria*, *Pittosporum*, *Pseudopanax* and *Coprosma*. New, undesirable combinations of provenance are being created by moving plant material from one part of the country or region to another, potentially mixing the genetic histories of the species (e.g. *Sophora* (Godley 1972); Spanish bluebells, hybrid elms, planes and spartina). And so rages a debate over eco-sourcing (Ferkins 2005); what should the boundaries be for wind- versus insect-pollinated species, or dust seed or spore-dispersed species?

Similar criteria need to be applied to imported animals. It is hard for continentals to imagine the extent of damage caused by introduced terrestrial mammals in lands that had none before human contact – although the interaction of American greys and English red squirrels is a boreal case in point.

In due course we should work out more efficient, sustainable and 'natural' ways of managing these cultural landscapes – the weeds, pests and competitive interactions. Functional integrity will be achieved when there is co-adaptation of plants and wildlife and indigenous species are in a steady state. Dunnet's pictorial meadows incorporate exotic forms and enhance invertebrate diversity. Again indigenous organisms may be beneficiaries (Graves and Shapiro 2003), but if an imbalance arises it may justify introduction of biocontrol agents thereby further integrating local and exotic species and creating new interdependencies.

Landscape perspectives

Recombinant vegetation is a component of new (cultural) landscapes. The structure and function of landscape has been drastically modified (Lindenmayer *et al.* 2008); there is less forest and more herbaceous vegetation, habitats are fragmented, have stronger edge effects and less connectivity, many communities are nearly monocultures, and as we have seen there are many introduced pests.

A new bio-diverse balance will have bits of everything in all combinations – all species forming viable populations within interbreeding range of one another (the meta-population). Most importantly it will have stasis – or rather dynamic equilibrium, which allows succession and evolution rather than constant human-imposed change at the whim of fashion. Over time hedgerows accumulate species and in NZ this has also begun, but the clock is continually wound back by the restless desire for change and the cheap technology and energy to satisfy these urges. And so history fails to materialise. We can however also accelerate the successional process, in keeping with an age of instant gratification – something that happened over centuries in Europe. Perhaps diverse hedges, as with lawns, will be enough to garner interest and desire to protect this diversity in otherwise mundane surroundings. To then allow evolution to continue at its own pace there has to be a reversal of the notions that history is bunk, continuity is stagnation, diversity is threatening, and the world should be reduced to some lowest common denominator. The key point here is that spatial ecological principals need to be applied to the future of landscape evolution (Meurk and Hall 2006; Breuste *et al.* 2008), otherwise globalised commerce will dictate 'efficiencies' of clearance and monoculture. The world is shrinking and diversity is likely to survive only by innovative planning and design (Pickett and Cadenasso 2008). The alternative is continual random loss. In urban environments, the required design is at a micro-scale. This presupposes a body of knowledge and practical experience. There is thus an urgent need for ecologists to enter the debate and provide evidence-based demonstrations of how biodiversity can be integrated and sustained in urban landscapes. It is not a matter of forcing the landscape but providing structural context and pattern that facilitates diversity and yet dynamism and evolution.

Conclusions

If one goal of urban ecology is to maintain or enhance biodiversity (Erfurt Declaration 2008; McDonnell *et al.* 2009) one must address the trends in the native-exotic balance and face the many dilemmas posed by the dynamics of mixed and invaded vegetation through some form of gradient management. Trends of globalisation on the one hand, and reassertion of indigenous heritage values and nature, on the other, have their advocates battling it out in local politics, magazines and actions (Lindenmayer *et al.* 2008).

We have recombinants whether we like it or not in both cultural and wild landscapes. I have described three modal types and an array of complexes according to their initial condition, proportion of indigenous and exotic species, and natural or human-induced trend. But what should our policy and approach be with respect to this phenomenon – passive neglect, active encouragement of ever more dispersal of species around the world to produce unified or whole new ecologies, or active constraint towards some clearly defined and manageable goal? This can really only be achieved through international convention and perhaps that is something worth promoting.

We have to go with the flow rather than fighting nature head on, make the best of it and derive something useful. There are fascinating new insights into the autecology of our indigenous species to be gained from pitching them into new environments and associations. But I do not go as far as those who regard all new species as exciting and desirable (cf. Marris 2009 'Ragamuffin Earth'). This represents one value usurping another (higher) one. Perhaps in the future we may develop technologies that can eradicate pest species from an area: we see this already in the ability of NZ wildlife managers to rid an 11 000 ha cliff-bound island in subantarctic waters of rats!

A new set of assembly (or decision support) rules will be needed which accommodate exotic species while providing for succession through to sustained presence of indigenous species (on every class of land) either passively in the form of building up seed sources or actively through micro-management. Gradient management is the most likely scenario to accommodate a wide range of species and attitudes. Natural areas have to be substantial in all environments to achieve viability, representativeness, baselines and the full Clementsian sere (pioneer to climax condition in his terminology) and, importantly, visibility. It will be the duty of urban ecology to provide the knowledge and of governments to implement it!

Summary of planning and management guidelines for urban environments:

- Identify and eliminate pathological species;
- Tolerate benign species;
- Encourage beneficial, especially indigenous species – all within their place;
- Maintain urban legibility and visibility (the full story) – sufficient for viability and sustainability, both ecologically and culturally;
- Maintain as many permutations as possible – along management gradients and within a stress-disturbance matrix. We cannot second guess what is the best or optimum environmental and management condition so we have to provide as many options as possible to ensure there is a high probability of something for every species – so that all can establish viable populations with minimal management. That is the aim – to expend the least amount of energy (and money) on maintaining these management states. Traditional tools such as grazing (with different animals), fire, herbicide, manual weeding, soil stripping, may all have to be employed – in many combinations.
- For successional species (Grubb 1977), build up populations, seed sources and seed banks and create punctuated disturbances so that indigenous and exotic species have an equal chance at being part of the mix and passing their propagules onto the next disturbed site – in dynamic mosaics of built environments.

There is a growing belief that urban biodiversity and urban green space are vital to the health and well-being of nature and people (Ernst and Monroe 2004; Mitchell and Popham 2009) but there is much to learn in terms of quantity, quality and configuration of green space and its effectiveness, multi-utilitarian value and cost efficiency (in terms of ecosystem services, maintenance costs and biodiversity specifically). Planning must take account of these various factors and have buy-in from community and business as well as bureaucrats (Breuste 2003; Baycan-Levent and Nijkamp 2009). The urban environment is a nexus for these concepts, values, debates and challenges and a catalyst for problem-solving. Solutions to ecological problems and conflicts need to be found in cities because this is where policies on everything from ecology to economics are generated.

Acknowledgements

Thanks to those who have shown me around the world. This only made it because of the almost infinite patience of Ian Douglas. Larry Burrows checked the story line at the last minute and was reassuring.

Bibliography

Acar, C., Acar, H. and Eroglu, E. (2007) 'Evaluation of ornamental plant resources in urban biodiversity and cultural changing: a case study of residential landscapes in Trabzon city (Turkey)', *Building and Environment*, 42(1): 218–29.

Baycan-Levent, T. and Nijkamp, P. (2009) 'Planning and management of urban green spaces in Europe: comparative analysis', *Journal of Urban Planning and Development*, 135(1): 1–12.

Breuste, J. (2003) 'Decision making, planning and design for the conservation of indigenous vegetation within urban development', *Landscape and Urban Planning*, 68(4): 439–52.

Breuste, J., Niemalä, J. and Snepp, R.P.H. (2008) 'Applying landscape ecological principles in urban environments', *Landscape Ecology*, 23(10): 1139–42.

Bridgewater, P.B. (1990) 'The role of synthetic vegetation in present and future landscapes of Australia', *Proceedings of the Ecological Society of Australia*, 16: 129–34.

Briske, D.D., Fuhlendorf, S.D. and Smeins, F.E. (2005) 'State-and-transition models, thresholds, and rangeland health: a synthesis of ecological concepts and perspectives', *Rangeland Ecology and Management*, 58(1): 1–10.

Buller, W.L. (1888) *A History of the Birds of New Zealand*, second edition, London: J.G. Koulemans.

Chapin, F.S. and Starfield, A.M. (1997) 'Time lags and novel ecosystems in response to transient climatic change in Alaska', *Climate Change*, 35(4): 449–61.

Crowl, T.A., Crist, T.O., Parmenter, R.R., Belovsky, G. and Lugo, A.E. (2008) 'The spread of invasive species and infectious disease as drivers of ecosystem change', *Frontiers in Ecology and the Environment*, 6(5): 238–46.

de Neef, D., Stewart, G.H. and Meurk, C.D. (2008) 'URban Biotopes of Aotearoa New Zealand (URBANZ) (III): Spontaneous urban wall vegetation in Christchurch and Dunedin', *Phyton (Horn, Austria)*, 48(1): 133–54.

Druce, A.P. (1957) 'Botanical survey of an experimental catchment, Taita, New Zealand', *Bulletin (New Zealand Department of Scientific and Industrial Research) 124*, Wellington.

Dunnett, N. and Hitchmough, J. (eds) (2004) *The Dynamic Landscape: Design, Ecology and Management of Naturalistic Urban Planning*, London/New York: Spon Press.

Erfurt Declaration (2008) online, available at: www.fh-erfurt.de/urbio/httpdocs/content/documents/ErfurtDeclarationUrbio200823.May2008eng.pdf.

Ernst, J.A. and Monroe, M. (2004) 'The effect of environment-based education on students' critical thinking skills and disposition toward critical thinking', *Environmental Education Research*, 10(4): 507–22.

Esler, A.E. and Astridge, S.J. (1987) 'The naturalisation of plants in urban Auckland, New Zealand, 2. Records of introduction and naturalisation', *New Zealand Journal of Botany* 25: 523–37.

Ferkins, C. (2005) *Ecosourcing: Code of Practice and Ethics*, second edition, Henderson: Waitakere City Council.

Flannery, T. (1994) *The Future Eaters*, Sydney: Reed Books.

Florgard, C. (2009) 'Preservation of original natural vegetation in urban areas: an overview', in M.J. McDonnell, A.K. Hahs and J.H. Breuste (eds), *Ecology of Cities and Towns: A Comparative Approach*, New York: Cambridge University Press.

Gilbert, O. (1991) *The Ecology of Urban Habitats*, London: Chapman and Hall.

Godley, E.J. (1972) 'Does planting achieve its purpose?', *Forest and Bird Journal*, 185: 25–6.

Graves, S.D. and Shapiro, A.M. (2003) 'Exotics as host plants of the California butterfly fauna', *Biological Conservation*, 110(3): 413–33.

Grime, J.P. (1979) *Plant Strategies and Vegetation Processes*, Chichester: Wiley.

Grove, J.M., Troy, A.R., O'Neil-Dunne, J.P.M., Burch Jr., W.R., Cadenasso, M.L. and Pickett, S.T.A. (2006) 'Characterization of households and its implications for the vegetation of urban ecosystems', *Ecosystems*, 9(4): 578–97.

Grubb, P.J. (1977) 'The maintenance of species-richness in plant communities: the importance of the regeneration niche', *Biological Reviews*, 52(1): 107–45.

Guthrie-Smith, H. (1921) *Tutira: The Story of a New Zealand Sheep Station*, London: William Blackwood & Sons, Ltd.

Hobbs, R.J., Higgs, E. and Harris, J.A. (2009) 'Novel ecosystems: implications for conservation and restoration', *Trends in Ecology and Evolution*, 24(11): 599–605.

Hobbs, R.J., Arico, S., Aronson, J., Baron, J.S., Bridgewater, P., Cramer, V.A., Epstein, P.R., Ewel, J.J., Klink, C.A., Lugo, A.E., Norton, D., Ojima, D., Richardson, D.M., Sanderson, E.W., Valladares, F., Montserrat, V., Zamora, R. and Zobel, M. (2006) 'Novel ecosystems: theoretical and management aspects of the new ecological order', *Global Ecology and Biogeography*, 15(1): 1–7.

Holmes, H. (2005) *Suburban Safari: A Year on the Lawn*, New York: Bloomsbury.

Hostetler, M. and Noiseux, K. (2010) 'Are green residential developments attracting environmentally savvy homeowners?' *Landscape and Urban Planning*, 94(3–4): 234–43.

Hulme, P.E., Pysek, P., Nentwig, W. and Vila, M. (2009) 'Will threat of biological invasions unite the European Union?' *Science*, 324(5923): 40–1.

Ignatieva, M., Meurk, C., van Roon, M., Simcock, R. and Stewart, G. (2008) 'How to put nature into our neighbourhoods', *Landcare Research Science Series* 35, Lincoln: Manaaki Whenua Press.

Jackson, S.T. and Hobbs, R.J. (2009) 'Ecological restoration in the light of ecological history', *Science*, 325(5940): 567–69.

Lee, W.G., Allen, R.B. and Johnson, P.N. (1986) 'Succession and dynamics of gorse (*Ulex europaeus* L.) communities in the Dunedin Ecological District South Island, New Zealand', *New Zealand Journal of Botany*, 24: 279–92.

Lee, W.G., Wood, J.R. and Rogers, G.M. (2009) 'Legacy of avian-dominated plant-herbivore systems in New Zealand', *New Zealand Journal of Ecology*, 34(1): 28–47.

Lehvävirta, S. and Hannu, R. (2002) 'Natural regeneration of trees in urban woodlands', *Journal of Vegetation Science*, 13(1): 57–66.

Lindenmayer, D.B., Fischer, J., Felton, A., Crane, M., Michael, D., Macgregor, C., Montague-Drake, R., Manning, A. and Hobbs, R.J. (2008) 'Novel ecosystems resulting from landscape transformation create dilemmas for modern conservation practice', *Conservation Letters* 1(3): 129–35.

Louv, R. (2005) *Last Child in the Woods: Saving our Children from Nature-Deficit Disorder*, Chapel Hill, NC: Algonquin Books.

Lugo, A.E. and Helmer, E. (2004) 'Emerging forests on abandoned land: Puerto Rico's new forests', *Forest Ecology and Management*, 190(2–3): 145–61.

McDonnell, M.J. and Pickett, S.T.A. (1990) 'Ecosystem structure and function along urban–rural gradients: an unexploited opportunity for ecology', *Ecology*, 71(4): 1232–7.

McDonnell, M.J., Hahs, A.K. and Breuste, J.H. (eds) (2009) *Ecology of Cities and Towns: A Comparative Approach*, New York: Cambridge University Press.

McGlone, M.S. (2006) 'Becoming New Zealanders: immigration and the formation of the biota', in R.B. Allen and W.G. Lee (eds), *Biological Invasions in New Zealand: Ecological Studies 186*, Berlin: Springer-Verlag.

McGlone, M.S. (2009) 'Postglacial history of New Zealand wetlands and implications for their conservation', *New Zealand Journal of Ecology*, 33(1): 1–23.

McGlone, M.S. and Clarkson, B.D. (1993) 'Ghost stories: moa, plant defences and evolution in New Zealand', *Tuatara*, 32: 2–22.

McGlone, M.S., Wilmshurst, J.M. and Leach, H.M. (2005) 'An Ecological and Historical Review of Bracken (*Pteridium esculentum*) in New Zealand, and its Cultural Significance', *New Zealand Journal of Ecology*, 29(2): 165–84.

McMahon, J.A. and Holl, K.D. (2002) 'Designer communities', *Conservation Biology. Practice* 3(1): 3–4.

Marris, E. (2009) 'Ragamuffin Earth', *Nature*, 460(7254): 450–3.

Mascaro, J., Beckland, K.K., Hughes, R.F. and Schnitzer, S.A. (2008) 'Limited native plant regeneration in novel, exotic-dominated forests on Hawai'i', *Forest Ecology and Management*, 256(4): 593–606.

Meurk, C.D. (2004) 'Beyond the forest: restoring the "herbs"', in I. Spellerberg and D. Given (eds), *Going Native*, Christchurch: Canterbury University Press.

Meurk, C.D. (2007) 'Implications of New Zealand's unique biogeography for conservation and urban design', *Proceedings of the Globalisation of landscape architecture: issues for education and practice conference, St Petersburg*, pp. 142–5.

Meurk, C.D. and Greenep, H. (2003) 'Practical conservation and restoration of herbaceous vegetation', *Canterbury Botanical Society Journal*, 37: 99–108.

Meurk, C.D. and Hall, G.M.J. (2006) 'Options for enhancing forest biodiversity across New Zealand's managed landscapes based on ecosystem modelling and spatial design', *New Zealand Journal of Ecology*, 30(1): 131–46.

Meurk, C.D. and Swaffield, S.R. (2000) 'A landscape ecology framework for indigenous regeneration in rural New Zealand-Aotearoa', *Landscape and Urban Planning*, 50(1–3): 129–44.

Meurk, C.D. and Swaffield, S.R. (2007) 'Cities as complex landscapes: biodiversity opportunities, landscape configurations and design directions', *New Zealand Garden Journal*, 10(1): 10–20.

Meurk, C.D., Zvyagna, N., Gardner, R.O., Forrester, G., Wilcox, M., Hall, G., North, H., Belliss, S., Whaley, K., Sykes, B., Cooper, J. and O'Halloran, K. (2009) 'Environmental, social and spatial determinants of urban arboreal character in Auckland, New Zealand', in M.J. McDonnell, A.K. Hahs and J.H. Breuste (eds) *Ecology of Cities and Towns: A Comparative Approach*, New York: Cambridge University Press.

Meyerson, L.A. and Mooney, H.A. (2007) 'Invasive alien species in an era of globalization', *Frontiers in Ecology and the Environment*, 5(4): 199–208.
Miller, J.R. (2005) 'Biodiversity conservation and the extinction of experience', *Trends in Ecology and Evolution*, 20(8): 430–4.
Miller, J.R. (2006) 'Restoration, reconciliation, and reconnecting with nature nearby', *Biological Conservation*, 127(3): 356–61.
Milton, S.J. (2003) 'Emerging ecosystems: a washing stone for ecologists, economists and sociologists?' *South African Journal of Science*, 99: 404–6.
Mitchell, R. and Popham, F. (2009) 'Effect of exposure to natural environment on health inequalities: an observational population study', *The Lancet*, 372(9650): 1655–60.
Molloy, B.P.J. (ed.) (1995) *Riccarton Bush: Putaringamotu*, Christchurch: Riccarton Bush Trust.
Mulcock, J. and Trigger, D. (2008) 'Ecology and Identity: a comparative perspective on the negotiation of "nativeness"', in D. Wylie (ed.), *Toxic Belonging? Identity and Ecology in Southern Africa*, Newcastle-upon-Tyne: Cambridge Scholars Publishing.
Nassauer, J.I. (ed.) (1997) *Placing Cature: Culture and Landscape Ecology*, Washington, DC: Island Press.
Norton, D.A. (2009) 'Species invasions and the limits to restoration: learning from the New Zealand experience', *Science* 325(5940): 569–71.
Pawson, E. and Brooking T. (eds) (2002) *Environmental Histories of New Zealand*, New York: Oxford University Press.
Pickett, S.T.A. and Cadenasso, M.L. (2008) 'Linking ecological and built components of urban mosaics: an open cycle of ecological design', *Journal of Ecology*, 96: 8–12.
Pollard, E., Hooper, M.D. and Moore, N.W. (1974) *Hedges*, London: Collins.
Rackham, O. (1986) *The History of the Countryside*, London: J.M. Dent & Sons Ltd.
Robinson, N. (1993) 'Place and planting design – plant signatures', *The Landscape*, 53: 26–8.
Rosenzweig, M.L. (2003) 'Reconciliation ecology and the future of species diversity', *Oryx*, 37(2): 194–205.
Saebo, A., Borzan, Z., Ducatillion, C., Hatzistathis, A., Lagerstrom, T., Supuka, J., Garcia-Valdecantos, J.L., Rego, F. and Van Slycken, J. (2005) 'The selection of plant materials for street trees, park trees and urban woodland', in C.C. Konijnendijk, K. Nilsson, T.B. Randrup, J. Schipperijn (eds), *Urban Forests and Trees*, Berlin/Heidelberg/New York: Springer.
Seastadt, T.R., Hobbs, R.J. and Suding, K.N. (2008) 'Management of novel ecosystems: are novel approaches required?' *Frontiers of Ecology and Environment*, 6(10): 547–53.
Song, I.-J., Hong, S.-K., Kim, H.-O., Byun, B. and Gin, Y. (2005) 'The pattern of landscape patches and invasion of naturalized plants in developed areas of urban Seoul', *Landscape and Urban Planning*, 70(3–4): 205–19.
Standish, R.J., Sparrow, A.D., Williams, P.A. and Hobbs, R.J. (2008) 'A state-and-transition model for the recovery of abandoned farmland in New Zealand', in R.J. Hobbs and K.N. Suding (eds) *New Models for Ecosystem Dynamics and Restoration*, Washington, DC: Island Press.
Stewart, G. and Wood, D. (1995). 'A second generation at Matawai Park', in M.C. Smale and C.D. Meurk (eds), *Proceedings of a Workshop on Scientific Issues in Ecological Restoration, Landcare Research Science Series* No. 14, Christchurch: Manaaki Whenua Press.
Stewart, G.H., Ignatieva, M.E., Meurk, C. and Earl, R.D. (2004) 'The re-emergence of indigenous forest in an urban environment, Christchurch, New Zealand', *Urban Forestry – Urban Greening* 2(3): 149–58.
Stewart, G.H., Ignatieva, M.E., Meurk, C.D., Buckley, H., Horne, B., and Braddick, T. (2009a) 'URban Biotopes of Aotearoa New Zealand (URBANZ) (I): Composition and diversity of temperate urban lawns in Christchurch', *Urban Ecosystems*, 12(3): 233–48.
Stewart, G.H., Meurk, C.D., Ignatieva, M.E., Buckley, H.L., Magueur, A., Case, B.S., Hudson, M. and Parker, M. (2009b) 'URban Biotopes of Aotearoa New Zealand (URBANZ) II: Floristics, biodiversity and conservation values of urban residential and public woodlands, Christchurch', *Urban Forestry and Urban Greening*, 8(3): 149–62.
Suding, K.N., Gross, K.L. and Houseman, G.R. (2004) 'Alternative states and positive feedbacks in restoration ecology', *Trends in Ecology and Evolution*, 19(1): 46–53.
Sullivan, J.J., Williams, P.A. and Timmins, S.M. (2007) 'Secondary forest succession differs through naturalized gorse and native kanuka near Wellington and Nelson', *New Zealand Journal of Ecology*, 31(1): 22–38.
Swaffield, S., Meurk, C. and Ignatieva, M. (2009) 'Urban biodiversity in New Zealand: Issues, challenges and opportunities', in P. Hedfors (ed.), *Urban naturmark i landskapet – en syntes genom landskapsarkitektur. Festskrift till Clas Florgard*, Rapportserien No. 3/09, Uppsala: Swedish University of Agricultural Sciences.

Turner, R.E., Redmond, A. and Zedler, J.B. (2001) 'Count it by acre or function? Mitigation adds up to net loss of wetlands', *National Wetlands Newsletter*, 23(6): 5–6, 14–16.

Van der Veken, S., Verheyen, K. and Hermy, M. (2004) 'Plant species loss in an urban area (Turnhout, Belgium) from 1880 to 1999 and its environmental determinants', *Flora*, 199(6): 516–23.

Walker, B., Holling, C.S., Carpenter, S.R. and Kinzig, A. (2004) 'Resilience, adaptability and transformability in social-ecological systems', *Ecology and Society*, 9(2): Article 5.

Williams, J.W. and Jackson, S.T. (2007) 'Novel climates, no-analog communities, and ecological surprises', *Frontiers of Ecology and Environment*, 5(9): 475–82.

Williams, N.S.G., Morgan, J.W., McCarthy, M.A. and McDonnell, M.J. (2006) 'Local extinction of grassland plants: the landscape matrix is more important than patch attributes', *Ecology*, 87(12): 3000–6.

Wilson, H.D. (1994) 'Regeneration of native forest on Hinewai Reserve, Banks Peninsula', *New Zealand Journal of Botany*, 9(3): 41–55.

Wilson, K.-J. (2004) *The Flight of the Huia*, Christchurch: Canterbury University Press.

Wittig, R. (2004) 'The origin and development of the urban flora of Central Europe', *Urban Ecosystems*, 7(4): 323–39.

Zerbe, S., Maurer, U., Peschel, T., Schmitz, S. and Sukopp, H. (2004) 'Diversity of flora and vegetation in European cities as a potential for nature conservation in urban-industrial areas – with examples from Berlin and Potsdam (Germany)', in Shaw *et al.* (eds) *Proceedings of 4th International Urban Wildlife Symposium*, Tucson, AZ: University of Tucson.

18

Creative conservation

Grant Luscombe and Richard Scott

Making a start

> What are the natural features which make a township handsome? A river, with its waterfalls and meadows, a lake, a hill, a cliff or individual rocks, a forest, and ancient trees standing singly. Such things are beautiful; they have a high use which dollars and cents never represent. If the inhabitants of a town were wise, they would seek to preserve these things, though at a considerable expense; for such things educate far more than any hired teachers or preachers, or any at present recognized system of school education.
>
> *(Thoreau 1861)*

Why creative conservation? Because the inhabitants have not been wise.

A 1997 report from the UK Government's Natural Environment Research Council on the changing face of nature conservation in Britain recognised that where nature conservation had begun to subsume nature preservation, some fifty years ago, notions of creative conservation had now similarly encouraged reassessment of the purpose and practice of wildlife-resource management (Sheail *et al.* 1997). The origins of this transition can be found two decades earlier in old urban centres across north-west England and the emerging New Town developments.

Ecological studies in Milton Keynes showed that relatively little was known about how to create new habitats for wildlife, and creative conservation was considered to be one aspect of ecological engineering, concerned with the exploration of opportunities to create new habitats and to rehabilitate degraded habitats, for the benefit of wildlife (Yoxon 1977).

Although most traditional conservationists find the creation of new ecological landscapes unnatural, the British landscape has undergone radical change in previous decades due to agricultural practices. Hedgerows introduced in the early nineteenth century were seen as alien features and eyesores, but are now cherished as part of the countryside. In fact, most of our prized and protected habitats are semi-natural, having arisen from obsolete farming practices. In towns and cities the changes have been even more dramatic, stimulating a new vision and understanding of nature conservation practice.

Some authors consider the spiritual home of man-made ecological landscaping in the UK resides somewhere between the Liverpool organisation Landlife, Warrington New Town and

Manchester University School of Landscape (Kendle and Forbes 1997). The bold stepping forwards from more limited ideas of conservation, taken by these pioneers, saw green deserts and derelict gap sites transformed using a multifunctional approach to deliver ecological, educational, recreational, aesthetic, health and economic benefits. This new thinking encompassed the core principle that creative conservation must never be a substitute for preserving existing habitats. It developed ideas about disturbance and opportunity in nature, and a wider definition of habitat as a place that supports or has the potential for supporting biodiversity.

Landlife, the first urban wildlife group, set out on this journey of discovery in 1975, driven by the energy, enthusiasm and optimism of local communities. This groundbreaking voluntary organisation (Baines 2007) and progenitor of the Groundwork Trusts (Nicholson-Lord 1987), focused on simple starting points rather than the complex end points that drive traditional nature preservation and conservation. It recognises the role of man in shaping the world, and nature's dynamic response. It is an approach that draws on ecological knowledge to initiate natural processes responsive to change, by incorporating stress and chaos rather than stability and routine. It considers the rural/urban divide to be an artificial construct in relation to biodiversity, creates new opportunities for wildlife to flourish, and engages with people to give nature a helping hand.

The concept of "people as a part of nature rather than apart from nature" was written into the first urban nature conservation strategy prepared by Landlife for the Nature Conservancy Council (Urban Wildlife Unit 1983). Throughout the 1980s similar policies were adopted in Birmingham, London, Bristol, Leicester and Sheffield. Early organisations replicating this approach included the Birmingham Urban Wildlife Group in 1979, and Operation Groundwork established in St Helens in 1981.

A set of guiding principles (Landlife and Urban Wildlife Partnership 2000), designed to encourage good practice on sites of little value to wildlife, emphasised the importance of survey work, as the ecological value of a site may not be immediately apparent. For example, it is possible that areas of little botanical interest or derelict appearance may be very important for local insect populations.

As the new millennium approached, ecological losses and the threat of climate change highlighted the importance of both conserving historic biodiversity, and reversing these trends through action-based programmes of habitat creation. In doing so, it raised the potential for new habitats to evolve, which may change established ecological maps. This approach still is one that many find difficult to accept. It is yet to find a proper place within UK biodiversity policy, where it can be integrated in such a way that its potential gains can be realised without cutting off the cultural engine of concern that drives conservation forward (Adams 1996a).

Taking the opportunity

Creative conservation applies good scientific methodology and lateral thinking to yield elegant solutions for a conservation benefit. Nature is about opportunity, so ecological theory linking species diversity to different environmental stresses and disturbances (Grime 1980) can be applied creatively. This means inert wastes considered to be low in nutrients and high pH can be used to good biodiversity benefit if the right cocktail of species is sown. Using such wastes establishes a virtuous circle in the locality, recycling materials as a landscape asset. This saves transport of waste materials away from sites, and the importation of costly topsoil with its associated problematic weed seeds.

In richer soil, productivity is greater. Such conditions favour domination by a few grass species unless checked by management or disturbance of some kind. Wildflower seed mixtures should be shallow sown at low rates without grass, as the grasses will colonise over time, once the wildflowers have established.

Practical examples on the ground illustrate the diverse nature of creative conservation. All these projects involve an imaginative engagement with the site and with the people who live nearby. Such sites do not have fixed outcomes and are designed to inspire people, because people copy things if they are inspired.

Colliery spoil

Standard reclamation of colliery sites seeks to create amenity grassland or restore it for low-grade agricultural purpose. The conversion of Bold Moss Colliery, St Helens – now known as Colliers' Moss – took an alternative approach by working with the existing substrates and local demolition materials. The Groundwork Trust created a mosaic of natural colonisation and introductions. The woodland creation was notable for the successful spread of bluebell and primrose populations introduced from seed.

Hay strewing

Since the 1980s, the University of Wolverhampton has been establishing a series of meadows utilising green hay strewing techniques from nature reserves outside the borough. The hay, spread over prepared bare soil, was inspired by project work at the Zuiderpark in The Hague. This resulted in a patchwork of meadows that compliment the work of the Wolverhampton based National Urban Forestry Unit.

Urban landfill

Pickerings Pasture was an industrial and household waste tip in Widnes from the 1950s until the 1980s. Between 1982–86 the site was made safe, capped with inert clay, and topsoiled by Halton

Figure 18.1 Pickerings Pasture, Halton.

Borough Council and Cheshire County Council. The site, one of the earliest wildflower sowings in such a prominent industrial location, comprises distinct spring and summer meadow areas, and is notable for displays of introduced cowslips, which have naturalised across the site.

Invertebrates

Sheffield University and the Sheffield City Wildlife Project merged applied ecology and landscape design with studies of garden wildlife, and green space management. Broad notions of ecological value were championed including the use of non-native species for aesthetic and invertebrate purposes. This work shattered many conservation garden myths by applying scientific method to what insects really find attractive.

Green roofs

The London Biodiversity Partnership responded to the *Biodiversity Action Plan* for black redstarts, in an imaginative way by establishing three mosaic habitats on new roof spaces at the Laban Dance Centre in Deptford. This included a sedum mat to give instant green cover, shingle areas for invertebrates and calcareous grassland to provide structured vegetation. Work by numerous groups has now linked together to form the European Federation of Green Roofs (see also Chapter 46 this volume).

Housing estates

In Hackney, the Grass Roof Company took on a contract for a council housing estate landscape, and totally changed the style of management. For example, it replaced former herbicide zones around fence lines and school boundaries with flower borders using a selection of annual and perennial wildflower species. The re-creation of one of these street corners was awarded a Silver Gilt medal at the 2007 Chelsea Flower Show.

Wastes

Inert wastes can provide the best opportunities for rich and attractive vegetation communities. In the past, this was known from natural colonisation. It was apparent that low fertility and lime rich substrates could be applied deliberately, either mixed with soils or used unadulterated. By applying theories of ecological stress, the low productivity and fresh start these substrates offer will result in sustainable landscapes, which though less productive in biomass are rich in biodiversity. These techniques are less reliant on management and can be created at low cost. Materials successfully demonstrated in Liverpool include crushed concrete, brick rubble, shell sand, cockleshells, quarry wastes and recycled clothing waste. The organization Buglife also promotes the sensitive design of brownfield sites and use of wastes to promote invertebrate conservation.

Roads

In the early 1990s experiments using wildflower seed mixtures along transport corridors were carried out, rather than implementing stand-alone engineering or topsoil solutions. This work, supported with a *Wildflower Handbook* (Department of Transport 1993) has now been widely adopted on roads feeding into the urban fabric.

Wetlands

Located on the banks of the River Thames in Barnes, south-west London, this 42 ha site comprises areas of standing open water, grazing marsh and reed bed created on the site of artificial reservoir basins constructed in 1886, which became redundant in 1989.

Whilst the whole site was of conservation value, the realisation of the centre was enabled by allowing development on 25 per cent of the site, thereby raising £11 million for the project through planning gain. This pragmatic approach by the Wildfowl and Wetlands Trust enabled wetland habitat creation to be initiated in 1995 and has resulted in one of the best wildlife habitats in the country, attracting more than 150 different bird species, twenty dragonfly and damselfly, six bat and over 300 butterfly and moth species each year. Some 150 volunteers contribute their time, and the 200,000 visitors each year have been rewarded with the return of the bittern to London.

Tumbling effects

There is strong evidence for the decline in species typical of habitats where nutrient levels are low, and a corresponding increase in species of habitats where nutrient levels are high and species more vigorous. These trends are likely to continue and some of the impacts can be addressed through urban-based creative conservation. For example, the loss of less competitive species is being addressed in urban areas by using inert wastes and developing techniques to expose subsoil.

One such initiative saw Landlife pioneer a soil inversion technique in the urban fringe – a simple intervention where a metre of soil is turned upside down, burying nutrients and weeds, and bringing low fertility subsoil to the surface. Best suited to lowland agricultural soils, the technique was cited in the UK's response to the *Global Plant Strategy* as a means of addressing eutrophication. It has now been applied across the UK with seventeen partners to create a variety of new habitats at a landscape scale (Luscombe *et al.* 2008).

Just 10 g of devil's bit scabious (*Succisa pratensis*) seed introduced onto a two hectare subsoil site in Huyton's outer estates, resulted in over half the area being covered in the plant ten years later. The significance of this scheme relates to the Marsh Fritillary butterfly (*Euphydryas aurinia*) whose larvae feed on this species. The lack of devil's bit scabious plants on the last site in Cumbria for this butterfly defeated officers in their desperate efforts to keep the species in this location. But with re-introduction programmes in hand, it is an option that has been used to conserve this butterfly in Devon.

Time for change

To paraphrase Darwin, 'It is not the strongest of species that survive nor the most intelligent, but the ones most responsive to change' (Anon).

Why is creative conservation important? Because climate is changing. In the wider context, creative conservation is vital to the way people think and act in the world. It provides a flexible framework that can respond to change and adapt to the inbuilt chaos of natural systems. Warmer winters may be one of the reasons behind the increase in frequency of Mediterranean species in the south of England, with London likely to have a climate similar to northern Portugal by 2071 (Kopf *et al.* 2007). By the 2080s, summer rain may reduce by 50–60 per cent and winter rain increase by 20–30 per cent (Defra 2002). This climate shift and associated extreme events is making it increasingly difficult to sustain existing fragmented habitats. Creative conservation interventions will be needed to build wildlife corridors that include urban conurbations.

In 2009, Natural England's *State of the Natural Environment* (Natural England 2009a) report concluded that England's natural environment is much less rich than fifty years ago, particularly outside protected sites. Defra's *New Atlas of British and Irish Flora* (2002) analysed nine million records over forty years; one of the conclusions was that plants are as mobile as birds and butterflies! The findings proved that species introduced in ancient times and plants with a northerly distribution have decreased in frequency, whilst species introduced in recent times (post-1500) and those with a Mediterranean or widespread distribution have increased. These changes reflect the impacts of climate change, which suggest that habitats are moving northwards at the rate of fifty to eighty kilometres per decade (DETR 1998). Nature conservation in Britain will have to change, and creative conservation techniques being developed in urban areas are leading the way.

How common is common

Plantlife's *Common Plants Survey* (McCarthy 2008) surveyed 524 sites, 121 of which contained none of the sixty-five common or familiar plant species. It is very often these species whose ranges have been declining most rapidly. The Joint Nature Conservation Committee recognises that once common species are under huge threat and that "the priorities for conserving Britain's wildflowers in future will need to be reordered, with more concern for commonplace plants" (JNCC 2005).

Creative conservation is focused on using common core species of native origin that occur widely across the country. It is an opportunity to put back simple habitats suited to soil type, for people to enjoy: the buttercup meadow, the poppy field and the cowslip bank. Such wildflower spectacles are supposed to be common, yet few have seen them, and fewer are there to be seen.

Introductions of small numbers of common wild plants in urban areas have evolved to resemble natural plant communities, delivering Biodiversity Action Plans that benefit wildlife. Working from an understanding of simple starting points, these living seed banks have, in the case of the Ascension Islands, created the "accidental rainforest" (Pearce 2004). Similar evolutionary drivers inform creative conservation approaches to sustaining future landscapes, be it urban or rural. On a two-hectare area of neglected grassland in Huyton, the topsoil was stripped off and sold to fund the project. Sown with sixteen wildflower species, Landlife recorded sixty-four new species recruited over eight years through natural colonisation. Grassland specialists thought the site resembled MG5/MG6 grassland (personal communication following visit by English Nature grassland specialists). This process may bring species new to the area, a process well documented on post-industrial land. In St Helens for example, glass waste supports maritime communities despite being twenty kilometres from the coast.

Introduced wildflower species can change in less than a decade through adaptive variation, resulting in local characteristics that reflect the dynamic nature of natural processes (Silvertown *et al.* 2006). This can be expected on new creative conservation habitats, as nature works to obliterate evidence of the hand of man. In time, future generations may well view such places as as natural as the countryside.

Creative conservation inclines more towards local distinctiveness as opposed to local provenance. Whilst localised sourcing can be used if ecologists wish to directly mimic regional flora characteristics, the movement of species northwards and uphill means that even this attempt to maintain historic character has limited sustainability.

Start points not end points

Attempts to improve sites for nature have often used the National Vegetation Classification to identify target habitats and specify seed mixes, often resulting in substantial acreages of grass. Nature is always in a state of flux and practitioners should perhaps be more concerned with creating the right foundations to enable a site to reach its full potential, rather than predetermining and fixating on a desired end point. If an end point is to be selected, which should be chosen, the meadows of 1926, just before the ice age, or yesterday?

Urban areas are home to new species on the move. Some leapt the garden wall, others jumped ship, and some hitched a ride with the Romans. Concerns on this front have put on hold new bluebell (*Hyacinthoides ssp.*) glades in the urban fringe. Problems distinguishing the Spanish and English varieties using DNA have resulted in researchers applying field markers to determine native and alien populations (Kohn *et al.* 2009). This methodology could see an increasingly subjective and purist approach applied to concepts of naturalness, and has darker societal undertones.

DNA mapping of species ignores gene flow within species and could stymie practical habitat creation efforts. Hard evidence from the Rothamsted Park grass experiment show changes occurring in the DNA of plants over a period of just ten years! Nature likes change, but many ecologists it seems, do not.

Creative conservation promotes conditions to maximise change, by incorporating edge effects. Each stone has a temperature and moisture gradient creating a varied microclimate for nature's evolutionary dance to take place. The edge is the meeting place between chaos and order, where nature wants to come to. People naturally gather at the edge of sea and sand, or the edge of woods; plants thrive at pond and woodland edges. The eye bounces off edges, whilst butterflies need edge effects to access grasses.

Creative conservation techniques reduce timescales involved and can provide a bedrock for lasting diversity. It is not just about survival of the fittest; it is also survival of the luckiest (Baker 2002). By incorporating an element of chance and disturbance into the system, nature periodically deals the opportunist a stronger hand.

Environmental justice

In the north-west of England 92 per cent of the land has little biodiversity value. People who have the most urgent environmental needs are those who live with the consequences of a poor quality of life, and live in depressing environments – close to polluted brooks, bland parks filled with litter and green deserts. Better off communities would not accept this, and those who have to endure such environments in which to bring up a family, or grow old, are being denied access to nature and environmental justice.

Bill Adams' book *Future Nature* (1996b) calls for a new vision for the conservation movement where people are placed high on the agenda when considering future landscapes. After all, the lack of open space is cited as one of the main reasons for leaving city areas, and it is most valued when it is natural in character.

For example, two-thirds of respondents in Kirkby tower blocks went out more often as a result of a wildflower meadow created in the vicinity (Cabe Space 2006). The wider impact of this Landlife project (Figure 18.2) is best illustrated by the following piece published on the web (*Kirkby Times* 2008):

> Of all the changes in Northwood, Kirkby, the wildflowers in the woods were perhaps the most eye-turning of all changes. The wildlife, and the people, made a return to the woods, with the wildflowers being a focal point, and a talking point.
>
> For a while at least the area became less troublesome – with local youths less inclined to wreck the place. A calming influence was witnessed by many of us. People began to use the

Figure 18.2 Cornfield annuals in Kirkby.

> area as it was intended – as a nice area to walk the dog, to take your children or for carers to accompany elderly people in wheelchairs, and for local residents to enjoy a nice walk or sit off for a while.

The natural environment is our natural health service; it can contribute to improved health outcomes (Dawe and Millward 2008). Physical activity and response to stress, contributory factors in causing circulatory disease, can be enhanced by access to green space, especially by exercise and walking in vegetated areas (see Chapters 32 and 33 this volume). Thirteen hundred extra deaths occur each year in the UK amongst lower income groups in areas where the provision of greenspace is poor (Natural England 2009b: 9). But what makes people want to continue to participate in health walks? For three out of five people the motivation to walk came more from observing changing seasons rather than doctor's orders. The numbers of deaths of people on both high and low incomes in green areas was half that compared with figures relating to built up areas. Creative conservation can contribute green space in urban settings and so help redress the balance.

Accessible nature

Natural England has called for the adoption of minimum standards for the provision of natural green space (Harrison *et al.* 1995; see also Chapter 41 this volume). These sites should be freely accessible in contrast to the restrictive access required by many nature reserves. To achieve this, new sites will need to be created in consultation with local communities. However, this process often engages people on a one-off basis at the outset of a project and then neglects them after the establishment phase.

Early wildflower projects in the heart of Knowsley and Bolton transformed areas of mown grass and derelict sites into visually stunning displays (Figure 18.2), which were the catalyst for new residents' groups through activities that included painting events, buttonhole days, aromatherapy and massage sessions, brass bands, kite flying, t'ai chi, story telling and teddy bears' picnics. Such events frequently celebrate local heritage that is often industrial in character. In some respects, creative conservation activity is part of a new urbanist approach linked to sustainable development, common ownership and the arts.

Every child is a nature lover and teachers have cited language development as the greatest single benefit resulting from exposure to nature. This fundamental benefit and nature's capacity to calm and stimulate curiosity has been recognised by the Danish education system, which engages children in nature exploration prior to school admission.

Routes to learning using creative conservation sites include early examples such as the William Curtis Ecological Park on a derelict site in London, and the more recent National Wildflower Centre, a focus for creative conservation set in a public park in Knowsley (Marren 2002).

New initiatives at the latter, aim to teach maths and physics using phenomena such as the Fibonacci sequence and Brownian motion, scientific concepts that can be readily demonstrated using wildflowers. Among inner city audiences there is a hunger to see wildflowers. In an effort to satisfy this demand, the Centre has trained over 2,500 people in how to make new wildflower landscapes – the seed corn for future urban green space.

Figure 18.3 New meadows: making townships handsome again.

Figure 18.4 The National Wildflower Centre, Knowsley.

Liberty in the system

Joan Ruddock MP, UK Parliamentary Under Secretary of State for the Environment acknowledged in 2008 that "we can restore some habitats" but was less optimistic about reversing historic biodiversity loss, arguing that "there is no way that we can just turn the clock back, in any comprehensive sense" (Ruddock 2008).

Halting biodiversity loss must not be an end-point for biodiversity conservation. It should go beyond this to enable growth in biodiversity into the future. Urban-based creative conservation offers a testing ground in less controversial situations, to trial the innovative approaches needed to address this challenge. The National Wildflower Centre is working with partners to establish creative conservation as part of a nationally recognised palette to rebuild biodiversity on a landscape scale.

In 1995 an English Nature officer commented:

> There is a chance in the next few years, not to set rigid rules for what habitats should be created where, but to evolve, criteria, procedures and practices that will help us and others to make the countryside a richer and more attractive place for wildlife and people.
>
> *(Holdgate 2003)*

However, the strictures on what can go where, continue to be constrained by Biodiversity Action Plans and the ongoing adherence to the National Vegetation Classification by many urban-based ecologists. Despite enhancing biodiversity semantics, policy remains firmly rooted in historic vegetation patterns. Habitat records mapped on paper can give the illusion that all is well on the ground; the reality can be shockingly different. In the light of Darwin's work, it should be obvious that nature is no respecter of the past.

Policy developments and action programmes over the past sixty years have failed to stem the alarming loss of all major habitat types. In 2004, English Nature was developing a new vision that did not seek a strategy to recreate an England of the past. Instead it sought to look to the future, take forward a landscape scale philosophy, and develop innovative approaches to deliver biodiversity gains (Townsend *et al.* 2004: 96). Urban creative conservation activists are delivering such innovation, vision and gains.

Researchers from Kent University undertaking a biodiversity policy review (Paisley and Swingland 2007: 19, emphasis in the original) wrote:

> Projects like Landlife, the Eden Project and a host of other creative conservation and re-wilding initiatives are achieving great things. *These initiatives are pursuing the Scottish notion of environmental justice: the access of all people to nature and the need to strengthen and celebrate the links between wildlife and local communities. Such projects should have full UK government support.*

When challenged about whether the target should be creating historic or new habitats in the face of climate change, John Rodwell, architect of the *National Vegetation Classification*, stated that what was needed, was "*some liberty in the system*" (Rodwell 2003, responding to Landlife question).

These expert opinions should be taken on board so that creative conservation is given its rightful place alongside genetic conservation, as a key tool in safeguarding our natural heritage for future generations in cities, towns, villages and hamlets.

Further reading

De La Bedoyere, C. (2007) *Starting out with Native Plants*, New Holland.
Koster, A. (2001) *Ecologisch groenbeheer*, Schuyt & Co.
Lloyd, C. (2004) *Meadows*, Cassell Illustrated.
Luscombe, G. and Scott, R. (2004) *Wildflowers Work*, Landlife.
Thompson, K. (2006) *No Nettles Required*, Eden Books.

References

Adams W.M. (1996a) "Creative conservation, landscape and loss", *Landscape Research*, 21(3): 265–76.
Adams W.M. (1996b) *Future Nature: A Vision for Conservation*, London: Earthscan.
Baines, C. (2007) "Wild Mersey", in I. Wray (ed.), *Mersey: The River That Changed The World*, Liverpool: The Bluecoat Press.
Baker, O. (2002) "Law of the jungle", *New Scientist*, 2329: 28.
Cabe Space (2006) *Making Contracts for Wildlife: How to Encourage Biodiversity in Urban Parks*, London: CABE.
Dawe, G. and Millward, A. (eds) (2008) *Statins and Green Spaces*, Manchester: UK MAB Urban Forum.
Defra (2002) *Climate Change Scenarios for the United Kingdom*, The UKCIP02 Briefing Report. London: Department for Environment, Food and Rural Affairs.
Department of Transport (1998), *The Wildflower Handbook*, London: DoT.
DETR (1998) *Climate Change Impacts in the UK*, London: Department for the Environment, Transport and the Regions.
Grime, J.P. (1979) *Plant Strategies and Vegetation Processes*, Chichester: Wiley.
Grime, J.P. (1980) "An ecological approach to management", in I.H. Rorison and R. Hunt (eds), *Amenity Grassland: An Ecological Perspective*, Chichester: John Wiley & Sons.

Harrison, C., Burgess, J., Millward, A. and Dawe, G. (1995) *Accessible Natural Greenspace in Towns and Cities: A Review of Appropriate Size and Distance Criteria*, English Nature Research Report No. 153, Peterborough: English Nature.
Holdgate, M. (2003) "The human stake in nature", *ECOS*, 24(1): 61.
Joint Nature Conservation Committee (2005) commentary on *Vascular Red Data List for Great Britain – Species Status No. 7*, online, available at: www.nhbs.com/the_vascular_plant_red_data_list_for_great_tefno_140371.html.
Kendle, T. and Forbes, S. (1997) *Urban Nature Conservation: Landscape Management in the Urban Countryside*, London: Taylor & Francis.
Kirkby Times (2008) "Fighting the scallies with flowers!" online, available at: www.kirkbytimes.co.uk.
Kohn, D.D., Hulme, P.E., Hollingsworth, P.M. and Butler, A. (2009) "Are native bluebells (Hyacinthoides non-scripta) at risk from alien congenerics? Evidence from distributions and co-occurrence in Scotland", *Biological Conservation*, 142(1): 61–74.
Kopf, S., Hallegatte, S. and Ha-Duong, M. (2007) "L'évolution climatique des villes européenes, Climat: Comment éviter la surchauffe?" *Dossier Pour la Science*, 54: 48–51.
Landlife and Urban Wildlife Partnership (2000) *Creative Conservation Guidelines and Principles*, Liverpool: Urban Wildlife Partnership.
Luscombe, G., Scott, R. and Young, D. (2008) *Soil Inversion Works*, Liverpool: Landlife.
Marren P. (2002) *Nature Conservation: A Review of The Conservation of Wildlife in Britain 1950–2001* (The New Naturalist series), London: Collins.
McCarthy, M. (2008) "Britain's most common wildflowers are latest species to disappear from countryside", *Independent*, 21 March.
Natural England (2009a) *The State of the Natural Environment*, Peterborough: Natural England.
Natural England (2009b) *Our Natural Health Service: The Role of the Natural Environment in Maintaining Healthy Lives*, Peterborough: Natural England.
Nicholson-Lord, D. (1987) *The Greening of the Cities*, London: Routledge and Kegan Paul.
Paisley, S. and Swingland, I. (2007) *Biodiversity Policy Review*, Canterbury: Durrell Institute of Conservation and Ecology, University of Kent.
Pearce F. (2004) "The accidental rainforest", *New Scientist*, 2465: 44–5.
Preston, C.D., Pearman, D.A. and Dines, T.D. (2002) *New Atlas of the British and Irish Flora*, Oxford: Oxford University Press
Rodwell, J. (1991–2000) *British Plant Communities, Volumes 1–5*, Cambridge: Cambridge University Press.
Rodwell, J. (2002) *Nature Conservation – Who Cares?* BANC/English Nature Conference, Market Harborough, October 2002.
Rodwell, J. (2003) "Human relationships with the natural world: an historical perspective", *Ecos*, 42(1): 10–16.
Ruddock, J. (2008) *House of Commons Select Committee on the Environment*, thirteenth report, Hansard Archives, London: House of Commons.
Sheail, J. Treweek, J.R. and Mountford J.O. (1997) *The UK Transition from Nature Preservation to "Creative Conservation"*, Huntingdon: Institute of Terrestrial Ecology.
Silvertown, J., Poulton, P., Johnston, E., Edwards, G., Heard, M. and Biss, P.M. (2006) "The Park Grass Experiment 1856–2006", *Journal of Ecology*, 94(4): 801–14.
Thoreau, H.D. (1861) *Journal*, 3 January, Camp Sherman OR: Thoreau Institute, online, available at: http://walden.org/Institute/thoreau/writings/Quotations/Conservation.
Townsend, D., Stace H. and Radley D. (2004) *State Of Nature: Lowlands – Future Landscapes for Wildlife*, Peterborough: English Nature.
Urban Wildlife Unit (1983) *Urban Nature Conservation Strategy (Liverpool and its fringes)* NCC contract no. HF3/05/115, Liverpool: Urban Wildlife Unit.
Yoxon, M. (1977) *Ecological studies in Milton Keynes*, Milton Keynes: Milton Keynes Development Corporation.

Part 3
The nature of urban habitats

Introduction

Ian Douglas

This section aims to set out the character of urban habitats from the buildings themselves to gardens, parks, woodlands and wetlands. The chapters review the main features of the habitat and the special combinations of plants and animals to be found in them. It also considers the way animals use these habitats and the characteristics of the adaptation of different animal species to urban living.

The construction of walls, courtyards, pavements and walkways also creates habitats for plants, even though such surfaces are usually dry and undergo extreme changes of temperature, warming up during the day and cooling rapidly at night. Irregular decay of the mortar in walls of stone or bricks provides the first footholds for plants, some of which extend their ranges dramatically. Ivy-leaved toadflax is almost exclusively found on walls in Britain. Wall plants include xerophytes, native cliff and rock plants and annuals, ferns, and garden escapes. Paving plants have to suffer compacted soil with consequent lack of aeration. Prostrate growth is common, with knotweed, plantains and pearlwort all showing this on paths. Pollution affects many of these surfaces, especially where exposed to vehicle emissions. Phil Wheater shows how stonework offers potentially important habitats for diverse communities of plants and animals. In addition, as such habitats are abundant, they provide opportunities for educational study and research.

The characteristics of our 'built' or urban environment reflect those of our ancestral cliff habitat and continue to provide opportunities for the same array of species with which we have been associated for almost one million years. Today the human habitat is dominated by rock outcrop organisms, and the more cities we build the more we create new habitats for this suite of organisms. People are spiritually attracted to cliffs because they have been sites of refuge and dwelling for most of our recent evolutionary history. *The Urban Cliff Hypothesis* argues that we are essentially the same 'cliff dwelling' or 'cliff dependent' species that arose more than a million years ago. Birds, such as peregrine falcons, are attracted to the cliff like features of tall buildings, especially those built adjacent to open expanses of water. Jeremy Lundholm explains how the framework presented by the Urban Cliff Hypothesis gives a context for understanding past human impacts and for envisioning better urban environments that benefit both people and the rest of nature.

Suburbia comprises a mosaic of contrived plant diversity, blatantly people-made, to which nature had added a variety of species of animals and plants that have adapted to urban life. The older, inner suburbs of cities contain widely different densities of housing and ratios of garden area to built-over area. Few are completely devoid of private or public gardens with mature trees. Many contain a relatively large portion of garden land which contains a rich habitat with a combination of mown grass, trees, and patches of flowers, shrubs and vegetables. These gardens respond rapidly to inputs of nutrients and usually tend to receive supplementary fertilizers and applications of water. Insects exploit the full variety of the gardens but may be the target of chemical sprays, though probably less intensively in gardens than areas of intense horticulture and arboriculture. Trees in gardens and parks are usually further apart than in natural woodland and have large canopy areas. They contribute to the reduction of particles in the air, but may dry the soils through high evapotranspiration in dry periods. Such gardens are important bird habitats and support a variety of animals, such as the squirrels and foxes now common in British suburbs. Within this category there are contrasts due to the age of the suburb. The new suburbs of cities, often built on former farmland, usually differ from older suburbs by having a smaller number and lower height of trees, and a greater proportion of lawn and mown grass than of shrubs and cultivated flower and vegetable gardens. The total plant biomass of such suburbs is much less than that of older inner suburban areas with consequently a less varied suburban fauna. The suburban mosaic also comprises other forms of greenspace, as Ian Douglas describes, including allotments, derelict and vacant land, golf courses, business parks, corridors and patches. Nowadays the 'urban green' is frequently interpreted as a set of sites of co-fabrication between humans and other city inhabitants, rather than as islands of nature in the city. The different kinds of spaces of ecological engagement include spaces associated with cultivation, such as allotments, community gardens and city farms; those associated with land restoration, such as derelict railway, mining, or industrial sites; and those associated with conservation, such as remnant woodlands, waterways and marshes.

Railway, road, canal and power-line reservations are major routes for plant migration in urban areas. Abandoned railway lines in urban areas can develop into major refuges for wildlife if they are not subject to heavy and frequent use by people. Weeds often spread unchecked along railways and canals. The regular mowing, or spraying of railway embankments restricts their biotic diversity, but such maintenance is not frequent and many complex plant associations develop. However, where railway lines have been abandoned, cuttings and embankments have undergone a succession of plant colonization whose nature depends on the surroundings of the segment of old railway track involved, the type of ballast used and the rate at which pioneer plants, can colonize the track, and the frequency and intensity of human use of the old track. While such areas are often unofficial playgrounds for children, they can also become used as a convenient place to dump rubbish and unwanted refrigerators and mattresses. In many instances, such areas have become part of footpath, cycle track and bridleway networks through cities. Ian Douglas and Jon Sadler show how corridors have six recognizable functions: habitat, conduit, filter, barrier, source and sink. In the habitat function, organisms can survive and reproduce in the corridor. In a conduit, organisms pass from one place to another, but do not reside in the corridor. In a filter, some organisms or materials can pass through the corridor, others cannot. A barrier means that organisms and materials cannot cross the corridor. If the corridor is a source, organisms or materials emanate from the corridor. If it is a sink, organisms or materials enter the corridor and are destroyed. These functions have to be addressed specifically in any debate about preserving, enhancing or creating wildlife corridors.

City parks range from highly formal, carefully managed gardens where people are not allowed to walk on lawns, to parks that are largely sports grounds, or parks that are basically remnants of common land or woodland. The character of urban parks reflects the ideas of their planners and designers, some of whom deliberately laid out formal gardens or planted exotic species in a formal manner. The range of ecological situations in city parks depends on the variety of management approaches. The extent to which insects can breed and multiply and form the basis of a food chain which supports a variety of avian and mammalian fauna depends in part on whether debris and plant litter are removed or allowed to decompose and how far insecticides and other forms of biocide are used to promote the growth of decorative plants. Where the management goal is the maintenance of lawns for active and passive games and recreation, the plant and animal diversity is low. Many city parks suffer such pressure of human activity that parts of them have become the equivalent of an overgrazed and overtrampled open woodland with relatively healthy trees, but the elimination of the ground-level flora and fauna. Martin Hermy examines the role of urban parks in terms of the ecopolis and sustainability concepts and critically assesses the types of management required for different aspects of urban parks, from water spaces to grasslands.

The grasslands of the redeveloped inner city areas are characteristic of a whole range of reclaimed, replanted areas in older cities. The substrate may often contain high concentrations of heavy metals, particularly of lead, about 50 cm below the surface. Such hangovers from former dwellings or industrial premises may influence the growth of roots of some species. Tony Kendle argues that alongside richer areas of planting or remnant natural habitats, the many mown grasslands that make up large almost featureless plains and are often scornfully referred to as 'green deserts', are not devoid of life or productivity. They may be better described as urban rangelands or savannah. Some diversity in the grasses may occur depending on the type of planting, but generally species diversity is low. Any increase in diversity over time depends on the use and management of the area. If mowing is infrequent or areas are untended, greater variety will occur.

Contaminated land is a widespread phenomenon in urban areas in industrialized countries. Much of it is a legacy of past activity, some of which ceased decades ago. Often records of the precise activities and their waste products no longer exist, making it difficult to trace the source and dispersal of any given contaminant. Adequate sampling is required to detect pockets of severe contamination and so reduce costs of rehabilitation. Good ecological understanding of species able to grow on soil forming materials placed over contaminated sites enables restoration to be effective. Michael Rivett and his co-authors explain the diversity of contaminated land derived from former industrial activities ranging from small garage workshops to a range of light/heavy manufacturing industries, from secret weapons research establishments to pesticide/chemical manufacturing plants, from large coking works to small dry-cleaner laundries, from railway sidings to cemeteries. Although diverse, each may pose significant risk and may assume a 'contaminated land' label. No two sites are the same. Even similar industrial types may have different combinations of contamination hazards, different contaminant transport scenarios possible, different and novel juxtapositions of urban habitat – receptors at risk and different pollution, economic, technical, regulatory and political drivers influencing assessment and remediation. However, high land values can enable costs of derelict land reclamation to be absorbed in development projects. Nonetheless, pockets of severe contamination occur around many old industrial areas and care has to be taken to contain the contamination and particularly to avoid the movement of contaminants in surface or groundwater bodies that may used for drinking water supplies.

Many cities contain areas of woodland that have become surrounded by urban growth. They are now key pieces of countryside landscape in the city. Examples include the woods of Hampstead Heath in London, Ecclesall Woods in Sheffield, the woodlands in Central Park, New York, the Bois de Boulogne in Paris, the peri-urban woodlands of Zurich and the Bois de la Cambre immediately south of Brussels. These urban woodlands contain many more species than the central city parks. In many cases these woodlands house important mammal and bird species, some of which migrate into gardens to seek food. C.Y. Jim shows how a cluster of small and dispersed woodland patches could denote remnants or relicts of a former continuous forest cover. Although the daughter patches share a common parental forest, they differ by topography, soil, hydrology, area, shape and human intervention. Depending upon the stress regime, such disjunct patches may experience differential degradation or progression. The divergent ecosystem development could generate assorted woodland patches in terms of species and structure.

Poorly drained, marshy ground occurs in small pockets in many urban areas. Some of it may be the undeveloped remnants of natural wetlands in flood plains or estuarine or coastal marshes. Other sites may be on the edges of waste dumps and derelict industrial land where natural drainage has been blocked. The plant and animal life of such wetlands depends on their nature, location size, surroundings, substrate and inputs. While all pools and marshy areas, temporary or permanent, on waste land are the habitat of insect life, perhaps the most medically significant are those in the ill-drained, poorly serviced, neglected areas of squatter settlements and temporary housing on the edges of most cities in low latitudes that are often the breeding grounds of mosquitoes and other disease vectors.

Animals are an integral part of the urban ecosystem, with major roles in species dispersal, habitat colonization and change, and varying responses to food sources and disturbance. People have varying responses to wild animals in the urban context, sometimes deliberately providing food, at other times accidentally leaving food out and encouraging the species that many people do not wish to have in or around their homes. Thus people have both positive and negatives (and often ambivalent) attitudes to urban wildlife. Whole businesses and municipal operations work to exterminate some species, such as wasps or rats. Yet animals are important cultural epiphenomena (in terms of education, mental health and enjoyment of urban greenspaces). Pets have to be seen as part of urban animal ecology. Cats are hyperpredators, occurring at many times their natural numbers in towns and cities. They affect bird populations, and sometimes whole species, deleteriously. Dogs can affect ground-nesting birds in similar ways. Nightingales – which used to be frequent on Richmond Common – have now disappeared due to dog-walkers. Yet there is also evidence that contact with dogs and cats improves the psychological and mental health, longevity, even, of human beings, some suffering from disturbances such as alcoholism. In some cases, e.g. certain autistic illnesses, they can lead to individuals 'breaking through' and communicating for the first time. Peter Jarvis, however, shows that animals succeeding in the urban environment are often opportunistic habitat and food generalists that can tolerate disturbance from and sometimes benefit from support by the public. Alternatively, they thrive in pockets of semi-natural habitat that provide similar or analogous conditions to that found in the wild. They are a significant component of community dynamics, and – with some exceptions – are generally viewed by people as a welcome addition to their concrete jungle.

Feral dogs and cats roam the streets of nearly every metropolitan city. The cat returns readily to a feral state if it has not been socialized properly in its young life. These cats, especially if left to proliferate, are frequently considered to be pests in urban areas, and may be blamed for devastating the bird, reptile and mammal populations. A local population of feral cats living in an urban area and using a common food source is sometimes called a feral cat colony. Feral animals affecting important wetland reserves around Sydney, Australia include cats, dogs, foxes, house mice, rats and starlings, all animals introduced to Australia by Europeans. They are attracted to gardens by food sources. Changes in rubbish collection bin systems can remove these food sources. Peter Jarvis notes that people have an ambivalent attitude towards feral animals in the urban environment, particularly in the case of the feral pigeon: many get pleasure out of feeding them, others rue the sight and cost of damage caused to buildings by the activities of these birds. He concludes that while feral pigeons are widespread and abundant in many towns, and are often viewed as pests, other feral animals are at most locally abundant, and issues relating to feral cats and dogs generally concern welfare. For animals that are often such highly visible components of the urban landscape their overall impact on the urban ecosystem tends to be rather low.

19

Walls and paved surfaces

Urban complexes with limited water and nutrients

C. Philip Wheater

Some of the most extreme urban habitats are artificial substrates comprising paving, road surfaces, courtyards, walkways, buildings and boundary walls. These environments (called "stonework" by Hadden 1978) are widespread in towns and cities, with estimates of 1 ha of vertical wall surface for every 10 ha of urban area (Darlington 1981). Such constructs can be hostile for plants and animals: new walls have steep, vertical, highly exposed faces retaining little nutrient material. Walls do, however, become more hospitable with age (as they decay) and in the absence of management and disturbance. Paving, too, despite the obvious impacts of trampling, provides habitats for plants and animals. Some groups can be well represented, with diverse lichen floras for example being found on stonework (e.g. Rosato 2006; Laundon 1967), and old walls can support diverse communities. Surveys of the Roman walls in Colchester recorded over 60 species of higher plant (including the nationally scarce *Clinopodium calamintha* – lesser calamint), 125 species of lichens including nationally rare species, and many invertebrates including notable species such as the bug *Asiraca clavicornis* (Jerry Bowdrey, Colchester and Ipswich Museum Service, pers. comm.). Older stonework may be relatively species-rich following decay, or because organisms have had longer to get there, or because they represent rural habitats (e.g. churchyard boundary walls) encapsulated by urban sprawl.

The similarity between the vegetation on walls and paving has long been recognized (e.g. Segal 1969). Lack of moisture and nutrients are major influences on colonization and establishment, and plants on stonework may have lower biomass, stunted stature, and lower fertility and productivity. The major differences between walls and paving are the composition of the building materials, how vertical the surfaces are (impinging on water retention and accumulation of organic matter), and the level of disturbance.

This chapter reviews work over the last 50 years on mural and paving ecology, examining the major drivers for the establishment and maintenance of communities on stonework. Although much early work was done in Britain and later work tends to be European, many of the principles apply to towns and cities across the world.

Colonization

Stonework floras vary geographically, partially reflecting climate. Payne (1978) compared walls in south-eastern England with previous mural studies and found similarities among common species with studies around London, but fewer similarities with walls in the north of England. Colonization can occur quickly with some mosses and lichens appearing within a few years of a wall being built. Subsequent development of vascular plant communities takes substantially longer, with mature communities usually being found only on walls over 100 years old (Darlington 1981; Segal 1969).

Stonework close to existing plant and animal communities tends to colonize quickly with species found locally: driveways and garden walls are often colonized by garden plants and weeds. Hence, there may be a relatively high abundance of non-indigenous species on walls (Simonová 2008) and paving (Muratet *et al.* 2008). Stonework tends to support relatively few common species, the majority being found only occasionally and resulting from local sources (e.g. dos Reis *et al.* 2006; Pavlova and Tonkov 2005; Duchoslav 2002). Isolated walls and paving develop more slowly than those forming part of networks of similar structures. Several studies suggest that stonework replicates natural rocky areas and record that plants typical of these environments predominate (e.g. Lundholm and Marlin 2006; Hruška 1987; Woodell 1979; see also Chapter 25). Other common wall species are eurytopic and hence common across urban habitats (e.g. Duchoslav 2002; Ignatieva *et al.* 2000; Lagiou *et al.* 1998).

The propagules of some stonework species may travel considerable distances, particularly in colonizing the centres of large urban areas. Mechanisms for dispersal include wind and other air currents, and through animal (including human) activity. Wind-borne seeds are deposited more on walls than surrounding areas, probably because walls develop strong updrafts around them (Darlington 1981). Pavement populations can be sparse, with plants reproducing by self-pollination and producing many wind-dispersed seeds (e.g. *Sagina procumbens* – procumbent pearlwort). Several common plants on walls are specialized wind-dispersers (e.g. many Asteraceae) and can be carried in vehicle slipstreams and by adhering to vehicles. Even woody plants such as *Buddleja* sp. (butterfly-bush), which produce large amounts of wind-borne seeds, can be found on walls and in pavement cracks.

Other species are carried by animals (e.g. birds and ants). Ants regularly scavenge on walls, frequently carrying seeds back to their nests (myrmecochory). Some seeds, especially those with oily surfaces (e.g. *Cheiranthus cheiri*, wallflower, and *Cymbalaria muralis*, ivy-leaved toadflax), are more attractive to ants than are others, and any lost along the way may germinate in cracks (Gilbert 1992). Benvenuti (2004) ascribed the presence of *Capparis spinosa* (caper) at heights on ancient walls of Pisa to movement by ants since this species does not have seeds adapted for wind-dispersal. The first colonizing animals are good dispersers such as winged insects (butterflies, moths, true bugs and some beetles) and small beetles and spiders that are carried on the wind.

Stonework communities

Several authors have described the factors that influence community development on stonework (e.g. Darlington 1981; Woodell 1979). There are two major routes to the colonization of bare surfaces (Lisci *et al.* 2003). On dry vertical surfaces, the primary colonizers are xerophytes such as lichens which, while particularly frequent on stonework, are often dependant on the substrate type (Figure 19.1). The second occurs where water collects (including on horizontal surfaces) and mosses are often the primary colonizers.

Cracks in stonework accumulate dead and decaying biotic matter, which contains nutrients and increases the moisture holding capacity. As organic matter accumulates in cracks, on ledges and where lichens, mosses and climbing plants grow, larger plants including grasses, ferns and flowering plants gain a foothold. These increase the accumulation of nutrient materials and can enlarge cracks enabling larger herbaceous and woody species to establish. Research on

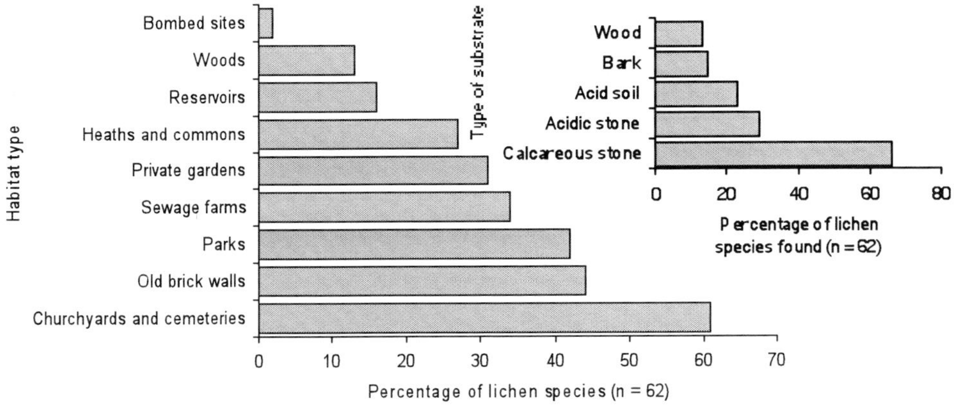

Figure 19.1 Percentage of lichen species found on different habitats and substrates (London, UK) (*source:* Laundon 1967).

artificial urban sites in Central Europe found clonal plants (self-replicating species) to be nearly twice as common as non-clonal plants (Prach and Pyšek 1994). Early colonizers included many annuals and some clonal species, with the latter spreading more rapidly initially then declining as communities developed.

Species with fast germination that exploit sudden rainfall may also be advantaged (Benvenuti 2004). Even large trees can establish on stonework. Jim (1998) found 30 tree species, some over 9 m tall, on stone retaining walls in Hong Kong, and Trocha *et al.* (2007) recorded *Betula pendula* (silver birch) on brick walls, often in association with symbiotic ectomycorrhizae (Figure 19.2). Walls can feature unusual communities with colonizers and woody species on the same

Figure 19.2 Silver birch growing from a domestic chimney.

site (e.g. Brandes 2004). Paving can also develop rich floras, although species are often at low frequencies. The dormant seeds of annual species often survive freezing in winter better than perennial species (except for those with deep roots). Many of the pavement plants that Chang and Kim (1990) recorded utilized C_4 carbon-fixation (an adaptation to the high temperatures, low nitrogen and drought conditions in which they were growing).

Vegetation on stonework provides habitats for animals. Steiner (1994) collected over 60,000 animals (194 taxa) from 216 moss samples on walls near Zürich. These included nematodes (47 taxa), tardigrades (13 taxa), rotifers and arthropods, of which mites and springtails were the most common (74 taxa) together with larger spiders, woodlice, centipedes, beetles and other insects. The pioneer stages in such communities are dominated by species such as mites, rotifers and snails that inhabit and consume lichens, algae and mosses and tolerate exposed surfaces. Snails are vulnerable to desiccation, sheltering in cracks in dry conditions, and prefer limestone or calcareous mortar (using the calcium for their shells). Some predators occur on bare surfaces; spiders such as Segestridae (jumping spiders) prey on small insects that alight to bask. Butterflies and lizards also bask on bare surfaces and Lepidoptera use crevices for over-wintering, to shelter from rain or wind, and for attaching pupae (Ruszczyk and Silva 1997). Ants explore open surfaces from nests away from the stonework, or in crevices, or under paving stones. Increased vegetation cover provides shelter for larger detritivores (e.g. woodlice and millipedes) and herbivores (e.g. true bugs). Woodlice shelter from dryness and aggregate under paving stones or in cracks on walls. Walls are important for taxa such as woodlice and spiders; over half of native British woodlice are recorded from walls (Harding and Sutton 1985), and Darlington (1981) considered that over 10 per cent of the British spider fauna were mural species.

Eventually stonework supports predators such as ground- and rove-beetles, harvestmen, and those orb-web spinning spiders that require established vegetation on which to spin webs. Nectar feeders such as bees and butterflies visit flowering plants. Wasps and bees, including mason wasps, bumblebees and especially mining bees, nest in gaps in stonework. Snakes and lizards hide in crevices and larger holes can form nests for birds and mammals (e.g. small mustelids and rodents). Stonework habitats are important in promoting biodiversity since there is often a paucity of animals such as bees and butterflies in urban areas (Eremeeva and Sushchev 2005).

Structure

The construction of walls or pavements determines whether plants can obtain and maintain a foothold. Concrete, tarmac and tight fitting bricks and stones provide few gaps for roots, or for accumulating water and organic matter. Most bricks, stones and mortar are alkaline (often only yellow bricks made of clays and stone such as granite are acidic). Limestone and lime-based mortar provide opportunities for calcicoles, whereas brick, rich in phosphorous, potassium, calcium and magnesium, supports different species (Gilbert 1989). The hardness of construction materials determines its resistance to pressure from roots and weathering.

Paving covers underlying soil, so pavement plants can suffer from lack of light and compacted soil with consequent deficiencies in aeration, water and nutrients. Unlike in urban walls, gaps between paving stones tend to be unfilled, providing access to soil (unless laid on concrete bases or weed membrane). Plants growing away from kerbs, and especially those near to walls, are sheltered, and may receive nutrients from rainwater runoff. Sunken areas in front of basements often maintain damp conditions and can support rich plant communities, including ferns (Figure 19.3).

The sides of buildings provide similar conditions to boundary walls, although those that are concrete or rendered do not have the nooks and crannies associated with bricks or stone. Many plants do not grow very high on buildings, and those on occupied buildings are often removed.

Figure 19.3 Plants concentrated in a sheltered basement frontage.

However, derelict buildings can develop complex communities similar to those on demolition sites (Gilbert 1989).

Decay

Eventually, even the hardest materials decay (e.g. Smith *et al.* 2008). Water freezes in winter forcing open small cracks, which roots can enlarge, providing space for other plants and animals. Mortar erodes at different rates from bricks or stones, creating crevices in which larger plants grow causing yet more erosion. Biodeterioration depends on the receptivity of the construction materials including the moisture availability on and in the structure (Warscheid and Braams 2000). Stones such as granite have lower porosity and poorer penetration of moisture than artificial materials such as concrete and brick, leading to more superficial biofilm development. Bioreceptivity depends on surface roughness (undamaged glazed bricks are rarely colonized) and nutrient availability (Guillitte 1998). Piervittori *et al.* (2004) review lichen deterioration of stonework. Conversely, micro-organisms are possible agents for remediation of building stone in removing sulphate crusts or producing protective layers via biomineralization of the surface of the stone (Webster and May 2006).

Colour

The colour of building materials influences surface temperatures; bare, dark surfaces heat rapidly during the day, readily losing moisture, then cool quickly at night. Dry, hot conditions are typified by xerophilous and thermophilous species. Under established vegetation or shade (for example north-facing walls, or those sheltered by buildings or trees) surface temperatures and moisture availability are less variable, enabling more species to survive. Similarly, the temperature of asphalt concrete is markedly higher than that of the surrounding soil, which itself is higher than that of adjacent vegetated areas (Yilmaz *et al.* 2008).

Runoff

The slope of paving and inclination of walls determines how much water runs off. Runoff can accumulate at wall bases and maintain damp conditions. Pavements adjacent to roads may be splashed by passing vehicles and can flood if drains are blocked. Floods can also be caused by localized subsidence. Flooding is problematic since many plants cannot withstand long periods of water-logging. Runoff increases where previously permeable areas are paved and this is increasingly an urban problem; about a quarter of front gardens in the north-east of England are now paved, while in London there are over 30 km^2 of paved front gardens with about 24,000 applications per year for lowering kerbs to accommodate off-road parking (RHS 2006). The problems of paving over gardens leading to lower biodiversity and increased runoff has also been recognized in the USA (Frazer 2005) and is now being controlled in England and Wales through the planning application process (Environment Agency 2008) (see also Chapters 13, 15 and 45 this volume).

Microclimate

The aspect of the site influences the amount of shade and insolation and hence the temperature and moisture retention of both walls and paving. Indeed, different species can be found on the south- and north-facing sides of the same wall, with the latter (in the northern hemisphere) having up to twice the species of vascular plants (and an even greater difference in moss coverage) than the former (Darlington 1981). Conversely, south faces may have more lichens. The decay of walls may be slower on north- compared to south-facing walls (Segal 1969). Microclimate is affected by the amount of shade, the proximity of buildings and trees, and exposure to wind. Canyon effects of surrounding buildings may increase the speed and gustiness of wind. Vegetation cover, including that of overhanging species ameliorates the microclimate and hence the growing conditions for other species (see also Chapters 10 and 11 this volume).

Pollution

Pollutants from atmospheric sources and building materials can accumulate in rainwater runoff. Kennedy and Gadd (2001) found that runoff from buildings in New Zealand was influenced by the composition of the construction materials and accumulation of organic matter in gutters and on roofs. They found higher pH, more suspended solids, organic matter, and heavy metals in runoff than in rain. The pH of pollutants can increase weathering, for example where acidic rain falls on alkaline materials including lime-based mortars. Historically, high levels of sulphur dioxide in urban air pollution reduced the lichen abundance in many industrialized cities. Subsequent reductions in sulphur dioxide levels have resulted in an increase in foliose lichen species (Darlington 1981).

Pollutants accumulate in soils (see also Chapter 15 this volume) and on walls alongside roads. The de-icing salts, sodium chloride and magnesium chloride (sometimes used near buildings because it is less corrosive to stonework), can lead to a decline in plant biomass (Cunningham *et al.* 2007). The levels of many vehicular pollutants (including heavy metals such as cadmium, lead, zinc and nickel associated with debris and emissions from vehicles) decline with the distance from the road.

Guano from perching and nesting birds, and dog urine and faeces increase nitrogen availability and affect the communities found on walls and paving. Nitrophilous algae (e.g. *Prasiola crispa* and *Desmococcus olivaceus*), lichens (e.g. *Lecanora dispersa*) and mosses (e.g. *Bryum capillare* – thread-moss) form communities low down on walls where urination is frequent (Gilbert 1989; Darlington 1981). Similarly, larger nitrophilous plants will grow at the base of such walls. Dog faeces can be widely distributed over urban streets: Rinaldi *et al.* (2006) found faeces in 98.6 per cent of sampling sites in Naples, southern Italy with a median of 25 per site (1 km × 700 m). Such levels can increase the presence of a variety of insects, especially Diptera.

Trampling

Trampling damages plants and increases soil compaction, leading to poor moisture retention and lack of aeration. Species subject to heavy trampling often exhibit adaptations such as prostrate growth either as rosettes (e.g. Plantaginaceae and some Asteraceae) or low growing mats (e.g. many mosses). Grasses on paving are often low growing and as the tops are damaged, the leaves grow from the base (the growth meristems are near the ground and are thus protected against damage). A very common pavement grass (*Poa annua*, annual meadow-grass) is highly tolerant to trampling and sets seed even when reduced to a very small plant. Where large gaps between paving stones accumulate moisture and nutrients, plant communities resemble those of heavily trampled gravel paths or, in sites with high nutrient levels, field gateways.

Management

Weeds on walls and buildings, at wall bases and especially on paving are often controlled to improve the aesthetics, reduce biodegradation and staining, and remove hazards to pedestrians. Organisms such as algae can discolour building materials; cement-rendered and limestone buildings are prone to this since they are porous (and hence retain moisture). Larger herbaceous and woody species (including climbing species) can physically damage walls and paving and make walking hazardous. Concern about pesticide use especially in public areas has led western cities to examine non-pesticide control (e.g. Rask and Kristoffersen 2007). Although it may be necessary to remove vegetation from stonework, small plants may do little or no damage. Indeed, some vegetation can bind walls together with their roots. Natural England (2007) gives guidance on when and how to manage mural vegetation.

Microhabitat

Ledges and joints provide microhabitats where material accumulates and moisture is conserved, especially at the junctions between different types of materials where differential movement creates opportunities for crack development. Darlington (1981) recorded that vertical walls were colonized by relatively few specialist mural species such as *Asplenium ruta-muraria* (wall rue), but where there are less severe angles, there are more ruderal species (e.g. *Taraxacum officinale* agg., dandelion). A range of microhabitats (e.g. Lisci *et al.* 2003) can be distinguished by the degree to which they retain moisture and substrate (Table 19.1). Generally, the number of plants decreases in microhabitats where there is little precipitation and relative humidity is low all year round.

Table 19.1 Characteristics of major microhabitats on walls

Microhabitats	*Features*	*Proportion of principal species[1] (%)*	*Comments*
Ground level	Rainwater retained; substrate accumulates	72	Roots may penetrate the ground; plants suffer less stress
Inclined surfaces	Dry, receiving wind-driven rain; more moisture than vertical surfaces	53	Offers possibilities for seeds to lodge
Between two types of building material	Dry, receiving only wind-driven rain	59	The greater the chemical difference between the two materials, the more nutrients are available.
Vertical faces of homogeneous material	Water limited to wind-driven rain	57	Can be hostile to plant growth
Horizontal surfaces	Water availability good; substrate accumulates	76	Plants supported by accumulation of substrate, moisture and seed deposits by ants and birds
Where vertical and horizontal surfaces meet	Rainwater retained; less substrate than wall bases	60	Plants supported by water retention
Where two vertical surfaces meet	Less moisture than where horizontal and vertical surfaces meet	24	Lack of moisture tends to restrict the number of species
Invaded by plants from adjacent soil	Wet where horizontal substrates absorb moisture or water seeps through	19	Retaining walls supporting trees, shrubs and rhizomatous and stoloniferous herbs

Source: Lisci *et al.* (2003).

Note
1 The proportion of principal species found on each microhabitat from those recorded on walls of historic interest.

Ledges on buildings are used for nesting and roosting by birds. Cliff-nesting species such as gulls have lower nesting densities and higher breeding success in towns than rural sites (Fisk 1978). The number and size of colonies of *Delichon urbica* (house martin), which build mud nests under the eaves of buildings, have been found to increase following a reduction in atmospheric pollution in Manchester, UK (Tatner 1978), and with tourism development (and hence more buildings providing potential nesting sites) in southern Spain (McCreery and Wheater unpublished data). *Falco tinnunculus* (kestrel) in Bratislava are found mainly on artificial structures (26 per cent on buildings under construction) and nest higher in the inner city (up to 50 m high) than at the edge (Darolová 1992).

Several authors recognize four zones to walls: base, middle, upper and top (e.g. Gilbert 1992) that are influenced differently by several factors (Table 19.2). Similarly, different heights on buildings and different distances from kerbs and walls on paving show differential influences. Wall bases are wetter (gaining moisture from runoff and capillary action from the soil), more shaded, and often accumulate organic matter, which is more neutral or acidic than the wall itself, sometimes becoming floristically diverse (Figure 19.4). Wall tops accumulate more material and are colonized by seeds dropped by perching birds and thus may also support diverse floras (e.g. Pavlova and Tonkov 2005). Although less well documented, animal species may also show zonation: the British web-spinning spider *Amaurobius similes* prefers dryer walls to *A. fenestralis*, which itself prefers drier conditions to *A. ferox*, which is often found at the base of old walls (Chinery 1977).

Conclusions

Stonework offers potentially important habitats for diverse communities of plants and animals. In addition, as such habitats are abundant, they provide opportunities for educational study and research (e.g. Jennings and Stewart 2000; Wheater 1999; Smith 1984). Increasingly the built environment is being managed to enhance biodiversity, including by modifying existing

Figure 19.4 Diverse vegetation growing at the base of a boundary wall.

Table 19.2 The influence of various factors on different zones of buildings, walls and paving on vegetation

Factors/zones on stonework	*Age*	*Construction*	*Slope/inclination*	*Aspect*	*Exposure/ shade*	*Wind*	*Moisture*	*Nutrients*	*Disturbance*	*Pollution*	*Animals*
Top of buildings, roofs, gutters, etc.	★	★	✶	★	★	★	★	★	✶	☆	✶
Ledges, sills, etc.	★	★	✶	★	★	★	★	★	✶	☆	✶
Top of boundary walls	★	★	✶	★	★	★	★	★	✶	☆	✶
Middle of walls	★	★	✶	★	★	✶	★	★	☆	☆	☆
Low down on walls	★	★	✶	★	✶	✶	★	★	☆	✶	✶
Ground at wall bases	★	★	✶	☆	☆	☆	★	★	☆	✶	✶
Paving away from walls and kerbs	✶	★	✶	✶	★	☆	★	★	★	✶	✶
Paving adjacent to kerb	✶	★	✶	✶	★	☆	★	★	★	★	✶

Notes
★ moderate impact
✶ major impact
☆ relatively minor impact.

Figure 19.5 Using vegetation to create green roofs and walls.

structures (e.g. adding ledges, bird and bat boxes), introducing species (such as *Falco peregrinus*, peregrine falcon), and designing new structures with wildlife in mind (URBED 2004). One initiative that is gaining in popularity in North America, and increasingly in Western Europe, is the use of green roofs and living walls to retain rainwater and promote biodiversity (Dunnett and Kingsbury 2008; Figure 19.5; Chapter 46 this volume). It is somewhat ironic that, whilst some authorities expend substantial funds and effort in removing vegetation from paving and walls, others are looking to exactly those opportunistic habitats in an attempt to 'green' our towns and cities.

References

Benvenuti, S. (2004) 'Weed dynamics in the Mediterranean urban ecosystem: ecology, biodiversity and management', *Weed Research*, 44(5): 341–54.

Brandes, D. (2004) 'Spontaneous flora of the old town centre of Metz (France)', Braunschweig: Working Group for Vegetation Ecology, Institute of Plant Biology, Technical University Braunschweig, pp. 1–4, online, available at: www.ruderal-vegetation.de/epub.

Chang, N.K. and Kim, E.A. (1990) 'Analysis of vegetation on the pavements and under the street trees in Seoul', *The Korean Journal of Ecology*, 13(4): 331–42.

Chinery, M. (1977) *The Natural History of the Garden*, London: Collins.

Cunningham, M.A., Snyder, E., Yonkin, D., Ross M. and Elsen, T. (2007) 'Accumulation of deicing salts in soils in an urban environment', *Urban Ecosystems*, 11(1): 17–31.

Darlington, A. (1981) *Ecology of Walls*, London: Heinemann Educational Books.

Darolová, A. (1992) 'Nesting of *Falco tinnunculus* (Linnaeus, 1758) in the urban agglomeration of Bratislava', *Biológica* (Bratislava), 47: 389–97.

dos Reis, V.A., Lombardi, J.A. and de Figueiredo, R.A. (2006) 'Diversity of vascular plants growing on walls of a Brazilian city', *Urban Ecosystems*, 9(1): 39–43.

Duchoslav, M. (2002) 'Flora and vegetation of stony walls in east Bohemia (Czech Republic)', *Preslia* (Praha), 74: 1–25.

Dunnett, N. and Kingsbury, N. (2008) *Planting Green Roofs and Living Walls*, London: Timber Press.

Environment Agency (2008) *Guidance on the Permeable Surfacing of Front Gardens*, Wetherby: Communities and Local Government Publications.

Eremeeva, N.I. and Sushchev, D.V. (2005) 'Structural changes in the fauna of pollinating insects in urban landscapes', *Russian Journal of Ecology*, 36(4): 259–65.

Fisk, E.J. (1978) 'The growing use of roofs by nesting birds', *Bird-Banding*, 49(2): 134–41.

Frazer, L. (2005) 'Paving paradise: the peril of impervious surfaces', *Environmental Health Perspectives*, 113(7): 457–62.

Gilbert, O.L. (1989) *The Ecology of Urban Habitats*, London: Chapman and Hall.

Gilbert, O.L. (1992) *Rooted in Stone: The Natural Flora of Urban Walls*, Peterborough: English Nature.

Guillitte, O. (1998) 'Bioreceptivity and biodeterioration of brick structures', in N.S.Baer, S. Fitz, R.A. Livingston and J.R Lupp (eds), *Conservation of Historic Brick Structures: Case Studies and Reports of Research*, Shaftesbury: Donhead Publishing, pp. 69–84.

Hadden, R.M. (1978) 'Wild flowers of London W1', *The London Naturalist*, 57: 26–33.

Harding, P.T. and Sutton, S.L. (1985) *Woodlice in Britain and Ireland: Distribution and Habitat*, Huntingdon: Natural Environment Research Council and the Institute of Terrestrial Ecology.

Hruška, K. (1987) 'Syntaxonomical study of Italian wall vegetation', *Vegetatio*, 73(1): 13–20.

Ignatieva, M., Meurk, C. and Newell, C. (2000) 'Urban biotypes: the typical and unique habitats of city environments and their natural analogues', in G.H. Stewart and M.E. Ignatieva (eds), *Urban Biodiversity and Ecology as a Basis for Holistic Planning and Design*, Proceedings of a workshop at Lincoln University, Lincoln University International Centre for Nature Conservation Publication No. 1, Christchurch: Wickliffe Press, pp. 46–53.

Jennings, V. and Stewart, D. (2000) 'Ecology and evolution of wall-dwelling organisms', *American Biology Teacher*, 62(6): 429–35.

Jim, C.Y. (1998) 'Old stone walls as an ecological habitat for urban trees in Hong Kong', *Landscape and Urban Planning*, 42(1): 29–43.

Kennedy, P. and Gadd, J. (2001) *Preliminary Examination of the Nature of Urban Roof Runoff in New Zealand*, Report prepared for the Ministry of Transport, Te Manatu Waka; North Shore City Council; Waitakere City Council; Metrowater, Wellington Regional Council; Canterbury Regional Council by Kingett Mitchell and Associates Ltd, Takapuna, North Shore.

Lagiou, E., Krigas, N., Hanlidou, E. and Kokkini, S. (1998) 'The vascular flora of the walls of Thessaloniki (N. Greece)', in I. Tsekos. and M. Moustakas (eds), *Progress in Botanical Research: Proceedings of the 1st Balkan Botanical Congress*, Dordrecht: Kluwer, pp. 81–4.

Laundon, J.R. (1967) 'A study of the lichen flora of London list of species', *The Lichenologist*, 3(3): 277–327.

Lisci, M., Monte, M. and Pacini, E. (2003) 'Lichens and higher plants on stone: a review', *International Biodeteriation and Biodegradation*, 51(1): 1–17.

Lundholm, J.T. and Marlin, A. (2006) 'Habitat origins and microhabitat preferences of urban plant species', *Urban Ecosystems*, 9(3): 139–59.

Muratet, A., Porcher, E., Devictor, V., Arnal, G., Moret, J., Wright, S. and Machon, N. (2008) 'Evaluation of floristic diversity in urban areas as a basis for habitat management', *Applied Vegetation Science*, 11(4): 451–60.

Natural England (2007) *Green Walls: An Introduction to the Flora and Fauna of Walls*, Natural England Technical Information Note TIN030, Peterborough: Natural England.

Pavlova, D. and Tonkov, S. (2005) 'The wall flora of the Nepal Tepe Architectural Reserve in the city of Plodiv (Bulgaria)', *Acta Botanica Croatica*, 64(2): 357–68.

Payne, R.M. (1978) 'The flora of the walls in south-eastern Essex', *Watsonia*, 12(1): 41–6.

Piervittori, R., Salvadori, O. and Isocrono, D. (2004) 'Literature on lichens and biodeterioration of stonework IV', *Lichenologist*, 36(2): 145–57.

Prach, K. and Pyšek, P. (1994) 'Clonal plants – what is their role in succession?' *Folio Geobotanica et Phytotaxonomica, Praha*, 29(2): 307–20.

Rask, A.M. and Kristoffersen P. (2007) 'A review of non-chemical weed control on hard surfaces', *Weed Research*, 47(5): 370–80.

RHS (2006) *Gardening Matters: Front Gardens*, Royal Horticultural Society Urban Series, Surrey: RHS.

Rinaldi, L., Biggeri, A., Carbone, S., Musella, V., Catelan, D., Veneziano, V. and Cringoli, G. (2006) 'Canine faecal contamination and parasitic risk in the city of Naples (southern Italy)', *BMC Veterinary Research*, 2(29): 1–6.

Rosato, V.G. (2006) 'Diversity and distribution of lichens on mortar and concrete in Buenos Aires Province, Argentina', *Darwiniana*, 44(1): 89–97.

Ruszczyk, A. and Silva, C.F. (1997) 'Butterflies select microhabitats on building walls', *Landscape and Urban Planning*, 38(1): 119–27.

Segal, S. (1969) *Ecological Notes on Wall Vegetation*, The Hague: Dr W. Junk.

Simonová, D. (2008) 'Alien flora on walls in southern and western Morovia', in B. Tokarska-Guzyk, J.H. Borck, G. Brundu, L. Child, C.C. Daehler and P. Pyšek (eds), *Plant Invasions: Human Perception, Ecological Impacts and Management*, Leiden: Backhuys Publishers, pp. 317–32.

Smith, B.J., Gomez-Heras, M. and McCabe, S. (2008) 'Understanding the decay of stone-built cultural heritage', *Progress in Physical Geography*, 32(4): 439–61.

Smith, D. (1984) *Urban Ecology*, Practical Ecology Series, London: George Allen and Unwin.

Steiner, W.A. (1994) 'The influence of air pollution on moss-dwelling animals: 1. methodology and the composition of flora and fauna', *Revue Suisse de Zoologie*, 101(2): 533–56.

Tatner, P. (1978) 'A review of house martins (*Delichon urbica*) in part of South Manchester, 1975', *Naturalist* 103: 59–68.

Trocha, L.K., Oleksyn, J., Turzanska, E., Rudawaska, M. and Reich, P.B. (2007) 'Living on the edge: ecology of an incipient *Betula*-fungal community growing on brick walls', *Trees*, 21(2): 239–47.

URBED (2004) *Biodiversity by Design; A Guide for Sustainable Communities*, London: Town and Country Planning Association.

Warscheid, T. and Braams, J. (2000) 'Biodeterioration of stone: a review', *International Biodeterioration and Biodegradation*, 46(4): 343–68.

Webster, A. and May E. (2006) 'Bioremediation of weathered-building stone surfaces', *TRENDS in Biotechnology*, 24(6): 255–60.

Wheater, C.P. (1999) *Urban Habitats*, London: Routledge.

Woodell, S.R.J. (1979) 'The flora of walls and pavings', in I.C. Laurie (ed.), *Nature in Cities*, Chichester: John Wiley and Sons.

Yilmaz, H., Toy, S., Irmak, M.A., Yilmaz, S. and Bulut, Y. (2008) 'Determination of temperature differences between asphalt concrete, soil and grass surfaces of the city of Erzurum, Turkey', *Atmósfera*, 21(2): 135–46.

20
Urban cliffs

Jeremy Lundholm

Introduction

Most accounts of urban ecosystems emphasize their harshness and inhospitability to native biodiversity: as one moves from the outskirts of a city to the core, the replacement of natural substrates with hard surfaces lowers the number of indigenous species (Kowarik 1990), weeding out all but "urban exploiters" (McKinney 2002) that are presumed highly adapted to this novel environment. The inner city features extremes of moisture availability and temperature which causes stress to plants (Spirn 1984; Whitlow and Bassuk 1988). The biological effects of urbanization on non-human nature are thought to represent a consequence of extreme novelty relative to the evolutionary history of the local biota (Hobbs *et al.* 2006): humans create new situations that many species have not experienced, causing extirpations or genetic evolution in response to these selection pressures. More recently, globalization has extended a homogenizing force into urbanization such that modern cities are increasingly self-similar throughout the globe (Pyšek 1998; McKinney and Lockwood 1999). In parallel, humans are viewed as the ultimate generalist: we can live anywhere, eat anything and we interact with others primarily by engineering ecosystems to suit our own needs.

This composite view of urban humanity is admittedly bleak, but contains some fundamental truths about the role of the built environment as an increasingly dominant force in shaping global ecosystems. Urbanization is proceeding rapidly in most of the world, and the built environment is profoundly different from many of the original, native habitats that it replaced. This chapter presents an alternative view of the process of urbanization that emphasizes the biological evolution of urban habitats, the similarities between built and natural forms, and the role of humans as ecological specialists. The contrarian view presented here is called the Urban Cliff Hypothesis and developed as part of a research program on naturally occurring cliffs and rock outcrops by a team of biologists (Larson *et al.* 2000). While cultural differences among human groups are an obvious consequence of cultural evolution, this account centralizes the biological similarities among human populations and their common evolutionary origins in Africa. It also presents a biologist's attempt to see urban habitats from the perspective of the other organisms that live there.

Outline of the urban cliff hypothesis

Landscape context of human evolution

Around 1.5 million years ago, hominids began making the transition from forests to open savanna, characterized by less tree cover and open grasslands (Tattersall 1998). The savanna was a biome on the rise at that time due to a general aridification of the climate (Bar-Yosef 1995). Several key developments in the evolution of our hominid ancestors may represent adaptations toward a more extreme climate characterized by more sun exposure: the loss of fur and the upright habitat would have protected our ancestors from the mid-day sun (Larson *et al.* 2004). The anthropological consensus is that our biome of origin over the last two million years is the east African savanna (Larson *et al.* 2004).

The savanna is characterized by grasslands and scattered trees, with pronounced seasonality and greater aridity than tropical forest biomes. Biomes contain within them many other types of ecosystems making up a mosaic of different "landscape elements" within the overall matrix that defines the biome. The east African savanna contains river valleys, marshlands and swamps, lakes, rock outcrops and cliffs that form important but less abundant elements of the landscape in a matrix of grassland and scattered trees (Larson *et al.* 2004). The savanna itself is a productive source of plant and animal food, but would have provided little in the way of protection for relatively weak bipeds, so the presence of rock outcrops and caves was essential in providing shelter and tools for early humans (Larson *et al.* 2004). Before the use of brush huts (a much later innovation), rock shelters would have provided shelter from the heat of the day and the cold nights on the elevated plateau of east Africa, safety from predators, as well as a place to protect food from scavengers (Larson *et al.* 2004). Rock outcrops would have also provided tool resources and represented permanent landscape features to aid in navigation and communication between hominid groups.

Once humans left Africa in migrations during the early Pleistocene, there is ample evidence that again, rock outcrops and caves were some of the main bases where populations were established. While rock outcrops are uncommon in most biomes, humans used them as shelter from which they could successfully forage into forested, grassland or montane landscapes (Larson *et al.* 2004). Cliffs and promontories provide what anthropologists have termed "prospect" and "refuge": the ability to see enemies and potential sources of food and other resources while remaining unseen or at least within easy access of shelter (Orians and Heerwagen 1992). The ancient association between humans and these kinds of landscapes appears to be quite ancient and possibly universal (Larson *et al.* 2000).

Habitat templates

The set of conditions in which a species finds its optimal survival, growth and reproduction represents a "habitat template" (Southwood 1977), and scientists classify organisms not only by their evolutionary relationships but also by these preferred habitats. Within the savanna biome, early humans exploited a distinct cliff or rock outcrop habitat template for shelter and other needs. Most people view *Homo sapiens* as a highly adaptable species that does not have a particular habitat template; we are the ultimate generalist or opportunist, we can exploit any environment. The Urban Cliff Hypothesis suggests that, despite our ability to exploit many different habitats to find food, we have an ancient association with a very particular habitat for shelter. Our needs for shelter could not have been met by other habitat templates within the savanna biome, for example, digging burrows in the soil large enough to fit people was not a possibility due to the structural properties of soil (Larson *et al.* 2004). *Homo sapiens* was a habitat specialist on caves and rock outcrops. From the perspective of the Urban Cliff Hypothesis, cities represent an extensive elaboration of this original human niche: we are still specialists on rock outcrop habitats and we build a version of them in every city.

Naturally occurring rock outcrop habitats

Rock outcrop habitats are rare on the surface of the earth: most biomes are dominated by habitats with substantial amounts of soil concealing the underlying bedrock (Larson *et al.* 2004). Rock outcrops can include vertical cliffs and escarpments (Figure 20.1), rock towers or mesas, talus or scree slopes, and exposed horizontal bedrock in the form of pavements (Figure 20.2). While outcrops can be composed of almost any type of bedrock, sedimentary rocks such as sandstone and limestone are most likely to contain the cave features sought by early humans (Larson *et al.* 2000). Each kind of rock outcrop habitat has distinct features that are exploited by different life forms (Table 20.1); what they have in common is relatively low productivity and high spatial heterogeneity, with many different kinds of microhabitats characterizing each feature (Larson *et al.* 2000).

Rock outcrops occur in all regions of the world and because of stark differences between their habitat templates and those of the surrounding biome, they function as a global refuge for biodiversity. First, because cliffs and other rocky habitats are less productive than surrounding habitats they are far less exploited for agriculture and forestry. In southern Ontario, the last old-growth trees, including some over 1000 years old, are found in a cliff face forest on the limestone

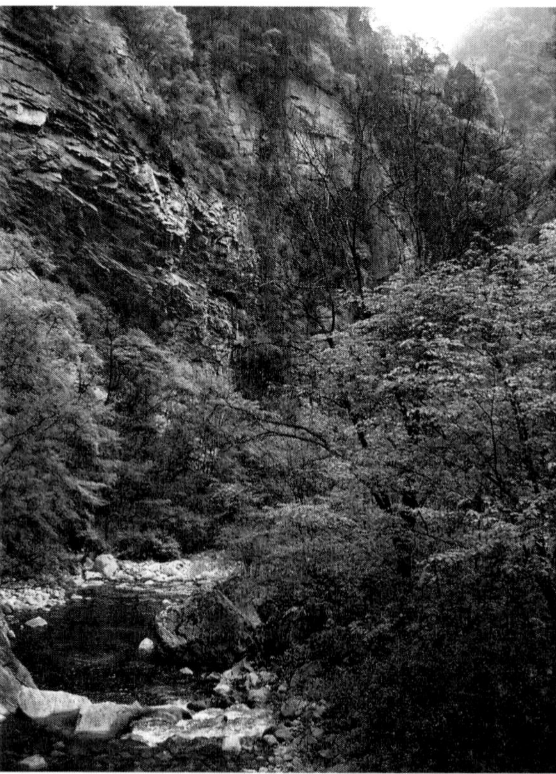

Figure 20.1 Natural rock outcrop showing cliff face and talus habitat (*photo:* J. Lundholm).

Figure 20.2 Natural high biodiversity limestone pavement (*photo:* J. Lundholm).

cliffs of the Niagara Escarpment (Larson and Kelly 1991). Second, cliffs are relatively inaccessible and can provide protection from predation or harsh environmental conditions. For example, tropical cliffs can shield organisms from the strongest sun around mid-day (Martorell and Patiño 2006). Third, the habitat templates in these areas represent a set of environmental conditions very different from the grasslands and forests that make up most terrestrial biomes (Larson *et al.* 2000), thus many species are restricted to rock outcrops, which are themselves uncommon on the landscape. In the tropics, limestone karst regions represent "arks of biodiversity" with many endemic species (Clements *et al.* 2006). The biodiversity protected by rock outcrops extends into the microscopic realm, with cliffs functioning as "evolution canyons" for soil microorganisms (Grishkan and Nevo 2008), and the rocks themselves containing an entire endolithic community of algae and fungi (Matthes-Sears *et al.* 1997).

Ecosystem engineering

The cave is considered to be the original and archetypical human dwelling, with some calling it the "paradigm for architecture" (Mezei 1974). The first evidence of independent structures (brush huts) does not occur until very late in the archaeological record (~20,000 BP), thus the first intentional buildings by human ancestors represent extensions and modifications of naturally occurring caves. Organisms which can exert large changes on their environment, beyond simply competing with or consuming other organisms are called "ecosystem engineers" (Jones *et al.* 1994). The key form of ecosystem engineering that shapes our current-day urban habitats revolves around our ability to create our own living spaces, to literally create niches ("hollows in the rock"). The Urban Cliff Hypothesis suggests that our built environment is historically connected to the original human habitat: buildings represent our attempts to improve the function of naturally occurring rock outcrops, while preserving many of their main features (Larson *et al.* 2004).

Hominids and early humans shared the rock outcrop habitat with a host of species, many of them found nowhere outside of this unique kind of place (Larson *et al.* 2000). When our ancestors learned how to extend caves and expand their shelters away from the cliff face, they unwittingly expanded the habitat itself, allowing these other species that were our cohabitants to expand as well.

Table 20.1 Brief summary of similar features in natural and urban rock outcrop habitats.

Natural rock outcrop feature	*Key features*	*Urban analog*	*Potential for explicit design and improvements*
Caves	Permanence; shelter from elements	Houses, rooms, basements, sheds garages	Improved by central heating, chimneys, lights, plumbing, internal divisions allowing separation of activities, drainage, alteration of surface properties (leveling, smoothing)
Cliff faces	Extreme heterogeneity (soil/no soil; wet/dry) hard surfaces; water sources; high biodiversity	Free standing walls, building walls, road cuts, quarry walls	Allow greater function as habitat for plants and animals by incorporating greater heterogeneity in microsite conditions, (e.g. eyries for birds of prey) (conversely, anti-pigeon structures are modifications of the urban cliff habitat in many cities), introduce species that are not colonizing spontaneously, harvest rainwater to support greater range of vegetation, green walls and facades
Talus slopes	Accumulation of organic matter and water; high biodiversity	Planters, rubble piles, base of exterior walls, gardens (pockets of fertile soil, protecting plants from competition by artificial structures or weeding)	Rubble roofs (brown roofs) for biodiversity, recognition of biodiversity value of rubble piles, incorporation of stormwater drainage into rain gardens, vegetable production in urban areas, species selection for aesthetics (semi-spontaneous urban habitats can be intentionally managed for aesthetics or other benefits such as provision of habitat for pollinators via floral resources)
Horizontal pavements	Extreme heterogeneity; high biodiversity	Vacant lots, gravel parking lots, brownfields, derelict land	Permeable pavements allow drainage, green roofs, incorporation of substrate heterogeneity to encourage greater biodiversity, add other components of rock outcrop biodiversity such as cryptogamic crusts to improve water retention, introduce native taxa not colonizing spontaneously, species selection for aesthetics

When people first constructed independent buildings of stone and other materials, the hard surfaces, crevices and other features were perceived by many organisms as indistinguishable or similar enough to their natural habitat that they followed us into the first built settlements. Many of these organisms would have been unwanted: mice, rats, and pigeons were all part of the original rock outcrop habitat mosaic and were some of the first species to become pests as they went after stored grain (Larson *et al.* 2004). Plants that colonized cracks in natural cliffs or the rubble beneath would have found many suitable habitats on the first buildings. There is much evidence that many of the species that were abundant around the first built settlements, including early domesticated species, both plants and animals, pests and weeds, were inhabitants of natural rock outcrop landscapes before people began creating their own replicas of these habitats (Larson *et al.* 2004).

Species with rock outcrop origins are far more abundant in urban and agricultural areas than expected by chance compared with species from other types of habitat, such as wetlands, forests and grasslands, when examined globally (Larson *et al.* 2004) or for a particular urban area (Lundholm and Marlin 2006). Spontaneously occurring plants in central European cities originate in several key natural habitats: floodplains, gravel or mud banks in rivers, dunes, rocky shorelines, talus slopes and rock outcrops (Wittig 2004). Most of these habitats have in common "openness" or lack of tree cover. Other inquiries into the origins of "weeds" opportunistic plants that frequently dominate urban settings, have found their original habitats to be permanently open habitats such as cliffs, riverbanks or dunes (Grubb 1976; Marks 1983). The built environment creates hard-surfaced, open habitats that represent good replicas of the original habitats of these plants and many other organisms that are now common in cities (Larson *et al.* 2004).

Biophilia

Accepting that humans evolved in a certain place (the savanna biome, as described above), several authors have proposed that modern humans, regardless of where they live, have an innate affinity for landscapes that resemble the east African savanna (Orians 1986). More recently, some have argued that the history of the built environment suggests that humans not only have innate preferences for landscapes, but that elements of rock outcrops such as cliffs and rock towers have cultural, even spiritual significance in many human groups (Larson *et al.* 2000, 2004). Sites of cultural importance are often related to natural cliffs or involve the recreation of rock features: the earliest monuments were monoliths discovered in hunter-gatherer sites in Turkey, dating 11,000 BP (Curry 2008), and the construction of stone pillars or standing stones as objects of spiritual power has been common to many human cultures since then. Reverence for natural cliffs and escarpments appears in west Africa, central America, the Himalayas, China and many other places and frequently extends to the burial of the dead in cliffs (Larson *et al.* 2004). Later developments include built structures surrounding holy sites in natural cave or cliff settings, such as Buddhist monasteries and statues carved from the rock, the stone churches of Ethiopia, and the Dome of the Rock in Jerusalem. An analysis of landscape images used in visual advertising shows that cliffs and other rock features are much more frequent than expected if people showed no preference for any particular landscape (Larson *et al.* 2000). The potential importance of this perspective is that built structures form not only our attempt to create appropriate physical housing for people, but also a profound recreation of the ancestral home of humanity.

Globalization of a rare habitat

A conventional view of urbanization sees the increasing destruction of natural habitats to make way for the built environment. The alternative view presented here adds nuance to this perspective by maintaining that the built environment still represents a kind of habitat, although not necessarily the original habitat on a particular site. The built environment of modern cities is a hardscape, where concrete, stone and metal form the backbone of the urban world. Humans construct a hardscape in response to current exigencies of commerce and culture but also because it provides for us a feeling of security that matches what our ancestors sought in natural cliffs and caves.

While natural rock outcrops remain rare elements of the landscape, the paradox of urbanization is that our artificial analogs of these habitats are becoming common. Many now question the homogenization of the natural world as a function of globalization and the transport of once indigenous and even endemic species to other parts of the world, where they can become pest species (La Sorte *et al.* 2007). Furthermore, artificial analogs of rock outcrop habitats are clearly not identical in physical characteristics or biodiversity with natural outcrops. As our technological prowess increased, buildings in some areas came to resemble and function less like the original stone shelters which were very close copies of cliff and cave environments, and more of the original biota from these habitats was shed. Urban habitats with cliff vegetation that appears natural are more likely to be found in older cities, where stone walls, the best artificial analog of natural cliffs, are still in abundance. The promise here is that, since the basic structure of the urban environment and its hardscapes still re-creates enough of the rock outcrop habitat template, we can improve the function of this habitat for us and for other species.

Comparisons with other frameworks

The "suburban savanna" hypothesis

The first efforts to link the form of artificial landscapes to prehistoric human habitat associations were based on psychological studies into landscape preference (Orians 1986). People were shown images of different landscapes and rated their liking for various habitats and individual features within them, such as the shapes of trees. The conclusions from this research were generally that people prefer landscapes that match the structure of the east African savanna: open vistas with scattered trees. These preferences were interpreted as an adaptive response to a landscape that provided essential elements for early human survival: openness to afford view of potential predators, competitors and food resources ("prospect"), and some trees to provide shelter and the means to escape predators (Orians 1986). Evaluation of the suburban landscape common in North America from the 1950s onward revealed that this same type of landscape has been recreated in the open lawns and scattered trees of suburbia (Orians and Heerwagen 1992).

One problem with this view is that it ignores the actual built structures in urban environments, emphasizing the vegetation and general form of the landscape (Lundholm 2006a). While the savanna offered a productive habitat in terms of food resources, hominids were dependent on rock shelters to survive in this biome (Larson *et al.* 2004), considering the buildings in the midst of the suburban savanna makes the picture complete. The Urban Cliff Hypothesis is thus a complement to this prior theory, and one that adds value in its explicit recognition of the habitat qualities of the built environment.

Humans as generalists

The Urban Cliff Hypothesis takes issue with the idea of the infinite flexibility and adaptability of *Homo sapiens*, instead arguing that we have set biological needs for very particular kinds of shelter. While the earliest exodus of human ancestors from Africa relied on the availability of naturally occurring rock shelters, later migrations took advantage of our increasing ability to construct our own shelters: brush huts and pit houses allowed us to colonize the far reaches of the globe. The variety of architectural forms visible once agricultural, and later industrial, civilizations became the norm belies the overwhelming similarities across continents in the dwellings constructed by hunter-gatherers. While cultural evolution provides the possibility for nearly infinite elaborations of visual arts, architecture and landscape design, all buildings that function well for humans use a set of design criteria that spring back to the biological needs of our species (Larson *et al.* 2004). In spite of our cultural elaborations, these consistent habitat features (Table 20.1) make the modern urban cliffs of the built environment amenable to this understanding of their deep relationship to our species' past. The hallmark of human evolution, from this perspective, is that innovations are built upon a highly conservative, biologically mandated form of ecosystem engineering.

The "no-analog" model

With the reality of human-caused changes to global biodiversity, some ecologists now argue that we are entering a "no-analog future," where biotic communities that once existed in relation to particular habitat templates are being replaced by combinations of species and environmental conditions that have not previously existed (Kaye *et al.* 2006; Fox 2007). These novel ecosystems require new approaches to management, as the factors that maintained native species assemblages are no longer operating in the same ways, and the movement of non-native species into most areas threatens historic communities (Seastedt *et al.* 2008).

Urban habitats in particular have long been approached with this view: ecologists believe that cities represent highly novel environments from the perspective of organisms native to the typical habitats of a particular region. These novel environments represent an evolutionary selective force that excludes many species from cities, and may alter the genetic structure of others (Wandeler *et al.* 2003; Hendry *et al.* 2008). Non-native species dominate in urban environments, either because they are generalists or opportunists that can rapidly adapt to human disturbances, or because they originate in regions, like Europe and western Asia where a long history of urbanization has resulted in the evolution of many new "urbanized" varieties which are best suited to colonizing cities worldwide (La Sorte *et al.* 2007).

The Urban Cliff Hypothesis provides a counterpart to these ideas in suggesting that our generation of ecological novelty is set within a matrix of ecological continuity with our original habitat template: the rock outcrop. There are undoubtedly novel environments being created by people in cities, witness the legacy of sites contaminated with toxic chemicals found nowhere in nature. But understanding urban environments requires us to also recognize that many of the habitat templates created by us are functionally analogous, not novel, from the perspective of the non-human denizens of cities (Lundholm 2006b). Understanding a no-analog future requires us to understand the ways in which novel habitats interact with habitats that are functionally similar to natural rock outcrops, cliffs, talus slopes and other natural hardscapes.

Reconciliation ecology

Existing nature reserves are only sufficient to protect approximately 5 percent of all biological species, so all other biodiversity will have to coexist with humans in highly altered landscapes consisting of forestry plantations, agroecosystems and the built environment (Rosenzweig 2003). Michael Rosenzweig argues that protecting the other 95 percent of species requires that we abandon the idea that we can maintain a pre-disturbance, pristine landscape and that we will have to use the highly altered, novel ecosystems created by humans to the mutual benefit of people and the rest of biodiversity. He argues that we should take the example of the many species that respond positively to human settlements and actively create habitats that would encourage even more species to coexist with humanity: "Reconciliation Ecology." A classic example is the recovery of peregrine falcon populations: peregrine falcons preferentially roost in urban canyons on tall buildings, echoing their preferred natural nest sites on rocky crags and cliff tops. Their favorite food, pigeon, is also abundant in cities, as a result of the same habitat preference for artificial rock outcrops.

The Urban Cliff Hypothesis adds an important dimension to reconciliation ecology in urban areas by describing the habitat template or backbone that we have created, and pointing out the many examples of species that already coexist with people because of the particular types of rock outcrop analog habitats we have created. We can then suggest ways in which we can alter the built environment to make it more compatible with biodiversity and to help this landscape regain its function as a biodiversity refuge (Table 20.1).

Practical applications

Cities represent a variety of habitats within a matrix of hardscape, the "Granite Garden" (Spirn 1984). Understanding the functions of these habitats is the key to improving the ecological value for both humans and the rest of biodiversity. We can recognize the potential of the hardscape to support biodiversity and actively build for nature. Current attempts to green building surfaces as green wall and roofs (Figure 20.3) represent the facilitation of species that may have lacked

Figure 20.3a Extensive (shallow substrate) green roof: urban analog of rock pavement or drought tolerant grassland over bedrock (*photo:* J. Lundholm).

Figure 20.3b Green wall incorporating vascular plants and bryophytes (*photo:* Gary Grant, used with permission).

Figure 20.3c Urban cliff landscape incorporating green facades, green roofs and other cliff and talus analogs (*photo:* J. Lundholm).

opportunities to colonize cities spontaneously and have great benefits not only for urban biodiversity (Kadas 2006; Brenneisen 2006), but also for ecosystem functioning (Oberndorfer *et al.* 2007). Many species used on green roofs indeed originate in natural rock outcrops (Lundholm 2006a). Green roofs themselves can mimic various habitats but the shallow-soil extensive roof systems that are the most economical to install have very similar properties to natural shallow-soil habitats such as rock pavements, cliffs and alpine environments. Some external green walls now incorporate water supply within the wall itself to sustain the vegetation (Figure 20.3a), mimicking the function of natural cliffs, which often feature a constant supply of groundwater seeping from the cliff face (Larson *et al.* 2000). Using native species from local examples of natural rock outcrops is also worth considering: this adds integrity to urban cliffs and enhances promotion of bioregional literacy and identity. The same approach is being used to facilitate other taxa such as birds, through the construction of eyries for birds of prey on the sides of buildings (Table 20.1).

Many of the environmental problems associated with urban habitats can be viewed as symptoms of an incomplete habitat replication project: we create hard surfaced environments because these provide for our shelter and transportation needs but real rock outcrop habitats are not completely "hard surfaced": rock pavements and especially talus slopes offer many opportunities for infiltration of runoff water, and natural cliff ecosystems offer many more potential niches for plants and animals than the modern glass office tower. The built environment has also neglected the role of vegetation in human well-being: our preference for landscapes that contain plants, not just hardscape may represent a latent understanding that vegetation represents resources and indicates a healthy ecosystem (Larson *et al.* 2004), as well as the more immediate psychological benefits of vegetation in urban environments (Kaplan 1995).

Educational resources

Humans are increasingly an urban species, and there is a long list of environmental problems associated with urbanization: habitat destruction, pollution, development of a car-based culture and the list goes on. There are also some potential benefits to the increasing concentration of people in smaller areas of land: denser settlements should give rise to resource use efficiencies as travel distances shrink, and high-density human habitations can also lead to energy efficiency. Whether these benefits are actually reaped, environmental educators are faced with a culture that denigrates urban environments and denies their status as habitats with much ecological value. Ecologists characterize species able to colonize urban centers as "exploiters" (McKinney 2002); less technical terms for urban habitats, especially in the urban core, are "wastelands," "abandoned lots" and their inhabitants are weeds, vermin, pests, or scavengers and are typically considered dirty or even sources of human diseases. In addition, it is a fact that non-native species are more abundant in urban areas than in more natural areas outside of cities, and non-native species, virtually by definition, cannot contribute ecological integrity under current definitions (Turner *et al.* 2005). For educators to take seriously urban habitats and their inhabitants, we face both a popular revulsion to urban species and an ecological rationale for emphasizing more "natural" habitats in our educational programs. How can we teach ecological literacy when we denigrate the very places and creatures we most frequently encounter? The Urban Cliff Hypothesis offers some resources toward reconciliation ecology in cities. First, this perspective forces us to confront the nature of the built environment as an evolved response to innate biological needs, and as a natural form of ecosystem engineering. This is in contrast to the widespread view that human niche construction inevitably represents an unnatural intrusion into the natural order of the world. Second, it allows us to recognize that many species benefit from urbanization as built forms replicate key features of natural habitats, and organisms respond positively to these habitat templates. The framework presented by the Urban Cliff Hypothesis gives a context for understanding past human impacts and for envisioning better urban environments that benefit both people and the rest of nature.

References

Bar-Yosef, O. (1995) "The role of climate in the interpretation of human movements and cultural transformations in western Asia," in E.S. Verba, G.H. Denton, T.C. Partridge and L.H. Burckle (eds.), *Paleoclimate and Evolution*, New Haven: Yale University Press.

Brenneisen, S. (2006) "Space for urban wildlife: designing green roofs as habitats in Switzerland," *Urban Habitats*, 4: 27–36.

Clements, R., Sodhi, N.S., Schilthuizen, M. and Ng, P.K.L. (2006) "Limestone Karsts of Southeast Asia: Imperiled Arks of Biodiversity," *BioScience*, 56(9): 733–42.

Curry, A. (2008) "Archaeology: Seeking the Roots of Ritual," *Science*, 319(5861): 278–80.

Fox, D. (2007) "Back to the no-analog future?" *Science*, 316(5826): 823–5.

Grishkan, I. and Nevo, E. (2008) "Soil microfungal communities of "Evolution Canyons" in Israel – extreme differences on a regional scale," *Biological Journal of the Linnean Society*, 93(1): 157–63.

Grubb, P.J. (1976) "A theoretical background to the conservation of ecologically distinct groups of annuals and biennials in the chalk grassland ecosystem," *Biological Conservation*, 10(1): 53–76.

Hendry, A.P., Farrugia, T.J. and Kinnison, M.T. (2008) "Human influences on rates of phenotypic change in wild animal populations," *Molecular Ecology*, 17(1): 20–9.

Hobbs, R.J., Arico, S., Aronson, J., Baron, J.S., Bridgewater, P., Cramer, V.A., Epstein, P.R., Ewel, J.J., Klink, C.A., Lugo, A.E., Norton, D., Ojima, D., Richardson, D. M., Sanderson, E.W., Valladares, F., Vila, M., Zamora, R. and Zobel, M. (2006) "Novel ecosystems: theoretical and management aspects of the new ecological world order," *Global Ecology and Biogeography*, 15(1): 1–7.

Jones, C.G., Lawton, J.H. and Shachak, M. (1994) "Organisms as ecosystem engineers," *Oikos*, 69(3): 373–86.

Kadas, G. (2006) "Rare invertebrates colonizing green roofs in London," *Urban Habitats*, 4(1): 66–86.

Kaplan, S. (1995) "The restorative benefits of nature: Toward an integrative framework," *Journal of Environmental Psychology*, 15(3): 169–82.

Kaye, J.P., Groffman, P.M., Grimm, N.B., Baker, L.A. and Pouyat, R.V. (2006) "A distinct urban biogeochemistry?" *Trends in Ecology & Evolution*, 21(4): 192–9.

Kowarik, I. (1990) "Some responses of flora and vegetation to urbanization in central Europe," in H. Sukopp, S. Hejny and I. Kowarik (eds.), *Urban Ecology*, The Hague: SPB Academic Publishing.

La Sorte, F.A., McKinney, M.L. and Pysek, P. (2007) "Compositional similarity among urban floras within and across continents: biogeographical consequences of human-mediated biotic interchange," *Global Change Biology*, 13(4): 913–21.

Larson, D. W. and Kelly, P. E. (1991) "The extent of old-growth Thuja occidentalis on cliffs of the Niagara Escarpment," *Canadian Journal of Botany*, 69: 1628–36.

Larson, D.W., Matthes, U. and Kelly, P.E. (2000) *Cliff Ecology*, Cambridge: Cambridge University Press.

Larson, D.W., Matthes, U., Kelly, P.E., Lundholm, J. and Gerrath, J. (2004) *The Urban Cliff Revolution*, Markham, Ontario: Fitzhenry & Whiteside.

Lundholm, J.T. (2006a) "Green roofs and facades: a habitat template approach," *Urban Habitats*, 4: 87–101.

Lundholm, J.T. (2006b) "How novel are urban ecosystems?" *Trends in Ecology and Evolution*, 21(12): 659–60.

Lundholm, J.T. and Marlin, A. (2006) "Habitat origins and microhabitat preferences of urban plant species," *Urban Ecosystems*, 9(3): 139–59.

Marks, P.L. (1983) "On the origin of the field plants of the northeastern United States," *The American Naturalist*, 122(2): 210–28.

Martorell, C. and Patiño, P. (2006) "Globose cacti (Mammillaria) living on cliffs avoid high temperatures in a hot dryland of Southern Mexico," *Journal of Arid Environments*, 67(4): 541–52.

Matthes-Sears, U., Gerrath, J.A. and Larson, D.W. (1997) "Abundance, biomass, and productivity of endolithic and epilithic lower plants on the temperate-zone cliffs of the Niagara Escarpment, Canada," *International Journal of Plant Sciences*, 158(4): 451–60.

McKinney, M.L. (2002) "Urbanization, biodiversity, and conservation," *BioScience*, 52(10): 883–90.

McKinney, M.L. and Lockwood, J.L. (1999) "Biotic homogenization: a few winners replacing many losers in the next mass extinction," *Trends in Ecology & Evolution*, 14(11): 450–3.

Mezei, A. (1974) "Book review of Joseph Rykwert's 'On Adam's house in Paradise: The idea of the primitive hut in architectural history'" *Leonardo*, 7: 376.

Oberndorfer, E., Lundholm, J., Bass, B., Coffman, R.R., Doshi, H., Dunnett, N., Gaffin, S., Kohler, M., Liu, K.K.Y. and Rowe, B. (2007) "Green roofs as urban ecosystems: ecological structures, functions, and services," *BioScience*, 57(10): 823–33.

Orians, G.H. (1986) "An ecological and evolutionary approach to landscape aesthetics," in E.C. Penning-Roswell and D. Lowenthal (eds.), *Landscape Meanings and Values*, London: Allen & Unwin.

Orians, G.H. and Heerwagen, J.H. (1992) "Evolved responses to landscape," in J.H. Barkow, L. Cosmides and J. Tooby (eds.), *The Adapted Mind*, Oxford: Oxford University Press.

Pyšek, P. (1998) "Alien and native species in Central European urban floras: A quantitative comparison," *Journal of Biogeography*, 25(1): 155–63.

Rosenzweig, M.L. (2003) *Win-Win Ecology: How the Earth's Species Can Survive in the Midst of Human Enterprise*, New York: Oxford University Press.

Seastedt, T.R., Hobbs, R.J. and Suding, K.N. (2008) "Management of novel ecosystems: are novel approaches required?" *Frontiers in Ecology and the Environment*, 6(10): 547–53.

Southwood, T.R.E. (1977) "Habitat, the templet for ecological strategies?" *Journal of Animal Ecology*, 46(2): 336–65.

Spirn, S.A. (1984) *The Granite Garden: Urban Nature and Human Design*, New York: Basic Books.

Tattersall, I. (1998) *Becoming Human*, San Diego, CA: Harcourt Brace.

Turner, K., Lefler, L. and Freedman, B. (2005) "Plant communities of selected urbanized areas of Halifax, Nova Scotia, Canada," *Landscape and Urban Planning*, 71(2–4): 191–206.

Wandeler, P., Funk, S.M., Largiader, C.R., Gloor, S. and Breitenmoser, U. (2003) "The city-fox phenomenon: genetic consequences of a recent colonization of urban habitat," *Molecular Ecology*, 12(3): 647–56.

Whitlow, T.H. and Bassuk, N.L. (1988) "Ecophysiology of urban trees and their management—The North American experience," *HortScience*, 23(3): 542–6.

Wittig, R. (2004) "The origin and development of the urban flora of Central Europe," *Urban Ecosystems*, 7(4): 323–39.

21

Suburban mosaic of houses, roads, gardens and mature trees

Ian Douglas

Mosaics are evident at all scales from the submicroscopic to the planet and universe. All mosaics are composed of spatial elements (Forman 1995). Cities develop mosaics of land use and land cover types, vegetation types, social conditions, surface conditions, heat absorption and emission as a function of human activities, particularly the subdivision of properties into smaller lots and then changes in the use of those lots. The simple paving of the yard in front of a house represents a change in the mosaic from vegetation to a hard surface. Essentially urban areas are fine-scale mosaic areas where jobs, homes, shops and recreation are close together with efficient transportation connecting built areas (Forman 1995). In the suburban mosaic, vegetation has a major role. In the densely built-up central areas of big cities, vegetation is less prominent, save in formal squares and parks.

The basic elements of the suburban mosaic are vegetation, buildings and hard, usually impermeable, surfaces, such as roads and car parks. Plants contribute to the spatial structure of urban systems both through their presence in parks, gardens and reserves, and through street trees and shade trees throughout the entire urban mosaic. In urban areas, the amounts, structure and condition of these three components (vegetation, buildings and surfaces) (Table 21.1) reflect human agency (Pickett and Cadenasso 2008).

Suburbia comprises a mosaic of contrived plant diversity, blatantly people-made, to which nature had added a variety of species of animals and plants that have adapted to urban life. These gardens respond rapidly to inputs of nutrients and usually tend to receive supplementary fertilizers and applications of water. Insects exploit the full variety of the gardens but may be the target of chemical sprays, though probably less intensively in gardens than areas of intense horticulture and arboriculture. Such gardens are important bird habitats and support a variety of animals, such as the squirrels and foxes now common in British suburbs. Within this category there are contrasts due to the age of the suburb. The new suburbs of cities, often built on former farmland, usually differ from older suburbs by having a smaller number and lower height of trees, and a greater proportion of lawn and mown grass than of shrubs and cultivated flower and vegetable gardens. The total plant biomass of such suburbs is much less than that of older inner suburban areas with consequently a less varied suburban fauna. Comparisons of urban bird populations in newly developed suburbs and in older well-established suburbs show a richer avifauna, in both diversity and population, in the older suburbs.

Table 21.1 Comparative data on urban land cover for cities in Europe and the USA

	Buildings % of area	*Impervious surfaces %*	*Grass %*	*Trees and shrubs %*	*Water %*
Brooklyn, NY, USA	34.5	32.8	20.8	11.4	0.5
Baltimore, MD, USA	–	–	27	20	–
Houston, TX, USA	30	30	21	18	1
Manchester, UK, high density	31	37	15	15	1
Manchester, UK, medium density	22	27	24	24	2
Manchester, UK, low density	14	18	26	39	1
Liverpool, UK high density	37.1	33.28	9.5	8.2	0
Liverpool, UK, low density	27.1	24.23	21	14.9	0.1
Munich, Germany	30	30	20	18	2
Stockholm, Sweden	–	–	–	26	13

Sources: Colding *et al.* 2006; Gill *et al.* 2008; Pauleit and Duhme 2000; Pauleit *et al.* 2005; Nowak *et al.* 2002.

New industrial estates and office parks often have similar large expanses of mown grass with a few trees or shrubs around the buildings. The same type of vegetation pattern, with abundant grass also surrounds new houses and blocks of flats in redeveloped inner city areas of older European and North American urban areas. Both sports grounds, typically with large areas of mown grass, and golf courses, combining mown grass with patches of more varied vegetation, form parts of the varied suburban mosaic. More specialized are the areas of allotments for food production and patches of derelict land recently colonized by both native and invasive species.

Urban land cover types

Characteristically, land within the suburban mosaic has moderate-to high-density, single-family housing with plot sizes of 0.1 to 1.0 ha. Lawns and gardens are common. Basic services, light industry, and multi-family housing are interspersed with the typical family dwellings (Marzluff *et al.* 2001). The older, inner suburbs of cities contain widely different densities of housing and ratios of garden area to built-over area. Few are completely devoid of private or public gardens with mature trees. Many contain a relatively large portion of garden land containing a rich habitat with a combination of mown grass, trees, and patches of flowers, shrubs and vegetables. Typically in the UK gardens occupy 18 to 27 per cent of urban areas (Loram *et al.* 2008). They can, collectively, form extensive, interconnected tracts of green space. However, the customary elements of domestic gardens, such as lawns, paved patios, cultivated borders for annuals, perennials, shrubs and trees vary greatly in their organization, layout, maintenance and variety from one garden to another according to the predilections and management efforts of individual householders. In a sample of five UK cities, patios of some description (concrete, paving stones or bricks) were present in more than 86 per cent of gardens in each city and cultivated border in more than 68 per cent, with at least 66 per cent of gardens possessing both features. More than 54 per cent also possessed a lawn. Approximately half of the gardens sampled also had trees taller than 3m, sheds and hard paths (Loram *et al.* 2008).

Great contrasts in the presence of vegetated surfaces occur between the centres of cities and the suburban surroundings. Trees in gardens and parks are usually further apart than in natural woodland and have large canopy areas. They contribute to the reduction of particles in the air, but may dry the soils through high evapotranspiration in dry periods. Analysis of impervious areas, those covered by buildings and other sealed surfaces such as roads and car parks, in the 147 km² River Croal catchment of Greater Manchester found the core town centre of only 0.9 km² was entirely impermeable, with 6.5 km² of the inner urban area being 82 per cent impermeable, but only 45 per cent of the remaining 28 km² of suburban residential areas was impervious (Douglas 1994). For Greater Manchester as a whole, 59 per cent of the 'urbanized' area consists of evapotranspiring (i.e. vegetated and water) surfaces. Residential areas, covering 48 per cent of 'urbanized' Greater Manchester, have a great impact on the ecology of the conurbation. The majority of residential areas form the 297 km² middle density category, with 48 km² classified as high density and 35 km² as low density and 48 km² as high covers from each other (Gill *et al.* 2008). In high density residential areas built surfaces (i.e. building and other impervious surfaces) cover about two-thirds of the area, compared to about half in medium density areas and one-third in low density areas. Tree cover is 26 per cent in low density areas, 13 per cent in medium density areas, and 7 per cent in high density areas (Gill *et al.* 2007). This can be compared with North American urban residential areas that have 15 to 40 per cent tree cover, typically 17 per cent in arid zone cities, 19 per cent in prairie cities and 40 per cent in humid regions (Oke 1989). The difference may reflect the contrasts in plot sizes, the generally larger American plots having more space for trees. Medium density residential areas in Greater Manchester have 32 per cent of all the evapotranspiring areas of the conurbation, with high density having 5 per cent and low density 3.5 per cent (Gill *et al.* 2008). This complex mosaic contains many varied elements. Although the emphasis has to be on the gardens that occupy most of the green patches of the mosaic (see Chapter 35), there are many aspects, such as green roofs (see Chapter 44) and walls (Chapter 19), as well as derelict land, allotments, golf courses, business parks, and linking corridors, that need attention.

Diversity of habitat with the suburban mosaic

Allotments

Allotments, often municipally owned but used by individuals for growing vegetables, may be significant, particularly in the higher density residential areas. Greater Stockholm, Sweden, has 126 allotment areas (mean size: 2.51 ha; median size: 1.47 ha) occupying 316 ha (Colding *et al.* 2006). The United Kingdom has some 300,000 allotments covering 10,000 (Royal Commission on Environmental Pollution 2007) to 12,150 ha (Buczacki 2007). The number had declined by 40 per cent between 1970 and 2005 (Royal Commission on Environmental Pollution 2007) but has subsequently begun to increase again with greater interest by urban residents in growing their own food. The National Trust in the UK, for example, has a target of 1,000 new allotments on its properties by 2012, planning to create 500 of them in 2010 (Jones 2010).

Derelict and vacant land

On average, 15 per cent of a United States city's land lies vacant. The land ranges from undisturbed open space to abandoned, contaminated brownfields (Pagano and Bowman 2000). England had 43,300 ha of derelict land in the early 1970s, 40,500 in the late 1980s and 39,600 in 1993. Derelict land in Scotland increased from 7,741 to 8,224 ha between 2002 and 2009. Within Scotland the increase or decrease in derelict land varied greatly, that in Glasgow declining from 767 to 712 ha (Scottish Government 2010) but derelict land still occupies 4.1 per cent of the city's area. Such declines are common in many urban areas of the UK where government policies have encouraged the use of brownfield sites for housing.

Derelict land can be colonized by many species. Three samples of derelict land in England had an average of 170 species, compared to 438 species found in a sample of gardens in Sheffield (Thompson *et al.* 2003). Across 50 derelict land sites in Birmingham, England, 379 species occurred in at least five sites (Angold *et al.* 2006).

Golf courses

Golf courses contain a variety of ground cover, with often quite large natural areas. Already 5,300 golf courses in Europe occupy 250,000 ha of land, not all within or adjacent to urban areas, with plans for more facilities to be built before 2010. Golf courses are also interspersed in the suburban mosaic, along with public parks and recreation grounds. In Greater Stockholm, Sweden, 24 golf courses (mean size: 61 ha; median size:57 ha) occupied 1,463 ha (Colding *et al.* 2006). These golf courses present opportunities for biodiversity (Stubbs 1996) and nurturing resilience in the overall landscape. Seasonal wetlands on golf courses increase amphibian biodiversity (Metts *et al.* 2001). Golf courses have the potential to function as corridors and buffer zones for biodiversity within built-up areas and to provide habitats for pollinators and species that control pests.

Typically, 70 per cent of a golf course is not used for playing, but comprises slopes, ponds, stream banks, grasslands, groups of trees and patches of woodland. For example, if the intensively managed playing surfaces are excluded, some 40 ha of a typical golf course in the Stockholm area consist of varying natural habitats (Colding *et al.* 2006). Biota on golf courses were of higher conservation value than those in green areas associated with other land uses in 64 per cent of a sample studied (Colding and Folke 2008). Greater values were also found in the majority of cases of comparisons based on measures of species richness, birds and insects, supporting the argument that any golf courses contribute to the conservation of fauna.

Business parks

Business parks can provide habitats for plant and animal species. They are often on the edge of urban areas in a strategic position on the urban-rural gradient (McDonnell and Pickett 1990), thereby forming a part of ecological corridors (Löfvenhaft *et al.* 2002). The flat roofs of office and warehouse buildings form a potential biotope for many plant and animal species (Grant *et al.* 2003). Vacant sites within these business parks provide opportunities for pioneer vegetation and pioneer animals such as the Natterjack toad (*Bufo calamita*) (Smit 2006). The night-time tranquillity, when business activity ceases, makes them ideal places for nocturnal species like amphibians and urban mammals (e.g. the urban fox, see Gloor *et al.* 2001). Dutch bird data showed that, generally, abundance and species richness of breeding birds were lower at business sites. However, business parks seemed to favour birds associated with early successional vegetation and open landscapes. Both coastal species and the Common Linnet, *Carduelis cannabina*, a Red List species, preferentially frequented urban business parks (Snep 2009).

Some business parks pride themselves on encouraging wildlife. For example, Hillington Business Park in Glasgow, Scotland claimed in November 2009 that the biodiversity value has moved from 'negligible' after the initial survey in 2007 to 'moderate and developing' in 2009. It invested £40,000 in new sustainable and native planting and changed the landscaping maintenance regime allowing meadow areas to grow, enabling flowering plants and grasses to flourish. A revised maintenance programme applied to many of the shrub beds led to unusual plant species such as Broad Leaved Helleborine, Orchids and Cuckoo Flower appearing (http://hillingtonbusinessparkglasgow.blogspot.com/ accessed 17 February 2010).

Corridors and patches

Most cities have many green corridors, some of which are extensive, some engineered and others fragments (see also Chapter 22 this volume). They range from small passages or pathways lined with vegetation from one suburban street to another, to wide river valleys that are opened up for public enjoyment and to provide refuges for wildlife. Transport corridors, such as urban freeways can have large numbers of trees, while railways have long been the means by which some plants have spread across the land.

In some arid cities former irrigation and water supply canals can provide green corridors. The linear nature of canals serves a number of environmental and ecological functions. Canals provide linkages between other canals, parks, streets, railways and bicycle paths. In this capacity they can serve as vital transportation corridors for commuter and recreational bicyclists and pedestrians, providing these canal users with an alternative to the use of motor vehicles. As a result, vehicular traffic and accompanying air pollution are reduced. Canal pathways can also create places for wildlife to find connections between areas of their natural habitat that are rapidly being eroded by urbanization. The preservation of biological diversity is an important goal of the environmental movement: the establishment of wildlife corridors has been suggested as a way to counteract the problems of habitat fragmentation that are occurring with urbanization. Advocates of greenways recommend that this type of linear corridor be established for ecological reasons such as encouraging species diversity (Smith 1993). As a human-made structure, however, a canal cannot equal a riparian waterway for the provision of animal habitat. Some even contend that it is possible for canals to have a negative impact on wildlife corridors by altering or interrupting an existing corridor (Blanton 1994).

Garden diversity

Origins and fashions in gardens

Gardens are cultural creations, made according to current tastes and conventions. The ancient Egyptians had papyrus and lotus gardens, while the ancient Greeks had enclosed villa gardens of asphodels. The Chinese and Japanese had gardens around temples and palaces with rock and stone formations that resembled wild nature and the step slopes of their mountains. In Islamic countries, from Arabia to Andalucia, there were water gardens with citrus trees. The Moguls established formal, private sanctuaries, with ornamental trees and shrubs that contrasted with the dry, stony landscape beyond the garden walls. For the Medici in Italy, nature, as reflected in their magnificent villa gardens, was a route to happiness, a model of divine order, which roughly held that God was in his heaven, humankind ranking somewhere below, and below humankind, nature, but each part related and invisible from the other (Mitchell 2002: 126).

The Ottoman Turks protected and shaded their beds of roses, scented flowers and twisting vines with walls and rows of trees. Suleiman I began to cultivate tulips in his gardens, sending bulbs as gifts to dignitaries in neighbouring countries (Mitchell 2002: 52). As the selection of plants began, exotics were cultivated, and animals were protected. Redistribution of seeds, bulbs, plants and animals from such gardens encouraged the increasing biodiversity of urban areas as settlements grew and wealth was accumulated by the elites of emerging states.

Garden biodiversity in the UK

Since 1990, with continuing urban growth, people have given more attention to the role of domestic (private) gardens in maintaining biodiversity in suburban and urban areas. Many UK

Local Biodiversity Action Plans specifically address urban green spaces (JNCC 2001). Their potential significance for biodiversity is demonstrated by:

i the high biodiversity that can be associated with the few domestic gardens in which detailed investigations have been made (e.g. Owen 1991; Miotk 1996);
ii the occurrence of nationally and regionally scarce species in domestic gardens (e.g. Owen 1991);
iii the high proportion of populations, high density or high productivity in suburban and urban gardens of some species that have become depleted in rural areas as a consequence of more intensive mechanized farming; and
iv the large total area of domestic gardens relative to the extent of other forms of suburban and urban green space such as urban parks and nature reserves (Gaston *et al.* 2005).

Today, in the UK, the size of the garden is the key factor influencing land cover within the garden, and especially the number of plants with a canopy of over two metres (Loram *et al.* 2008).

Detailed study of urban gardens in Birmingham and Sheffield, UK, has revealed an astonishing total richness a total of 438 species being found in Sheffield, of which 48.6 per cent occurred just once (Thompson *et al.* 2003). This almost certainly reflects two factors. First, there is a very large pool of plants available to gardeners – the 2010 *Royal Horticultural Society Plant Finder* (Cubey 2010) lists over 70,000 taxa (*c.*14,000 distinct species) available from UK nurseries. Second, owing to active management and maintenance by gardeners, garden plants have a highly 'unnatural' ability to persist at remarkably low population sizes. Compared to semi-natural habitats, gardens are much less homogeneous, i.e. individual gardens may be largely open, grassy or deeply shaded, while larger gardens may contain all these habitats. However, while gardens share this property with urban derelict land, in most respects derelict land is similar to semi-natural habitats, implying that management and the size of the species pool are the key unique features of gardens.

Intercontinental movement of garden tastes

Tastes and fashions in garden design have been transported from one continent to another, with gardens being created as an escape from the reality of the natural environment. People create lawns in arid areas, plant palms in places subject to frost, and import alpine species to temperate western European lowlands. This can lead to the displacement and even elimination of native species. In Christchurch, New Zealand, of 139 species in 350 sampled lawns, the great majority were naturalized plants originating in the northern hemisphere; 87 species from Europe, 81 from the Middle East, Turkey and the Mediterranean, 42 from Asia, 35 from India and 13 from North America. Just 16 species were native to New Zealand. Some of these, including *Pseudopanax aboreus* and *Coprosma repens* will not survive mowing.

Arid areas pose particular problems for urban greenspace. People moving into arid areas from wetter regions often prefer to have well-watered green areas around their houses. However in some cities and in individual developments there may be covenants that require the planting of front gardens with desert vegetation more suited to a dry climate. These covenants may sometimes be irksome to householders. A Phoenix, Arizona, USA survey found a distinct incongruence between conforming desert landscape layouts in front yard landscapes and the occupiers' actual landscape preferences. Many homeowners preferred a garden design that mixed desert-adapted plants, mesic plant species and lawns. The increasing popularity of desert landscaping in cities like Phoenix (Martin *et al.* 2003) and Tuscon (McPherson and Haip 1989) gives diversity to the suburban landscape, but is essentially a top-down phenomenon, directed by public and private interest groups, persuading homeowners to change their attitudes and practices.

Insects in gardens

Insects exploit the full variety of the gardens but may be the target of chemical sprays, though probably less intensively in gardens than areas of intense horticulture and arboriculture. About ten species or functionally related groups account for 63 to 97 per cent of the arthropod pests found in North American gardens. While common plants are just as likely to be attacked by arthropods as rare plants, plants of the *Rosacaeae* family are more likely to have arthropod pests than those of other families (Raupp and Shrewsbury 2007). Applications of cover sprays to residential landscapes in Maryland, USA, over prolonged periods of time have the potential to exacerbate problems with many species of scale insects on suburban trees (Raupp *et al.* 2001). Repetitive cover sprays could selectively remove natural insect enemies from suburban areas, allowing scale populations to increase. The application of residual pesticides to the canopies of oak trees significantly reduced the abundance of several taxa of natural enemies.

In detail, the distribution of individual insect species may be determined by their ability to adapt to particular habitats. While a rural-urban gradient has been noted in butterfly distributions, some species of blow flies, in London at least, appear to be affected by the opportunities in the suburban mosaic. While more urban-adapted blow fly species, such as *Calliphora vicina, Lucilia sericata and L. illustris*, may spread across most of London and form a more or less continuous distribution, the asynanthropic species, such as *C. vomitoria, L. ampullacea and L. caesar*, seem to be confined to semi-natural habitats having more limited and disjunctive distribution range (Hwang and Turner 2005). The grassy and wooded parklands in London are their major habitats, with private gardens as connection corridors. The dispersal ability of each species and the connectivity between suitable habitats determine the isolation of flies in different parts of London.

The suburban mosaic in rapidly growing low-latitude cities

While much of the literature deals with the suburban mosaics of North America and Europe, most of the rapidly growing cities are in the tropics. Here too, urban gardens and other parts of the urban mosaic provide important habitats. Private gardens of traditional cities of Central America may play an important role in wildlife conservation. Here, in the city of Léon in Nicaragua, trees cover 38.5 per cent of the suburban mosaic, in strong contrast to the large monoculture agricultural fields beyond the city limits. One of the endangered species of the region, the black spiny-tailed iguana (*Ctenosaura similes*), is locally surviving where high trees provide daytime shelter away from people and gaps in garden fences allow them to move between gardens to find the types of tree they prefer (González-García *et al.* 2009). Instead of burrowing into sand banks, iguanas use a variety of human-made structures, such rubbish tips, for their burrows, only rarely digging directly into garden soil.

Developing, tropical countries have the largest human growth rates, most of the Earth's biodiversity, and rapid expected rates of urbanization, yet studies of the impacts of urbanization in the tropics are exceedingly rare. Current investigations are just beginning address the complex effects of human settlement in the tropics (Brooks and Begazo 2001). Bird observations in the south-western portion of Mexico City have shown that the contrasts in bird number and species between green areas and built-up areas of cities further north and in Europe also apply in one of the world's largest cities (Ortega-Álvarez and MacGregor-Fors 2009). In Concepción, Chile, also, the impacts of urban sprawl on biodiversity differ little from the world-wide situation: native ecosystems are replaced by pavement and buildings and what is left of the natural soil is covered with green areas, which are dominated by non-native ornamental species.

The new and expanded urban areas of Asia offer particularly good examples of new forms of suburb, with an emphasis on high-rise apartment buildings interspersed with managed greenspaces. Singapore has developed an island-wide park connector network designed to meet the perceived growing need for a range of alternative recreational facilities. The highly urbanized island has planned its parks and open spaces as part of a network system to optimize the use of limited land resources. Green corridors at least 20 m wide link these open spaces. They provide jogging and cycling tracks but also serve as green linkages for birds and other fauna to move from one park to another, thereby enhancing the natural elements in the environment (Tan 2006). As these connectors mature over time and their vegetation becomes more attractive, the bird species in them come to resemble those of well-wooded nearby parks (Sodhi *et al.* 1999) thus resembling the character of larger gardens near woodlots in temperate cities such as Vancouver, Canada. In many ways this suburban mosaic of high-rise apartments with managed interconnected spaces has an alternative path towards the habitat variety and conservation value of the gardens of the low rise suburbs of Europe and North America. Similar approaches can be seen in some of the newer urban developments in China (Kong *et al.* 2010) and have been advocated for Beijing in terms of a green network system of green wedges, parks and green corridors (Li *et al.* 2005). In a way, this combination of greenspaces between buildings connected to a series of parks resembles some of the idealized plans Le Corbusier presented for his 'City of Tomorrow' (Le Corbusier 2002). In the Singapore context it works and is well used.

Conclusion

While it can be argued that scientists working on urban ecology issues face a challenge in making their complex, and to some degree uncertain, science accessible to policy makers, it is import to recognize that local people have for generations made use of and understood the landscape variety and biodiversity of the suburban mosaic in many different ways. Nowadays the 'urban green' is frequently interpreted as a set of sites of co-fabrication between humans and other city inhabitants, rather than as islands of nature in the city. The different kinds of spaces of ecological engagement include spaces associated with cultivation, such as allotments, community gardens and city farms; those associated with land restoration, such as derelict railway, mining or industrial sites; and those associated with conservation, such as remnant woodlands, waterways and marshes. In each case, various people have intimate knowledge of the history and culture of these ecological sites and communities that consists of more than scientific ways of knowing (Whatmore and Hinchcliffe 2003). Their skills and know-how contribute to sustaining and managing the biodiversity and landscapes that make up the ever-changing urban ecology of the suburban mosaic.

References

Angold, P.G., Sadler, J.P., Hill, M.O., Pullin, A., Rushton, S., Austin, K., Small, E., Wood, B., Wadsworth, R., Sanderson, R. and Thompson, K. (2006) 'Biodiversity in urban habitat patches', *Science of the Total Environment*, 360(1–3): 196–204.

Blanton, J. (1994) *Landscape Impacts Created by Human-Made Barriers: A Case Study of the Hayden Rhodes Aqueduct*, Master of Environmental Planning thesis, Arizona State University, Tempe, AZ.

Brooks, D. and Begazo, A. (2001) 'Macaw abundance in relation to human population density', in J.M. Marzluff, R. Bowman, and R. Donnelly (eds) *Avian Ecology and Conservation in an Urbanizing World*, Norwell, MA: Kluwer Academic, pp. 429–39.

Buczacki, S. (2007) *Garden Natural History*, London: Collins.

Colding, J. and Folke, C. (2008) 'The Role of Golf Courses in Biodiversity Conservation and Ecosystem Management', *Ecosystems*, 12(2): 191–206.

Colding, J., Lundberg, J. and Folke, C. (2006) 'Incorporating green-area user groups in urban ecosystem management', *Ambio*, 35(5): 237–44.

Cubey, J. (editor-in-chief) (2010) *RHS Plant Finder 2010-2011*, London: Dorling Kindersley.

Douglas, I. (1994) 'Human settlements' in W.B. Meyer and B.L. Turner (eds), *Global Land Use/Cover Change*, Cambridge: Cambridge University Press, pp. 149–69.

Forman, R.T.T. (1995) 'Some general principles of landscape and regional ecology', *Landscape Ecology*, 10(3): 133–42.

Gaston, K.J., Smith, R.M., Thompson, K. and Warren, P.H. (2005) 'Urban domestic gardens II: experimental tests of methods for increasing biodiversity', *Biodiversity and Conservation*, 14(2): 395–413

Gill, S.E., Handley, J.F., Ennos, A.R. and Pauleit, S. (2007) 'Adapting cities for climate change: the role of the green infrastructure', *Built Environment*, 33(1): 115–33.

Gill, S.E., Handley, J.F., Ennos, A.R., Pauleit, S., Theuray, N. and Lindley S.J. (2008) 'Characterising the urban environment of UK cities and towns: a template for landscape planning', *Landscape and Urban Planning*, 87(3): 210–22.

Gloor, S., Bontadina, F., Hegglin, D., Deplazes, P. and Breitenmoser, U. (2001) 'The rise of urban fox populations in Switzerland', *Mammalian Biology*, 66: 155–64.

González-García, A., Belliure, J., Gómez-Sal, A. and Dávila, P. (2009) 'The role of urban greenspaces in fauna conservation: the case of the iguana *Ctenosaura similis* in the "patios" of León city, Nicaragua', *Biodiversity and Conservation*, 18(7): 1909–20.

Grant, G., Engleback, L. and Nicholson, B. (2003) *Green Roofs: Their Existing Status and Potential for Conserving Biodiversity in Urban Areas*, English Nature Report 498, Peterborough: English Nature.

Hwang, C. and Turner, B.D. (2005) 'Spatial and temporal variability of necrophagous Diptera from urban to rural areas', *Medical and Veterinary Entomology*, 19(4): 379–91.

Jones, S. (2010) 'National Trust creates 300 new allotments: conservation charity hopes to have 1,000 allotments by 2012', *Guardian*, Friday 19 February 2010.

JNCC (2001) *Sustaining the Variety of Life: 5 Years of the UK Biodiversity Action Plan*, London: Department of the Environment, Transport and the Regions.

Kong, F., Yin, H., Nakagoshi, N. and Zong, Z. (2010) 'Urban green space network development for biodiversity conservation: identification based on graph theory and gravity modeling', *Landscape and Urban Planning*, 95(1–2): 16–27.

Le Corbusier (2002) '*from* The City of Tomorrow and its Planning', in G. Bridge, and S. Watson (eds), *The Blackwell City Reader*, Oxford: Blackwell, pp. 20–9.

Li, F., Wang, R., Paulussen, J. and Liu, X. (2005) 'Comprehensive concept planning of urban greening based on ecological principles: a case study of Beijing, China', *Landscape and Urban Planning*, 72(4): 325–36.

Löfvenhaft, K., Bjorn, C. and Ihse, M. (2002) 'Biotope patterns in urban areas: a conceptual model integrating biodiversity issues in spatial planning', *Landscape and Urban Planning*, 58(2): 223–40.

Loram, A., Thompson, K., Warren, P.H. and Gaston, K.J. (2008) 'Urban domestic gardens (XII): the richness and composition of the flora in five cities', *Journal of Vegetation Science*, 19(3): 21–30.

McDonnell, M.J. and Pickett, S.T.A. (1990) 'Ecosystem structure and function along urban-rural gradients: an unexploited opportunity for ecology', *Ecology*, 71(4): 1232–7.

McPherson, E.G. and Haip, R.A. (1989) 'Emerging desert landscape in Tucson', *Geographical Review*, 79(4): 435–49.

Martin C.A., Petersen, K.A. and Stabler, L.B. (2003) 'Residential landscaping in Phoenix, Arizona, U.S.: Practices and preferences relative to covenants, codes and restrictions', *Journal of Arboriculture*, 29(1): 9–17.

Marzluff, J.M., Bowman, R. and Donnelly R. (2001) 'A historical perspective on urban bird research: trends, terms, and approaches', in J.M. Marzluff, R. Bowman, and R. Donnelly R. (eds), *Avian Ecology and Conservation in an Urbanizing World*, Norwell, MA: Kluwer Academic: 1–17.

Metts, B.S., Scott, D.E. and Gibbons, J.W. (2001) *Enhancing Amphibian Biodiversity on Golf Courses through the use of Seasonal Wetlands*, Far Hills, NJ: US Golf Association, online, available at: www.uga.edu/srel/Reprint/3093.htm.

Miotk P. (1996), 'The naturalized garden – a refuge for animals? – first results', *Zoologischer Anzeiger*, 235: 101–16.

Mitchell, J.H. (2002) *The Wildest Place on Earth: Italian Gardens and the Invention of Wilderness*, Washington, DC: Counterpoint.

Nowak, D.J., Crane, D.E., Stevens, J.C. and Ibarra, M. (2002) *Brooklyn's Urban Forest*, General Technical Report NE-290, Newtown Square, PA: US Department of Agriculture, Forest Service, Northeastern Forest Experiment Station.

Oke, T.R. (1989) 'The micrometeorology of the urban forest', *Philosophical Transactions Royal Society London, B* 324(1223): 335–48.

Ortega-Álvarez, R. and MacGregor-Fors, I. (2009) 'Living in the big city: effects of urban land-use on bird community structure, diversity, and composition', *Landscape and Urban Planning*, 90(3–4): 189–95.

Owen J. (1991) *The Ecology of a Garden: The First Fifteen Years*, Cambridge: Cambridge University Press.

Pagano, M.A. and Bowman, A.O'M. (2000) *Vacant Land in Cities: An Urban Resource*, Washington DC: Brookings Institution Survey Series.

Pauleit, S. and Duhme, F. (2000) 'Assessing the environmental performance of land cover types for urban planning', *Landscape and Urban Planning*, 52(1): 1–20.

Pauleit, S., Ennos, R. and Golding, Y. (2005) 'Modeling the environmental impacts of urban land use and land cover change—a study in Merseyside, UK,' *Landscape and Urban Planning*, 71(2–4): 295–310.

Pickett, S.T.A. and Cadenasso, M.L. (2008) 'Linking ecological and built components of urban mosaics: an open cycle of ecological design', *Journal of Ecology*, 96(1): 8–12.

Raupp, M.J. and Shrewsbury, P. (2007) 'Landscape ornamentals', in D. Pimental (ed.), *Encyclopedia of Pest Management, Volume 2*, London: Taylor and Francis, pp. 298–301.

Raupp, M.J., Holmes, J.J., Sadof, C., Shrewsbury, P. and Davidson, J.A. (2001) 'Effects of cover sprays and residual pesticides on scale insects and natural enemies in urban forests', *Journal of Arboriculture*, 27(4): 203–14.

Royal Commission on Environmental Pollution (2007) *The Urban Environment (26th Report)*, London: The Commission.

Scottish Government (2010) *Scottish Vacant and Derelict Land Survey 2009*, Edinburgh: National Statistics for Scotland.

Smit, G.F.J. (2006) 'Urban development and the natterjack toad (*Bufo calamita*): Implementation of the habitat directive in the Netherlands', in M. Vences, J. Köhler, T. Ziegler and W, Böhme (eds), *Herpetologia Bonnensis II. Proceedings of the 13th Congress of the Societas Europaea Herpetologica*, pp. 167–70.

Smith, D.S. (1993) 'An overview of greenways: their history, ecological context and specific functions', in D.S. Smith and P.C. Hellmund (eds), *Ecology of Greenways: Design Functions of Linear Conservation Areas,* Minneapolis, MN: University of Minnesota Press, pp. 1–22.

Snep, R. (2009) *Biodiversity Conservation at Business Sites: Options and Opportunities*, Alterra Scientific Contributions 28, Wageningen: Alterra.

Sodhi, N.S., Briffett., C., Kong, L. and Yuen, B. (1999) 'Bird use of linear areas of a tropical city: implications for park connector design and management', *Landscape and Urban Planning*, 45(2–3): 123–30.

Stubbs, D. (1996) *An Environmental Management Programme for Golf Courses*, Dorking: European Golf Association Ecology Unit.

Tan K.W. (2006) 'A greenway network for Singapore', *Landscape and Urban Planning*, 76(1–4): 45–66.

Thompson, K., Austin, K.C., Smith, R.H., Warren, P.H., Angold, P.G. and Gaston, K.J. (2003) 'Urban domestic gardens I: putting local garden diversity in context', *Journal of Vegetation Science*, 14(1): 71–8.

Whatmore, S. and Hinchliffe, S. (2003) 'Living cities: making space for nature', *Soundings: Journal of Politics and Culture*, 22: 37–50.

22

Urban wildlife corridors

Conduits for movement or linear habitat?

Ian Douglas and Jonathan P. Sadler

In his excellent description of places to see wildlife in London, David Goode (1986) describes railway lines were often the best place to see foxes in London. He also extolled the ecological interest of walks along abandoned railways, such as that from Mill Hill to Edgware. Without specifically using the word 'corridor' he was advocating the virtues of linear habitats in cities (Figure 22.1). In a similar book on Belfast, published almost 20 years later, Robert Scott (2004) describes the Lagan Valley as a two-way corridor, enticing wild creatures upstream from the marine environment of Belfast Lough and downstream from the surrounding countryside into the heart of the city; a pattern mirrored in other large cities with riparian corridors (e.g. Newcastle and the Tyne valley). Connective features such as green networks and corridors have been influential in guiding planning policies in many areas of the world (Turner 2006; von Haaren and Reich 2006), but are also subject to considerable debate and confusion (Hess and Fischer 2001). They have long been seen as providing connectivity, linking greenspaces, and minimizing the potential effects of fragmentation on wildlife (Jongman *et al.* 2004), while providing important recreational, leisure and nature experience possibilities for people (Gobster and Westphal 2004). The idea of urban wildlife corridors is now firmly established in natural history and has become part of urban environmental planning.

Urban greenways have developed from the cleaning up of rivers such as the American River in Sacramento California, damaged by gold mining and the Willamette River in Portland, Oregon damaged by paper mill effluent (Bauer 1980). River valley greenways in these cities were initiated in the 1960s, initially to provide riverside parkland for people to enjoy close to the city centre. Over time the emphasis of the plans for these greenways has changed to give greater importance to their role as wildlife habitat and potential wildlife corridors.

By 1995, more than 500 communities, in North America alone, had greenway projects under way (Searns 1995). Modern greenways set out to be multi-purpose, going beyond recreation and beautification to address such areas as habitat needs of wildlife, promoting urban flood damage reduction, enhancing water quality, and providing a resource for outdoor education (Figure 22.2). Some greenway projects specifically recognize the needs of wildlife and aim to provide corridors for wildlife migration. Following a report recommending that Tuscon, Arizona, set up and maintain a system of interconnected open spaces with adequate water and vegetation to sustain wildlife, the city adopted an 'environmental resource zone'

Figure 22.1 Abandoned railway line at Stamford Brook, Trafford, Greater Manchester, UK, forming a potential wildlife corridor.

Figure 22.2 Sustainable urban drainage landscaping at Stamford Brook, Trafford, Greater Manchester, UK, forming a potential wildlife corridor.

ordinance help protect areas essential to wildlife, including migration corridors (Schwab 1994).

Many cities incorporate a concept of green network or set of open spaces in their local plans. One of them is Telford New Town, some 50 km to the west of Birmingham in the West Midlands region of England. A planned new town comprising a number of existing villages which, from the early 1960s onwards, have been extended and linked by major new housing, industrial and commercial areas. Greenspace development in the town involved combining natural regeneration of old industrial areas with high-quality mass planting involving around six million trees and ten million shrubs (Douglas and Box 2000). Thirty-eight per cent of Telford's area is now open space, including much of the Ironbridge Gorge, a World Heritage site as the home of some of the earliest stages of the Industrial Revolution, but now also an important wildlife corridor (see also Chapter 41 this volume).

Some 80 km north of Telford, the old industrial areas around Liverpool and Manchester in the UK, for decades a landscape dominated by signs of former mining and heavy industry, are now gaining new woodlands and areas of natural vegetation as land is restored and tree-planting is carried out (Figure 22.3). Some of the greenspaces so created are being incorporated into new systems of linear parks, especially along river valleys. Such linear systems have begun to provide green corridors from the heart of the cities to the open countryside beyond the suburbs (Barker 1997). Two of the agencies involved, the Mersey Forest and the Red Forest, work with partners to develop the Green Infrastructure for this part of north-west England. The vision for Green

Figure 22.3 Woodland planted by the Red Rose Forest at the peri-urban Dainewell Woods, Greater Manchester, UK. The trackway is the Trans Pennine Trail from Liverpool to Hull, used by horse riders, cyclists and walkers (*photo:* Maureen Douglas).

Infrastructure (GI) is to bring together a coherent network of components, such as open spaces, green corridors and woodlands, for the benefit of people and wildlife. This will be achieved through adopting a more systematic approach to planning and managing these components across local authority boundaries and all spatial planning levels. In this way GI will become as important a consideration in the growth and development of communities as transport, grey infrastructure, built infrastructure and social infrastructure. GI differs from conventional approaches to open space planning by considering the multiple functions and benefits of green infrastructure alongside land development, growth management and built infrastructure planning issues. As such it is integral to the management of growth and development and essential for building well-designed and sustainable communities.

Many other industrial and urban regions are thinking similarly. A report on the English Midlands plateau notes:

> The potential for creating wildlife corridors and multi-functional green networks in the Midlands Plateau, via non-statutory site linkage, is great. By improving the wildlife value of buildings, gardens, local authority-owned open spaces, canals, disused railway lines, and road verges, a connectivity can be achieved.
>
> *(Middlemarch Environmental 1997: 15)*

Central to this vision of green infrastructure is the idea of linkages and connectivity between patches of open spaces and woodlands, between diverse types of wildlife habitats in urban and peri-urban areas. The connections are characterised by the words 'green corridors', a notion of strips of vegetated land running through a matrix of urban and industrial development. In Birmingham the wildlife conservation strategy was explicitly built around the concept when it was published in 1997 (BCC 1997). Since then management of wildlife in the city has relied heavily on corridors as a strategic planning tool. This chapter:

1 Examines the concept of the corridor in wildlife terms;
2 Considers how the concept has been applied in urban areas;
3 Outlines research that has tested whether or not green corridors have delivered the links, and connectivity expected in terms of conservation value, and;
4 Suggests some practical benefits that have emerged from urban wildlife corridors.

The wildlife corridor concept: what's in a name?

The idea of wildlife corridors in the landscape has long been discussed. Forman and Gordon (1986) devote a chapter to the topic in their book on landscape ecology. For them plant and animal movement across the landscape is a key element of good environmental management.

The term corridor has been used in a variety of ways. Beier and Noss (1998: 1242) define a corridor as

> A linear habitat embedded in a dissimilar matrix, that connects two or more larger block of habitat and that is proposed for conservation on the grounds that it will enhance or maintain the viability of specific wildlife populations in the habitat blocks

Bennett (2003) uses the term link, or linkage, in the sense of an arrangement of habitat (not necessarily linear or continuous) that enhances the movement of animals or the continuity of ecological processes through the landscape. He also uses the term 'habitat corridor' to describe a linear strip of vegetation that provides a continuous (or near continuous) pathway between to habitats. This term has no implications about its relative use by animals.

The key function of the corridor or link is the provision of connectivity. Landscape patterns that promote connectivity for species, communities and ecological processes are a key element of nature conservation in environments modified by human impacts. The concept of connectivity is used to describe how the spatial arrangement and the quality of elements in the landscape affect the movement of organisms among habitat patches (Bennett 2003). At the landscape scale, connectivity has been defined as 'the degree to which the landscape facilitates or impedes movement among resource patches' (Taylor *et al.* 1993).

It is argued that corridors of greenspace linking habitat islands help to overcome barriers to migration, gene exchange and species turnover, and provide avenues for colonization. In this way they may mitigate local extinctions of resident populations because of the replacement of individuals and/or genes into existing, but declining populations (Harris and Scheck 1991).

Corridors have six recognisable functions: habitat, conduit, filter, barrier, source and sink (Figure 22.4). In the habitat function, organisms can survive and reproduce in the corridor (Hess and Fischer 2001). In a conduit, organisms pass from one place to another, but do not reside in the corridor. In a filter, some organisms or materials can pass through the corridor, others

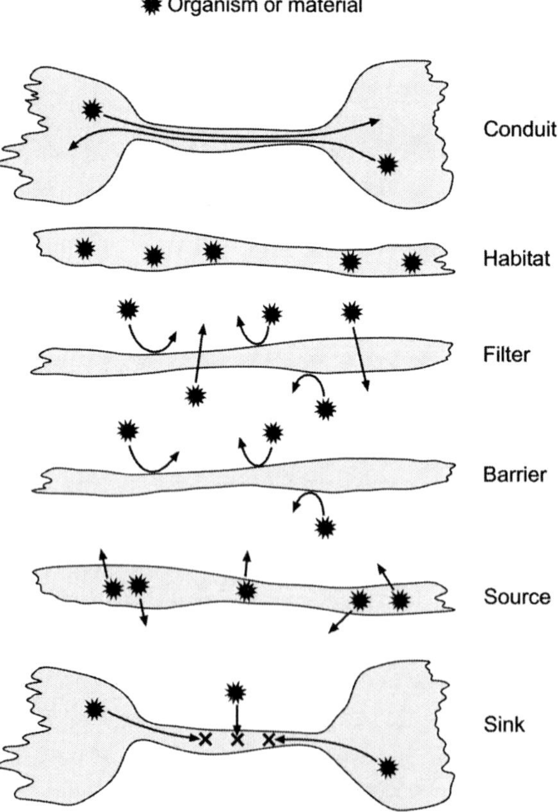

Figure 22.4 Diagram showing the patterns of movement associated with the six key functions of wildlife corridors.

cannot. A barrier means that organisms and materials cannot cross the corridor. If the corridor is a source, organisms or materials emanate from the corridor. If it is a sink, organisms or materials enter the corridor and are destroyed. These functions have to be addressed specifically in any debate about preserving, enhancing or creating wildlife corridors.

The explosion of interest in corridors has not been without scepticism, criticism and debate and the conservation benefits gained from corridors have become a contentious issue. Criticisms have centred around three points (Bennett 2003):

- Whether sufficient scientific evidence is available to demonstrate the potential conservation benefits of corridors;
- Whether the potential negative effects of corridors may outweigh any conservation value;
- Whether corridors are a cost effective option in comparison with other ways of using scarce conservation resources.

It is critical to recognize that a landscape is perceived differently by different species and communities, so the level of connectivity varies between communities and between species (Dawson 1994). A particular landscape or region may, at the same time, provide high connectivity for some organisms, such as mobile wide-ranging birds, and low connectivity for others such as snails and small sedentary reptiles (Bennett 2003).

The careful review by Beier and Noss (1998) concluded that evidence from well-designed investigations generally supported the value of corridors and a conservation tool. Almost all the reports suggested that corridors provide benefits to, or are used by, animals in real landscapes. However design defects in many studies meant that only about 12 of those examined allowed meaningful inferences of conservation value. Ten of those studies offered persuasive evidence that corridors created sufficient connectivity to improve the viability of populations in habitats connected by corridors (Figures 22.5 and 22.6).

Figure 22.5 The Bridgewater Canal at Brooklands, Trafford, Greater Manchester, UK, with broad vegetated areas on both banks offering a diversity of cover for wildlife.

Figure 22.6 The Bridgewater Canal at Sale, Trafford, just one kilometre north of Brooklands (Figure 22.5), where both banks of the canal have little vegetation, disrupting the connectivity for terrestrial fauna.

While a wealth of scientific work has accumulated, the scientific basis of wildlife corridors remains inconclusive, with ecologists, planners and conservationists sharing doubts about their efficacy. In the English West Midlands, research into planning politics showed that wildlife corridors played a powerful role in mediating between ecological and developmental interests in the city, while scientific data concerning the actual usage of corridors by animals and plants in the city suggested they were largely irrelevant for species dispersal (Angold *et al.* 2006; Evans and Randalls 2008). The greenway and green infrastructure movement may perhaps have established a momentum and mythology of its own, that becomes detached from the way organisms actually behave. One argument is that the emphasis on habitat connectivity has diverted attention from discovering how it is that so many species find their way through the city and colonize so many diverse urban habitats (Evans 2007). Much of the mystery and confusion over the utility and efficacy of corridors has come from conflation, these functions and a lack of clarity in their use in policy documents and academic studies (cf. Beier and Noss 1998).

To find out whether that is really the case, urban wildlife corridors have to be examined more closely.

Urban wildlife corridors

Understanding the role of connectivity and linkage is a major challenge to urban ecology. Even in non-urban areas only a few empirical studies have illustrated the successful role of corridors as conduits for species movement, and most of these derive from carefully controlled landscape scale experiments (Beier and Noss 1998; Haddad and Baum 1999; Haddad *et al.* 2003). This is a vastly different situation from most urban areas where green corridors are multifunctional spaces, with a lot of habitat variability, frequently adjacent to a wealth of other habitats. Gardens, which are both habitats in their own right and potential linking habitats, have been intensively studied in the UK (e.g. Smith *et al.* 2005, 2006) but less so elsewhere (Daniels and Kirkpatrick 2006). The efficacy of linkages via corridors (Angold *et al.* 2006), permeability of matrix and habitat isolation are not well documented within cities. There are also important linkages within the rural fringe (the so-called peri-urban zone) that require fuller evaluation (Snep *et al.* 2006).

Depending on the organisms involved, urban wildlife corridors can take many forms. For some birds, wooded streets can be important. In Madrid, 14 bird species were found in wooded streets, compared to 24 species in parks. The wooded street species were only tree/tree hole, ground/tree hole, and ground/tree, with just one ground/rock species. Actual occurrence of particular birds depended on the amount of foraging cover and the nature of the street vegetation (Fernández-Juricic 2000). This pattern is not repeated for similarly mobile species such as bats. Oprea *et al.*'s (2009) detailed study of connectivity of wooded streets in Mexico City indicated only sporadic and limited use of them by commuting bats.

Investigations of birds in relation to the many woodland patches in Stockholm, Sweden, found that the nutcracker and the honey buzzard, the least common species in the survey, had the largest area requirements and certain demands on the quality of their habitat. The occurrence of these species was affected by connectivity in the landscape in the form of amount of suitable habitat in the surroundings. Thus, these two species appeared to be favoured by the connectivity properties of the green space corridors (Mörtberg and Wallentinus 2000).

In the Los Angeles area of southern California, bobcats and coyotes occurred frequently in corridors as habitats, but use of corridors as conduits occurred much less often. Only coyotes used corridors to travel from one habitat fragment to another. Bobcats may have used corridors less than coyotes because they were more likely to confine their movement to one fragment, probably due to their greater sensitivity to human development. Although corridors were not always used as conduits, their use was sufficient to suggest that they are significant in maintaining connectivity (Tigas *et al.* 2002).

For some invertebrates, potential movement corridors may be irrelevant. In the English West Midlands, analysis of the carabid fauna on derelict industrial sites found no evidence to support the hypothesis that sites away from railway corridors are more impoverished in their carabid fauna than sites on corridors (Small *et al.* 2006). This was probably because derelict carabid species are generally good dispersers rather than that these railways are poor corridors.

Urban greenways do not function as conduits enhancing the movement of species, rather they have a role as a sequence of linked (sometimes poor quality) habitats (Angold *et al.* 2006). Corridors are species specific, one size does not fit all. Corridors do provide habitat and it is the quality of that habitat which is of central importance to urban biodiversity.

Green corridors may make little difference to the diversity of plants and beetles found in our towns and cities by virtue of their function as corridors. They do often provide valuable habitat, especially on river corridors and railway land. However, the plant communities of derelict sites show no greater similarity to each other if adjacent or close to designated urban corridors, suggesting that the corridors do not increase connectivity. Nevertheless, sites within areas with a high local density of derelict sites were more similar, suggesting greater inter-site dispersal of propagules when sites are less isolated. In the West Midlands study no evidence was found that wetland specialist beetle diversity is greater on or near the green corridors, and no evidence that corridors are necessary for dispersal of butterflies (Wood and Pullin 2002). For the invertebrates in particular, habitat quality appears to be the significant factor, and the habitat of designated 'green corridors' in the Black Country is rather a chain of habitats of different type and/or quality rather than a linear continuous habitat.

The West Midlands research (Angold *et al.* 2006) found no evidence that plants or invertebrates use urban greenways for dispersal. Their importance for these groups is rather as a chain of different habitats permeating the urban environment. However, it is possible that small- and medium-sized mammals, may use corridors. The practical implication of this is that urban greenspace planners and managers should identify a target species or group of species for urban greenways intended as dispersal routeways rather than as habitat in their own right.

Streams and rivers form potential wildlife corridors through urban areas, but the small first and second order stream routes, in particular, often become the site of rubbish dumping, both in the stream and on valley sides. The presence of a stream corridor can thus indicate the potential for extended human influence into the interior of a site. In addition, trails may become established on slightly higher ground, giving access to the wetland. Small (2002) examined the distribution of wetland beetles associated with a riverine corridor (the River Cole), which bisects the southern part of Birmingham. Comparisons of wetland spider and carabid assemblages at sites both on- and off-corridor showed no clear community differences indicating the lack of a clear functional conduit role for these organisms. In contrast, recent unpublished data from bird survey work using wetland bird survey techniques and ringing data (Bodnar and Sadler unpubl.) indicates that river corridors permit the movement of some wetland bird species (e.g. reed buntings) into the city. Blakely *et al.*'s (2006) study of the effect of culverts and road bridges on oviposition and dispersal flight in river macroinvertebrates (caddis, may and stoneflies) clearly indicates that for these taxa river corridors are important dispersal corridors. Moreover, it strongly highlighted the significant reduction that river culverts can have on upstream dispersal of macroinvertebrates along urban rivers.

Underpasses and stormwater drains as corridors

Roads, especially wide freeways or motorways, can have significant impacts on wildlife movement and survival, particularly on wide-ranging species, such as mammalian carnivores (Ng *et al.* 2004), but also less mobile organisms such amphibians (Eigenbrod *et al.* 2009). For example in the Sunrise Corridor area of south-east Portland, Oregon, approximately half the existing wildlife habitat and movement corridors are vulnerable to future planned and potential development as a result of current zoning and land use ordinances (Trask 2007). Existing commercial and residential development already constricts the main wildlife corridor, and wildlife access between the remaining habitat patches in the area will be severed if further zoned development occurs. These impacts can be reduced if animals have over- or under-passes to cross highways.

Several studies in southern California have examined how connectivity can be maintained between wildlife habitats in an area dissected by many freeways. Underpasses for predators and ungulates to move under highways that bisect urban wildlife corridors between habitat patches have been effective, but the size of animal using the underpasses varies according to the dimensions and location of the underpass. To encourage the larger predators (coyotes and bobcats) to use them, the underpasses need to be away from areas of high road density and residential development (Haas 2001).

A year-long monitoring study of animals using 15 different passages and culverts, ranging in cross-sectional area from 2 to 282 m^2, under freeways to the north, west and south of the Simi Hills, just west of the San Fernando Valley, California, found that the wild mammals passing through one or more passages at least once included: deer mice (*Peromyscus spp.*), woodrats (*Neotoma spp.*), ground squirrels (*Spermophilus beecheyi*), cottontail rabbits (*Sylvilagus audubonii*), opossums (*Didelphis virginianus*), striped skunks (*Mephitis mephitis*), spotted skunks (*Spilogale putorius*), raccoons (*Procyon lotor*), coyotes (*Canis latrans*), bobcats (*Lynx rufus*), mountain lions (*Puma concolor*), and mule deer ( ) (Ng *et al.* 2004). Mule deer used only three shorter passages of large cross-section. Bobcats and coyotes tended to avoid passages surrounded largely by developed habitat. Mid-sized mammals (raccoons, opossums and skunks) tended to use the longer and smaller passages. Cats and dogs used passages also used by humans and their movements were not affected by the habitat conditions at either end of the passage. For wild animals habitat disturbed by human activity close to passage entrances appeared to deter use of the passage (Ng *et al.* 2004). This study showed the benefits of underpasses for wildlife but also noted that there may be adverse impacts as well. Underpasses are used by house cats which could have harmful impacts on other native species. For example, in areas of high human density, domestic animals, particularly house cats, have been associated with the decline and extinction of bird and small mammal populations in fragmented habitats (Crooks and Soulé 1999).

Stormwater management facilities, particularly detention ponds, have become significant landscape features in most new urban and suburban residential developments. The variety of aquatic and terrestrial habitats attracts much urban wildlife seeking refuge from the inhospitable urban matrix (Adams *et al.* 1985; Vermonden *et al.* 2009). However, because most of these habitat 'islands' are isolated, they do not encourage animal movements and constrain nutrient and energy flows. Changes to stormwater design could enhance opportunities for wildlife, but such gains can be much greater if they are part of a regional landscape ecological planning approach that recognizes these facilities as important landscape elements – patches and corridors – in an urban landscape matrix (Forman and Gordon 1986). Such a matrix could be built around a regional landscape network of greenways forming a set of interconnected nodes, patches, corridors and multiuse modules (MUMS) in an urban context (Cook 1991). The network would link stormwater patches and corridors to other typical urban greenspace, such as parks, cemeteries and remnant woodlands, using stream and riparian corridors, abandoned railway routes, and roadsides. Where possible, the network would link to a larger, undisturbed and protected species source pool, such as a nature reserve or national park, from which species could migrate and spread (McGuckin and Brown 1995).

Urban corridors as dispersal routes for introduced species

An unexpected and significant issue that has been identified in recent studies of linear corridor features such as roads and rivers in urban areas is their association with the presence and spread of non-native or introduced species. Cilliers and Bredenkamp's (2000) study of roadside verges in Potchefstroom indicated that rather than facilitate movement of indigenous species into the city from the surrounding landscape, the verges enhanced the possibilities for dispersal of introduced species out of the urban centre. This process seems prevalent in other regions where regional biota is dominated by indigenous and endemics species, such as islands like Tenerife (Canary Islands). Kalwij *et al.* (2008) concluded in their study, however, that although road verges in towns provided the opportunity for invasive species migration they host such species rather than facilitate their spread. This is a little surprising as von der Lippe and Kowarik's (2007; 2008) detailed studies of traffic enhanced movement of propagules along roads in Berlin showed that long-distance dispersal occurred significantly more frequently in seeds of non-native (mean share 38.5 per cent) than native species (mean share 4.1 per cent). These data suggest that propagule dispersal of non-native species along roads is a constant, not sporadic, occurrence that may well accelerate plant invasions and induce rapid changes in biodiversity patterns.

Similarly, Angold *et al.* (2006) report that of a total of 433 plant species recorded in their work in the West Midlands only ten were associated with riparian corridors ($p < 0.005$). Only one of these, the introduced species Himalayan Balsam, was >20 per cent more frequent than might have been expected by chance alone. In a later study, Maskell *et al.* (2006) show that the presence of this exotic species has a negative impact on the abundance of native plant species ($p < 0.05$), indicating the loss of one native species per quadrat as Balsam cover increased. The spread of the Himalayan Balsam also reduced invertebrate and beetle abundance on riparian marginal habitats ($p < 0.05$) (Figure 22.7).

Practical applications and tools

Examples of planned urban wildlife corridors include the 'Green Links' scheme in Greater Vancouver. It has three demonstration sites, two based on utility corridors (Coquitlam and Surrey), and one running through a matrix of residential development (Burnaby) (Schaefer 1998). The 8 km long and 100 m wide Coquitlam corridor occupying 128 ha on land largely owned by the city through which run electricity power lines. It links five ecosystem fragments including, fields and a marsh, an arboretum containing a representative of every tree species known to grow in British Columbia, a large municipal park containing a remnant forest and small lake with bog habitat and two ravines cutting back into an escarpment and plateau. This utility corridor supports 121 species of higher plants and 51 species of birds. Over 6,000 trees have planted and invasive species (Scotch Broom, Himalayan Blackberry and Purple Loosestrife) are regularly removed. Community satisfaction with the corridor is high, with strong support for the habitat improvement work (Schaefer 1998).

A model, such as that of McGuckin and Brown (1995), able to predict spatial patterns of stormwater management facilities under different simulations and provide measures of landscape ecological structure and function, such as connectivity and porosity, would help urban planners, drainage engineers and decision makers to select, plan and design stormwater detention and conveyance systems that would also be highly effective in increasing the ecological integrity of the urban landscape.

There are two primary considerations in corridor design: (1) including corridor attributes that can contribute to corridor quality; (2) tailoring corridors to the needs of the species it was designed to serve (Fleury and Brown 1997). The design needs to consider the critical corridor attributes of: matrix, patch, network connectivity, barrier, length, width, shape, edge, structure

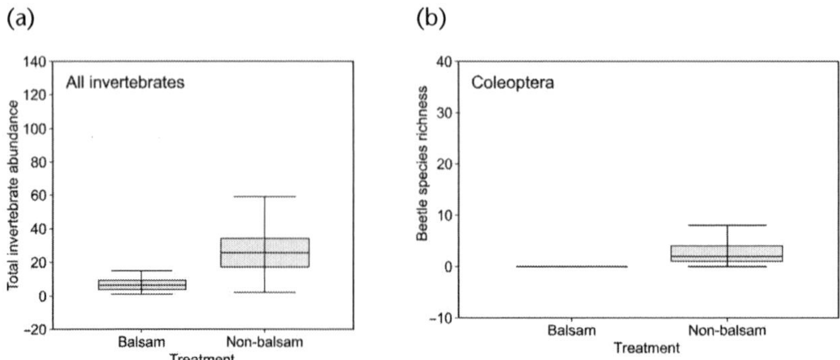

Figure 22.7 The abundance of terrestrial (a) invertebrates and (b) beetles (Coleoptera) in contrasting plots of Balsam rich and Balsam poor areas on the Rivers Cole, Rea and Tame, Birmingham, UK (*source:* unpublished data J.P. Sadler).

Table 22.1 Corridor design table (after Hess and Fischer 2001)

Function	*Primary uses*	*Other uses*	*Potential problems*
Societal goals			
Habitat			
Conduit			
Barrier			
Filter			
Source			
Sink			

and composition. Because of the differences in the way various species use corridors, as discussed above, the target species for a particular corridor have to be considered. One way is to group species into guilds. A guild may be defined as 'groups of species that exploit the same class of environmental resources in a similar way' (Fleury and Brown 1997: 173). One such grouping could be into (1) insects, (2) reptiles and amphibians, (3) birds, (4) small mammals, (5) medium mammals, and (6) large mammals, but equally, as the Madrid street tree example above shows, a corridor might be designed for a specific group of birds. It would be possible to have different types of corridor for different species in a single city's green infrastructure.

However, it is also important to recall the six functions of wildlife corridors: habitat, conduit, filter, barrier, source and sink (Hess and Fischer 2001). A simple corridor design table (Table 22.1) that details desired functions, primary uses, other uses and potential problems can be used to help discussion. By requiring explicit definition of the proposed corridor's functions, such an approach increases the visibility of important issues and questions, and makes potential conflicts apparent.

In applying wildlife corridor theories to urban green infrastructure it is extremely important to use both the available science and local knowledge (Evans 2007). There are no design rules to readily specify how a wildlife corridor should be developed. Experience of the sort gained through Creative Conservation (see Chapter 18) may help. Such experience may be available locally.

Conclusion

Because greenways are increasingly popular as a means of improving urban and suburban living environments and may themselves serve as important habitat, biologists should work with urban planners and community groups to design greenways that contribute to urban wildlife conservation and education, rather than arguing whether greenways function as corridors (Rosenberg *et al.* 1997). Nevertheless, there is a danger that enthusiasm for greenways and green infrastructure will allow ecologists and planners to slip into a tendency to think that wildlife corridors are essential for the movement of wildlife through cities. Anyone who has seen urban foxes crossing suburban streets at night knows that the city is almost completely permeable to many mammal species. Clearly species that are not adjusted to urban foraging will benefit from routes across urban features, such as highways, in order to move from one piece of remnant countryside to another. Yet for many groups of organisms the urban landscape may well prove to be extremely permeable. The key here seems to be habitat quality. Recent studies of urban birds emphasize the importance of local habitat quality (Tratalos *et al.* 2007; Evans *et al.* 2009); a pattern clearly shown in data from invertebrates (Sadler *et al.* 2006). Unpublished data from a recent bird ringing study in Birmingham has shown movements in one year of birds over many kilometres within the city, with one male great tit making a 60 kilometre movement.

In a way the development of urban wildlife corridors has almost outstripped the science. These corridors, or greenways, are good for many reasons. They are extremely pleasant for people to walk along, to ride horses and bicycles along and to hear birds. They are good for urban biodiversity and bring all the other benefits of urban vegetation. However, there are not enough patient observations over long enough periods, of the movements of animals for firm statements to be made about their benefits for most forms of wildlife. Short of encouraging more ecologists to pay greater attention to the ecological consequences of small-scale urban environmental change, the change that most directly affects individual people and animals, urban greenspace managers and conservation groups have to think of the multiple benefits and design different types of wildlife corridor to target various types of mammals and birds. Variety of opportunity for wildlife will bring all the other benefits as well.

References

Adams, L.W., Dove, L.E. and Franklin, T.M. (1985) 'Use of urban stormwater control impoundments by wetland birds', *Wilson Bulletin*, 97(1): 120–2.

Angold, P.G., Sadler, J.P., Hill, M.O., Pullin, A., Rushton, S., Austin, K., Small, E., Wood, B., Wadsworth, R., Sanderson, R. and Thompson, K. (2006) 'Biodiverisity of urban habitat patches', *Science of the Total Environment*, 360 (1–3): 196–204.

Barker, G. (1997) *A Framework for the Future: Green Networks with Multiple Uses in and around Towns and Cities*, English Nature Research Reports, No. 256, Peterborough: English Nature.

Bauer, W.S. (2000) *A Case Analysis of Oregon's Willamette River Greenway Program*, PhD thesis, Corvallis, OR: Oregon State University.

Beier, P. and Noss, R.F. (1998) 'Do habitat corridors provide connectivity?' *Conservation Biology*, 12(6): 1241–52.

Bennett, A.F. (2003) *Linkages in the Landscape: The Role of Corridors and Connectivity in Wildlife Conservation*, Gland/Cambridge: IUCN.

BCC (1997) *A Nature Conservation Strategy for Birmingham*, Birmingham: Birmingham City Council.

Blakely, T.J., Harding, J.S., McIntosh, A.R. and Winterbourn, M.J. (2006) 'Barriers to the recovery of aquatic insect communities in urban streams', *Freshwater Biology*, 51(9): 1634–45.

Cilliers, S.S. and Bredenkamp, G.J. (2000) 'Vegetation of road verges on an urbanisation gradient in Potchefstroom, South Africa', *Landscape and Urban Planning*, 46(4): 217–39.

Cook, E.A. (1991) 'Urban landscape networks: an ecological planning framework', *Landscape Research*, 16(3): 7–16.

Crooks, K.R. and Soulé, M.E. (1999) 'Mesopredator release and avifaunal extinctions in a fragmented system', *Nature*, 400: 563–6.

Daniels, G.D. and Kirkpatrick, J.B. (2006) 'Comparing the characteristics of front and back domestic gardens in Hobart, Tasmania, Australia', *Landscape and Urban Planning*, 78(4): 344–52.

Dawson, D. (1994) *Are Habitat Corridors Conduits for Animals and Plants in a Fragmented Landscape?* English Nature: Peterborough.

Douglas, I. and Box, J. (2000) *The Changing Relationship between Cities and Biosphere Reserves*, Peterborough: UK MAB Urban Forum.

Eigenbrod, F., Hecnar, S.J. and Fahrig, L. (2009) 'Quantifying the road-effect zone: threshold effects of a motorway on anuran populations in Ontario, Canada', *Ecology and Society*, 14(1): Articel 24, online, available at: www. ecologyandsociety.org/vol14/iss1/art24/.

Evans, J. and Randalls, S. (2008) 'Geography and paratactical interdisciplinarity: Views from the ESRC–NERC PhD studentship programme', *Geoforum*, 39(2): 581–92.

Evans, J.P. (2007) 'Wildlife Corridors: an urban political ecology', *Local Environment*, 12(2): 129–52.

Evans, K.L., Newson, S.E. and Gaston, K.J. (2009) 'Habitat influences on urban avian assemblages', *Ibis*, 151(1): 19–39.

Fernández-Juricic, E. (2000) 'Avifaunal use of wooded streets in an urban landscape', *Conservation Biology*, 14(2): 513–21.

Fleury, A.M. and Brown, R.D. (1997) 'A framework for the design of wildlife conservation corridors with specific application to southwestern Ontario,' *Landscape and Urban Planning*, 37(3–4): 163–86.

Forman, R.T.T. and Gordon, M. (1986) *Landscape Ecology*, Chichester: Wiley.

Gobster, P.H. and Westphal, L.M. (2004) 'The human dimensions of urban greenways: planning for recreation and related experiences', *Landscape and Urban Planning*, 68(2–3): 147–65.

Goode, D. (1986) *Wild in London*, London: Michael Joseph.

Haas, C.D. (2001) *Responses of Mammals to Roadway Underpasses Across an Urban Wildlife Corridor, the Puente-Chino Hills, California*, UC Davis, CA: Road Ecology Center, online, available at: www.escholarship.org/uc/item/26m9g40j.

Haddad, N.M. and Baum, K.A. (1999) 'An experimental test of corridor effects on butterfly densities', *Ecological Applications*, 9(2): 623–33.

Haddad, N.M., Bowne, D.R., Cunningham, A., Danielson, B.J., Levey, D.J., Sargent, S. and Spira, T. (2003) 'Corridor use by diverse taxa', *Ecology*, 84(3): 609–15.

Harris L.D. and Scheck, J. (1991) 'From implications to applications: the dispersal corridor principle applied to the conservation of biological diversity', in D.A. Saunders and R.J. Hobbs. (eds) *Nature Conservation 2: The Role of Corridors*, Chipping Norton, NSW: Surrey Beatty and Sons, pp. 189–220.

Hess, G.R. and Fischer, R.A. (2001) 'Communicating clearly about conservation corridors', *Landscape and Urban Planning*, 55(3): 195–208.

Jongman, R.H.G., Kulvik, M. and Kristiansen, I. (2004) 'European ecological networks and greenways', *Landscape and Urban Planning*, 68(2–3): 305–19.

Kalwij, J.M., Milton, S.J. and McGeoch, M.A. (2008) 'Road verges as invasion corridors? A spatial hierarchical test in an arid ecosystem', *Landscape Ecology*, 23(4): 439–51.

Maskell, L.C., Firbank, L.G., Thompson, K., Bullock, J.M. and Smart, S.M. (2006) 'Interactions between non-native plant species and the floristic composition of common habitats', *Journal of Ecology*, 94(6): 1052–60.

McGuckin, C.P. and Brown, R.D. (1995) 'A landscape ecological model for wildlife enhancement of stormwater management practices in urban greenways', *Landscape and Urban Planning*, 33(1–3): 227–46.

Middlemarch Environmental (1997) *The Midlands Plateau: Natural Area Report*, Peterborough: English Nature.

Mörtberg, U. and Wallentinus, H.G. (2000) 'Red-listed forest bird species in an urban environment – assessment for green space corridors', *Landscape and Urban Planning*, 50(4): 215–26.

Ng, S.J., Dole, J.W., Sauvajot, R.M., Riley, S.P.D. and Valone, T.J. (2004) 'Use of highway undercrossings by wildlife in southern California', *Biological Conservation*, 115(3): 499–507.

Oprea, M., Mendes, P., Vieira, T.B. and Ditchfield, A.D. (2009) 'Do wooded streets provide connectivity for bats in an urban landscape?' *Biodiversity and Conservation*, 18(9): 2361–71.

Rosenberg, D.K., Noon, B.R. and Meslow, E.C. (1997) 'Biological corridors: form, function, and efficacy', *BioScience*, 47(10): 677–87.

Sadler, J.P., Small, E.C., Fiszpan, H., Telfer, M.G. and Niemelä, J. (2006) 'Investigating environmental variation and landscape characteristics of an urban-rural gradient using woodland carabid assemblages', *Journal of Biogeography*, 33(6): 1126–38.

Schaefer, V. (1998) 'Green links and urban connectivity: an experiment', *Urban Nature Magazine*, 4: 100–4.

Schwab, J. (1994). *Planning for Wildlife Mitigation Corridors: Environment and Development*, Chicago, IL: American Planning Association.

Scott, R. (2004) *Wild Belfast: On Safari in the City*, Belfast: Blackstaff Press.

Searns, R.M. (1995) 'The evolution of greenways as an adaptive urban landscape form', *Landscape and Urban Planning*, 33(1–3): 65–80.

Small, E.C. (2002) *Biodiversity and persistence of carabid beetles (Coleoptera: Carabidae) in fragmented urban spaces*, Birmingham: Unpubl. PhD thesis, School of Geography and Environmental Sciences, University of Birmingham.

Small, S. Sadler, J.P. and Telfer, M. (2006) 'Do landscape factors affect brownfield carabid assemblages?' *Science of the Total Environment*, 360(1–3): 205–22.

Smith, R.M., Gaston, K.J., Warren, P.H., and Thompson, K. (2005) 'Urban domestic gardens (V): relationships between landcover composition, housing and landscape', *Landscape Ecology*, 20(2): 235–53.

Smith, R.M., Warren, P.H., Thompson, K., and Gaston, K.J. (2006) 'Urban domestic gardens (VI): environmental correlates of invertebrate species richness', *Biodiversity and Conservation*, 15(8): 2415–38.

Snep, R.P.H., Opdam, P.F.M., Baveco, J.M., WallisDeVries, M.F., Timmermans, W., Kwak, R.G.M. and Kuypers, V. (2006) 'How peri-urban areas can strengthen animal populations within cities: A modeling approach', *Biological Conservation*, 127(3): 345–55.

Taylor, P.D., Fahrig, L., Henein, K. and Merriam, G. (1993) 'Connectivity is a vital element of landscape structure', Oikos, 68: 571–3.

Tigas, L.A., Van Vuren, D.H. and Sauvajot, R.M. (2002) 'Behavioral responses of bobcats and coyotes to habitat fragmentation and corridors in an urban environment', *Biological Conservation*, 108(3): 299–306.

Trask, M. (2007) *Limitations to Wildlife Habitat Connectivity in Urban Areas*, UC Davis, CA: Road Ecology Center, online, available at: http://escholarship.org/uc/ item/0sf780k9.

Tratalos, J., Fuller, R.A., Evans, K.L., Davies, R.G., Newson, S.E., Greenwood, J.J.D. and Gaston, K.J. (2007) 'Bird densities are associated with household densities', *Global Change Biology*, 13(8): 1685–95.

Turner, T. (2006) 'Greenway planning in Britain: recent work and future plans', *Landscape and Urban Planning*, 76(1–4): 240–51.

Vermonden, K., Leuven R.S.E.W., van der Velde, G., van Katwijk, M.M., Roelofs, J.G.M. and Hendriks, J. (2009) 'Urban drainage systems: An undervalued habitat for aquatic macroinvertebrates', *Biological Conservation*, 142(5): 1105–15.

Von der Lippe, M. and Kowarik, I. (2007) 'Long-distance dispersal by vehicles as driver in plant invasions', *Conservation Biology*, 21(4): 986–96.

Von der Lippe, M. and Kowarik, I. (2008) 'Do cities export biodiversity? Traffic as dispersal vector across urban–rural gradients', *Diversity and Distributions*, 14(1): 8–25.

von Haaren, C. and Reich, M. (2006) 'The German way to greenways and habitat networks', *Landscape and Urban Planning*, 76(1–4): 7–22.

Wood, B.C. and Pullin, A.S. (2002) 'Persistence of species in a fragmented urban landscape: the importance of dispersal ability and habitat availability for grassland butterflies', *Biodiversity and Conservation*, 11(8): 1451–68.

23
Landscaped parks and open spaces

M. Hermy

Introduction

Landscaped parks and open spaces in (sub)urban areas vary greatly in structure, extent, functions, history and management. They range from highly formal, carefully managed areas where access by people often is restricted, to parks that are largely sports grounds, or parks that are basically remnants of common land, woodland or open space. The character of urban parks also reflects their history, and thus also the ideas of their planners and designers, some of whom deliberately laid out formal areas or planted exotic species in a formal manner. They usually owe their societal importance to their location, close to densely populated areas. For people, parks and open spaces are oases of green, tranquility, relaxation and pleasure. Their size, which usually is a proxy for their habitat heterogeneity, makes them an essential component of urban green structure. They usually contain a considerable number of trees, but elements such as lawns, pastures, ponds, garden elements and related infrastructure are more dominant, while forest stands are often limited or even absent. So habitat variety often is high. Life forms range from the visually dominant trees to invertebrates and fungi. Plant and animal life are a mixture of naturally occurring and deliberately introduced species; an assemblage that changes in space and time. All this makes parks complex ecosystems that are difficult to comprehend, even more difficult to manage in an optimal way and to treat in a concise way as is the goal of this chapter.

Integrative concepts

Given the complexity of landscaped parks and open spaces, and their multifunctionality, integrative and adaptive concepts and approaches are required (see Gustavsson *et al.* 2004). One example of a highly integrative approach arises from the Ecopolis-concept of Tjallingii (1995), which is based on ecological and sustainability principles. According to the Ecopolis-concept, urban parks (UP) may be regarded as eco-device models with inputs and outputs – just like ecosystems. They may be presented as a triangle in which the three corners refer to areas, flows and actors. Each of these themes is managed according to a certain motto. The motto for the areas could then be the Living Urban Park (LUP), for the flows the Responsible UP, and for the actors the Participating UP (Figure 23.1). The Living UP refers to ecosystem biodiversity in the broad

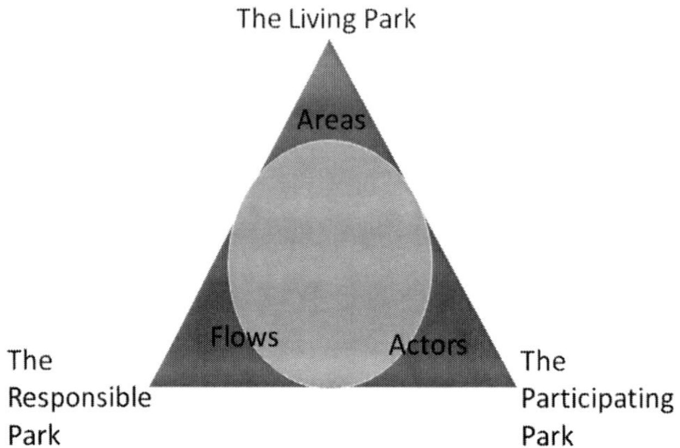

Figure 23.1 The ECOPOLIS concept (Tjallingii 1995) of urban parks.

sense, a complex system of which the number of species increases with the area and may easily run into thousands of species (see further). RUP management regulates flows, e.g. water flow, traffic and visitors, export of products (such as green waste), and this should be achieved in a responsible way. The third theme (Participating UP) refers to all users of the UP. This includes the main management actors, including UP managers and public authorities, visitors and other interest groups. All of these should participate in one way or another in order to meet both the multitude of functions and services and the need for communication of values, preferences and knowledge. In this way a sustainable system may be achieved.

The Ecopolis-concept thus strongly links to sustainability. One key question is therefore, how sustainable are parks? This is particularly important in terms of management, as ultimately the services and functions provided by parks will depend on it. Thus it is important to provide an overview of a possible, more sustainable, management system, before focusing on the nature of parks and thus mainly on LUP.

Developing sustainable management

What is sustainable urban park management?

Until recently, only little attention was given to the sustainable management of UP ecosystems. Of importance in that respect is the fact that management of urban parks has changed considerably since 1900. Higher labour costs, increasing throughout the twentieth century, have led to much more extensive management. Temporary green is used less frequently and variegated collections of plants, so typical for the nineteenth century gardens and park styles, have almost completely disappeared. The decrease in management intensity gave more opportunities for natural developments. Consequently, it is not surprising that the ecological aspects of urban greenspace management received much more attention after 1960.

The concept of sustainable development is usually described as a development that provides for the needs of the current generation without compromising the needs and possibilities of future generations (e.g. WCED 1987). Despite this definition, the implementation of the sustainability concept remains difficult. Translated to urban parks, sustainable management of UP could be described as aimed at long-term management of these complex habitats in order to perform their multiple functions also in the future. But what does that basically mean in practice?

Adaptation to local conditions

In order to maintain a desired park situation or 'scenery', a particular management is needed. Vegetation succession in a temperate climate leads to closed, forested landscapes in which trees dominate, and thus open habitats will need management if we want to keep them open. Furthermore, a natural forest may be far from the most desired given the multiple functions expected from urban parks and open spaces. In any case, long-term management is best guaranteed when current and future funding is secured. Management with minimal costs assumes that habitats can spontaneously exist and develop without much human help. As a result, the aim could be 'stable' communities where spontaneous processes of establishment of both animals and plants regulate community composition and only need minimal care. In general these communities will be well balanced, adapted to the local environmental conditions. They usually are characterised by a variety of native species that are part of fully developed local food webs. However, many non-native (plant) species can survive and even regenerate without much human help as long as they are adapted to local climate (e.g. frost hardy) and soil conditions. Exceptionally, they can even be the most suitable for urban conditions. Yet their food webs are less developed, as indicated by the lower number of insect species associated with exotic tree species (Kennedy and Southwood 1984). Exotic species may also cause great problems due to strong competitive ability, although this has so far been considered less of a problem in Europe than for instance in North America or New Zealand. But under climate change some of the exotics may become more aggressive and pest, disease and weed incidence may increase (for overview see Bisgrove and Hadley 2002).

The greatest biological diversity of plants, fungi and animals is associated with diverse vegetation which generally reflects a diverse environment. Indeed, particularly along gradual environmental gradients (e.g. light conditions or soil conditions) a variety of species-rich communities develop. Plant species composition then gradually changes and offers great opportunities for a variety of wildlife. This often contrasts strongly with the actual situation with sharp boundaries in UP. The steep banks of lakes and ponds only offer limited space for wildlife. Environmental diversity increases significantly if a gradual gradient from open water to land is created. For the same reasons, gradual transitions from grassland, via shrub to forest also offer a high biological diversity (Bradshaw *et al.* 1986). Furthermore the probability of long term survival of taxa is expected to be much higher when environmental gradients change slowly compared to steep gradients that offer little space and opportunities for biological diversity and adaptation.

Using natural processes and conditions

Ideally management should in a conscious way use the natural processes of spontaneous development of flora and fauna, and eventually take measures to increase the potential through directed, environmental interference. In practice, however, spontaneous development is not always feasible or desirable. For example, it may take decades before natural regeneration of trees will occur and develop in a way attractive to visitors. A high proportion of luck may be required in terms of desirable plants entering the area first (dispersal limitation). Additionally it may be that seed sources, essential to regeneration, are not even present nearby. As a consequence, sowing or planting may be required for regeneration or establishment of species. But native and locally indigenous genetic material should be preferred when possible as this probably offers the best 'fit' to local conditions. If one prefers exotic or non-indigenous species, it is essential to match the choice of species as closely as possible with these local environmental conditions (see Hansen and Stahl 1993). If this is not possible (e.g. with large collections of species on limited space) it is preferred to adapt environmental conditions, i.e. mainly soil, but also light regimes so that the long term survival of exotic species is possible under minimal management. Those management measures taken to create good starting conditions are grouped under the heading of habitat creation and/or restoration (e.g. Buckley 1989).

Management continuity

Sustainable management also requires *continuity*. If management measures change continuously, the vegetation – mainly composed of perennial plant species – has no time to adapt to the new conditions and it will probably not reach the targets set by management plans, particularly in systems which include living organisms of great longevity (i.e. trees). Only few species adapt to frequent disturbance in combination with stress (Grime *et al.* 1988). Disturbance is considered here as any measure which destroys part of the whole vegetation and thus lowers the biomass. Stress is defined in this concept as any event causing a shortage or suboptimal condition of growth (e.g. shortage of nutrients or light).

In this view, management measures are considered as disturbances in as far as they destroy biomass. Annual ploughing or digging only allows ruderal plant species to persist, such as weeds on arable lands. Under conditions of low disturbance and high nutrients competitive species dominate (e.g. *Urtica dioica*), as in tall herb vegetation on river banks. Annual mowing of grassland with litter removal directs succession to the stress-tolerant and to some extent to the ruderal part of the spectrum. Thus the CSR-theory (Competetive-Stress tolerant- Ruderal) enables us to understand the effect of various management measures as well as the important vegetation processes of dominance and species coexistence (see Grime *et al.* 1988) in parts of urban parks where (semi-)natural communities are aimed at.

In order to achieve well-developed vegetation, it is important to maintain a chosen management regime or measure. Addition of species (e.g. bulbous perennials in grasslands, or forested parts) through planting may increase diversity and attractiveness if they manage to maintain viable populations (see e.g. Lloyd 2004). Usually this results in an increase of species diversity and stability of the community, both of which are favourable for achieving a more sustainable ecosystem. The management principle of continuity also assumes an a priori developed long-term vision or strategy for the UP management of a particular site.

Use of chemicals

The use of chemicals in managing vegetation should be avoided as much as possible, as serious objections to it can be brought forward from the perspectives of the environment, public health as well as wildlife. As many chemicals are not selective, they also affect other plants and/or animals than those aimed at. Wildlife (and in particular insects) may also suffer from the loss of food plants. Some specific pesticides may be needed – special situations (e.g. for the management of pavements or ruins), but in general they are not part of sustainable UP management.

Water and recreation management

Public urban parks serve recreational purposes, including walking, sports, and relaxing. Different recreational needs demand different types of infrastructure and may thus be difficult to jointly accommodate (e.g. horse riding, walking, mountain biking).

In order to meet recreational demands, at least minimal recreational facilities should be provided and maintained. The higher the recreational pressure, the more open pavements can be, as trampling is a primary factor in controlling vegetation development. For a review of the ecological effects of recreation see Liddle (1997). Open or half-open hardening also enables precipitation to percolate. Run-off increases when the soil-surface sealing increases. Management respecting or enhancing water infiltration and retention is important particularly in urban environments. Water management in which infiltration, retention and restrictive water use facilities are sought is another integral part of UP management (cf. Tjallingii 1995; see Dunnett and Clayden 2007). All too often, water problems are 'exported' to neighbouring areas.

Biodiversity

Biodiversity has a multiscale content going from genes, species to habitat and ecosystems; biodiversity also contains structural and functional attributes at these four levels of organisation. Most often however, the biodiversity is approached at the species level. Then it can be measured as species richness or species diversity. The species diversity of urban and suburban habitats, including parks and gardens, can be extremely high (e.g. Loram *et al.* 2008), as is also illustrated convincingly by Owen (1991). Of the total species number for Great Britain (i.e. 1032 species) 21 per cent was observed over a 15 year period in a young 741 m^2 garden. Given this high biodiversity it is not easy to monitor it.

Papers looking at the biodiversity of urban parks are therefore often restricted to specific species groups, for example bats (Kurta and Teramino 1992), mammals (Chernousova 1996), arthropods (Natuhara *et al.* 1994) or birds (Morneau *et al.* 1999; Fernández-Juricic 2001).

Therefore it is justifiable to examine variation in habitats and habitat diversity in urban parks. This has the advantage, if defined accurately, of being easier to apply. Also habitats perhaps incorporate better some threats to biodiversity, such as fragmentation, habitat loss and habitat change through management (Maddock and Du Plessis 1999).

Habitat units as templates for biodiversity

To determine habitat diversity in an efficient way, a list with all possible habitat units in parks is essential (Table 23.1). Although made for the parks in Flanders, Table 23.1 is useful to other parts of the temperate world. Each unit has to be unambiguously defined leaving no confusion about its delineation in the field. Because each habitat unit is measured differently and it impacts the architectural design of an urban park in a different way, a distinction should be made between punctual, linear and planar elements (Hermy and Cornelis 2000). The planar elements include inter alia different types of woodland, plantings, grassland and gardens, ponds and buildings. Alleys and different types of hedges, verges, rides, watercourses and riverbanks are the most important linear elements. The punctuating elements include single trees, pools, icehouses and some other infrastructure elements.

Geographical Information Systems (GIS) combined with aerial photographs, or even remote sensing, are powerful tools for quantifying the amount (e.g. number, area, length) of these habitat units, to be used in diversity measurements and for displaying purposes such as the distribution of certain habitat units and/or species, eventual overlaying with knowledge system building, management, and planning. In a study of 15 suburban parks in Flanders we found a positive correlations between habitat diversity and the number of plant taxa, breeding birds and amphibian species, but not with butterfly species richness (Figure 23.2a) (Cornelis and Hermy 2004). It also became clear that total park area is not necessary positively correlated with species richness (Figure 23.2b).

Table 23.1 List of habitat units distinguished in (sub)urban parks in Flanders; for coding purposes each element has a unique number (in brackets)

1 Planar elements

1.1 Forest stand: unit composed of a more or less natural forest vegetation
  1.1.1 deciduous wood: forest stand of deciduous trees
    1.1.1.1 coppice: forest stand of regularly cutted thickets (1)
    1.1.1.2 coppice with standards: forest stand of regularly cutted thickets and upper trees (2)
    1.1.1.3 park wood: forest stand of single trees with ligneous undergrowth (3)
    1.1.1.4 leafy, regular high forest: forest stand of regular high deciduous trees (4)
  1.1.2 coniferous wood: forest stand of conifers (5)
  1.1.3 mixed wood: forest stand of deciduous and coniferous trees (6)
1.2 Plantation: unit composed of planted trees
  1.2.1 orchard: enclosed unit planted with fruit trees (7)
  1.2.2 forest grassland: grassland planted with forest trees (8)
  1.2.3 tree gallery: linear plantation of trees without undergrowth (9)
  1.2.4 arboretum: plantation of different tree species with an educational function (10)
  1.2.5 forest plantation: plantation of forest trees (<3 m) (11)
1.3 Labyrinth: unit composed of close hedges in labyrinth form (12)
1.4 Shrub plantation: unit composed of shrubs (13)
1.5 Grassland: unit composed of grass species
  1.5.1 lawn: frequently mown grassland (14)
  1.5.2 sports field: frequently mown grassland used as sports ground (15)
  1.5.3 hay meadow: grassland used to make hay (16)
  1.5.4 pasture: grassland grazed by animals (17)
  1.5.5 hay-pasture: grassland that is grazed after hay-making (18)
1.6 Tall herb vegetation: unit composed of rough herbs, inclusive reed vegetation (19)
1.7 Heathland: unit composed of dwarf shrubs (20)
1.8 Agricultural area: unit composed of arable crops (21)
1.9 Fallow land: temporary unit composed of fallow ground (22)
1.10 Garden: enclosed unit composed of vegetables, fruit or ornamental plants
  1.10.1 kitchen garden: garden composed of vegetables and fruit (23)
  1.10.2 herb garden: garden composed of medicinal herbs (24)
  1.10.3 rose garden: garden composed of roses (25)
  1.10.4 ornamental garden: garden composed of other ornamental plants (26)
1.11 Ornamental plantation: non-enclosed unit composed of ornamental plants (27)
1.12 Water feature: unit composed of water
  1.12.1 castle-moat: water feature round a historical building (28)
  1.12.2 pond: water feature free from each building (29)
1.13 Building: unit composed of buildings, inclusive the limited space between the buildings (30)
1.14 Car park: unit composed of parking places for vehicles
  1.14.1 half-hardened: parking with a hardening that is not completely sealed (31)
  1.14.2 not hardened: parking without any hardening (32)

2 Linear elements

2.1 Alley: double or four-double row of trees, including the verges (33)
2.2 Tree row: row of trees (34)
2.3 Hedge: linear woody vegetation
  2.3.1 sheared hedge: hedge that is regularly sheared (35)
  2.3.2 non-sheared hedge: hedge that is not sheared (36)
  2.3.3 woody embankment: hedge on an embankment created by humans (37)
2.4 Road verge: non-hardened strip along a road (38)

2.5 Bank: strip of land on each side of a water feature or a watercourse
  2.4.1 bank of a water feature: bank of a castle-moat or pond
    2.4.1.1 natural: bank not consolidated by humans (39)
    2.4.1.2 semi-natural: bank consolidated by humans where vegetation is still possible (40)
  2.4.2 bank of a watercourse: bank of a ditch, brook or river
    2.4.2.1 natural: bank not consolidated by humans (41)
    2.4.2.2 semi-natural: bank consolidated by humans where vegetation is still possible (42)
2.6 Watercourse: linear element used for the discharge of water
  2.6.1 ditch: watercourse with a width of max. 1 m that may contain water (43)
  2.6.2 brook: watercourse with a width of max. 3 m that always contains water (44)
  2.6.3 river: watercourse with a width of more than 3 m (45)
2.7 Road infrastructure: strip used and prepared for pedestrians and service traffic
  2.7.1 road: road infrastructure with a width of more than 2 m
    2.7.1.1 half-hardened: road with a hardening that is not completely sealed (46)
    2.7.1.2 not hardened: road without any hardening (47)
  2.7.2 sunken road: sunken road infrastructure, including the verges (48)
  2.7.3 path: road infrastructure with a width of less than 2 m
    2.7.3.1 half-hardened: path with a hardening that is not completely sealed (49)
    2.7.3.2 not hardened: path without any hardening (50)
2.8 Wall: linear masonry used as enclosing (51)

3 Punctual elements
3.1 Single tree or shrub: tree or shrub not surrounded by other trees or shrubs (52)
3.2 Pool: small, shallow, stagnant water ≤ 100 m² (53)
3.3 Icehouse: house where ice was kept (54)
3.4 Tumulus: burial mound (55)
3.5 Infrastructure element: human construction (well, fountain, kiosk, chapel, monument, statue, bridge, aviary…) (56)

Source: Hermy and Cornelis 2000.

Species diversity

As species diversity may be great even in urban habitats (see Table 23.2), monitoring needs to be restricted to some groups. Many possibilities are open here. Yet all will probably include (vascular) plant species as the latter determines the architecture and structural variation in urban parks, they are usually also well known and they offer a variety of ecological niches to animal species and services to people. Additionally urban parks traditionally have high botanical and/or horticultural values. Lists should therefore preferably include indigenous, naturalised and exotic species, including intraspecific taxa and varieties. But most plants have relative long generation times and therefore often possess a time lag in responding to environmental changes. The latter makes them less suitable for monitoring purposes.

As water plays an important landscape role in parks and attracts wildlife, water ponds should be included e.g. by studying amphibians or invertebrate populations. Through their amphibian lifestyle many of these animals both incorporate terrestrial and aquatic habitat qualities of the park. Amphibians are very sensitive to environmental stress factors. As invertebrates easily take up more than 50 per cent of the total species richness on earth some groups should be included for monitoring purposes. They usually have short life cycles and high habitat specificity, which makes them potentially sensitive for monitoring changes. Butterflies are such a group. They quickly respond to changes in the structure, configuration and botanical composition of habitats, their taxonomy is well known and they can be monitored relatively easily. As with all more

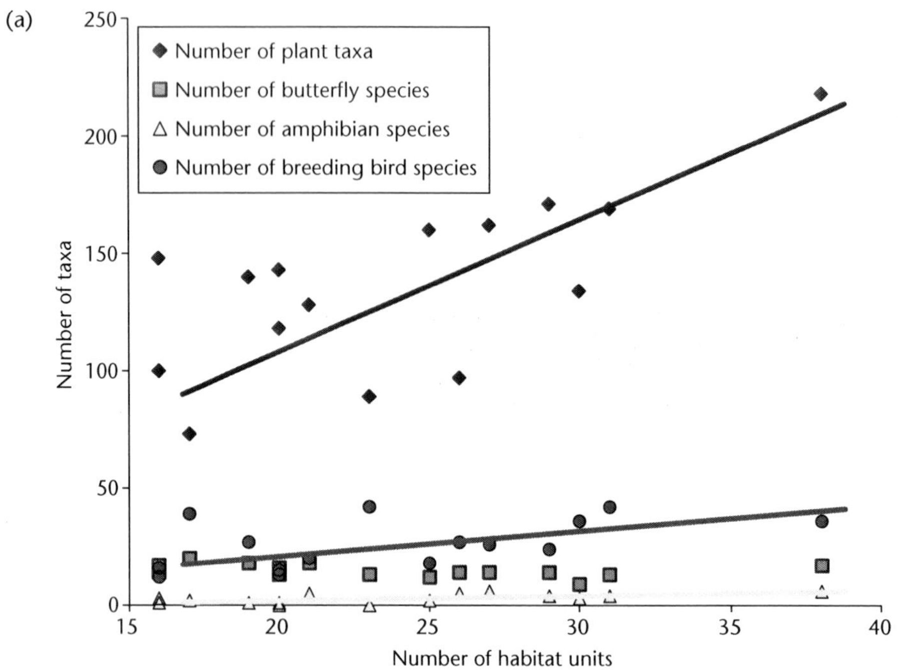

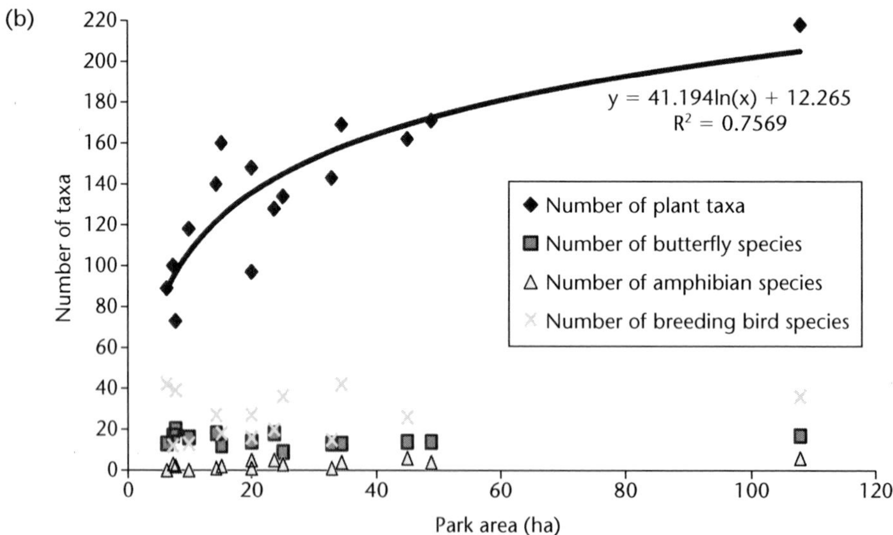

Figure 23.2 Diversity relationships in 15 (sub)urban parks in Flanders (Belgium). (a) the relationship between habitat diversity (number of habitat units, cf. Table 23.1) and number of species in vascular plant species ($y = 5.6238x$, with x number of habitat units; $R^2 = 0.40$), breeding bird species ($y = 1.0837x$; $R^2 = 0.25$), number of amphibian species ($y = 0.2053x–2.0331$; $R^2 = 0.40$) and number of butterfly species (not sign.correlation); (b) the relationships between park area and species richness in various taxa (equation for plant species = $y = 41.194\ln(x) + 12.265$; $R^3 = 0.76$).

Table 23.2 Park area and biodiversity characteristics for 15 (sub)urban parks in Flanders (calculated from Cornelis and Hermy 2004)

Characteristic	*Average (SD)*	*Cumulative total*	*Saturation %*[1]
Park area (ha)	27.9 (±24.12)	418	–
Number of plant taxa	136.7 (±35.12)	497[2]	29.9[3]
Number of butterfly species	14.8 (±2.65)	82	48.5
Number of amphibian species	3.1 (±1.84)	28	38.9
Number of breeding bird species	26.2 (±9.91)	8	61.5

Notes
1 ratio between total observed number over all parks and the total number known from Flanders, multiplied by 100.
2 including introduced species; native and naturalised spp. = 393.
3 only for native and naturalised spp.

mobile species groups, they may not only respond to changes within urban habitats, but also to changes beyond these habitats. Birds integrate a number of features of urban parks (tranquillity, structure, age of trees, management, habitat heterogeneity). This group also contains a number of top-of-the-food-chain species (e.g. owls) and migratory songbirds.

Ecologically sound management measures

Urban woodland, tree-rich parts of parks, and shrub areas

Sustainable management should focus on maintaining a variety of functions, from timber production to nature conservation and not in the least recreation. Creating and maintaining an attractive and diverse woodland image for recreational purposes will mostly be the main goal. It should be kept in mind that with spontaneous forest development investment and cost are at their lowest. But amenity values may be lower as well, at least during part of the forest development cycle. Moreover, in case of arboreta, and also for many parks, spontaneous development is not possible or desirable. Yet even here, more nature-like forms of woodland management will keep costs at a minimum.

Apart from non-intervention, woodlands and wooded elements may be managed in a variety of ways. It has to be remembered, however, that radical changes in woodland management usually have large consequences, as many of the typical woodland species are typical for steady-state communities. Changes in woodland management practices therefore can only be implemented over longer periods of time, and consequently also here management continuity is an important principle. It also means, however, that traditionally managed parts, such as coppice or coppice-with-standards (Buckley 1994) or high forest should be maintained or only gradually changed. This is particularly true where these traditional forms of management have resulted in valuable plant communities, often indicated by abundant carpets of spring flowers (e.g. *Scilla non-scripta, Anemone nemorosa*). Coppice cycles usually are in the order of 10–20 years, while shorter and longer cycles are possible. Cut lots can be distributed in time and space over the forested area. This will generate a variety of regeneration stages, each of which will be associated with typical wildlife, and which is motivated by aesthetic reasons as well. Cycles for standards are usually longer than 100 years. For an overview we refer to Buckley (1994).

When aesthetical, social or ecological values of coppice or coppice-with-standard systems are low, it is better to convert these woodlands to some form of high forest that requires less cost and is closer to nature. Most sustainable forest management mimics nature.

Urban parks usually have no production function, although the value of timber may contribute to compensating management costs. Yet old and dead trees support a wide variety of wildlife; heavy wood debris in particular is highly valuable for many animal species and fungi, and old trees serve bird species and even bats. Old trees thus should be maintained as long as possible if the safety of visitors allows this.

Water and bank vegetation

Many urban woodlands and parks contain bodies of water. These serve a variety of functions, including water and fish production and providing amenity values. Sustainable management should aim to develop a gradual transition from water to land, thus creating smooth gradients from inundated and wet to drier conditions (see above). Large fish or waterfowl populations do not match large populations of hydrophytes. Management of the terrestrial part of the water 'ecotone' may involve an annual or biannual mowing regime. Moss (2010) has provided an overview of fresh water ecology. Even if hard edges cannot be removed, some form of mitigation is usually possible, for example through stepping stones from the water or shallow water.

If reed marshes are present, management will usually involve a winter or autumn mowing regime with litter removal. This will usually result in vital reed marshes with considerable plant species richness. Autumn mowing can also be combined with bird conservation, as it yields shelter through reed re-growth in spring. Mowing from the ice during winter – where winter freezing occurs – may be the only easy way of working; fortunately reed needs not to be managed every year. Wet meadows (e.g. with *Caltha palustris*) or tall herb communities (e.g. *Filipendula ulmaria*, *Epilobium hirsutum*) usually are best mown every 2–5 years, in late summer or early autumn. During this time of year the ground water table is usually at its lowest. If not managed these areas will develop into willow shrub and/or ultimately into some form of black alder (*Alnus glutinosa*) forest.

Hydrophytes such as *Nymphaea alba* and *Nuphar lutea* may generate impressive 'pictures' greatly enjoyed by people. But apart from these flagship species many other smaller species may be present as well. Water plants will only thrive when they obtain sufficient light, so shade-covered ponds are usually poor in wildlife. Managers of urban woodlands and parks should also take all necessary measures to ensure the best water quality possible. This may involve the creation of helophyte water purification systems. As time passes all water bodies – and particularly shallow ones – will gradually turn into land. Succession will ultimately result in willow and alder marshes. With this in mind water bodies need cleaning from time to time. This will reverse succession into the open water phase.

Grasslands

Grasslands, including lawns, are often an important part of urban parks and woodlands, and therefore they need specific attention. However, lawns that are not frequently used, apart from the edges along paths, are best replaced by meadows. These are mown once or – in case of heavily fertilised soils – twice a year during early summer and a second time late summer or early autumn. A properly managed meadow flowers, apart from a short period after mowing, almost continuously from spring to autumn (Ash *et al.* 1992). As hay or freshly cut grass is removed, the regeneration niche and ultimately the coexistence of many plant species is assured (cf. Grime *et al.* 1988). To reach greater amenity values naturalised – originally non-native – plant species may be introduced (often geophytes e.g. *Narcissus* spp., *Camassia* spp., *Crocus* spp.). Most of the spring flowering species are found on richer soils, so for a sustainable result selection should follow local environmental conditions. Hill *et al.* (1999) and Hansen and Stahl (1993) provide comprehensive reviews of the habitat requirements of plant species. Excellent examples include the meadows of Great Dixter in Sussex, England (Lloyd 2004). In order to replace species-poor grassland on nutrient-rich soils litter should always be removed and no fertilising allowed. Through mowing with litter removal the number of more stress-tolerant species increases gradually, allowing for greater coexistence of a large number of plant species (Grime *et al.* 1988). Hay-making is preferred to the immediate removal of the litter as it enables larger seed dissemination, although direct removal of litter may be less expensive. The application of this mowing regime will also considerably reduce the annual biomass production and thus will reduce the cost.

Particularly around the edges of forested parts it can be essential to mow the grassland only once every 2–5 years. This will allow for the development of specific woodland edge vegetation, creating valuable habitats for plants but also for butterflies and other invertebrates. Structurally this allows for a more gradual transition towards forest habitats and thus creating a gradual environmental gradient, where light intensity but also air humidity and other factors slowly change.

In this concept lawns are limited to gardens, to the edge of paths or to places frequently used by visitors for activities such as picnicking and playing. Along paths lawns make gradients of height more gradual. But even on lawns spontaneous processes of germination and establishment may occur. If lawns are mown regularly, litter is removed each time and no fertilisation is applied, gradually more species and particularly more herbs such as *Bellis perennis*, *Veronica filiformis*, *Trifolium repens*, and *Hypochoeris radicata* will establish.

More competitive species occur along the edges of forests, in the transition from grassland to woodland, on open, not frequently managed areas, along ditches, rivers and ponds. Often vegetation with large herbaceous plant species is flower-rich (e.g. *Hypericum perforatum* and *Teucrium scorodonia* on nutrient-poor and dryer sites). Through their structure, these species are highly significant for wildlife: as nectar or food source for many insect species, as refuge for small animals, hibernation sites for invertebrates, as food sources for singing birds, and so forth. Unwanted species may be of great importance for invertebrates. *Urtica dioica* is an important food plant for butterfly species such as *Vanessa atalanta*, *Inachis io*, *Araschnia levana*, *Polygona c-album* and *Aglais urticae*. To maintain tall herb vegetation, mowing is required once every 2–5 years. Mowing intensity is lowest on poor soils. Most of the tall herb species – all related to the competitive plant strategy – are not well adapted to annual mowing regimes, at least during summer. When they are mown annually with removal of litter, grasses and other smaller plant species will take over (cf. Grime *et al.* 1988) and plant species diversity will increase. Mowing within urban parks is best distributed in time and space. This will yield seasonal peaks and an outdrawn flowering season to enjoy, through various phases of re-growth, and thus also shelter and food for a variety of wildlife.

Further reading

Kendle, T. and Forbes, S. (1997) *Urban Nature Conservation: Landscape Management in the Urban Countryside*, London: E&FN Spon.

Dunnett, N. and Hitchmough, J. (eds) (2004) *The Dynamic Landscape: Design, Ecology and Management of Naturalistic Urban Planting*, London: Taylor & Francis.

References

Ash, H.J., Bennett, R. and Scott, R. (1992) *Flowers in the Grass: Creating and Managing Grasslands with Wild Flowers*, Peterborough: English Nature.

Bisgrove, R. and Hadley, P. (2002) *Gardening in the Global Greenhouse: The Impacts of Climate Change on Gardens in the UK: Technical Report*, Oxford: UKCIP.

Bradshaw, A.D., Goode, D.A. and Thorpe, H.P. (eds) (1986) *Ecology and Design in the Landscape*, Oxford: Blackwell Scientific Publications.

Buckley, G.P. (ed.) (1989) *Biological Habitat Reconstruction*, London: Belhaven Press.

Buckley, G.P. (ed.) (1994) *Ecology and Management of Coppice Woodlands*, London: Chapman & Hall.

Chernousova, N.F. (1996) 'Effects of urbanisation on communities of small mammals in park-forests in a large industrial centre', *Russian Journal of Ecology*, 27(4): 278–83.

Cornelis, J. and Hermy, M. (2004) 'Biodiversity relationships in urban and suburban parks in Flanders', *Landscape and Urban Planning*, 69(4): 385–401.

Dunnett, N. and Claydon, A. (2007) *Rain Gardens: Managing Water Sustainably in the Garden and Designed Landscape*, Portland OR: Timber Press.

Fernández-Juricic, E. (2001) 'Avian spatial segregation at edges and interiors of urban parks in Madrid, Spain', *Biodiversity and Conservation*, 10(8): 1303–16.

Grime, J.P., Hodgson, J.G. and Hunt, R (1988) *Comparative Plant Ecology: An Ecological History*, London: Unwin Hyman.

Gustavsson, R., Hermy M., Konijnendijk C. and Steidle-Schwahn A. (2004) 'Management of urban woodland and parks – searching for creative and sustainable concepts', in C. Konijnendijk, K. Nilsson, T.B. Randrup and J. Schipperijn (eds), *Urban Forests and Trees: A Reference Book*, Berlin: Springer, pp. 269–397.

Hansen, R. and Stahl, F. (1993) *Perennials and Their Garden Habitats*, Cambridge: Cambridge University Press.

Hermy, M. and Cornelis, J. (2000) 'Towards a monitoring method and a number of multifaceted and hierarchical biodiversity indicators for urban and suburban parks', *Landscape and Urban Planning*, 49(3–4): 149–62.

Hill, M.O., Mountford, J.O., Roy, D.B. and Bunce, R.G.H. (1999) *Ellenberg's Indicator Values for British Plants: Ecofact Research Report Series 2b*, Huntingdon: Institute of Terrestrial Ecology.

Kennedy, C.E.J. and Southwood, T.R.E. (1984) 'The number of species of insects associated with British trees: a re-analysis', *Journal of Animal Ecology*, 53(2): 455–78.

Kurta, A. and Teramino, J.A. (1992) 'Bat community structure in an urban park', *Ecography*, 15(3): 257–61.

Liddle, M.J. (1997) *Recreation Ecology: The Ecological Impact of Outdoor Recreation and Ecotourism*, London: Chapman and Hall.

Lloyd, C. (2004) *Meadows*, London: Cassell Illustrated.

Loram, A., Thompson, K., Warren, P.H. and Gaston, K.J. (2008) 'Urban domestic gardens (XII): the richness and composition of the flora in five UK cities'. *Journal of Vegetation Science*, 19(3): 321–30.

Maddock, A. and Du Plessis, M.A. (1999) 'Can species data only be appropriately used to conserve biodiversity?' *Biodiversity and Conservation*, 8(5): 603–15.

Morneau, F., Décarie, R., Pelletier, R., Lambert, D., DesGranges, J.-L. and Savard, J.-P.L. (1999) 'Changes in bird abundance and diversity in Montreal Parks over a period of 15 years', *Landscape and Urban Planning*, 44(2–3): 111–21.

Moss, B. (2010) *Ecology of Fresh Water: A View for the Twenty-First Century*, fourth edition, London: Wiley-Blackwell.

Natuhara, Y., Imai, C. and Takeda, H. (1994) 'Classification and ordination of communities of soil arthropods in an urban park of Osaka City', *Ecological Research*, 9(2): 131–41.

Owen, J. (1991) *The Ecology of a Garden: The First Fifteen Years*, Cambridge: Cambridge University Press.

Tjallingii, S. (1995) *Ecopolis: Strategies for Ecologically Sound Urban Development*, Leiden: Backhuys Publishers.

WCED (1987) *Our Common Future: From One Earth to One World*, Oxford: Oxford University Press.

24

Grassland on reclaimed soil, with streets, car parks and buildings but few or no mature trees

Tony Kendle

Introduction

Urban greenspace is a complex mosaic of different landscape and habitat types and different origins. As other chapters have explained, fragments of land of high conservation value can even survive within the urban framework, often undergoing shifts in ecology but retaining much of the original plant and animal communities.

This chapter however is about redeveloped land – the spaces that are created by people with more less-intensive intervention. The communities that are found on them are therefore derived from either recent natural colonisation or from deliberate introduction. There is a widespread assumption that the only species that will be found living there will be common, opportunistic, synanthropic, possibly even invasive and therefore of no great interest. There are many exceptions, such as where fragments of relatively undisturbed land still exist within the matrix so that a legacy of site history remains. But even where it is true that nothing 'unusual' is found, there is still a lot of interest for those who bother to look.

The survival of these sites often results from government policies that aim to provide and protect a given proportion of greenspace in urban settings for the health and enjoyment of residents. Target levels of greenspace are frequently applied in contemporary town and city planning, ranging from 15–40 per cent. The latter sounds high but it is surprisingly achievable in locations where high rise tower blocks are separated by expanses of open space. Other major sources of land can be road verges and spaces between roads and temporary spaces awaiting development.

However it is often the case that the bodies that designate and create these spaces have neither the capital nor operational funds to make them interesting, diverse or detailed. So alongside richer areas of planting or remnant natural habitats are many mown grasslands that make up large, almost featureless, plains, often scornfully referred to as 'green deserts'. They are not devoid of life or productivity though and may be better described as urban rangelands or savannah.

Whatever the terminology, we have the unusual situation of a species investing energy and resources to create a habitat that it seems to find uninviting and not especially useful. Of course it is true that not all functions of these grassland are dependent on our direct use. They can play important roles in issues such as climate regulation or rainwater absorption even if no one sets foot on them.

Mostly, though, the grasslands are intended for multiple uses and in theory they are there for recreation and amenity. As with any other animal, in practice the extent and patterns of human activity on these sites depend on aspects such as shelter, accessibility, location including proximity to homes or workplaces, exposure, height of grass, and drainage, and the density of large dogs as well as territoriality and aggression in human social groups. In some countries the grasslands may contain other species that are threatening, such as the snakes or harmful spiders that can be found in Australia.

As a consequence of these factors some grasslands receive high density and regular use in good weather, especially for children's games or sport or for lunchtime relaxation for office workers. Some are dominated by use types that are not entirely compatible with other people – team sports, golf practice, dog exercising. Other sites are rarely visited by anyone (although if they are not too overlooked this may give them a higher value for activities that need a degree of privacy, not all of which are criminal).

One of the groups least likely to be found on these grasslands is ecologists and naturalists, most of whom are put off by apparent 'sterility' of these grasslands reinforced by the uniformity of the management applied, especially mowing. But for those who are interested to look closer, what will be found is a dynamic mosaic of different species responding to different environmental conditions, with the dance of competition and collaboration between them. These rangelands remind us that many ecological processes and relationships are constant and manifest in a thousand different ways all around us.

These habitats also give us a lesson in observation. The most common and accessible landscapes for most of us who live in towns and cities have much to show us if we only bother to look. Maybe the 'sterility and lack of interest' lies more in our minds than in reality?

Ecological factors

There are three main factors that determine the initial nature of the plant and animal communities found in urban rangelands – soil, sowing and mowing. (N.B. the species referred to in this chapter are those found in the UK, but there are clear analogues in other countries). We refer to those commonly found, but in every case there is a long 'tail' of diversity, made up of species that are only found in a few locations. A survey of 52 lawns (Thompson *et al.* 2004) found 159 species, of which 60 were recorded only once.

The influence of soils

The soils of urban areas can vary enormously. They can encompass almost any of the typical soil types found in rural areas, as well as substrates derived from building rubble and post-industrial wastes that can show extreme chemical and physical properties (see Chapter 15 this volume). Not surprisingly the plants and animals that survive in these different materials vary greatly. Some sites have substrates derived from materials that are the legacy of industrial activities, such as mining spoils. The extreme chemical and physical nature of these naterials can give rise to unusual plant and animal assemblages. As a general rule, the longer sites take to revegetate, the more interesting the flora and fauna that develops (Ash *et al.* 1991).

The factor that unites most redevelopment sites is that the soil will have undergone some form of disturbance. Where soils have been moved during demolition and construction work sudden boundaries between different soil types can sometimes occur, as can many extreme chemical and physical gradients. In some situations this contributes to the development of more diversity in the plant and animal communities. However, unless it is extreme, this variation is also easily masked, especially if a layer of topsoil is spread, grasses are sown and uniform mowing applied.

It has been widely recognised that it is easy to establish vegetation on crushed building waste from demolition sites. Rubble and finer material with a high proportion of mortar is free draining and alkaline, rich in some nutrients but low in organic matter and hence low in nitrogen. This can be addressed by the use of fertiliser or by sowing nitrogen-fixing species such as clovers (Bradshaw and Chadwick 1992). These materials have physical characteristics that are not very different to some natural substrates such as glacial tills. In these settings the growth of smaller herbs and grasses is partly determined by the percentage of fine particles, but it is not uncommon to find that trees and woody plants find a blocky open substrate perfectly acceptable, and certainly easier to grow in than clay-rich soils that have been heavily compacted by construction.

Soil compaction is a frequent hazard on demolition and construction sites, and paradoxically the most fertile soils in their natural state, such as heavy loams, can become the most hostile once denatured. They exhibit problems of drainage, poor aeration (which can happen even on dry soils) and impedance of root growth. Depending on topography and climate these patches are either seasonally or almost permanently wet. The grasslands can develop into distinctive communities that include higher density of creeping buttercup (*Ranunculus repens*), docks (*Rumex* spp.) and different grasses from those normally found, such as Yorkshire fog (*Holcus lanatus*) and rough-stalked meadow grass (*Poa pratensis*). In some cases extensive mats of moss dominate the sward (e.g. *Brachythecium rutabellum*).

Where soil movement by machinery has occurred, compacted layers can impede drainage at depth. A second, less severe but more widespread factor is surface compaction from people walking, especially along desire lines. The species that do well in these situations have some tolerance of poor drainage and low aeration in winter but must also contend with drying out and poor water infiltration in summer, as well as direct physical abrasion. Successful species include creeping bent (*Agrostis stolonifera*) and plantain (*Plantago major*).

Contamination from heavy metals and hydrocarbons dumped or spilled onto soil or aerial deposited pollution (less common now in post-industrial economies) can make soils toxic. However, unless the toxicity is extreme, the effect is a general decrease in vigour and density of cover that hardly gets noticed.

Close to the edges of roads and car parks the use of de-icing salt can sometimes lead to significant soil salinisation (see also Chapter 15 this volume). In extreme cases a community of halophytic plants can develop in a small strip running the entire length of the road kerbings. Common in the UK are *Puccinellia distans* and *Spergularia marina*.

Plants that are not clearly recognised as halophytes are also found in these boundaries and clearly have some tolerance of the conditions e.g. *Matricaria perforata*. Actually the plants have to be tolerant of multiple stresses as salinity is not the only problem that these strips have – they are frequently compacted and have very shallow soil profiles because of the concrete used to set the kerbs in place.

Colonisation or sowing/planting and succession

Natural colonisation

One of the major factors controlling the composition of the grassland is whether it is sown or turfed or develops though natural colonisation.

The general trend for sites that vegetate through natural invasion is that the early successional stages are dominated by wind dispersed mobile and usually ephemeral plants mixed with a range of the propagules that can emerge from re-used and displaced soils. The actual communities can be highly variable with factors such as the nature of the substrate and the distance from sources of colonising plants having a major influence.

These ephemerals give way to taller herbaceous clump-forming species, in turn giving way to grasses where the broadleaved components maintain a presence by vegetative spread more than reproduction from seed. However there are many exceptions to these rules, for example it is common to find that scrub species can colonise in the earliest stages alongside annual and perennial herbs. Many locally distinct stands of exotic species can arise representing unique combinations of site, colonisation sources and chance. Gilbert (1989) reviews many of these in detail.

However there are a consistent selection of species that become familiar to urban naturalists and also (if not nameable) to those children who enjoy free range play in illicit places. These include:

annual meadow grass (*Poa annua*)
false oat grass (*Arrhenatherum elatius*)
Yorkshire fog (*Holcus lanatus*)
Oxford ragwort (*Senecio squalidus*)
purple toadflax (*Linaria purpurea*)
bind weeds (*Convulvulus arvenisis, Calystegia sepium*)
docks (*Rumex obtusifolia*)
hogweed (*Heracleum sphondylium*)
wild carrot (*Daucus carota*)
yarrow (*Achillea millefolium*)
creeping thistles (*Cirsium arvense*)
Michaelmus daisies (*Aster hybrids*)
Japanese knotweed (*Reynoutria japonica*)
goat willow (*Salix caprea*)
butterfly bush (*Buddleia japonica*)

Pockets of redistributed top soil usually brings persistent perennial garden weeds such as nettles (*Urtica dioica*), couch grass (*Elymus repens*), ground elder (*Aegopodium podagraria*) and horsetail (*Equisetum arvense*).

If this natural colonisation process begins but the community is then mown the dominance is restricted to a mix of rosette forming or creeping prostrate herbs and grasses that eventually coalesce to something that superficially has the visual uniformity of a sown grassland, but is likely to remain much more diverse in reality.

Herbivorous invertebrates can be quite rapid colonists of these communities especially winged ones, such as grasshoppers and caterpillars, especially those associated with early urban colonising plants such as the cinnabar moth (*Tyria jacobae*) that feeds on ragwort. Adult stage butterflies can be abundant feeding on some of the flowers especially the late summer flowering plants like *Aster*.

Spiders are usually amongst the earliest colonists including money spiders (*Linyphiidae*) that can blow in but also larger species such as zebra jumping spiders (*Salticus scenicus*). They clearly appreciate the open but complex physical structure of rubbles and stony ground – situations that are becoming recognised in the UK as Inland Rock habitats (Tucker *et al.* 2005). Slugs and snails can come in rapidly as well, depending on the proximity of nearby colonisation sources such as gardens, or pockets of garden soil incorporated in the new habitat. Pockets of buried organic rubbish can support high densities of decomposer organisms.

Sometimes highly unusual species and species assemblages have been found in these early successions, especially of invertebrates, and some of these have been regarded as being of conservation significance (Eyre *et al.* 2002). The problem is that we have almost no baseline of survey data to evaluate the real scarcity of these species, nor do we understand their dispersal patterns well enough to judge how far they travel. The identification of ephemeral colonisers of disturbed land that appear to be rare causes something of a conservation conundrum. Should conservation resources be dedicated to maintaining their survival? This could be achieved by, for example, periodic re-disturbance of early successional sites, as sometimes proposed (Gemmell and Connell 1984). In practice it is hard to find examples of this approach being initiated and maintained except on small scales within urban nature parks or similar (Emery 1986).

Lichens and mosses can also be rapid colonists, especially calcicole species that appreciate the calcium, magnesium, potassium and other nutrients found in rubbles. Every layer rapidly develops a food web that often goes unnoticed. For example Gilbert (1976) showed how a concrete post supported lichens such as *Lecanorion dispersae*, which were grazed on by mites such as *Ameronothrus maculatus* which in turn are predated by bugs *Temnosthethus pusillus*, all of which ultimately fed a decomposer community of collembids.

Once established the grasslands can lose the open niches that support these early colonists, but they do become home for burrowing invertebrates including sometimes large populations of those that feed on plant roots. In a garden context this latter group can be regarded as lawn pests and controlled by insecticides, but in the less manicured settings of urban rangelands they are frequently unnoticed or forgiven. The best known in that latter category are 'leatherjackets' or crane fly larvae (*Tipula paludosa*) that can over-winter in large numbers underground. As the site ages and higher levels of organic matter begin to accumulate there are more opportunities for decomposer organisms such as earthworms (*Lumbricus terrestris*).

Another burrowing invertebrate group are the mining bees such as the Tawny Mining Bee (*Andrena fulva*) and are found in short turf, especially where the ground is warm and dry. As is the way of these things, each of these species can provide food for predators, parasites, and disease organisms such as the common bee fly (*Bombylius major*). Of course, these invertebrates provide the food for many of the birds seen on the grasslands.

The colonisation and distribution of fungi in urban grasslands of this type is largely one big wonderful mystery. Mycorrhizal species that are associated with transplanted trees are often found fruiting in the planting beds and adjacent grasslands. It is highly likely that transplanted trees and shrubs carry the associations with them when they are planted. Where there are scattered young trees, the open climate often means that leaves are dispersed and do not accumulate (except in shop doorways), but where the trees are grouped an organic leaf litter soon develops that is home to diverse decomposer species. Decomposer species are also found associated with buried organic debris.

Away from woody plants fungal fruiting bodies are still sometimes found, such as inkcaps (*Coprinus* spp.) and field mushrooms (*Agaricus campestris*), but discerning the important factors is difficult. They are usually localised and may come from mycelium in redistributed soils rather than colonising from spores.

The extent to which wild herbaceous plants are really dependent on obligate symbiosis with fungi is not well understood. However there are some species for which this is known to be true, notably terrestrial orchids such as *Dactylorhiza* spp. and the occasional huge populations of these plants that can develop on post-industrial land (Ash *et al.* 1991) hints at the possibility that there is unusually strong dominance by fungal species that are held in check in normal soil communities.

Sown or turfed grasslands

The range of turf grasses used for amenity and sports use in the UK contains the following species:

Agrostis tenuis (browntop)
Festuca rubra litoralis (slender creeping red fescue)
Festuca rubra rubra (creeping red fescue)
Festuca rubra commutata (chewings fescus)
Poa pratensis (smooth stalked meadow grass)
Lolium perenne (perennial rye grass)

Lolium perenne tends to dominate in practice, because many landscape managers appreciate its reputation of tolerating wear, but it can have a higher maintenance demand than some of the others and also is less tolerant of stress conditions. In some situations where soil conditions are not favourable it can fail to maintain its dominance. More sophisticated assessment of conditions and needs for redeveloped areas usually ends up leading to mixes dominated by fescues. In situations where low nitrogen levels are anticipated then white clover (*Trifolium repens*) is included in the mix.

Planted trees and shrubs

New tree and shrub plantings around these grasslands vary in composition depending on the prevailing fashion in urban landscape design. An early to mid-twentieth century interest in exotic diversity and flowering and fruiting shrubs gradually moved to a dominance of evergreen species such as *Aucuba japonica* or *Prunus laurocerasus*, which were believed to have a low maintenance need compared to the long season of'interest', with the dense and year round shade helping with weed control. These have in turn fallen out of favour, being recognised as having in fact a long season of no-interest. Current fashion favours deciduous species with a much greater focus on native species, such as alders (*Alnus glutinosa*) and birches (*Betula pendula*), or those trees that have a long history of success in urban conditions such as sycamore (*Acer pseudoplatanus*).

The motivation for favouring natives sometimes include a belief that they are better adapted to UK conditions and support more wildlife, neither of which is likely to be true in urban areas and on disturbed ground. However native trees do help maintain an appropriate 'sense of place' for the landscape, alongside those few non-natives that give particular character to some urban areas such as the London Plane (*Platanus x hispanica*).

Young trees and shrubs support a much less diverse range of organisms than a mature, or especially over-mature, specimen of the same species would do. Unsurprisingly the missing groups are decomposers and species that find shelters and niches within the increasingly complex architecture of the tree form. However those species that can exploit young trees can be present in high numbers and contribute to a food web that is simplified compared to older habitats but still important. Perhaps the most dominant group are leaf and young-stem feeding invertebrates such as the sycamore aphid (*Drepanosiphum platanoidis*) (Dixon 1973).These in turn feed predatory species – birds and other insects such as ladybirds (*Adalia bipunctata*), lacewings (*Chrysapa* spp.) and hoverflies (*Syrphus* spp.) as well as parasites such as the wasp *Aphelinus flavus*. Meanwhile, ants feed on the honeydew residue and sooty moulds develop and the web of life grows.

Mowing and other management

Not mown at all

There are often pockets of land in the urban fabric where because of lack of clarity about responsibility, accessibility, topography or the nature of the substrate no mowing is done. If the grassland community is established before woody plants invade, and the soil is relatively fertile a dense grass mat will develop that can be surprisingly resistant to invasion and further succession. These can stand for years before any significant woody plant invasion occurs.

The most effective colonisers of these mats are frequently plants with heavy animal dispersed seed such as elders (*Sambucus nigra*) and especially those that can also spread vegetatively, such as brambles (*Rubus fruticosus*). As these grow and shade out the herbaceous mat they can create colonisation opportunities for other woody species that have shade tolerant seedlings, such as oak.

Stands of competitive clump-forming herbaceous plants, such as *Urtica* or *Aegopodium*, that have become established before a closed grass canopy, can persist. If the soil is less fertile so that the grassland community is less competitive, or if disturbance gaps are maintained for some reason, or if grasses are not sown allowing time for other plants to establish then more diversity can be found. Other species can become significant components including those that spread by wind-blown seed but that form strong aggressive clumps when established such as *Aster* spp.

In some cases where the substrate is made up of chemically or physically extreme materials more unusual plant communities can build up, giving fascinating but poorly understood glimpses of interactions between the physiological tolerances of the plant, the reproductive mechanisms and the site. For example substrates that are made up of large particles of acidic (or at least neutral) stone can become colonised by dense stands of foxglove (*Digitalis purpurea*) and evening primrose (*Oenothera biennis*). Even these gradually disappear as organic matter builds up and grasses establish.

Periodically mown

Grasslands that have periodic rather than continual mowing can support different species complements. The most common occurrence of this pattern in an urban context is when a grassland has been planted with bulbs such as daffodils (*Narcissus* hybrids) which are not mown until the bulb foliage dies down.

Such bulbs are often referred to as 'naturalised' but they may be made up of persistent planted populations that reproduce only vegetatively or not at all. The periods without cutting can provide opportunities for other species to flower such as mouse-ear hawkweed (*Pillosella officinarum*), *Primula* spp. or Lady's Smock (*Cardamine pratensis*).

Sometimes more adventurous landscape managers will experiment with relaxing mowing at different times of year to favour peak flowering of other grassland species such as knapweed (*Centaurea nigra*) and yarrow (*Achillea millefolium*).

Regularly mown

Most urban grasslands are mown on a cycle of typically 10–14 days and to a height of 10–15 cm, They can look devoid of broadleaved species at first glance but in many cases they have diverged significantly away from any grasses that were sown and there is a rich population of prostrate and rosette forming herbs that are tolerant of close cropping (Thompson *et al.* 2004). The species found include white clover (*Trofilium repens*), daisy (*Bellis perennis*) creeping speedwell (*Veronica filiformis*), selfheal (*Prunella vulgaris*), dove's foot cranesbill (*Geranim molle*), creeping buttercup

(*Ranunculus repens*) and creeping thistle (*Cirsium arvense*). In some cases surprising populations of herbs can persist that people do not normally realise can be cut so close and still survive, such as *Primula vulgaris* and some grassland orchids.

Herbicides and fertilisers

There is a common assumption that these green expanses are sterile spaces partly because they are dosed regularly in herbicides by the parks department. Actually although it may have been a common practice in the twentieth century, this now hardly ever occurs. Public reaction to spraying is one disincentive, but the most significant are budget pressures – one of the few positive contributions accountants may have made to maintaining species diversity. Most greenspace managers have little reason to create species uniformity in extensive grasslands as long as the visual appearance and functional characteristics are adequate. (High performance sports turf is of course different.)

Herbicides are often used around the bases of street furniture and sometimes trees, essentially locations that mowers cannot access easily. The use of well-chosen herbicides at tree bases is less threatening to the tree than the use of strimmers, the alternative management device, as strimmer damage to tree bark can rapidly reach a lethal level.

Diversity can also be reduced by high soil fertility. Again however it is increasingly rare to find greenspace managers fertilising grasslands except in very high wear, high performance areas. The extensive urban rangelands that this chapter addresses rarely see any such inputs. However, high fertility can arise from topsoiling with nutrient rich soil. The disturbance caused by soil lifting and spreading can cause a nutrient flush in the short term that increases the likelihood of a dense competitive sward establishing. Fertility can also be increased by high levels of excreta from dogs or other animals.

Sometimes where soil nitrogen is low, but other resources are adequate, a population of nitrogen fixing herbs gets established that can form a dense sward that is again resistant to further invasion. This is much more likely where seed of these herbs has been sown, but even when establishment is patchy individual plants of species such as *Trifolium repens* can expand to form a colony several metres across.

Animal use and interactions

The prevalence and behaviour of a wide range of animal species on urban rangelands depends very much on the distance to and diversity of surrounding landscape. A few phytophageous and soil-dwelling species find their whole needs met within the grassland. Others use the grasslands only part of the time.

Dogs (*Canis lupus familiaris*) use urban rangelands for a range of territorial and social bonding behaviour including mating. The extent to which cats (*Felis catus*) use these sites is less clear as mostly they avoid open spaces that are well lit, unless in transit. It would be unsurprising if some degree of hunting takes place, especially in urban areas undersupplied with private gardens. Most other urban mammals are at least semi-nocturnal but, like cats, are bound to be making some use of these spaces, including rodents, foxes (*Vulpes vulpes*) and badgers (*Meles meles*).

Birds are the most visible animals to use these grasslands, and a wide range usually can be seen, although extensive urban rangelands are often not as well frequented by songbirds as garden turf, almost certainly because of the absence of shelter and food diversity, although free range large dogs are probably also a factor.

The birds that are seen exhibit a diverse range of ecological behaviours but again it feels worth distinguishing between those that make casual use of the grasslands and those for whom it provides an integral component of what they need. The former group includes those that have become colonists in urban areas because of their tendency to live off detritus and waste human food, notably in the UK pigeons and gulls. Herring gulls (*Larus argentatus*) and Lesser black-backed gulls (*Larus fuscus*) have urban populations in the UK that are together over a hundred thousand pairs (Rock 2005) and growing. They are often seen on grasslands no doubt supplementing their diet or sometimes mating, but in small numbers. They gather in larger numbers mostly around places where lots of waste food can be found such as landfill sites or outdoor 'food courts' in shopping malls.

These gulls have become synanthropic over a period of less than 100 years, colonising towns from their more natural habitats. Urban feral pigeons (*Columba livia*) have a longer assocation with urbanisation. They may have returned to a wild state after having been domesticated. They are very gregarious and gather in large populations in many urban areas, often in close association with people, and sometimes are found both in small and large groups spilling onto grasslands.

House sparrows (*Passer domesticus*) are the third common species that will sometimes feed off discarded food and the animals seem happy to follow people into areas without any vegetation cover. However there is no doubt that the food supply is supplemented by food gathered from grasslands and shrubs. This species has undergone decline in urban areas in the UK and Europe in recent years, but in many parts of the world it is still increasing in numbers.

The European starling (*Sturnus vulgaris*) can roost in extraordinary numbers in some urban areas (50,000 reported as living in one abandoned pier in Brighton). They can sometimes be seen in flocks feeding on urban grasslands, but usually fly out to suburban areas.

The birds that clearly have a higher dependence on grasslands include many songbirds that hunt for invertebrates in short turf or exposed soil such as blackbirds (*Terdus merula*), robins (*Erithacus rubecula*), carrion crows (*Corvus cornone*) and magpies (*Pica pica*).

The other major cohort of note comprises the vegetation feeders. In grasslands near to water this would include ducks and swans but most dramatic of all must be the Canada geese (*Branta canadensis*) that can sometimes be seen roosting and grazing on urban grasslands.

All of these animal species interact, and vary in their tolerance of each other. Sometimes there can be indirect factors that change the nature of the habitats and hence what lives there. For example the bases of trees can be subjected to high levels of dog excretion and urination, leading to a shift in the lichen flora and in extreme cases dominance by the alga *Prasiola crispa*. Dog excreta patches also provide opportunities for slugs and flies to feed and breed and doubtless add significantly (but not pleasantly) to the invertebrate diversity and biomass of urban areas. The same must be true for fungi as well, with doubtless many species colonising unnoticed.

The tendency of dogs to focus on any one specific tree obviously goes up where trees are isolated and especially if they are adjacent to footpaths and benches. Bird populations will also often favour specific trees and considerable quantities of bird excreta can collect on grassland and soil beneath corvid roosts and starling roosts. For some reason the floral and faunal impact of these deposits does not attract much research attention.

Conclusions

The species and communities of urban rangelands may be deadened in many people's eyes by over-familiarity and an assumption of bland uniformity, but actually we have, on our doorsteps, superb examples of the irrepressible nature of life. Fascinating but overlooked food webs surround us, and in every direction we would see the constant shimmer of the dynamic dance between colonisation, competition and collaboration between species if only we were prepared to invest in the core skill of naturalists – the ability and willingness to observe.

References and further reading

Aldous, D. and Chivers, I. (2002) *Sports Turf and Amenity Grasses: A Manual for Use and Identification*, Collingwood,Victoria: Landlinks Press.

Ash, H.J., Gemmell, R.P. and Bradshaw, A.D. (1991) 'The introduction of native plant species on industrial waste heaps: a test of immigration and other factors affecting primary succession', *Journal of Applied Ecology*, 31(1): 74–8.

Bradshaw, A.D. and Chadwick, M.J. (1992) *The Restoration of Land. The Ecology and Reclamation of Derelict and Degraded Land*, Cambridge: University of Cambridge Press.

Dixon, A.F.G. (1973) *Biology of Aphids*, London: Edward Arnold.

Emery, M. (1986) *Promoting Nature in Cities and Towns: A Practical Guide*, Chichester: Packard Publishing.

Eyre, M.D., Luff, M.L. and Woodward, J.C. (2002) 'Rare and notable Coleoptera from post-industrial and urban sites in England', *Coleopterist*, 11(3): 91–101.

Gemmell, R.P. and Connell, R.K. (1984) 'Conservation and creation of wildlife habitats on industrial lands in Greater Manchester', *Landscape Planning*, 11(3): 175–86.

Gilbert, O.L. (1976) 'A lichen-arthropod community', *Lichenologist*, 8(1): 96.

Gilbert, O.L. (1989) *The Ecology of Urban Habitats*, London: Chapman and Hall.

Rock, P. (2005) 'Urban gulls: problems and solutions', *British Birds*, 98(7): 338–55.

Thompson, K, Hodgson, J.G., Smith, R.M., Warren, P.H. and Gaston K.J. (2004) 'Urban domestic gardens (III): Composition and diversity of lawn floras', *Journal of Vegetation Science*, 15(3): 373–8.

Tucker, G., Ash, H. and Plant, C. (2005) *Review of the Coverage of Urban Habitats and Species Within the UK Biodiversity Action Plan. English Nature Research Report 651*, Peterborough: English Nature.

25

Urban contaminated land

Michael O. Rivett, Jonathan P. Sadler and Bob C. Barnes

Introduction

Urban contaminated land is an international environmental problem. Decades of industrial activity with little environmental consideration have led to significant legacies of urban chemical contamination. The EA (Environment Agency for England and Wales) estimates that approximately 2 per cent of England and Wales (300,000 hectares) was affected by industrial activity and potentially contaminated (EA 2009). Although significant capability now exists to manage the problem, contaminated land remains an expensive, technically challenging proposition. Contaminated land management has been driven by the protection of human health and water resources (Rivett *et al.* 2002). Although ecological concerns about its conservation value (Gibson 1998) have not necessarily been ignored, it is only in quite recent times that they have figured more prominently in legislation and published guidance, for example the European Water Framework Directive (2000/60/EEC) and the EA guidance *An Ecological Risk Assessment Framework for Contaminants in Soil* (EA 2008a).

This chapter reviews the nature of urban contaminated land relevant to urban ecology; how such land is assessed with specific emphasis on ecological risk assessment (ERA); and key issues relevant to the remediation of contaminated land and associated ecological habitat enhancement. The focus is upon the assessment of the ecological risk posed by contaminated land in that the ERA process fundamentally provides both a detailed descriptor of the specific nature of the contaminated land through the conceptual site model (CSM) as well as a quantified understanding of risks posed by the land and need for remedial actions necessary to develop ecological habitat. The chapter has a UK emphasis, but is internationally relevant, especially to areas that are undergoing rapid and recent industrialization, such as India and China. It is built around the aforementioned ERA framework proposed by the EA (EA 2008a), named 'ERA 1' by the EA and its supporting guidance, namely ERA 2a on CSMs (EA 2008b), ERA 2b on soil screening values (EA 2008c), ERA 2c on bioassays (EA 2008d), ERA 2d on ecological surveys (EA 2008e) and ERA 2e on attribution of cause and effect (EA 2008f). These works are an excellent example of current best practice that is legislatively driven and provide internationally relevant guidance on the assessment and management of ecological risks associated with contaminated land. The reader is duly referred to that extensive documentation of which the later sections of this chapter provide a precis.

The nature of urban contaminated land

The highly diverse nature of urban contaminated land and its ecological relevance are illustrated in Figure 25.1. The immense variety of industrial types ranges from small garage workshops to a range of light/heavy manufacturing industries, from secret weapons research establishments to pesticide/chemical manufacturing plants, from large coking works to small dry-cleaner laundries, from railway sidings to cemeteries. Although diverse, each may pose significant risk and may assume a 'contaminated land' label. No two sites are the same. Even similar industrial types may have different combinations of contamination hazards, different contaminant transport scenarios possible, different and novel juxtapositions of urban habitat – receptors at risk and different pollution, economic, technical, regulatory and political drivers influencing assessment and remediation. Still, there remains much commonality in the general nature and approach taken to contaminated land, even internationally.

Figure 25.1 Diverse examples of contaminated land: (a) tar wells excavated at a former gasworks; (b) LNAPL oil emerging from a bankside; (c) DNAPL coal-tar pool on the riverbed (pale blocks are house bricks); (d) colliery spoil heap with chemical disposal pit beneath the non-vegetated area; (e) erosion channel in a spoil-tip; (f) water course choked with run-off sediment; (g) leachate from demolition spoil on a former factory site; (h) nutrient poor capping waste on a former landfill site; (i) acid ballast from old railway tippings; (j) a classic brownfield (derelict) 'urban common' exhibiting great floral diversity.

Statutory meaning

Contaminated land is frequently used as a generic term, but may hold statutory meaning. For example, Part 2A (Section 78A) of the Environmental Protection Act 1990 introduced a statutory regime for the identification and control of contaminated land in England and Wales (EA 2009) and states that:

> 'Contaminated land' is any land which appears to the local authority in whose area it is situated to be in such a condition, by reason of substances in, on or under the land, that – significant harm is being caused or there is a significant possibility of such harm being caused; or pollution of controlled waters is being, or is likely to be, caused ... where 'harm' is defined as: harm to the health of living organisms or other interference with the ecological systems of which they form a part, and in the case of man includes harm to his property.

The definition's broadness and specific inclusion of ecological systems are noteworthy. However, Part 2A 'Ecological harm' is restricted to specified, key ecological, receptor locations, e.g. a site of special scientific interest (SSSI), a national nature reserve, a marine nature reserve. As such, it typically fails to include much urban ecological habitat. Other ecological regulatory drivers, however, also exist including the European Habitats Directive (92/43/EEC) and statutory land planning guidance that underpins much urban re-development. The UK's Planning Policy Statement (PPS) 23: Planning and Pollution Control that states: "*Land contamination, or the possibility of it, is a material planning consideration in the preparation of development plan documents and in taking decisions on individual planning applications*". The ERA framework developed for Part 2A can be appropriately used to assess the possible risks to nature conservation interests when potentially polluting activities are proposed, or the assessment and remediation of historic contamination (EA 2008a).

The definition above does nevertheless highlight a significant and perhaps counter intuitive issue. Certain contaminants on urban sites impact ecological systems by halting processes such as ecological succession. This creates a mosaic of 'early successional' habitats that are hotspots of biodiversity, especially for invertebrate species (Small *et al.* 2003; Eyre *et al.* 2003). As a result planning decisions need to take into account a broader range of ecological issues, not merely statutory requirements.

Historic influence

Urban contamination has often arisen from historic activity, possibly decades previous. Modern-day contaminant release is much reduced due to increased environmental awareness, operational controls and anticipated liabilities. Unfortunately detailed knowledge of past industrial activity often may be limited, good site records of waste disposal activities and releases prior to the 1970s being rare. Masses of chemical contaminant released to ground are rarely known, even in modern times, as are the precise release locations, their timings and durations. Identification of the polluter, or 'responsible party', is often highly contentious, particularly in historic cases.

Waste disposal practices performed in much of the twentieth century may now be banned, including disposal to landfill 'dumps' that allowed plumes to 'dilute and disperse' into the subsurface, pouring of waste volatile solvents on the ground; waste sludge applications for dust suppression and chemical wastes used for weed control. Although now banned, their legacy may remain for decades. Hydrophobic organic contaminants such as PCBs (polychlorinated biphenyls) used in electrical transformers; PAHs (polycyclic aromatic hydrocarbons) present in coal tars; toxic metals in sludges and PFA (pulverized fly/fuel ash) can be highly sorbing and remain in soils, land, dumps and surface-water sediments decades after their original deposition. Liquid organic chemicals often released to ground as light or dense oils, referred to as light/dense non-aqueous phase liquids (L/DNAPLs), have likewise proven to be highly persistent subsurface sources of contamination generating both vapour and groundwater plumes (Mackay and Cherry 1989).

Contaminant source zones

Identification of contaminant 'source zones' that act as on-going sources of contamination to the surrounding environment is critical in site assessment. Such zones often constitute the contamination 'hot spots' where significant remedial attention may be warranted via source-zone removal, reduction or containment to allow land development or re-establishment of habitat and protection of receptors. Source zones may be obvious, for example an existing leaky chemical/fuel storage tank/facility, a wastewater soakaway, a solid waste dumping ground. Alternatively, where the contributing activity was historic and associated infrastructure long gone, the source may be much less obvious, for instance, unrecorded chemical storage and waste-disposal areas, buried former fuel tanks or gasworks tar wells, unknown leakage/effluent points to ground, concealed illegal disposal activities, accidental, non-reported or forgotten spillages.

Site investigation with soil/water sampling for chemical analysis is the typical way such sources are identified. Obvious stresses to what ecological habitat is present may also reveal hot spot source areas. For example discrete areas of elevated phytotoxic heavy metals may manifest as areas of impoverished vegetation. Deeper rooting phreatic trees may become stressed by underlying contaminated groundwater.

Interaction with the wider environment

The label 'contaminated land' can be misleading and focus attention on shallow site contamination. This narrow interpretation is not implied in the Part 2A definition, nor within contaminated land investigations that assess interactions with the wider environment via risk assessment frameworks involving evaluation of various individual source – pathway – receptor 'pollutant linkages' and generation of a conceptual site model (CSM) (EA 2008b). This wider understanding of contaminated land impact critically underpins the effective assessment of risks posed to receptors. It is paramount to recognize that receptors, including ecological habitats, some distance from a contaminated site may be (or become) impacted by that site.

Contamination spreads into the wider environment in various ways, for instance, via windblown transport of particulate matter, volatilization of vapours to the atmosphere or soil gas, leaching (or dissolving) of contamination into water infiltrating through the source. Suspended particulate, vapour and dissolved-phase contaminant plumes above or below ground may threaten both immediate and remote receptors, possibly kilometres away. Volatile organic compound (VOC) plumes are widespread in urban groundwaters (Mackay and Cherry 1989). The atmospheric wet/dry deposition of contaminants from stack emissions has led to metal-rich 'acid rain' on areas downwind of poorly controlled metal refining/smelting operations. Subsurface migration of landfill-gas plumes may cause stressed overlying vegetation and dry/cracked soils.

Contaminant interactions with sediments in ecologically rich surface waters may be important. Direct effects of poorly sited chemical/waste use/disposal adjacent to surface waters may be visually obvious and ecologically damaging (De Jonge *et al.* 2008), but influences of contaminated groundwater baseflows comparatively unobtrusive yet significant. Many wetlands, lakes, rivers, riparian zones and associated habitats and ecologies are critically dependent on good quality groundwater that may be derogated by remote contaminated land. Some 15 per cent of the 191 chlorinated solvent groundwater plumes studied by McGuire *et al.* (2004) discharged to rivers. VOCs were prevalent in baseflow to the River Tame in Birmingham, UK (Ellis and Rivett 2007) and an additional factor contributing to the already poor ecological status of rivers in the city (Beavan *et al.* 2001).

Ecological Risk Assessment (ERA)

ERA is one component of a much broader contaminated land risk assessment process (EA 2004). Specific ecological guidance, such as that outlined in the EA's ERA framework (EA 2008a) to assess risks to ecology from contaminated land under Part 2A legislation, is a relatively new phenomenon. Similar ERA frameworks have recently been developed elsewhere, including Canada (Gaudet *et al.* 1995) and USA (US EPA 1997). The EA outlines a three-tiered ERA framework founded on a pollutant linkage-based CSM that explicitly considers: 'Tier 1' – a screening step comparing chemical analyses of site soils with a soil screening value (SSV); 'Tier 2' – uses a choice of ecological survey and biological testing tools to gather evidence for any harm to ecological receptors; 'Tier 3' – seeks to attribute the harm to the chemical contamination. This framework is set out in Figure 25.2 and seeks to establish whether pollutant linkages are likely to

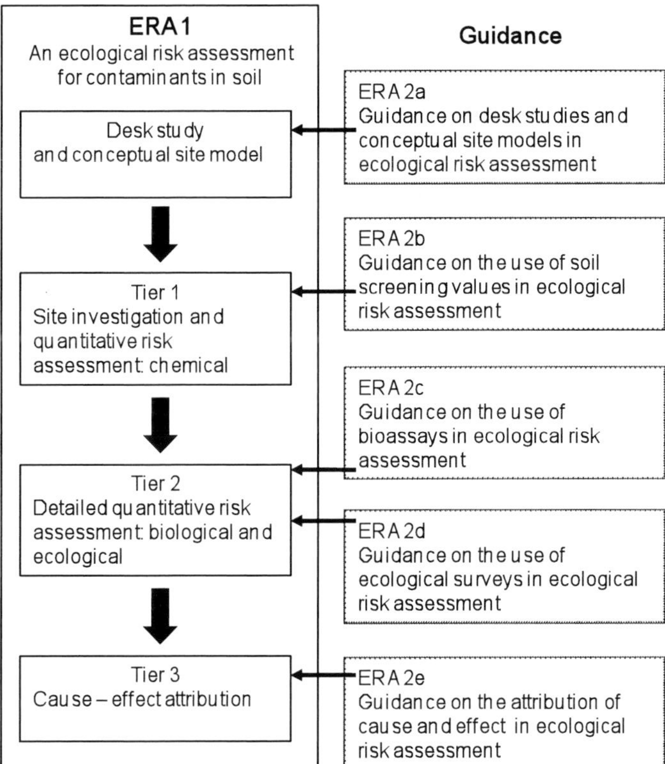

Figure 25.2 Schematic representation of the Environment Agency's Ecological Risk Assessment framework (EA 2008a).

exist between the contamination and the designated ecological receptors and to gather sufficient information for making decisions regarding whether harm to these receptors is occurring or could occur. Its best practice basis permits the ERA framework to be used in other contexts including conservation regulations, and planning and pollution control.

Development of the Conceptual Site Model (CSM)

Prior to embarking on the formal ecological risk assessment tiers, a CSM is developed for the contaminated land site that is based on existing data to postulate the source – pathway – receptor pollutant linkages appropriate for the site and to establish if any of these have reasonable potential to cause harm to ecological receptors (EA 2008b). If these linkages are deemed to be possible, then the ERA is progressed to Tier 1.

Tier 1: soil screening values

Tier 1 involves intrusive site investigation with soil (and water) sampling for chemical analysis that is targeted in the area that supports the ecological receptors. It seeks to establish whether sufficient contamination is present to pose a risk (cause harm) to ecological receptors by comparing sampled concentrations with Soil Screening Values (SSVs) (EA 2008c). The latter have been developed from a suite of ecotoxicological tests using soil organisms for a range of contaminants to generate a 'Predicted No Effect Concentration for soil' ($PNEC_{soil}$). For reliable assessment of effects on soil organisms such toxicity test data need to represent primary producers (plants), consumers (e.g. invertebrates) and decomposers (e.g. micro-organisms). Unfortunately toxicity data for soil compartments are sparse and deriving PNECs challenging. EA (2008c) indicates four scenarios, with the first two deemed insufficient for inclusion in the ERA:

- When no soil organism toxicity data are available, the equilibrium partitioning method based on aquatic toxicity data can be applied;
- When only one test result with soil-dwelling organisms is available, a large assessment factor (AF) is usually applied, as the degree of protection is not fully characterized;
- When toxicity data are available for a producer, a consumer and/or a decomposer, the PNEC is calculated using an appropriately low AF (<50);
- When sufficient data are available, statistical extrapolation techniques, or Species Sensitivity Distributions (SSDs) are used with a robust PNEC derived with an AF <2.

The ERA list of SSVs (EA 2008c) is hence brief, comprising (mg/kg): benzo(a)pyrene 0.15; cadmium 1.15 (0.09); chromium 21.1; copper 88.4 (57.8); lead 167.9; mercury 0.06; nickel 25.1 (20.3); pentachlorobenzene 0.029; pentachlorophenol 0.6; tetrachloroethene 0.01; toluene 0.3; zinc 90.1(72.5). Bracketed values indicate SSVs dependent on soil conditions (e.g. pH, organic matter content) or for cadmium, a secondary poisoning value allowing for potential bioaccumulation/magnification in higher organisms. In the absence of specific contaminant data, there is encouragement, to 'borrow' values from other international sources. Ecological Soil Screening Levels (Eco-SSLs; www.epa.gov/ecotox/ecossl) developed by the US EPA are intentionally conservative and protective of ecological receptors commonly in contact with soil, or ingested by soil biota. Eco-SSLs are based on the geometric mean of selected chronic toxicity data for plants, invertebrates, mammals and birds. RIVM (National Institute of Public Health and the Environment), Netherlands, has developed Serious Risk Concentrations for ecosystems (SRCeco) (www.rivm.nl/bibliotheek/rapporten/711701023.pdf).

SSVs provide an early indication of potential risk, but should not be used to determine whether there are ecological risks from soil contamination. They are set conservatively to avoid assessing a site as safe when there are risks. Assessing a site as posing risks when it is in fact safe will be more common meaning that decisions just based on SSV exceedance may lead to unnecessary remediation. Rather, exceedance with feasible pollutant linkages dictates progression to a higher tier.

Tier 2: ecological surveys

Tier 2 ecological surveys and/or biological testing (bioassays) are used to establish whether identified ecological receptors are subject to significant harm or the significant possibility of significant harm (EA 2008a). Part 2A descriptions that apply to protected locations are:

> Harm which results in an irreversible adverse change, or in some other substantial adverse change, in the functioning of the ecological system within any substantial part of that location;
>
> Harm which affects any species of special interest within that location and which endangers the long-term maintenance of the population of the species at that location;

and

> in the case of a protected location which is a European Site (or a candidate Special Area of Conservation or a potential Special Protection Area), harm which is incompatible with the favourable conservation status of natural habitats at that location or species typically found there.

So-called 'ecosurveys' aim to gather spatial and temporal ecological data on habitat presence and status and species occurrence. They are required to identify whether there is any evidence of significant harm occurring now or potentially in the future. Activities include surveys of population size, density and age structure as well as composition, i.e. particular species presence, overall diversity. Hill *et al.* (2005) and Sutherland (1996) give detailed methodologies and with EA (2008e) provide an overview relevant to contaminated land and associated case studies on an industrial site colonized by nesting plovers with PAH/PCBs present and a disused metal smelter site with song-thrush receptors present. These studies illustrate the great effort required to establish whether ecological harm is occurring and significant.

Many challenges arise. Ecosystem functions are variable and may relate to many factors that are independent of contamination levels, but difficult to isolate from chemical impacts. The ecosurvey search radius is influenced by both contaminant movement away from a site and the potential movements of ecological receptors. Some bat species may forage over 30 km and be affected by distant contamination, whereas a great crested newt may only be impacted if present within a site or within 500 m of a contaminated site. Typically a radius is set 1–2 km from the study site boundary, however, the above warrants some searches be much extended.

A weight of evidence for, or against risk of significant harm occurring is developed by: considering the potential impacts on 'favourable conservation status', 'favourable condition status' and integrity of the receptor sites; comparison with historical datasets or 'baseline' uncontaminated sites to assess adverse change; and a robust statistical treatment of the data. Impacts are assessed with reference to whether they are beneficial, or adverse and their magnitude, extent, duration, reversibility, timing and frequency (EA 2008e).

Tier 2: bioassays

Bioassay biological testing measures the toxicity of a contaminant/sample via exposure of a specific organism and measuring a life cycle parameter. Controlled laboratory conditions avoid confounding environmental factors with a null hypothesis test approach. If significant differences are observed (typically 95 per cent confidence), the 'effect' is deemed worthy of further consideration. Bioassays may hence help evaluate whether soil contamination may cause harm to identified organisms or 'Receptors of Potential Concern' (EA 2008d). Bioassays used in the EA's ERA are: solid-phase Microtox®; bait lamina; nitrogen mineralization; earthworm survival and reproduction test; earthworm lysosomal stability test; collembolan survival and reproduction test; plant seedling emergence and vegetative vigour test. Table 25.1 describes strengths and weaknesses of three test methods. Selection of bioassays appropriate to a relevant pollutant linkage is critical; however, there is no single, perfect bioassay suitable for all cases, rather, each test contributes to the weight of evidence approach.

Tier 3: cause-effect attribution

Tier 3 determines whether observed impacts can be attributed to the contaminants via significant correlations and/or measured chemical gradients, or other indicative patterns of relationship (EA 2008f). All data are assessed and links between Tier 1 contamination and Tier 2 harm to receptors are considered. The critical question is does the weight of evidence point to the ecological impacts being caused by the contaminant source and associated pollutant linkage? The design of the Tier 2 ecological survey is of paramount importance in urban settings due to the wide range of contamination sources making attribution of cause and effect especially problematic. A balance of carefully designed field sampling, supplemented with laboratory studies is preferred

Table 25.1 Example bioassays (EA 2008d)

Method	*Strengths and weaknesses*
Microtox®: Commercial test based on the bioluminescent bacterium, *Vibrio fischeri*, which produces light as a respiration by-product that is compared to control bacteria. The degree of respiration suppression is a surrogate expression of contaminant effects.	*Strengths:* Fast; sensitive to a range of contaminants; small sample size; standardized. *Weaknesses:* unrepresentative (marine bacterium); sensitivity variation; sample treatment.
Bait lamina: Test assesses the feeding activity of soil organisms by measuring breakdown of organic material set into plastic strips inserted into the soil.	*Strengths:* In situ; process focused; ecologically relevant. *Weaknesses:* Low interpretive power; physicochemical parameter sensitivity; spatially limited; unsuitable for all soils.
Plant seedling emergence: Plant tests to assess soil quality based on seedling emergence, vegetative vigour and growth.	*Strengths:* Sensitive to a range of contaminants; indicative of higher level effects; standardized. *Weaknesses:* Physico-chemical parameter sensitivity; plant-specific differences.

(Underwood 1996). If the evidence suggests that impacts are occurring then appropriate risk management and remedial action need to be implemented to achieve adequate risk reduction. It is important to re-examine the extent to which non-chemical hazards may have caused observed harm to receptors, perhaps through disease, human disturbance, competition from invasive species and longer term environmental variables (EA 2008f). This is challenging, but necessary to avoid expensive and inappropriate remedial works.

Case studies

Case studies that carefully document the ERA process at contaminated land sites support both wider practitioner practice and the underpinning science. The reader is referred to case studies in EA (2008a) and US EPA (2009). A case study condensed from EA (2008a) is provided in Box 25.1.

Box 25.1 Ecological Risk Assessment (ERA) of a gasworks site (adapted from EA 2008a)

Conceptual site model (CSM): The case concerns a former gasworks adjacent to a Local Nature Reserve (LNR) designated (and surveyed) in 1964 due to its continuous woodland supporting songbird species (firecrests) and dormice. 'Contaminants of Potential Concern' (CoPCs) included aromatic hydrocarbons, phenols, cyanides and metals. The ERA progressed to Tier 1 as several pollutant linkages were possible in the CSM, including direct contaminant uptake by plants; direct contact/incidental contaminant ingestion by wildlife; ingestion of contaminants contained in plants.

Tier 1: Intrusive site investigation was undertaken that demonstrated contamination in the LNR 'Zone 1' closest to the gasworks with made ground evident supporting anecdotal evidence that wastes had been deposited within the LNR during gas works decommissioning (late 1960s). The ERA progressed to Tier 2 as CoPC concentrations exceeded SSVs, or CoPCs did not have a proposed SSV, or other agreed substitute.

Tier 2: Ecosurveys were undertaken to assess woodland plant communities, invertebrate populations, bird populations and small mammal populations. Deterioration was shown in the Zone 1 plant community since the 1964 survey with habitat less suitable for dormice and birds (foraging and nesting). Bioassays used as relevant measurement endpoints, included plant (cabbage, tomato, wheat) tests, collembola and earthworm reproduction and nitrogen mineralization. Comparison of samples from Zone1 with contaminant-free areas indicated significant adverse effects relating to plant growth, nitrogen mineralization and collembolan reproduction. Although bioassays and ecosurveys did not indicate contaminants were having a direct toxic effect on species of special interest, the food resources on which they depend, plants and seeds (dormice) and insects (firecrests) were likely impacted. Assessment progressed to Tier 3.

Tier 3: Evidence was reviewed and it was considered whether the contamination was responsible for the adverse effects observed in the measurement endpoints that inferred harm to the ecological receptors. A Hill's Causal Criteria approach (EA 2008f) identified that the criteria for strength of association, ecological gradient, plausibility, coherence, temporality and specificity were satisfied. It was concluded that the adverse effects observed during the ecosurveys and bioassays were caused by the presence of the contamination.

Remediation

Remediation (including site assessment) of urban contaminated land can be incredibly expensive. Costs for relatively small sites are in the \$50–500,000 range, large sites \$1–100 million and megasites >\$100 million. England's Department of the Environment, Food and Rural Affairs (Defra) recently estimated average remedial costs at £0.5 million per hectare (Defra 2007). High land values may, however, absorb such costs in development projects. Remediation may be required to mitigate against a combination of risks posed to human health, water resources, infrastructure integrity and ecological habitats. The latter may be a minor or major remedial programme component. A general goal of many remedial actions or broader urban regenerations, fortunately, is to improve ecological habitat regardless of there being specific ecological risks originally posed. This is negotiated, however, on a site by site basis by site developers and city planning departments. This may lead to inconsistencies locally and regionally and planning requirements may be overturned on appeal, especially where developments involve large sums of money and large tracts of land.

Complete site restoration to the natural pre-industrial baseline is not a viable option, both from technical and economic standpoints – some contamination will always remain. Effective risk management necessitates remedial actions are targeted to pollutant linkages that pose unacceptable threats. Action may be targeted at any of the source, pathway or receptor to moderate the risk to an acceptable level and 'break' the linkage. Solutions are diverse and range from contaminant source removal to landfill, coverage by a car park, or perhaps a groundwater pump-and-treat scheme to intercept contamination, or alternatively to even relocate the receptor – housing may be purchased and people at risk relocated, or water courses diverted around problem areas with relocation of associated habitat.

Development and enhancement of ecological habitat associated with urban contaminated land is non-trivial. Whilst engineering-based objectives of a programme are possibly rapidly realized, establishment of mature ecological habitats generally requires many years and needs to be planned carefully from the onset of the development. Relationships between contaminants and ecological receptors are rarely straightforward. Some land deemed as contaminated from one perspective, e.g. underlying groundwater pollution risk, may actually contain contamination that is beneficial to certain ecological assemblages or species of special interest. In surface waters contaminants can persist in sediments long after the initial aqueous contaminants have been removed from the water (Hutchinson and Rothwell 2008). Poorly conceived remedial works failing to recognize relationships such as this could do more ecological harm than good.

Careful remediation planning is required to understand and effectively manage the impacts of remedial works that may involve much heavy plant machinery and the bringing to surface of severe contamination perhaps concealed for decades. Unless carefully managed, the active remediation phase may pose significant risk to surrounding ecological habitats via windblown particulate matter, vapours, disturbance of contaminated surface-water sediments, or the foraging of wildlife on exposed contamination. There has been significant development of more sustainable '*in-situ* remediation' technologies in recent decades where contamination is managed *in situ* reducing such risks. *In-situ* techniques include bioremediation, soil stabilization, phytoremediation, permeable reactive barriers, thermal methods and chemical treatments (Genske 2003). Whilst some methods may have detrimental impacts on soil-based ecologies in the treatment areas during the active phase, other more 'green technologies' may contribute to the establishment of ecological habitat, for example the phytoremediation of soils and groundwater contamination by willow and poplar species or the use of sustainable urban drainage (SUDs) technologies to intercept run-off toxins.

Ecological habitat development objectives have to nestle alongside other objectives for the urban land including industrial/commercial land requirements and housing, infrastructure and leisure developments. Habitat development may be moulded by such and by the original nature and distribution of habitat and contamination present. Some areas of a site will more easily lend themselves to habitat re-establishment than others and attract wider support from other interested parties. Whilst proven unacceptable risks to ecological habitats shown through the above ERA process will have to be dealt with, just how they are dealt with may be open to significant and potentially prolonged debate with perhaps no easy solutions. Without a robust underpinning ERA process, ecological habitat development needs may be potentially 'squeezed out' compared to more apparently profit-making parts of a development. Greater opportunity is perhaps afforded for urban ecology where there is more widespread urban regeneration works as opposed to rather piecemeal, individual site works. This may be afforded by the redevelopment of large industrial sites, or else local authority lead programmes to redevelop a particular urban district, e.g. Salford Quays, Manchester, and Paddington Basin, London (Raco *et al.* 2007). Establishment of greater urban ecological habitat coordinated and interlinked with, for instance, the establishment of urban habitat corridors is hence permitted (Angold *et al.* 2006).

Conclusions

The need to develop urban ecological habitats associated with contaminated land is widely appreciated, but the challenges are significant. Although ecological habitat figures more prominently within contaminated land programmes and relevant legislation and ecological risk assessment frameworks are in place, it is just one part of a complex urban problem. Ecological considerations are many and varied. They may prove to be time consuming, awkward and expensive with consequences of remedial actions often not apparent until years after those works. Much research remains to further underpin site practice, including measurement of ecotoxicological data to support regulatory criteria (standards), development of more diverse and rapid bioassays, greater documentation of robust case studies, and improvement in modelling tools to better predict risks and remedial actions required.

References

Angold, P.G., Sadler, J.P.,Hill, M.O., Pullin, A., Rushton, S., Austin, K., Small, E., Wood, B., Wadsworth, R., Sanderson, R. and Thompson, K. (2006) 'Biodiversity in urban habitat patches', *Science of the Total Environment*, 360(1–3): 196–204.

Beavan, L., Sadler, J. and Pinder, C. (2001) 'The invertebrate fauna of a physically modified urban river', *Hydrobiologia*, 445(1–3): 97–108.

Defra (Department for Environment, Food and Rural Affairs) (2007) *Soil Framework Directive: Initial Regulatory Impact Assessment*, London: Defra.

De Jonge, M., Van der Vijver, B., Blust, R. and Verbotes, L. (2008) 'Responses of aquatic organisms to metal pollution in a lowland river in Flanders: A comparison of diatoms and macroinvertebrates', *Science of the Total Environment*, 407(1): 615–29.

EA (Environment Agency) (2004) *Model Procedures for the Management of Land Contamination: Contaminated Land Report (CLR) 11*, Bristol: Environment Agency.

EA (Environment Agency) (2008a) *An Ecological Risk Assessment Framework for Contaminants in Soil: Environment Agency Science Report SC070009/SR1*, Bristol: Environment Agency.

EA (Environment Agency) (2008b) *Guidance on Desk Studies and Conceptual Site Models in Ecological Risk Assessment: Environment Agency Science Report SC070009/SR2a*, Bristol: Environment Agency.

EA (Environment Agency) (2008c) *Guidance on the use of soil screening values in ecological risk assessment, Environment Agency Science Report: SC070009/SR2b*, Bristol: Environment Agency.

EA (Environment Agency) (2008d) *Guidance on the Use of Bioassays in Ecological Risk Assessment: Environment Agency Science Report SC070009/SR2c*, Bristol: Environment Agency.

EA (Environment Agency) (2008e) *Guidance on the Use of Ecological Surveys in Ecological Risk Assessment: Environment Agency Science Report SC070009/SR2d*, Bristol: Environment Agency.

EA (Environment Agency) (2008f) *Guidance on the Attribution of Cause and Effect in Ecological Risk Assessment: Environment Agency Science Report SC070009/SR2e*, Bristol: Environment Agency.

EA (Environment Agency) (2009) *Dealing with Contaminated Land in England and Wales*, Bristol: Environment Agency.

Ellis, P.A. and Rivett, M.O. (2007) 'Assessing the impact of VOC-contaminated groundwater on surface-water at the city scale', *Journal of Contaminant Hydrology*, 91(1–2): 107–27.

Eyre, M.D., Luff, M.L. and Woodward, J.C. (2003) 'Beetles (Coleoptera) on brownfield sites in England: An important conservation resource?', *Journal of Insect Conservation*, 7(4): 223–31.

Gaudet C.L., Power E.A., Milne D.A., Nason T.G.E. and Wong M.P. (1995) 'A framework for ecological risk assessment at contaminated sites in Canada', *Human and Ecological Risk Assessment*, 1(2): 43–115.

Genske, D.D. (2003) *Urban Land: Degradation, Investigation, Remediation*, Berlin/ Heidelberg/New York: Springer.

Gibson, C.W.D. (1998) *Brownfield: Red Data, The Values Artificial Habitats Have for Uncommon Invertebrates*, Peterborough: English Nature.

Hill, D., Fasham, M., Tucker, G., Shewry, M. and Shaw, P. (2005) *Handbook of Biodiversity Methods: Survey, Evaluation and Monitoring*, Cambridge: Cambridge University Press.

Hutchinson, S.M. and Rothwell, J.J. (2008) 'Mobilisation of sediment-associated metals from historical Pb working sites on the River Sheaf, Sheffield, UK', *Environmental Pollution*. 155(1): 61–71.

Mackay, D.M.and Cherry, J.A. (1989) 'Groundwater contamination: limits of pump-and-treat remediation', *Environmental Science and Technology*, 23(6): 630–6.

McGuire, T.M., Newell, T.J., Looney B.B., Vangelas, K.M. and Sink, S.H. (2004) 'Historical analysis of monitored natural attenuation: A survey of 191 chlorinated solvent sites and 45 solvent plumes', *Remediation*, 15(1): 99–112.

Raco, S., Henderson, S. and Bowlby, S. (2007) 'Delivering brownfield regeneration: Sustainable community-building in London and Manchester', in T. Dixon, M. Raco, P. Catney and D.N. Lerner (eds), *Sustainable Brownfield Regeneration*, Oxford: Wiley – Blackwell.

Rivett, M.O., Petts, J., Butler, B. and Martin, I. (2002) 'Remediation of contaminated land and groundwater: Experience in England and Wales', *Journal of Environmental Management*, 65(3): 251–68.

Small, E.C., Sadler, J.P. and Telfer, M.G. (2003) 'Carabid beetle assemblages on urban derelict sites in Birmingham, UK', *Journal of Insect Conservation*, 6(4): 233–46.

Sutherland, W.J. (1996) *Ecological Census Techniques: A Handbook*, Cambridge: Cambridge University Press.

Underwood, A.J. (1996) 'Spatial and temporal problems', in G.E. Petts and P. Callow (eds), *River Restoration*, Oxford: Blackwell Science, pp. 182–204.

US EPA (1997) *Ecological Risk Assessment Guidance for Superfund: Process for Designing and Conducting Ecological Risk Assessments – Interim Final, June 1997* (OSWER Publication Number 9285.7–25; NTIS Order Number PB97–963211), Washington, DC: US Environmental Protection Agency.

US EPA (2009) *Case Studies for Selected Region 5 Superfund Sites*, online, available at: www.epa.gov/region5superfund/ecology/html/casestudiestoc.html.

26

Urban woodlands as distinctive and threatened nature-in-city patches

C.Y. Jim

Introduction

In urban greening, larger and more permanent plants such as trees and shrubs are emphasized. Urban ecological studies tend to focus on woody vegetation (Kunick 1987; Tregay 1979). Urban woodlands refer to the most complex arboreal vegetation in urban areas dominated by the tree growth form in close proximity to each other, with the soil and field-layer vegetation not influenced by active management. They are differentiated from anthropogenic sites with trees in urban parks which receive rather intensive and regular horticultural treatments. The connotation usually conjures up rather natural vegetated areas developed with little human interference. The signature characters to laypersons are the generally wild and non-manicured landscape, and the elaborate combination of plants of different forms, sizes and species to generate a feeling of naturalness.

To the experts and professionals, urban woodlands are subsumed under a tree-dominated biotope situated in urban areas. They are composed of an assemblage of plants of varied species, size, density, growth form and structural complexity, in association with the faunal and abiotic components. The commonality is manifested by trees that constitute the overarching members and biomass framework to impart the woodland ambience. Vegetation with such inherent traits is commonly found in wilderness and rural areas. The urban woodland biotope owes its significance to its unusual occurrence in cities. The contradistinction and incongruity between nature and culture, embedded within the built-up matrix, bestow a special status on urban woodlands.

Few urban woodlands are truly pristine. Many originated from natural forests and were subject to decades if not centuries of continual human interventions. Forests situated near villages were often managed for productive purposes when the sites were still rural. Timber, poles, firewood, bark (for tannin), fruits, herbal medicine, mushrooms and game were commonly tapped under the traditional rural economy. Trees were often selectively removed with or without planting of new trees to serve productive or utilitarian purposes. Once urbanized, which means inclusion into the city's ambit, such resource-tapping activities would usually dwindle or disappear. Instead, recreational incursion imposes a different suite of changes. The human biophilia instinct (Wilson 1984) would gravitate towards the green enclaves. Whether in the rural or urban mode, such human interference could gradually modify the habitat conditions and constituent flora and fauna (Sukopp 2004). Some urban woodlands are unrelated to former natural forests. They are entirely planted, or are developed spontaneously on abandoned green fields or brown fields (Kowarik 2005).

Some cities are fortunate to have been bequeathed by design or default fine and sizeable urban woodlands. The outstanding examples, mainly in large metropolises, include:

1 Paris: Bois de Boulogne (Figure 26.1) and Bois de Vincennes;
2 Berlin: Grunewald;
3 Brussels: Bois de la Cambre;
4 Vienna: Wienerwald;
5 London: Holland Park (Figure 26.2), Hampstead Heath, Richmond Park and Wimbledon Common;
6 Dublin: some parts of Phoenix Park (Figure 26.3);
7 Tokyo: Yoyogi Park around Meiji Jingu (Figure 26.4; Ono 2005);
8 Nanjing: Zijin Mountain (Jim and Chen 2003);
9 Guangzhou: Baiyun Mountain (Chen and Jim 2006); and
10 Warrington New Town (UK): Birchwood, with intimate integration of woodland and residential development providing a special case of created woodland on a sprawling abandoned industrial site (Jorgensen *et al.* 2005).

The fortuitous combination of vision, political will, enlightened leadership, benign neglect or a dosage of luck, has created the infrequent circumstances for their preservation. It is indeed a remarkable achievement to have escaped from the relentless and destructive march of urbanization. They are truly the pride of the community and excellent object lessons for developing cities.

Figure 26.1 The Bois de Boulogne offers a sprawling urban natural area in the western fringe of Paris.

Figure 26.2 The small remnant urban woodland inside Holland Park is encapsulated in the core of London.

Figure 26.3 Some parts of the huge Phoenix Park in Dublin are managed as urban woodlands.

Figure 26.4 The Yoyogi Park is a 70 ha urban woodland planted in the early twentieth century around the Meiji Jingu in the heart of Tokyo.

With half of humanity dwelling in cities, reaching 70 percent in Western Europe, urban residents are increasingly detached from nature (Berry 2008). Urban green spaces, often over-designed and manicured, present a low-caliber surrogate and emulation of nature (Jim 2002; Jim and Chen 2008a). With increasing awareness of the important physical and psychological benefits and ecosystem services of nature in cities (Chen and Jim 2008; Jim and Chen 2008b), the ecological or naturalistic approach has been advocated as an alternative for urban green areas. Nature in cities could be inherited or created. Existing pockets of nature are increasingly protected in the course of urban expansion (Tyrväinen *et al.* 2005).

Urban woodlands offer the best natural ingredients to deserve conservation. Often harboring rare or endangered species and communities, they contribute to conservation of biodiversity and special biotopes (Kowarik 1990; Sachse *et al.* 1990). Some ancient woodlands with several hundred years of tenure carry both ecological and cultural heritage value (Spencer and Kirby 1992; Thomas *et al.* 1997). Sacred woods associated with old villages, temples or shrines (Mishra *et al.* 2004; Chandrakanth *et al.* 2004) are occasionally inherited by cities. Their increasing recreational use conflicts with conservation objectives, calling for innovative site and visitor management (Jim 1989a). The deleterious urban effects, ranging from modification to annihilation, need to be ameliorated.

This paper explores the concept of urban woodlands by surveying comprehensively their distinctive and diagnostic traits. They are evaluated with reference to the convergence and divergence of abiotic, biotic, and geometric characteristics. The role of human mediation and intervention is considered as the almost ubiquitous factor of woodland changes. The management of valuable urban nature could be informed by knowledge of their ecological intricacies and human-nature interactions.

Diagnostic abiotic traits

Location and area

Due to the reality of urban land economics, urban woodlands can hardly compete with the high bids of built-up uses, hence they are seldom preserved in city centers. Small sites cover only a few hectares, whereas large sites such as the Bois de Boulogne or the Grunewald cover hundreds of hectares. Some small sites are associated with religious establishments or traditional beliefs. Most large sites are located in the urban fringe, close enough to the population concentration and yet somewhat detached from the stifling city core. A notable exception is the relatively extensive 70 ha Yoyogi in the heart of Tokyo, presenting a vivid contrast to the compact built-up environs (Figure 26.4). It represents the world's largest city-center urban woodland, planted entirely in the early twentieth century to commemorate Emperor Meiji's reign. Sites separated from built-up areas by rural lands should be subsumed under rural or countryside woodlands. Despite inclusion within the city boundary, their detachment from urban influence disqualifies them as urban woodlands.

A series of urban woodlands in terms of location may be found in some cities in a spatial sequence: countryside, rural-suburban, urban fringe, and urban core. The individual sites are interspersed in the developed land matrix. Despite the apparent increase in development intensity and hence urban effect along the apparent chain, the gradation may not represent a discernible gradient of ecological changes in flora, fauna, soil or human disturbance regime (Porter *et al.* 2001; Hedblom and Söderström 2008). Also, it may not denote a progression in terms of woodland succession or other vegetation features.

Topography

Rugged terrain with steep or unstable slopes unsuitable for development, such as hills, ridges, cliffs, escarpments, or ravines, is sometimes encountered in urban expansion. Mining and quarrying could degrade the natural topography by cutting and filling. Water bodies such as streams, ponds, lakes and reservoirs and their associated terrestrial fringes are commonly excluded from development. Their high-quality and attractive natural scenery and setting could be actively enlisted as landscape-recreational resources for the host city. The physical obstacles may be overcome at a cost in terms of money or time. More often, they are avoided to leave perforations or gaps in the urban matrix, providing pleasant pockets of inherited nature to punctuate the cityscape.

Hillocks are often left occluded within cities to provide remnant habitats and persist as isolated islets of nature. Association with religious entities such as temples or shrines enhances the chance of preservation. Sometimes, such sites are put aside as land reserve with low development priority. Such serendipitous existence of nature could be construed as property-development opportunities. Their green and natural qualities command a premium for high-grade residential projects. With changing circumstances, they may become cost-effective for urbanization in the future. It is necessary to guard the precious natural islands from undue infilling pressure.

Flat and filled urban sites are seldom devoted to woodland preservation or planting. The abandonment of urban-industrial sites in some western cities, however, could provide opportunities for woodland development on flat surfaces. Urban expansion could identify relatively flat lands with mature forest for inclusion in the new city areas (Coughlin *et al.* 1988).

Soil profile and properties

The undisturbed woodlands inherited from natural forests tend to keep a natural and mature soil profile, with the least sealed surface amongst different urban land cover types (Pauleit and

Duhme 2000). In a given climatic zone, the woodland presents the zonal soil type in terms of profile development and capability to support primary production. The original woodland soil is endowed with actively recycled and rich organic matter and nutrient contents, strong and stable structure, and a balanced distribution of porosity to facilitate infiltration, drainage, aeration, and storage of plant-available moisture (Jim 2003a). The site's history, especially regarding human use and management regime, would impose pedological changes. Even if the impacts have stopped, the residual effects will linger because soil formation and recovery are sluggish in comparison with vegetation changes (Jim 2003b; Jim and Chan 2004).

The surface litter layer (O horizon) is the most sensitive to disturbance, followed by the underlying mineral-organic topsoil (A horizon). Trampling could comminute the organic litter which is susceptible to loss by accelerated decomposition and erosion. O-horizon degradation would reduce the cushioning and protective function of litter on the A horizon. It also trims the supply of organic matter and nutrient in humus form to the A horizon. Recreational trampling could compact and seal the exposed soil, and induce soil structural damage and erosion by wind and running water (Jim 1987). A trampled and compacted soil could not sustain a healthy population of soil organisms including the decomposers. Overall, it becomes a poor medium for plant growth (Jim 1993).

If the antecedent land use of the woodland site was urban-industrial, the human impacts on the soil are more drastic and protracted (Weiss *et al.* 2005). The original soil could be truncated, removed, mixed with imported fills or buried in the course of site formation and construction. The soil surface is often paved or sealed by impermeable materials such as concrete or asphalt. Infiltration of precipitation water into the soil is drastically curtailed. Pollutants generated by industry and mining could contaminate the soil (see also Chapter 15 this volume). Demolition or disintegration of old buildings could accelerate pollutant release. The dissolution and dispersion of calcareous and alkaline construction rubbles could raise soil pH considerably. The large rubble fragments reduce the capability of the soil to store water and nutrients, and restrict root growth (Jim 1998a, 1998b). In some mining sites, the spoil heaps and tailings could be seriously contaminated with heavy metals, acids or acid precursors.

Not all human impacts are negative. At some localized urban-industrial spots, the soil could have been improved by beneficial management, including enrichment by organic matter, fertilizers, and other amendments. Physical manipulation by plowing could improve soil structure as well as water-and nutrient-holding capacity.

After abandonment of urban-industrial sites, the disturbed soil could evolve without human intervention. Such soil changes would be slow, and spontaneous vegetation succession would be retarded. For relatively sterile sites, it could mean prolonging the pioneer stage dominated by herbaceous species adopting the r-strategy (a strategy often adopted by pioneer herbaceous species in unstable habitats to reproduce quickly by delivering an abundant quantity of small seeds that can disperse widely, and completing the life cycle in a short generation time). The progression to advanced seral communities composed of equilibrium woodland species is literally arrested until the substrate has been adequately improved. An intervention approach is adopted in some sites to improve the soil, or replacing it with a prepared mix to facilitate woody plant growth.

Hydrology and microclimate

A small urban woodland has a limited effect on overall urban hydrology. An aggregate urban woodland coverage exceeding 25 percent of the city area has some significant effects. It ameliorates some ill effects of impermeable urban surfaces on water flows and pathways in the city. It increases canopy interception, temporary storage on soil surface, infiltration, percolation, and groundwater recharge. By reducing surface runoff, it lowers the occurrence of urban floods (see also Chapter 13 this volume). By increasing the evapotranspiration component from the vegetation and the soil, it cools the air by latent heat extraction. Overall, in terms of the hydrograph response, it helps to delay and suppress the peak storm runoff, and extend the discharge duration.

A woodland literally creates its own microclimate. The shading effect of the canopy reduces solar penetration to the woodland floor, creating a dim light regime that is exploited by shade-tolerant species (skiophytes) with low compensation point and high photosynthetic efficiency. Shading plus the strong passive evaporative cooling lower the intra-woodland air temperature. Compared to bare sites, a significant proportion of the incident solar radiation is consumed by latent heat absorption, and much less is converted into longwave convective sensible heat. Biomass and soil temperature are correspondingly depressed.

Living plant tissues, including the foliage, branches and trunks, contain a notable quantity of liquid water. The distribution of biomass water at different heights enhances the woodland's thermal capacity. More heat energy is needed to warm up the tissues, and more heat energy is stored to delay the cooling down process. The net result is raising the minimum temperature and lowering the maximum to contract the diurnal range. The higher moisture content in the subcanopy air and soil supplements the temperature dampening effect. The relatively less mobile subcanopy air, trapped in the dense foliage-cum-branches of the main canopy stratum and the undergrowth, serves as a thermal insulation blanket to keep the warm air within the woodland.

The center of a dense woodland with a closed canopy of interlocking tree crowns experiences the maximum microclimatic effect. The periphery interfaces with the external climate has a subdued edge effect. Woodlands that are large with dense tree cover generate significant microclimatic benefits. In downwind areas, the cooling effect could spillover to benefit neighboring residential areas. It could also flush out the warmer, drier, and more polluted air in the adjacent developed areas with cool, fresh, clean, and aromatic air. A woodland situated near urban areas could serve as an airshed for the sustainable supply of health-enhancing air. The biogenic aerosols generated by a woodland, however, could induce allergic reactions in the skin and respiratory system of vulnerable victims.

Diagnostic biotic traits

Biomass structure and floristic composition

The biomass structure of urban woodlands is contingent upon species composition, tree age, and disturbance regime. The mature woodland with minimal urban effect offers the tallest, densest, and most complex structure analogous to the zonal climax forest. It is characterized by a continuous main canopy composed of tall trees with interlocking crowns. In the vertical dimension, the biomass is stratified into the ground surface, undergrowth and main tree layers. An even-age structure is expected of a stable and healthy woodland, with representatives in different age groups that include a good cohort of regeneration stems. Disturbed natural or young plantation woodlands have a different degree of departure from the maximum structural state.

Floristic composition is a fundamental woodland trait that has been intensively studied with reference to urban effect and conservation assessment. Besides individual species, some studies explore the affinity between members to identify the occurrence of natural assemblages in the form of biotic communities (Bertin *et al.* 2005; Jim and Chen 2008a). Species diversity is a cardinal concern, which is more informative if it is compared with the mature natural woodland of the climatic zone. Less disturbed woodlands register a high similarity index vis-à-vis natural counterparts. Human impacts tend to suppress species richness and Simpson and Shannon-Wiener diversity indices. Species dominance or importance could be expressed as the size of a species population (abundance), or density expressed as basal area/ha. Dominance by natives could be diluted by exotic plantation or invasion due to urban effects. Most urban woodlands suffer from biotic pauperization (McBride and Jacobs 1986). The exceptional sites have biotic augmentation due to enrichment planting, and habitat heterogeneity of edge and ecotone. The woodland flora contributes significantly to urban biodiversity and its conservation, helping some cities to accommodate a higher floristic diversity than their surrounding countryside (Stewart *et al.* 2004; Alvey 2006).

Two critical measures of spatial variations in species composition, namely alpha (intra-patch) and beta (inter-patch) diversities, are important to tropical forest studies (Duivenvoorden *et al.* 2002; Wright 2002). They have been sparingly applied in urban woodland studies, which so far have been concentrated in developed countries in temperate lands. Almost all the published research comes from Europe and North America, plus some from Australia and New Zealand. The alpha-beta concept could help to decipher the exceptionally high spatial heterogeneity in species composition in tropical urban woodlands.

Rare species deserve special attention, interpreted according to their vicissitudes under the interplay of natural and urban effects. Rarity could be inherent, or denote relicts of a previously larger population, or indicate new arrivals that have not had the time and chance yet to increase their presence. The decline of equilibrium woodland species, which serve as indicator species, echoes fundamental deterioration in the biotic or abiotic conditions. Rarity could also signify the sporadic insertion of cultivated plants into the natural site. Some unusual species seek refuge in the remnant woodland site, as the required habitat conditions no longer exist in the locality. The diagnosis and prognosis of the rarity symptom could throw light on the past site history and future trend of urban woodlands to inform ecological management decisions.

Recruitment and regeneration dynamics

Regeneration dynamics are a crucial aspect of woodland ecology. A vigorous woodland should harbor a healthy population of tree seedlings and saplings in its undergrowth stratum. A continual process of recruitment to fill canopy gaps is pertinent to ascertaining continuity of the woodland ecosystem (Lehvävirta and Hannu 2002). The supply of propagules from resident plants and nearby ecosystems denotes a long-term lifeline of the natural area. The generosity of the seed rain relies on seed production as well as dispersal efficacy. The soil seed bank needs to be continually refurbished to sustain the regenerative process. A species with a balanced age-structure of young, middle-aged and old trees represents a progressive population that connotes stability. A species with an age-structure heavily biased towards old trees denotes a retrogressive population beset by further decline.

The arrival of diaspores at a site does not guarantee successful germination and establishment. The high mortality of seedlings and the differential susceptibility to losses determine the species make-up of the recruits. Predation and especially herbivory plays a crucial role in seedling survival. The arrival may not be of the preferred species. Native species are usually chosen in planting programs to maintain the naturalness of the woodlands. Nevertheless, the disturbed state within and in the environs of the woodlands would often favor the invasion of exotics. Native young trees are usually shade-tolerant, whereas invasive exotics often thrive under open conditions. Human disturbance tends to create more open sites to favor exotics at the expense of natives.

The incursion by aggressive exotics could trigger drastic and long-term floristic changes. Once a foothold is secured, loci of spread would be formed to facilitate if not accelerate the invasion (Robinson and Handel 2000). Native recruits could be outcompeted and decline to modify the composition, structure and character of the woodland. Enrichment planting commonly emulates the natural spreading phenomenon. Seedlings of preferred species could be planted in a woodland in the form of scattered loci to trigger the natural infilling process.

Human visitors to urban woodlands could contribute to seed dispersal (Mount and Pickering 2009). Our clothing, socks, shoes and bags could take inadvertent seed passengers into woodlands and so alter their species make-up. Seeds picked up from a diverse range of sources could be brought into wooded sites. The nature of the source area influences the species provenance of the propagules. Roadside or disturbed sites in general contribute more non-natives which include invasive elements. Intact natural sites offer more natives to recipient sites.

Floristic provenance and synanthropic influence

An important aspect of woodland ecology is the geographical provenance of its constituent species (Pyšek 1998; Honnay *et al.* 1999). A systematic classification of the urban synanthropic flora is presented in Figure 26.5. Most studies focus on woody species, especially trees. A clear distinction is drawn between natives (indigenous species) and exotics (introduced or alien species).

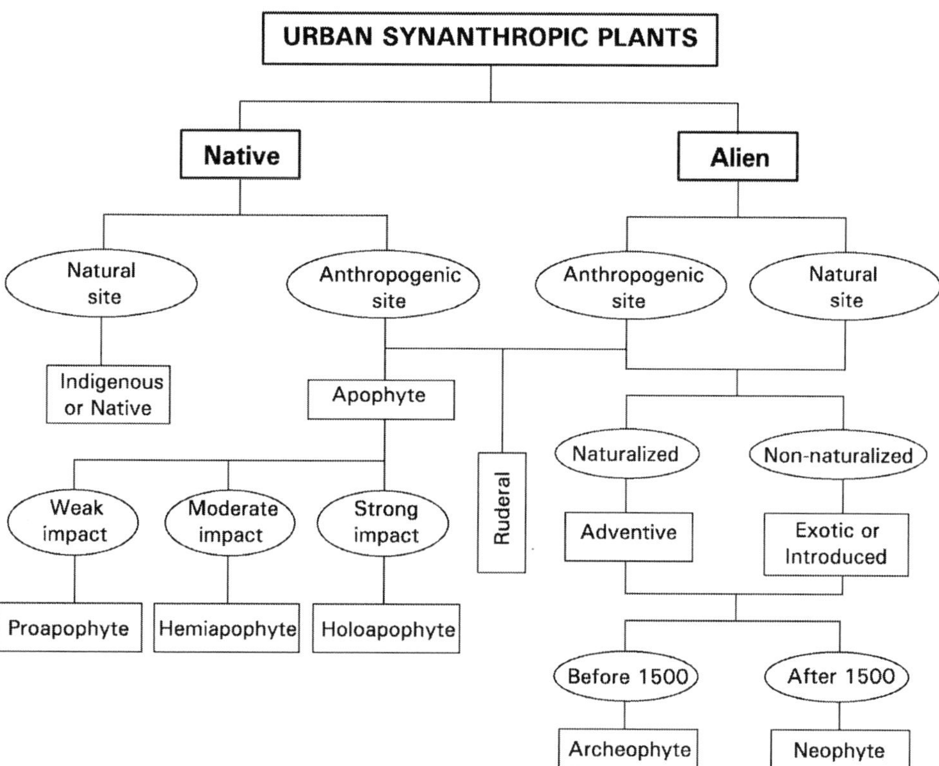

Figure 26.5 Classification of urban synanthropic plants based on the native-alien dichotomy and human mediation.

Disturbed woodlands carry the telltale symptom of a strong exotic presence. In the extreme case, they could out-compete the natives to form new communities and become dominants. Exotics could become highly successful in their adopted homes as naturalized species or adventives, some of which could enter urban woodlands. European studies divide exotics into archeophytes that arrive at a site before 1500, and neophytes afterwards. In ancient woodlands with centuries of lineage and historical botanical records, the changing flora could be tracked in detail to depict the fluxes in ecological and anthropogenic influences.

Plants that grow on human-modified or anthropogenic sites are ruderals, of which the natives are apophytes. Based on the hemeroby concept (degree of human impacts on the site), proapophytes (weak), hemiapophytes (moderate) and holoapophytes (strong) can be differentiated (Kowarick 1990). As local denizens, the apophytes commonly join the exotics to colonize brownfields, including the abandoned industrial-mining sites. In the post-industrial landscape of some European cities, perforations have appeared in the urban matrix due to the abandonment of developed lands (Henne 2005). Such sites, increasingly produced by the urbanization reversal process (Berry 2008), could be converted into a new genre of urban woodlands that begin their tenure on anthropogenic substrates (Tischew and Lorenz 2005; Weiss *et al.* 2005). Rather than remaining as urban blights, the surplus lands could become welcome solution spaces. The new woodlands could enhance the quantity and quality of the urban green space stock and enhance urban ecology.

Adopting the spontaneous growth approach, natural woodland succession could lead to rather natural woodlands in terms of composition and structure. Allowing nature to run its own course will take several decades for constituent seral communities to culminate in the climax stage. At different points along the successional continuum, the early, middle and late species arrays are sequentially displayed. Particularly difficult and sterile sites, however, could arrest the succession progression. Human intervention could pump-prime or accelerate woodland formation by sowing seeds and planting seedlings or saplings. This emerging new form of urban wildland has been labeled as secondary nature to distinguish it from primary nature.

Synanthropic imprints are commonly found in urban woodlands. Due to past or extant disturbances, most sites embody a mixture of natives and exotics. The relative proportion of the two contesting groups varies from the pure natives end member to the pure exotics (plantation woodland). The species make-up, abundance and vigor of the exotics league provide an indicator of human influences. Some brownfield sites are spatially heterogeneous in substrate, hydrological and microclimatic conditions, leading to elaborate habitat differentiation, which in turn fosters floristic differentiation into phytosociological alliances.

Anthropogenic processes impose direct and indirect effects on the urban woodland flora. Site fragmentation raises the edge-to-core area ratio. The urban influence at the edge is correspondingly increased to favor exotics that are better adapted to disturbance. Some landscape plants, both natives and exotics, cultivated in parks and gardens near urban woodlands, could intrude as garden escapees. Existing species considered as undesirable could be deliberately removed or thinned. Silvicultural treatments could unintentionally but preferentially favor some species whilst suppressing others. The choice of species for afforestation, plantation and enrichment-planting could directly impinge on species composition (Jim 1993, 2001). Like the ornamental counterparts, the species choice follows certain fashions or fads that tend to shift with time.

Diagnostic geometric traits

Spatial pattern and connectivity

Urban woodlands could be characterized by a host of geometric properties. Adopting landscape ecology concepts, an urban woodland denotes a patch embedded in an urban matrix. Woodlands could be linked by corridors or greenways. More often than not, however, urban woodlands exist in isolation in the form of nature islets.

Applying the equilibrium theory of island biogeography (MacArthur and Wilson 1967), an urban woodland that is large and connected to other woodlands can support more species and has a lower probability of suffering from species extinction. Some urban woodlands are extensions of the countryside woodland into the city, in which case habitat continuity is preserved. Plants, animals, diaspores and associated genes could literally flow from the countryside into the city. Such woodlands could theoretically accommodate a more diversified species composition.

The proximity to built-up areas would incur urban effects to suppress floristic and ecological developments of the urban woodland extension. Thus green tongues, fingers, or spokes of woodland could penetrate an urban area, in which case only the elongated daughter extensions could be considered as urban woodlands. Wider and shorter natural extensions are green wedges. A series of such woodland extensions interspersing in a developed area creates a desirable interpenetration spatial pattern. They are often preserved in spatial conjunction with linear topographic features such as spurs, valleys, or river courses. The vast parent countryside woodland lying beyond the urban sphere of influence, analogous to the mainland connected to the peninsula, should not be subsumed under urban woodlands.

Spatial continuity or connectivity between woodland patches, a key geometric feature of urban woodlands, could be characterized by canopy configuration interpreted from aerial photographs. By analyzing the coverage, shape and connectivity of the urban forest canopy, three basic types could be divided into nine subtypes (Jim 1989b, 1989c):

1 isolated: dispersed, clustered, clumped;
2 linear: rectilinear, curvilinear, annular; and
3 connected: reticulate, ramified, continuous.

The pattern of woodland patches indicates the underlying areal associations. They are strongly influenced by topography, and antecedent and contemporary land use, reflecting the resultant of actions and reactions between urban and natural landscapes.

Fragmentation and edge effect

Fragmentation encompasses two related phenomena: (1) the breaking up of a species population and its habitat patch; and (2) reduction in connectivity among habitat patches. In highly developed regions such as Western Europe, the eradication of forests and breaking up of the remnant patches have occurred since historical times (Gkaraveli *et al.* 2001; Zerbe and Brande 2003). Whereas past forest losses were mainly linked to rural activities, fragmentation of urban woodlands signifies a continuation of this process under novel urban influences (Thomas *et al.* 1997; Atkinson *et al.* 1999). Rural woodland patches are commonly fragmented, and even more so in cities to yield mainly small, isolated and scattered patches.

A cluster of small and dispersed woodland patches could denote remnants or relicts of a former continuous forest cover. Although the daughter patches share a common parental forest, they differ by topography, soil, hydrology, area, shape, and human intervention. Depending upon the stress regime, such disjunct patches may experience differential degradation or progression. The divergent ecosystem development could generate assorted woodland patches in terms of species and structure. The small patches have a limited capacity to accommodate equilibrium species populations. Some species, especially natives, could be driven into critically small populations threatened by local extinction (Smale and Gardner 1999).

Small sites have a higher edge-to-area ratio to increase the proportion of the perimeter belt at the expense of core. The edge provides an alternative ecotone habitat as a transition between woodland and the matrix envelope (Godefroid and Koedam 2003). The light, wind, microclimate, and human-disturbance regimes of the edge differ from the core. It is often continually disturbed to favor species with pioneer or adventives characters. Some species are adapted to grow in this hybrid living environment. Where conditions permit, the combination of core and ecotone habitats could increase the total species diversity.

The exceptionally high alpha and beta species diversities in tropical forests demand a different interpretation of urban woodlands. Each small isolated patch could harbor relatively high species richness due to alpha diversity. Moreover, individual patches could accommodate different species assemblages due to beta diversity, which is betrayed by the low species similarity indices between patch pairs. Even small urban woodlands could hold uncommon or rare species unique to individual sites to bestow conservation value.

Each small patch subpopulation has a limited gene pool. If the inter-patch distance is short enough to be spanned by normal dispersal means, gene flow between the proximal patches could continue albeit at a low frequency. If the distance exceeds a traversable threshold, gene exchanges are seriously curtailed and often limited to occasional jump dispersals with low frequency of occurrence and low probability of success. Detachment of daughter gene pools from the parent pool could bring long-term repercussions on the viability of the daughter populations.

Habitat heterogeneity

Urban woodlands with a high degree of naturalness tend to accommodate a heterogeneous array of microhabitats or microsites. The habitat diversity, expressed in horizontal and vertical dimensions has a bearing on biotic composition and diversity (Rudnicky and McDonnell 1989). The fine mosaic of living conditions increases the chances for plants and animals with different requirements to live in the woodland. A natural, mature, and undisturbed woodland could offer a wide complement of microhabitats, encompassing dense woodland, sparse woodland, old woodland, young woodland, open glade, shrubland, grassland, bare patch, rocky patch, marshy patch, sunny patch, shaded patch, standing deadwood, fallen deadwood, ponds, and streams. The interfaces between microhabitats, with different width, could offer ecotone conditions for a suite of transitional species to further augment the aggregate biotic diversity of the woodland.

Overzealous woodland management has a notorious tendency to suppress or even eliminate the fine subdivision of the woodland ecosystem. The inappropriate urban-park management mindset, emphasizing neatness, order, and human dominance, could be mistakenly applied to urban woodlands near built-up areas. The plantation strand of silvicultural practices could also find their way into urban woodlands. The field and undergrowth vegetation, in particular, could be subject to unnecessary disturbance, clearance, and sanitization. Besides unnecessary manicure operations, excessive care should be avoided. In particular, the indiscriminate removal of natural stresses, which are often needed to maintain diversity, could inadvertently eliminate key or rare species. Some delicate microhabitats are easy to disrupt but exceedingly sluggish to recover. The preferred urban woodland management calls for an ecologically sensitive approach to maximize the opportunities for habitat-cum-species diversities. The urge to simplify, regularize, tame, and homogenize should be resisted.

Conclusion

With relentless urban growth and environmental degradation, the nature deficit of cities could continue to aggravate especially in developing countries. Increasing awareness of the need to contact nature has highlighted the importance of protecting, enhancing and creating natural areas in cities. Urban woodland offers the highest caliber nature, the value of which is accentuated by proximity to urban populations. As gems of nature in human settlements, their cherished status owes very much to their location and rarity in the urban context, as well as their intrinsic natural virtues. The esteem is due to association with the urban hinterland, to which its fate is closely tied. The attitude and behavior of the surrounding people constitute the principal determinants of the vicissitude and destiny of urban woodlands.

Some sites attract heavy patronage of city dwellers seeking salubrious outdoor recreation. Some sites contain important habitats and biotic communities worthy of conservation. As remnants or islands of nature in urbanized realms, they provide pertinent ecosystem services to the host city. The conflict between the principal uses demands custom-designed and innovative management solutions. The common pitfall of adopting the urban-park or silvicultural overkill approaches contradicts the unique needs of the threatened resource base. Neither would benign negligence be acceptable.

A nature-cum-people oriented strategy, based firmly on research findings on the composition, structure, function, urban effects, and ecological prognosis, could help to sustain the precious urban-nature enclaves. If valuable nature in the extra-urban realm can be earnestly designated as protected areas, an equal treatment could be extended to precious urban nature. The inclination to ignore or belittle backdoor and familiar conservation areas in cities needs to be weaned. Urban woodlands present worthy candidates for conservation and proactive management. It is high time that urban planning should fully embrace urban natural areas as integral ingredients of a truly sustainable city.

References

Alvey, A.A. (2006) 'Promoting and preserving biodiversity in the urban forest,' *Urban Forestry and Urban Greening*, 5(4): 195–201.

Atkinson, D., Smart, R.A., Fairhurst, J., Oldfield, P., and Lageard, J.G.A. (1999) 'A history of woodland in the Mersey Basin,' in E.F. Greenwood (ed.), *Ecology and Landscape Development: A History of the Mersey Basin*, Liverpool: Liverpool University Press and National Museums and Galleries on Merseyside, pp. 91–107.

Berry, B.J.L. (2008) 'Urbanization,' in J.M. Marzluff, E. Schulenberger, W. Endlicher, M. Alberti, G. Bradley, C. Ryan, C. ZumBrunnen, and U. Simon (eds.), *Urban Ecology: An International Perspective on the Interaction between Humans and Nature*, New York: Springer, pp. 25–48.

Bertin, R.I., Manner, M.E., Larrow, B.F., Cantwell, T.W., and Berstene, E.M. (2005) 'Norway maple (*Acer platanoides*) and other non-native trees in urban woodlands of central Massachusetts,' *Journal of the Torrey Botanical Society*, 132(2): 225–35.

Chandrakanth, M.G., Bhat, M.G., and Accavva, M.S. (2004) 'Socio-economic changes and sacred groves in South India: Protecting a community-based resource management institution,' *Natural Resources Forum*, 28(2): 102–11.

Chen, W.Y. and Jim, C.Y. (2006) 'Valuation of nature: economic contribution of peri-urban protected areas in Guangzhou Baiyun Mountain Scenic Area,' in C.Y. Jim and R.T. Corlett (eds.), *Sustainable Management of Protected Areas for Future Generations*. Gland, Switzerland: World Conservation Union (IUCN), pp. 59–75.

Chen, W.Y. and Jim, C.Y. (2008) 'Evaluation and valuation of the diversified ecosystem services provided by urban forests,' in M.M. Carreiro, Y.C. Song, and J.G. Wu (eds.), *Ecology and Management of Urban Forests: An International Perspective*, New York: Springer, pp. 53–83.

Coughlin, R.E., Mendes, D.C., and Strong, A.L. (1988) 'Local programs in the United States for preventing the destruction of trees on private land,' *Landscape and Urban Planning*, 15(1–2): 165–71.

Duivenvoorden, J.F., Svenning, J.C., and Wright, S.J. (2002) 'Beta diversity in tropical forests,' *Science*, 295(5555): 636–7.

Gkaraveli, A., Williams, J.H., and Good, J.E.G. (2001) 'Fragmented native woodlands in Snowdonia (UK): assessment and amelioration,' *Forestry*, 74(2): 89–103.

Godefroid, S. and Koedam, N. (2003) 'Distribution pattern of the flora in a peri-urban forest: an effect of the city-forest ecotone,' *Landscape and Urban Planning*, 65(4): 169–85.
Hedblom, M. and Söderström, B. (2008) 'Woodlands across Swedish urban gradients: status, structure and management implications,' *Landscape and Urban Planning*, 84(1): 62–73.
Henne, S.K. (2005) 'New wilderness as an element of the peri-urban landscape,' in I. Kowarik and S. Körner (eds.), *Wild Urban Woodlands*, Berlin: Springer, pp. 247–62.
Honnay, O., Endels, P., Vereecken, H., and Hermy, M. (1999) 'The role of patch area and habitat diversity in explaining native plant species richness in disturbed suburban forest patches in northern Belgium,' *Diversity and Distribution*, 5(4): 129–41.
Jim, C.Y. (1987) 'Trampling impacts of recreationists on picnic sites in a Hong Kong country park,' *Environmental Conservation*, 14(2): 117–27.
Jim, C.Y. (1989a) 'Visitor management in recreation areas,' *Environmental Conservation*, 16(1): 19–32, 40.
Jim, C.Y. (1989b) 'Tree canopy characteristics and urban development in Hong Kong,' *Geographical Review*, 79(2): 210–25.
Jim, C.Y. (1989c) 'Tree canopy cover, land use and planning implications in urban Hong Kong,' *Geoforum*, 20(1): 57–68.
Jim, C.Y. (1993) 'Soil compaction as a constraint to tree growth in tropical and subtropical urban habitats,' *Environmental Conservation*, 20(1): 35–49.
Jim, C.Y. (1993) 'Ecological rehabilitation of disturbed lands in the humid tropics,' in W.E. Parnham, P.J. Durana, and A.L. Hess (eds.), *Improving Degraded Lands: Promising Experiences from South China*, Honolulu, Hawaii: Bishop Museum Bulletin in Botany 3, Bishop Museum, pp. 217–29.
Jim, C.Y. (1998a) 'Soil characteristics and management in an urban park in Hong Kong,' *Environmental Management*, 22(5): 683–95.
Jim, C.Y. (1998b) 'Urban soil characteristics and limitations for landscape planting in Hong Kong,' *Landscape and Urban Planning*, 40(4): 235–49.
Jim, C.Y. (2001) 'Ecological and landscape rehabilitation of a quarry site in Hong Kong,' *Restoration Ecology*, 9(1): 85–94.
Jim, C.Y. (2002) 'Heterogeneity and differentiation of the tree flora in three major land uses in Guangzhou City, China,' *Annals of Forest Science*, 59(1): 107–18.
Jim, C.Y. (2003a) 'Conservation of soils in culturally protected woodlands in rural Hong Kong,' *Forest Ecology and Management*, 175(1–3): 339–53.
Jim, C.Y. (2003b) 'Soil recovery from human disturbance in tropical woodlands in Hong Kong,' *Catena*, 52(2): 85–103.
Jim, C.Y. and Chan, M.W.H. (2004) 'Assessing natural and cultural influence on soil in remnant tropical woodlands,' *Area*, 36(1): 6–18.
Jim, C.Y and Chen S.S. (2003) 'Comprehensive greenspace planning based on landscape ecology principles in compact Nanjing city, China,' *Landscape and Urban Planning*, 65(3): 95–116.
Jim, C.Y. and Chen, W.Y. (2008a) 'Pattern and divergence of tree communities in Taipei's main urban green spaces,' *Landscape and Urban Planning*, 84(3–4): 312–23.
Jim, C.Y. and Chen, W.Y. (2008b) 'Assessing the ecosystem service of air pollution removal by urban vegetation in Guangzhou (China),' *Journal of Environmental Management*, 88(4): 665–76.
Jorgensen, A., Hitchmough, J., and Dunnett, N. (2005) 'Living in the urban wildwoods: a case study of Birchwood, Warrington New Town, UK,' in I. Kowarik and S. Körner (eds.), *Wild Urban Woodlands*, Berlin: Springer, pp. 95–116.
Kowarik, I. (1990) 'Some responses of flora and vegetation to urbanization in Central Europe,' in H. Sukopp, S. Hejný, and I. Kowarik (eds.) *Urban Ecology: Plants and Plant Communities in Urban Environments*, The Hague: SPB Academic, pp. 75–97.
Kowarik, I. (2005) 'Wild urban woodlands towards a conceptual framework,' in I. Kowarik and S. Körner (eds.), *Wild Urban Woodlands*, Berlin: Springer, pp. 1–32.
Kunick, W. (1987) 'Woody vegetation in settlements,' *Landscape and Urban Planning*, 14(1): 57–78.
Lehvävirta, S. and Hannu, R. (2002) 'Natural regeneration of trees in urban woodlands,' *Journal of Vegetation Science*, 13(1): 57–66.
MacArthur, R.H. and Wilson, E.O. (1967) *The Theory of Island Biogeography*, Princeton, NJ: Princeton University Press.
McBride, J.R. and Jacobs, D.F. (1986) 'Presettlement forest structure as a factor in urban forest development,' *Urban Ecology*, 9(3–4): 245–66.
Mishra, B.P., Tripathi, O.P., Tripathi, R.S., and Pandey, H.N. (2004) 'Effects of anthropogenic disturbance on

plant diversity and community structure of a sacred grove in Meghalaya, northeast India,' *Biodiversity and Conservation*, 13(2): 421–36.

Mount, A. and Pickering, C.M. (2009) 'Testing the capacity of clothing to act as a vector for non-native seed in protected areas,' *Journal of Environmental Management*, 91(1): 168–79.

Ono, R. (2005) 'Approaches for developing urban forests from the cultural context of landscapes in Japan,' in I. Kowarik and S. Körner (eds.) *Wild Urban Woodlands*, Berlin: Springer, pp. 221–30.

Pauleit, S. and Duhme, F. (2000) 'Assessing the environmental performance of land cover types for urban planning,' *Landscape and Urban Planning*, 52(1): 1–20.

Porter, E.E., Forschner, B.R., and Blair, R.B. (2001) 'Woody vegetation and canopy fragmentation along a forest-to-urban gradient,' *Urban Ecosystems*, 5(2): 131–51.

Pyšek, P. (1998) 'Alien and native species in central European urban floras: a quantitative comparison,' *Journal of Biogeography*, 25(1): 155–63.

Robinson, G.R. and Handel, S.N. (2000) 'Directing spatial patterns of recruitment during an experimental urban woodland reclamation,' *Ecological Applications*, 10(1): 174–88.

Rudnicky, J.L. and McDonnell, M.J. (1989) 'Forty-eight years of canopy change in a hardwood-hemlock forest in New York City,' *Bulletin of the Torrey Botanical Club*, 116(1): 52–64.

Sachse, U., Starfinger, U., and Kowarik, I. (1990) 'Synanthropic woody species in the urban area of Berlin,' in H. Sukopp, S. Hejný, and I. Kowarik (eds.) *Urban Ecology: Plants and Plant Communities in Urban Environments*, The Hague: SPB Academic, pp. 233–43.

Smale, M.C. and Gardner, R.O. (1999) 'Survival of Mount Eden Bush, an urban forest remnant in Auckland, New Zealand,' *Pacific Conservation Biology*, 5: 83–93.

Spencer, J.W. and Kirby, K.J. (1992) 'An inventory of ancient woodland for England,' *Biological Conservation*, 62: 77–93.

Stewart, G.H., Ignatieva, M.E., Meurk, C.D., and Earl, R.D. (2004) 'The re-emergence of indigenous forest in an urban environment, Christchurch, New Zealand,' *Urban Forestry and Urban Greening*, 2(3): 149–58.

Sukopp, H. (2004) 'Human-caused impact on preserved vegetation,' *Landscape and Urban Planning*, 68(4): 347–55.

Thomas, R.C., Kirby, K.J., and Reid, C.M. (1997) 'The conservation of a fragmented ecosystem within a cultural landscape: the case of ancient woodland in England,' *Biological Conservation*, 82(3): 243–52.

Tischew, S. and Lorenz, A. (2005) 'Spontaneous development of peri-urban woodlands in lignite mining areas of Eastern Germany,' in I. Kowarik and S. Körner (eds.), *Wild Urban Woodlands*, Berlin: Springer, pp. 163–80.

Tregay, R. (1979) 'Urban woodlands,' in I.C. Laurie (ed.), *Nature in Cities*, New York: Wiley, pp. 267–95.

Tyrväinen, L., Pauleit, S., Seeland, K., and de Viries, S. (2005) 'Benefits and uses of urban forests and trees,' in C.C. Konijnendijk, K. Nilsson, T.B. Randrup, and J. Schipperijn (eds.), *Urban Forests and Trees: A Reference Book*, Berlin: Springer, pp. 81–114.

Weiss, J., Burghardt, W., Gausmann, P., Haag, R., Haeupler, H., Hamann, M., Leder, B., Schulte, A., and Stempelmann, I. (2005) 'Nature returns to abandoned industrial land: monitoring succession in urban-industrial woodlands in the German Ruhr,' in I. Kowarik and S. Körner (eds.), *Wild Urban Woodlands*, Berlin: Springer, pp. 143–62.

Wilson, E.O. (1984) *Biophilia: The Human Bond with other Species*, Cambridge, MA: Harvard University Press.

Wright, S.J. (2002) 'Plant diversity in tropical forests: a review of mechanisms of species coexistence,' *Oecologia*, 130(1): 1–14.

Zerbre, S. and Brande, A. (2003) 'Woodland degradation and regeneration in Central Europe during the last 1,000 years: a case study in NE Germany,' *Phytocoenologia*, 33(4): 683–700.

27

Wetlands in urban environments

Joan G. Ehrenfeld, Monica Palta, and Emilie Stander

Wetlands provide a unique resource in urban environments. They offer ecological services, including habitat for animals and plants, the storage of floodwaters, the maintenance of stream flow, and the capacity to remove and sequester pollutants, and they also give urban residents access to nature. While considerable research has addressed the ecological function of urban wetlands, much remains to be done. Below, we survey the current state of knowledge.

Wetlands in the urban context

Wetlands are the product of three interacting sets of factors – hydrology, including the source(s), temporal patterns, directions, and velocities of water flow; the vegetation, consisting of plants adapted to the rigors of saturation and flowing water, and soils, which have unique biogeochemical properties resulting from the extended presence of water (Mitsch and Gosselink 2000). In urban environments, however, although these three elements remain the key to their structure and function, other factors unique to urban environments assume equal importance (Figure 27.1), because they modify hydrology, soils, and vegetation. For example, the hydrogeomorphic setting, the combination of geology and topography that determines wetland hydrology, is often highly modified in urban areas by reshaping of the land surface for development, and replacement of soil by concrete or construction debris. Wetlands are fragmented by ditching, draining, or filling. The wetland biota must cope with lack of connection among habitats, including adjacent upland, dispersal barriers, and exotic species from landscaped yards. Water quality is degraded by nutrients and toxic substances, with effects on all components of the biota. Finally, adjacent human populations may directly disturb the interior through cutting trails and dumping trash, while also valuing the wetland for a variety of practical and aesthetic reasons. Together, the basic properties of hydrology, soils, and vegetation as modified by the anthropogenic features of urban landscapes determine the range and quality of ecosystem services available from urban wetlands.

Distribution of wetlands in urban regions

Because there is no information about the extent of wetlands in urban areas, we undertook a brief survey of wetlands in urban areas of the United States, using the spatial data available

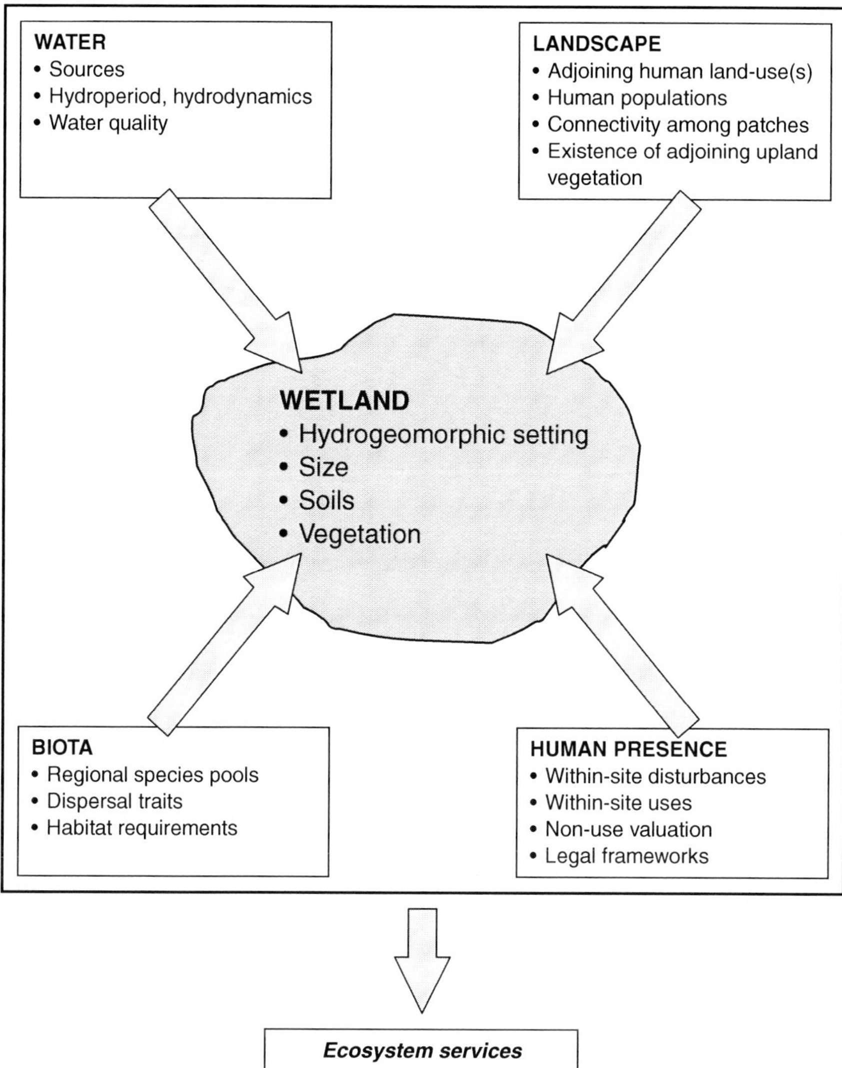

Figure 27.1 Anthropogenic factors modifying wetland ecology in urban regions.

from the National Wetlands Inventory (NWI; www.fws.gov/wetlands), in order to assess how well wetlands have survived in urban settings. We chose the largest city within each of five broad geographic regions of the county, and obtained NWI data on wetlands for the nine US Geological Survey quadrangles (7.5 × 7.5 minute) covering the city center and adjacent lands (15,627 km^2 area). We then arbitrarily chose for comparison nine quadrangles as close to the city as possible but with minimum population densities and representing as similar topographies as possible.

Surprisingly, three of the five cities examined have more wetland area than nearby rural areas (Table 27.1). Although we did not analyze land-use data, we speculate that in these regions, intensive agriculture in the non-urban areas explains the low amounts of wetland, similar to

Table 27.1 Wetland area within and outside of large American cities. Amounts of wetland (km^2) in each class are for a 1,736.4 km^2 region around each city center ('urban') and for a similar area in a rural region within the same geographic section of the country ('rural'). Total area is computed for both all wetland types ('total') and for freshwater only ('FW') to control for the coastal locations of many cities. Deepwater habitats, as classified by Cowardin *et al.* (1978) are excluded from the analysis. Percentages are given for total wetlands ('Percent') and as percent of only-freshwater wetlands (%FW) for locations with estuarine-marine wetlands present. Population (millions of people) are based on data from the 1998 census and include only the areas falling within the quadrangles used.

City	*Region (population × 10^6)*	*Units*	*Estuarine and marine*	*Emergent marsh*	*Forested and shrub*	*Pond*	*Lake*	*Riverine*	*Other*	*Total (FW)*
Chicago, IL	Urban (3.8)	Area	0	8.9	5.5	6.1	10.6	15.6	0	46.6
		Percent (%FW)	–	19.1	11.8	13.1	22.7	33.5	0	–
	Rural (0.014)	Area	0	10.4	47.4	2.0	243.6	10.1	0	313.5 (313.5)
		Percent (%FW)	–	3.3	15.1	0.6	77.7	3.2	0	–
Houston, TX	Urban (2.1)	Area	19.7	13.0	72.6	5.3	54.8	9.9	0.7	176.0 (156.4)
		Percent (%FW)	11.2	7.4 (8.3)	41.3 (46.5)	3.0 (3.4)	31.1 (35.0)	5.6 (6.3)	0.4 (0.5)	-
	Rural (0.06)	Area	0	21.2	7.1	11.5	6.3	35.2	0.6	81.9
		Percent (%FW)	–	25.9	8.6	14.0	7.7	43.0	0.8	–
Jacksonville, FL	Urban (0.8)	Area	2,689.7	5,058.8	259,082.2	2,295.8	3.9	342.1	107.4	269,579.9 (266,890.2)
		Percent (%FW)	1.0	1.9 (1.9)	96.1 (97.1)	0.9 (0.9)	<0.1 (<0.1)	<0.1 (<0.1)	0.1 (0.1)	–
	Rural (0.08)	Area	129.6	53.1	848.2	3.4	0.5	107.7	<0.1	1,142.5 (1,012.9)
		Percent (%FW)	11.4	4.7 (5.2)	74.2 (83.7)	0.3 (0.3)	<0.1 (<0.1)	9.4 (10.6)	<0.1 (<0.1)	–
New York, NY	Urban (9.5)	Area	45.0	4.1	4.4	2.4	2.4	4.2	0.2	62.5 (17.5)

		Percent (%FW)	72.0	6.6 (23.5)	7.0 (25.1)	3.8 (13.5)	3.8 (13.5)	6.6 (23.7)	0.2 (0.8)	–
	Rural (0.02)	Area	0	9.6	40.2	3.5	9.7	55.3	<0.1	118.3
		Percent (%FW)	–	8.1	34.0	2.9	8.2	46.7	<0.1	–
San Francisco, CA	Urban (1.8)	Area	49.4	1.7	1.0	1.8	8.6	0.4	0	62.9 (13.4)
		Percent (%FW)	78.6	2.7 (12.7)	1.5 (7.2)	2.8 (13.3)	13.6 (63.7)	0.7 (3.2)	0	–
	Rural (0.02)	Area	5.7	3.5	5.1	0.4	0.2	0	11.6	26.4 (20.8)
		Percent (%FW)	21.5	13.4 (17.0)	19.2 (24.4)	1.6 (2.0)	0.6 (0.8)	0	43.8 (55.8)	–

findings of Thibault and Zipperer (1994). In contrast, New York City and Chicago have the expected smaller amount of remaining wetland than nearby rural areas. All of the large cities have a coastal location (Chicago on Lake Michigan, the others on oceans), and they retain large areas of coastal wetlands. Notably, riverine wetlands are more poorly represented within most cities (all but Chicago) than in nearby rural areas, despite the attention given to greenways along streams as public open space. This preliminary survey suggests that urban regions may actually support more wetland area than nearby rural regions. It also suggests that the representation of different types of wetlands among cities, and between cities and their adjacent rural areas are quite varied, and probably reflect regional geology as well as historical land-use patterns.

Hydrology and water quality in urban wetlands

Hydrology

As land surfaces are altered by urban development in ways that change topography and especially the partitioning of precipitation between infiltration and surface run-off, the hydrologic forcing factors that shape wetland structure and function will be altered (Brinson 1993; Bedford 1996). The channeling of urban stormwater runoff from impervious surfaces directly to streams results in rapid development of peak flows, and often at higher discharge levels, followed by more rapid recessions, decreases in base flow, larger discharge variations, changes in seasonal patterns, and alteration and deepening of channel morphology (Azous and Horner 2001; Paul and Meyer 2001; Groffman *et al.* 2003; Booth *et al.* 2004; Walsh *et al.* 2005). In addition, ditches, berms, and drainage for mosquito and flood control alter water tables and surface flow patterns (Ehrenfeld and Schneider 1991; Ehrenfeld *et al.* 2003) and create novel hydrologic regimes (Kentula *et al.* 2004). The effect of these hydrological alterations is to lower water tables, so that wetlands are dry for extended periods, and have 'flashy,' rapidly changing hydrographs (Figure 27.2). Furthermore, waste water and storm runoff inputs, from septic fields, road runoff, and leaking sewer pipes affect both stream flow and the loading of nutrients and pollutants (Bernhardt *et al.* 2008).

Nutrients and pollutants in urban wetland environments

Urban wetlands are exposed to a variety of sources of pollution in both surface and groundwater, resulting in elevated amounts of nutrients, heavy metals, sediment, pathogens, pesticides, and pharmaceuticals. Urban wetlands are assumed to improve water quality, through removal and/or sequestration of nutrients and pollutants, but there is surprisingly little specific study of their pollutant removal capacity. Most emphasis has been placed on nitrogen removal, which depends on anaerobic conditions to promote denitrification (Seitzinger *et al.* 2006). However, the hydrologic alterations described above inhibit nitrate removal capacity (Groffman *et al.* 2002; Stander and Ehrenfeld 2009) when wetland soils become aerobic and/or when nitrate-laden groundwater flows beneath the active zone of the surface soils. Some urban wetlands display high levels of nitrate retention, if soils have prolonged wet periods (Hanson *et al.* 1994; Stander 2007). Changes in wetland nitrogen processing can also result from ancillary changes in soil properties; Zhu and Ehrenfeld (1999, 2000) found that higher pH and sand content in the soil due to storm water inputs into Atlantic white-cedar swamps led to decreased ability of peat soils to retain added N and increased net nitrification rates. These studies together suggest that alteration of wetland hydrology and soils in urban areas may decrease N removal function if soils become predominantly aerobic.

Removal of other nutrients and pollutants has been less studied in urban wetlands. Phosphorus concentrations in soils and plant tissues are often higher than elsewhere. Several studies, however,

(a)

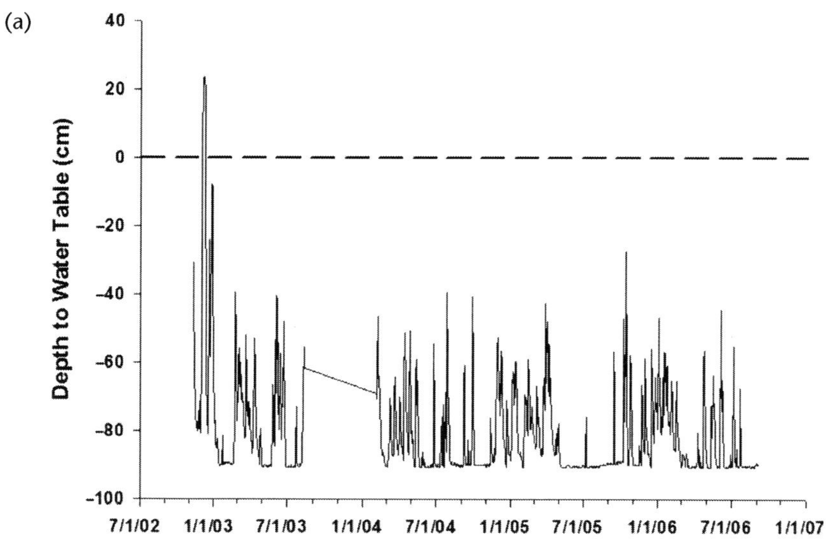

(b)

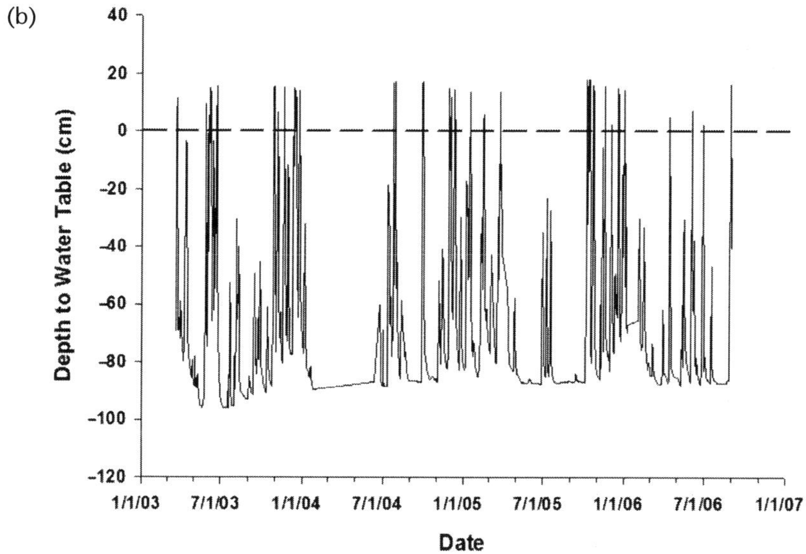

Figure 27.2 Examples of disturbed hydrologic patterns in urban forested wetlands (a) wetland with lowered water table due to downcut adjacent strear; (b) wetland with 'flashy' hydroperiod. Dashed line indicates the ground surface (*source:* Stander 2007).

have found that elevated P occurs at intermediate levels of urbanization (Horner *et al.* 2001, Hogan and Walbridge 2007). Higher sediment deposition in newly urbanizing watersheds that have high erosion rates from construction sites, versus lower sediment mobilization coupled to hydrologic disconnections in long-urbanized sites may explain this curvilinear relationship.

Much less is known about urban wetlands' ability to remove other pollutants contained in urban stormwater runoff. Like P, heavy metals are often transported to urban wetlands via adsorption to sediment particles, so sediment trapping and infiltration mechanisms are the most effective removal mechanisms. Urban-associated metals (Pb, Cu, Zn, Cr, Cd, and Ni) and organic contaminants accumulate in wetlands receiving high inputs of sediments (Sanger *et al.* 1999a, 1999b; Kimbrough and Dickhut 2006; Lee *et al.* 2006). Toxic substances may be taken up by plants, with potential movement through the food web (e.g. Windham *et al.* 2001; Weis *et al.* 2002, 2003). Root-associated microbes in coastal urban wetlands are capable of dehalogenation (Ravit *et al.* 2005; Launen *et al.* 2008), thereby potentially ameliorating toxic pollution. However, most studies of toxic substances in urban wetlands focus on estuarine wetlands (Kennish 2002), and there is little knowledge of these substances in freshwater wetlands.

Plant and animal communities

Communities of plants and animals reflect both the intrinsic characteristics of wetlands as habitat, and the influence of adjacent and regional urban land use. Below, we briefly survey patterns of response of biotic communities to urban environments.

Plant communities

Not surprisingly, wetlands within heavily-urbanized regions are frequently invaded by non-native species (e.g. Magee *et al.* 1999; Azous and Horner 2001; Lavoie *et al.* 2003; Moffatt *et al.* 2004; Wolin and MacKeigan 2005; Alston and Richardson 2006, among many others). Exotic species are strongly associated with physical disturbance (D'Antonio *et al.* 1999), a natural feature of riparian wetlands that may be exacerbated in urban regions, and an anthropogenic feature of non-riparian wetlands. However, lower levels of invasion similar to non-urban wetlands have also been recorded (Ehrenfeld 2005; Leck and Leck 2005). These studies mostly involve comparisons of urban wetlands with non-urban wetlands; very few studies have examined whether different types of urban land-use affect the rates of exotic invasion (Cutway and Ehrenfeld 2009), or whether different types of wetlands within urban areas are more susceptible to invasion. Little is known about the relative importance of complete anthropogenic disturbance, human-modified natural disturbance regimes (e.g. changes in flood regimes), and natural disturbance characteristics of wetlands in general.

Exotic species often become a larger fraction of the flora because the number of native species declines (Moffatt *et al.* 2004). Other environmental factors, such as salinity in coastal wetlands, may override the effects of urbanization (Lavoie *et al.* 2003). Ehrenfeld and colleagues (Ehrenfeld *et al.* 2003; Ehrenfeld 2005, 2008; Cutway and Ehrenfeld 2009) found both a wide range of exotic species invasion, in both absolute and relative numbers, in a diverse population of wetlands, and that riparian wetland sites with coarser-textured soils, and soils with high nutrients (N and P) tended to have higher representation of exotics. The presence of residential, rather than industrial-commercial adjoining land use increased the presence of exotics, a pattern also found by Vidra and Shear (2008). However, exotic species have variable and individualistic responses to environmental factors (Ehrenfeld 2008; Oneal and Rotenberry 2008). Other studies have also associated enhanced invasion into urban wetlands with increased nutrient availability (Panno *et al.* 1999; Wolin and MacKeigan 2005; Vidra *et al.* 2006; Aguiar *et al.* 2007) and drier conditions (Pyle 1995; Groffman *et al.* 2003; Predick and Turner 2008).

Disturbance associated with direct human presence (e.g. trails, dumping of trash, vegetation destruction) may enhance invasion by increasing light and other resources (Moffatt and McLachlan 2004; Moffatt *et al.* 2004; Aguiar *et al.* 2007; Burton and Samuelson 2008; Oneal and Rotenberry 2008). However, not all studies have found a clear association with human activities and disturbance (Cutway 2004; Ehrenfeld 2004; Houlahan *et al.* 2006; Ehrenfeld 2008). Thus, exotic invasion does not increase with urban disturbance and human population size in a simple, linear way. Rather, as McDonnell and Hahs (2008) have demonstrated, urban land use creates a complex gradient that affects different species in varying ways dependent on local context.

Vegetation structure in urban wetlands is equally variable. Often, urban forested wetlands have more large trees, and fewer small trees than in non-urban environments. This may reflect human management, or it may reflect the fact that the wettest areas have resisted draining and development, and therefore have older plant communities (Burton and Samuelson 2008; Pennington *et al.* 2008). There may be more dead trees in urban riparian zones than in non-urban sites. However, metrics of vegetation structure are sometimes comparable to those of non-urban wetlands of the same general types (e.g. Ehrenfeld 2005).

Plant communities are also influenced by wetland area. In southern Ontario, Houlahan *et al.* (2006) found that wetland size was strongly correlated with overall plant diversity, although less strongly related to the richness of individual functional groups (e.g. native and exotic species, annuals vs. perennials), a finding similar to other studies. Fragmentation, and thus small patch size, is characteristic of urban wetlands (e.g. Pyle 1995; Moffatt *et al.* 2004; Predick and Turner 2008; Wang *et al.* 2008). For example, 50 percent of 262 wetlands surveyed in northeastern New Jersey, were less than 10 ha (Ehrenfeld 2000).

There has been little examination of urban wetland vegetation with respect to biogeochemical functions or other ecosystem functions, despite the frequent invocation of vegetation in urban regions for amelioration of temperature, noise, and air pollution stressors, and many other functions. Some economic analyses of wetland value to residents have examined whether different types of vegetation (e.g. marsh, forest, shrubs) are differentially valued; not surprisingly, there is no consensus (Doss and Taff 1996; Mahan *et al.* 2000; Brander *et al.* 2006). The role of plant community characteristics and functions in determining functions and values needs further exploration.

Animal communities of urban wetlands

Animal communities show many of the same patterns as do plant communities: increased presence (richness and abundance) of exotic, cosmopolitan, and/or tolerant species, and decreases in native species, especially those dependent on large intact areas of natural habitat (Knutson *et al.* 1999; Miller *et al.* 2003; Lussier *et al.* 2006; Price *et al.* 2006; Villagran-Mella *et al.* 2006; Hamer and McDonnell 2008; Pillsbury and Miller 2008). Many of these changes reflect alterations of the physical environment within the wetland, such as habitat patches (Willson and Dorcas 2003; Miller *et al.* 2007), as well as changes in water quality (Lougheed *et al.* 2008) and loss of adjacent upland habitat (Knutson *et al.* 1999).

Bird communities are similarly affected by urbanization (Chace and Walsh 2006). The vertical structure of the plant community and the connectivity to other habitat types emerge as primary variables affecting bird community composition and abundance (Germaine *et al.* 1998; Rodewald and Bakermans 2006; Mason *et al.* 2007; Luther *et al.* 2008; Palmer *et al.* 2008, Pennington *et al.* 2008). Sundell-Turner and Rodewald (2008), for example, found that the composition of the breeding bird community in Ohio wetlands was most strongly related to the amount of forested land within a 1 km radius of a site, and secondarily to the area of roads within that radius. Similar patterns have been documented in the Pacific northwest (Hennings and Edge 2003), and the semi-arid mid-west (Miller *et al.* 2003). As with plant species, any given species and/or guild of birds responds individualistically to the combination of vegetation structure and landscape structure, and some species or groups of birds are more abundant in the urban-affected wetlands than in rural ones. Augmentation of bird abundances may reflect bird-feeder-based food subsidies and/or warmer winter temperatures (heat island effect) (Atchison and Rodewald 2006; Leston and Rodewald 2006). Overall, disturbance-intolerant species are less common, or missing entirely, from urban wetlands, exotic species are more common, and other species vary in their population responses to urbanization. Also, while urban wetlands may provide poor habitat for some nesting species, they are important as stop-over sites during migrations (Pennington *et al.* 2008). Finally, the representation of exotic and disturbance-tolerant birds in the fauna is positively correlated with the abundance of exotic plant species in the flora, suggesting important linkages among the multiple components of the wetland biota.

Wetland management in urban landscapes

Large wetland complexes are now widely recognized as critical sources of ecosystem services and non-use values, as shown by growing amounts of scientific research, popular press articles, and money dedicated to preservation and restoration of urban wetlands. Two examples will illustrate this trend.

In the New York area, the Hackensack Meadowlands of New Jersey (Figure 27.3a) originally covered about 112 km^2, and included a variety of saline and freshwater communities. Many efforts to drain the area for agriculture and commerce, to build dams, roads and railroads across it, and to extract resources, fragmented the wetland, drastically altered hydrology and soils, and eliminated most forest vegetation (Quinn 1997; Marshall 2004). Today, the wetland occupies about 78 km^2, most of which is heavily dominated by mono-specific stands of *Phragmites australis*. Despite the long history of degradation, the area supports a high diversity of plants (275 species), birds (334 species), fish (51 species), and mammals (24 species) (Kiviat and MacDonald 2004; Hackensack Meadows Initiative 2007). Paradoxically, some of this diversity results from the topographic heterogeneity afforded by numerous landfills. As a result of greatly enhanced public attention during the 1990s, the region has now become the target of a comprehensive restoration and protection effort (Hackensack Meadows Initiative 2007), with strong local political and social support (www.njmeadowlands.gov). However, in addition to the presence of invasive species, the restoration effort needs to cope with contamination with nutrients, heavy metals, PCBs (polychlorinated biphenyls), dioxins, and other contaminants (e.g. Hackensack Meadows Initiative 2007; Barrett and McBrien 2007; Tsipoura *et al.* 2008).

In China, the government has established 26 'national urban wetland parks' plus another 38 'national wetland parks,' at least some of which are in urban areas. The best developed is the XiXi National Wetland Park (Fig. 27.3b) in Hangzhou (Zhejiang province), a metropolis of nearly 6.5 million people. The park, covering nearly 10 km^2, is established within a wetland landscape whose cultural importance dates back at least to 223 CE. Historic buildings are interspersed with conservation areas, and a large education center offers programming that focuses on the functions and values of wetlands. The park receives thousands of visitors each year (www.xixiwetland.com.cn/en/main.do).

Research directions

Despite a rapidly increasing amount of interest by urban residents, planners and policymakers in maintaining wetlands within urban regions, there has been surprisingly little research on the functions and services of wetlands in urban areas, and therefore little guidance to help managers. Indeed, the potential for conflict and the need for trade-offs in providing multiple functions in urban regions is high (Ehrenfeld 2004). We offer the following suggestions about research

(a)

(b)

Figure 27.3 Examples of urban wetland systems (a) Hackensack Meadowlands, NJ (the New York City skyline in visible in the background); (b) XiXi National Wetland Park, Hangzhou, China.

directions that could improve the conservation, management and restoration of wetlands in urban environments.

1 Research is needed on all aspects of wetland and urban ecology on a greater range of wetland types, over a greater range of climates, hydrologic regimes, urban land uses, and socioeconomic systems than currently exists. The small number of regions and wetland types that have been extensively researched make broad generalizations about urban wetlands unwarranted at present.
2 Despite a wealth of research on urban hydrology, there is little research on the hydrologic responses of non-riparian, groundwater and precipitation-fed wetlands to urbanization. A more detailed and mechanistic linking of urban hydrology to these other kinds of wetlands is urgently needed.
3 As with hydrology, a better knowledge of nutrient and pollutant retention mechanisms in wetlands other than riparian sites is needed. Source-sink relationships of nutrients, metals, and toxic hydrocarbon compounds in urban wetlands are urgently needed to quantify the water-quality-maintenance functions, and to support wetland management to reliably produce this service.
4 Specific quantification of a wide range of ecosystem services in urban regions should be a high priority for research. Before wetlands can be managed to provide ecosystem services, there needs to be clear documentation of the conditions under which specific services are provided, at what rates, and how the integrity of these ecosystems can be protected in a market environment.

References

Aguiar, F.C., Ferreira, M.T., Albuquerque, A., and Moreira, I. (2007) 'Alien and endemic flora at reference and non-reference sites in Mediterranean-type streams in Portugal,' *Aquatic Conservation-Marine and Freshwater Ecosystems*, 17(4): 335–47.

Alston, K.P. and Richardson, D.M. (2006) 'The roles of habitat features, disturbance, and distance from putative source populations in structuring alien plant invasions at the urban/wildland interface on the Cape Peninsula, South Africa,' *Biological Conservation*, 132(2): 183–98.

Atchison, K.A. and Rodewald, A.D. (2006) 'The value of urban forests to wintering birds,' *Natural Areas Journal*, 26(3): 280–8.

Azous, A.L. and Horner, R.R. (eds.) (2001) *Wetlands and Urbanization: Implications for the Future*, Boca Raton, FL: Lewis Publishers.

Barrett, K.R. and McBrien, M.A. (2007) 'Chemical and biological assessment of an urban, estuarine marsh in northeastern New Jersey, USA,' *Environmental Monitoring and Assessment*, 124(1–3): 63–88.

Bedford, B. (1996) 'The need to define hydrologic equivalence at the landscape scale for freshwater wetland mitigation,' *Ecological Applications*, 6(1): 57–68.

Bernhardt, E.S., Band, L.E., Walsh, C.J., and Berke, P.E. (2008) 'Understanding, managing, and minimizing urban impacts on surface water nitrogen loading,' *Annals of the New York Academy of Sciences No. 1134, Year in Ecology and Conservation Biology 2008*, pp. 61–96.

Booth, D.B., Karr, J.R., Schauman, S., Konrad, C., Morley, S.A., Larson, M.G., and Burges, S.J. (2004) 'Reviving urban streams: land use, hydrology and human behaviour,' *Journal of the American Water Resources Association*, 40(5): 1351–65.

Brander, L.M., Florax, R.J.G.M., and Vermaat, J.E. (2006) 'The empirics of wetland valuation: a comprehensive summary and a meta-analysis of the literature,' *Environmental and Resource Economics*, 33(2): 223–50.

Brinson, M.M. (1993) *A Hydrogeomorphic Classification for Wetlands*, Vicksburg, MS: U.S. Army Engineer Waterways Experiment Station.

Burton, M.L. and Samuelson, L.J. (2008) 'Influence of urbanization on riparian forest diversity and structure in the Georgia Piedmont, US,' *Plant Ecology*, 195(1): 99–115.

Chace, J.F. and Walsh, J.J. (2006) 'Urban effects on native avifauna: a review,' *Landscape and Urban Planning*, 74(1): 46–69.

Cowardin, L.M., Carter, V., Golet, F.C., and LaRoe, E.T. (1979) *Classification of wetlands and deepwater habitats of the United States*, Washington, DC: U.S. Department of the Interior, U.S. Fish and Wildlife Service, Office of Biological Services.

Cutway, H.B. (2004) *Ecology, Evolution, and Natural Resources*, New Brunswick, NJ: Rutgers University.

Cutway, H.B. and Ehrenfeld, J.G. (2009) 'Exotic plant invasions in forested wetlands: effects of adjacent urban land use type,' *Urban Ecosystems*, 12(1): 371–90.

D'Antonio, C.M., Dudley, T.L., and Mack, M.C. (1999) 'Disturbance and biological invasions: direct effects and feedbacks,' in L.R. Walker (ed.), *Ecosystems of Disturbed Ground*, Amsterdam: Elsevier, pp. 413–452.

Doss, C.R. and Taff, S.J. (1996) 'The influence of wetland type and wetland proximity on residential property values,' *Journal of Agricultural and Resource Economics*, 21(1): 20–129.

Ehrenfeld, J.G. (2000) 'Evaluating wetlands within an urban context,' *Urban Ecosystems*, 4(1): 69–85.

Ehrenfeld, J.G. (2004) 'The expression of multiple functions in urban forested wetlands,' *Wetlands*, 24(4): 719–33.

Ehrenfeld, J.G. (2005) 'Vegetation of forested wetlands of urban and suburban landscapes in New Jersey,' *Journal of the Torrey Botanical Society*, 132(2): 262–79.

Ehrenfeld, J.G. (2008) 'Exotic invasive species in urban wetlands: environmental correlates and implications for wetland management', *Journal of Applied Ecology*, 45(4): 1160–9.

Ehrenfeld, J.G. and Schneider, J.P. (1991) '*Chamaecyparis thyoides* wetlands and suburbanization: effects of nonpoint source water pollution on hydrology and plant community structure,' *Journal of Applied Ecology*, 28(2): 467–90.

Ehrenfeld, J.G., Cutway, H.B., Hamilton IV, R., and Stander, E. (2003) 'Hydrologic description of forested wetlands in northeastern New Jersey, USA – an urban/suburban region,' *Wetlands*, 23(4): 685–700.

Germaine, S.S., Rosenstock, S.S., Schweinsburg, R.E., and Richardson, W.S. (1998) 'Relationships among breeding birds, habitat, and residential development in Greater Tucson, Arizona,' *Ecological Applications*, 8(3): 680–91.

Groffman, P.M., Bain, D.J., Band, L.E., Belt, K.T., Brush, G.S., Grove, J.M., Pouyat, R.V., Yesilonis, I.C., and Zipperer, W.C. (2003) 'Down by the riverside: urban riparian ecology,' *Frontiers in Ecology and the Environment*, 1(6): 315–21.

Groffman, P.M., Boulware, N.J., Zipperer, W.C., Pouyat, R.V., Band, L.E., and Colosimo, M.F. (2002) 'Soil nitrogen cycle processes in urban riparian zones,' *Environmental Science and Technology*, 36(21): 4547–52.

Hackensack Meadows Initative (2007) *Preliminary Conservation Planning for the Hackensack Meadowlands, Hudson and Bergen Counties, New* Jersey, Pleasantville, NJ: US Fish and Wildlife Service, New Jersey Field Office.

Hamer, A.J. and McDonnell, M.J. (2008) 'Amphibian ecology and conservation in the urbanising world: A review,' *Biological Conservation*, 141(10): 2432–49.

Hanson, G.C., Groffman, P.M., and Gold, A.J. (1994) 'Symptoms of nitrogen saturation in a riparian wetland,' *Ecological Applications*, 4(4): 750–6.

Hennings, L.A. and Edge, W.D. (2003) 'Riparian bird community structure in Portland, Oregon: Habitat, urbanization, and spatial scale patterns,' *Condor*, 105(2): 288–302.

Hogan, D.M. and Walbridge, M.R. (2007) 'Best management practices for nutrient and sediment retention in urban stormwater runoff,' *Journal of Environmental Quality*, 36(2): 386–95.

Horner, R.R., Cooke, S.S., Reinelt, L.E., Ludwa, K.A., and Chin, N.T. (2001) 'Water quality and soils,' in A.L. Azous and R.R. Horner (eds.), *Wetlands and Urbanization: Implications for the Future*, Boca Raton, FL: Lewis Publishers, pp. 237–254.

Houlahan, J.E., Keddy, P.A., Makkay, K., and Findlay, C.S. (2006) 'The effects of adjacent land-use on wetland species richness and community composition,' *Wetlands*, 26(1): 79–96.

Kennish, M.J. (2002) 'Environmental threats and environmental future of estuaries,' *Environmental Conservation*, 29(1): 78–107.

Kentula, M.E., Gwin, S.E., and Pierson, S.M. (2004) 'Tracking changes in wetlands with urbanization: Sixteen years of experience in Portland, Oregon, USA,' *Wetlands*, 24(4): 734–43.

Kimbrough, K.L. and Dickhut, R.M. (2006) 'Assessment of polycyclic aromatic hydrocarbon input to urban wetlands in relation to adjacent land use,' *Marine Pollution Bulletin*, 52(11): 1355–63.

Kiviat, E. and MacDonald, K. (2004) 'Biodiversity Patterns and Conservation in the Hackensack Meadowlands, New Jersey,' *Urban Habitats*, 2(1): 28–61.

Knutson, M.G., Sauer, J.R., Olsen, D.A., Mossman, M.J., Hemesath, L.M., and Lannoo, M.J. (1999) 'Effects of landscape composition and wetland fragmentation on frog and toad abundance and species richness in Iowa and Wisconsin, USA,' *Conservation Biology*, 13(6): 1437–46.

Launen, L.A., Dutta, J., Turpeinen, R., Eastep, M.E., Dorn, R., Buggs, V.H., Leonard, J.W., and Haggblom, M.M. (2008) 'Characterization of the indigenous PAH-degrading bacteria of Spartina dominated salt marshes in the New York/New Jersey Harbor,' *Biodegradation*, 19(3): 347–63.

Lavoie, C., Jean, M., Delisle, F., and Létourneau, G. (2003) 'Exotic plant species of the St. Lawrence River wetlands: a spatial and historical analysis,' *Journal of Biogeography*, 30(4): 537–49.

Leck, M.A. and Leck, C.F. (2005) 'Vascular plants of a Delaware River tidal freshwater wetland and adjacent terrestrial areas: Seed bank and vegetation comparisons of reference and constructed marshes and annotated species list,' *Journal of the Torrey Botanical Society*, 132(2): 323–54.

Lee, S.Y., Dunn, R.J.K., Young, R.A., Connolly, R.M., Dale, P.E.R., Dehayr, R., Lemckert, C.J., McKinnon, S., Powell, B., Teasdale, P.R., and Welsh, D.T. (2006) 'Impact of urbanization on coastal wetland structure and function,' *Australian Ecology*, 31(2):149–63.

Leston, L.F.V. and Rodewald, A.D. (2006) 'Are urban forests ecological traps for understory birds? An examination using Northern cardinals,' *Biological Conservation*, 131(4): 566–74.

Lougheed, V.L., McIntosh, M.D., Parker, C.A., and Stevenson, R.J. (2008) 'Wetland degradation leads to homogenization of the biota at local and landscape scales,' *Freshwater Biology*, 53(12): 2402–13.

Lussier, S.M., Enser, R.W., Dasilva, S.N., and Charpentier, M. (2006) 'Effects of habitat disturbance from residential development on breeding bird communities in riparian corridors,' *Environmental Management*, 38(3): 504–21.

Luther, D., Hilty, J., Weiss, J., Cornwall, C., Wipf, M., and Ballard, G. (2008) 'Assessing the impact of local habitat variables and landscape context on riparian birds in agricultural, urbanized, and native landscapes,' *Biodiversity and Conservation*, 17(8): 1923–35.

Magee, T.K., Ernst, T.L., Kentula, M.E., and Dwire, K.A. (1999) 'Floristic comparison of freshwater wetlands in an urbanizing environment,' *Wetlands*, 19(3): 517–34.

Mahan, B.L., Polasky, S., and Adams, R.M. (2000) 'Valuing urban wetlands: A property price approach,' *Land Economics*, 76(1): 100–13.

Marshall, S. (2004) 'The Meadowlands Before the Commission: Three Centuries of Human Use and Alteration of the Newark and Hackensack Meadows,' *Urban Habitats*, 2(1): 4–27.

Mason, J., Moorman, C., Hess, G., and Sinclair, K. (2007) 'Designing suburban greenways to provide habitat for forest-breeding birds,' *Landscape and Urban Planning*, 80(1–2): 153–64.

McDonnell, M.J. and Hahs, A.K. (2008) 'The use of gradient analysis studies in advancing our understanding of the ecology of urbanizing landscapes: current status and future directions,' *Landscape Ecology*, 23(10): 1143–55.

Miller, J.E., Hess, G.R., and Moorman, C.E. (2007) 'Southern two-lined salamanders in urbanizing watersheds,' *Urban Ecosystems*, 10(1): 73–85.

Miller, J.R., Wiens, J.A., Hobbs, N.T., and Theobald, D.M. (2003) 'Effects of human settlement on bird communities in lowland riparian areas of Colorado (USA),' *Ecological Applications*, 13(4): 1041–59.

Mitsch, W. and Gosselink, J.G. (2000) *Wetlands*, third edition, New York, NY: Van Nostrand Reinhold.

Moffatt, S.F. and McLachlan, S.M. (2004) 'Understorey indicators of disturbance for riparian forests along an urban-rural gradient in Manitoba,' *Ecological Indicators*, 4(1): 1–16.

Moffatt, S.F., McLachlan, S.M., and Kenkel, N.C. (2004) 'Impacts of land use on riparian forest along an urban-rural gradient in southern Manitoba,' *Plant Ecology*, 174(1): 119–35.

Oneal, A.S. and Rotenberry, J.T. (2008) 'Riparian plant composition in an urbanizing landscape in southern California, USA,' *Landscape Ecology*, 23(5): 553–67.

Palmer, G.C., Fitzsimons, J.A., Antos, M.J., and White, J.G. (2008) 'Determinants of native avian richness in suburban remnant vegetation: Implications for conservation planning,' *Biological Conservation*, 141(9): 2329–41.

Panno, S.V., Nuzzo, V.A., Cartwright, K., Hensel, B.R., and Krapac, I.G. (1999) 'Impact of urban development on the chemical composition of ground water in a fen-wetland complex,' *Wetlands*, 19(1): 236–45.

Paul, M.J. and Meyer, J.L. (2001) 'Streams in the urban landscape,' *Annual Review of Ecology and Systematics*, 32: 333–66.

Pennington, D.N., Hansel, J., and Blair, R.B. (2008) 'The conservation value of urban riparian areas for landbirds during spring migration: Land cover, scale, and vegetation effects,' *Biological Conservation*, 141(5): 1235–48.

Pillsbury, F.C. and Miller, J.R. (2008) 'Habitat and landscape characteristics underlying anuran community structure along an urban-rural gradient,' *Ecological Applications*, 18(5): 1107–18.

Predick, K.I. and Turner, M.G. (2008) 'Landscape configuration and flood frequency influence invasive shrubs in floodplain forests of the Wisconsin River (USA),' *Journal of Ecology*, 96(1): 91–102.

Price, S.J., Dorcas, M.E., Gallant, A.L., Klaver, R.W., and Willson, J.D. (2006) 'Three decades of urbanization: Estimating the impact of land-cover change on stream salamander populations,' *Biological Conservation*, 133(4): 436–41.

Pyle, L.L. (1995) 'Effects of disturbance on herbaceous exotic plant species on the floodplain of the Potomac River,' *American Midland Naturalist*, 134(2): 244–53.

Quinn, J. (1997) *Fields of Sun and Grass: An Artist's Journal of the New Jersey Meadowlands*, New Brunswick, NJ: Rutgers University Press.

Ravit, B., Haggblom, M., and Ehrenfeld, J.G. (2005) 'Salt marsh rhizosphere affects microbial transformation of the widespread halogenated contaminant tetrabromobisphenol A (TBBPA),' *Soil Biology and Biochemistry*, 37(6): 1049–57.

Rodewald, A.D. and Bakermans, M.H. (2006) 'What is the appropriate paradigm for riparian forest conservation?' *Biological Conservation*, 128(2): 193–200.

Sanger, D.M., Holland, A.E., and Scott, G.I. (1999a) 'Tidal creek and salt marsh sediments in South Carolina coastal estuaries: I. Distribution of trace metals,' *Archives of Environmental Contamination and Toxicology*, 37(4): 445–57.

Sanger, D.M., Holland, A.E., and Scott, G.I. (1999b) 'Tidal creek and salt marsh sediments in South Carolina coastal estuaries: II. Distribution of organic contaminants,' *Archives of Environmental Contamination and Toxicology*, 37(4): 458–71.

Seitzinger, S., Harrison, J.A., Bohlke, J.K., Bouwman, A.F., Lowrance, R., Peterson, B., Tobias, C., and Van Drecht, G. (2006) 'Denitrification across landscapes and waterscapes: A synthesis,' *Ecological Applications*, 16(6): 2064–90.

Stander, E.K. (2007) *The Effects of Urban Hydrology and elevated Atmospheric Deposition on Nitrate Retention and Loss in Urban Wetlands*, New Brunswick, NJ: PhD thesis, Rutgers University.

Stander, E.K. and Ehrenfeld, J.G. (2009) 'Rapid assessment of urban wetlands: Do hydrogeomorpic classification and reference criteria work?' *Environmental Management*, 43(4): 725–42.

Sundell-Turner, N.M. and Rodewald, A.D. (2008) 'A comparison of landscape metrics for conservation planning,' *Landscape and Urban Planning*, 86(3–4): 219–25.

Thibault, P.A. and Zipperer, W.C. (1994) 'Temporal changes of wetlands within an urbanizing agricultural landscape,' *Landscape and Urban Planning*, 28(2–3): 245–51.

Tsipoura, N., Burger, J., Feltes, R., Yacabucci, J., Mizrahi, D., Jeitner, C., and Gochfeld, M. (2008) 'Metal concentrations in three species of passerine birds breeding in the Hackensack Meadowlands of New Jersey,' *Environmental Research*, 107(2): 218–28.

Vidra, R.L. and Shear, T.H. (2008) 'Thinking locally for urban forest restoration: A simple method links exotic species invasion to local landscape structure,' *Restoration Ecology*, 16(2): 217–20.

Vidra, R.L., Shear, T.H., and Wentworth, T.R. (2006) 'Testing the paradigms of exotic species invasion in urban riparian forests,' *Natural Areas Journal*, 26(4): 339–50.

Villagran-Mella, R., Aguayo, M., Parra, L.E., and Gonzalez, A. (2006) 'Relationship between habitat characteristics and insect assemblage structure in urban freshwater marshes from central-south Chile,' *Revista Chilena De Historia Natural*, 79(2): 195–211.

Walsh, C.J., Roy, A.H., Feminella, J.W., Cottingham, P.D., Groffman, P.M., and Morgan, R.P. (2005) 'The urban stream syndrome: current knowledge and the search for a cure,' *Journal of the North American Benthological Society*, 24(3): 706–23.

Wang, X.L., Ning, L.M., Yu, J., Xiao, R., and Li, T. (2008) 'Changes of urban wetland landscape pattern and impacts of urbanization on wetland in Wuhan City,' *Chinese Geographical Science*, 18(1): 47–53.

Weis, J.S., Windham, L., and Weis, P. (2003) 'Patterns of metal accumulation in leaves of the tidal marsh plants Spartina alterniflora Loisel and Phragmites australis Cav. Trin Ex Steud. over the growing season,' *Wetlands*, 23(2): 459–65.

Weis, P., Windham, L., Burke, D.J., and Weis, J.S. (2002) 'Release into the environment of metals by two vascular salt marsh plants,' *Marine Environmental Research*, 54(3–5): 325–29.

Willson, J.D. and Dorcas, M.E. (2003) 'Effects of habitat disturbance on stream salamanders: Implications for buffer zones and watershed management,' *Conservation Biology*, 17(3): 763–71.

Windham, L., Weis, J., and Weis, P. (2001) 'Lead uptake, distribution and effects in two dominant salt marsh macrophytes, *Spartina alterniflora* (Cordgrass) and *Phragmites australis* (Common Reed),' *Marine Pollution Bulletin*, 42(10): 811–16.

Wolin, J.A. and MacKeigan, P. (2005) 'Human influence past and present – Relationship of nutrient and hydrologic conditions to urban wetland macrophyte distribution,' *Ohio Journal of Science*, 105(5): 125–32.

Zhu, W.-X. and Ehrenfeld, J.G. (1999) 'Nitrogen mineralization and nitrification in urbanized Atlantic white cedar wetlands,' *Journal of Environmental Quality*, 28(2): 523–29.

Zhu, W.-X. and Ehrenfeld, J.G. (2000) 'Nitrogen retention and release in Atlantic white cedar wetlands,' *Journal of Environmental Quality*, 29(2): 612–20.

28
Urban animal ecology

Peter J. Jarvis

Introduction

Animals require food, shelter for every-day activities, and appropriate places for breeding. In towns and cities they are often opportunistic in where they live and what they eat. They have to be flexible and adaptable in their requirements, and generally have broad food ranges (food generalists and scavengers). They can often tolerate disturbance of various kinds, particularly from human activity, and generally live in closer proximity to humans than their equivalents in non-urban areas. Successful mammals tend to be nocturnal.

There is almost certainly some kind of size threshold for the sensitivity of animal communities to urbanization and town size below which the impact on community attributes is trivial, but above which can be dramatic, as suggested by Garaffa *et al.* (2009) in their comparison of bird communities in different-sized settlements in Argentina.

In their overview of the effects of urbanization on bird communities (much of which can be extrapolated to other animal groups), Chace and Walsh (2006) report that omnivorous and granivorous feeders tend to be particularly successful. Many animal groups respond to vegetation composition and structure, and areas that retain native vegetation generally retain more native animal species. This point has been demonstrated in Melbourne, Australia, for example: overall abundance and species richness were lower in exotic and recently-developed street landscapes than in parks and patches with native vegetation, and parks had the greatest number of foraging guilds (including granivory, insectivory, nectarivory and omnivory) (White *et al.* 2005).

Fecundity in urban areas is a reflection of species-specific adaptability to urban resources, and to levels of predation. Increased urbanization typically leads to an increase in animal biomass but a reduction in species richness. Bird species density increases almost concentrically away from central London, as does average bird size (despite the abundance of feral pigeons *Columba livia* in the central area), but in land snails it is the larger species (particularly *Helix pomatia*) and individual size that are associated with inner London, probably because water relations, critical for snail survival, are more suitable here (Cousins 1982).

Habitat use by urban animals

Blair (1996) has grouped city animals into urban exploiters (adept at exploiting the ecosystem changes caused by urban growth), urban avoiders (sensitive to anthropogenic changes), and suburban adaptable species (able to exploit additional resources such as ornamental vegetation).

Animals in towns may be found:

- On or in humans and their pets, for example head lice, fleas and intestinal worms; or associated with building interiors, for instance clothes moths and silverfish.
- Primarily associated with technological landscapes: a number of arthropods (McIntyre 2000; Pacheco and Vasconcelos 2007) and feral pigeons in concrete-dominated landscapes, for example, and gulls on roof tops and landfill sites (though all are clearly also found elsewhere). In 1997, gull colonies on buildings hosted over 8 per cent and around 4 per cent of the British population of herring gull *Larus argentatus* and lesser black-backed gull *L. fuscus*, compared to 0.6 per cent for herring gull in 1976, and even less for the black-backed, demonstrating the rapidly increasing importance of the built environment for these seabirds (Monaghan 1982; Raven and Coulson 1997).
- In substitute or analogue urban habitats, sometimes using constructed features such as buildings (analogous to cliffs, see Chapter 20) or radio masts (equivalent to trees), but also relying on green habitats such as parks and gardens, especially for food. Gardens are a critical habitat for many urban animals (Owen 1991; Bland *et al.* 2004; Cannon *et al.* 2005; Daniels and Kirkpatrick 2006; and see Chapter 26). So-called brownfield sites (derelict or post-industrial landscapes) are also important for many animal groups, for example carabid ground beetles (Small *et al.* 2003).
- Species which rely on a wide dispersed habitat range. They may be primarily rural, but they exploit urban habitats temporarily at some stage of their lives, for example during winter.
- Species which have no particular affinity for or tolerance of urban conditions, but which have become incorporated into the urban fauna within patches of land which are similar or analogous to rural conditions (for example woodland or so-called wasteland), and large enough and sufficiently undisturbed to give them refuge.

Spatial patterns in animal distribution

Many studies have demonstrated that there often exists a gradient of change in species diversity and abundance along an urban-suburban-rural gradient, with peaks in suburbia, supporting the intermediate disturbance hypothesis. At Palo Alto, California, for example, species richness and diversity of both birds and butterflies peak at intermediate levels of urban development; bird abundance was also greatest in suburbia, but butterfly abundance declined with increasing building development (Blair 1996, 1999; Blair and Launer 1997). Similar trends have been noted for birds in other countries, for example in Dunedin, New Zealand (Van Heezik *et al.* 2008), and for other animal groups, for example rodents in Buenos Aires, Argentina (Cavia *et al.* 2009) and carabid ground beetles in Debrecen, Hungary (Magura *et al.* 2008). Lizard abundance and species number in Tucson, Arizona, also decline along a five-point scale of increasing housing density (Germaine and Wakeling 2000). Generalist species tend to be more abundant in more urbanised areas.

Seasonal effects must also be considered, many birds for example making broader use of the urban landscape in winter, even in Mediterranean towns, such as Montpellier, France, where the urban heat island is not especially pronounced (Caula *et al.* 2008).

Frequency of nest predation can be dependent on position along an urban gradient. Predation pressure in Palo Alto exhibits an overall decline from natural to urban sites, suggesting that urban environments have relatively low predation pressures (Gering and Blair 2006), a situation also noted in Quebec, Canada, and Rennes, France (Clergeau *et al.* 1998), and Finland (Jokimäki and Huhta 2008). Predatory relaxation in urban environments may partially explain the greater abundance of some species in urban environments, for example starlings *Sturnus vulgaris*, house sparrows *Passer domesticus* and feral pigeons.

Melles *et al.* (2003), however, while finding that species richness of birds declined with increasing urbanization in Vancouver and Burnaby, British Columbia, argue that both local and landscape-scale resources are important in determining urban bird distributions, with local-scale features such as large trees, berry-producing shrubs and streams being particularly important in elevating bird abundance.

Habitat fragmentation, size reduction, increased edge effect and isolation might all generally be predicted to reduce diversity and have an adverse effect on population dynamics, while green patches might be expected to increase animal species diversity and abundance, and habitat corridors accentuate such trends by increasing opportunities for dispersal (Adams *et al.* 2006). For many invertebrates, however, dispersal is generally not a limiting factor in presence or persistence. However, small- and medium-sized mammals may use habitat corridors, and water-based specialists such as otters *Lutra lutra* and water voles *Arvicola terrestris* certainly rely on appropriate linear features (Angold *et al.* 2006).

Food in the urban habitat

In many habitats found in parts of the townscape, such as woodland and tall herb communities, there are few differences between urban and non-urban patterns of feeding behaviour, though environmental factors such as the urban heat island and flooding of channelized streams may modify what is available in the way of food.

In the actual built environment, a few animal species depend entirely on humans providing food, for example feral pigeons and some urban waterfowl. Humans often importantly supplement food, for example providing bird feeders in gardens. Most successful urban animals are opportunist feeders, sometimes adopting innovative approaches to acquiring food, for example the piercing of milk bottle tops by birds such as blue tits *Parus caeruleus*, first reported from near Southampton, southern England in 1921, and its subsequent geographical spread (Fisher and Hinde 1949), possibly by copying (Sherry and Galef 1984), and seen as a form of 'cultural' transmission. Bottle opening originated at several independent sites, and has spread through an accelerating process that could have included direct and/or indirect social influences (Lefebvre 1995).

Animals can also predict regular fluctuations in the food supply, for example pigeons at humans' lunch time, or swans *Cygnus olor* in city areas such as Krakow, Poland, and London (Jozkowicz and Gorska-Klek 1996; Sears 1989).

Some urban animals are carnivorous, for example birds of prey, feral cats, amphibians and an array of invertebrates. Many animals in urban habitats are omnivores. Many essentially plant-eating birds require some insect food at the fledgling stage. Herbivores generally take a wide range of plants or plant products such as seeds and berries, i.e. they are polyphagous.

Oligophages (animals with a small dietary range) or monophages (those relying on a single kind of food) must have a guaranteed and predictable supply of their food plant(s) over time. Scavenging involves the use of waste material, storage areas, roadkill etc. Scavengers include foxes *Vulpes vulpes* and feral cats, as well as mice and rats, cockroaches and silverfish. Invertebrates such as woodlice and millipedes live on detritus, and there is also a sizeable xylophagous (wood-eating) fauna.

Some animals require different habitats to provide food for both adults and young, for example butterflies and amphibians.

Shelter and microclimate

Many urban structures provide shelter for animals, sometimes literally providing a roof over their heads. A variety of microclimates, for instance, are provided by gardens. As well as a heat island, towns also provide a number of open habitats that provide warm environments, for example much early succession-stage wasteland includes bare or thinly-vegetated ground, which can heat up quickly. Similarly, on sites used by humans for recreation, where feet and wheels have destroyed the vegetation cover, warm, exposed soil may attract a range of insects and even some snakes and lizards. If such habitats attract a range of insects, they in turn will attract insectivores such as hedgehogs *Erinaceus europaeus* and shrews *Sorex* species as well as a variety of birds.

Social behaviour and population dynamics

The social structure of many higher animals, particularly mammals and birds, may differ in urban compared to non-urban areas. Flocking may be more common in birds, for example, and large roosts might exploit limited nocturnal resting sites, for instance the 350–500 pied wagtails *Motacilla alba* that each year roost in just three trees in central Wolverhampton (Jarvis and Clifford 1998).

Starlings *Sturnus vulgaris* were a common sight in many British city centres in the mid-twentieth century, commuting to the suburbs or surrounding countryside during the day to eat, but returning at dusk in huge numbers in order to roost. Birmingham, UK, for instance, had a population of some 30,000 starlings during the 1960s and 1970s (Shirley 1996) when numbers suddenly crashed. The majority of British starlings are migrants from Eastern Europe and Russia, and a series of cold winters in the 1970s had caused a dramatic population decline.

Starling numbers have only really begun to recover during the twenty-first century. Now many starlings seem to stay in the city centre and compete with pigeons for scraps of food. Elsewhere in Europe, however, the story has been different. In the 1950s, for instance, a few thousand starlings used Perpignan in south-western France as a stopover during their autumn migration from northern Europe to North Africa. In the 1970s their numbers doubled, in the 1990s they doubled again, and this trend has continued. As well as damaging trees the Perpignan starlings, now numbering perhaps two million birds, leave two metric tonnes of droppings each day (Lichfield 1997: 37). One of the advantages that starlings have in adapting to the urban environment is their behavioural plasticity and flexibility in habitat selection, characteristics indeed of many successful urban animals. Clergeau and Quenot (2007), for example, demonstrate this for roost-site selection by urban starlings in Rennes, France.

Territorial behaviour may also differ, though some animals may be more tolerant of boundary incursions, others more possessive of scarce territorial resources than in the countryside. Social structure may have to accommodate a more dispersed population. Territorial display, including song, may be altered in some bird species. Songs in urban great tits *Parus major* are shorter and faster than their woodland counterparts: results from ten European cities suggest that this has been a response to anthropogenic noise, and is an indication of a behavioural plasticity that may be critical for living in towns (Slabbekoorn and den Boer-Visser 2006).

Urban animals may show signs of stress that could be deleterious to survival, but Partecke *et al.* (2006) have shown that blackbirds *Turdus merula* born in Munich have lower stress levels than forest conspecifics, a genetically-determined response, suggesting that urbanization can change the stress physiology of animals to allow them to cope.

Population dynamics may be different in towns. Above a certain level of availability, food is rarely a limiting factor. Animals may have lower life expectancies, especially having to withstand greater casualty numbers from sources such as motor vehicles; electrocution, and impact on structures such as TV masts and power lines; drowning in canals and ornamental pools; poisoning; and confinement in litter, netting etc.

Disease is a little understood aspect of mortality in urban animals. Spread of disease is affected by population density, rates of contact, feeding behaviour, breeding, genetics, general health and age profiles, and these may be very different for urban animals compared to their rural cousins.

Status

A few animal species found in towns are rare and endangered, for example the black redstart *Phoenicurus ochros*, whose distribution in Britain is almost entirely urban. Status changes over time. Red kite *Milvus milvus*, for example, a common scavenger in medieval England disappeared during early modern times, and is now found scattered in a few rural areas of Britain. The decline in starling numbers noted above led to a reduction in competition for nest holes in urban trees which in turn led to an increase in great spotted woodpecker *Dendrocopos major* numbers in British towns (Smith 2005).

The house sparrow is another example. This bird became very common in late Victorian and early Edwardian Britain, in large part because of the growing amount of horse-drawn transport: horses were fed grain, and the sparrows ate both spillage and grain that had been excreted by the horses. With motor transport taking over from the horse, sparrow numbers declined dramatically. During the last 25 years sparrow numbers have halved and the species has become Red Listed; reasons are unclear, but might involve a reduction in both seed and invertebrates (the latter required by fledglings) in manicured, pesticide-swamped gardens.

Public attitudes to urban wildlife

In the last half century people have generally become not just more tolerant to urban wildlife but also get a good deal of pleasure from observing it, and in many cases have become pro-active by feeding animals in parks and gardens. There remain a number of animal species, of course, that are undisputedly pests, for example, cockroaches, Argentine ants *Linepithema humile* (*Iridomyrmex humilis*), rats and mice. A national overview of the ten vertebrates of greatest concern in American cities identified, in order of greatest to least magnitude of damage, raccoons *Procyon lotor*, coyotes *Canis latrans*, skunks *Mephitis* species, beavers *Castor canadensis*, deer, geese, squirrels, opossums *Didelphis virginiana*, foxes and red-winged blackbirds *Agelaius phoeniceus* (Adams *et al.* 2006). Many of these animals are attracted to towns because of the availability of food and shelter, and a relative absence of predators. It is right that attempts are made at controlling pests. Many animals, however, are viewed with mixed feelings (DeStefano and Deblinger 2005), for example foxes, raccoons and even domestic cats.

The urban red fox

The red fox *Vulpes vulpes* is an example of an animal whose status in urban areas has changed over the last 100 years. The key elements of the fox's life cycle are much the same in towns as in the countryside, with mating in mid-winter, females acquiring a den in February and giving birth in March. Cubs emerge in April and remain with the mother until October, when a mass dispersal is noted. Young adults take over vacant territories in November. Dispersed animals born near the centre of Bristol, UK, lived significantly longer than those born nearer the city edge (Harris and Trewhella 1988).

Foxes are found in a number of British towns, though not in every one. Densities vary, and are locally high (five fox family groups/km^2), but average densities are much lower, ranging from >2 families/km^2 in Poole, Dorset, to 0.188 families/km^2 in Wolverhampton.

By looking at the urban morphology of 157 towns in England and Wales, Harris and Rayner (1986) explained distribution of urban foxes as follows: the 1930s saw a change in housing policy that led to low-density, owner-occupied residential suburbs – ideal habitat for foxes. Foxes moved in from the surrounding countryside or were 'engulfed' by suburban expansion. Once the suburbs had been colonized foxes then moved further into less favoured urban habitats, maintaining themselves there at lower densities. The rarity of urban foxes in most northern and eastern towns in England may be explained by the greater incidence there of high-density council-rented housing and of industry. Where they are found, preferred habitats are gardens associated with detached and semi-detached housing, woodland, parkland and tall-herb sites such as wasteland and railway embankments.

Sarcoptic mange is endemic in many urban fox populations, occasionally leading to a population crash. Mange has spread through Bristol's fox population since the mid-1990s, and is now the most significant mortality factor, possibly exacerbated by the high density of foxes and the presence of dogs as an additional host for the ectoparasitic mite. In 2004, Bristol's fox population was only 15 per cent of that in 1994 (Soulsbury *et al.* 2007).

About a third of the fox's food is scavenged from rubbish tips, dustbins, bird tables and litter from fast-food outlets. Urban foxes eat fruits and berries, hunt birds, rabbits and a variety of small mammals, and consume a variety of invertebrates, especially earthworms (Macdonald 1989). Questionnaire surveys by the author suggest a shift in public opinion since the 1970s, when around 40 per cent of respondents viewed urban foxes as welcome, to the twenty-first century when a 75 per cent positive rating has been found, largely a consequence of knowing more about the animal through television programmes.

The urban raccoon

In many ways the habits of the raccoon and public attitudes towards this species in the USA echo those of the fox in Britain. This is another omnivore and scavenger whose activities lead to complaints of damage, noise and it being a vector of rabies, and it has consistently been ranked highest in perceived vertebrate pest problems, for example by 26 per cent of home-owners in Chicago (Gehrt 2004). Urban home ranges are 5–79 ha compared to the 50–300 ha of rural areas, and density in towns range from 55/km^2 up to 333/km^2 in parts of Washington DC (Riley *et al.* 1998). Perhaps less rehabilitated in the mind of the public than the fox, the raccoon is nevertheless welcomed by a sizeable minority of urban dwellers, and indeed food is often deliberately left out for these animals.

Urban cats and their impact on wildlife

Of the eight million pet cats in the UK (Pet Food Manufacturers Association 2010) and up to 65 million in the USA (Center for Information Management 1997) the majority live in towns and cities. In both of these countries cat numbers are increasing by about 13 per cent per annum. Australia has seen a decline in pet cat numbers from 3.2 million in 1988 to 2.4 million cats in 2005 (Baldock *et al.* 2003; Kendall 2006), but again a majority are found in towns (unlike the large number of ferals that live in the countryside).

Cats are predators, so it is not surprising that the effects of predation on urban wildlife have been debated. Two studies that have suggested wide-scale slaughter are methodologically unsound. The estimation by May (1988) that 100 million birds and small mammals were killed each year in Great Britain is based on inappropriately extrapolating data on feline predation in one village (Churcher and Lawton 1987; Jarvis 1990). Similarly, a study by Britain's Mammal Society used questionnaire responses by a self-selected and (in a number of critical ways) non-random sample to identify prey type and number (Woods *et al.* 2003), concluding that some 250 million mammals, birds, reptiles and amphibians were killed each year. Extrapolating results from a more scientific study in a part of Bristol, UK, would provide estimates of prey numbers an order of magnitude lower (Baker *et al.* 2005).

Whatever the actual figures, a great deal of urban wildlife is killed by cats. The key question, though, is whether cats prey on individuals that would have died in any case (from other predators, starvation, disease or weather) – the 'doomed surplus' hypothesis – or whether cat predation is additive – the 'hapless survivor' hypothesis. The Bristol study showed that certain bird species were particularly vulnerable. Cat predation on house sparrow was equivalent to 45 per cent of the combined total of pre-breeding density and annual productivity; equivalent figures for dunnock *Prunus modularis* and robin *Erithacus rubecula* were both 46 per cent. With young birds being particularly vulnerable, results suggest that 80–91 per cent of productivity were lost, results that are certainly not trivial. Sims *et al.* (2008), looking at data from a number of British towns, similarly found negative relationships between cat densities and number of breeding bird species, though correlations with individual bird species were varied and equivocal.

However, Beckerman *et al.* (2007) argue that the mere presence of cats can lead to animals avoiding otherwise preferred habitat and in this way reduce productivity. The effects of predation can be measured by looking at trade-offs between maximizing instantaneous survival in the presence of predators and acquiring resources for long-term survival or reproduction (Cresswell 2008). Even a small reduction in fecundity due to sub-lethal effects (<1 offspring per annum per cat) could result in marked decreases (up to 95 per cent) in bird abundance.

Conclusions

Animals succeeding in the urban environment are often opportunistic habitat and food generalists that can tolerate disturbance from and sometimes benefit from support by the public. Alternatively, they thrive in pockets of semi-natural habitat that provide similar or analogous conditions to that found in the wild. They are a significant component of community dynamics, and – with some exceptions – are generally viewed by people as a welcome addition to their concrete jungle.

References

Adams, C.E., Lindsey, K.J. and Ash, S.J. (2006) *Urban Wildlife Management*, Boca Raton, FL: Taylor & Francis.

Angold, P.G. and 10 others (2006) 'Biodiversity in urban habitat patches', *Science of the Total Environment*, 360(1–3): 196–204.

Baker, P.J., Bentley, A.J., Ansell, R.J. and Harris, S. (2005) 'Impact of predation by domestic cats *Felis catus* in an urban area', *Mammal Review*, 35(3–4): 302–12.

Baldock, F.C., Alexander, L. and More, S.J. (2003) 'Estimated and predicted changes in the cat population of Australian households from 1979 to 2005', *Australian Veterinary Journal*, 81(5): 289–92.

Beckerman, A.P., Boots, M. and Gaston, K.J. (2007) 'Urban bird declines and the fear of cats', *Animal Conservation*, 10(3): 320–5.

Blair, R.B. (1996) 'Land use and avian species diversity along an urban gradient', *Ecological Applications*, 6(2): 506–19.
Blair, R.B. (1999) 'Birds and butterflies along an urban gradient: surrogate taxa for assessing biodiversity', *Ecological Applications*, 9(1): 164–70.
Blair, R.B. and Launer, A.E. (1997) 'Butterfly diversity and human land use: species assemblages along an urban gradient', *Biological Conservation*, 80(1): 113–25.
Bland, R.L., Tully, J. and Greenwood, J.J.D. (2004) 'Birds breeding in British gardens: an underestimated population?' *Bird Study*, 51(2): 97–106.
Cannon, A.R., Chamberlain, D.E., Toms, M.P., Hatchwell, B.J. and Gaston, K.J. (2005) 'Trends in the use of private gardens by wild birds in Great Britain 1995–2002', *Journal of Applied Ecology*, 42(4): 659–71.
Caula, S., Marty, P. and Martin, J.-L. (2008) 'Seasonal variation in species composition of an urban bird community in Mediterranean France', *Landscape and Urban Planning*, 87(1): 1–9.
Cavia, R., Cueto, G.R. and Suárez, O.V. (2009) 'Changes in rodent communities according to the landscape structure in an urban ecosystem', *Landscape and Urban Planning*, 90(1–2): 11–19.
Center for Information Management (1997) *US Pet Ownership and Demographic Sourcebook*, Shaumberg, IL: American Veterinary Medical Association.
Chace, J.F. and Walsh, J.J. (2006) 'Urban effects on native avifauna: a review', *Landscape and Urban Planning*, 74(1): 46–69.
Churcher, P.B. and Lawton, J.H. (1987) 'Predation by domestic cats in an English village', *Journal of Zoology*, 212(3): 439–55.
Clergeau, P. and Quenot, F. (2007) 'Roost selection flexibility of European starlings aids invasion of urban landscape', *Landscape and Urban Planning*, 80(1–2): 56–62.
Clergeau, P., Savard, P., Savard, J.-P., Mennechez, G. and Falardeau, G. (1998) 'Bird abundance and diversity along an urban-rural gradient: a comparative study between two cities on different continents', *Condor*, 100(3): 413–25.
Cousins, S.H. (1982) 'Species size distributions of birds and snails in an urban area', in R. Bornkamm, J.A. Lee and M.R.D. Seaward (eds), *Urban Ecology*, Oxford: Blackwell Scientific, pp. 99–109.
Cresswell, W. (2008) 'Non-lethal effects of predation in birds', *Ibis*, 150(1): 3–17.
Daniels, G.D. and Kirkpatrick, J.B. (2006) 'Does variation in garden characteristics influence the conservation of birds in suburbia', *Biological Conservation*, 133(3): 326–35.
DeStefano, S. and Deblinger, R.D. (2005) 'Wildlife as valuable natural resources vs. intolerable pests: a suburban wildlife management model', *Urban Ecosystems*, 8(2): 1573–642.
Fisher, J. and Hinde, R.A. (1949) 'The opening of milk bottles by birds', *British Birds*, 42: 347–57.
Garaffa, P.I., Filloy, J. and Bellocq, M.I. (2009) 'Bird community responses along urban-rural gradients: does the size of the urbanized area matter?' *Landscape and Urban Planning*, 90(1–2): 33–41.
Gehrt, S.D. (2004) 'Ecology and management of striped skunks, raccoons, and coyotes in urban landscapes', in N. Fascione, A. Delach and M.E. Smith (eds), *People and Predators: From Conflict to Coexistence*, Washington DC: Island Press, pp. 81–104.
Gering J.C. and Blair, R.B. (2006) 'Predation on artificial bird nests along an urban gradient: predatory risk or relaxation in urban environments?' *Ecography*, 22(5): 532–41.
Germaine, S.S. and Wakeling, B.F. (2000) 'Lizard species distributions and habitat occupation along an urban gradient in Tucson, Arizona, USA', *Biological Conservation*, 97(2): 229–37.
Harris, S. and Rayner, J.M.V. (1986) 'Models for predicting urban fox (*Vulpes vulpes*) numbers in British cities and their application for rabies control', *Journal of Animal Ecology*, 55(2): 593–603.
Harris, S. and Trewhella, W.J. (1988) 'An analysis of some of the factors affecting dispersal in an urban fox (*Vulpes vulpes*) population', *Journal of Applied Ecology*, 25(2): 409–22.
Jarvis, P.J. (1990) 'Urban cats as pests and pets', *Environmental Conservation*, 17(2): 169–71.
Jarvis, P.J. and Clifford, K. (1998) 'Winter roosting wagtails in Wolverhampton', *Urban Nature Magazine*, 4: 98.
Jokimäki, J. and Huhta, E. (2008) 'Artificial nest predation and abundance of birds along an urban gradient', *Condor*, 102(4): 838–47.
Jozkowicz, A. and Gorska-Klek, L. (1996) 'Activity patterns of the mute swan *Cygnus olor* wintering in rural and urban areas: a comparison', *Acta Ornithologica*, 31(1): 45–51.
Kendall, K. (2006) 'Cat ownership in Australia: barriers to ownership and behavior', *Journal of Veterinary Behavior: Clinical Applications and Research*, 1(1): 5–16.
Lefebvre, L. (1995) 'The opening of milk bottles by birds: evidence for accelerating learning rates, but against the wave-of-advance model of cultural transmission', *Behavioural Processes*, 34(1): 43–53.

Lichfield, J. (1997) 'City of the dreadful flight', *Independent on Sunday* 16 January, pp. 36–7.
Macdonald, D. (1989) *Running with the Fox*, London: Unwin Hyman.
Magura, T., Tóthmérész, B. and Molnár, T. (2008) 'A species-level comparison of occurrence patterns in carabids along an urbanisation gradient', *Landscape and Urban Planning*, 86(2): 134–40.
May, R.M. (1988) 'Control of feline delinquency', *Nature (London)*, 332: 392–3.
McIntyre, N.E. (2000) 'Ecology of urban arthropods: a review and call to action', *Annals of the Entomological Society of America*, 93(4): 825–35.
Melles, S., Glenn, S. and Martin, K. (2003) 'Urban bird diversity and landscape complexity: species-environment associations along a multiscale habitat gradient', *Conservation Ecology*, 7(1): Article 5, online, available at: www.consecol.org/vol7/iss1/art5.
Monaghan, P. (1982) 'The breeding ecology of urban nesting gulls', in R. Bornkamm, J.A. Lee and M.R.D. Seaward (eds), *Urban Ecology*, Oxford: Blackwell Scientific, pp. 111–21.
Owen, J. (1991) *The Ecology of a Garden: The First Fifteen Years*, Cambridge: Cambridge University Press.
Pacheco, R. and Vasconcelos, H.L. (2007) 'Invertebrate conservation in urban areas: ants in the Brazilian cerrado', *Landscape and Urban Planning*, 81(3): 193–9.
Partecke, J., Schwabl, I. and Gwinner, E. (2006) 'Stress and the city: urbanization and its effects on the stress physiology in European blackbirds', *Ecology*, 87(8): 1945–52.
Pet Food Manufacturers Association (2010) *Pet Population 2010*, online, available at: www.pfma.org.uk/overall/pet-population-figures-htm.
Raven, S.J. and Coulson, J.C. (1997) 'The distribution and abundance of *Larus* gulls nesting on buildings in Britain and Ireland', *Bird Study*, 44(1): 13–34.
Riley, S.P.D., Hadidian, J. and Manski, D.A. (1998) 'Population density, survival, and rabies in raccoons in an urban national park', *Canadian Journal of Zoology*, 76(6): 1153–64.
Sears, J. (1989) 'Feeding activity and body condition of mute swans *Cygnus olor* in rural and urban areas of a lowland river system', *Wildfowl*, 40: 88–98.
Sherry, D.F. and Galef Jr, B.G. (1984) 'Cultural transmission without imitation: milk bottle opening by birds', *Animal Behaviour*, 32: 937–8.
Shirley, P. (1996) *Urban Wildlife*, London: Whittet.
Sims, V., Evans, K.L., Newson, S.E., Tratalos, J.A. and Gaston, K.J. (2008) 'Avian assemblage structure and domestic cat densities in urban environments', *Biodiversity Research*, 14(2): 387–99.
Slabbekoorn, H. and den Boer-Visser, A. (2006) 'Cities change the songs of birds', *Current Biology*, 16: 2326–31.
Small, E.C., Sadler, J.P. and Telfer, M.G. (2003) 'Carabid beetle assemblages on urban derelict sites in Birmingham, UK', *Journal of Insect Conservation*, 6(4): 233–46.
Smith, K.W. (2005) 'Has the reduction in nest-site competition from starlings *Sturnus vulgaris* been a factor in the recent increase of great spotted woodpecker *Dendrocopos major* numbers in Britain?' *Bird Study*, 52(3): 307–13.
Soulsbury, C.D., Iossa, G., Baker, P.J., Cole, N.C., Funk, S.M. and Harris, S. (2007) 'The impact of sarcoptic mange *Sarcoptes scabiei* on the British fox *Vulpes vulpes* population', *Mammal Review*, 37(4): 278–96.
Van Heezik, Y., Smyth, A. and Methieu, R. (2008) 'Diversity of native and exotic birds across an urban gradient in a New Zealand city', *Landscape and Urban Planning*, 87(3): 223–32.
White, J.G., Antos, M.J., Fitzsimons, J.A., and Palmer, G.C. (2005) 'Non-uniform bird assemblages in urban environment: the influence of streetscape vegetation', *Landscape and Urban Planning*, 71(2–4): 123–35.
Woods, M., McDonald, R.A. and Harris, S. (2003) 'Predation of wildlife by domestic cats *Felis catus* in Great Britain', *Mammal Review*, 33(2): 174–88.

29

Feral animals in the urban environment

Peter J. Jarvis

Introduction

Feral animals are domesticated or tame species where either they or their antecedents have reverted to the wild, and where they have been able to establish breeding populations. They are not the same as wild animals simply habituated to human proximity.

The commonest urban feral animal is the feral pigeon *Columba livia*, but pet animals such as cats *Felis catus*, dogs *Canis familaris*, and species of parakeet and terrapin can also sustain populations in the urban environment.

Feral pigeons

People have an ambivalent attitude towards feral animals in the urban environment, particularly in the case of the feral pigeon: many get pleasure out of feeding them, others rue the sight and cost of damage caused to buildings by the activities of these birds (Simms 1979; Jarvis 1996; Johnston and Janiga 1995).

History and distribution

Feral pigeons originated from escaped domestic pigeons, themselves derived from wild rock doves around 5000 years ago. The domestic pigeon's origins probably lie in the Neolithic, though the first evidence of domestication comes from Mesopotamian figurines dating from *c.*4500 BCE (Simms 1979). This bird was quickly adopted by many cultures and spread westward throughout Europe (probably arriving in Britain with the Romans) and eastward into India and China. When domesticated pigeons first reverted to the wild is not known, but the second-century writer Plautus reported that feral pigeons were very tame and lived on the rooftops in Rome.

Feral pigeons were known in London by the late fourteenth century, for the Bishop of London complained that the building of nests on St Paul's Cathedral had led to people throwing stones which broke windows and statues (Simms 1979). Pigeons were well-established in London by the seventeenth century (Lever 1987), and by the mid-nineteenth century were commonly nesting elsewhere in urban Britain and indeed Europe. While pigeons were probably widely distributed actual numbers sharply increased during the late nineteenth and early twentieth centuries.

Until the early twentieth century, pigeons benefited from spillage of grain used to feed the horses that drew wagons and carriages; this important source of food disappeared with the arrival of motor vehicles. During the twentieth century, the genetic stock of feral pigeons has been enriched through breeding with escaped racing birds. In many parts of Europe numbers fell during the Second World War because of a lack of food, then picked up again during the 1950s.

Urban feral pigeons are common in much of south and south-east Asia and parts of the Far East. Domesticated pigeons were introduced to South Africa in 1654 and some went feral shortly afterwards. Such a story was repeated in Australia (domestic pigeons introduced in 1788, ferals noted in the late nineteenth century); New Zealand (1850s); Latin America (various dates); Canada (1606); and the USA (probably in the 1820s) (Lever 1987).

Natural history

Flocks of feral pigeons are stable in composition, flock size and reproductive success both related to the amount of food that can be found. Flocks show little group cohesion: individuals simply use the same area. While young birds may disperse, there is a very low level of dispersion among adults (0–0.8 per cent per annum in one Polish study) due to a strong philopatry (faithfulness) towards their breeding sites (Hetmanski 2007). They have relatively small home ranges. In Basel, Switzerland, for example, over 32 per cent of pigeons remained within 0.3 km of their roosting sites, and only 7.5 per cent occasionally flew distances greater than 2 km (Rose *et al.* 2006).

In temperate latitudes pigeons can generally breed throughout the year, although not all birds remain in reproductive condition over winter. At Salford Docks near Manchester, UK, only a third of the population attempted to nest, a reflection of availability of food rather than of nesting sites (Murton *et al.* 1972). The number of young birds leaving the nest is usually greatest in spring and summer. Ewins and Bazely (1995), however, report on a colony in Toronto where reproductive success peaked in winter (November–April), attributed to a superabundance of food provided by humans and a lack of disturbance.

Roosting is normally communal; protection on one side against wind and rain seems to be the key in deciding where to rest, hence their abundance on ledges. Preference is given to spaces on or in buildings and structures where they can roost or breed as a group. Ferals may rest in trees but only exceptionally nest there.

Feral pigeons cluster around places where the public deliberately feeds them or they scavenge on litter. These sources of food are often a limiting factor, but pigeons also use rubbish tips, gardens, parks, playing fields and wasteland. Cereal is the key dietary requirement whether as seed or in the form of bread or cake, but these birds will opportunistically take in a variety of foods.

After a night fast, hungry birds feed in the early morning, storing food in their crops; birds rest while this food is digested, and feeding is resumed in the late afternoon. While this pattern has been observed, for example, by Murton *et al.* (1972) in an environment (grain storage areas) where food is abundant, a similar bimodal feeding schedule was found in Montreal where food was patchy and unpredictable (Lefebvre and Giraldeau 1984): a single area could only provide food irregularly and in small amounts; sufficient amounts could only be achieved by the flock breaking up and individual birds or small groups visiting different feeding areas. The bimodal schedule for the group was an integration of the different usage schedules on different sites.

Pigeons have little impact on the urban ecosystem, though they do provide an important source of food for feral cats and raptors such as peregrine falcon *Falco peregrinus*. In Warsaw, for example, pigeons formed 32 per cent of the peregrine's diet, reaching 40 per cent in winter and summer (Lukasz 2001). Pigeons might provide competition for other birds of urban squares and sidewalks, such as house sparrows *Passer domesticus*.

Pest status and the control of urban feral pigeons

Humphries (2008) suggests that contact with people has sharpened this 'superdove's' survival skills. When originally housed in dovecotes they were free to mate and forage for themselves, so feral descendents retain many of the behaviours originally shaped in the wild. Their adaptability and opportunism, however, have led to such great numbers that in many cities they are viewed as nuisances.

They harbour around 60 different human pathogenic organisms. The commonest disease-causing organisms are the bacterium *Chlamydophila psittaci*, which can lead to respiratory psittacosis in humans; the fungus *Cryptococcus neoformans* (leading to cryptococcosis which can cause meningitis); and the pigeon tick *Argas reflexus*, which can cause anaphylactic shock (Haag-Wackernagel 2005). However, the risk of these to humans is very low, even for those whose work brings them into close contact with the birds: there were only 176 documented transmissions of disease from feral pigeons to humans reported between 1941 and 2003 (Haag-Wackernagel and Moch 2003).

Pigeon droppings deface and corrode buildings, monuments and statuary, particularly those made of limestone, because excrement mixed with water forms a weak acid, and because contact with water encourages mould to grow on the faeces, the mycelia (whose metabolic products include acids) then entering the stone. Excreta also damage paintwork and foul pavements. An accumulation of excreta, feathers and nest debris can clog gutters, drains and ventilators.

A pigeon produces around 11.5 kg of excrement each year, and cleaning costs are massive. In 1996, it cost £14,000 for cleaning statuary and nearly £91,000 for cleaning pavements in London's Trafalgar Square (Harris 1996). In 2000, Railtrack reported an annual cleaning cost of £12,000 for just one bridge at Balham, London, with an unsuccessful attempt at pigeon-proofing the bridge with wire mesh having cost £9000 (*[London] Independent*, 1 August 2000). Some $1.1 billion is spent each year in the USA in cleaning urban bird mess, much of which comes from the feral pigeon (*New York Times Magazine*, 15 October 2006).

Costs of prevention are also high, yet many efforts have been ineffective (Haag-Wackernagel 2000). Trapping is time-consuming and can involve cage-trapping, catching birds by hand on night roosts or catapult-netting. Shooting using air rifles has been undertaken around some factories and warehouses. Stupefactants such as alpha-chloralose, mixed in with seed, have been widely used in the USA, Scandinavia, Germany, France and Australia.

Use of chemicals such as flake naphthalene and calcium chloride has deterred pigeons from some sites, but these birds have a poor olfactory sense and therefore only respond, if at all, to regular and expensive applications. Physical repellents include spikes, netting or wire mesh, and plastic gels placed on ledges which, giving way under the bird's weight, make it feel insecure and inclined to move elsewhere. High-frequency sound waves, bright (including strobe) lighting, continuous or irregular loud noises and electrification of roosting ledges have all also been of little value.

Use of trained hawks to deter pigeons is inefficient and expensive. Using a pair of hawks to scare the pigeons of Trafalgar Square in London between 2003 and 2006 cost £226,000. Some 2500 out of 3500 pigeons are estimated to have disappeared, but at an average cost of £90 each not considered to be value for money.

Anti-cholesterol compounds and other chemicals have been adopted as contraceptives in a number of cities, from Rome and Venice in the 1970s to Hollywood, Denver and St Paul, Minnesota, in the twenty-first century. Since 2007, these three American cities have used OvoControl P on rooftop feeders. The compound prevents eggs from hatching by impeding development of the layer between yolk and egg white. This method has also recently been promoted in New York where commuters using a recently renovated terminal of the Staten Island ferry have complained not only about pigeon droppings but also about falling maggots that had been feeding on the birds' feces. However, pest controllers cannot ensure that each pigeon consumes the required 5 g of OvoControl every day over a number of months, and there are risks that non-target bird species will eat the doctored food pellets.

All of the above methods, however, either simply move the birds and the problem elsewhere or provide no more than short-term reductions in pigeon numbers. In Basel an intensive campaign of culling pigeons initially reduced the population of some 20,000 birds by 80 per cent (Haag-Wackernagel 1995). However, after only a few weeks flocks had regained their previous size or were even larger than before. Vacancies left by the removed individuals had quickly been filled by young birds.

A breeding female can produce up to 12 young each year. Because annual adult mortality is only around 11 per cent there is strong competition for food. Killing adult pigeons benefits juvenile birds which would otherwise have no chance of survival – they usually suffer a 90 per cent mortality, the main cause of death being starvation.

Haag-Wackernagel's work showed that a permanent population reduction can only be achieved by reducing the ecological capacity of the urban ecosystem, and it is food, not adult mortality, that is the limiting factor. In 1988 Pigeon Action was founded with the aim of providing an ethically responsible and ecologically sound means of clearing Basel's pigeon population. With overpopulation having been caused by the large amount of feeding by the public, the aim was to combine culling with effecting a change in people's behaviour. An information campaign promoted the idea that feeding harmed the birds by leading to overcrowding, poor living conditions and disease. Many people, however, enjoy feeding pigeons, so supervised pigeon lofts and designated human-pigeon 'encounter areas' were created as homes to a small but healthy population but where eggs were regularly culled. The effectiveness of this approach is seen by the fact that the initial 20,000 birds were reduced to 10,000 within 50 months (Haag-Wackernagel 1995).

The success of the pigeon control programme in Basel has been followed elsewhere, for example in Paris, Aachen and Augsburg, and it has been proposed for New York.

Parakeets

Three species of parakeet that have escaped or have been released from cages and aviaries have established themselves in the urban wild with generally low but often sustained populations: ring-necked (or rose-ringed) parakeets *Psittacula krameri*, monk (or quaker) parakeets *Myiopsitta monachus*, and canary-winged parakeets *Brotogeris versicolorus*.

Afro-Asian ring-necked parakeets have escaped in sufficient numbers to breed in a number of places in England (first noted in 1969); Cologne, Bonn, Hamburg, Wiesbaden and elsewhere in Germany (from 1967); Antwerp and Brussels in Belgium (1970s); The Hague, Rotterdam and Amsterdam in the Netherlands (possibly 1968); Barcelona, Spain; Tokyo, Japan (by the mid-1980s) (Eguchi and Amano 2004); and Los Angeles, New York and a few other cities in the USA (Lever 1987).

Feral ring-necks were breeding in south-eastern England by the mid-1970s (Sharrock 1976). By 1984 they had also been reported from north-western England (Lack 1986), but this population had died out by 1996, by which time the south-east population had spread further westwards and southwards, and sightings had also been reported from parts of East Anglia (Gibbons *et al.* 1996). A census in 1996 showed maximum numbers in October of 1508 individuals, of which 1123 were counted in the west London area (Pithon and Dytham 1999). In 2008 the RSPB estimated the British breeding population to be 4300 (www.rspb.org.uk/wildlife/birdguide/name/r/ringneckedparakeet). These birds are generally sedentary, and sightings in Wales and the Scottish borders are almost certainly single escapees.

These long-lived birds start breeding at three years old, and nesting success can be quite high (figures range from 0.8 to 1.9 fledging per nest), but while the populations west of London increased quite substantially during the 1990s, others grew more slowly or not at all. By 1992 populations in the Greater London area were increasing at a rate of around 30 per cent per annum, while in Thanet the increase was 15 per cent per annum. Parakeets were expanding their range at about 0.4 km/yr in the Greater London area (Butler 2003).

No other European population has increased at such a rapid rate as around London with the possible exception of the one in Brussels, now over 5000 birds (Rabosée 1995); the population in Cologne was stable, while the larger Wiesbaden population had increased slightly (Ernst 1995; Zingel 1997).

Population expansion is unlikely to have been limited by food as these birds eat a wide variety of plant material in the wild and they also commonly visit gardens. They might be limited to areas with a mild climate and indeed may not be able to survive British winters without using garden feeders, so this would limit their distribution to urban and suburban locations. Pithon and Dytham (2002) argue that dispersal may be inhibited by communal roosting behavior: for parakeets to colonize new areas benefits should outweigh such costs as increased predation risk and reduced information on food sources, so dispersal is most likely when high densities make foraging areas near roost sites saturated.

To date these birds seem to have little impact on the native avifauna, although as nest-hole breeders they potentially compete for breeding sites with such native birds as woodpeckers. Some fruit growers are concerned that these birds might take buds to an extent that they become pests.

The South American Monk parakeet is a commonly-kept cage bird in the USA, where the earliest reports of free-birds were from the New York metropolitan area in 1967, where eight nests were found three years later. Details of other flocks sighted in the USA are given in Lever (1987): by the early 1970s the species was established in seven states, and by 1995 it had spread to seven more (www.monkparakeet.com). As one of the few temperate-zone parrots, monk parakeets are better able than most to survive cold climates, and as well as in New York City, small colonies exist as far north as Chicago, Cincinnati, and urban areas in Rhode Island, Connecticut and south-west Washington State.

The brightly-coloured monk parakeets are communal breeders and, uniquely for parrots, build their nests of twigs often a metre in both width and height. They will nest in trees but in the built environment also commandeer a variety of spaces such as on buildings, girders and utility poles. Numbers have generally remained low. Flocks in the New York/New Jersey area appear to have been the largest, with at least 200 birds noted by 1973. In that year a concern about the possible impacts of alien birds generally, together with a fear that the parakeets might be instrumental in spreading ornithosis and Newcastle disease, together led to the implementation of control measures, and numbers were certainly reduced, and flocks elsewhere disappeared. Monk parakeets have also been found in Spain, being a common sight in parks in Barcelona (often as numerous as the feral pigeon), Madrid and Cadiz.

Canary-winged parakeets, also from South America, are again important cage birds in the USA. Fifty free-flying birds were noted in Miami in 1969, while flocks of several hundred were reported there in 1976. Other major sightings have been from Los Angeles, San Francisco and Long Island, New York, where in the early 1970s groups of up to 50 birds successfully overwintered (Bull 1973). These birds depend on seed and fruit in their native range, but feral populations have added flowers and nectar to their diet. Fears of competition with native birds for food or nest-holes are obviated by low numbers and high mortality rates.

There are a few other localized feral parakeets in urban North America, for example red-masked parakeets *Aratinga erythrogenys* in San Francisco.

All of these feral urban parakeets eat a variety of foods, but their survival over winter in some of their more northerly US locations may – like the ring-necked parakeet in Britain – depend on using suburban bird feeders.

Feral cats

Estimates of cat population sizes are hedged with uncertainties. Pet cat ownership in the USA during the 1990s was probably 59–65 million (Center for Information Management 1997), with anything between ten to 65 million additional animals being stray or feral (Mahlow and Slater 1996; Patronek 1998). Tabor (1983) suggested that the British cat population was 3.5–5 million, of which 1.5 million were feral. The Pet Food Manufacturers Association (2010) estimate in 2010 was eight million pet cats in Britain, but the ratio of ferals is probably now lower than that suggested by Tabor because of animal welfare programs, so an informed guess might be 1.5–2.5 million.

Most feral cats live in rural or wilderness areas but many towns have sometimes sizeable pockets of ferals. Originating from escaped or abandoned animals these loosely-structured colonies now have a history of many feral generations, so that members have never had or have lost any socialization with humans.

Rome probably has the largest urban feral cat population in the world with 250,000–350,000 animals associated with around 2000 separate colonies. Aggressive (agonistic) interactions within each colony demonstrate a linear dominance hierarchy in the males, though this does not correlate with copulatory success. In domestic cat groups females generally optimize mating by copulating with a number of males, while males try to monopolize and guard the female. In a study of feral cats in central Rome, however, while females did indeed mate polygamously males did not attempt a monopoly, possibly because of a physiological mechanism that ensures fertilization by the dominant males, but perhaps because where cats are living at unnaturally high densities (12.5 adults/ha), competition based on agonistic interactions could be costly, in terms of fight injury and because non-interacting males might exploit the situation and effect a successful mount (Natoli and De Vito 1991). Where densities are lower, as studied in a park in Lyon, France, where there were 7–9 adults/ha, males can secure larger home ranges (average 0.8 ha) which overlap with a number of smaller female home ranges (0.2 ha), and a male can dominate copulations within his own range (Say and Pontier 2004). Similar behaviour was noted by Dards (1983) in Portsmouth Naval Dockyard where feral animals lived at a density of 2 adults/ha, but where over half of their food came from deliberate feeding by dockyard workers. In a study at Avonmouth Docks, Bristol, where density was much lower (0.10–0.15 adults/ha) home range sizes were similar between the sexes (10–15 ha) and were much smaller than expected from the low density, and cats were mostly solitary rather than group living (Page *et al.* 1992).

Because of their association with public parks and squares, and factories, warehouses and docklands feral cats probably have less impact on urban wildlife than domestic cats. With a diet in which rats and mice play a large part they are arguably more beneficial than they are pests. Most issues surrounding feral cats are to do with welfare. During the 1960s the British model

Celia Hammond began a programme of rescuing, neutering and re-homing stray and feral cats at a time when euthanasia was the usual method of control. She opened a sanctuary for the many cats that could not be returned to their original environment, and in 1986 set up the Celia Hammond Animal Trust. Many other welfare organizations throughout the world have followed this lead: capturing, testing for and vaccinating against infectious diseases, neutering, then returning cats to the capture site (Remfry 1996; Slater 2005).

Feral dogs

As with cats it is important to distinguish between strays, which have had human association at some time in their life, and truly feral dogs. The former group often atavistically form loosely-structured packs, and can certainly become nuisances or even dangers to humans, pets and wildlife (Beck 1973). In Petrozavodsk, density varies according to the availability of secluded areas and the extent of control by the municipal authorities (Ivanter and Sedova 2008).

Truly feral dogs are generally associated with rural rather than urban areas, though packs of ferals are associated with parts of the built environment where human activity is limited, for example airports and military areas. These are opportunistic animals with a catholic diet. Dens are often found under abandoned buildings and equipment. Packs have a rigid social hierarchy from which stray dogs are generally excluded (Daniels and Beckoff 1989; Pal *et al.* 1998). Nevertheless, with pup survival rates of only 5 per cent after one year (Beck 1973) free-ranging dogs have difficulty in sustaining populations without continuous recruitment.

Feral pigs

Feral pigs *Sus scrofa* have become common particularly in parts of south-eastern USA: with the spread of urban development there have been increasing reports of these animals causing damage to suburban gardens, fencing and outbuildings, parks and golf courses (Brown 1985). Wild boar can cross-fertilize with the domestic form, and can cause problems even in such major cities as Berlin where over 5000 such animals live in the parks and suburbs. Feral pigs and wild boar can transmit diseases such as hemorrhagic leptospirosis to humans (Jansen *et al.* 2006).

Terrapins

Some species of terrapin, commonly kept as pets particularly in North America and parts of Europe, have escaped or have been released into ponds. A boom in keeping the Mississippi Basin red-eared terrapin *Trechemys scripta elegans* was associated in the US and UK with the fleeting popularity of the cartoon mutant ninja turtles; the British Chelonia Group estimated that during the late 1970s and early 1980s around 33,000 such turtles were introduced each year. Owners finding that the hatchling that fitted into a fish tank took only two or three years to grow into a 0.3 m-long 2 kg animal consequently illegally released these pets. While they overwinter by hibernating at the bottom of often urban ponds and shallow lakes, summer temperatures are generally too low for this species to breed in Britain, eggs requiring an incubation temperature of 25°C for 60 days. However, they can live for over forty years and grow to a large size. They have a voracious appetite, feeding (when temperatures are above 16–18°C) on vegetation, insects, small fish, frogs and frogspawn; they compete for this food with native fauna; and a habit of basking on waterside nests disrupts breeding: their impact on the wildlife of local ponds and lakes could be deleterious (Beebee and Griffiths 2000). Substantial populations have built up in a number of urban ponds particularly in the London area (from Kensington Gardens to Epping Forest), and in 2008 colonies were also reported from the Bournemouth-Poole area on the south coast, and Coventry and Swindon in the Midlands (www.nonnativespecies.org). Red-necked terrapins have also been introduced to many parts of south-east Asia, particularly in temple ponds, public parks and reservoirs (www.ecologyasia.com/verts/turtles/red-eared_terrapin.htm).

Conclusions

While feral pigeons are widespread and abundant in many towns, and are often viewed as pests, other feral animals are at most locally abundant, and issues relating to feral cats and dogs generally concern welfare. For animals that are often such highly visible components of the urban landscape their overall impact on the urban ecosystem tends to be rather low.

References

Beck, A.M. (1973) *The Ecology of Stray Dogs: A Study of Free-Ranging Urban Animals*, Purdue, IN: Purdue University Press.

Beebee, T.J. and Griffiths, R.A. (2000) *Amphibians and Reptiles: A Natural History of the British Herpetofauna*, London: HarperCollins.

Brown, L.N. (1985) 'Elimination of a small feral swine population in an urbanizing section of central Florida', *Florida Scientist*, 48(2): 120–3.

Bull, J. (1973) 'Exotic birds in the New York City area', *Wilson Bulletin*, 85(4): 501–5.

Butler, C.J. (2003) *Population Biology of the Introduced Rose-Ringed Parakeet Psittacula krameri in the UK*, Unpublished DPhil thesis, University of Oxford, online, available at: http://biology.uco.edu/PersonalPages/CButler/thesis.pdf.

Center for Information Management (1997) *US Pet Ownership and Demographic Sourcebook*, Shaumberg, IL: American Veterinary Medical Association.

Daniels, T.J. and Bekoff, M. (1989) 'Spatial and temporal resource use by feral and abandoned dogs', *Ethology*, 81(4): 300–12.

Dards, J.L. (1983) 'The behaviour of dockyard cats: interactions of adult males', *Applied Animal Ethology*, 10(1–2): 133–53.

Eguchi, K. and Amano, H.E. (2004) 'Invasive birds in Japan', *Global Environmental Research*, 8(1): 29–39.

Ernst, U. (1995) 'Afro-asiatische Sittiche in einer mitteleuropäischen Grossstadt: Einnischung und Auswirkungen auf die Vogelfauna', *Jahrbuch für Papageienkunde*, 1: 23–114.

Ewins, P.J. and Bazely, D.R. (1995) 'Phenology and breeding success of feral rock doves, *Columba livia*, in Toronto, Canada', *Canadian Field-Naturalist*, 109: 426–32.

Gibbons, D.W., Reid, J.B. and Chapman, R.A. (1996) *The New Atlas of Breeding Birds in Britain and Ireland: 1988–1991*, Calton: Poyser.

Haag-Wackernagel, D. (1995) 'Regulation of the street pigeon in Basel', *Wildlife Society Bulletin*, 23(2): 256–60.

Haag-Wackernagel, D. (2000) 'Behavioural responses of the feral pigeon (Columbidae) to deterring systems', *Folia Zoologica*, 49(2): 25–39.

Haag-Wackernagel, D. (2005) 'Parasites from feral pigeons as a health hazard for Humans', *Annals of Applied Biology*. 147(2): 203–10.

Haag-Wackernagel, D. and Moch, H. (2003) 'Health hazards posed by feral pigeons', *Journal of Infection*, 48(4): 307–13.

Harris, E.C. (1996) *Report for the Control of the Pigeon in Trafalgar Square*, unpublished report to Department of National Heritage.

Hetmanski, T. (2007) 'Dispersion asymmetry within a feral pigeon *Columba livia* population', *Acta Ornithologica*, 42(1): 23–31.

Humphries, E. (2008) *Superdove: How the Pigeons took Manhattan … and the World*, London: HarperCollins.

Ivanter, E.V. and Sedova, N.A. (2008) 'Ecological monitoring of urban groups of stray dogs: an example of the city of Petrozavodsk', *Russian Journal of Ecology*, 39(2): 105–10.

Jansen, A., Nöckler, K., Schönberg, A. Luge, E., Ehlert, D. and Schneider, T. (2006) 'Wild boars as a possible source of hemorrhagic leptospirosis in Berlin, Germany', *European Journal of Clinical Microbiology and Infectious Diseases*, 25(8): 544–6.

Jarvis, P. (1996) 'Pigeons and people', *Urban Nature Magazine*, 2: 141–3.

Johnston, R.F. and Janiga, M. (1995) *Feral Pigeons*, Oxford: Oxford University Press.

Lack, P. (1986) *The Atlas of Wintering Birds in Britain and Ireland*, Calton: Poyser.

Lefebvre, L. and Giraldeau, L.-A. (1984) 'Daily feeding site use of urban pigeons', *Canadian Journal of Zoology*, 62(7): 1425–8.

Lever, C. (1987) *Naturalised Birds of the World*, London: Longman Scientific & Technical.

Lukasz, R. (2001) 'Feeding activity and seasonal change in prey consumption of urban peregrine falcons *Falco peregrinus*', *Acta Ornithologica*, 36: 165–9.

Mahlow, J.C. and Slater, M.R. (1996) 'Current issues in the control of stray and feral cats', *Journal of the American Veterinary Medical Association*, 209(12): 2016–20.

Murton, R.K., Thearle, R.J.P. and Thompson, J. (1972) 'Ecological studies of the feral pigeon (*Columba livia* var.). I. Population, breeding biology and methods of control', *Journal of Applied Ecology*, 9(3): 835–54.

Natoli, E. and De Vito, E. (1991) 'Agonistic behaviour, dominance rank and copulatory success in a large multi-male feral cat, *Felis catus* L., colony in central Rome', *Animal Behaviour*, 42(2): 227–41.

Page, R.J.C., Ross, J. and Bennett, D.H. (1992) 'A study of the home ranges, movements and behaviour of the feral cat population at Avonmouth Docks', *Wildlife Research*, 19(3): 263–77.

Pal, S.K., Ghosh, B. and Roy, S. (1998) 'Agonistic behaviour of free-ranging dogs (*Canis familiaris*) in relation to season, sex and age', *Applied Animal Behaviour Science*, 59(4): 331–48.

Patronek, G.J. (1998) 'Free-roaming and feral cats – their impacts on wildlife and human beings', *Journal of the American Veterinary Medical Association*, 212(2): 218–26.

Pet Food Manufacturers Association (2010) *Pet Population 2010*, online, available at: www.pfma.org.uk/overall/pet-population-figures-htm.

Pithon, J. and Dytham, D. (1999) 'Census of the British population of ring-necked parakeets *Psittacula krameri* by simultaneous roost counts', *Bird Study*, 46(1): 112–15.

Pithon, J.A. and Dytham, C. (2002) 'Distribution and population development of introduced ring-necked parakeets *Psittacula krameri* in small introduced populations in southeast England', *Bird Study*, 49(2): 110–17.

Rabosée, D. (1995) *Atlas des oiseaux nicheurs de Bruxelles, 1989–1991*, Brussels: Société d'Études Ornithologiques Aves.

Remfry, J. (1996) 'Feral cats in the United Kingdom', *Journal of the American Veterinary Medical Association*, 208(4): 520–3.

Rose, E., Nagel, P. and Haag-Wackernagel, D. (2006) 'Spatio-temporal use of the urban habitat by feral pigeons (*Columba livia*)', *Behavioral Ecology and Sociobiology*, 60(2): 242–54.

Say, L. and Pontier D. (2004) 'Spacing pattern in a social group of stray cats: effects on male reproductive success', *Animal Behaviour*, 68(1): 175–80.

Sharrock, J.T.R. (1976) *The Atlas of the Breeding Birds in Britain and Ireland*, Calton: Poyser.

Simms, E. (1979) *The Public Life of the Street Pigeon*, London: Hutchinson.

Slater, M.R. (2005) 'The welfare of feral cats', in I. Rochlitz (ed.) *The Welfare of Cats*, Dordrecht: Springer, pp. 41–75.

Tabor, R. (1983) *The Wild Life of the Domestic Cat*, London: Arrow Books.

Zingel, D. (1997) 'Zum Verhalten von Halsbandsittich und Alexandersittich *Psittacula krameri* und *Psittacula eupatria* im Schlosspark Wiesbaden-und in ihren Heimatländern', *Ornithologische Mitteilungen*, 49(6): 143–66.

Part 4

Ecosystem services and urban ecology

Introduction

Ian Douglas

This section examines the importance of urban ecosystems for human society. Beginning with an overview of the many ways in which nature in the built environment contributes to a good urban life and a consideration of the different ways various groups of people may view nature and natural areas, the section then examines specific benefits in terms of implications for health and well-being. It then looks at the many contributions vegetation offers, from the benefits of street trees to gardens and biodiversity.

The human need for nature shows itself as an urge for contact with other forms of life besides our own. Yet for most people such contact has to take place on our own terms, as part of the ordered, regulated conditions of our daily lives. Urban nature cannot be permitted to overwhelm our lives or the artefacts that are so important to them. Thus we both fight pests in our houses and gardens, but at the same time care for domestic animals and nurture flowers and shrubs, both indoors and outdoors. Yet many people find wandering in relatively natural woodland and along river banks stimulating and relaxing. Some enjoy the space for jogging or walking the dog; others exploit it for fishing or water sports. Most outdoor activities in urban greenspaces involve physical exercise that is beneficial for our physical and mental health. Yet those greenspaces can be key elements in urban environmental management and in helping cities adapt to climate change. A 10 per cent increase in the amount of greenspace in built-up centres would reduce urban surface temperatures by as much as 4°C in north-west Europe. Floodplain parks, sports grounds and golf courses can all be part of floodbasins that can temporarily store extreme flood flows that would otherwise overtop banks and inundate built-up areas. Tree planting, especially along roads and highways can not only absorb CO_2 but can also filter out part of the airborne pollutant load and also attenuate traffic noises and other sounds. Urban nature can stimulate children, encourage community togetherness and provide relaxation for elderly citizens.

In separate chapters David Nicholson-Lord and Rachel Kaplan take independent views of how people value nature in cities and their reactions to it. By the mid-nineteenth century, the simple benefits of the countryside had become popularly associated with nature itself. These associations inherited from European romanticism spread in North America through thinking about how nature, through attunement with, contemplation of it, and immersion in it, uplifted the spirit. This strongly influenced the design of public greenspaces and parks. Yet uncultivated naturalness was undesirable. People expected parks to cater for a whole variety of interests, from being pleasure grounds full of alternative recreational opportunities, from bandstands to rowing boats on a lake. For them nature was simply a backdrop or frame within which other activities took place. After 1900 parks began to be organized to provide regulated recreation under park leaders, play directors and efficiency minded experts. To a certain extent they saw parks as a mechanism of social control: a moral defence against the abundance of chaos perceived to stem from the greater free time brought about by shorter working hours. People's attitudes to nature changed until the spirit of the right to roam outside the cities across open countryside stimulated a wider demand for national parks and access to mountains and moorlands, rivers and waterfalls. Gradually such accessibility and the idea of accessible urban nature took hold. Places to watch birds, see wildlife, and escape from urban stress and noise became valued. Even so, this was not for everyone. Many people were always concerned about unruly behaviour and vandalism in natural areas and nature reserves in and around cities. Many would not venture into shady wooded areas, were put off by signs of broken fences, damaged plants and carelessly dumped rubbish. In the worst cases, paths along well vegetated urban stream valleys became seen as locations where drug dealing and other unpleasant activities took place. The intrinsic and aesthetic values of urban nature are complex. Many people still like to have that element of control and safety.

High quality green spaces go a long way to encouraging people to pursue healthier lifestyles through exercise such as walking, cycling and active children's play. Urban green space encourages physical activity across a broader range of the community than those likely to use other avenues for physical activity such as gyms. In particular, access to green spaces tends to be free in contrast to other centres for leisure activity. A diversity of jogging routes, nature trails, cycle paths and bridleways in urban greenspaces, such as river valley areas, encourages a variety of forms of physical activity. It is also important to identify the beneficial elements in aiding recovery from specific conditions. Jenna Tilt finds that though there is strong evidence that vegetation, particularly high levels of vegetation and mature trees, contribute to one's physical activity and/or health, the pattern and maintenance of this high level of vegetation is also extremely important. Perceptions and fears about safety and a general dislike for poorly maintained urban nature spaces might limit the positive effects of the nearby natural space on physical activity and health considerably.

Looking at green space for quite short time periods has an impact on the nervous system – lowering blood pressure and reducing stress. Regular access to restorative, natural environments can halt or slow processes that negatively affect mental and physical health. Walking in natural areas provides opportunities for social interaction that are particularly beneficial for the elderly. Exposure to natural scenes reduces stress. Trees play an important social role in easing tensions and improving psychological health. People feel better living around trees. Houses surrounded by nature help to raise children's attention capabilities. Thus living in areas with trees helps to reduce anger and violence and improve the ability to concentrate and work effectively. Landscapes can be used in this way to help develop programmes designed to help recovery from specific conditions. Rod Matsuoka and William Sullivan review the mounting evidence regarding the profound and systemic benefits that contact with nature provides urban dwellers. Taken together, the studies reveal that in a wide variety of settings – parks, work, home, school, and hospitals – the benefits of being exposed to nature are available to urban residents. Both passive engagement with nature (e.g. viewing nature through a building window) and more direct interactions with nature (e.g. climbing a tree, planting a vegetable garden) produce an array of benefits. The benefits are far from trivial. For individuals, they include an enhanced capacity to concentrate and pay attention, greater ability to cope with life's stressors, higher levels of life satisfaction, and increased levels of psychological well-being. For neighbours and communities, they include stronger ties among neighbours, lower levels of incivilities, fewer instances of aggressive behaviour, reduced levels of violence, and fewer reported crimes.

The inclusion of street trees and other streetscape features on median strips and roadsides in dual carriageways may actually reduce crashes and injuries on urban roadways. Trees at the edges of roads offer a safer walking environment for pedestrians. Trees take up water, encourage greater evapotranspiration and help to reduce the volume of water going to drains and thus the cost of drainage infrastructure. Trees provide shade and help keep streets cool in hot weather. Trees close to the traffic absorb more pollutants than those further away. Trees improve the aesthetic appeal of streets as places in which to walk. Trees probably have a calming effect on both pedestrians and drivers, perhaps lowering the likelihood of 'road rage'. In discussing street trees it is important to recognize that different groups of people and different institutions place contrasting values on street trees, ranging from the insurance industry, through gardeners, house purchasers, and people working in offices. Often trees are largely argued for in terms of their quantitative impact on particulate interception, climate amelioration, and argued against in economic terms, namely of cracks they may/may not make in house walls, ceilings etc. rather than in other terms such as their value as 'inter-generational objects' and objects of simple inspiration in peoples' lives. Gerald Dawe considers the ecological and anthropocentric views of street trees held by different groups of people and argues that the highly sensitive issues involve need further consideration and consultation in order to achieve a greater, more sustainable, consensus on the future of street trees.

Gardens are major contributors to biodiversity. In Sheffield, UK, 61 gardens studied in detail contained nearly as many plants as the native flora of the British Isles and 786 species of invertebrates were found in them. The location of a garden within Sheffield did not appear to affect its biodiversity. The presence of alien or native plants also did not appear to affect the biodiversity. Over 50 per cent of the entire UK populations of House Sparrow and Starling are associated with gardens, with 25 per cent of Song Thrushes and 10 per cent of Bullfinches. The common frog has increased in numbers in urban Britain thanks to fringed garden ponds. Kevin and Sian Gaston conclude that there is little likelihood that the majority of explicitly wildlife gardening actions have a marked local negative or otherwise undesirable impact. However, at least under some circumstances, there are some issues of potential concern, and which warrant much further work than has been done to date. These include the role of domestic gardens as potential sources of alien species in the wider landscape, the potential for alien populations to increase at the expense of local populations of native species, the potential for genetic contamination of wild populations from garden populations, and the impact of the production or extraction of resources for wildlife gardening in the regions from which these are obtained.

30

Intrinsic and aesthetic values of urban nature

A journalist's view from London

David Nicholson-Lord

The interplay between cities and nature is one of the most powerful themes of contemporary culture. This is not only because they are "where we live" – or, in the case of nature, where many of us *used to* live. It is also because they represent opposite polarities of the human mind. The overlap between the spatial and the psychological – between cities and "nature" as physical places and as countervailing cultural and psychic forces and processes – is one of the things that makes analysis of their relationship so fascinating and yet so complex.

The last three centuries have seen these tensions – present but latent throughout the previous six or seven millennia, since the rise of the first cities in the fertile crescent of Mesopotamia – emerge on a striking new scale, playing themselves out in ways that bear directly on human occupancy of the planet – not least what form it takes and how well it survives. It is a story with a number of elements: the increase in human numbers, affluence and impact; the rise of urbanization as a global force, dominating the Earth – culminating in the moment (in 2007) when the world's population became predominantly urban; the shrinkage of "nature" and loss of the wild; the emergence, in the second half of the twentieth century, of a new political and social movement – environmentalism; and, partly as a consequence, a profound shift in attitudes towards nature and cities. These were the deep social and cultural patterns that underlay and influenced the more obvious changes in aesthetics – defined here as the design language we use for our settlements, in particular those parts of our settlements that we leave unsettled.

Settlement and its opposite, indeed, is one aspect of the polarity of city and nature. As will become clear later in this chapter, a key property of nature is its "unsettlement" – the fact that it is both unsettled and, potentially, unsettling. Cities, by contrast, archetypally represent order, control, predictability – and thus, in one sense, liberation from the threat posed by the subversive, anarchic forces of the wild. This was the case for most of history, from the Greek *polis* to the medieval city – where, it was said, *stadt luft macht frei* (city air makes you free) – and it was a fundamental characteristic of the world in which nature, in its scope and mystery, simply outweighed humanity: it remained, largely, unexplored and unexplained. Wildness was unregenerate, pagan and dangerous: it was something to be abolished, rescued or reclaimed.

The language we use to describe the nature that existed within the boundaries of settlements of this period, therefore, crystallizes around the concept of formality. From the hanging gardens of Babylon to the Elizabethan knot garden, nature, to a modern sensibility, seems tamed and tidied, organized and subjugated. That there was a need for a connection with it, on the part of human beings, is not in doubt. According to the Jewish historian Josephus, Nebuchadnezzar, in constructing the gardens at Babylon, "rendered the prospect an exact resemblance of a mountainous country" in order to please his queen, who had been brought up in a remote region of Asia "and was fond of a mountainous situation". And since the loftiest point in the gardens was at the same height as the city walls, it must have been possible to stand in this paradise – as gardens were then known – and gaze out over rural Mesopotamia. The survival of greenery within settlements throughout the millennia that followed is testimony to the human urge for a connection with nature beyond the merely functional, yet for the most part it was a connection that had to be firmly controlled – like nature itself.

The last three centuries have witnessed a sea-change in attitudes, at least in the developed world, and a corresponding revolution in nature aesthetics. Much of the evidence for this comes from the eighteenth and early nineteenth centuries, which saw the spread of cities, the birth of Romanticism, the rise – in the poetry of Wordsworth and his contemporaries, for example – of a recognizably modern nature paganism. In north America the exploration and colonization of a pristine continent by "civilized" man was the catalyst for the development of a wilderness movement, best exemplified in the writings of Thoreau and Emerson, and an appreciation of the American "sublime". In the second half of the eighteenth century – around 1760–1770, according to the cultural critic Raymond Williams, attitudes to the city reached a turning-point. Before this period, cities represented manners and refinement; after it, they came to be viewed as oppressive (Williams 1983). In John Ash's *New and Complete Dictionary of the English Language*, meanwhile, published in 1775, the word "civilization" was defined as the "state of being civilized" as well as the act of civilizing – in other words, as an (achieved) condition, not merely a process (Nicholson-Lord 1987).

Cities can be seen as the spatial or physical expression of the cultural state we describe as civilization, and both have soured in the last two centuries. The environmental psychologist Y.F. Tuan argues that an inversion of long-established polarity has taken place – that wilderness has become sacred and settlements profane (Tuan 1974). Tastes in landscape reflected this – notably the new appetite for uplands. At the start of the eighteenth century, for example, Daniel Defoe was still describing the English Lake District as "barren and frightful": less than a century later Wordsworth and his fellow Lake poets were discovering there the joys of solitude and "tranquillity". In the intervening period a new aesthetic had arisen which was to dominate Western thinking about the design of larger landscapes into the twentieth century.

The English landscape movement associated with the names of William Kent, Humphry Repton and, above all, Lancelot "Capability" Brown was many things: a revolt against formality, seen as a Continental (European) phenomenon associated with absolutist monarchies – not least the geometrics of Versailles; an expression of dissatisfaction with enclosure and with the near-at-hand; an aspiration towards its opposite – the far away, that which lay outside; and a new union with nature, now seen as embodying these qualities of "outsideness". The poet Alexander Pope, a landscapist himself, used a striking phrase to describe the new school of landscape design – "calling in the country" – and one equally striking technical innovation expressed this perhaps more than any other. This was the ha-ha, or sunken ditch, an "invisible" boundary that removed the sense of separation between viewer and viewed, creating a psychological connection between humans, assumed to inhabit the civilized foreground, and nature – assumed, tellingly, to have been lost but recovered yet still, somehow, remaining "outside".

The ability of the eye to sweep uninterrupted from the near-at-hand to the distant was a crucial element in the sense of psychological freedom that these new eighteenth century landscapes conferred. So too was the informality and "naturalness" of the landscapes themselves

– rolling green vistas, dotted or clumped trees, a body of water. The Brown style, developed at Stowe in Buckinghamshire after 1741 and achieving the royal imprimatur when he became Master Gardener to George III, was widely copied throughout Europe and the United States and gradually assumed the status of an international landscape language. It was no surprise, therefore, that when the spread of cities accelerated in the nineteenth century and new parks were created as a counterweight to the perceived horrors of urban life, the landscape movement of the previous century should provide the template for their design. The Victorian city, with its densely packed masses, was seen, by those in power at least, as not merely a breeding ground of disease and immorality but as a potential hotbed of revolution. Nature – embodied in the surrogate countryside of the urban park – was an all-purpose antidote: calming, diverting, health-giving.

In their heyday – the mid Victorian era when urban densities were at their peak – the parks were enormously popular. In Victoria Park, the first major public park in London, 25,000 bathers plunged into the open-air lakes before 8 AM; on one Sunday in June shortly after it opened, the park received 118,000 visitors. The authorities responded with measures of crowd-control, banning games, forbidding contact with the grass, employing park police, replacing grassy walks with paving or gravel. Floral "entertainment" became de rigueur – brightly-coloured ornamental beds, specimen shrubs or trees, horticultural "features" such as rockeries. Bandstands were ubiquitous; later organized sports were introduced. But by the second half of the twentieth century their glory days seemed over. Cities were thinning out, rambling had become a mass pastime, the car, in particular, was providing a passport to the "real" countryside beyond. A downward spiral set in: as visible usage declined, hard-pressed local authorities increasingly begrudged the funds needed for their upkeep. Maintenance regimes lapsed, staffing was reduced, an air of neglect prevailed. Parks, notoriously, became the "Cinderella" service.

Behind this institutional decline, however, lay a deeper cultural narrative. By the later twentieth century, it seemed, the inspiration for parks had gone. They embodied an outdated, defunct vision of nature – or, perhaps, no vision at all. One reason was that nature had, yet again, retreated – to the national parks, for instance, "created" (as legal entities) in the UK in the aftermath of the Second World War. Here you could find both natural features – mountains, forests – unmediated by human design *and* the solitude in which to appreciate them. Far from serving as a connection to this distant nature, as the eighteenth century landscape had done, the late twentieth century park seemed to be a mockery of it. A century or more of usage had taken its toll: the parks had in some sense become encrusted with the detritus of design, of piecemeal intervention. They had become, themselves, exercises in control, bureaucracy, perhaps even formality – and, what was worse, a *failed* exercise.

Several broader lessons offered themselves. First, it seems, "nature" – or our understanding of it – changes from generation to generation, and the aesthetic – the design language – we employ when we incorporate it into our settlements needs to change too. Second, a process of design creep appears to be at work in which the original vision of a natural – and thus, by implication, *undesigned* – world is overlain by visible accretions of human agency, tarnishing and ultimately destroying the sense of naturalness. And third, both these processes are driven or influenced by social, cultural or economic trends in society at large.

Under the influence of counter-urbanization, for example, cities from the 1960s onwards throughout much of the developed world began to empty dramatically, opening up possibilities for green space creation on a scale not seen since the nineteenth century. The aesthetic behind these new green spaces drew much of its force from environmentalism and ecology: it thus placed much emphasis on biodiversity, on (re)creating ecosystems and habitats, on "untidiness". This went hand in hand with an intensified appetite for naturalness, driven partly by an awareness of the loss of the natural in the world at large but also by the experience of the late twentieth century urbanite – of a life lived increasingly indoors, inside man-made habitats, of distancing and disconnection from nature.

For the first time, too, a concerted scientific analysis of human-nature interactions was attempted. By and large, the nineteenth century had created parks without a "scientific" rationale: nature was assumed to be health-giving and improving. This was partly a function of Romanticism, partly a reaction to the perceived awfulness of the new industrial city. But there was also a sense in which both the policy and the spatial context was looser, more flexible: cities were younger, smaller, institutions fewer, newer, less monolithic. By the late twentieth century, although cities had opened up spatially somewhat under the influence of population exodus, the policy context remained relatively unforgiving: in a crowded, complex, competitive world – crowded institutionally as well as demographically – policies had to *prove* themselves. Hunch and gut feeling were no longer enough.

The research into why nature might be "good for us" (and therefore worth public investment), and what makes nature, or the countryside, different from towns and cities, has taken us into new and uncertain territory. In 1978 the Nature Conservancy Council asked people involved in four urban conservation projects why they enjoyed the experience of nature in cities. They spoke of escape, freedom, adventure, discovery; of the sense of a world apart – a "timeless" world, a "paradise", an "oasis"; of the rediscovered richness of once-ordinary sensations. One schoolboy talked about "fun with dirt". Others dwelt on fresh air, the "feel" of flowers, the crackle of ice, above all, perhaps, on smells – "smells", as one Londoner said, that "you wouldn't smell anywhere else. Your whole senses are alive" (Nicholson-Lord 1987).

Later research by University College, London, with residents of Greenwich has suggested that people see nature in cities as a "gateway to a better world" – one that is uncommercialized, rich in sensory impressions and, most important, alive. People feel "part of a living word in which plants, insects, birds, water, mud, birdsong and earthy smell all have their place," the researchers concluded (Nicholson-Lord 1987).

Cities, by contrast – or more specifically, the built environment – are typically seen as dead. A "sensory mapping" exercise in an American town found that four-fifths of its best-loved places were natural landscapes; the most disliked parts were "constructed-urban." Three-quarters of the most memorable sensory experiences cited by residents were linked with "primitive-natural" landscapes. When, in another study, psychologists asked adults – not, it should be noted, country people – to describe the most significant places in their childhood, they drew a sketch of trees, rocks or bushes – in other words, somewhere out of doors. University students shown photographs of urban and rural scenes found that the natural scenes made people friendlier, more playful, less nervous, more content; the urban ones made them depressed and aggressive (Nicholson-Lord 1987).

A dominant theme of such studies is not only that the physical shades over into the psychological but that the two often cannot be disentangled. We react to nature with body and mind: and the two kinds of response feed off and enrich each other. A Countryside Commission study in 1996 found that the feature people most appreciated about the countryside was the sense of relaxation and well-being, followed by "fresh air" and peace and quiet. But, significantly, 93 per cent of people benefit from "just knowing it is there" – merely the thought of it is a comfort (Nicholson-Lord 2003). American student campers, asked what they enjoyed most about nature, put the natural environment top of the list, followed by "cognitive freedom" – the freedom to control one's thoughts, actions, use of time (Nicholson-Lord 1987).

More recent research on "tranquillity" – the most common reason for visiting the countryside – has shown that the absence of people – *other* people – is its most crucial ingredient. As to what constitutes tranquillity, people surveyed mention openness, greenery, "natural places", water (and particularly its sounds), views and horizons, wildlife. One poll of BBC Radio 4 listeners felt tranquillity was encapsulated in sounds like waves or wind through trees (CPRE 2005).

Much of the evidence of the powerful symbolic meanings represented by nature has come in studies by psychologists. In 1994 a study for English Nature reviewed over 250 of these and came to some intriguing conclusions. Nature, it found, offers a "sense of coherence" – in contrast to the confusion of the man-made world. It is mysterious – provoking awe and wonder, a sense of the sublime, encouraging contemplation and "effortless attention" yet resisting explanation. It is largely devoid of "negative feedback" – it does not, in other words, carry a burden of human meaning, or rejection – and thus reinforces self-esteem (Nicholson-Lord 2003).

With concepts such as coherence, mystery and freedom, we are into challenging terrain. Cities, as we have seen, were once associated with freedom: how, and why, have roles been reversed? One reason, clearly, is that the urban freedom of medieval times was political, to do with emancipation from serfdom. No doubt there was also a sense of liberation from what Marx called the "idiocy of rural life". The freedom that nature confers today, by contrast, has more of a psychic dimension to it – the freedom of a world from which people, their rules and hierarchies and interfering ways, have been excluded. A world in which nature is seen as free is a world in which human society has become – or so it seems to many people – oppressive, invasive and intrusive.

Behind such responses there is a long cultural history – a history of associations that derive from art, fiction, religion and mythology and have coloured our attitudes at a level below conscious thought. Nature is rich in such meanings, from the prelapsarian idylls of a Golden Age, of which the Biblical Garden of Eden is one example – the term Paradise derives from the Avestic (ancient Persian) word for enclosure or park – to the role of forests and wilderness. Throughout myth, legend and literature, as authorities such as Joseph Campbell have pointed out, forests are not only places of awe, mystery and fearfulness – places inhabited by wild men and beasts. They are also places where quests begin and adventures follow – places of escape, loss of self and subsequent finding of self (Campbell 1968).

The questing knights of the Grail, for instance, enter the forest "where it is thickest". In Shakespeare's Forest of Arden, wrongs are righted, the world-weary refreshed, the world turned upside down. From Robin Hood to the Zapatistas of Chiapas in Mexico, forests are home to outlaws: to subversion, revolution and world-changing. Indeed it is hard to avoid the conclusion that a vital part of ourselves lies in forests, or at least in the rich yet unknown space they represent, and that if the forests and the wilderness die, this part of ourselves will die too – or, perhaps worse, atrophy and turn septic. Some such logic helps to explain the paradox that increasing numbers of comfortable, affluent Westerners are now actively courting danger, walking across continents or rowing round the world – activities that former ages would have deemed inexplicable.

Many of these psychological responses, of course – mystery, awe, redemption – have long had religious associations and there is much evidence that nature, for many people, now serves as a spiritual focus, rivalling or replacing that of organized religion. While church attendances have been falling, secular religions such as paganism and witchcraft have undergone a resurgence. Movements such as creation spirituality and green Christianity – the latter stressing man's *stewardship* of the planet as opposed to his *dominion* over it – have emerged out of Christian orthodoxy. New Age beliefs – the product of a new distaste for the disenchantments of science, a new openness to mysticism and mysteries – have proliferated.

In the main, however, the new nature-based spirituality has not organized itself or codified its beliefs, preferring to remain private, celebratory, free of ideology. Its public face is the sprawling confederation of green NGOs and pressure groups now referred to as the environmental movement, which has grown explosively since the 1960s and is estimated to number, in the

UK, between four and five million people. In their defence of wilderness and resistance to development and "pollution" can be glimpsed a much older sense of what is sacred, profane and taboo.

Origins of nature spirituality

That environmentalism has become a form of secular religion would surprise nobody who has seen tree-huggers protesting against bulldozers or heard deep ecologists telling us to "think like a mountain". Nor would it surprise historians or philosophers of religion. A century ago the psychologist William James collected scores of accounts of life-changing or life-enhancing experiences – semi-mystical moments that submerge the ego and give a new sense of life and hope – for his classic work *The Varieties of Religious Experience*. James was struck by the number of cases that occurred out of doors. "Certain aspects of nature," he wrote, "seem to have a peculiar power of awakening such mystical moods." Religious awe was "the same organic thrill we feel in a forest at twilight or in a mountain gorge" (James 1902).

Such moments are not as uncommon as one might imagine. Freud labelled them "oceanic"; the American psychologist, Abraham Maslow, called them "peak experiences". One survey found that 36 per cent of British people (and 42 per cent of Americans) own up to them (Nicholson-Lord 1987). And it is clear not only that they are the raw material of religion – the emotional charge that generates a belief in divinity – but that nature is a potent source of them.

This may be not so much because nature is "beautiful" as because it is mysterious, awe-inspiring, endlessly fascinating. According to the German philosopher Rudolph Otto, author of *Das Heilige* (The Idea of the Holy), a sense of the sacred involves a recognition of a power which is *ganz andere* – wholly other. Otto distinguished two chief components of this perception – the *mysterium tremendum* and the *mysterium fascinans*. The first can be translated as "fearful majesty", the second approximates to a sense of "plenitude of being" – the richness and diversity of life (Otto 1923). Charles Darwin experienced something of both, it seems, on his first encounter with a tropical forest in Brazil in 1832. "Wonder, astonishment and sublime devotion fill and elevate the mind," he wrote afterwards.

Nature's potency, in other words, lies in its *otherness* – the fact that it is fundamentally and inalienably different from man and his works. The historian Mircea Eliade, attempting to analyse what it was that led "primitive" cultures to worship the vital force they believed nature to express, chose a slightly different formulation – "real existence". Nature was mysterious, awe-inspiring, certainly; more important, it was real, in a way humans were not (Eliade 1968).

Whatever term is used, attempts to capture this quality of mystery about nature are a dominant theme of myth, religion, art and literature – poetry, in particular. Yet different cultures have gone about the task in different ways. As Eliade has shown, older, more earthbound cultures – North American Indians, Pacific Islanders – thought the otherness was immanent. In other words, it was *within* nature – a vital, indwelling force permeating living things which they called *wakanda* or *mana*. By contrast, theologies of transcendence, such as Christianity, moved God "outside" nature.

Many Christians thus believed that in celebrating the beauty of nature they were celebrating the glory of God, whose handiwork it was. As the poet William Cowper put it, "God made the country and man made the town". Yet Eliade also showed that it is part of the natural cycle of religious belief for monotheistic Gods to grow remote and unloved – at which point there arises a desire among their former worshippers for a more vital and immediate contact with "real existence". Eliade calls this a "fall into life" – and if the diagnosis holds true for the twentieth and twenty-first centuries, it carries some far-reaching implications.

First it means that from the slow collapse of Christian monotheism a new spiritual quest is emerging – for a direct and unmediated relationship with nature, no longer camouflaged by theological doctrine or confused with divinity. If that is the case, then the presence, or absence, of nature in cities becomes, in effect, a question of religious rights – of freedom to worship. For growing numbers of people, nature is their church – a sacred place, a place of "otherness", and thus a source of inspiration, illumination, comfort and celebration. To deny them this outlet is tantamount to religious persecution.

Second, nature in cities is a vital ingredient of spiritual health as well as physical and psychological health. In a self-avowedly secular society, this may seem a strange observation but it is clear that the decay of Christian belief has left many people spiritually stranded. It is also clear, from evidence such as the World Values Survey, which has investigated changing patterns of belief over the last three decades, that the spread of affluence and the satisfaction of material needs in the developed world has brought with it a slow but seemingly inexorable increase in "post-materialism" – the search for a meaning and pattern to human existence.

Over the last two decades evidence has accumulated about the health benefits of contact with nature (see Chapters 32 and 33 this volume). Studies have linked it variously with: faster post-operative recovery; relief of anxiety; better all-round health; reduced muscle tension, blood pressure and heart rate; greater happiness; reduced stress and aggression; enhanced mental functioning and ability to concentrate; and relief of hyperactivity. Green neighbourhoods – parks, tree-lined streets, even a view of nature – have been associated with less crime, better community relationships, greater longevity, and more maturity, concentration and self-discipline amongst teenage girls (Nicholson-Lord 2003).

These are somatic effects with their origins, apparently, in the psyche. Yet advances in understanding of the immune system in recent years has demonstrated how "inner", psychological health feeds through into outer, bodily, health. Studies have shown that people with a deep religious faith tend to be more optimistic; that spiritual activity vies with exercise as the most successful strategy for coping with anxiety and depression. It may be, however, that they are the same thing – that going for a country walk is, for many people, a form of spiritual activity, refreshing the soul just as the "fresh air and exercise" refreshes the body. Initiatives such as Health Walks and the Green Gym have used nature as a way of getting people to exercise and lose weight. But surveys of those taking part have demonstrated clearly that without the nature component, the idea would not work. The most important element in encouraging people to walk – cited by 80 per cent – was "to be in the countryside/green space". Sixty per cent cited "watching the seasons change". Only 10 per cent mentioned losing weight (Nicholson-Lord 2003).

The potential consequences for the future of cities are profound. In T.S. Eliot's poem, *The Waste Land*, the narrator watches a crowd crossing Westminster Bridge and laments: "So many – I had not thought death had undone so many". If only nature is "real" and "other", cities need nature to survive. Without it, they may be efficient as places to work, shop, eat and drink, they may even boast a role as cultural centres, but they will remain, like Eliot's "unreal city", psychologically and spiritually desiccated. And in pursuit of that missing dimension – in pursuit of emotional and spiritual energy and fulfilment – people will continue to leave them, in search of a better life outside.

Both major efforts to green the city – the creation of parks in the nineteenth century and the ecologically-based greening initatives of the late twentieth – sought to make urban living more tolerable by incorporating the vision of freedom, beauty and spiritual satisfaction that nature had come to embody. Many of the qualities that cities had traditionally embodied – refinement, order, predictability, human design – now seemed tarnished or irrelevant. It was a sign, perhaps, of a new planetary condition – one species dominating its living space so completely it had grown tired of its own reflection. To be free was to escape that reflection, to lose oneself in otherness. Humanity, on the brink of urbanizing the planet, sensed itself increasingly a prisoner of its own urban creation.

Against such a background, the need to green the city, more imaginatively and comprehensively than ever before, might appear the one positive way forward – and, indeed, a broad policy consensus around the need for greener cities has taken shape in the UK since the late 1990s. As with so much of urban history, however, powerful forces are pushing in the opposite direction. Population growth and housing need are raising urban densities. A new planning orthodoxy favours "compact" – tightly settled – cities, arguing that they are more carbon-efficient and therefore more sustainable. The rise of an anxiety culture, centring on worries about crime, health and safety, claims for insurance or compensation, is proving inimical to many natural landscape features – notably larger trees and denser stands of vegetation – which are seen to constitute risk.

Partly in response to such pressures, there is evidence that urban design has moved away from the enthusiasm of a previous generation for ecologically-based "design with nature" and back towards its default position – self-advertising human design. Cities, and the green spaces within them, thus face new problems of crowding, usage, denaturing, a larger, more intense and more complex human footprint. Recent studies, meanwhile, have suggested that the growth of computer-based "videophilia" and the loss of contact between city-dwellers and nature could have serious implications for humanity's future management of the planet (Nicholson-Lord 2006). The green city vision, despite well-nigh universal endorsement, remains as elusive as ever.

References

Campbell, J. (1968) *The Masks of God, Vol. 4, Creative Mythology*, London: Secker and Warburg.

CPRE (Campaign to Protect Rural England)/Countryside Agency (2005) *Mapping Tranquillity*, London: CPRE.

Eliade, M. (1968) *Myths, Dreams and Mysteries*, London: Collins.

James, W. (1902) *The Varieties of Religious Experience*, New York: Modern Library.

Nicholson-Lord, D. (1987) *The Greening of the Cities*, London: Routledge & Kegan Paul.

Nicholson-Lord, D. (2003) *Green Cities – And Why We Need Them*, London: New Economics Foundation.

Nicholson-Lord, D. (2006) "Cities for biophiliacs", in J. Norman (ed.) *Living for the City*, London: Policy Exchange, pp. 21–33.

Otto, R. (1923) *Das Heilige*, transl. J.W. Harvey as *The Idea of the Holy*, Oxford: Oxford University Press [first published in German in 1917].

Tuan, Y.F. (1974) *Topophilia*, Englewood Cliffs, NJ: Prentice-Hall.

Williams, R. (1983) *Keywords*, London: Fontana.

31

Intrinsic and aesthetic values of urban nature

A psychological perspective

Rachel Kaplan

Nature is beautiful; nature is peaceful; nature is messy; nature is scary. These are among many perspectives of the urban natural environment. To some there is no nature in the city; for many urban people, by contrast, the urban environment is enriched by many kinds of nature. Whose definition of nature is appropriate? Are people's values of urban nature too idiosyncratic to permit meaningful discussion?

The quick answer to the second question is "no," but more about that in the third section of the paper which draws on insights gained from research on environmental preference. In the first section we examine the multitude of approaches to defining *urban nature*. The differing views of nature are based, in part, on diverse experiences and training; they are also guided by the purposes or functions that "nature" is intended to serve. That is the focus of the second section. Going beyond the diversity of perspectives, inclinations, and purposes, the final section touches on the dual goals of ecologically sound practices and people's preferences and well-being.

What to name it? Whose definition?

"Urban nature" has been questioned as an oxymoron by some and used to signify "wild" nature by others. In several American cities (e.g. San Francisco, Seattle, Denver) organizations have been devoted to urban nature programs to increase the availability and viability of biodiversity in the urban framework, often through ecological restoration projects. Such groups have no doubt that nature does and must exist in the city and, furthermore, that it must be sustained for future generations. In other contexts, urban nature efforts have been referred to as "open space" or "green infrastructure" (Benedict and McMahon 2006).

While there would be little debate about urban parks as examples of urban nature, there might be less agreement if the park serves mostly as a facility for a variety of sports, as is quite common in American towns and cities. Are lawns part of urban nature? A great deal of urban land is consumed by such turf in sizes varying from a bit of grass leading up to a private residence to vast expanses surrounding corporate facilities (not to mention golf courses). What about gardens? Public gardens and arboreta are likely to be widely accepted as urban nature, but private gardens and community gardens perhaps less so.

The dictionary cannot answer the quest for defining "nature" to encompass the diversity of ways in which the word is used. We raise the question of "whose definition" to highlight that the defining qualities of urban nature depend on many issues. In the context of the discussion here, a particularly salient consideration is the background or experience of the person who would be answering the question. Ecologists, planners, managers, and restorationists may define nature in different terms. Their perspectives, in turn, may contrast in important ways from public notions of what constitutes nature. In other words, the answer to what constitutes "nature" is strongly influenced by the kind of training and experience one has had.

The issue of what to call it and whose characterization to use are not a matter of semantics or of getting agreement on a formal definition. People's notion of what does or does not constitute "nature" (or, more specifically "urban nature") has ramifications that reflect deeply felt differences in the values associated with the nearby natural environment. Intense controversies and strong emotions surrounding some restoration efforts offer vivid examples. For example, the "Chicago restoration controversy" (Gobster 2000) can be viewed as public opposition to restoration practices, although what outraged many residents was the removal of healthy trees in their viewshed (Shore 1997). The emotional storm led to a moratorium on all restoration activities in the Forest Preserve District of Cook County (Illinois) with some restrictions sustained for a decade. Embedded in the diverse views of "nature" is far more than professional understanding of restoration ecology. As Nelson *et al.* (2008) observed, societal values must also be understood if the intentions of managing the urban natural resources are to be achieved.

The focus in this chapter is on these societal values or social factors, or put another way, on the perspectives of "nature" that characterize those with no particular professional training. The empirical evidence that has informed our understanding of these perspectives has had quite a few surprises. Perhaps least expected was the consistency across diverse groups and cultures. For the public, as we shall see in the next few pages, the notion of urban nature is broad and inclusive as opposed to purist. At the same time, however, it does not subsume any and all vegetation.

For what purposes or functions?

Closely related to the different perspectives of what is subsumed by "urban nature" are the differing orientations to its functions. As other chapters in this volume have shown, the functions of the land can be described from numerous perspectives. Consider a particular parcel of land at the urban edge, perhaps ten hectares of old fields and forest that are being considered for development. Now think of the different ways such a property might be described. What might an ecologist say about it? How about a planner, realtor, or landscape architect? The same parcel might be considered "derelict" land or "a prime business location." The presence of particular plant species might be highlighted or totally ignored. The trees on the land may be considered "no problem to bulldoze" or "the jewels to preserve."

For many individuals who have no particular training related to the environment the functions or purposes are often related to how they might use the setting. But "use" is an awkward term when considering many of the benefits derived from nature experiences that are often subtle, not necessarily conscious, and possibly entail little physical activity. *Active uses* are the ones most readily considered when talking about forms of involvement or uses of the natural environment. Yet the role that nature plays in these uses can vary from backdrop (e.g. the "nature" behind a tennis court) to being the focus of attention (e.g. a botanic garden). Two other dimensions of involvement with nature deserve consideration: A great deal of nature appreciation comes from *observation*. The contexts for observing nature can vary greatly in terms of physical activity, yet the mental activity may be substantial. Another form of involvement with nature is even less active, but closely aligned with nature's intrinsic values. We have called this more conceptual kind of use "*thereness*" (Kaplan 1985) as it is not dependent on the presence of the natural setting. While the active, observation, and thereness forms of involvement are distinguishable, they often occur together (e.g. in the context of gardening or taking neighborhood walks).

Observing nature

Observation is central to the appreciation of urban nature. Some observation is deliberate; nature photography captures many such moments. At other times, however, observation is less purposeful. One may not realize that the last few minutes were spent looking at a tree until a bird suddenly appears to catch one's gaze.

Perhaps the most pervasive interactions with urban nature occur when one is not even in the nature setting. Whether or not one is cognizant of it, the view from the window can have significant impacts. The powerful role of seeing nature through a window has been shown in numerous contexts. Moore's (1981) study, in the unlikely context of a huge prison, showed that inmates whose cells face the farm fields beyond the facility drew on health services significantly less often than those whose cells viewed other inmates. Ulrich (1984) found greater speed of recovery of hospital patients with nature views. In his analysis of 100 high school campuses Matsuoka (2010) documented that the schools with greater potential for viewing nearby nature, especially from the cafeteria window, had superior student performance (after controlling for socioeconomic variables). Research in the "home" context includes dormitories (Tennessen and Cimprich 1995), public housing (Kuo and Sullivan 2001), and multiple-family housing arrangements (e.g. Kaplan 2001). In the work context, employees with nature views reported fewer ailments, greater patience, and higher job satisfaction (Kaplan 1993). Across these many studies the views varied widely. In some cases they entailed little more than a tree or two or some flowers; in other cases the view may have been of a larger expanse of nature. The psychological benefits that were assessed also differed greatly. Collectively, however, the studies help to substantiate the enormous intrinsic values that the natural setting provided.

"Thereness"

Nature observation can be described as being in the *eyes* of the beholder. Even in its physical absence, however, nature can also play a role in the *minds* of beholders. It does this quite often and easily: We imagine places we know or would like to know, we think about familiar places in other seasons (for example, pondering what to plant in the garden when spring reappears), we try to describe to someone the awesome glow of the full moon amidst the trees. None of these require that the environment is present; all rely on the vast capacity humans have for drawing on mental models and images.

Such "uses" of nature exemplify its intrinsic values. They also help explain why people can so passionately work to protect natural areas they may have never seen and grieve the loss of natural environments they do not plan to visit (Windle 1992). Low user counts of a local park may be taken as justification for reallocating the land without recognizing that the knowledge of the availability of the resource (its "thereness") may be more important to nearby residents than its actual use.

The vital role of knowing a nature area is available is readily evidenced in the negative outcries when its thereness is threatened. The same strong feelings also help to explain citizens' willingness to pay for land conservation initiatives even when it is not likely to benefit them personally. During the 20 years that the Trust for Public Land has maintained its LandVote database (in the US), it has documented that more than 75 percent of "conservation finance ballot measures have passed in urban and rural communities in nearly all 50 states" (www.tpl.org/tier3_cd.cfm?content_item_id=12010andfolder_id=2386/). Not only is this percentage noteworthy, but it has resulted in 1678 passed measures!

The intrinsic values of urban nature are thus far-reaching, subtle as well as blatant, accruing with or without our cognizance, and even whether or not nature is physically present.

What do nature preferences tell us?

One would expect that people's preferences for natural settings would differ. After all, people differ in backgrounds, capabilities, experiences, and penchants. At the same time, however, people also have shared experiences and inclinations and these can help us understand why, despite the great diversity, there are some common dimensions to the importance played by nearby nature.

A fundamental human commonality is our dependence on information. Information is the source of bonds and battles, secrets and slights, delights and dismays. It turns out that an examination of some characteristics of information is useful in understanding common themes in the role nature plays in values and preferences. As used here, information is not restricted to the written or spoken word, nor to the information millions access on television or through the Internet. All the settings, contexts, and circumstances in our lives are sources of information. Humans extract information through all sense modalities and they store vast quantities of information in readily accessible form. Both what surrounds us and what is stored in our heads affects our perceptions of subsequent information.

The preference matrix

All well and good, yet how this relates to preferences may not be self-evident. Research on environmental preferences led us (Kaplan and Kaplan 1978) to grasp that what people like in the environment has a great deal to do with their perceived ability to function in that setting. When the information provided by the environment seems confusing or threatening, the environment is less likely to be preferred. Environments that are more supportive of people's informational needs are likely to be favored. In other words, rather than an idiosyncratic whim, preference is an adaptive response based on the available (or inferred) information. Without necessarily realizing it, we depend on rapid assessments of what the environment – physical, virtual, or conceptual – might afford. These insights, as well as the emerging results from numerous studies, led us to propose the preference matrix (Kaplan and Kaplan 1989).

We have come to see understanding and exploration as pivotal informational needs (Table 31.1). The failure to understand can severely undermine the capacity to function effectively. People dislike being lost; they want to be able to quickly extract the information that is critical to making sense of the situation. At the same time, however, people readily become restless when things are very predictable. They yearn for new challenges and even want to venture beyond their comfort zone. Thus although both the familiar and the unknown are major vectors; neither

Table 31.1 Preference matrix

	Understanding	*Exploration*
Immediate	Coherence	Complexity
Inferred	Legibility	Mystery

by itself is sufficient. Similarly, the four quadrants of the matrix point to qualities that are all desired if the environment is to satisfy our needs.

As the matrix shows, coherence and complexity (or diversity) both entail qualities that can be rapidly assessed. An environment that is low in coherence would be difficult to make sense of; it would be hard to quickly distinguish the salient components. Coherence offers a sense of predictability. An environment that is low in complexity, by contrast, lacks diversity, giving the sense that there is nothing to explore. Both qualities are important; the lack of either may reduce preference. It can be easy to confuse their separate contribution. For example, lower preference for a setting that appears a jumble, might (erroneously) be attributed to too much complexity. The elements comprising the jumble, however, would not appear excessive if they were organized in a coherent way. In other words, the reason for the confusion is more likely to be a lack of coherence than excessive complexity. Similarly, a highly coherent environment may not be preferred if it lacks complexity. Characterizing it as boring is attributable to the low complexity, rather than the assumption of too much coherence.

Legibility and mystery, as the matrix shows, both require some inference. In a physical environment, one can think of these properties as giving the setting three dimensions, or depth. Legibility is enhanced by landmarks and other cues that would help one find one's way through the setting. It is reassuring to have clarity not only about reaching a destination, but to expect that the return trip will also be straightforward. Mystery, by contrast, involves cues that invite further exploration; it is conveyed by suggestions that one could learn more if one can go deeper into the setting. A classic means of inviting such exploration is a bend in the path. Other cues that do not fully obstruct one's view, yet also do not permit seeing all there is to see, can be equally compelling in engaging one's curiosity about what lies ahead.

Dozens of studies of environmental preferences show that each of the preference predictors plays an important role. The important issues in the present context are not so much the relative importance of these predictors, but the insights that the preference research provides with respect to intrinsic and aesthetic values of urban nature. Urban nature was the context for most of the studies (see Kaplan and Kaplan (1989) for overviews of many earlier studies). Preference – people's indication of how much they like a scene – is certainly an important component of aesthetics. At the same time, the critical role played by understanding and exploration speaks to vital intrinsic factors. It would seem reasonable that preferences, rather than being idiosyncratic expressions of a momentary whim, are closely related to one's perception of safety and the expectations of one's ability to cope.

By the same token, it is useful to consider the preference matrix in other contexts. Dependence on information is not limited to urban nature. It applies broadly to the settings and circumstances people constantly encounter. Some of these entail the physical environment, both indoors and outdoors, or perhaps even the environment created by the computer monitor. The many approaches we all depend on for acquiring information – maps, guides, books, presentations, signage – can all be examined in terms of their capacity to facilitate understandability and exploration. How often do we find these situations and informational sources falling short with respect to these concerns? It is not difficult to think of presentations that were confusing – lacking coherence and legibility. Nor is it difficult to think of examples where one's sense of exploration is undermined.

Urban nature – wild and tame

The chapter began with a description of perceptions of nature as beautiful as well as messy, peaceful as well as scary. Are these clashing perspectives reflections of failures of the preference predictors or are there additional factors that need to be considered? In particular, it is important to explore changes in the urban landscape from the prevalence of smooth ground texture characterized by many urban parks and widespread use of lawns to the increasing recognition of the negative environmental consequences of such manicured landscapes.

Lawns and their care are a major industry and preoccupation in many parts of the world. By some estimates lawns consume about 128,000 square kilometres (nearly 32 million acres) of land in the US (Milesi *et al.* 2005), and some 80 percent of US households are said to have private lawns. Lawn maintenance, however, has come under considerable attack for its many health and environmental impacts. Not only are lawns expensive to maintain, but they place sizable demands on water; they depend on massive amounts of fuel both in mowing and fertilizer production, and maintenance practices pollute the air from mowing as well as the groundwater from chemicals spread on the lawn. In response, many American and Canadian municipalities have imposed restrictions on pesticide and fertilizer use for lawns as well as limiting use of water for lawn maintenance. Clearly "the obsessive quest for the perfect lawn" (Steinberg 2006) and the substantial industry behind it are being challenged (see also Chapters 23 and 24 this volume).

While devotion to lawns can be considered an indication of preference for them, it is interesting that the research that generated the preference predictors discussed earlier does not offer strong support for the preference of large expanses of lawn. In fact, large lawn areas are likely to be lacking in each of the predictors given the low complexity of the monocrop, as well as lack of landmarks or mystery. At the same time, however, the preference predictors clearly point to smooth ground texture as an effective way to signal ease of locomotion and support exploration.

The apparent incongruity may relate to the scale of the smooth ground texture. Small focal areas need not compromise ecological health while providing the qualities that speak to people's preferences. Negotiating the conflicting vectors between advocates of naturalistic landscapes and those who find them distasteful may be informed by such considerations. Research examining the perceptions and preferences for ecological restoration and naturalistic landscapes has been notable for its international flavor as well as for offering a mixed picture often showing that rougher textures are less well liked (Kaplan 2007). However, the juxtaposition of the rougher textures with a path or smooth-textured edge can enhance the potential for exploration and make the setting more inviting.

What this analysis would suggest, then, is that the cues suggested by the preference matrix can go a long way toward making naturalistic landscapes not only acceptable but liked. At the same time it is important to recognize that both understandability and exploration are necessarily influenced by experience. After all, the mental models we carry with us result from experiences. Our familiarity with different kinds of environments is not only the result of previous exposures but has a persistent effect on how we perceive new settings as well as our comfort with them. Thus as naturalistic landscapes become more common, and there is a more widely shared understanding of their benefits, one would expect greater acceptance of the rougher textures and increasing appreciation of such approaches. That is not to say that the smoother textures will lose favor, but rather, as Özgüner and Kendle's (2006) study shows, both design approaches can be appreciated in their own right.

Urban nature and gardening

If grass is a common characteristic of urban vegetation, so are gardens. Gardens too can be distinguished in terms of their scale, public vs. private ownership, and the degree to which they are perceived as an attraction or blight. However, despite their prevalence as well as the important social and psychological roles they play both private and community gardens are often overlooked in the context of urban nature. Here we only briefly highlight some ways in which such gardens exemplify many of the values that make urban nature such a vital source of human well-being.

One purpose of gardens is for growing food. The various kinds of community and allotment gardens, as well as gardens maintained at private residences, devote much of the precious urban land resource to food production. Furthermore, increasing the availability of local food is an important consideration for urban sustainability. However, whether or not the garden is devoted to edibles, it also fulfills other benefits and these are closely related to the themes of this chapter. The American Community Gardening Association, founded 30 years ago, includes several in their list of "Benefits of community gardens" (www.communitygarden.org/learn). In particular, these facilities have played major roles as a community resource; they have enhanced the sense of community and triggered other neighborhood projects that increase residents' social interactions across generations and cultures, pride in their community, and self-reliance.

The tangible benefits (such as food production) are certainly important to gardeners. However, our research on the psychological benefits of gardening (Kaplan 1973; Kaplan and Kaplan 1989) shows that the ability of the garden to afford tranquility and the many ways it fosters fascination are even greater sources of satisfaction. Checking on the plants, working close to nature, and even planning the garden are included among such fascinations. Lewis' (1996) keen observations of residents of public housing apartment complexes show many parallels in a context that included no edible plants, namely the flower competition sponsored by the New York City Housing Authority. The intensity of their valuing of their flower gardens is evident in the imaginative steps residents took to guard them from vandalism.

Urban nature provides many kinds of satisfactions and is valued for a multitude of reasons. When asked why they appreciate nature study participants will often say because it is "green." Much of the year in many places nature is not green. Rather green serves as a shorthand for many dimensions of nature that are difficult to articulate and subtle, yet deeply felt and treasured.

Can we combine ecological and people considerations?

One does not argue over the things that do not matter. Nearby nature, however, matters greatly to many people. The more valued it is, the more easily it can be grounds for controversy. The threat of development where there is a patch of open space, the removal of invasive species if they served as a nature view from one's window, a neighbor's naturally landscaped yard rather than kempt front lawn, the decision to change an area that attracts particular birds to a groomed park – all these can be hotly contested among citizens who had previously seemed calm and respectful. The emotions are barometers of how much it matters.

Conflicting views, however, need not be considered insoluble problems. Many times a solution need not be a matter of winners and losers. Many times the warring parties fail to realize the strong commonalities in their values. All too often the divisiveness and conflict are the outcomes of a failed process for seeking common ground.

Phalen (2009), for example, contrasts two ecological restoration projects. One of them, the so-called Chicago controversy mentioned earlier, met with substantial outcries and disagreements; in the other case – in the same city at about the same time – disparate groups found ways to resolve their concerns. While the process in the latter instance required the willingness of many individuals to spend the time, such commitment is neither requisite to successful outcomes nor a guarantee that it will happen. The realization of common values can also result from far simpler approaches. Kearney *et al.* (2008), for example, show how the results of a photo-questionnaire uncovered stronger commonalities than the opposing groups would have suspected.

To ecologists and planners it may appear that citizens' values are at odds with sound management, leading to the assumption that any proposed solution will be disputed. To citizens, by contrast, it often appears that the managers are dismissive of their concerns. Dismay and protest may serve a purpose, but they readily obscure the fact that the many groups and parties share in caring deeply about the availability of urban nature. They want it to be managed in a way that is sustainable, ecologically healthy, and supportive of people's values. No doubt other chapters of this volume provide insights into ways the much valued urban nature can be designed and managed to meet these challenges.

Acknowledgments

The chapter represents much more than my own work. The benefits of a life-long collaboration go far beyond "acknowledgment." Stephen Kaplan's contributions, like so many of the values discussed in this chapter, are vast, pervasive, and far-reaching. Our many students and their students have also been important influences. The financial support provided by the US Department of Agriculture, Forest Service Northern Research Station, is also gratefully acknowledged.

References

Benedict, M.A. and McMahon, E.T. (2006) *Green Infrastructure: Linking Landscapes and Communities*, Washington, DC: Island Press.

Gobster, P H. (2000) "Restoring nature: Human actions, interactions, and reactions," in P.H. Gobster and R.B. Hull (eds.), *Restoring Nature: Perspectives from the Social Sciences and Humanities*, Washington DC: Island Press, 1–19.

Kaplan, R. (1973) "Some psychological benefits of gardening," *Environment and Behavior*, 5(2): 145–52.

Kaplan, R. (1985) "Nature at the doorstep: Residential satisfaction and the nearby environment," *Journal of Architectural and Planning Research*, 2(2): 115–27.

Kaplan, R. (1993) "The role of nature in the context of the workplace," *Landscape and Urban Planning*, 26(1–4): 193–201.

Kaplan, R. (2001) "The nature of the view from home: Psychological benefits," *Environment and Behavior*, 33(4): 507–42.

Kaplan, R. (2007) "Employees' reactions to nearby nature at their workplace: The wile and the tame," *Landscape and Urban Planning*, 82(1–2): 17–24.

Kaplan, R. and Kaplan, S. (1989) *The Experience of Nature: A Psychological Perspective*, New York: Cambridge University Press (republished by Ann Arbor, MI: Ulrich's 1995.)

Kaplan, S. and Kaplan, R. (eds.) (1978) *Humanscape: Environments for People*, North Scituate, MA: Duxbury (republished by Ann Arbor, MI: Ulrich's, 1982.)

Kearney A.R., Bradley, G.A., Petrich, C.H., Kaplan, R., Kaplan, S, and Simpson-Colebank, D. (2008) "Public perception as support for scenic quality regulation in a nationally treasured landscape," *Landscape and Urban Planning*, 87(2): 117–28.

Kuo, F.E. and Sullivan, W.C. (2001) "Aggression and violence in the inner city – effects of environment via mental fatigue," *Environment and Behavior*, 33(4): 543–71.

Lewis, C.A. (1996) *Green Nature Human Nature: The Meaning of Plants in our Lives*, Urbana, IL: University of Illinois Press.

Matsuoka, R.H. (2010) "Student performance and high school landscapes: Examining the links," *Landscape and Urban Planning*, 97(4), 273–282.

Milesi, C., Running, S.W., Elvidge, C.D., Dietz, J.B. Tuttle, B.T., and Nemani, R.R. (2005) "Mapping and modeling the biogeochemical cycling of turf grasses in the United States," *Environmental Management*, 36(3): 426–38.

Moore, E.O. (1981) "A prison environment's effect on health care service demands," *Journal of Environmental Systems*, 11(1): 17–34.

Nelson, C.R., Schoennagel, T., and Gregory, E. (2008) "Opportunities for academic training in the science and practice of restoration within the United States and Canada," Restoration Ecology, 16(2): 225–30.
Örgünzer, H. and Kendle, A.D. (2006) "Public attitudes towards naturalistic versus designed landscapes in the city of Sheffield (UK)," *Landscape and Urban Planning*, 74(2): 139–57.
Phalen, K.B. (2009) "An invitation for public participation in ecological restoration: The reasonable person model," *Ecological Restoration*, 27(2): 178–86.
Shore, D. 1997 "The Chicago Wilderness and its critics. II. Controversy erupts over restoration in Chicago area," *Restoration and Management Notes*, 15(1): 25–31.
Steinberg, T. (2006) *American Green: The Obsessive Quest for the Perfect Lawn*, New York: W.W. Norton and Co.
Tennessen, C.M. and Cimprich, B. (1995) "View to nature: Effect on attention," *Journal of Environmental Psychology*, 15(1): 77–85.
Ulrich, R.S. (1984) "View through a window may influence recovery from surgery," *Science*, 224(4647): 420–1.
Windle, P. (1992) "The ecology of grief," *BioScience*, 42(5): 363–6.

32

Urban nature and human physical health

Jenna H. Tilt

The act of walking is to the urban system what the speech act is to language.

Michel de Certeau (2002)

Hardly a day goes by without a news story on obesity, diet and/or exercise. But despite this media attention we are currently losing the "battle with the bulge"; obesity rates continue to rise in industrialized countries around the world (World Health Organization 2006). As the research evidence builds for the link between lack of physical activity and chronic diseases (e.g. coronary heart disease, hypertension, non-insulin-dependent diabetes, osteoporosis, colon cancer, and anxiety and depression) (Pate *et al.* 1995), public health researchers, urban planners and other researchers are beginning to investigate how the urban environment can create, enhance, or limit opportunities for humans to participate in physical activities such as walking, jogging, and biking. This chapter will discuss how urban nature spaces, such as parks and highly vegetated neighborhoods, can contribute positively to physical activity levels and other health measures.

The new emerging transdisciplinary field of "active community environment studies" combines the dual foci of urban planning and public health to investigate the relationship between the urban environment and physical activity using a social-ecological approach (Stokols 1992; Sallis and Owen 1999). This approach recognizes that the physical environment, in addition to other factors, such as demographic factors, can influence behavior and behavioral change. Several active community environment studies have shown that features of the built urban environment such as residential density (Saelens *et al.* 2003; Frank *et al.* 2005; Leslie *et al.* 2005; Moudon 2005), close proximity to a variety of stores and services (Humpel *et al.* 2004; Hoehner *et al.* 2005; Lee and Moudon 2006; Pikora *et al.* 2006), street connectivity (Duncan and Mummery 2005; Frank *et al.* 2005; Leslie *et al.* 2005) and perceptions of safety (Booth *et al.* 2000; Doyle *et al.* 2006) are related to patterns in physical activity behavior, particularly walking for transportation or recreation purposes. These associations hold true even after considering and controlling for the effects of personal and social factors such as socio-economic status and self-efficacy.

Aesthetic features accompanying the walking area or route have also been shown to be associated with human physical activity (Humpel *et al.* 2002; Pikora *et al.* 2003; Lee and Moudon 2004; Owen *et al.* 2004; Tilt *et al.* 2007) and urban nature spaces such as parks, street trees and overall residential greenness are increasingly being recognized as an important feature in creating healthy and walkable urban environments. This chapter will discuss some of the relevant research regarding how two types of urban nearby natures – parks and neighborhood vegetation – may contribute to the physical health of urban residents. Key issues to consider when incorporating nearby nature into urban settings for the purpose of fostering physical activity will be explored. In addition, implications and future research regarding the relationship between nearby nature and physical health will also be examined.

Parks and physical activity

Urban parks have long been considered the "lungs" of the city. The Olmsted family of landscape architects and other renowned architects of the early twentieth century designed parks as a place for the working man and his family to escape from the ills of the city (Olmsted 1979; Cranz 1982). At that time, parks were envisioned to be places of passive recreation: places to picnic, promenade and to relax from the high physical demands of an early labor-intensive industrializing country. Nowadays, the vast majority of Americans, and citizens from other post-industrial countries, spend their working week sitting passively. Not only are we passive in our jobs, but we come home to televisions, game stations and other entertainment systems that allow us to continue to be passive in our leisure time as well. Thus, more than 60 percent of Americans do not achieve the CDC's recommended weekly physical activity levels (CDC 1996). Parks today are still the lungs of the city, but now are viewed by public health professionals as critical spaces that have the capacity to encourage and foster physical activity for a wide range of populations beyond the passive enjoyment of yesterday. Public health and urban planning studies focusing on the relationship between parks and physical activity have found two key components in analyzing this association: (1) park proximity from one's home; and (2) specific park features.

Park proximity

Being able to access a park easily is fundamental to park use and increasing physical activity levels for a wide variety of populations (Kahn *et al.* 2002; Cohen *et al.* 2007). Living close to a park allows people to take an alternative form of transportation to arrive at that park setting, such as walking or biking, and thus increasing their overall physical activity. In addition, those, especially children and adolescents, who arrive at parks on foot or by bike are more active in the park setting than those arriving by other means (Grow *et al.* 2008). Older adults spend more active time in a park setting when they perceive it to be close to their home (Mowen *et al.* 2007). Ethnic minorities and low-income populations also reap physical health benefits and are more physically active when living within walking distance from a park (Cohen *et al.* 2007). However, historically, ethnic minority and lower-income residential urban areas suffer from poor access to parks (Wolch *et al.* 2005).

As more research confirms the connection between park proximity and physical health, the call for equitable access to parks has become part of the larger environmental justice movement and a key concern for several local and national environmental and sustainability non-profit organizations such as Trust of Public Lands (Gies 2006). Parks provide a space for physical activities free of charge, unlike other locations for physical activities such as private clubs and gyms. Exactly what constitutes "good" park proximity or distance from a home to the park is debated by public health and urban professionals, but is generally thought to be within 0.5 mile or 1 km distance from the home to the park (Cohen *et al.* 2007). Although the distance may vary slightly, more important is the directness of the route to the park. Street connectivity (e.g. low preponderance of "dead-end" streets), sidewalk availability, low traffic and crosswalks all aid in creating an easily accessible route to the park, increasing the frequency of walking and biking trips to the park (Troped *et al.* 2003; Frank *et al.* 2005).

Park features and users

Once inside the park, specific park features have been related to physical activity levels for a variety of populations. Park audit tools such as System for Observing Play and Recreation in Communities (SOPARC) (McKenzie *et al.* 2006) and physical activity tracking devices such as accelerometers have helped researchers examine how different areas and features of a park may contribute to physical activity. Not surprisingly, researchers found that young children were most active on playgrounds (Potwarka *et al.* 2008). Ball fields and courts, such as soccer fields and basketball courts, are areas of intense physical activity, especially for adolescent males. Observational research showed that teenagers use parks to meet friends and observe peers of the opposite sex rather than engaging in physical activities (Loukaitou-Sideris 1995; Marcus *et al.* 1998). Observational and ethnographic studies report that children who have outgrown the playgrounds are virtually absent from parks, especially young girls ages 6–12 (Tuttle 1997); while other researchers have observed this "tween" group as more interested in the natural surroundings for imaginative play (Marcus *et al.* 1998).

Racial and ethnic groups tend to use parks in different ways. African Americans preferred recreation and park facilities (Taylor 1993; Payne *et al.* 2002) and put a high priority on the quality of park facilities, particularly maintenance and cleanliness (Gobster 2002). Hispanic ethnic groups tend to use parks more often than other groups and for more passive uses (primarily picnicking) (Hutchinson 1987; Gobster 1992; Loukaitou-Sideris 1995; Gobster 2002; Payne *et al.* 2002; Tinsley *et al.* 2002). Asian groups tend to come to parks as part of an organized group, such as a church group (Gobster 2002; Tinsley *et al.* 2002); mainland Chinese, in particular, prefer walking and swimming activities in park settings (Zhang and Gobster 1998). Women park users of all backgrounds tend to be more passive users than men, primarily participating in caretaking duties like watching young children play at the playground (Hutchinson 1994; Krenichyn 2004) or pushing strollers around the park (Gobster 1992). In addition, women park users were most often seen in parks in the middle of the day and on weekends when use levels are the highest (Franck and Paxson 1989; Hutchinson 1994).

Despite the differences discussed above, some commonalities in park use and preferences do exist between these various user groups. Several studies have shown that the aesthetic appeal of the park is related to park use and physical activity (Tinsley *et al.* 2002). In particular, water features, wooded areas, and landscaping are important reasons in choosing to use a park setting for physical activities (Tucker *et al.* 2007). Not only are these features important to choosing a park setting, they are also related to more active use within a park. For example, adults and seniors tend to be more physically active within parks with paved pathways, wooded areas and trails (Kaczynski and Henderson 2008; Kaczynski *et al.* 2008).

These natural features within a park setting can also mitigate much of the effect of close proximity of parks on physical activity levels (Giles-Corti *et al.* 2005; Kaczynski *et al.* 2008). In other words, the quality of the park features, particularly its natural features, may be more important in obtaining higher levels of physical activity within a park than simply living within close proximity to a park. These parks that tend to have more and a wider diversity of these natural features are generally larger in physical size and are often referred to as "regional" or "central parks" of cities and counties. A wider range of visitors may access these parks by various means – car, on foot, public transportation – and may come specifically to engage in physical activities.

However, even though these large parks may be important for diverse physical activities, smaller neighborhood parks or "pocket parks" can also play a critical role in promoting physical activity among many groups of people. It is the combination of (1) having a park within an easy walking distance that is direct and free of barriers, plus (2) having, once at the park, some space in which to continue to be active that is aesthetically pleasing, enjoyable and that facilitates physical activity. A regional or large park may already have many natural features that can encourage more active use in a park setting, but not everyone can live within walking distance of such parks. An alternative strategy is to design or re-design existing smaller neighborhood parks to incorporate more spaces that encourage active use to a wider range of people. For example, ball fields offer opportunities for vigorous physical activities, but only a few groups (primarily adolescent and adult males) use these fields. Placing a paved pathway around the perimeter of the field along with trees and landscaping may encourage active use of the space for other groups (Joseph and Zimring 2007). A group of researchers in Missouri found that these types of trails were particularly useful in getting women and people in lower socioeconomic status groups to increase their physical activity levels (Brownson *et al.* 2000). Women in lower socioeconomic groups have been shown to be at a particular risk of not getting enough physical activity (Crespo *et al.* 2000; Wilcox *et al.* 2000). These ideas and others, such as providing on-site day care or physical activity leaders and classes (Floyd *et al.* 2008; Kaczynski and Henderson 2008; Potwarka *et al.* 2008), may help to encourage new park users and to help passive park users to become more physically active park users.

The variety of research collected on park use – both active and passive – point to the importance of creating or preserving nearby nature in a park setting. Nearby nature, such as a small wooded area with a paved path to walk or bike on, or a pond with benches surrounding it, can appeal to a wide variety of park users. Creating more parks, providing some nearby nature, such as trees, landscaping, water features, and providing opportunities to actively engage with those features, such as pathways, may prove to be effective in increasing physical activity levels for many diverse groups of people. This is particularly important for low-income and minority areas, which disproportionately have access to fewer parks within their neighborhoods.

Neighborhood vegetation and physical activity

During the late nineteenth and early twentieth centuries, ensuring adequate public health was an important issue in US and European cities. Cholera, yellow fever and other disease outbreaks were regular phenomena in crowded, unsanitary urban neighborhoods. As a result, the profession of planning devoted much effort to understanding the linkages between urban space and its impacts on health (Frumkin *et al.* 2004; Frank and Engelke 2005). Ebenezer Howard and other early figures in planning and design saw the importance of separating industrial uses from living areas and the importance of green spaces to improve the health of urban residents (MacMaster 1990). These planning efforts to separate uses and create healthier public spaces also created unintended consequences. Infectious disease outbreaks are rare these days and sanitation conditions of our urban areas have improved greatly. However, decades of plans and strict zoning laws designed to separate residential, commercial and industrial land uses have also created a homogenous landscape of low-density residential communities where the occupants must drive to work, school and for most errands (Frumkin *et al.* 2004).

Active community environments studies have shown that neighborhoods with high accessibility to stores and services (Lee and Moudon 2006), high residential density and high street connectivity are neighborhoods most supportive of walking and other physical activities (Frank *et al.* 2005). In addition, aesthetics have also been shown to play a key role in understanding the physical environmental characteristics of streets and neighborhoods that create more walkable places. Reviews of active community environments studies (Humpel *et al.* 2002; Pikora *et al.* 2003; Lee and Moudon 2004; Owen *et al.* 2004) point to aesthetics as a key physical environment variable associated with walking behavior, including studies that examine walking for recreation (Ball *et al.* 2001; Hoehner *et al.* 2005; Pikora *et al.* 2006; Joseph and Zimring 2007) and transportation (Suminski *et al.* 2005; Tilt *et al.* 2007). Although "aesthetics" in these studies is ill-defined, the concept is mainly related to the natural features and vegetation found in the neighborhood or park environment. The most thorough operational definition of aesthetics in active community environment research is from a qualitative study that found trees, water features and birds had a highly positive influence on physical activity behavior in a park setting (Corti *et al.* 1996).

Neighborhood aesthetics, or greenness, has been measured subjectively via residential perceptions (Takano *et al.* 2002; Tilt 2007; Tilt *et al.* 2007; Sugiyama *et al.* 2008) and objectively using the Normalized Difference Vegetation Index (NDVI) (Liu *et al.* 2007; Tilt *et al.* 2007; Bell *et al.* 2008), which is a remotely sensed spectral vegetation index derived from satellite-mounted sensors that measures photosynthetic light. Other objective measures of neighborhood vegetation used in active community environment studies include the percentage of neighborhood canopy coverage (Tilt 2007), land use (de Vries *et al.* 2003) and neighborhood street tree counts (Giles-Corti and Donovan 2003; Hoehner *et al.* 2005; Pikora *et al.* 2006). Though the methodologies differ from study to study, this body of research generally shows a consistent positive relationship between neighborhood greenness or vegetation and walking or general physical health. Key findings include a link between body mass index (BMI) and greenness (NDVI) in children and youth (Bell *et al.* 2008) and an interaction between high accessible neighborhoods, BMI and greenness (NDVI) in adults (Tilt *et al.* 2007). Other studies have shown high levels of vegetation (e.g. canopy coverage and residential perceptions) to be related to walking trip frequency to a variety of neighborhood destinations (Tilt 2007) and for neighborhood recreation walking (Kaczynski 2010). Research has also found that neighborhood greenness is a factor in perceived physical (Maas *et al.* 2006; Mitchell and Popham 2007; Sugiyama *et al.* 2008) and mental (Sugiyama *et al.* 2008) health and, in one study, the longevity of seniors (Takano *et al.* 2002).

In some of my recent research, I used a photo-questionnaire of black and white neighborhood walking scenes (Figure 32.1) illustrating varying levels of vegetation to assess how respondents perceived their own neighborhood greenness (e.g. *"How similar is the scene to where they walk in their neighborhood"*) and what respondents preferred in a neighborhood walking environment (e.g. *"How much they would like to walk in this neighborhood"*). Similar to the literature reviewed above, I found that those respondents in the Seattle, Washington urban and suburban areas that perceived their neighborhood to be similar to the high vegetation scenes made more walking trips to a variety of destinations, after controlling for a range of socio-demographic variables and built environment factors such as access to destinations (Tilt 2007, 2010).

But why does having and preferring highly vegetative neighborhood environments facilitate physical activities and better physical health? In the same photo-questionnaire, I also asked respondents to briefly list the reasons behind their preference scores for the last three scenes, shown here in Table 32.1. These reasons give considerable insight into why urban nature contributes to physical activities. Themes emerging from a content analysis completed on the reasons for preference for these three scenes revealed that the high vegetation scene was preferred because its high level of vegetation provided interest to the walking area (Table 32.1). Other themes showed that the pattern and maintenance of the vegetation also influence preference ratings. For example, the pattern of the high vegetation scene (Figure 32.1a) is very enclosed, limiting the visual access of the walking area. Some respondents mentioned this pattern as a positive

(a)

(b)

(c)

Figure 32.1 Individual and factor preference for (a) high, (b) medium, and (c) low vegetation scenes.

Table 32.1 Overall stated reasons for preferences of the three vegetation scenes

Meta-theme	*Theme*	*High vegetation*	*Medium vegetation*	*Low vegetation*
Vegetation	Lots of vegetation (+)	36% (221)[1]	6% (36)	–
	Some vegetation (+/–)	25% (154)	34% (208)	1% (5)
	No vegetation (–)	–	2% (13)	45% (275)
Exposure	Private, protected (+)	13% (79)	1% (3)	–
	Too isolated (–)	18% (111)	2% (15)	–
	Too open, exposed (–)	–	2% (10)	17% (104)
Interesting	Interesting (+)	21% (127)	7% (41)	1% (4)
	Boring (–)	–	4% (24)	33% (206)
Maintenance	Poor upkeep (–)	6% (38)	22% (137)	–
Safety	Not safe (–)	10% (60)	–	–

Note

1 *n* = 617 total respondents from Seattle, Washington and suburban areas, % of respondents mentioning each theme.

characteristic of the scene, but most did not and also perceived the walking environment to be unsafe – someone or something could be lurking behind that dense vegetation.

Based on these results, it appears that the high vegetation walking scene created a polarization of preferences. Respondents who preferred this highly vegetative walking environment mentioned that the high level of vegetation offered a diverse and interesting walking environment. On the other hand, respondents who did not prefer the highly vegetated walking environment scene perceived the high level of vegetation as contributing to a sense of being enclosed and isolated within the walking space, decreasing the visual and physical access of the walking environment and leading to concerns about safety. These seemingly conflicting findings show that while high levels of vegetation are associated with walking trips in this study and others listed above, and additionally have been related to overall physical health, the pattern of these highly vegetative neighborhoods could possibly prove to be even more influential in its relationship to physical activity and physical health.

Issues to consider when incorporating nature into urban settings for the purpose of fostering physical activity

Safety

The evidence here strongly makes the case that nearby nature, particularly high levels of vegetation including mature trees, is related to physical activities, particularly walking, within a park and neighborhood setting. However, there are several issues to consider before incorporating more vegetation into parks and neighborhoods for the purpose of fostering physical activities. The most important issue is that of perceived safety of where the physical activity is taking place. As shown above in the example from my research, high levels of vegetation do not necessarily equate with more physical activity. The pattern of this vegetation is critical; if vegetation is planted and allowed to grow in such a way that it limits the ability to visually and physically access one's immediate surroundings, then fears of being in a potentially unsafe environment will most likely override any motivation to use the space for physical activities.

The qualitative reasons related to exposure in this study relate to Appleton's prospect and refuge theory (Appleton 1975), a key early finding in environmental psychology. Open environments with unimpeded opportunities to see are referred to as prospect and environments where one has an opportunity to hide are called refuge. Appleton states that landscapes that have *both* opportunities to see and to hide are aesthetically more pleasing (Appleton 1975). For example, a bench on a hill, partially hidden from the view of others by vegetation, yet pervious enough to those sitting on the bench to be able to look down upon a pathway below, would be considered an ideal location for both prospect and refuge. Compare this mental image of a bench on a hillside partially hidden by vegetation to a bench in the middle of a football field. On the football field, there is no cover or refuge; yes, one can see out clearly from the bench, but also many can see in creating feelings of vulnerability.

However, when moving through an environment, being able to constantly survey one's immediate surroundings, thus having high prospect, becomes more important than do spaces of refuge. Prospect can be blocked by trees, thick shrubs, walls or other features and may evoke a sense of fear since one cannot see what is behind the thick layers of vegetation. For example, Nasar and Jones (1997) found that concealed or partially hidden areas created a sense of fear for female college students when walking in these areas at night. Many of these concealed areas were created by the vegetation patterns of the area – thick shrubs and trees that blocked visual access. Similarly in the qualitative analysis of preference for walking environments discussed above, many of these same respondents feared that someone or something may be hiding behind the vegetation that they could not see into or over in the scene. On the other hand, having some vegetation in the neighborhood may help to limit crime and civil disorder as shown by researchers studying the effects of vegetation on residents of Section-8 housing projects in Chicago. These researchers found that housing complexes surrounded by trees had fewer incidences of crime and civil disorder (Kuo 2001). However, the vegetation pattern around these projects was a combination sparse grass and widely spaced trees that did not limit visual or physical accessibility.

Many active community environment studies point to safety concerns as a factor limiting women's use of urban parks of many different ethnicities (Westover 1986; Carr *et al.* 1992; Burgess 1998). Fear for personal safety is a barrier to neighborhood adult walking (Loukaitou-Sideris 2006) and has also been negatively related to a child's physical activity level (Molnar *et al.* 2004; Beets and Foley 2008) and to adults' readiness to encourage children's physical activity (Miles 2008). On the other hand, perceptions of a safe neighborhood environment have been associated with walking and/or physical activity in many studies (Hawthorne 1989; Ross 1993; Bauman *et al.* 1996; Weinstein *et al.* 1999; Booth *et al.* 2000; Chandola 2001; Kirtland *et al.* 2003; Saelens *et al.* 2003; Doyle *et al.* 2006).

Together these findings point to a complex relationship of having a vegetated, yet safe park and neighborhood environment to walk or participate in other physical activities. Highly vegetated neighborhoods and parks can provide environments that promote physical activities, *if* the vegetation is managed in such a way that it does not limit visual or physical access of the space where the activity is taking place. Having mature trees which have been pruned to raise the tree canopy; and offering low-level shrubs and flowers which are colorful, but organized in a coherent manner, allows visual interest and access to the area where the physical activity is occurring, without leading to safety concerns. In addition, trees should be well spaced and not be planted near a path corner or intersection, since these are points where visual access is already limited and where fear for personal safety is high (Nasar and Jones 1997).

Maintenance of the neighborhood and civil disorder

Not only can the pattern of vegetation affect physical activity behaviors because of potential safety issues, but the maintenance of the natural area can also cause safety concerns and limit people's overall desire to be in a park or neighborhood sidewalk environment. Nassauer (1995) found that natural spaces that have "cues of care" such as mowed edges, flowering plants, and trimmed shrubs create a sense of familiarity and can give a place an orderly frame. This orderly frame helps to increase preference of the environment. Cues of physical disorder such as graffiti, littering and untamed natural areas have been negatively related to parents' encouragement of children to use local playgrounds (Miles 2008). Similarly, in the qualitative assessment given above, negative perceptions of maintenance were related to lower preferences for walking environments (Table 32.1). Signs of physical disorder or visual cues of park neglect, such as overgrown and weedy vegetation, and broken equipment have also been associated with neighborhoods with lower health status (Coen and Ross 2006) and have low levels of use for vulnerable populations such as African American adolescents (Ries *et al.* 2008). These signs of disorder and lack of maintenance can also lead to perceptions of being in an unsafe environment.

While having trees and landscaping in park and neighborhood environments will most certainly add to aesthetic value and possibly promote higher levels of physical activity for those living and visiting these areas, urban planners and policy makers also need to consider and account for the maintenance costs associated with this high level of greenness. This vegetation needs a certain level of upkeep, such as regular pruning, weed and litter removal, to help create and maintain an aesthetically appealing and safe environment.

Implications and future research

The implications of the connection between urban nature and physical activity are many. Understanding and providing concrete evidence regarding the relationship of vegetation level and pattern on physical activity and health has far-reaching policy implications for neighborhood planning. Planning and zoning laws across the nation are slowly starting to shift towards creating neighborhoods with a renewed focus on walking and accessibility to services and transit. For example, the city of Cornelius, North Carolina adopted a new land development code to emphasize pedestrian travel (Design 2006). However, changes to planning ordinances and codes are incremental and often slowed by political and economic forces. Incorporating more vegetation, particularly trees in appropriate patterns as discussed above, may provide a way for communities to make small improvements to their walking environments and parks and enhance the health of its citizenry. Such small improvements could be made more immediately rather than waiting for larger planning policies, such as multi-use zoning or increased street connectivity, to be implemented. In addition, linking vegetation to physical well-being could be thought of as another "ecosystem service" provided by urban nature and could be used as a way to leverage more funds for urban green spaces.

However, there is still much work that needs to be done in investigating the link between urban nature and physical activity and health. For example, future research needs to address questions such as: does the level of vegetation in the neighborhood influence route choice to destinations or walking routes around the neighborhood? Although many studies have culminated together to provide strong evidence linking physical activity to greenness, future studies should concentrate on describing in greater detail the mechanisms behind vegetation's influence on physical activity behavior and general health. Another avenue for future research that may provide valuable information is to explore the effect of vegetation on physical activity and health in different geographic areas or seasons.

Conclusions

This chapter has explored the research linking urban nature, particularly nearby natural spaces such as vegetation found along neighborhood sidewalks and within parks, to physical activity levels and overall health. Though there is strong evidence that vegetation, particularly high levels of vegetation and mature trees, contribute to one's physical activity and/or health, the pattern and maintenance of this high level of vegetation is also extremely important. Perceptions and fears about safety and a general dislike for poorly maintained urban nature spaces may considerably limit the positive effect of the nearby natural space on physical activity and health.

In an effort to tie this chapter to the others in this volume, two important final points should be made: (1) urban planners, policy makers and urban forestry professionals should list physical health as another "ecosystem service" that trees and urban nature provide as they advocate for increasing natural spaces within urban areas; and (2) greater research effort should be made to establish linkages between urban nature, physical health and mental well being. For example, Stephens (1988) found that physical activities done for leisure (such as taking a walk on a neighborhood park trail) were correlated with psychological health (mood, general well-being) and negatively correlated with depression measures (Stephens 1988). He suggests that when looking at physical activity, not only is the energy expenditure important, but also the "perception" or "meaning" that might be derived from the physical activity (Stephens 1988). Conceivably, this perception or meaning is part of the mental restoration benefits that are often found in natural environments discussed here in this handbook by Matsuoka and Sullivan (Chapter 33). Together, these chapters provide more evidence that we as researchers, students, and advocates need to think more holistically about how urban planning and land use change can facilitate or limit access and availability of urban natures and the possible impacts these changes have on urban ecosystem health, including human physical and mental health.

References

Appleton, J. (1975) *The Experience of Landscape*, London: Wiley.

Ball, K., Bauman, A., Leslie, E., and Owen N. (2001) "Perceived environmental aesthetics and convenience and company are associated with walking for exercise among Australian adults," *Preventive Medicine*, 33(5): 434–40.

Bauman, A., Wallner, F., Miners, A., and Westley-Wise, V. (1996) *No Ifs No Buts: Illawarra physical activity project: Baseline research report*, Warrong, NSW: Commonwealth Department of Health and Family Services.

Beets, M.W. and Foley, J.T. (2008) "Association of father involvement and neighborhood quality with kindergartners" physical activity: A multilevel structural equation model," *American Journal of Health Promotion*, 22(3): 195–203.

Bell, J.F., Wilson, J.S., and Liu, G.C. (2008) "Neighborhood greenness and 2-year changes in body mass index of children and youth," *American Journal of Preventive Medicine*, 35(6): 547–53.

Booth, M.L., Owen, N., Bauman, A., Clavisi, O., and Leslie, E. (2000) "Social-cognitive and perceived environment influences associated with physical activity in older Australians," *Preventive Medicine*, 31(1): 15–22.

Brownson, R. C., Housemann, R., Brown, D.R., Jackson-Thompson, J., King, A.C., Malone, B.R., and Sallis, J.F. (2000) "Promoting physical activity in rural communities: Walking trail access, use, and effects," *American Journal of Preventive Medicine*, 18(3): 235–41.

Burgess, J. (1998) "But is it worth taking the risk? How women negotiate access to urban woodland: A case study," in R. Ainley (ed.), *New Frontiers of Space, Bodies and Gender*, New York, NY: Routledge, pp. 115–28.

Carr, S., Francis, M., Rivlin, L.G., and Stone, A.M. (1992) *Public Space*, New York, NY: Cambridge University Press.

CDC (1996) *Physical Activity and Health: A Report of the Surgeon General*, Atlanta, GA: U.S. Department of Health and Human Services, Centers for Disease Control and Prevention, National Center for Chronic Disease Prevention and Health Promotion.

Chandola, T. (2001) "The fear of crime and area differences in health," *Health and Place*, 7(2): 105–16.
Coen, S.E. and Ross, N.A. (2006) "Exploring the material basis for health: Characteristics of parks in Montreal neighborhoods with contrasting health outcomes," *Health and Place*, 12(4): 361–71.
Cohen, D.A., McKenzie, T.L., Sehgal, A., Williamson, S., Golinelli, D., and Lurie, N. (2007) "Contribution of public parks to physical activity," *American Journal of Public Health*, 97(3): 509–14.
Corti, B., Donovan, R., and Holman, C.J. (1996) "Factors influencing the use of physical activity facilities: results from qualitative research," *Health Promotion Journal of Australia*, 6(1): 16–21.
Cranz, G. (1982) *The Politics of Park Design*, Cambridge, MA: MIT Press.
Crespo, C.J., Smit, E., Andersen, R.E., Carter-Pokras, O., and Ainsworth, B.E. (2000) "Race/ethnicity, social class and their relation to physical inactivity during leisure time: Results from the Third National Health and Nutrition Examination Survey, 1988–1994," *American Journal of Preventive Medicine*, 18(1): 46–53.
de Certeau, M. (2002) *The Practice of Everyday Life*, Berkeley, CA: University of California Press.
de Vries, S., Verheij, R.A., Groenewegen, P.P., and Spreeuwenberg, P. (2003) "Natural environments–healthy environments? An exploratory analysis of the relationship between greenspace and health," *Environment and Planning A*, 35(10): 1717–31.
Design, A.L.B. (2006) *Staying Connected in Cornelius through Responsible Community Design*, San Diego, CA: Active Living Research.
Doyle, S., Kelly-Schwartz, A., Schlossberg, M., and Stockard, J. (2006) "Active community environments and health," *Journal of the American Planning Association*, 72(1): 19–31.
Duncan, M.J. and Mummery, W.K. (2005) "Psychosocial and environmental factors associated with physical activity among city dwellers in regional Queensland," *Preventive Medicine*, 40(4): 363–72.
Floyd, M.F., Spengler, J.O., Maddock, J.E., Gobster, P.H., and Suau, L.J. (2008) "Park-based physical activity in diverse communities of two U.S. cities," *American Journal of Preventive Medicine*, 34(4): 299–305.
Franck, K.A. and Paxson, L. (1989) "Women and urban public space," in I. Altham and E. Zube (eds.), *Public Places and Spaces*, New York: Plenum Press, pp. 121–46.
Frank, L.D. and Engelke, P.O. (2005) "Multiple impacts of the built environment on public health: Walkable places and the exposure to air pollution," *International Regional Science Review*, 28(2): 193–216.
Frank, L.D., Schmid, T.L., Sallis, J.F. Chapman, J., and Saelens, B.E. (2005) "Linking objectively measured physical activity with objectively measured urban form: Findings from SMARTRAQ," *American Journal of Preventive Medicine*, 28(2) Supplement 2: 117–25.
Frumkin, H., Frank, L.D., and Jackson, R. (2004) *Urban Sprawl and Public Health: Designing, Planning, and Building for Healthy Communities*, Washington, D.C.: Island Press.
Gies, E. (2006) *The Health Benefits of Parks: How Parks Help Keep Americans and their Communities Fit and Healthy*, San Francisco, CA: The Trust for Public Lands.
Giles-Corti, B. and Donovan, R.J. (2003) "Relative influences of individual, social environmental, and physical environmental correlates of walking," *American Journal of Public Health*, 93(9): 1583–9.
Giles-Corti, B., Broomhall, M., Knuiman, M., Collins, C., Douglas, K., Ng, K., Lange, A., and Donovan, R.J. (2005) "Increase walking: How important is distance to, attractiveness, and size of public open space?" *American Journal of Preventive Medicine*, 28(2): 169–76.
Gobster, P.H. (1992). "Urban park trail use: An observational approach," *Proceedings of the 1991 Northeastern Recreation Research Symposium*, Radnor, PA: USDA, Forest Service, Northeastern Forest Experiment Station, General Technical Report NE-160, pp. 215–24.
Gobster, P.H. (2002) "Managing urban parks for a racially and ethnically diverse clientele," *Leisure Sciences*, 24(2): 143–59.
Grow, H.M., Saelens, B.E., Kerr, J., Durant, N.H., Norman, G.J., and Sallis, J.F. (2008) "Where are youth active? Roles of proximity, active transport, and built environment," *Medicine and Science in Sports and Exercise*, 40(12): 2071–9.
Hawthorne, W. (1989) *Why Ontarians walk, why Ontarians don't walk more: A study of the walking habits of Ontarians*, Toronto: Energy Probe Research Foundation.
Hoehner, C.M., Brennan Ramirez, L.K., Elliott, M.B., Handy, S.L., and Brownson, R.C. (2005) "Perceived and objective environmental measures and physical activity among urban adults," *American Journal of Preventive Medicine*. 28(2): 105–15.
Humpel, N., Owen, N., and Leslie, E. (2002) "Environmental factors associated with adults" participation in physical activity," *American Journal of Preventive Medicine*, 22(3): 188–99.
Humpel, N., Owen, N., Iverson, D., Leslie, E., and Bauman, A. (2004) "Perceived environment attributes, residential location, and walking for particular purposes," *American Journal of Preventive Medicine*, 26(2): 119–25.

Hutchinson, R. (1987) "Ethnicity and urban recreation: Whites, blacks, and hispanics in Chicago's public parks," *Journal of Leisure Research*, 19(3): 205–22.

Hutchinson, R. (1994) "Women and the elderly in Chicago's public parks," *Leisure Sciences*, 16(4): 229–47.

Joseph, A. and Zimring, C. (2007) "Where active older adults walk: Understanding the factors related to path choice for walking among active retirement community residents," *Environment and Behavior*, 39(1): 75–105.

Kaczynski, A.T. (2010) "Neighborhood walkability perceptions: Associations with amount of neighborhood-based physical activity by intensity and purpose," *Journal of Physical Activity and Health*, 7(1): 3–10.

Kaczynski, A.T. and Henderson, K.A. (2008) "Parks and recreation settings and active living: A review of associations with physical activity function and intensity," *Journal of Physical Activity and Health*, 5(4): 619–32.

Kaczynski, A.T., Potwarka, L.R., and Saelens, B.E. (2008) "Association of park size, distance, and features with physical activity in neighborhood parks," *American Journal of Public Health*, 98(8): 1451–6.

Kahn, E.B., Ramsey, L.T., Brownson, R.C., Heath, G.W., Howze, E.H., Powell, K.E., Stone, E.J., Rajab, M.W., and Corso, P. (2002) "The effectiveness of interventions to increase physical activity: A systematic review," *American Journal of Preventive Medicine*, 22(4S): 73–107.

Kirtland, K.A., Porter, D.E., Addy, C.L., Neet, M.J., Williams, J.E., Sharpe, P.A., Neff, L.J., Kimsey Jr., C.D., Ainsworth, B.E. (2003) "Environmental measures of physical activity supports: Perception versus reality," *American Journal of Preventive Medicine*. 24(4): 323–31.

Krenichyn, K. (2004) "Women and physical activity in an urban park: Enrichment and support through an ethic of care," *Journal of Environmental Psychology*, 24(1): 117–30.

Kuo, F.E. (2001) "Environment and crime in the inner city: Does vegetation reduce crime?" *Environment and Behavior*, 33(3): 343–67.

Lee, C. and Moudon, A.V. (2004) "Physical activity and environment research in the health field: Implications for urban and transportation planning practice and research," *Journal of Planning Literature*. 19(2): 147–81.

Lee, C. and Moudon, A.V. (2006) "Correlates of walking for transportation or recreation purposes," *Journal of Physical Activity and Health*, 3(Suppl 1): S77–S98.

Leslie, E., Saelens, B.E., Frank, L., Owen, N., Bauman, A., Coffee, N., and Hugo, G. (2005) "Residents' perceptions of walkability attributes in objectively different neighborhoods: a pilot study," *Health and Place*, 11(3): 227–36.

Liu, G.C., Wilson, J.S., Qi, R., and Ying, J. (2007) "Green neighborhoods, food retail and childhood overweight: Differences by population density," *American Journal of Health Promotion*, 21(4S): S317–25.

Loukaitou-Sideris, A. (1995) "Urban form and social context: Cultural differentiation in the uses of urban parks," *Journal of Planning Education and Research*, 14(2): 89–102.

Loukaitou-Sideris, A. (2006) "Is it safe to walk? Neighborhood safety and security considerations and their effects on walking," *Journal of Planning Literature*, 20(3): 219–32.

Maas, J., Verheij, R.A., Groenewegen, P.P., Sjerp de Vries.S., and Spreeuwenberg, P. (2006) "Green space, urbanity, and health: how strong is the relation?" *Journal of Epidemiology and Community Health*, 60(7): 587–92.

MacMaster, N. (1990) "The battle for Mousehold Heath 1857–1884: 'popular politics' and the Victorian public park," *Past and Present*, 127(1): 117–54.

Marcus, C.C., Watsky, C.M., Insley, E. and Francis. C. (1998) "Neighborhood parks," in C.C. Marcus and C. Francis (eds.), *People Places: Design Guidelines for Urban Open Space*, second edition, New York, NY: John Wiley & Sons Inc, pp. 85–148.

McKenzie, T.L., Cohen, D.A., Sehgal, A., Williamson, S., and Gonilelli, D. (2006) "System for observing play and recreation in communities (SOPARC): Reliability and feasibility measures," *Journal of Physical Activity and Health*, 3(S1): S208–S222.

Miles, R. (2008) "Neighborhood disorder, perceived safety, and readiness to encourage use of local playgrounds," *American Journal of Preventive Medicine*, 34(4): 275–81.

Mitchell, R. and Popham, F. (2007), "Greenspace, urbanity and health: relationships in England," *Journal of Epidemiology and Community Health*, 61(8): 681–3.

Molnar, B., Gortmaker, S., Bull, F.C., and Buka, S.L. (2004) "Unsafe to play? Neighborhood disorder and lack of safety predict reduced physical activity among urban children and adolescents," *American Journal of Health Promotion*, 18(5): 378–86.

Moudon, A.V. (2005) "Active living research and the urban design, planning, and transportation disciplines," *American Journal of Preventive Medicine*, 28(2 Supp 2): 214–15.

Mowen, A., Orsega-Smith, E., Payne, L., Ainsworth, B., and Godbey, G. (2007) "The role of park proximity and social support in shaping park visitation, physical activity, and perceived health among older adults," *Journal of Physical Activity and Health*, 4(2): 167–79.

Nasar, J.L. and Jones, K.M. (1997) "Landscapes of fear and stress," *Environment and Behavior*, 29(3): 291–323.

Nassauer, J.I. (1995) "Messy ecosystems, orderly frames," *Landscape Journal*, 14(2): 161–70.

Olmsted, F. (ed.) (1979) *Civilizing American Cities: A Selection of Frederick Law Olmsted's Writings on City Landscapes*, Cambridge, MA: MIT Press.

Owen, N., Humpel, N., Leslie, E., Bauman, A., and Sallis, J.F. (2004) "Understanding environmental influences on walking: Review and research agenda," *American Journal of Preventive Medicine*, 27(1): 67–76.

Pate, R.R., Pratt, M., Blair, S.N., Haskell, W.L., Macera, C.A., Bouchard, C., Buchner, D., Ettinger, W., Heath, G.W., and King, A.C. (1995) "Physical activity and public health: A recommendation from the Centers for Disease Control and Prevention and the American College of Sports Medicine," *Journal of American Medical Association*, 273(5): 402–7.

Payne, L.L., Mowen, A.J., and Orsega-Smith, E. (2002) "An examination of park preferences and behaviors among urban residents: The role of residential location, race, and age," *Leisure Sciences*, 24(2): 181–98.

Pikora, T., Giles-Corti, B., Bull, F., Jamrozik, K., and Donovan, R. (2003) "Developing a framework for assessment of the environmental determinants of walking and cycling," *Social Science and Medicine*, 56(8): 1693–703.

Pikora, T., Giles-Corti, B., Knuiman, M.W., Bull, F.C., Jamrozik, K., and Donovan, R. (2006) "Neighborhood environmental factors correlated with walking near home: Using SPACES," *Medicine and Science in Sports and Exercise*, 38(4): 708–14.

Potwarka, L.R., Kaczynski, A.T., and Flack, A.L. (2008) "Places to play: Association of park space and facilities with healthy weight status among children," *Journal of Community Health*, 33(5): 344–50.

Ries, A.V., Gittelsohn, J., Vorhees, C.C., Roche, K.M., Clifton, K.J., and Astone, N.M. (2008) "The environment and urban adolescents" use of recreational facilities for physical activity: A qualitative study," *American Journal of Health Promotion*, 23(1): 43–50.

Ross, C.E. (1993) "Fear of victimization and health," *Journal of Quantitative Criminology*, 9(2): 159–75.

Saelens, B.E., Sallis, J.F., Black, D.B., and Chen, D. (2003) "Neighborhood-based differences in physical activity: An environment scale evaluation," *American Journal of Public Health*, 93(9): 1552–8.

Sallis, J.F. and Owen, N. (1999) *Physical Activity and Behavioral Medicine*, Thousand Oaks, CA: Sage Publications.

Stephens, T. (1988) "Physical activity and mental health in the United States and Canada: Evidence from four population surveys," *Preventive Medicine*, 17(1): 35–47.

Stokols, D. (1992) "Establishing and maintaining healthy environments: Toward a social ecology of health promotion," *American Psychologist*, 47(1): 6–22.

Sugiyama, T., Leslie, E., Giles-Corti, B., and Owen, N. (2008) "Associations of neighbourhood greenness with physical and mental health: do walking, social coherence and local social interaction explain the relationships?" *Journal of Epidemiology and Community Health*, 62(5): e9, doi:10.1136/jech.2007.064287.

Suminski, R.R., Poston, W.S.C., Petosa, R.L., Stevens, E., and Katzenmoyer, L.M. (2005) "Features of the neighborhood environment and walking by U.S. adults," *American Journal of Preventive Medicine*, 28(2): 149–55.

Takano, T., Nakamura, K., and Watanabe, M. (2002) "Urban residential environments and senior citizens' longevity in megacity areas: The importance of walkable green spaces," *Journal of Epidemiology and Community Health*, 56(12): 913–18.

Taylor, D.E. (1993) "Urban park use: race, ancestry and gender," in P. Gobster (ed.), *Managing Urban High-Use Recreation Settings*, St. Paul, Minnesota: U.S.D.A. Forest Service North Central Forest Experiment Station, General Technical Report NC-163, pp. 82–6.

Tilt, J.H. (2007) *Neighborhood Vegetation and Preferences: Exploring Walking Behaviors in Urban and Suburban Environments*, PhD thesis Seattle, WA: University of Washington, College of Forest Resources, Social Sciences.

Tilt, J.H. (2010) "Walking trips to parks: Exploring demographic, environmental factors, and preferences for adults with children in the household," *Preventive Medicine*, 50(S1): S69–S73.

Tilt, J.H., Unfried, T.M., and Roca, B. (2007) "Using objective and subjective measures of neighborhood greenness and accessible destinations for understanding walking trips and BMI in Seattle, Washington," *American Journal of Health Promotion*, 21(4): 371–9.

Tinsley, H.E.A., Tinsley, D.J., and Croskeys, C.E. (2002) "Park usage, social milieu, and psychological benefits of park use reported by older urban park users from four ethnic groups," *Leisure Sciences*, 24(2): 199–218.

Troped, P.J., Saunders, R.P., Pate, R.R., Reininger, B., and Addy, C.L. (2003) "Correlates of recreational and transportation physical activity among adults in a New England community," *Preventive Medicine*, 37(4): 304–10.

Tucker, P., Gilliland, J., and Irwin, J.D. (2007) "Splashpads, swings, and shade: Parents' preferences for neighbourhood parks," *Canadian Journal of Public Health*, 98(3): 198–202.

Tuttle, C. (1997) *Being Outside: How High and Low Income Residents of Seattle Perceive, Use and Value Urban Open Space*, PhD thesis, Seattle: University of Washington, Interdisciplinary Program in Urban Design and Planning.

Weinstein, A., Felgley, P., Pullen, P., Mann, L., and Redman, L. (1999) "Neighborhood safety and the prevalence of physical inactivity–selected states," *Journal of American Medical Association*, 281(5): 1373.

Westover, T.N. (1986) "Park use and perception: Gender differences," *Journal of Park and Recreation Administration*, 4: 1–8.

Wilcox, S., Castro, C., King, A.C., Housemann, R., and Brownson, R.C. (2000) "Determinants of leisure time physical activity in rural compared with urban older and ethnically diverse women in the United States," *Journal of Epidemiology and Community Health*, 54(9): 667–72.

Wolch, J., Wilson, J.P., and Fehrenbach, J. (2005) "Parks and park funding in Los Angeles: An equity-mapping analysis," *Urban Geography*, 26(1): 4–35.

World Health Organization (2006) *Obesity and Overweight Factsheet*, online, available at: www.who.int/mediacentre/factsheets/fs311/en/index.html [accessed March 6, 2009].

Zhang, T. and Gobster, P.H. (1998) "Leisure preferences and open space needs in an urban Chinese American community," *Journal of Architectural and Planning Research*, 15(4): 338–55.

33

Urban nature

Human psychological and community health

Rod Matsuoka and William Sullivan

Introduction

Over forty years of research has produced a substantial body of evidence revealing a variety of benefits from having contact with nature in urban settings. Although the evidence base is not complete, the findings of the overwhelming majority of the studies demonstrate that urban inhabitants benefit to a considerable degree, and in a variety of ways, from having everyday contact with nature. It has become clear that everyday exposure to nearby nature is associated with improved psychological health and functioning and increased community health of urban dwellers.

Urban nature in large and small doses has been shown to benefit individuals and communities. Large expanses of nature, such as those found in urban parks, produce a variety of benefits for people. But so, too, do relatively small patches of trees, lawns, and other forms of vegetation found along streets, at schools and corporate campuses, on civic landscapes, residential sites, and on hospital and prison grounds. So-called accidental nature—plants growing on neglected urban sites (e.g. abandoned industrial and residential sites, along derelict railroad tracks or in neglected stream corridors)—has also been shown to provide benefits to folks who have frequent contact with it.

It is interesting to note that the benefits from having contact with nature are associated with both direct and indirect contact. Direct contact involves activities such as gardening, or walking in a park. Indirect contact includes viewing natural settings through a window or looking at a painting of a landscape.

In this chapter, we concentrate on the evidence of the benefits of nature culled from recent studies, specifically those conducted since 1990. These studies were found by initially using online academic databases (e.g. ISI Web of Knowledge, PsycINFO, Google Scholar) to search for relevant published papers and review articles. The key studies referenced by the authors of those papers were then examined. Finally, of these collected studies, those that satisfied the criteria presented in Table 33.1 were cited in this review.

We begin by categorizing the benefits of having contact with nature in specific settings. Next, we describe the mechanisms that convey these benefits. We end by examining the interrelationships among the benefits and their underlying mechanisms.

Table 33.1 Criteria for studies being included in this review concerning the benefits of urban nature

Included	*Not included*
Urban nature benefits	
• Psychological health and well-being • Community benefits (e.g. social interactions, reduction in incivilities)	• Physical health and activity • Benefits of indoor plants • Benefits of artificially created nature (e.g. computer-generated imagery)
Settings	
• Workplace • Home • Hospitals • Schools • Community gardens • Parks • Greater neighborhood • Citywide regions • Laboratory studies (e.g. photo surveys) concentrating on urban settings	• Rural locales (e.g. farms) • National and wilderness parks located outside of cities • Laboratory studies (e.g. photo surveys) not concentrating on urban settings
Study type	
• Empirical studies (e.g. surveys, interviews, on-site observations, laboratory studies)	• Non-empirical studies (e.g. research proposals, editorials) • Review papers
Publications	
• Studies published from 1991 onwards • Studies published in English	• Studies published before 1991 • Studies published in non-English languages

Categorizing the benefits

In this chapter, the benefits provided by contact with nearby nature are categorized as accruing to individuals or communities (see Figure 33.1). We focus on the psychological health aspect of individual benefits, and both the social interaction and reduction in incivilities facets of community benefits. The physical health gains related to contact with nature are addressed above in Chapter 32.

Psychological well-being

Contact with urban nature has been linked with overall psychological health among members of all age groups. Researchers have uncovered a variety of benefits that can be categorized as:

- Enhanced capacity to pay attention
- Greater ability to cope with life stressors and crises
- Improvements in overall psychological well-being
- Greater satisfaction with their neighborhoods and their lives.

Evidence of these benefits has come from a wide variety of real world settings: in the workplace, at home, in parks, neighborhood spaces, community gardens, hospitals, and schools (see Table 33.2).

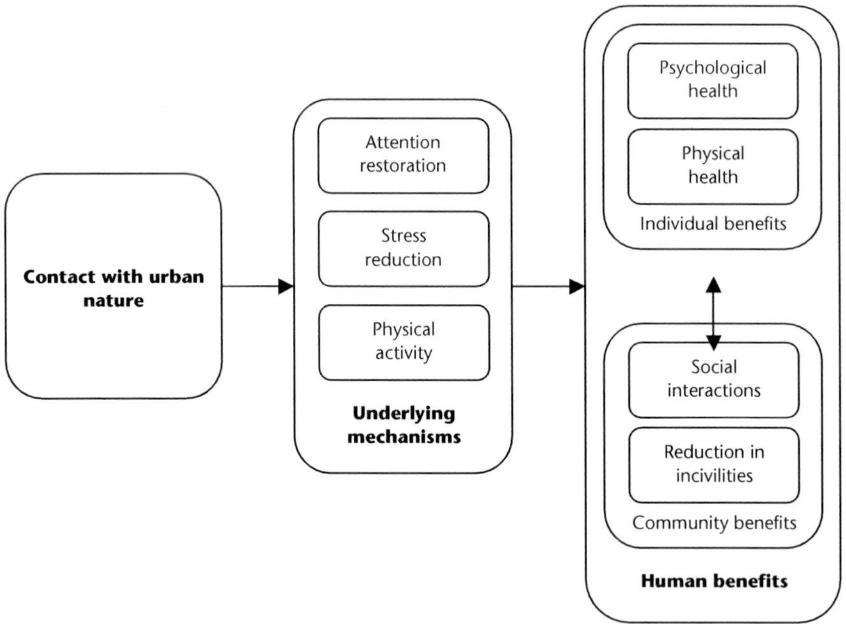

Figure 33.1 The human benefits,[1] involving both individual and community benefits, of urban nature contact.

Note

1 The benefits of contact with urban nature accrue to individuals and communities via a variety of underlying mechanisms.

Workplace

Being exposed to nature while at work has a variety of far-reaching consequences. In recent investigations, views of nature from workplace settings (e.g. office buildings, factories) have been associated with increased employee productivity (Heerwagen and Wise 1998; Heschong Mahone Group 2003b), reduced levels of job-related frustration and stress (R. Kaplan 1993), greater psychological well-being (Dravigne *et al.* 2008; Heerwagen and Wise 1998; R. Kaplan 1993; Leather *et al.* 1998), enhanced feelings of job satisfaction (Dravigne *et al.* 2008; Heerwagen and Wise 1998; R. Kaplan 1993; Leather *et al.* 1998), and higher levels of life satisfaction (R. Kaplan 1993).

Home

Exposure to nature at home has profoundly positive benefits for individuals. Views of, and direct exposure to, nearby nature has been repeatedly shown to increase resident's cognitive functioning (Kuo 2001; Rappe and Kivela 2005; Tennessen and Cimprich 1995), sense of well-being (Day 2008; R. Kaplan 2001; Rappe and Kivela 2005; Rappe *et al.* 2006), sense of peace and quiet (Day 2008; Yuen and Hien 2005), and satisfaction with home and neighborhood (R. Kaplan 2001; Kearney 2006; Lee *et al.* 2008; Talbot and Kaplan 1991). Individuals experience a higher quality of life when they have nature near their residences. The evidence suggests that exposure to trees is especially important.

Table 33.2 Real world settings in which the studies were conducted, and the corresponding benefits

Study setting	*Individual benefits: psychological health*	*Community benefit #1: social interactions*	*Community benefit #2: reduction in incivilities*
Workplace	Increases in: • Effectiveness • Job satisfaction • Life satisfaction • Productivity • Quality of life • Sense of well-being • Work spirit Reductions in: • Fatigue • Frustration • Stress		
Home	Increases in: • Cognitive functioning • Improved mood • Mental effectiveness • Neighborhood satisfaction • Peace of mind • Powers of concentration • Sense of peace and quiet • Sense of well-being Reductions in: • Attention-deficit hyperactivity disorder symptoms • Impact of stressful life events • Symptoms of depression	Increases in: • Opportunities for social interaction • Sense of community • Social activity • Social support • Social ties Reductions in: • Feelings of loneliness	Increases in: • Sense of safety Reductions in: • Aggressive behaviors • Incivilities • Levels of fear • Property crime • Violent behaviors • Violent crime
Hospitals	Increases in: • Attentional abilities • Morale • Psychological restoration • Sense of well-being Reductions in: • Behavioral problems • Emotional distress • Emotional pain • Stress	Increases in: • Talking with others Reductions in: • Emotional distress	
Schools	Increases in: • Cognitive play	Increases in: • Positive changes to social hierarchy of play behavior	

continued

Table 33.2 continued

Study setting	*Individual benefits: psychological health*	*Community benefit #1: social interactions*	*Community benefit #2: reduction in incivilities*
	• Plans to attend college • Graduation rates • Inventive play • Sense of quality of life • Test scores Reductions in: • Impulsive behaviors • Inattentive behaviors	Reductions in: • Antisocial behaviors • Bullying • Hyperactive behaviors	
Community gardens	Increases in: • Life satisfaction • Optimism • Psychological health • Quality of life • Self-esteem • Sense of well-being • Zest for life	Increases in: • Community cohesion • Community development • Social health • Social networks	
Parks	Increases in: • Directed-attention abilities • Feelings of revitalization • Mental capacity restoration • Positive emotions • Positive feelings • Sense of tranquility • Sense of well-being Reductions in: • Anger • Anxiety • Attention-deficit hyperactivity disorder symptoms • Depression • Stress	Increases in: • Friendships across cultures • Interracial relations • Social relations • Strong family ties	
Greater neighborhood	Increases in: • Positive feelings • Psychological healing Reductions in: • Stress		
Citywide			Increases in: • Feelings of social safety Reductions in: • Incidents of crime

The benefits of nearby nature at home have been shown to impact individuals across the lifespan. For children, there is evidence that nature near the home can buffer the impact of life stressors (Wells and Evans 2003) and enhance cognitive functioning (Taylor *et al.* 2002; Wells 2000). In older adults, regular access to natural environments has been associated with enhanced powers of concentration (Jansen and von Sadovszky 2004; Ottosson and Grahn 2005; Rappe and Kivela 2005), feelings of peacefulness and refreshment (Jansen and von Sadovszky 2004; Rappe and Kivela 2005), and feelings of well-being (Rappe and Kivela 2005).

Hospitals

Views and physical access to green spaces have been shown to benefit individuals who are being treated in, or who work in, hospitals and other therapeutic settings. Perhaps the most famous study regarding the impact of nearby nature on human health was Ulrich's (1984) work in a Pennsylvania hospital.[1] His study examined the impact of having either a view of a green space with trees or a brick wall outside the rooms of individuals who had gallbladder surgery. Patients with a view of the green space took fewer self-administered painkillers and were released from the hospital sooner than their counterparts who were assigned to rooms that had a view of the wall.[2]

In recent years, a number of other studies have uncovered evidence of the impact of exposure to therapeutic landscapes or healing gardens. Access to these sites has been positively associated with self-reported well-being (Barnhart *et al.* 1998; Curtis *et al.* 2007) quality of life (Detweiler *et al.* 2008; Detweiler and Warf 2005), morale (Mather *et al.* 1997), mood and restoration (Whitehouse *et al.* 2001), and reductions of distress and emotional pain for both patients and staff (Sherman *et al.* 2005). Having a view of a green space is clearly supportive of a person who is sick or recovering from illness. The evidence shows that such views also have wonderful impacts on caregivers.

Schools

There is growing evidence that exposure to green spaces on school grounds has benefits for students. A handful of studies have directly investigated how natural features of the campus landscape can influence student behavior. Most have examined the effects on students of alternative types of playgrounds in preschool, kindergarten, and elementary school. In these contexts, natural playscapes have been found to benefit children's creative play, and emotional and cognitive development (Lindholm 1995; Malone and Tranter 2003; Mårtensson *et al.* 2009). The number of even these studies, though, is small (Herrington and Studtmann 1998; Neville 1994).

A few studies have investigated the extent to which views of nature from primary and secondary school buildings affect student performance. The findings show that views of nature from classroom and cafeteria windows are associated with enhanced performance on standardized tests, higher graduation rates, higher percentages of students planning to attend a four-year college, and more pro-social student behaviors (Heschong Mahone Group 2003a; Matsuoka 2010). Moreover, two studies conducted on college campuses have linked greater ecodiversity (Ogunseitan 2005) and greater use of green spaces (McFarland *et al.* 2008) with higher levels of self-reported quality of life. Although these studies are still relatively few in number, the evidence shows that having easy access to nature from schools have consistent and systematically positive impacts on learning and student behavior.

Community gardens

The opportunities to directly interact with nature in community gardens have been linked with enhancements in feelings of overall well-being and quality of life (Armstrong 2000; Boyer *et al.* 2002; Milligan *et al.* 2004; Teig *et al.* 2009; Waliczek *et al.* 2005). These enhancements have been tied to diverse sources, involving the creation of mutually supportive social environments (Boyer *et al.* 2002; Glover 2004; Glover and Parry 2005; Milligan *et al.* 2004; Saldivar-Tanaka and Krasny 2004; Shinew *et al.* 2004; Teig *et al.* 2009; Wakefield *et al.* 2007), sense of participation and accomplishment (Milligan *et al.* 2004), and people's perceptions that the act of gardening will improve their mental health (Armstrong 2000; Wakefield *et al.* 2007).

Parks

The experience of nature in urban parks is related to improved mental health in a variety of ways. Direct and indirect access to urban parks has been linked to improved cognitive functioning—that is, people score higher on tests that measure their capacity to pay attention and to concentrate (Berman *et al.* 2008; Fuller *et al.* 2007; Hartig *et al.* 2003; Korpela *et al.* 2008; Krenichyn 2006). Access to urban parks has also been linked to having an opportunity to get away from the stresses and pressures of urban life (Guite *et al.* 2006; Macnaghten and Urry 2000; Tyrväinen *et al.* 2007), to enhanced positive feelings (Chiesura 2004; Hartig *et al.* 2003), to an increase in self-reported well-being (Milligan and Bingley 2007), and a reduction in levels of stress, anxiety, and depression (Bodin and Hartig 2003).

Greater neighborhood

Greater levels of vegetation in one's daily surroundings have been linked with a greater sense of psychological well-being. Researchers have attributed these benefits from nature contact to the promotion of positive feelings (English *et al.* 2008; Korpela and Ylén 2007) and the reduction of stress (English *et al.* 2008; Nielsen and Hansen 2007).

Perhaps most strikingly, children diagnosed with attention-deficit hyperactivity disorder (ADHD) have been found to benefit from exposure to nature in urban parks and near their homes. In a series of studies, such access has been consistently linked with a reduction in ADHD symptoms (Kuo and Taylor 2004; Taylor and Kuo 2009; Taylor *et al.* 2001). Taylor and Kuo report that children with ADHD concentrated better after the walk in the park than after the downtown walk or in a neighborhood (2009).

Summary

Taken together, these studies demonstrate that views to and experiences in urban nature provide a variety of benefits to psychological health. When individuals are exposed to urban nature on a regular basis, they reliably show an enhanced capacity to concentrate and pay attention, greater ability to cope with the stressors in their lives, higher levels of life satisfaction, and increased levels of psychological well-being.

The studies that produced these findings were conducted under a great diversity of conditions. The settings varied from places along the rural-urban fringe to urban public housing neighborhoods. The participants varied in gender and ethnicity and represent individuals from most stages across the lifespan. The methods of investigation varied among experiments, field experiments, observation and mapping, face-to-face interviews, and mail-back questionnaires.

Even the operational definitions of nature varied. That this diversity should lead to consistent findings demonstrating the salutary benefits of urban nature gives us considerable confidence in their validity. It is clear that exposure to nature where people work, reside, and recreate, as well as where individuals are hospitalized or educated, leads to a variety of psychological benefits that the vast majority of individuals would welcome in their own lives.

Contact with urban nature builds communities

A number of recent studies demonstrate the positive impacts that access to urban nature can have on the amount of social interaction, and ultimately, the strength of social ties, among neighbors. These are important findings because social integration and the strength of social ties are important predictors of well-being and, for older adults, of longevity. Green urban spaces appear to attract people outdoors, increasing opportunities for casual social encounters among neighbors and fostering the development of stronger neighborhood social ties.

There is now a compelling body of evidence demonstrating that access to urban green spaces is reliably related to:

- Enhanced social contacts and increased levels of social support among neighbors.
- More acquaintances among people of different age groups, racial, ethnic, and cultural backgrounds.
- Increased sense of community among neighbors.

These benefits accrue to individuals who gain exposure to nature in residential, park, hospital, and school settings, and in community gardens.

Home

Green spaces promote social interaction among neighbors in urban and suburban areas. Among residents of an inner-city public housing neighborhood, researchers found that higher levels of nearby vegetation (e.g. trees, grass) are associated with greater use of outdoor spaces (Coley *et al.* 1997), and higher levels of social activity in the neighborhood (Kuo *et al.* 1998; Kweon *et al.* 1998; Sullivan *et al.* 2004). Furthermore, residents of buildings with more trees and grass report that they know their neighbors better, socialize with them more often, have stronger feelings of community, and feel safer and better adjusted than do residents of more barren, but otherwise identical buildings (Kuo *et al.* 1998).

The social benefits of green neighborhood spaces are not restricted to individuals who live in urban public housing neighborhoods. Studies conducted in new urbanist communities and traditional suburban developments show that increases in natural features or perceived greenness in these neighborhoods were associated with higher levels of social contact and increased feelings of social support among neighbors (Kim and Kaplan 2004; Maas, van Dillen *et al.* 2009).

Hospitals

Green spaces are reliably associated with social interactions on hospital grounds. Hospital gardens and landscapes are used by both staff and patients for privacy and socializing (Barnhart *et al.* 1998; Sherman *et al.* 2005; Whitehouse *et al.* 2001). One study found that staff and patients selected open, natural settings for passive behaviors (i.e. sitting, viewing scenery), and more enclosed settings for active behaviors (i.e. walking, socializing) (Barnhart *et al.* 1998).

Schools

On school grounds, more natural playscapes are associated with more active, social play among children (Dyment and Bell 2008; Fjørtoft 2004; Herrington and Studtmann 1998). Studies have found that social interactions among children improved when natural features were introduced (Herrington and Studtmann 1998). The findings of one investigation revealed that these interactions became based more on cognitive, social, and emotional skills rather than primarily physical prowess (Herrington and Studtmann 1998).

Community gardens

The establishment of community gardens in residential neighborhoods has been linked with increased social interaction leading to improved social networks among neighbors (Armstrong 2000), increased community building among nearby neighbors (Teig *et al.* 2009), and enhanced interpersonal relationships among juvenile offenders (Cammack *et al.* 2002).

Parks

In light of these findings, it may not be surprising to learn that public urban green spaces, including recreation and nature parks, have been found to play an important role in increasing social activity among community members (Chiesura 2004). These activities include neighbors making contact with individuals of different racial, ethnic, or cultural backgrounds (Gobster 1998; Seeland *et al.* 2009).

Summary

Green settings in neighborhoods are associated with greater social cohesion among neighbors. The presence of trees and grass in neighborhood spaces increases the use of those spaces and the number of individuals involved in social interactions within them. By increasing face-to-face contact and the number of individuals involved in social interactions, trees and grass in inner-city common spaces contribute to the social cohesion and vitality of a neighborhood.

The findings from these studies show that social interaction and the development of supportive relationships among urban residents can be fostered and enhanced by providing nature contact in a wide variety of settings.

Green places may reduce incivilities, aggression, and violence

A smaller number of studies have investigated the possibility that contact with urban nature is associated with fewer occurrences of incivilities, aggression, and violence. Evidence in support of this possibility has been uncovered through field experiments, surveys, and a review of neighborhood crime statistics.

Home

Research in Chicago neighborhoods indicates that residents who live in 'greener' surroundings report lower levels of fear (Kweon *et al.* 1998; Kuo *et al.* 1998) and fewer incivilities (Brunson 1999). Another study compared levels of aggression and violence in an urban public housing neighborhood where residents were randomly assigned to levels of greenness. Levels of aggression and violence were systematically lower for individuals living in green surroundings than for individuals living in barren surroundings (Kuo and Sullivan 2001a). These findings were reinforced by research using police crime reports for ninety-eight Chicago apartment buildings with varying levels of nearby vegetation. Results indicated that although residents were randomly assigned to different levels of nearby vegetation, the greener a building's surroundings, the fewer crimes reported. Furthermore, this pattern held for both property crimes and violent crimes (Kuo and Sullivan 2001b).

Citywide

Recent studies conducted at an urban scale show that greater everyday contact with nature is associated with lower levels of incivilities. For example, one study found that the presence of more green space in a person's living environment is associated with enhanced feelings of safety, except in very dense urban areas (Maas, Spreeuwenberg *et al.* 2009). Another study discovered that areas of a major city with less than the average level of greenness have been associated with an increased amount of reported crime (Snelgrove *et al.* 2004).

How does urban nature produce these results?

How might exposure to green urban spaces produce such a range of wonderful outcomes? By what mechanism or pathways do these results occur? The literature suggests three possible means through:

- Enhanced mental restoration
- Reductions in stress
- Increases in social interaction and social support.

The two most widely cited pathways for the connections between nature and psychological health involve Kaplan and Kaplan's Attention Restoration Theory and Roger Ulrich's Psycho-Evolutionary Theory. Many researchers have cited one or both mechanisms to explain their findings (see Table 33.3).

Attention Restoration Theory recognizes that humans have two modes of attending to things. Some objects, ideas, settings, and situations are effortlessly engaging and require no effort as we take them in. Kaplan and Kaplan call this mode *involuntary attention* (1989). Other stimuli and settings are not effortlessly engaging and thus require us to pay attention – or as the Kaplans say, to *direct attention*.

Our capacity to direct our attention is subject to fatigue. That is, just like muscles that require rest after a period of intense use, our capacity to deliberately direct attention declines with use. The costs of attentional fatigue (what many call mental fatigue) are considerable. An individual experiencing mental fatigue will have a reduced ability to concentrate, and may also become more antisocial, irritable, distractible, impulsive, and accident prone. Thus, it is not surprising then that mentally fatigued individuals are more likely to have trouble meeting their goals.

Kaplan and Kaplan observed that settings and stimuli that draw primarily on involuntary attention give directed attention a chance to rest and restore. They note that contact with nature should assist in the recovery from mental fatigue because green settings draw primarily on involuntary attention (Kaplan and Kaplan 1989; S. Kaplan 1995). The studies described above demonstrate that when individuals have greater exposure to green spaces, they are better at managing social situations, avoiding or resolving conflicts, making progress on their goals, and

Table 33.3 Researchers citing attention restoration, psycho-evolutionary, or both theories as underlying mechanisms to explain their findings

	Attention restoration theory	*Psycho-evolutionary theory*	*Both theories*
Individual Benefits: Mental Health			
Workplace	R. Kaplan 1993	–	Heschong Mahone Group 2003b; Leather *et al.* 1998
Home	Jansen and von Sadovszky 2004; Kaplan 2001; Kuo 2001; Taylor *et al.* 2001, 2002; Tennessen and Cimprich 1995; Wells and Evans 2003	Rappe and Kivela 2005	Day, 2008; Lee *et al.* 2008; Ottosson and Grahn 2005; Rappe *et al.* 2006; Wells 2000
Hospitals	–	Barnhart *et al.* 1998; Whitehouse *et al.* 2001	Detweiler *et al.* 2008; Detweiler and Warf 2005; Sherman *et al.*, 2005
Schools	Ogunseitan 2005	–	Heschong Mahone Group 2003a; Mårtensson *et al.* 2009; Matsuoka 2010
Community gardens	Waliczek *et al.* 2005	–	–
Parks	Berman *et al.* 2008; Bodin and Hartig 2003; Fuller *et al.* 2007; Kuo and Taylor 2004; Taylor and Kuo 2009; Taylor *et al.* 2001; Tyrväinen *et al.* 2007	–	Hartig *et al.* 2003; Krenichyn 2006;
Greater neighborhood	–	Nielsen and Hansen 2007	Korpela and Ylén 2007;
Community Benefit #1: Social Interactions			
Home			Coley *et al.* 1997
Hospitals	Barnhart *et al.* 1998	–	Sherman *et al.* 2005
Schools	Dymentand Bell 2008	–	–
Community Gardens	Teig *et al.* 2009	–	–
Parks	–	–	Korpela *et al.* 2008
Community Benefit #2: Reduction in Incivilities			
Home	Kuo and Sullivan 2001a, 2001b	–	–
Citywide	Snelgrove *et al.* 2004	–	–

coping with the stresses of life. Given all the benefits that grow from improved capacity to pay attention, it may not be surprising to find that individuals who report greater exposure to green spaces also report higher levels of life satisfaction and psychological well-being.

The Psycho-Evolutionary Theory proposes an alternative pathway for describing the relationship between contact with nature and some of the positive outcomes described above. This theory deals with the role that emotion and physiology play in the recovery from stressful events.

A person feels stress in response to a situation that is perceived as threatening. This feeling of stress produces negative emotions and some short-term arousal of the autonomic nervous system – that is, signs of physiological stress. The Psycho-Evolutionary framework holds that the cost of these negative emotions and physiological responses is fatigue. Exposure to green spaces that produce feelings of mild to moderate interest, pleasantness, and calmness should help a person recover from the feeling of stress. That is, through exposure to non-threatening green spaces, an individual's psychological and physiological arousal is returned to more moderate levels, which fosters an overall sense of well-being (Hartig *et al.* 1991; Ulrich *et al.* 1991).

Finally, researchers have noted that social interaction may be a mechanism underlying the connection between exposure to urban nature and some of the psychological and social benefits described above (Note: social interaction is displayed as a community benefit in Table 33.1). The mechanism here is clear. Green spaces act as gathering places or spaces where neighbors are likely to linger. Such settings support frequent, informal interaction among neighbors and nurture the development of neighborhood social ties. Neighborhood social ties are the glue that transforms a collection of unrelated neighbors into a neighborhood. When people know their neighbors, they are more likely to feel a comforting sense of support and community. In addition, they feel more satisfied with their neighborhood, better adjusted, and safer (Kuo *et al.* 1998).

A number of scholars have proposed this mechanism to explain findings that show contact with nature is associated with enhanced levels of psychological well-being in the home environment (Day 2008; Wells and Evans 2003) and in community gardens (Boyer *et al.* 2002; Milligan *et al.* 2004; Teig *et al.* 2009).

Summary and conclusions

There is mounting evidence regarding the profound and systemic benefits that contact with nature provides urban dwellers. In this chapter, we have emphasized the empirical evidence published since 1990 that describes and examines these benefits. Taken together, the studies reveal that in a wide variety of settings – parks, work, home, school, and hospitals – the benefits of being exposed to nature are available to urban residents. Both passive engagement with nature (e.g. viewing nature through a building window) and more direct interactions with nature (e.g. climbing a tree, planting a vegetable garden) produce an array of benefits.

The benefits are far from trivial. For individuals, they include an enhanced capacity to concentrate and pay attention, greater ability to cope with life's stressors, higher levels of life satisfaction, and increased levels of psychological well-being. For neighbors and communities, they include stronger ties among neighbors, lower levels of incivilities, fewer instances of aggressive behavior, reduced levels of violence, and fewer reported crimes.

Still, much work needs to be done. We know little about the dose-response curve describing the relationship between contact with nature and the benefits described in this chapter. How much exposure is necessary to produce benefits? Does greater exposure lead to linear increases in benefits? How frequently should exposure occur? For what duration? With what concentration of nature? The opportunities for future research are plentiful and rich.

Attention Restoration Theory and the Psycho-Evolutionary Theory have been widely cited by researchers to explain their findings. A number of researchers cite both of these underlying mechanisms (see Table 33.3). In addition, social interaction, a benefit in itself, has been proposed as an underlying mechanism to explain the psychological health benefits of nature contact (see Figure 33.1).

Urban planners, designers, ecologists, and the public can help create cities that bring out the best in people by providing abundant opportunities for people to have contact with nature. In addition to urban parks, such contact can occur along tree-lined streets, and on school campuses, hospital grounds, civic centers, and in public housing neighborhoods.

Based on the empirical evidence reviewed here, it is clear that having a green space nearby is more than just an aesthetic amenity. Such spaces are a critical part of healthy urban habitats.

Acknowledgments

This study was supported by a research joint venture agreement with The USDA Forest Service Northern Research Station, and by the Department of Landscape Architecture at the University of Illinois, Urbana-Champaign.

Notes

1 Although published before 1991, this study is discussed because of its importance in this particular category and setting of human benefits.

2 The health impacts of the green space outside the Pennsylvania hospital were clearly mediated through a psychological process or psychological processes. Although no data are available for us to further examine the extent to which one psychological process or another was responsible for those benefits, that a psychological process was at work is clear because the only difference in the experience of the two groups was the content of the view—which could only have acted on the patients through a psychological process.

References

Armstrong, D. (2000) 'A survey of community gardens in upstate New York: implications for health promotion and community development,' *Health and Place*, 6(4): 319–27.

Barnhart, S. K., Perkins, N. H., and Fitzsimonds, J. (1998) 'Behaviour and outdoor setting preferences at a psychiatric hospital,' *Landscape and Urban Planning*, 42(2–4): 147–56.

Berman, M. G., Jonides, J., and Kaplan, S. (2008) 'The cognitive benefits of interacting with nature,' *Psychological Science*, 19(12): 1207–12.

Bodin, M. and Hartig, T. (2003) 'Does the outdoor environment matter for psychological restoration gained through running?' *Psychology of Sport and Exercise*, 4(2): 141–53.

Boyer, R., Waliczek, T. M., and Zajicek, J. M. (2002) 'The Master Gardener program: do benefits of the program go beyond improving the horticultural knowledge of the participants?' *HortTechnology*, 12(3): 432–6.

Brunson, L. (1999) *Resident appropriation of defensible space in public housing: Implications for safety and community*, unpublished doctoral dissertation, University of Illinois, Urbana-Champaign.

Cammack, C., Waliczek, T. M., and Zajicek, J. M. (2002) 'The Green Brigade: The psychological effects of a community-based horticultural program on the self-development characteristics of juvenile offenders,' *HortTechnology*, 12(1): 82–6.

Chiesura, A. (2004) 'The role of urban parks for the sustainable city,' *Landscape and Urban Planning*, 68(1): 129–38.

Coley, R. L., Kuo, F. E., and Sullivan, W. C. (1997) 'Where does community grow? The social context created by nature in urban public housing,' *Environment and Behavior*, 29(4): 468–94.

Curtis, S., Gesler, W., Fabian, K., Francis, S., and Priebe, S. (2007) 'Therapeutic landscapes in hospital design: a qualitative assessment by staff and service users of the design of a new mental health inpatient unit,' *Environmental Planning C: Government and Policy*, 25(4): 591–610.

Day, R. (2008) 'Local environments and older people's health: dimensions from a comparative qualitative study in Scotland,' *Health and Place*, 14(2): 299–312.

Detweiler, M. B., Murphy, P. F., Myers, L. C., and Kim, K. Y. (2008) 'Does a wander garden influence inappropriate behaviors in dementia residents?' *American Journal of Alzheimer's Disease and Other Dementias*, 23(1): 31–45.

Detweiler, M. B. and Warf, C. (2005) 'Dementia wander garden aids post cerebrovascular stroke restorative therapy: a case study,' *Alternative Therapies in Health and Medicine*, 11(4), 54–8.

Dravigne, A., Waliczek, T. M., Lineberger, R. D. and Zajicek, J. M. (2008) 'The effect of live plants and window views of green spaces on employee perceptions of job satisfaction,' *HortScience*, 43(1), 183–7.

Dyment, J. E. and Bell, A. C. (2008) 'Grounds for movement: green school grounds as sites for promoting physical activity,' *Health Education Research*, 23(6): 952–62.

English, J., Wilson, K., and Keller-Olaman, S. (2008) 'Health, healing and recovery: therapeutic landscapes and the everyday lives of breast cancer survivors,' *Social Science and Medicine*, 67(1): 68–78.

Fjørtoft, I. (2004) 'Landscape as playscape: the effects of natural environments on children's play and motor development,' *Children, Youth and Environments*, 14(2): 21–44.

Fuller, R. A., Irvine, K. N., Devine-Wright, P., Warren, P. H., and Gaston, K. J. (2007). 'Psychological benefits of greenspace increase with biodiversity,' *Biological Letters*, 3(4): 390–4.

Glover, T. D. (2004) 'Social capital in the lived experiences of community gardeners,' *Leisure Sciences*, 26(2): 143–62.

Glover, T. D. and Parry, D. C. (2005) 'Building relationships, accessing resources: mobilizing social capital in community garden contexts,' *Journal of Leisure Research*, 37(4): 450–74.

Gobster, P. H. (1998) 'Urban parks as green walls or green magnets? Interracial relations in neighborhood boundary parks,' *Landscape and Urban Planning*, 41(1): 43–55.

Guite, H. F., Clark, C., and Ackrill, G. (2006) 'The impact of the physical and urban environment on mental well-being,' *Public Health*, 120(12): 1117–26.

Hartig, T., Evans, G. W., Jamner, L. D., Davis, D. S., and Gärling, T. (2003) 'Tracking restoration in natural and urban field settings,' *Journal of Environmental Psychology*, 23(2): 109–23.

Hartig, T., Mang, M. M., and Evans, G. W. (1991) 'Restorative effects of natural environment experiences,' *Environment and Behavior*, 23(1): 3–26.

Heerwagen, J. H. and Wise, J. A. (1998) 'Green building benefits: differences in perceptions and experiences across manufacturing shifts,' *Heating/Piping/Air Conditioning Magazine*, 70 (February): 57–63.

Herrington, S. and Studtmann, K. (1998) 'Landscape interventions: new directions for the design of children's outdoor play environments,' *Landscape and Urban Planning*, 42(2–4): 191–205.

Heschong Mahone Group (2003a) *Windows and Classrooms: A Study of Student Performance and the Indoor Environment. Technical Report P500–03–082-A-9*, Sacramento, CA: California Energy Commission.

Heschong Mahone Group. (2003b) *Windows and Offices: A Study of Office Worker Performance and the Indoor Environment. Technical Report P500–03–082-A-9*, Sacramento, CA: California Energy Commission.

Jansen, D. A. and von Sadovszky, V. (2004) 'Restorative activities of community-dwelling elders.' *Western Journal of Nursing Research*, 26(4): 381–99.

Kaplan, R. (1993) 'The role of nature in the context of the workplace,' *Landscape and Urban Planning*, 26(1–4): 193–201.

Kaplan, R. (2001) 'The nature of the view from home – psychological benefits,' *Environment and Behavior*, 33(4), 507–42.

Kaplan, R. and Kaplan, S. (1989) *The Experience of Nature: A Psychological Perspective*, New York, NY: Cambridge University Press.

Kaplan, S. (1995) 'The restorative benefits of nature – toward an integrative framework,' *Journal of Environmental Psychology*, 15(3), 169–82.

Kearney, A. R. (2006) 'Residential development patterns and neighborhood satisfaction – impacts of density and nearby nature,' *Environment and Behavior*, 38(1): 112–39.

Kim, J. and Kaplan, R. (2004) 'Physical and psychological factors in sense of community: new urbanist Kentlands and nearby Orchard Village,' *Environment and Behavior*, 36(3): 313–40.

Korpela, K. M. and Ylén, M. (2007) 'Perceived health is associated with visiting natural favourite places in the vicinity,' *Health and Place*, 13(1): 138–51.

Korpela, K. M., Ylén, M., Tyrväinen, L., and Silvennoinen, H. (2008) 'Determinants of restorative experiences in everyday favorite places,' *Health and Place*, 14(4): 636–52.

Krenichyn, K. (2006) '"The only place to go and be in the city": women talk about exercise, being outdoors, and the meanings of a large urban park,' *Health and Place*, 12(4): 631–43.

Kuo, F. E. (2001) 'Coping with poverty – impacts of environment and attention in the inner city,' *Environment and Behavior*, 33(1): 5–34.
Kuo, F. E. and Sullivan, W. C. (2001a) 'Environment and crime in the inner city – does vegetation reduce crime?' *Environment and Behavior*, 33(3): 343–67.
Kuo, F. E. and Sullivan, W. C. (2001b) 'Aggression and violence in the inner city – effects of environment via mental fatigue,' *Environment and Behavior*, 33(4): 543–71.
Kuo, F. E. and Taylor, A. F. (2004) 'A potential natural treatment for attention-deficit/hyperactivity disorder: evidence from a national study,' *American Journal of Public Health*, 94(9): 1580–6.
Kuo, F. E., Bacaicoa, M., and Sullivan, W. C. (1998) 'Transforming inner-city landscapes: trees, sense of safety, and preference,' *Environment and Behavior*, 30(1): 28–59.
Kweon, B.-S., Sullivan, W. C., and Wiley, A. R. (1998) 'Green common spaces and the social integration of inner-city older adults,' *Environment and Behavior*, 30(6): 832–58.
Leather, P., Pyrgas, M., Beale, D., and Lawrence, C. (1998) 'Windows in the workplace: sunlight, view, and occupational stress,' *Environment and Behavior*, 30(6): 739–62.
Lee, S.-W., Ellis, C. D., Kweon, B.-S., and Hong, S.-K. (2008) 'Relationship between landscape structure and neighborhood satisfaction in urbanized areas,' *Landscape and Urban Planning*, 85(1): 60–70.
Lindholm, G. (1995) 'Schoolyards – the significance of place properties to outdoor activities in schools,' *Environment and Behavior*, 27(3): 259–93.
Maas, J., Spreeuwenberg, P., Van Winsum-Westra, M., Verheij, R. A., de Vries, S., and Groenewegen, P. P. (2009) 'Is green space in the living environment associated with people's feelings of social safety?' *Environment and Planning A*, 41(7): 1763–77.
Maas, J., van Dillen, S. M. E., Verheij, R. A., and Groenewegen, P. P. (2009) 'Social contacts as a possible mechanism behind the relation between green space and health,' *Health and Place*, 15(2): 586–95.
Macnaghten, P. and Urry, J. (2000) 'Bodies in the woods,' *Body and Society*, 6(3–4), 166–82.
Malone, K. and Tranter, P. (2003) 'Children's environmental learning and the use, design and management of schoolgrounds,' *Children, Youth, and Environments*, 13(2), online, available at: http://colorado.edu/journals/cye/13_2/Malone_Tranter/ChildrensEnvLearning.htm [accessed October 18, 2009].
Mårtensson, F., Boldemann, C., Söderström, M., Blennow, M., Englund, J.-E., and Grahn, P. (2009) 'Outdoor environmental assessment of attention promoting settings for preschool children,' *Health and Place*, 15(4): 1149–57.
Mather, J. A., Nemecek, D., and Oliver, K. (1997) 'The effect of a walled garden on behavior of individuals with Alzheimer's,' *American Journal of Alzheimer's Disease and Other Dementias*, 12(6): 252–7.
Matsuoka, R. H. (2010) 'Student performance and high school landscapes,' *Landscape and Urban Planning*, 97(4): 273–82.
McFarland, A. L., Waliczek, T. M., and Zajicek, J. M. (2008) 'The relationship between student use of campus green spaces and perceptions of quality of life,' *HortTechnology*, 18(2): 232–8.
Milligan, C. and Bingley, A. (2007) 'Restorative places or scary spaces? The impact of woodland on the mental well-being of young adults,' *Health and Place*, 13(4): 799–811.
Milligan, C., Gatrell, A., and Bingley, A. (2004). '"Cultivating health": therapeutic landscapes and older people in northern England,' *Social Science and Medicine*, 58(9): 1781–93.
Neville, S. J. (1994) *Infants' Sensorimotor Play in Two Yards: A Traditional Play Yard and the Infant Garden*, unpublished master's thesis, University of California, Davis.
Nielsen, T. S. and Hansen, K. B. (2007) 'Do green areas affect health? Results from a Danish survey on the use of green areas and health indicators,' *Health and Place*, 13(4): 839–50.
Ogunseitan, O. A. (2005) 'Topophilia and the quality of life,' *Environmental Health Perspectives*, 113(2): 143–8.
Ottosson, J. and Grahn, P. (2005) 'A comparison of leisure time spent in a garden with leisure time spent indoors: on measures of restoration in residents in geriatric care,' *Landscape Research*, 30(1): 23–55.
Rappe, E., and Kivela, S.-L. (2005) 'Effects of garden visits on long-term care residents as related to depression,' *HortTechnology*, 15(2): 298–303.
Rappe, E., Kivela, S.-L., and Rita, H. (2006) 'Visiting outdoor green environments positively impacts self-rated health among older people in long-term care,' *HortTechnology*, 16(1): 55–9.
Saldivar-Tanaka, L. and Krasny, M. E. (2004) 'Culturing community development, neighborhood open space, and civic agriculture: the case of Latino community gardens in New York City,' *Agriculture and Human Values*, 21(4): 399–412.
Seeland, K., Dübendorfer, S., and Hansmann, R. (2009) 'Making friends in Zurich's urban forests and parks: the role of public green space for social inclusion of youths from different cultures,' *Forest Policy and Economics*, 11(1): 10–17.

Sherman, S. A., Varni, J. W., Ulrich, R. S., and Malcarne, V. L. (2005) 'Post-occupancy evaluation of healing gardens in a pediatric cancer center,' *Landscape and Urban Planning*, 73(2–3): 167–83.
Shinew, K. J., Glover, T. D., and Parry, D. C. (2004) 'Leisure spaces as potential sites for interracial interaction: community gardens in urban areas,' *Journal of Leisure Research*, 36(3): 336–55.
Snelgrove, A. G., Michael, J. H., Waliczek, T. M., and Zajicek, J. M. (2004) 'Urban greening and criminal behavior: A geographic information system perspective,' *HortTechnology*, 14(1): 48–51.
Sullivan, W. C., Kuo, F. E., and DePooter, S. F. (2004) 'The fruit of urban nature – vital neighborhood spaces,' *Environment and Behavior*, 36(5): 678–700.
Talbot, J. F. and Kaplan, R. (1991) 'The benefits of nearby nature for elderly apartment residents,' *International Journal of Aging and Human Development*, 33(2): 119–30.
Taylor, A. F. and Kuo, F. E. (2009) 'Children with attention deficits concentrate better after walk in the park,' *Journal of Attention Disorders*, 12(5): 402–9.
Taylor, A. F., Kuo, F. E., and Sullivan, W. C. (2001) 'Coping with ADD – the surprising connection to green play settings,' *Environment and Behavior*, 33(1): 54–77.
Taylor, A. F., Kuo, F. E., and Sullivan, W. C. (2002) 'Views of nature and self-discipline: evidence from inner city children,' *Journal of Environmental Psychology*, 22(1–2): 49–63.
Teig, E., Amulya, J., Bardwell, L., Buchenau, M., Marshall, J. A., and Litt, J. S. (2009) 'Collective efficacy in Denver, Colorado: strengthening neighborhoods and health through community gardens,' *Health and Place*, 15(4): 1115–22.
Tennessen, C. M. and Cimprich, B. (1995) 'Views to nature: effects on attention,' *Journal of Environmental Psychology*, 15(1): 77–85.
Tyrväinen, L., Mäkinen, K., and Schipperijn, J. (2007) 'Tools for mapping social values of urban woodlands and other green areas,' *Landscape and Urban Planning*, 79(1): 5–19.
Ulrich, R.S. (1984) 'View through a window may influence recovery from surgery,' *Science*, 224(4647): 420–1.
Ulrich, R. S., Simons, R. F., Losito, B. D., Fiorito, E., Miles, M. A., and Zelson, M. (1991) 'Stress recovery during exposure to natural and urban environments,' *Journal of Environmental Psychology*, 11(3): 201–30.
Wakefield, S., Yeudall, F., Taron, C., Reynolds, J., and Skinner, A. (2007) 'Growing urban health: community gardening in South-East Toronto,' *Health Promotion International*, 22(2): 92–101.
Waliczek, T. M., Zajicek, J. M., and Lineberger, R. D. (2005) 'The influence of gardening activities on consumer perceptions of life satisfaction,' *HortScience*, 40(5): 1360–5.
Wells, N. M. (2000) 'At home with nature – effects of "greenness" on children's cognitive functioning,' *Environment and Behavior*, 32(6): 775–95.
Wells, N. M. and Evans, G. W. (2003) 'Nearby nature – a buffer of life stress among rural children,' *Environment and Behavior*, 35(3): 311–30.
Whitehouse, S., Varni, J. W., Seid, M., Cooper-Marcus, C., Ensberg, M. J., Jacobs, J. R., and Mehlenbeck, R. S. (2001) 'Evaluating a children's hospital garden environment: utilization and consumer satisfaction,' *Journal of Environmental Psychology*, 21(3): 301–14.
Yuen, B. and Hien, W. N. (2005) 'Resident perceptions and expectations of rooftop gardens in Singapore,' *Landscape and Urban Planning*, 73(4): 263–76.

34

Street trees and the urban environment

Gerald F.M. Dawe

Summary

Few can dispute that cities with street trees are much more attractive than those without. Nevertheless, street trees are virtually ignored in modern urban planning. There is firstly a short characterisation of their physical and biological nature. Benefits of street trees are then reviewed under the following headings: temperature reduction and shade, reduction in windspeed, hydrological effects, improvement in air quality, via pollutant interception, gaseous pollutants (O_3, NO_x, SO_2 and CO_2), carbon sequestration and oxygen (O_2) production, particulate interception, noise reduction, health benefits, cultural and community benefits. The disbenefits are then described under two headings: 'old' ecocentric (tree-centred) problems such as disease and minor nuisance value, and 'new' anthropocentric (built environment and danger to people) problems, such as damage to housing, buildings and services. Evidence from the popular media on recent street tree felling is provided. In the discussion section two methods of ensuring street tree survival, which sum up some of the above factors, are looked at: (1) the economic/quantitative perspective; and (2) the political/sustainability perspective. Often trees are largely argued for in terms of their quantitative impact on particulate interception, climate amelioration or economic value, and argued against also in economic or health and safety terms, namely of cracks they may/may not make in house walls, ceilings etc. or injury to people, rather than in other terms such as their value as 'inter-generational objects' and objects of simple inspiration in people's lives.

Introduction

Street trees are individual trees, planted along streets or in gardens abutting onto streets. As such they are not urban woodlands or urban forests. Nevertheless when seen from above mature town or city street trees can resemble a woodland or a forest with housing scattered among them. They make a significant contribution to vegetation of cities. It is estimated that four times as much is spent on 'street trees' as on 'park trees' in the USA (Tschantz and Sacamano 1994).

Street trees provide benefits because of their value to wildlife, their aesthetic beauty and visual amenity, and for their effects at intercepting airborne pollutants (gases, particulates and noise), and ameliorating the extremes of the urban climate. They also provide demonstrable benefits to human health and culture. Set against these 'services' (*sensu* Spray 2002) are occasional disease, 'nuisances' and impacts on the built environment, e.g. house or building foundations, pavements or sewerage systems. Many street trees were planned in England by pre-Victorian and Victorian improvers. In central London a substantial number were planted around 1800 (Pitt *et al.* 1979). At that time dangers came not from trees but from public health and law and order. Trees were considered beneficial. Over time the perceived negative influences have won out, with the result that many cities are losing their trees, so important for their beauty and for the well-being of people. This paper looks at benefits, disbenefits and changing attitudes towards street trees.

Four types can be distinguished:

1 privately owned trees of individuals, in gardens adjoining properties (often at their highest numbers in low-density residential neighbourhoods (Britt and Johnston 2008))
2 privately owned street trees of industrial or commercial areas (e.g. industrial estates, central business districts)
3 publicly owned street trees (e.g. along roadsides, in parks fringing roads, or in shopping centres)
4 community-planted or managed trees (e.g. initiated by neighbourhood land trusts, etc.)

Street trees, because they are planted in avenues, have much in common with urban riverside or railway trees.

Street trees, although conferring multiple benefits on the urban environment, often exist under environmental stress themselves. Dewers (1981) recounts of how 95 species of woody plants were trialled in San Antonio, Texas, but only 22 of them – a mixture of 'natives' and 'exotics' – survived after two growing seasons. Street tree decline in relation to construction damage is documented by Hauer *et al.* (1994).

Physical and biological characterisation of street trees

Normal street tree planting recommendations assume distances of between 12 to 15 m in the USA (Miller 1997: 206–7), but municipal ordinances and concerns about interactions with pavements and built structures are increasingly dictating that smaller trees are planted, despite the often larger size of house lots, at least in parts of the US (Kalmbach and Kielbaso 1979). An estimate of the number of street trees in the USA is 100 million assuming a spacing of 17.6 m and 620,000 miles of municipal streets. The figure implies 114 trees per km of street (Barker 1976).

Street tree lifespans are often surprisingly short, and in cities such as Beijing, 29 per cent of trees were found to be 'in poor, critical, dying and dead condition' (Yang *et al.* 2005). In Jersey City, New Jersey, London Plane (*Platanus* x *acerifera*) averaged a mere 39 years of age and Norway Maple (*Acer platanoides*) 48 years. Collectively these trees made up 55 per cent of the trees in the city (Miller 1997: 220). Elsewhere, street tree longevity has been put at ten to 15 years (Nilsson *et al.* 2001: 348). In Beijing measurement of tree diameters revealed a skewed distribution towards small trees. In older historic city centres, more mature trees are present: in central London, 9.8 per cent were saplings, 49 per cent were young, 33.1 per cent were mature (0.35 to 1.8 m girth at shoulder-height) and 7.6 per cent were 'over mature' (over 1.8 m and also trees with dead branches or in need of support). The main species here were London Plane (*Platanus* x *acerifolia*) and Lime (*Tilia* x *europaea* and *T. platyphyllos*). Approximately 75 per cent of London Planes were mature or 'over mature' (Plummer and Shewan 1992), or possibly even 'dying'. The language used to describe street trees is critical. Old and ancient trees are often in need of preserving for the valuable wildlife they contain and for their own sake, but they may not be when they are in the street.

Street trees are biologically diverse and contain rarities. In a survey of 3 per cent (4,700 ha) of the Greater London area 240 species from 82 genera were recorded (Matthews *et al.* 1993). In Melbourne, Australia, 1,127 'taxa' (some unknown) were recorded (Frank *et al.* 2006) and Beijing was found to contain 170 tree species used in 'landscape planting' (Yang *et al.* 2005), though significantly, only a few tree species made up the majority of those found in the samples. This generality can probably be extended to other cities: that despite biologically diverse street tree floras, relatively few species make up the majority of street tree cover.

Benefits of street trees

This section identifies 'services' (*sensu* Spray 2002) that trees provide to the urban population. Benefits to wildlife have been eschewed here in favour of a concentration on more human-related areas. Due to the physical proximity of street trees to people, it is their benefit *for something* (e.g. 'instrumental value') that has been much more widely written about, than the birds, insects, lichens or mosses which may inhabit them.

Cities are complex areas, differing in their building make-up, presence of car parks, courtyards, parks and other open spaces, and they also have variable effects on the human experience of comfort within them.

The urban climate and urban air pollution in the absence of trees can be disadvantageous to urban residents, because of two factors: (1) relative drought and dryness—coupled with sometimes high temperatures—due to dominance of built structures and lack of shade; and (2) relatively high pollutant levels, arising mainly from vehicular traffic. Around houses, street trees can help to ameliorate such conditions, and they may also impact on air-conditioning systems in warmer climates.

Climate and effects on atmosphere

Temperature reduction and shade

Cities are hotter and drier places than their environs, giving rise to the 'urban heat island' phenomenon. Comparing the pre-urban to the urban situation, evapotranspiration reduces from 40 per cent to 25 per cent of total rainfall, due to the extensive presence of hard surfaces, with consequent effects on temperature and humidity (Hough 1984: 72). Parks and street trees can both counter these effects, leading to less runoff, more natural watercourses in urban areas, and a lesser likelihood of flooding. Reports of temperature reduction due to the presence of urban trees follow:

Parks

A detailed study in 1965 scientifically measured the effect of a well-treed park on temperature in London. Traverses across Hyde Park which recorded a temperature decrease of 1.3°C compared to its surroundings (Chandler 1965). Other more recent notable records for parks include that of Rock Creek Park in Washington DC, which was 3°C cooler than its immediate environs (Landsberg 1981). Cooling is attributable to three factors: the distance from buildings which would radiate out heat, together with heat generated by traffic, versus evapotranspiration from within the park.

Street trees

No temperature differences were found between trees on city streets and nearby residential blocks in Syracuse, New York (Miller 1997: 59, citing Hiesler and Herrington 1976). Cooling of between 0.7°C and 1.3°C was found under urban tree canopies in Bloomington, Indiana (Souch and Souch 1993), similar to that of the 1.6°C cooling found in Beijing due to the presence of a tree/shrub cover of 16.4 per cent (Yang *et al.* 2005).

In hotter climates temperature reductions can be more marked under trees: an average reduction in air temperature of 3.6°C was recorded beneath the canopy of a tree in Miami, Florida (Parker 1989). In a detailed study of Tel-Aviv, Israel, cooling effects due to street trees were found to be an average of 0.5°C with an average cooling of 3°C at noon (Shashua-Bar and Hoffman 2000).

Some have found that night temperatures are higher under trees than in the open and that the differential can range from 5 to 8°C (Grey and Deneke 1978, citing Federer 1971). These temperature variations are explained both by inherent urban heterogeneity, by different climatic zones, and by trees acting as barriers to incoming radiation but also doing the opposite: providing barriers to radiation emitting from pavements and the built environment during the night. To put this in context, heavy urban traffic raises urban temperatures by around 2°C (Swaid 1993).

There is no doubt that sitting under trees in shade in the summer gives pleasure and comfort to urban residents, whether this is due to temperature reduction, or simple blocking of sunlight.

Reduction in windspeed

Reduction of wind around buildings via street-tree plantings can save energy use, and therefore, indirectly, carbon dioxide (CO_2) emissions, which are implicated in climate change. It can also provide a better environment for people. 'Narrow shelterbelts' with a moderate porosity (50 per cent) are the best options to reduce windspeeds (Coppin and Richards 1990: 245). Without the porosity, the wind on the downward side of the barrier remains highly disturbed. The ideal conditions approximate to the planting of an avenue of street trees. Open belts give the greatest shelter downwind, the effect lasting for around nine times the height of the trees. However, dense belts give protection over a greater distance, but have the disadvantage of producing eddying immediately downwind. Nevertheless, they demonstrably reduce wind-loading on properties. In open windswept areas such as the USA's Great Plains area, windspeed may reduce from two to five times the height of the tallest trees in front of the barrier and extend to distances 30 to 40 times height on the leeward side (Grey and Deneke 1978: 52).

Hydrological effects

To sum up some practical aspects of the 'urban heat island' and street trees' role in ameliorating this, the hydrological nature of cities has an overarching role: in the pre-urban situation, 40 per cent of rainfall goes back into the atmosphere via evapotranspiration, 10 per cent forms surface runoff and 50 per cent goes into groundwater. In the post-urban situation 25 per cent forms evapotranspiration, 30 per cent plus 13 per cent from built areas (43 per cent in total) runs off and 32 per cent goes into groundwater (Hough 1984: 72). The much higher surface runoff has deleterious effects on urban watercourses, meaning that many are canalised and almost devoid of vegetation. In a typical urban environment, rainfall infiltration capacity is reduced and runoff becomes so severe as to limit recharging of soil water and groundwater reserves (Bruijnzeel 2001). Urban forestry, and street trees in particular have a significant role in reducing this urban runoff, by holding back water and allowing it to return to the atmosphere via evapotranspiration.

Improvement in air quality by pollutant interception

The paucity of data and directed science in this area is notable. Yang *et al.* 2005, in a study of pollution interception by the urban forest of Beijing point to several problems: first, cities are far from being homogeneous and neither are air pollutants, both in their spread and concentration, and in their occurrence over time, in the urban atmosphere. The problem is further compounded by difference in tree species' abilities to intercept pollutants, and by morphological factors such as tree size which can make a tree, of a given species, vary in its ability to intercept pollutants by at least tenfold. The following approximate ratio of tonnages of major pollutants intercepted by trees is derived from Yang *et al.* 2005: PM_{10} (Particulate Matter of 10 μm or less in diameter) stand at three to four times that of ozone (O_3) (when it is near to maximum level), and nitrogen dioxide (NO_2) and sulphur dioxide (SO_2) are intercepted at much lower levels still, around one sixth or a tenth of PM_{10}.

Negative effects: VOCs

On the negative front, street trees are able to contribute to emissions of Volatile Organic Compounds (VOCs) (Jarvis and Fowler 2001) which assist in the generation of low-level O_3 pollution though this is only likely to be a fractional amount compared to emissions from vehicular traffic. The term Biogenic VOCs (or BVOCs) has been coined to distinguish VOCs arising from living plants compared to those of commercial/industrial origin. An estimate for those BVOC chemical species making a major contribution to O_3 production by trees in Beijing was 205 tonnes of isoprene and 36 tonnes of monoterpene – together it was estimated that these could go on to generate a theoretical 1,984 tonnes of O_3 (Yang *et al.* 2005). *Populus, Robinia* and *Salix* have high BVOC production rates (Benjamin and Winer 1998). Overall, however, when set against this, the benefits of street trees are considerable.

Gaseous pollutants (O_3, NO_x, SO_2 and CO_2)

O_3, nitrogen oxides (NO_x) and SO_2 are mostly associated with vehicular emissions (and O_3 partly with BVOCs), though SO_2 also with power stations. CO_2 is a gas which has recently risen to prominence due to its role in increasing the rate of climate change. It is a product of burning fossil fuels, via vehicular emissions but also residential and industrial activity within urban areas. CO_2 is also involved in entirely natural biological processes, e.g. photosynthesis, where O_2 is given off – e.g. by street trees – and C is converted into more complex molecules used by the tree, ultimately to lignins (e.g. wood).

A few conifer species are intolerant to air pollutants, especially SO_2, and are therefore unsuitable for planting as street trees (Landsberg 1981). This is less relevant now because of the rapid decline of SO_2 across many urban areas due to fewer city-located coal-fired power stations. Increases in ambient SO_2 in rapidly expanding industrial areas of China and South East Asia, are of most concern here. NO_2 is readily taken up by leaves, and O_3 can cause individual tree and forest damage, as well as impact on human health, though surface levels of pre-industrial O_3 are not thought to be harmful (Jarvis and Fowler 2001). It is estimated that in Auckland, New Zealand, 1,230 tonnes of NO_x and 1,990 tonnes of O_3 are taken up by the urban forest (Cavanagh and Clemons 2006). To give an idea of their relative magnitude, O_3, NO_2, and SO_2 were estimated as being 20 per cent, 11 per cent and 8 per cent of the pollutants removed by trees in Beijing (a total of 39 per cent), PM_{10} accounting for the remaining 61 per cent.

Carbon sequestration and oxygen production

Some have argued that there are local variations in the CO_2:O_2 ratio within cities, ranging from 1:3000 in pure air to 1:1000 alongside many city highways (Brown 1983). Street trees, as part of the urban forest, do make a substantial contribution to the CO_2:O_2 balance. The 61 m tonnes of O_2 estimated to be produced by the US urban forest collectively, is enough to offset O_2 consumption by approximately two-thirds of the US population (Nowak *et al.* 2007). The earliest description of urban parks as 'lungs of the metropolis' may be that of Brezina and Schmidt in 1937 (cited by Landsberg 1981). Their hypothesised effects, at that time, were that the excesses of carbon dioxide (CO_2) would be offset by oxygen (O_2) produced in parks. This relates, some 70 years on, to current climate change concerns and the potentially beneficial role of trees in carbon sequestration (e.g. Kohlmaier *et al.* 1998). As well as direct C sequestration, street trees assist in the avoidance of production of CO_2 by secondary effects on residential buildings The total carbon stored and avoided by the 100 million estimated trees of the US urban forest (363 million tC) is <1 per cent of the estimated amount of carbon emitted in the USA over the same 50-year period. Increasing fuel efficiency of passenger automobiles by 0.5 km/l over 50 years would also produce the same carbon effects as the 100 million trees (Nowak and Crane 2002). Another study showed that a building in full sun required 2.6 times as much electricity for cooling than one in full shade, the context for this being Alabama, USA (Laband and Sophocleus 2009). Again, this would imply indirect savings in net CO_2 emissions due to the presence of street trees. C sequestration by street trees (direct and indirect) needs to be seen in a well-rounded context, that of the global C cycle and the fact that C sinks may eventually become C sources (Jarvis and Fowler 2001; Nowak's perspective is also useful here – see e.g. Nowak and Crane 2002).

Particulate interception

Particulates are small-sized relatively inert bodies associated with fossil-fuel combustion and vehicular traffic (especially diesel). There are actually very few scientific studies of the interception of particulates by street trees. Dochinger (1980), estimated that overall reduction in 'dustfall' by urban trees was 27 per cent, though conifers were more effective, at 38 per cent. A notable study by Beckett *et al.* 2000, describes particulate (e.g. around 10 µm in diameter, so called PM_{10}) interception by five different urban tree species. The amount of particulates intercepted by the trees varied from 2,936 g per tree for a London-based Plane tree (*Platanus* x *hispanica*) to only *c.*40 g for Planes at The Level, in Brighton. The majority were within the 10 µm (PM_{10}) or smaller size range. The disparity between trees can be explained by the fact that most particulates are emitted from vehicular traffic and the London site was within a 100 metres of Marble Arch. Also, the trees themselves differed in size and thus their potential for intercepting pollution varied. Elms (*Ulmus*) were also found to be significant particulate interceptors whereas Ash (*Fraxinus excelsior*), European Lime (*Tilia* x *europaea*) and Cherry (*Prunus avium*) were found to be generally intermediate in performance. Larger trees with greater leaf surface areas were more effective at intercepting particulates, but young trees, because of their relatively high canopy density were also found to be effective interceptors of particulates. It has been estimated that in Auckland, New Zealand, 1,320 tonnes of particulates are taken up by the urban forest (Cavanagh and Clemons 2006) and that in Beijing, 784 tonnes of PM_{10} were taken up by trees (Yang *et al.* 2005).

Noise reduction

Noise attenuation varies from 1.5 dB to 30 dB per 100 m. OECD's guidance is 2–3 dB(A) per 100 m to that of 5–10 dB(A) for thick coniferous plantations: 'in practice, however, different tree species do not appear to differ greatly in their ability to reduce traffic noise, though evergreens are best when year-round screening is required' (Coppin and Richards 1990: 83). Brown (1983) gives figures for trees and vegetation (compared to pavements) of 5 dB per *c.*30 m, 10 dB/*c.*60 m, and 14 dB/*c.*115 m. (That dB is a logarithmic scale needs bearing in mind.) In fact, there are more subtleties about the perceived noise reduction due to the presence of street trees: 'the whisper of pines, the rustle of oak leaves, or the quaking of aspens' (Grey and Deneke 1978: 76) plus the presence of birds can distract from undesirable urban noise.

Health benefits

There are a myriad series of interactions between green open spaces and human health (Sanesi *et al.* 2006; see also Chapters 32 and 33 this volume). The interaction of street trees with pollution and climate, as outlined above, is significant. To give examples of health effects, particulate pollution 10 μm in size (PM_{10}) can be damaging to health, and the 2.5 μm ($PM_{2.5}$) fraction has been highlighted because of its potential to get deep inside lung tissue (Beckett *et al.* 2000). O_3 and SO_2 and high urban temperatures have also been implicated in exacerbating lung and coronary disease. These are specific examples of direct injury to health and the role of street trees in ameliorating it, but there is evidence aplenty of the broader health benefits of green vegetated spaces (Dawe and Millward 2008) and an integral part of this is that trees, as the largest structures within natural vegetation, have multiple benefits. Street trees often cannot be teased out from the evidence but, at the street-tree scale, community gardens have multiple benefits (Armstrong 2000; Burls 2008; Elmendorf 2008). Street trees themselves are sometimes a direct product of community organisation (Armstrong 2000) and in turn, may foster the wider involvement of participants in urban problems, a concept termed 'embracement' by Burls and Caan (2004). It has also been found that views from hospital windows including trees and soft landscape enhances recovery from surgery (Ulrich 1984). Physical involvement with green urban landscapes also promotes health (Guite *et al.* 2006). In a wide ranging review of benefits of greenspaces to psychological and mental health, Douglas (2008) prefaces his paper with a quote from Roszak (1996) which refers to the role of trees in indigenous cultures, in helping to 'heal troubled souls'.

Cultural and community benefits of street trees

Street trees are not only the background to living in an attractive place, people are almost always more directly involved with them. This may involve either (1) direct use of street trees, or (2) mental inspiration or other psychological benefits from street trees.

Direct use of street trees – In one Milwaukee neighbourhood, Lime or Lindens (*Tilia* spp.) are planted because people make wine from the flowers (Miller 1997: 235). Tree prunings in London could be used in a biomass fuel supply for the capital, albeit that data did not include that of street trees of private gardens (Bright *et al.* 2001). Tree values are international. Urban forests in Ibadan, Nigeria, supply both 'timber and medicinal plants' to the local population, according to 36 per cent of responses to a survey (Popoola and Ajewole 2001). Other uses may be for forest products or for a 'spirit of sacredness', or simply because of a regard for folklore (Bass 2001).

Mental inspiration: Numerous people have been inspired by trees. They form a pleasant backdrop to their lives, cementing them with nature and beyond. These contributions are probably unquantifiable: 'A Nightingale Sang in Berkeley Square', a 1940 song by Manning Sherwin and Eric Maschwit, is what is meant here but in lay terms: simply the spirit of relaxing under, or enjoying trees, whether or not books are written or music composed as a result.

Schroeder *et al.* (2006) discuss the 'intangible benefits and values' of street trees, identifying two high scoring factors among US respondents: 'increases sense of home and family' and 'increases sense of community'. These, together with enabling better privacy for the individual, show just how intimate is the relationship between street trees and people.

There is sometimes a more negative side to community engagement with street trees, which poses difficulties for tree managers. Street trees can sometimes be perceived as things 'for setting fire to', 'for lurking behind' and this, together with the tendency of some buildings to have street trees planted around them which are sufficiently small that a person cannot hide behind them are some of the factors which arise here (Spray 2002; see also Douglas 2008).

The visual appearance of trees, combined with pollution and climatic benefits and their cultural uses, are what constitutes their 'aesthetic' value. This has recently been defined in terms of the sustainability/political dimension of urban forest management.

Disbenefits: ecocentric versus anthropocentric

Two types of 'problems' are often expressed around street trees.[1] The 'older' problems about street tree pathology and damage to street trees have increasingly been replaced by 'new' concerns, primarily relating to negative effects of street trees on urban built structures and people. The 'new' does not necessarily replace the 'old' in chronological terms, but it reflects two completely different orientations which have become evident in street tree management (Table 34.1).

'Old' problems: tree-oriented (ecocentric)

These 'problems' are mainly centred on tree pests or disease, nuisance or tree health (Table 34.2).

I have deliberately excluded work to do with salt damage to tree roots and also matters to do with utilities (e.g. 'Rights-of-Way' (ROW)) since these are perennial.

Table 34.1 How street-tree problems are changing

	'Old' street tree problems	*'New' street tree problems*
Orientation:	Trees at the heart	Built environment and people at the heart
Known as:	Ecocentric	Anthropocentric
Description:	Tree pathology and damage to trees due to built environment (e.g. stresses due to drought, soil compression) is the central orientation	Damage to built environment or to people is the core orientation. Concerns are around minimising impact of street trees on built structures, and to people ('health and safety') which in turn may be driven by the insurance industry
Implications:	• perception of trees as needing to be cared for	• perception of trees as problem
	• concentration on the tree	• concentration on the built environment and on dangers to people

Table 34.2 Examples of 'Old' problems of street trees – ecocentric – tree-oriented

Species	*Problem*	*Geographic area*	*Reference*
Silver Maple (*Acer saccharinum*)	• Carpenter ants nesting, causing limb breakage • aphid honeydew (associated with the above)	New Jersey, USA	Fowler and Parrish 1982
Sugar Maple (*Acer saccharum*)	• compressed urban soils • nutrient conditions resulting in decline	Amherst/Northampton, Massachusetts, USA	Dyer and Mader 1986
Ash Trees (*Fraxinus* spp.)	• pest – Emerald Ash Borer (*Agrilus planipennis*)	Toledo, Ohio, USA	Heimlich *et al.* 2008
Tulip Tree (*Liriodendron tulipifera*)	Multiple: • pavement planting • air pollution • aphid honeydew	Berkeley, California, USA	Dreistadt and Dahlsten 1986
Honeylocust (*Gleditsia triacanthos*)	• pest – *Homadaula anisocentra* aided by urban heat island	Northern USA	Hart et al. 1986
	• drought	USA	Potts and Herrington 1982
Norway Maple (*Acer platanoides*)	• multiple	USA	Marion 1981

'New' problems: built-environment and people oriented (anthropocentric)

These are 'problems' resulting from street tree contact with, and effects on the built environment. Dominating over all of them are 'health and safety' concerns, which in turn are driven by the insurance industry. 'New' tree problems began, in the UK, with National House-Building Council, 1974, and a subsequent Department of the Environment (DoE now Defra) report on 'foundation movement' associated with street trees (see Reynolds 1980 for a review). A series of papers by Reynolds, in the 1980s, followed. At the time, it was emphasised that 'heave' or 'settlement' of houses in relation to tree roots was still 'a matter of opinion' (Reynolds and Alder 1980). In the early 1980s a survey of London's trees revealed that: 'Issues arising from this consultation included: ... the pressures on big trees and trees growing in streets arising from conflicts with buildings and services' (Matthews *et al.* 1993). Further warning bells were sounded for the street tree-mycorrhiza system, via: 'allowing urban structures (roads, paths, car-parks) to impinge too closely onto tree root systems [together with] the common occurrence of harsh and unnecessary pruning regimes...' (Green 2002). For examples of 'new' problems see Table 34.3.

Many reports of investigations of tree stability or rot, prompted at least in part by 'new' 'problems', have been published (to give just a few examples: Catena 2003; Ouis 2001; Schwarze 2001). The tables hint that 'new' problems are more prevalent within the UK than the USA, due to inherently more dense street and housing layouts (Schroeder *et al.* 2006), but that increased urbanisation may predict their arrival in almost ubiquitous fashion. However, their exact origins are badly in need of exploration. The 'old' ecocentric and 'New' anthropocentric problems are summarised in Figure 34.1.

Table 34.3 Examples of 'New' street tree – anthropocentric – problems

Species	*Problem*	*Geographic area*	*Reference*
Populus x *canadensis, Salix alba, S. fragilis*, etc.	• root intrusion and damage to sewer pipes	Sweden	Stål 1998
Betula, Populus and Salix	• root intrusion and damage to sewer pipes	Denmark	Randrup 2000
Leyland Cypress (*Cupressocyparis leylandii*) Richardson 2002	• visual intrusion and neighbour disputes between private gardens	UK	Evans 2002a Richardson 2002
Wild Cherry (*Prunus avium*) rootstocks	• pavement damage	Sheffield, UK	Nicoll and Armstrong 1998
Broad range	• private houses –root damage by trees	UK	National House-Building Council 1974
Broad range (veteran trees)	• roads, paths and car-parks impinging on veteran trees	UK	Green 2002
Wide species range	• conflicts with buildings and services	London	Matthews *et al.* 1993
Wide species range	• subsidence under private houses and other buildings	UK	Biddle 1998a, 1998b
Wide species range	• cars receiving damage due to collisions with trees	USA	Wolf and Bratton 2006
Wide species range	• interference with Wi-Fi computer networks	California, USA	Lacán and McBride 2009

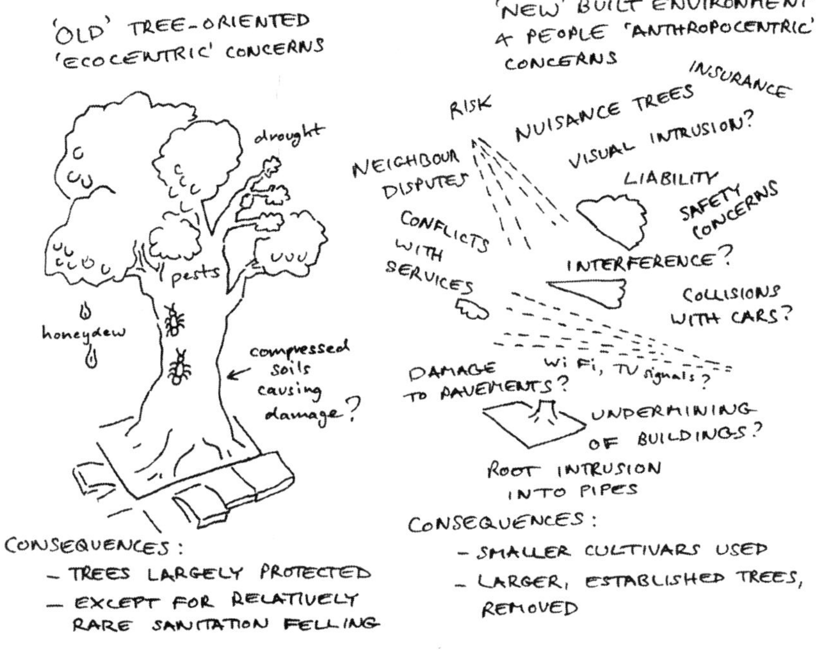

Figure 34.1 Cartoon of the contrasts between 'Old' and 'New' street tree problems.

Are street trees increasing or decreasing?

There is one final crucial question to answer, in relation to both 'benefits' and 'disbenefits'. As a result of this constantly changing balance of arguments, are street trees actually in decline? There are many effective street tree planting programmes leading to increases in street tree numbers, for example in London, where 48,000 trees were planted while 40,000 were lost (Anon 2007), and in the 105 neighbourhoods of Portland, Oregon (Mike Houck, *pers. comm.*). Nevertheless, there is also some evidence for decline, and certainly for mature street tree felling (Booth and Montague 2007; Britt and Johnston 2008; Cowan and Harbinson 2009; in Bristol see Anon 2009a; and in London, see Anon 2007). An important caveat is that in some instances (e.g. Anon 2007) while actual *numbers* of street trees do not appear to diminish, mature trees may actually be being lost (Johnson 2007). As Brown (1983) says: 'Mature trees will outlast two or three human generations and may assume massive size and scale; yet they are easily lost and destroyed.'

How then, can the social consequences be derived? A sequential listing of popular news media from the Nexis® UK 'all English' news database (Appendix 1) has produced results from which some tentative conclusions can be drawn. The list shows that large or mature trees are still disappearing and that such trees have:

1 powerful emotive significance with communities;
2 sometimes trees are removed without apparent notice being given to those communities within which they live;
3 often the 'pruning' or 'felling' is more extensive than was originally advertised;
4 sometimes the community is critically aware of the long life of trees and the potential difficulty of substituting for ones that have been removed; and
5 sometimes the community would prefer to retain tree(s) even when they 'have to be removed' by the local authority for primarily anthropocentric reasons.

Inevitably, a sequential listing from 'popular news media' will be subject to criticism, partly because this very media is known to 'sensationalise' findings and events, and partly because the 'sample' is not random and hence cannot be generalised either spatially or into the longer term. However, the first 'sensationalised' reason is precisely why the media is such a relevant source: the media impinges directly onto the consciousness of both tree officers and local people, and indeed onto other urban 'actors'. Whilst it is certainly far from a 'random sample' showing whether mature street trees are actually in decline, what it illustrates, very directly, is something of the emotive significance of mature street tree removal, and its consequences for both local government and for the local community. This removal, together with its impact on communities, and how to minimise both, is something badly in need of more exploration.

Discussion: the battle for street trees

A first important question to be asked about street trees is whether they are considered to be significant individuals, 'street furniture', or an integral part of the broader urban ecosystem. Such views govern how we react to them, and how they may or may not prosper in the urban landscape.

It is interesting to note that one of the driving forces for removing trees, the insurance industry, did not always have this attitude. 'Casualty' losses of trees and appropriate insurance for this were certainly once a factor in the US urban forest (Grey and Deneke 1978: 161). Something has caused the switch, from 'old' to 'new' street tree problems. Probably because of 'health and safety', 'subsidence', and 'heave' concerns (the subtext is insurance and insurance claims), combined sometimes with smaller-sized household plots, and worries about crime, the fears of many local government officials have now led to a decrease in street trees, and in their physical diminution via the use of new cultivars. For example, the American Elm (*U. americana*) routinely grew to 30 m tall (More and White 2003). The Dutch Elm Disease (DED)-resistant variety 'New Harmony' reaches only 21 m at maturity (United States National Arboretum 2009). Lime or Basswood (*Tilia* spp.) normally reaches 35 m maximum height (Mitchell 1996). New cultivars are between 10–15 m in height (Morgan 2003). This is often despite residents' wishes to see larger trees planted, at least in the US (Kalmbach and Kielbaso 1979).

This is unlike the 'high day of the street tree' in the eighteenth and nineteenth century when attitudes to cracked pipes and falling tree branches had not assumed the same importance as they do today. I am not going to deal with risk in detail, but it clearly is a balance between the severity of outcome and the likelihood of it happening. If taken to its logical conclusion, all risk and death minimised, it would probably not only be trees which would vanish from our streets, but cars too.

Sometimes residents will even defend street trees beyond expert opinion in favour of felling (e.g. Heimlich *et al.* 2008) (Appendix 34.1). This is something which gives rise to a little-discussed dilemma at the heart of street tree management: should it be the tree 'expert' or the community which decides to fell or not to fell, and what then will be the consequences? This is a serious, but little discussed, emotional and scientific conflict. A positive way forward would be to allow contradictions between tree experts and communities to co-exist, and investigate them further, rather than suppress one side's reality over another's: 'it shouldn't be surprising to find cases where two or more radically different and indeed incommensurable sets of beliefs have equal practical adequacy' (Parker 2001: 258). Parker argues against the violent suppression of one or other opposing view. However, here we are not simply discussing 'beliefs' per se. These are beliefs with very practical consequences for street trees and, as has been shown, sometimes trust has been lost by the public, due to precipitate actions involving the drastic pruning or removal of street trees.

Several different orientations to valuing street trees bring at least some of the above factors together:

1 the economic/quantitative perspective; and
2 the political/sustainability perspective.

Economic/quantitative perspective

The presence of street trees is associated with residential districts of higher socio-economic status (Handley 1983; Tratalos *et al.* 2007), and residents have a very positive attitude towards them, albeit that their perceived need for street tree planting varies (Schroeder *et al.* 2006). There is evidence that the presence of street trees can lift residential price values. At 1971 prices street trees were able to raise the price of residential lots by \$10 per foot of front garden (e.g. \$90/m^2), and individual street trees were valued at \$300 in 1956 prices. This compares to today's valuations, which put them at hundreds of thousands of pounds or dollars per individual tree. Increases in property value of 20 per cent with averages of between 5 to 10 per cent, or \$3,000 to \$7,000 were also estimated from the late 1970s (Grey and Deneke 1978: 159–60). An average of \$8,870 per house was recorded in a recent study (Donovan and Butry 2010). The value of street trees in two streets in the City of London were estimated at £201,144 and £36,144 using the Helliwell system, and £111,769 and £23,478 using the Swiss system (1987 prices: Plummer and Shewan (1992: 145) derived from

Helliwell (1967) and Radd (1976)). This worked out at figures of £227.54, £126.43, £69.51 and £45.15 per linear metre. There is not space to go into the methodology here, but it does include a number of factors which are subjective, but which have been further developed by Neilan (2008) whose report is illustrated by two street trees worth '£150,000 and £250,000'. Because they are formulaic and give a result, such systems have been thought worthy of use. Economic valuation does give street trees 'value', particularly in relation to surrounding house prices. However, there is a necessary corollary: that such values need to be connected with the wider political processes of urban tree management systems, thereby making their value transparent to residents, as well as to tree managers and local government politicians. Without such generally perceived value, there may be problems. As Plummer and Shewan (1992: 150) say: 'Cost benefit is a useful exercise but here again there are problems of assessing benefits due to their intangible nature.' To give an example of just how dangerously arbitrary such results can be, especially when disconnected from tree management systems, 12 trees, one of them around 30 years old, were valued at only £2,500 – *c.* £200 per tree – in the popular press in 2009 (Anon 2009b).

To give another example of the multiple ways in which tree valuation is used, very often 'costs and benefits' have been regarded by local authority tree people in personal career terms, in terms of tree officers' benefits, rather than those which ought to accrue to a broader general public. For example, Land Use Consultants (1994: 15) record the 'costs and benefits' of tree strategies – as opposed, admittedly, to the planting or preservation of street trees – as being: (1) to raise the profile of tree and tree management 'with other departments in the authority, the general public and other organisations'; (2) to promote 'the role of tree officers'; (3) 'to provide coherent tree management'; and (4) 'to aid in the procurement, maintenance and management of funds'.

Political/sustainability perspective

The value of economic analysis itself needs questioning. It is an attempt to give political value to street trees. In local government, the normal power relationships may be useful for preserving street trees, irrespective of any economic analysis (Table 34.4). However, this argument needs setting in the wider context of local and national governmental planning policies.

Table 34.4 Normal power relationships in local government, and their consequences for preservation or diminution of street tree stock (▶ = implication)

Local people	*Elected politicians*	*Local government employees*	*Consequences*
Unaware or unconcerned about street trees ▶	▶ Indifferent or slightly concerned about street trees	▶ Take the lead from politicians: hence indifference, and possible over-concern with safety issues	▶ Street trees have indifferent political status, and may diminish – there is no concern about their removal
Aware and vitally concerned about street tree growth and health ▶	▶ Likely to take a keen interest, if campaigning from the public is strong enough	▶ Become sensitive to street tree preservation and enhancement: now a 'political issue'	▶ Street trees have political status, and care is taken with them, possibly at the expense of the built environment

These broader environmental-political issues will increasingly include 'sustainability', in other words consideration of street trees as 'intergenerational' objects of value, rather than as disposable parts of the 'soft'-landscape (*sic*) (Raison *et al.* 2001). Green (2002) has argued that arborists themselves should 'have a central role' in educating the public about veteran trees. It is doubtful, however, if they could be in at the start of a political process, advocating for the appreciation of street trees among local people (since, almost by definition, they work as servants of local government). More utilitarian aspects such as keeping tree inventories may also offer a better solution but at a more mundane level. They will not deal with the fundamental problem. The tree officers are in a corner, fighting their cause. This is because the strong political commitment to having trees prominently in our streets is all too often just not there (though there are rare exceptions, e.g. work in some London boroughs and certain US cities and the commitment of London's Mayor, Boris Johnson to spend £4 million over four years for 10,000 new street trees in the neediest areas and to support programmes to raise funds for tree planting projects). Everything is now about tree damage and health and safety, not aesthetics, emotions, beauty, nor indeed, even quantifiable benefits.

Far better conciliation between tree officers' concerns and communities fighting to protect trees offers one very positive way forward, and also forms a core part of the tradition of sustainability, inasmuch as 'participative decision-making', whatever its drawbacks, is at the heart of the process. It can also feed powerfully into the political argument outlined above. Lest emotion, on the part of local people in favour of preserving trees, seem too controversial: 'Bureaucratic measures – such as Planning Policy Guidances (PPGs), Tree Preservation Orders (TPOs), etc. – step aside from evaluating the significance of trees' (Evans 2002b: 257).

> I want to argue that a failure to understand emotion ascriptions to Nature implies a lack of understanding of that emotion. It is through these emotions, and particularly through moods which may not have any cause or effect at all, that Nature opens to consciousness and gives consciousness something to be conscious of. Trees do this wonderfully, magically, powerfully.
>
> *(Evans 2002a: 256)*

Spray (2002: 263) says: 'a new ethical relationship with trees may be emerging, which is more balanced than our predominantly rational, objective one'. Schroeder 1987 (cited by Spray) argues that tree managers may commit 'grievous' blunders with the public if they fail to recognise their emotional attachment to trees.

Conclusion

Street tree conservation and increase must be argued through reference to:

1 their utilitarian function (e.g. climate and pollution interception)
2 their cultural value (e.g. as sources of products and stimulation or enjoyment)
3 their economic value, but only when tied in to tree management plans or to the wider cultural context
4 local politics of street trees, allied to sustainability considerations
5 human emotion (set apart from science)

(or (1) to (5) collectively, arguing for street trees as part of the whole urban ecosystem).

To return to the question asked at the beginning of this paper, just how are street trees perceived?

1 As single valued entities (in which case emotion, politics or local campaigning may be more effective than benefit-cost analysis in preserving them),
2 As 'pieces of street furniture' (in which case benefit-cost analysis may be more effective, but the trees may be vulnerable to anthropocentric views taken of them), or
3 As integral parts of the urban ecosystem (in which case they may prevail, provided the public, local government officers and elected politicians have an informed understanding of such an ecosystem).

These are categorical distinctions.

The ecological argument that trees are individuals which co-exist with other species and ourselves is, just maybe, the most important perspective. Trees are an important part of the community. There is a tendency to view urban or street trees as of less significance because of their 'numerous problems'. However these need to be tolerated and the built environment, literally built around them. Instead, the 'war of attrition' continues, both by removal of substantial trees, and the selection of smaller cultivars.

Gasson and Cutler (1998) ask 'Can we live with trees in our towns and cities?' For street trees the answer must still be: 'the jury is still out', but it has been out for far too long a time. I believe that the greatest efforts need to be put into areas (1) to (5) above and in tandem to give some chance of street trees' positive survival, even *their fostering at the expense of the built environment.* Comery (2000: 240) shows how a paved path can be curved around a street tree, saving it from damage in the process. Mitchell (1996: 272), arguing for the Oriental Plane (*Platanus orientalis*) rather than the London Plane as a street tree says: 'That the squares might be severely restricted in their open places and the streets impassable to traffic, need not concern us'. There are germs of ideas here, which challenge – head-on – the anthropocentric perspective. They imply that the built environment should not always have primacy over street trees, but rather, that street trees ought – sometimes – to have primacy over the built environment. One way in which this might happen is by both communities and local government officials interacting more effectively in participative and political processes. These ideas urgently need further exploration.

Acknowledgement

I thank Paul Evans for helping with some philosophical parts of this paper, and also editors Ian Douglas, David Goode and Mike Houck for their help.

Note

1 Here, it is not intended to be critical of the research itself, but more, the context in which it takes place.

Appendix 34.1 Street tree felling, and reaction to it[1]

	Headline	*Paper and date*	*Location*	*Issue*	*Concerns by local people for trees Yes/No*[2]	*Age, no., type of tree(s)*
1	India: Tree felling confirmed at Tataguni	*Right Vision News*, February 17, 2010	Bangalore, India	Felling – Illegal: 5 trees	NK	–
2	Farmer's widow awaits tree felling; 'Another horror story waiting to happen'	*The Southland Times* (New Zealand), February 16, 2010	Queenstown, New Zealand	Death due to tree: 72 now to be felled	No: in favour of more felling	One 20 m high Poplar (*Populus* sp.). 72 trees to be felled
3	Tree felling continues	*Guelph Tribune*, February 11, 2010	Guelph, Ontario, Canada	NK	NK	–
4	Public views needed on tree felling plan	*North Devon Journal*, February 11, 2010	Barnstaple, Devon, UK	Fear of crime: 15 trees to be felled	Yes: consultation needed	–
5	Nigeria; Niger Campaigns Against Tree Felling	*Africa News*, February 10, 2010	Niger State, Nigeria	Campaign by govt. against illegal tree-felling	NK	–
6	Tree-felling fuels cost row over Irish Embassy	*The Irish Times*, February 8, 2010	Ottawa, Canada	Felling in grounds	Yes: residents' association	Felled 4 'mature beeches' (*Fagus*)
7	Road shut for tree felling	*Aberdeen Press and Journal*, February 2, 2010	Aberdeen to Inverurie road, UK	Road closure. Felling for 'safety reasons'	NK	–
8	Township fury over school's tree felling	*Hobart Mercury* (Australia), January 30, 2010	Swansea, Australia	Felling for 'safety reasons'	Yes: 'people in tears' and removal 'total surprise'	–
9	Tree-felling to reveal views of cathedral	*The Northern Echo*, January 15, 2010	Durham Cathedral, Durham, UK	Felling to clear views, plus safety	NK	–
10	Gosnells tree felling applauded	*Weekend Courier* (Perth, Australia), January 15, 2010	Rockingham City, Australia	Felling to deal with 'dangerous' trees	No: some in favour Yes: some protesting	20 'big gum trees' (*Eucalyptus*)

continued

Appendix 34.1 continued

11	Anger at tree felling in Brockworth	*Gloucestershire Echo*, January 8, 2010	Brockworth, Glos, UK	Felling – trees' debris causing floods	Yes: 'uproar' – bigger trees not left standing	14 mature trees
12	Residents angered by tree felling; A tree screen at Camborne separating houses at Treswithian from the busy A30 has been chopped down	*The West Briton*, January 7, 2010	Camborne, Cornwall, UK	Felling	Yes: 'very angry', told mature trees would be pruned, not removed	12 mature trees
13	India: Panel begins to dispose of cases on tree felling	*Right Vision News*, January 4, 2010	Shimla, India	Government attempts to rationalise tree-felling	Y/N: Hazards to housing, developers felling too many	–
14	Approval for tree felling	*South Wales Echo*, January 1, 2010	Hensol Castle, Vale of Glamorgan, UK	Felling: approval of	NK	–
15	India: Dhaka mulls law for illegal tree felling	*Daily the Pak Banker*, December 24, 2009	Dhaka, Bangladesh	Legislation by govt. against illegal tree-felling	NK	–
16	Czech Republic: New Legislation Regarding Tree Felling	*Mondaq Business Briefing*, December 14, 2009	Czech Republic	Legislation by govt. regarding tree-felling	NK	–
17	Indiscriminate tree felling for projects opposed to ecology: Sheila	*UNI (United News of India)*, December 5, 2009	India	Indiscriminate tree-felling 'creates an ecological imbalance in the world'	NK	–
18	Tree felling causes much local anger	*The Argus* (Ireland), December 2, 2009	County Kerry, Ireland	Felling due to road-widening	Yes: 'unaccceptable face of development'	–
19	Action threat over tree felling	*Hobart Mercury* (Australia), December 2, 2009	Dodges Ferry, Australia	Felling due to 'power cuts'	Yes: protest by resident standing in front of tree	*Eucalyptus viminalis* trees

20	Horticulturist condemns tree-felling	*Scunthorpe Evening Telegraph*, November 28, 2009	Scunthorpe, UK	Felling '100 dead, dying or dangerous trees' to make way for £2 million leisure complex	Yes: man 89 years old, who for more than 30 years had responsibility for many of Scunthorpe's trees in parks and gardens, 'sad' at development	100 trees
21	Greens call for probe into city's tree felling: party plans assembly motion	*Canberra Times*, November 11, 2009	Canberra, and Australia more generally	'The number of residents surprised and outraged by the tree removals' suggests process is not working as it should do	Yes: 'surprised and outraged' by fellings	–
22	Residents blast tree felling on prime site	*Sunshine Coast Daily* (Queensland), November 11, 2009	Buderim, Sunshine Coast, Australia	Developer consent for removing trees granted by local authority	Yes: 'furious' that nothing done to stop 'carnage'	17 'well-established' Figs (*Ficus*), 28 Pines (*Pinus*) and possibly 16 Camphor Laurels (*Cinnamomum camphora*)
23	Urgent need for federal input on tree felling in the capital	*Canberra Times*, November 10, 2009	Canberra, Australia	Tree replacement programme: 282 trees 'in poor health' coming down, replaced by 588 saplings	Yes: once cut down 'trunks and main branches show no visible signs they are ailing' nor is there evidence of risk to public health	–
24	Tree-felling on hold: fury wins reprieve	*Canberra Times*, November 8, 2009	Griffith, Narrabundah and Manuka, Australia	Felling suspended after protest from residents	Yes: 'heated community debate' over tree-removal programme	Felling of 'landmark trees' suspended
25	Inquiry after tree felling at The Old Croft, Knutsford	*Knutsford Guardian*, November 4, 2009	Knutsford, Cheshire, UK	Legality of felling around private home to be investigated: 'ruining' character	NK: local authority concerned	30 trees

continued

Appendix 34.1 continued

26	Nigeria; Aliero Set for War With PHCN Over Tree Felling in Abuja	*Africa News*, November 2, 2009	Abuja, Nigeria	Indiscriminate felling in capitai city, by power company	NK: local authority concerned	1,000+ trees felled
27	Tree felling site visit	*South Wales Evening Post*, November 2, 2009	Plas Gwernfadog, Wales	Assessment visit	NK	–
28	Tree felling work held up	*The Northern Echo* (Newsquest Regional Press), October 30, 2009	Northallerton, North Yorkshire, UK	Felling of 2 'diseased' trees on children's play area	Yes: 'peaceful protest' resident stood protecting tree from felling	Ash (*Fraxinus excelsior*) estimated at 200 years old
29	Tree felling along road 22N94	*US Fed News*, October 21, 2009 12:30 PM	Feather River Ranger District, Plumas National Forest, CA, USA	Felling of 'hazard trees'	NK	–
30	Ainslie residents dispute tree-felling	*Canberra Times*, October 3, 2009	Ainslie, Canberra, Australia	Felling of tree 'in poor condition'	Yes: 'pine plantation mentality' for street trees, and no notification of felling, say protestors	80 year old White gum (*Eucalyptus mannifera*) felled
31	Anger at earl's tree felling bid; plans to fell nearly 400 trees on the outskirts of Exeter have angered residents	*Express and Echo* (Exeter), September 30, 2009	Exeter, Devon, UK	Felling needed near to road, some of trees 'in poor condition'	Yes: 'only a handful needed to be felled' –Tree Preservation Orders sought for many of them	Nearly 400 trees implicated
32	Residents oppose tree felling plan	*Chelmsford Weekly News*, September 28, 2009	Galleywood Common, Chelmsford, Essex	Felling needed for road-widening	Yes: 'mature trees' are going to be completely destroyed	Mature oak (*Quercus* spp.) and other species
33	Aberdeen City Council: Gray Street closed during tree-felling work	*M2 PressWIRE*, September 28, 2009	Aberdeen	Felling will cause street closure	NK	–
34	250 sign petition to stop cliff tree felling	*North Devon Journal*, September 17, 2009	Torridge, Devon	Felling near to a manor, development related	Yes: 250 people signed a petition against the felling	–

35	Tree felling to proceed	*The Courier Mail* (Australia), September 8, 2009	Toowoomba, Queensland, Australia	Felling for road-widening	Yes: locals 'stunned' and 'prepared to take action'	–
36	Outraged at trust's planned tree felling	*Leatherhead Advertiser*, August 20, 2009	Boxhill, Dorking	Felling for restaurant and car park extension	Yes: letter written	–
37	Tree felling plan backed	*Morpeth Herald*, August 7, 2009	Morpeth, Northumberland	Felling to prevent damage to wall	NK	Sycamore (*Acer pseudoplatanus*), ash (*Fraxinus excelsior*) and two conifers
38	Upset over tree felling in estate's upgrading programme	*The Straits Times* (Singapore), July 31, 2009	Singapore	Felling: 'massive tree felling exercise' which is 'questionable'	Yes: letter written in which respondent questions replacement of 'mature trees' 90 per cent of which 'were healthy'	mature trees
39	Homebush Bay tree felling for V8 Supercars gets black flag	*Sydney Morning Herald* (Australia), July 31, 2009, Third Edition	Homebush Bay, Sydney, Australia	Felling to enable V8 Supercar race to go ahead. Removal of 147 mature trees, replacement with 440 trees. The event 'needed a wide straight without trees'.	Yes: complaint about removing the hard won 'feeling of nature [in] urban areas'	147 'mature trees' removed – planted prior to games in 2000
40	Council delay over tree felling is 'lousy service'; A 'nuisance tree' in Crediton is still standing, six months after Mid Devon District Council agreed to remove it.	*Mid Devon Gazette*, July 28, 2009	Crediton, Devon, UK	Felling of 'nuisance tree'	No: complaint about delay in felling –	

continued

Appendix 34.1 continued

41	Tree felling makes M20 noise worse	*Kent Messenger*, July 24, 2009	Aylesford, Kent, UK	Felling of row of trees after sound barrier installed along motorway	Yes: residents 'fought' to save trees, thought that success was achieved, then they were felled	–
42	Bristol council officer's tree-felling ban challenge fails	*Bristol Evening Post*, July 22, 2009	Bristol, Avon, UK	Felling of trees adjacent to Blackberry Hill	Y/N: Yes: 100 objections received No: tree officer grants permission for trees to be felled	–
43	Police appeal over £2,500 tree felling	*Penarth Times*, July 11, 2009	Penarth, Wales	Felling –Illegal, done overnight. Trees worth estimated £2,500. Police investigating	NK	30-year-old oak (*Quercus* sp.) and Ash (*Fraxinus excelsior*). 12 trees felled
44	Council blames flying foxes for tree felling	*Newcastle Herald* (Australia), July 6, 2009	Singleton, Upper Hunter, Australia	Trees due to be felled 'plagued' with a protected species of flying fox: now dead or a 'risk to the public'	NK	At least 10 *Jacaranda* spp. and pine (*Pinus* spp.) trees. Some trees 'well over' 100 years old
45	Remind home owners of rule banning tree felling	*The Straits Times* (Singapore), July 6, 2009	Singapore	Felling of 3 trees results in fine of $6,000 to homeowner	No: letter writer expresses empathy with homeowners' case for tree felling	–
46	Lawyers condemn tree-felling	*The Statesman* (India), June 17, 2009	Jalpaiguri, India	Felling of trees in the court compound, some planted by the judiciary	Yes: 'mindless' cutting condemned, but who authorised it remains a mystery	Young trees: 5 years old. Included several Sishu and Debdaru (*Polyalthia* sp.)
47	Port tree-felling concern	*Aberdeen Press and Journal*, June 6, 2009	Invergordon, UK	Felling resumed as 'part of the landscaping at the entrance to the service base'	Yes: are birds nesting? Following a survey continuance of felling was given the okay	–

48	Tree-felling at cemetery called act of vandalism	*The Daily News* (New Plymouth, New Zealand, June 6, 2009	Omata, New Zealand	Felling to create space for new carpark in cemetery	Yes: act of 'vandalism', 'intimacy' lost, and complaints about non-consultation	30 Matipo trees felled
49	Resident's anger over tree-felling	*Kentish Gazette*, June 4, 2009	Faversham, Kent, UK	Felling –proposed – of 'big' trees by supermarket	Yes: one resident managed to save two of them, but says, of the remainder: 'it will take a generation for them to come back'	10 trees, some 'very big' and have been here for 'some time'
50	Inquiry into tree felling on Sudbury estate	*Suffolk Free Press*, May 28, 2009	Sudbury, Suffolk, UK	Felled trees on new housing estate, seemingly without approval	Yes: residents 'furious' and town council 'feels exactly the same'	–

Notes

1 This listing is from 17 February 2010 to 28 May 2009 in reverse sequential order using Nexis® UK 'all English news' database, and searching for 'tree felling'. The download took place on 18 February 2010. All material presented is based on the articles. A search through the first 124 on the database revealed 50 concerning street trees, which are listed above. Circa 40 were repetitions or variants of previous articles. Therefore street trees represented 50 out of 80 or around 60 per cent of the news content.

2 'Y/N' indicates 'Yes' and 'No' in different interest groups, 'NK' means Not Known.

References

Anon (2007) *Chainsaw Massacre: A Review of London's Street Trees*, London: Greater London Assembly, online, available at: http://legacy.london.gov.uk/assembly/envmtgs/2007/envjun12/item09a.pdf.

Anon (2009a) Internet download: www.bristolstreettrees.org/Information.html.

Anon (2009b) 'Police appeal over £2,500 tree felling', *Penarth Times*, 11 July 2009. [See Appendix 1: example 43].

Armstrong, D. (2000) 'A survey of community gardens in upstate New York: Implications for health promotion and community development', *Health and Place*, 6(4): 319–27.

Barker, P.A. (1976) *Planting strips in street Rights-of-Way: a key public resource. Trees and Forests for Human Settlements*, Toronto: University of Toronto.

Bass, S. (2001) 'The importance of social values', in J. Evans (ed.), *The Forests Handbook, Volume 1*, Oxford: Blackwell Science, pp. 363–71.

Beckett, K.P., Freer-Smith, P.H. and Taylor, G. (2000) 'The capture of particulate pollution by trees at five contrasting urban sites', *Arboricultural Journal*, 24(2–3): 209–30.

Benjamin, M.T. and Winer, A.M. (1998) 'Estimating the ozone forming potential of urban trees and shrubs', *Atmospheric Environment*, 32(1): 53–68.

Biddle, P.G. (1998a) *Tree Root Damage to Buildings. Volume 1. Causes, Diagnosis and Remedy*, Wantage: Willowmead Publishing.

Biddle, P.G. (1998b) *Tree Root Damage to Buildings. Volume 2. Patterns of Soil Drying in Proximity to Trees on Clay Soils*, Wantage: Willowmead Publishing.

Booth, R. and Montague, B. (2007) 'Timber! Councils fell urban trees', *The Sunday Times*, 5 August, online, available at: www.timesonline.co.uk/tol/news/uk/article2199239.ece.

Brezina, E. and Schmidt, W. (1937) *Das künstliche Klima in der Umgebung des Menschen*, Stuttgart: Ferdinand Enke Verlag, pp. 196–9.

Bright, I., Hesch, R., Bentley, N. and Parrish, S. (2001) 'A study of the potential for developing a biomass fuel supply from tree management operations in London', *Arboricultural Journal*, 25(3): 255–88.

Britt, C. and Johnston, M. (eds) (2008) *Trees in Towns II. A New Survey of Urban Trees in England and Their Condition and Management. Research for Amenity Trees No. 9*, London: Department for Communities and Local Government.

Brown, M. (1983) 'Design of planting and paved areas and the role in the city', in A.B. Grove and A.B Cresswell (eds), *City Landscape: A contribution to the Council of Europe's European Campaign for Urban Rennaissance*, London: Butterworths, pp. 87–124.

Bruijnzeel, L.A. (2001) 'Forest hydrology', in J. Evans (ed.), *The Forests Handbook, Volume 1*, Oxford: Blackwell Science, pp. 301–43.

Burls, A. (2008) 'Meanwhile wildlife gardens, with nature in mind (DVD presentation)', in G. Dawe and A. Millward (eds), *'Statins and Greenspaces': Health and the Urban Environment*, London: UK MAB Urban Forum, pp. 36–9.

Burls, A. and Caan A.W. (2004) 'Social exclusion and embracement: a helpful concept?' *Primary Health Care Research and Development*, 5(3): 191–2.

Catena, A. (2003) 'Thermography reveals hidden tree decay', *Arboricultural Journal*, 27(1): 27–42.

Cavanagh, J.-A.E. and Clemons, J. (2006) 'Do urban forests enhance air quality?' *Australasian Journal of Environmental Management*, 13(2): 120–30.

Chandler, T.J. (1965) *The Climate of London*, London: Hutchinson and Co.

Comery, W.R. (2000) 'Tree roots versus sidewalks and sewers', in J.E. Kuser (ed.), *Handbook of Urban and Community Forestry in the Northeast*, New York: Kluwer Academic, pp. 227–42.

Coppin, N.J. and Richards, I.G. (1990) *Use of Vegetation in Civil Engineering*, London: Construction Industry Research and Information Association (CIRIA) and Butterworths.

Cowan, A. and Harbinson, C. (2009) 'Stemming the tide: urban tree decline, revelations from Trees in Towns II', *Essential* ARB, online, available at: http://frontpage.woodland-trust.org.uk/ancient-tree-forum/atfnews/images/TreesInTowns.pdf [accessed 14 May 2010].

Dawe, G.F.M. and Millward, A. (eds) (2008) *'Statins and Greenspaces': Health and the Urban Environment*. London: UK MAB Urban Forum, online, available at: www. ukmaburbanforum.co.uk/Downloadable%20Pubs/StatinsandGreenspaces.pdf.

Dewers, R.S. (1981) 'Evaluation of native and exotic woody plants under severe environmental stress', *Journal of Arboriculture*, 7(11): 299–302.

Dochinger, L.S. (1980) 'Interception of airborne particles by tree plantings', *Journal of Environmental Quality*, 9(2): 265–8.

Donovan, G.H. and Butry, D.T. (2010) 'Trees in the city: valuing street trees in Portland, Oregon', *Landscape and Urban Planning*, 94(2): 77–83.

Douglas, I. (2008) 'Psychological and mental health benefits from nature and urban greenspace', in G. Dawe and A. Millward (eds), *'Statins and Greenspaces': Health and the Urban Environment*, London: UK MAB Urban Forum, pp. 12–22.

Dreistadt, S.H. and Dahlsten, D.L.(1986) 'A problem-prone street tree – a case study of the Tulip Tree in Berkeley, California', *Journal of Arboriculture*, 12(6): 146–9.

Dyer, S.M. and Mader, D.L. (1986) 'Declined urban sugar maples: growth patterns, nutritional status and site factors', *Journal of Arboriculture*, 12(1): 6–13.

Elmendorf, W. (2008) 'The importance of trees and nature in community: a review of the relative literature', *Arboriculture and Urban Forestry*, 34(3): 152–6.

Evans, P. (2002a) 'A night of dark trees', *Arboricultural Journal*, 26(3): 249–56.

Evans, P. (2002b) 'The care of trees in the built environment', *Arboricultural Journal*, 26(3): 257–61.

Federer, C.A. (1971) 'Effects of trees in modifying urban microclimate', in *Trees and Forests in an Urbanizing Environment: University of Massachusetts Cooperative Extension Service Monograph No. 17*, pp. 23–8.

Fowler, H.G. and Parrish, M.D. (1982) 'Urban shade trees and carpenter ants', *Journal of Arboriculture*, 8(11): 281–4.

Frank, S., Waters, G., Beer, R. and May, P. (2006) 'An analysis of the street tree population of Greater Melbourne at the beginning of the 21st century', *Arboriculture and Urban Forestry*, 32(4): 155–63.

Gasson, P.E. and Cutler, D.F. (1998) 'Can we live with trees in our towns and cities?' *Arboricultural Journal*, 22(1): 1–10.

Green, E. (2002) 'Arborists should have a central role in educating the public about veteran trees', *Arboricultural Journal*, 26(3): 239–48.

Grey, G.W. and Deneke, F.J. (1978) *Urban Forestry*, New York: Wiley.

Guite, H.F., Clark, C. and Ackrill, G. (2006) 'The impact of the physical and urban environment on mental wellbeing', *Public Health*, 120(12): 1117–26.

Handley, J. (1983) 'Nature in the urban environment', in A.B. Grove and R.W. Cresswell (eds), *City Landscape: A contribution to the Council of Europe's European Campaign for Urban Rennaissance*, London: Butterworths, pp. 47–59.

Hart, E.R., Miller, F.D. and Bastian, R.A. (1986) 'Tree location and winter temperature influence on mimosa webworm populations in a northern urban environment', *Journal of Arboriculture*, 12(10): 237–40.

Hauer, R.J., Miller, R.W. and Oumet, D.M. (1994) 'Street tree decline and construction damage', *Journal of Arboriculture*, 20(2): 94–7.

Heimlich, J., Sydnor, D., Bumgardner, M. and O'Brien, P. (2008) 'Attitudes of residents toward street trees on four streets in Toledo, Ohio, U.S. before removal of Ash trees (*Fraxinus* spp.) from Emerald Ash Borer (*Agrilus planipennis*)', *Arboriculture and Urban Forestry*, 34(1): 47–53.

Helliwell, R. (1967) 'The amenity value of trees and woodlands', *Journal of the Arboricultural Association*, 1: 128–31.

Hiesler, G.M. and Herrington, L.P. (1976) 'Selection of trees for modifying metropolitan climates', in *Better Trees for Metropolitan Landscapes*, Radnor, PA: USDA-Forest Service, Northeastern Forest Research Station, General Technical Report NE-22, pp. 31–7.

Hough, M. (1984) *City Form and Natural Process: Towards a New Urban Vernacular*, London: Croom Helm.

Jarvis, P.G. and Fowler, D. (2001) 'Forests and the atmosphere', in J. Evans (ed.), *The Forests Handbook. Volume 1*. Oxford: Blackwell Science, pp. 229–81.

Johnson, D. (2007) 'Chair's foreword', in Anon, 2007.

Kalmbach, K.L. and Kielbaso, J.J. (1979) 'Resident attitudes toward selected characteristics of street tree planting', *Journal of Arboriculture*, 5(6): 124–9.

Kohlmaier, G.H., Weber, M. and Houghton, R.A. (eds) (1998) *Carbon Dioxide Mitigation in Forestry and Wood Industry*, Berlin and Heidelberg: Springer.

Laband, D.N. and Sophocleus, J.P. (2009) 'An experimental analysis of the impact of tree shade on electricity consumption', *Arboriculture and Urban Forestry*, 35(4): 197–202.

Lacán, I. and McBride, J.R. (2009) 'City trees and municipal Wi-Fi networks: compatibility or conflict?' *Arboriculture and Urban Forestry*, 35(4): 203–10.

Landsberg, H.E. (1981) *The Urban Climate*, New York: Academic Press.

Land Use Consultants (1994) *Urban Tree Strategies. Research for Amenity Trees No. 3*, London: Department of the Environment.

Marion, P.D. (1981) 'Norway Maple decline', *Journal of Arboriculture*, 7(2): 38–42.

Matthews, R., Mottram, N. and Ward, K. (1993) *Action for London's Trees: Investing in a Leafy Capital*, Walgrave: Task Force Trees/Countryside Commission, CCP 433.
Miller, R.W. (1997) *Urban Forestry: Planning and Managing Urban Greenspaces*, New Jersey: Prentice Hall.
Mitchell, A. (1996) *Alan Mitchell's Trees of Britain*, London: HarperCollins.
More, D. and White, J. (2003) *Cassell's Trees of Britain and Northern Europe*, London: Cassell.
Morgan, D.L. (2003) 'Lovely, lonesome lindens', *American Nurseryman*, April 15, pp. 20–2.
National House Building Council (NHBC) (1974) *Practice Note 3: Root Damage by Trees*, London: NHBC.
Neilan, N. (2008) *Capital Asset Value for Amenity Trees (CAVAT): Full Method, Users' Guide*, London: London Tree Officers' Association, online, available at: www.ltoa.org.uk.
Nicoll, B.C. and Armstrong, A. (1998) 'Development of *Prunus* root systems in a city street: pavement damage and root architecture', *Arboricultural Journal*, 22(3): 259–70.
Nilsson, K., Randrup, T.B. and Wandall, B.M. (2001) 'Trees in the urban environment', in J. Evans (ed.), *The Forests Handbook, Volume 1*, Oxford: Blackwell Science, pp. 347–61.
Nowak, D.J. and Crane, D.E. (2002) 'Carbon storage and sequestration by urban trees in the USA', *Environmental Pollution*, 116(3): 381–9.
Nowak, D.J., Hoehn, R. and Crane, D.E. (2007) 'Oxygen production by urban trees in the United States', *Arboriculture and Urban Forestry*, 33(3): 220–6.
Ouis, D. (2001) 'Detection of rot in standing trees by means of an acoustic technique', *Arboricultural Journal*, 25(2): 117–52.
Parker, J. (1989) 'The impact of vegetation on air conditioning consumption', in *Controlling Summer Heat Islands: Proceedings of the Workshop on Saving Energy and Reducing Atmospheric Pollution by Controlling Summer Heat Islands*, Berkeley: University of California, pp. 45–52.
Parker, J. (2001) 'Social movements and science: the question of plural knowledge systems', in J. López and G. Potter (eds), *After Postmodernism: An Introduction to Critical Realism. Continuum Studies in Critical Theory*, London: Athlone Press, pp. 251–9.
Pitt, D., Soergell II, K. and Zube, E. (1979) 'Trees in the city', in I.C. Laurie (ed.), *Nature in Cities*, Chichester: Wiley, pp. 205–29.
Plummer, B. and Shewan, D. (1992) *City Gardens: An Open Spaces Survey in the City of London*, London: Belhaven Press.
Popoola, L. and Ajewole, O. (2001) 'Public perceptions of urban forests in Ibadan, Nigeria: implications for environmental conservation', *Arboricultural Journal*, 25(1): 1–22.
Potts, D.F. and Herrington, L.P. (1982) 'Drought resistance adaptations in urban honeylocust', *Journal of Arboriculture*, 8(3): 75–80.
Radd, A. (1976) 'Trees in towns and their evaluation', *Arboricultural Journal*, 3(1): 2–26.
Raison, R.J., Brown, A.G. and Flinn, D.W. (2001) *IUFRO Research Series 7: Criteria and Indicators for Sustainable Forest Management*, Wallingford: CABI Publishing.
Randrup, T. (2000) 'Occurrence of tree roots in Danish municipal sewer systems', *Arboricultural Journal*, 24(4): 283–306.
Reynolds, E.R.C. (1980) 'Tree roots and buildings, 7: Tree roots and built developments – a review', *Arboricultural Journal*, 4(1): 31.
Reynolds, E.R.C. and Alder, D. (1980) 'Trees and buildings, 6: A matter of opinion – settlement or heave of houses and the role of tree roots', *Arboricultural Journal*, 4(1): 24–30.
Richardson, J. (2002) 'High hedges – new laws', *Arboricultural Journal*, 26(1): 55–64.
Roszak, T. (1996) 'The nature of sanity (mental health and the outdoors)', *Psychology Today*, 29(Jan./Feb.): 22–6.
Sanesi, G., Lafortezza, R., Bonnes, M. and Carrus, G. (2006) 'Comparison of two different approaches for assessing the psychological and social dimensions of green spaces', *Urban Forestry & Urban Greening*, 5(3): 121–9.
Schroeder, H.W. (1987) 'Psychological value of urban trees: measurement, meaning and imagination', in A.F. Phillips and D.J. Gangloff (eds), *Proceedings of the Third National Urban Forestry Conference*, Orlando, FL.
Schroeder, H., Flannigan, J. and Coles, R. (2006) 'Residents' attitudes toward street trees in the UK and U.S. communities', *Arboriculture and Urban Forestry*, 32(5): 236–46.
Schwarze, F.W.M.R. (2001) 'Development and prognosis of decay in the sapwood of living trees', *Arboricultural Journal*, 25(4): 321–37.
Shashua-Bar, L. and Hoffman, M.E. (2000) 'Vegetation as a climatic component in the design of an urban street: an empirical model for predicting the cooling effect of urban green areas with trees', *Energy and Buildings*, 31(3): 221–35.

Souch, C.A. and Souch, C. (1993) 'The effect of trees on summertime below canopy urban climates: a case study Bloomington, Indiana', *Journal of Arboriculture*, 19(5): 303–12.

Spray, M. (2002) 'What are trees for?' *Arboricultural Journal*, 26(3): 263–79.

Stål, Ö. (1998) 'The interaction of tree roots and sewers: the Swedish experience', *Arboricultural Journal*, 22(4): 359–67.

Swaid, H. (1993) 'Urban climate effects of artificial heat sources and ground shadowing by buildings', *International Journal of Climatology*, 13(7): 797–812.

Tratalos, J., Fuller, R.A., Warren, P.H., Davies, R.G. and Gaston, K.J. (2007) 'Urban form, biodiversity potential and ecosystem services', *Landscape and Urban Planning*, 83(4): 308–17.

Tschantz, B.A. and Sacamano, P.L. (1994) *Municipal Tree Management in the United States*, Kent, OH: Davey Tree Expert Co.

Ulrich, R.S. (1984) 'View through a window may influence recovery from surgery', *Science*, 224(4647): 420–1.

United States National Arboretum (2009) *Ulmus americana* 'Valley Forge' *and* 'New Harmony', online, available at: www.usna.usda.gov/Newintro/american.html.

Wolf, K.L. and Bratton, N. (2006) 'Urban trees and traffic safety: considering U.S. roadside policy and crash data', *Arboriculture and Urban Forestry*, 32(4): 171–9.

Yang, J., McBride, J., Zhou, J. and Sun, Z. (2005) 'The urban forest in Beijing and its role in air pollution reduction', *Urban Forestry & Urban Greening*, 3(2): 65–78.

35
Urban gardens and biodiversity

Kevin J. Gaston and Sian Gaston

Introduction

Domestic or private gardens/yards (those associated with domestic dwelling) make highly variable contributions to the greenspace of urban areas. In many towns and cities, particularly in much of the developing world, they are often small in size, small in number (relative to the number of dwellings), and/or given over almost entirely to impermeable surfaces and largely devoid of life. However, in substantial parts of western Europe, North America, Australia, and elsewhere, domestic gardens are frequently more substantial, numerous, and in large part vegetated. Here they may comprise high proportions of urban greenspace. Thus, for example, such gardens have variously been estimated to make up 16.4 percent of the central parts of Stockholm County, Sweden, and *c*.25 percent (UK) and 36 percent (New Zealand) of the overall surface area, and >35 percent (UK) and 96 percent (Nicaragua) of the greenspace, of individual cities (Gaston *et al.* 2005a; Colding *et al.* 2006; Loram *et al.* 2007; Mathieu *et al.* 2007; González-García and Sal 2008). The vast differences in the number and coverage of domestic gardens across the globe reflect geographic, historical, cultural, and socioeconomic factors.

Where the coverage of domestic gardens and their contribution to greenspace is high they have the potential to make a substantial impact on the biodiversity of urban areas, and may for many people provide the primary (and perhaps sole) arena for interactions with wildlife (at a time when such interactions are in global decline). Where the coverage of domestic gardens is low they may still be important, particularly if they continue to contribute substantially to urban greenspace, or provide resources that are otherwise lacking, including linkages between other kinds of greenspace. In this chapter we provide a broad overview of the relationship between urban gardens and biodiversity.

Whilst an attempt has been made to identify genuinely general principles, and draw on a wide range of examples, there is an inevitable bias towards the developed regions (and particularly Britain) on which the majority of the literature is focused. At the very least this highlights the pressing need for studies of the ecology of domestic gardens in a much wider spectrum of countries, environments and cultures.

Biodiversity of gardens

Although there are some older works (e.g. Barnes and Weil 1944, 1945; Morley 1944; Guichard and Yarrow 1948; Tutin 1973; Mathias 1975), it is only relatively recently that much attention has been paid to the biodiversity that is associated with domestic gardens. Studies tend to have taken two forms. The first have investigated the biodiversity associated with individual gardens (e.g. Owen 1991; Miotk 1996). This has had the advantage of providing often very detailed information. However, the gardens chosen have arguably been somewhat atypical, and thus the generalizations that can be drawn have been rather limited. The second set of studies has investigated the biodiversity associated with multiple gardens, sometimes using volunteers to help assemble data for hundreds or even thousands of sites (e.g. Davis 1979; Saville 1997; Thompson *et al.* 2003, 2004; Cannon *et al.* 2005; Daniels and Kirkpatrick 2006; Smith *et al.* 2006a, 2006b, 2006c; Fetridge *et al.* 2008; Loram *et al.* 2008a). Such work has identified broad patterns and quantified the aggregate contribution of different numbers of gardens. But, this has inevitably been at the cost of the depth of information that could be gathered, resulting in a focus often on just a single or very few higher taxa (most commonly birds).

Obviously, most progress will be achieved through a combination of the two approaches, particularly where detailed studies are conducted of gardens that are also included in broader analyses.

Species richness

Contrary to earlier assertions that domestic gardens were akin to biological deserts (Elton 1966), studies have revealed the occurrence of a surprisingly high diversity of species in gardens, including those in urban settings. Perhaps most famously, Owen (1991) documented the occurrence of a minimum of 2204 species in a single garden in Leicester (UK) between 1972–86, including 422 species of plants, 1602 insects, 121 other invertebrates, and 59 vertebrates. In a similar vein, but without particularly intensive sampling, Miotk (1996) recorded more than 700 species of animals in a domestic garden on the edge of a village in Germany.

Other studies have documented:

- i 159 vascular plant species in the lawns of 52 urban gardens (UK; Thompson *et al.* 2004) – obviously a high proportion of plants (both native and alien) in domestic gardens are planted;
- ii 1166 vascular plant species (30 percent native), *c.*80 lichen species, 68 bryophyte species (in samples) in 61 urban gardens (UK; Gaston *et al.* 2004);
- iii 293 vascular plant species (38 percent native) in 96 urban garden patios (Venezuela; González-García and Sal 2008);
- iv 110 bee species in 21 urban gardens (US; Fetridge *et al.* 2008);
- v 54 leaf-mining insect species in 56 urban gardens (UK; Smith *et al.* 2006c);
- vi 40 bird species in 214 urban back and front gardens (Australia; Daniels and Kirkpatrick 2006).

Comparisons between the species richness of domestic gardens and other kinds of land cover are surprisingly scarce. However, Thompson *et al.* (2003) and Loram *et al.* (2008a) showed that species accumulation curves for native floras in gardens in the UK were, for a given area, not dissimilar from those for a range of other land covers. More generally, along rural-urban gradients, the species richness of some major taxa (particularly plants and birds) commonly exhibits a peaked pattern (e.g. Tratalos *et al.* 2007; McKinney 2008; Sanford *et al.* 2008). Where rural areas have low productivity or have experienced marked habitat modification, particularly as a consequence of intensive agriculture, species richness is often low, rises towards suburban areas, and then declines again with increasing levels of urbanization. The peak in suburban areas results in large part from greater habitat heterogeneity, much of which is contributed by domestic gardens (and the often complex patterns of land cover within them).

In some regions at least, a substantial proportion of species, particularly of plants, in domestic gardens may be alien in origin (and often planted; e.g. see Smith *et al.* 2006b; Loram *et al.* 2008a; Marco *et al.* 2008). The implications that this has for native biodiversity in other groups remain poorly explored. Whilst it is clear that gardens with high numbers of alien plant species may none the less contain large numbers of native animal species, and these may use alien plants for resources, there is some evidence both that biodiversity is nonetheless suppressed and that less use is made of these plants (e.g. Day 1995; French *et al.* 2005; Daniels and Kirkpatrick 2006; Burghardt *et al.* 2008).

Populations

Whilst large numbers of native species may be recorded in an individual domestic garden, it is clear that only a rather small proportion of these could maintain viable populations therein. The importance of gardens for populations thus lies in the aggregate numbers of individuals that they can sustain through part or the entirety of their life cycles, or the contribution that they make to maintaining populations that may depend equally, or more heavily, on other habitat types.

Species use domestic gardens for a wide variety of reasons. Amongst the more sedentary, excluding vagrants, it is a matter of suitable habitat being available for fulfilling their life cycles (or recolonization from elsewhere). Amongst the more mobile species, gardens may be used as sources of water, food, shelter, breeding sites, and wintering sites. This may involve moving into individual gardens (or indeed any gardens) on a daily, seasonal or annual basis. Thus, for example, the birds occurring in the city of Morelia (Mexico) include some that mostly live in the oak-pine forest outside and travel into the city to use parks and really large gardens (e.g. Gray silky-flycatcher *Ptilogonys cinereus*, Red warbler *Ergaticus ruber*, Slate-throated redstart *Myioborus miniatus*, Elegant euphonia *Euphonia elegantissima*, Rufous-capped warbler *Basileuterus rufifrons*, Cassin's vireo *Vireo cassinii*), one that invades the city during the breeding season (Canyon wren *Catherpes mexicanus*), and others that occur in residential areas year-round (e.g. Sharp-shinned hawk *Accipiter striatus*, Berylline hummingbird *Amazilia beryllina*, Spotted wren *Campylorhynchus gularis*, Blue mockingbird *Melanotiscae rulescens*, Blue grosbeak *Passerina caerulea*, Swainson's thrush *Catharus ustulatus*, Greater peewee *Contopus pertinax*, Common raven *Corvus corax*, Rufous-backed robin *Turdus rufopalliatus*) (J.E. Schondube, I. MacGregor-Fors, and L. Morales-Pérez, pers. comm.)

The majority of the organisms in urban gardens tend to belong to species that until recently at least have been abundant and widespread. Particularly in developed regions that have experienced lengthy periods of extensive and intensive agriculture many such species are in broad decline (e.g. Donald *et al.* 2001; Murphy 2003; Conrad *et al.* 2004, 2006; Gregory *et al.* 2004). Here, a high proportion of the remaining populations may in some cases apparently reside in urban areas, and especially suburban gardens, and their numbers may be more stable or declining at a slower rate (e.g. in the UK potential examples include bumblebees, the Common Frog *Rana temporaria*, Song Thrush *Turdus philomelos*, and Hedgehog *Erinaceus europaeus*; Swan and Oldham 1993; Doncaster 1994; Gregory and Baillie 1998; Mason 2000; Carrier and Beebee 2003; Osborne *et al.* 2008).

More generally, species of conservation concern, and rare and restricted species can occur in urban gardens, particularly in regions in which much natural and semi-natural habitat has been lost, and that which remains has become highly fragmented (e.g. Owen 1991; Cannon 1999; Burghardt *et al.* 2008). However, the role of domestic gardens in this regard remains poorly understood, and a situation that could usefully be resolved in many regions.

Finally, like many other land uses, urban gardens almost certainly remain home to much as yet entirely undocumented biodiversity. Even in regions which have been relatively well studied, undescribed species can still be found. For example, moss on one tarmac garden path in Sheffield (UK) supported what is likely a new, miniscule species of *Macentina* lichen (Gilbert 2001).

Factors influencing garden biodiversity

Although doubtless rooted in general ecological principles, understanding of the factors which influence levels of biodiversity in gardens remains in its infancy, and has focused on a rather small range of global variation in garden form and context. The factors which have been identified to date include some that are intrinsic to individual gardens (i.e. are characteristics of the gardens themselves) and others that are extrinsic (i.e. that characterize the broader context in which the garden lies).

Intrinsic factors

Area

Garden size can vary substantially both between and within regions. Although it is not clear how representative these samples are in each case, Rapoport (1993), reporting the results of surveys of the numbers of cultivated plants and weeds in gardens, gives garden areas of $11.9 \pm 25.7\,m^2$ (mean $\pm$ SD) for London (n = 65), $17.4 \pm 6.2\,m^2$ for Szczecin (n = 8), $35.8 \pm 54.4\,m^2$ for Mexico City (n = 19), $40.8 \pm 16.1\,m^2$ for Warsaw (n = 9), $53.4 \pm 133.3\,m^2$ for Buenos Aires (n = 21), $55.3 \pm 86\,m^2$ for Gdansk (n = 7), and $115.4 \pm 259.9\,m^2$ for six cities in Poland (n = 34). Other estimates include means of $155.4–253.0\,m^2$ for urban gardens in each of five cities in the UK (Loram *et al.* 2007), $1000–2000\,m^2$ for Hamilton (New Zealand; Day 1995), and $500–600\,m^2$ for Hobart (Australia; Daniels and Kirkpatrick 2006).

Larger gardens typically comprise greater areas of less formal land cover types, and a greater diversity of land cover types (Smith *et al.* 2005; Loram *et al.* 2008b). In consequence, biodiversity within gardens exhibits classical species-area relationships, with species numbers increasing with garden size (e.g. Rapoport 1993; Daniels and Kirkpatrick 2006; Smith *et al.* 2006b; Loram *et al.* 2008a; van Heezik *et al.* 2008). This is typically true both of those components that have been intentionally introduced (i.e. planted flora) and those that have not. The occurrence of particular species within gardens may also frequently be correlated with their size (e.g. Chamberlain *et al.* 2004; Daniels and Kirkpatrick 2006; Baker and Harris 2007; González-García *et al.* 2009).

The typical size of domestic gardens is likely to be increasing in rather few regions. It is, however, certainly decreasing in some, with pressure for residential dwellings resulting in increased housing densities, infilling and backland developments (Syme *et al.* 2001; Gaston and Evans 2010). Such changes will reduce the biodiversity associated with individual gardens. This trend will be further exacerbated in some regions by the increased hard surfacing of garden space for off-road parking (Alexander 2006).

Vegetational complexity

At least in temperate regions, one of the best predictors of the animal biodiversity occurring in gardens has been found to be the structural complexity of the vegetation that they contain (Day 1995; Daniels and Kirkpatrick 2006; Smith *et al.* 2006c; van Heezik *et al.* 2008; González-García *et al.* 2009). This may be reflected in direct measures of such complexity or, say, in the occurrence, numbers, height, and the clumping of trees that are present.

Age

In regions in which the creation of new housing developments tends to involve stripping the landscape of most if not all pre-existing vegetation and other biodiversity features, the age of a domestic garden may play a significant role in predicting the level of biodiversity that it contains. Thus, relationships between the species richness of particular higher taxa and age have been documented on some occasions (e.g. Smith *et al.* 2006c).

Extrinsic factors

Greenspace coverage

At least for more mobile taxa, a major determinant of the biodiversity occurring in individual gardens is the overall extent of coverage by greenspace in the surrounding area (e.g. Chamberlain *et al.* 2004; Smith *et al.* 2006c; Baker and Harris 2007). This may include rural land covers (farmland, woodland, etc.), parks, green landscaping, road verges, and, of course, other domestic gardens. The most speciose gardens are typically those surrounded by large amounts of such greenspace. Equally, domestic gardens may be important in providing habitat linkages between other kinds of greenspace (e.g. Rudd *et al.* 2002), and their proximity may enhance species richness in other greenspaces (Chamberlain *et al.* 2007).

Habitat and resource occurrence

In a similar vein, the biodiversity of individual gardens may often be higher in those which are surrounded by a high density of a particular key habitat or resource. For example, the numbers of species recorded from ponds in urban gardens has been found to be positively correlated with the density of ponds in the surrounding landscape (Gledhill *et al.* 2008). This effect may be enhanced by the tendency for marked spatial autocorrelation in gardening practices as a consequence of societal and cultural pressures to conformity. Equally, these forces mean that key habitats and resources may not normally become more widespread if they are already scarce in the gardens of a region. However, they may potentially be harnessed to generate neighborhood level changes in garden structures which directly benefit biodiversity (Warren *et al.* 2008).

Gardening for biodiversity

In some regions of the world the principal objective of managing semi-natural and natural biota in gardens is to ensure its eradication (especially where it may include species that are directly or indirectly hazardous or at least perceived as such; e.g. mosquitoes, scorpions, snakes, rodents). Cannon (1999) termed this 'ungardening', and in global terms it is doubtless widespread.

In other regions, even in urban areas, domestic gardens are widely used to grow fruit and vegetables for consumption or sale. Here, the biodiversity of semi-natural and natural biota is often also low, particularly where it is perceived as competing with or preying upon these products and is treated accordingly.

Finally, in yet other regions, biota, and especially vegetation is predominantly managed with a view to its amenity value. In such regions, however, the prevalence of gardening culture has also been accompanied by the development of the notion of actively gardening for biodiversity (or at least components thereof). Indeed, 50 percent or more of garden owners may participate in some form of wildlife gardening, although much of this participation takes the form simply of feeding wild birds, and the percentages may be inflated by those who do so rather infrequently (e.g. Lepczyk *et al.* 2004; Gaston *et al.* 2007a; Davies *et al.* 2009). Such management may be sufficient markedly to reduce some of the negative influences of small garden areas and low coverage of greenspace in surrounding neighborhoods.

In some parts of the world there is a large and rich vein of popular media (including books, magazine articles, radio and television programs, web sites) suggesting simple changes that can be made to improve the value of domestic gardens for biodiversity (e.g. Gibbons and Gibbons 1988; Baines 2000; Moss and Cottridge 1998; Packham 2001). Indeed, in some cases the take up of wildlife gardening has increased perhaps partly as a result (e.g. Chamberlain *et al.* 2005). Unfortunately, the origins of many of these recommendations are often obscure or unclear, and evidence of their effectiveness remains largely anecdotal or based solely on observational data, with the associated difficulties of determining causality. Observational studies and replicated experiments to evaluate the frequency with which particular management actions have a measurable influence on garden biodiversity on an appropriate time scale are largely wanting. However, those that have been conducted have principally reported positive effects (Gaston *et al.* 2005b; Daniels and Kirkpatrick 2006; Baker and Harris 2007; Fuller *et al.* 2008).

There is little likelihood that the majority of explicitly wildlife gardening actions have a marked local negative or otherwise undesirable impact (Gaston *et al.* 2007b). However, at least under some circumstances, there are some issues of potential concern, and which warrant much further work than has been done to date. These include the role of domestic gardens as potential sources of alien species in the wider landscape, the potential for alien populations to increase at the expense of local populations of native species, the potential for genetic contamination of wild populations from garden populations, and the impact of the production or extraction of resources for wildlife gardening in the regions from which these are obtained (e.g. Groves *et al.* 2005; Whelan *et al.* 2006; Dehnen-Schmutz *et al.* 2007a, 2007b).

In conclusion

Domestic gardens can make important contributions to urban biodiversity, through the provision of greenspace and resources. Depending on the region and taxon of concern, they may variously provide additions to the space and resources provided by other greenspaces (including through enhancing connectivity), key habitat and other resources at particular times of day, season or year, or space and resources for resident individuals. These influences can be improved both through planning policy which determines the broad structure of domestic gardens, and through the management of individual spaces which determine how well they realize their potential.

Acknowledgments

We are grateful to M.G. Gaston, P. Johnson, and J.E. Schondube for generous assistance.

Further reading

Baines, C. (2000) *How to Make a Wildlife Garden*, London: Frances Lincoln. (The benchmark volume on wildlife gardening, at least in the UK.)

Gaston, K.J., Smith, R.M., Thompson, K., and Warren, P.H. (2004) "Gardens and wildlife–the BUGS project," *British Wildlife*,16(1): 1–9. (A summary of the first phase of perhaps the most comprehensive study of garden biodiversity to be undertaken to date.)

Lewington, R. (2008) *Guide to Garden Wildlife*, Gillingham: Wildlife Publishing. (A benchmark field guide to garden biodiversity.)

Owen, J. (1991) *The Ecology of a Garden: The First Fifteen Years*, Cambridge: Cambridge University Press. (The classic study of garden biodiversity.)

Toms, M. (2004) *The BTO/CJ Garden Birdwatch Book*. Thetford: British Trust for Ornithology. (The findings of an extraordinary example of citizen science.)

References

Alexander, D.A. (2006) "The environmental importance of front gardens," *Municipal Engineer*, 159(4): 239–44.

Baines, C. (2000) *How to Make a Wildlife Garden*, London: Frances Lincoln.

Baker, P.J. and Harris, S. (2007) "Urban mammals: what does the future hold? An analysis of the factors affecting patterns of use of residential gardens in Great Britain," *Mammal Review*, 37(4): 297–315.

Barnes, H.F. and Weil, J.W. (1944) "Slugs in gardens: their numbers, activities and distribution, Part1," *Journal of Animal Ecology*, 13(2): 140–75.

Barnes, H.F. and Weil, J.W. (1945) "Slugs in gardens: their numbers, activities and distribution, Part 2," *Journal of Animal Ecology*, 14(2): 71–105.

Burghardt, K.T., Tallamy, D.W., and Shriver, W.G. (2008) "Impact of native plants on bird and butterfly diversity in suburban landscapes," *Conservation Biology*, 23(1): 219–24.

Cannon, A. (1999) "The significance of private gardens for bird conservation," *Bird Conservation International*, 9(4): 287–97.

Cannon, A.R., Chamberlain, D.E., Toms, M.P., Hatchwell, B.J., and Gaston, K.J. (2005) "Trends in the use of private gardens by wild birds in Great Britain 1995–2002," *Journal of Applied Ecology*, 42(4): 659–71.

Carrier, J. and Beebee, T.J.C. (2003) "Recent, substantial, and unexplained declines of the common toad *Bufo bufo* in lowland England," *Biological Conservation*, 111(3): 395–9.

Chamberlain, D.E., Cannon, A.R. and Toms, M.P. (2004) "Associations of garden birds with gradients in garden habitats and local habitat," *Ecography*, 27(5): 589–600.

Chamberlain, D.E., Gough, S., Vaughan, H., Vickery, J.A. and Appleton, F. (2007) "Determinants of bird species richness in public greenspaces," *Bird Study*, 54(1): 87–97.

Chamberlain, D.E., Vickery, J.A., Glue, D.E., Robinson, R.A., Conway, G.J., Woodburn, R.J.W., and Cannon, A.R. (2005) "Annual and seasonal trends in the use of garden feeders by birds in winter," *Ibis*, 147(3): 563–75.

Colding, J., Lundberg, J., and Folke, C. (2006) "Incorporating green-area use groups in urban ecosystem management," *Ambio*, 35(5): 237–44.

Conrad, K.F., Warren, M.S., Fox, R., Parsons, M.S., and Woiwod, I.P. (2006) "Rapid declines of common, widespread British moths provide evidence of an insect biodiversity crisis," *Biological Conservation*, 132(3): 279–91.

Conrad, K.F., Woiwod, I.P., Parsons, M., Fox, R., and Warren, M.S. (2004) "Long-term population trends in widespread British moths," *Journal of Insect Conservation*, 8(2–3): 119–36.

Daniels, G.D. and Kirkpatrick, J.B. (2006) "Does variation in garden characteristics influence the conservation of birds in suburbia?" *Biological Conservation*, 133(3): 326–35.

Davies, Z.G., Fuller, R.A., Loram, A., Irvine, K.N., Sims, V., and Gaston, K.J. (2009) "A national scale inventory of resource provision for biodiversity within domestic gardens," *Biological Conservation*, 142(4): 761–71.

Davis, B.N.K. (1979) "The ground arthropods of London gardens," *London Naturalist*, 58: 15–24.

Day, T.D. (1995) "Bird species composition and abundance in relation to native plants in urban gardens, Hamilton, New Zealand," *Notornis*, 42(3): 175–86.

Dehnen-Schmutz, K., Touza, J., Perrings, C., and Williamson, M. (2007a) "A century of the ornamental plant trade and its impact on invasion success," *Diversity and Distributions*, 13(5): 527–34.

Dehnen-Schmutz, K., Touza, J., Perrings, C., and Williamson, M. (2007b) "The horticultural trade and ornamental plant invasions in Britain," *Conservation Biology*, 21(1): 224–31.

Donald, P.F., Green, R.E., and Heath, M.F. (2001) "Agricultural intensification and the collapse of Europe's farmland bird populations," *Proceedings of the Royal Society, London B*, 268(1462): 25–9.

Doncaster, C.P. (1994) "Factors regulating local variations in abundance: field tests on hedgehogs (*Erinaceus europaeus*)," *Oikos*, 6(2): 182–92.

Elton, C.S. (1966) *The Pattern of Animal Communities*, London: Methuen.

Fetridge, E.D., Ascher, J.S., and Langellotto, G.A. (2008) "The bee fauna of residential gardens in a suburb of New York city (Hymenoptera: Apoidea)," *Annals of the Entomological Society of America*, 101(6): 1067–77.

French, K., Major, R., and Hely, K. (2005) "Use of native and exotic garden plants by suburban nectarivorous birds," *Biological Conservation*, 121(4): 545–59.

Fuller, R.A., Warren, P.H., Armsworth, P.R., Barbosa, O., and Gaston, K.J. (2008) "Garden bird feeding predicts the structure of urban avian assemblages," *Diversity and Distributions*, 14(1): 131–7.

Gaston, K.J. and Evans, K.L. (2010) "Urbanisation and development," in N. Maclean (ed.), *Silent Summer the State of the Wildlife in Britain and Ireland*, Cambridge: Cambridge University Press, pp. 72–83.

Gaston, K.J., Cush, P., Ferguson, S., Frost, P., Gaston, S., Knight, D., Loram, A., Smith, R.M., Thompson, K., and Warren, P.H. (2007b) "Improving the contribution of urban gardens for wildlife: some guiding propositions," *British Wildlife*, 18(3): 171–7.

Gaston, K.J., Fuller, R.A., Loram, A., MacDonald, C., Power, S., and Dempsey, N. (2007a) "Urban domestic gardens (XI): Variation in urban wildlife gardening in the UK," *Biodiversity and Conservation*, 16(11): 3227–38.

Gaston, K.J., Smith, R.M., Thompson, K., and Warren, P.H. (2004) "Gardens and wildlife – the BUGS project," *British Wildlife*, 16(1): 1–9.

Gaston, K.J., Smith, R.M., Thompson, K., and Warren, P.H. (2005b) "Urban domestic gardens (II): Experimental tests of methods for increasing biodiversity," *Biodiversity and Conservation*, 14(2): 395–413.

Gaston, K.J., Warren, P.H., Thompson, K., and Smith, R.M. (2005a) "Urban domestic gardens (IV): the extent of the resource and its associated features," *Biodiversity and Conservation*, 14(14): 3327–49.

Gibbons, B. and Gibbons, L. (1988) *Creating a Wildlife Garden*, London: Hamlyn.

Gilbert, O.L. (2001) "Wildlife reports – Lichens," *British Wildlife*, 12(6): 440–1.

Gledhill, D.G., James, P., and Davies, D.H. (2008) "Pond density as a determinant of aquatic species richness in an urban landscape," *Landscape Ecology*, 23(10): 1219–30.

González-García, A. and Sal, A.G. (2008) "Private urban greenspaces or "patios" as a key element in the urban ecology of tropical central America," *Human Ecology*, 36(2): 291–300.

González-García, A., Belliure, J., Gómez-Sal, A., and Dávila, P. (2009) "The role of urban greenspaces in fauna conservation: the case of the iguana *Ctenosaura similes* in the 'patios' of León city, Nicaragua," *Biodiversity and Conservation*, 18(7): 1909–20.

Gregory, R.D. and Baillie, S.R. (1998) "Large-scale habitat use of some declining British birds," *Journal of Applied Ecology*, 35(5): 785–99.

Gregory, R.D., Noble, D.G., and Custance, J. (2004) "The state of play of farmland birds: population trends and conservation status of lowland farmland birds in the United Kingdom," *Ibis*, 146(Suppl. 2): 1–13.

Groves, R.H., Boden, R., and Lonsdale, W.M. (2005) *Jumping the Garden Fence: Invasive Garden Plants in Australia and their Environmental and Agricultural Impacts*, CSIRO report prepared for WWF-Australia, Sydney, NSW: WWF-Australia.

Guichard, K.M. and Yarrow, J.H.H. (1948) "The Hymenoptera Aculeata of Hampstead Heath and the surrounding district, 1832–1947," *London Naturalist*, 27: 81–111.

Lepczyk, C.A., Mertig, A.G., and Liu, J. (2004) "Assessing landowner activities related to birds across rural-to-urban landscapes," *Environmental Management*, 33(1): 11–125.

Loram, A., Thompson, K., Warren, P.H., and Gaston, K.J. (2008a) "Urban domestic gardens (XII): The richness and composition of the flora in five cities," *Journal of Vegetation Science*, 19(3): 321–30.

Loram, A., Tratalos, J., Warren, P.H., and Gaston, K.J. (2007) "Urban domestic gardens (X): the extent and structure of the resource in five major cities," *Landscape Ecology*, 22(4): 601–15.

Loram, A., Warren, P.H., and Gaston, K.J. (2008b) "Urban domestic gardens (XIV): the characteristics of gardens in five cities," *Environmental Management*, 42(3): 361–76.

Marco, A., Dutoit, T., Deschamps-Cottin, M., Mauffrey, J.-F., Vennetier, M., and Bertaudiere-Montes, V. (2008) "Gardens in urbanizing rural areas reveal an unexpected floral diversity related to housing density," *Comptes Rendus Biologies*, 331(6): 452–65.

Mason, C.F. (2000) "Thrushes now largely restricted to the built environment in eastern England," *Diversity and Distributions*, 6(4): 189–94.

Mathias, J.H. (1975) "A survey of amphibians in Leicestershire gardens," *Transactions of Leicester Literary and Philosophical Society*, 69: 28–41.

Mathieu, R., Freeman, C., and Aryal, J. (2007) "Mapping private gardens in urban areas using object-oriented techniques and very high-resolution satellite imagery," *Landscape and Urban Planning*, 81(3): 179–92.

McKinney, M.L. (2008) "Effects of urbanization on species richness: a review of plants and animals," *Urban Ecosystems*, 11(2): 161–76.

Miotk, P. (1996) "The naturalized garden – a refuge for animals? – first results," *Zoologischer Anzeiger*, 235(1–2): 101–16.

Morley, B.D.W. (1944) "A study of the ant fauna of a garden, 1934–42," *Journal of Animal Ecology*, 13(2): 123–7.

Moss, S. and Cottridge, D. (1998) *Attracting Birds to Your Garden*, London: New Holland.

Murphy, M.T. (2003) "Avian population trends within the evolving agricultural landscape of eastern and central United States," *The Auk*, 120(1): 20–34.

Osborne J.L., Martin A.P., Shortall C.R., Todd, A.D., Goulson, D., Knight, M.E., Hale, R.J., and Sanderson, R.A. (2008) "Quantifying and comparing bumblebee nest densities in gardens and countryside habitats," *Journal of Applied Ecology*, 45(3): 784–92.

Owen J. (1991) *The Ecology of a Garden: The First Fifteen Years*, Cambridge: Cambridge University Press.

Packham, C. (2001) *Chris Packham's Back Garden Nature Reserve*, London: New Holland.

Rapoport, E.H. (1993) "The process of plant colonization in small settlements and large cities," in M.J. McDonnell and S.T.A. Pickett (eds.), *Humans as Components of Ecosystems: The Ecology of Subtle Human Effects and Populated Areas*, New York: Springer, pp. 190–207.

Rudd, H., Vala, J., and Schaefer, V. (2002) "Importance of backyard habitat in a comprehensive biodiversity conservation strategy: a connectivity analysis of urban greenspaces," *Restoration Ecology*, 10(2): 368–75.

Sanford, M.P., Manley, P.N., and Murphy, D.D. (2008), "Effects of urban development on ant communities: implications for ecosystem services and management," *Conservation Biology*, 23(1): 131–41.

Saville, B. (1997) *The Secret Garden: Report of the Lothian Secret Garden Survey*, Edinburgh: Lothian Wildlife Information Centre.

Smith, R.M., Gaston, K.J., Warren, P.H., and Thompson, K. (2005) "Urban domestic gardens (V): Relationships between land cover composition, housing and landscape," *Landscape Ecology*, 20(2): 235–53.

Smith, R.M., Gaston, K.J., Warren, P.H., and Thompson, K. (2006a) "Urban domestic gardens (VIII): environmental correlates of invertebrate abundance," *Biodiversity and Conservation*, 15(8): 2515–45.

Smith, R.M., Thompson, K., Hodgson, J.G., Warren, P.H., and Gaston, K.J. (2006b) "Urban domestic gardens (IX): Composition and richness of the vascular plant flora, and implications for native biodiversity," *Biological Conservation*, 129(3): 312–22.

Smith, R.M., Warren, P.H., Thompson, K., and Gaston, K.J. (2006c) "Urban domestic gardens (VI): Environmental correlates of invertebrate species richness," *Biodiversity and Conservation*, 15(8): 2415–38.

Swan, M.J.S. and Oldham, R.S. (1993) *Herptile Sites Volume 1: National Amphibian Survey Final Report*, English Nature Research Report No. 38, Peterborough: English Nature.

Syme, G.J., Fenton, D.M., and Coakes, S. (2001) "Lot size, garden satisfaction and local park and wetland visitation," *Landscape and Urban Planning*, 56(3): 161–70.

Thompson, K., Austin, K.C., Smith, R.H., Warren, P.H., Angold, P.G., and Gaston, K.J. (2003) "Urban domestic gardens (I): Putting small-scale plant diversity in context," *Journal of Vegetation Science*, 14(1): 71–8.

Thompson, K., Hodgson, J.G., Smith, R.M., Warren, P.H., and Gaston, K.J. (2004) "Urban domestic gardens (III): Composition and diversity of lawn floras," *Journal of Vegetation Science*, 15(3): 371–6.

Tratalos, J., Fuller, R.A., Evans, K.L., Davies, R.G., Newson, S.E., Greenwood, J.J.D., and Gaston, K.J. (2007) "Bird densities are associated with household densities," *Global Change Biology*, 13(8): 1685–95.

Tutin, T.G. (1973) "Weeds of a Leicester garden," *Watsonia*, 9: 263–7.

van Heezik, Y., Smyth, A., and Mathieu, R. (2008) "Diversity of native and exotic birds across an urban gradient in a New Zealand city," *Landscape and Urban Planning*, 87(3): 223–32.

Warren, P.S., Lerman, S.B., and Charney, N.D. (2008) "Plants of a feather: spatial autocorrelation of gardening practices in suburban neighbourhood," *Biological Conservation*, 141(1): 3–4.

Whelan, R.J., Roberts, D.G., England, P.R., and Ayre, D.J. (2006) "The potential for genetic contamination vs. augmentation by native plants in urban gardens," *Biological Conservation*, 128(4): 493–500.

Part 5
Methodologies

Introduction

Ian Douglas

This section introduces some of the methods used in urban ecology to analyse urban habitats, map those habitats, study invasive and introduced species, examine nutrient cycles and biogeochemical fluxes, and to analyse urban metabolism. The section does not, and cannot, provide an overview of all the ecological methods likely to be needed in understanding urban ecosystems, but reference is made to the authoritative sources where those generic methods can be found.

Urban habitats are extremely varied and offer a wide variety of opportunities for vegetation to become established, for the protection of natural vegetation and the establishment and maintenance of gardens and other planted landscapes. This diversity has been subdivided in terms of vegetation types, biotic urban ecosystem characteristics and urban land use types. Ian Douglas re-examines these and compares them with the European EUNIS habitat classification.

In Britain, Phase 1 surveys provide a standardized system for classifying and mapping wildlife habitats, providing, relatively rapidly, a record of the semi-natural vegetation and wildlife habitat over large areas, including urban areas. The habitat classification is based on vegetation, with topographic and substrate features included where appropriate, and is augmented by as full a species list as was possible in and at the time of survey. The mapping is based on field visits and is usually plotted at 1:10,000. Surveys also need to identify potential patches and corridors in built-up areas and may require more detailed mapping. Urban land-cover and land-use mapping by means of satellite remote sensing has been investigated since the early 1970s. Increasingly more powerful data and more sophisticated evaluation methods have been introduced. As soon as reliable results on vegetation cover or sealing degrees could be obtained for small reporting units like city blocks, remote-sensing techniques became interesting for operational use in the framework of city planning. In 1997 the European Union tested Russian high-resolution photographic satellite data (KVR 1000) for land-use purposes. Land-use classes at a certain (but not on the highest) level could be obtained for statistical blocks both in Athens and in Berlin. Automated classification of urban vegetation using remote sensing and geographical information systems can subdivide trees, grass and lawn effectively. False colour IKONOS images are useful for establishing urban land cover types, urban vegetation health and can be used for detailed evaluation of water use by urban vegetation. Peter Jarvis assesses the validity of various types of mapping, including habitat and biotype mapping, selective surveys and tree cover mapping. He notes that throughout the UK, habitat surveys have also provided information for the many local and regional biodiversity action plans that were produced in the latter part of the 1990s and the first few years of the twenty-first century. There remain a number of limitations concerning the acquisition, storage, cartographic representation, analytical modelling, interpretation and use of spatial data appropriate to planning for a green urban environment. Nevertheless a great deal of progress has been made, particularly since 1980.

Urban areas contain large numbers of deliberately and inadvertently introduced plant and animal species. Long records exist of some plants that escaped from botanic gardens or were dumped into ponds after domestic use. Activities as seemingly benign as the planting of exotic trees and shrubs in parks and along byways disrupt the distribution of natural components of biodiversity. These activities combine to decrease habitat area and disturb the equilibrium between extinction and immigration amongst the remaining natural habitats. Domestic pets have escaped and become feral. Animals such as mink have been released on the fringes of cities. These introductions and invasions change urban ecosystems, but climate change is adding more. Globally, many species are already being affected. Ian Douglas shows how pest distributions are beginning to change and diseases of plants and animals are spreading, not merely because of alterations to the climate, but through the combined effect of the whole range of human impacts on the environment and ecosystems. Many pathogens of terrestrial and marine taxa are sensitive to temperature, rainfall and humidity, creating synergisms that could affect biodiversity. Climatic warming may increase pathogen development and survival rates, disease transmission and host susceptibility.

Many animals have shown adaptations to the urban environment. In Manchester in the nineteenth century, insects adapted, with the melanic form of the peppered moth gradually replacing other forms after 1848. By 1895 black peppered moths accounted for 98 per cent of the total peppered moth population. Urbanization typically leads to a turnover in species composition such that species that cope well with urban conditions replacing those that occurred in the pre-urban habitat. Many species do not live in cities and do not breed close to highways, and indeed the birds of urbanized areas are highly similar: the same few species become common everywhere, while the area's original species variety is lost. Studies of the songs of the great tit (*Parus major*), a successful urban-dwelling species, in the centre of ten major European cities, including London, Prague, Paris and Amsterdam, compared to those of great tits in nearby forest sites showed that for songs important for mate attractions and territory defense, the urban songs were shorter and sung faster than the forest songs. The urban songs also showed an upshift in frequency that is consistent with the need to compete with low-frequency environmental noise, such as traffic noise.

Urban biogeochemistry should not be regarded as distinct from that of other ecosystems because the physical and chemical laws that govern biogeochemical reactions are universal. However, constant physicochemical laws do not enable a simple transfer of biogeochemical models to urban ecosystems because the drivers of biogeochemical reactions are under human control. Although human control is complex, the drivers are linked to three classes of human activity: engineering, urban demographic trends and household-scale actions. A biogeochemical analysis has to examine the combination of the natural circulation of chemicals with the deliberate shifting and storing of materials in the city, the emissions of gaseous, liquid and solid wastes and the accumulation of materials in temporary sinks such as landfills. Nancy Grimm emphasizes that a major challenge in scaling urban biogeochemistry to the globe is the heterogeneity of cities. Many variables interact to determine the biogeochemical fluxes to and from an individual city. There are significant biogeochemical differences between cities in developed and developing countries due to the differences in municipal services (e.g. sewage and solid waste removal) and available resources for technology (e.g. wastewater treatment, recycling). Ultimately, the challenge is to integrate human choices and ecosystem dynamics into a seamless, transdisciplinary model of biogeochemical cycling in urban ecosystems. Such a model will inform the nature and location of significant habitat changes that will influence urban wildlife and human activity in built-up areas.

Urban ecosystems have a unique metabolism that we can characterize and measure in terms of stocks and flows of materials and energy that move between ecosystems and socio-economic systems. Cities cycle and transform raw materials, fuel and water into the urban built environment, and into human biomass, consumable products and waste. Studies of the mass balance of cities have provided important insights about the role that urban dwellers play in the cycling of chemicals. They are beginning to gain information on the mechanisms by which other factors – demographics, human activities and wastewater infrastructure – affect nutrient cycling in urban ecosystems. Such analyses also incorporate the impact of urban consumption on other ecosystems, both in terms of the urban ecological footprint and in terms of land use competition and soil and water resource degradation. Huang and Lee note that most current research on socio-economic metabolism ignores energy flow because of the difficulty of comparing materials and energy with the same units. Aggregating material flows according to their mass neglects the relative contribution associated with the values of materials with different qualitative contents. They use energy and transformity introduced by H.T. Odum to show how they obtained an overview of the socio-economic metabolism of Taiwan material flows and energy flows by incorporating an emergy synthesis.

These specific analyses could be augmented by an assessment of greenspace quality. Good quality greenspace plays a vital role in enhancing the quality of urban life. In many cities the quality of greenspace declined during the second half of the twentieth century. After 2000 greater emphasis on the role of open air recreation in improving health and the role of green areas in adapting to climate change led to signs of improvement in some countries. Qualitative assessment of greenspace improvement is usually made by assessing the views of users and managers, but those who live close by, but do not use the greenspaces, are seldom surveyed and brought into focus groups. More quantitative assessments of greenspace quality are made using indicators of various functions, such as biodiversity indicators for the nature conservation function or user statistics (e.g. number of persons playing football per week) for the recreational functions. There is little debate about the best way to examine quality of multiple functions and multiple benefit provisions. Correcting this is a task for the future.

36
Urban habitat analysis

Ian Douglas

Introduction

Maintenance of urban biodiversity and greenspace for both the residents and for its natural value in the face of the growing world population and expanding cities requires a better integration of ecological knowledge into urban planning. To achieve this goal ecological patterns and processes in urban ecosystems have to be understood. Thus to start with we have to know the kind of organisms, ecosystems and habitats we have in our cities. The first step in the necessary urban ecological research is to find out what kind of nature exists in cities. The second is to comprehend the workings of ecological processes affecting those organisms and their habitats (Niemelä 1999). This task involves the subject of this chapter: analysing urban habitats.

The term habitat is used here when generally referring to the physical, chemical and biological components of a defined geographical area. The term 'habitat type' is employed for specific kinds of habitats that have been described as separate from other such entities in habitat classification systems.

Habitats are characterized by a typology relating the various habitats to a specific classification and a given habitat patch to a specific type, where each type has a set of defining characteristics. The texture of habitats concerns the number of patches for each habitat type and the size distribution of habitat patches. The structure of habitats is given by the spatial structure or layout of the patches and the geographical relationships between the patches. Most often, the typology, texture and structure of habitats is described on habitat maps, showing the patches of different types. The spatial structure may also be described by a variety of spatial statistics or indices (e.g. fragmentation indices, landscape metrics). Finally, each habitat patch can be characterized by its internal properties, i.e. various aspects of habitat quality (Lengyel *et al.* 2008).

Classification of urban habitats

Urban areas harbour diverse nature ranging from semi-natural habitats to wastelands, parks and other highly human-influenced biotopes with their associated species assemblages. Analysing this dynamic heterogeneity remains a challenge for urban ecologists. Often in the past, people classified urban habitats in terms of two broad categories: vestiges of the natural world and the man-made wild (Goode 1986). Within these categories, several landscape types are described in terms of their use of function as part of the human ecosystem, such as derelict industrial land. Different authors devise a classification to suit the nature of the city or area they are dealing with (Table 36.1).

Urban habitats are extremely varied and offer a wide variety of opportunities for vegetation to become established, for the protection of natural vegetation and the establishment and maintenance of gardens and other planted landscapes. This diversity has been subdivided in terms of vegetation types (Shimwell 1983), types of biotic zone (Dorney 1979), urban ecosystem (Duvigneaud 1974) and urban land use types (Douglas 1983) (Table 36.2). Automated classification of urban vegetation using remote sensing and geographical information systems can subdivide trees, grass and lawn effectively (Ridd 1995; Shirai *et al.* 1998). False colour IKONOS images are useful for establishing urban land cover types and urban vegetation health and can be used for detailed evaluation of water use by urban vegetation (Herrold *et al.* 2002). The various mapping methods and the use of remotely sensed data are further discussed in the following chapter on urban habitat type mapping (see Chapter 37 this volume).

A classic view (Marren 2002) is that there are two main kinds of wildlife habitat in urban areas. One is based on built structures and landscape features resulting from human activity: roofs and streets, railway sidings and canal banks, urban parks and churchyards and derelict land. The other comprises the remnants of the countryside caught up in the growth of the city, the pieces of woodland, wetland and even farmland that are now surrounded by urban growth. However, today so much reclamation of old mining, quarrying and derelict industrial land has gone on that many of the key wildlife areas are now as much created by human endeavour as the buildings and roads to the urban area. In central London, close to Kings Cross Station, an old coal yard has been converted into Camley Street Natural Park, with a large pond and reed beds fringed by willow and birch trees and has become a haven for both wading birds and people taking a break from city routines (Goode 1986). Such examples show that we need to think of emerging ecosystems in urban and peri-urban areas in terms of relict rural systems, adventitious exploiters and colonizers of the built environment and deliberately created ecosystems, through land reclamation or the creation of areas of wild vegetation, parks and gardens (see Chapter 18 this volume).

The Second Land Utilisation Survey of Britain

The Second Land Utilisation Survey of Britain introduced a land use classification for urban areas that made it possible to map both types of human urban and industrial activities, as well as vegetation and crop types (Coleman and Maggs 1965). In built-up districts a great many forms of land use are crowded into a small area as compared with the countryside, and there is relatively little space in which to represent them. It was decided, therefore, to associate together as a grey-coloured background all those forms of residential and commercial use which are common to all or most settlements, and vary more according to the settlements' size than to any other factor. This group includes housing, shops, hospitals, churches, business and administrative offices, places of entertainment and so on. Conspicuous against this grey background are brighter colours representing forms of use more individual to particular settlements, e.g. industry, transport, open greenspace or allotment gardens (Coleman 1961).

This could serve as a basis for classifying urban land cover types, but would not correspond to ecological habitats. Nevertheless it enabled the continuum of vegetated land between high-biomass climax forest and low-biomass mown lawns to be subdivided (Coleman and Shaw 1980) and provided the most comprehensive land-use typology covering all the gradations of greenspace introduced up to that time.

Table 36.1 Urban habitats described by various authors (compiled for this volume by Ian Douglas)

Gilbert	*Goode*	*Scott*	*Dickman And Doncaster*	*Schulte* et al.	*Nicol And Blake*
Urban commons	Heaths and commons				
	–	Scrub	–	Agricultural land	
Woodland	Woodlands	Woodlands	Woodland	–	Woodlands
	Meadows and hedgerows	Meadows	Long grass	–	Grassland
	–	–	Orchard	–	–
	Downland	–	Short grass	Border	Arable
	–	Wetlands	–	–	Wetlands
	–	Waterways	–	–	–
Rivers, canals, lakes, ponds and water mains					
	River and marsh	Ponds and lakes	–	Tenant garden	Land left to nature
Allotments and leisure gardens	–	Gardens	Allotment, detached house garden	Front garden	Allotments
	–	–	Semi-detached house garden	–	–
Roads					
	–	–	–	Path	–
Railways	Railways	Coastal areas	–	Garage roof	–
	Waterworks and reservoirs	Hills and mountains	–	–	–
	Disused sand and gravel pits				
	–	–	–	–	
Industrial areas	Derelict industrial sites and railways sidings				
	–	–	Waste land	Unmanaged land	
Cemeteries					
	Cemeteries	–	–	–	–
City parks	Parks	Parks and open spaces			
	–	Lawn	Formal open space		
City centres	City squares	Inner city	–	Street area	–

Table 36.2 Types of urban and peri-urban land cover and ecological conditions (after Douglas 2008)

Habitat type (Douglas 1983)	*Habitat type used in Iowa Nature Mapping*	*Substrate type used in urban LTER, USA*	*Biotic zone (after Dorney 1979)*	*Type of urban ecosystem (after Duvigneaud 1974)*	*Urban vegetation type (Shimwell 1983)*
Paved roofed, densely urban complexes devoid of vegetation and water bodies	Urban high-density commercial/industrial	Asphalt +/– concrete	Cliff/organic detritus	Anthropogenic	Pioneer, open communities of rocks, walls, roofs, pavements and other metalled or trodden surfaces, dominated by cryptograms or other specifically adapted rock plants (lithophytes)
Suburban mosaic of houses, roads, gardens and mature trees	Urban high-density gardens	Tile roofs + granitic soil +/– woody vegetation +/– grass +/– water; asphalt + concrete + soil +/– metal roofs +/– grass woodland	Old urban savanna	Urbanophile	Managed mown grasslands, managed urban savanna woodland
Corridor zones of wild plants	Urban railway right-of-way	Grass + woody vegetation +/– soil +/– asphalt	Grassland/weed complex	Peripheral	Rank, perennial, tall grass and tall herb (usually >70 cm) communities of embankments road verges, abandoned sewage beds and damp swamp and marsh margins
Landscaped parks and open spaces	Urban maintained parkland	Grass + water +/– sandy soil	Mown grassland	Anthropogenic and modified natural	Managed, mown grasslands and weedy, perennial herb grass communities
Derelict land construction sites			Abiotic weed complex	Urbanophile	Therophyte-dominated communities of derelict brick-rubble, cinder and fuel-ash tips, etc. (includes communities with large number, or even mostly, perennials)

New suburbs, devoid of mature trees and a high grassland: cultivated land ratio	Urban low-density gardens	Dry soil + grass +/– concrete	New urban savanna	Urbanophile	Managed, mown grasslands
Grassland on reclaimed soil, with streets, car parks and buildings but few or no mature trees	Urban low-density commercial	Asphalt + concrete + soil +/– metal roofs	New urban savanna/ mown grassland	Anthropogenic/ urbanophile	Open communities of low growing annuals of ornamental park borders and open communities of low growing annuals of ornamental park borders and roadsides/regularly mown grasslands
Small woodland and rural areas within the city	Rural grassland/ woodland	Grass + woody vegetation +/– soil +/– asphalt	Remnant ecosystems/ natural islands	Relict	Deciduous and evergreen woodland, greater than 5 m in height with a closed canopy/ grasslands on a variety of natural habitats
Water bodies	Pond/impoundment; open water (reservoir/ lake); river/stream	Water	Lake-stream/aquatic complex	Modified natural	Emergent tall, swamp communities of rivers canals and lake margins
Wetlands and much-modified water bodies	Pond/impoundment	Grass + water +/–	Derelict/weedy grassland and aquatic complex	Urbanophile	

The New York urban land use inventory

In the 1980s an inventory of land use, vegetative and natural characteristics of urban areas was a basic and essential component for the development of an urban wildlife programme for the State of New York, USA (Matthews *et al.* 1988). The programme was seen as the forerunner of other state and federal wildlife programmes geared towards the needs of urban residents. The inventory identified, defined, categorized and quantified potential urban wildlife habitat within the seven urbanized areas of the state. The survey recognized two categories of land cover in built-up areas, vegetated cover and built-over areas. It obtained existing vegetation cover types and land use from aerial photographs and subdivided the study areas into 31 vegetation cover types and 27 land uses. Cover types within an area were determined according to the characteristics of the surrounding vegetation and land for each particular parcel. The smallest area interpreted was approximately 0.3 ha in size for upstate areas.

Examining the cover types for the city of Troy (Table 36.3), 88 per cent of the city area is made up of ten cover types, three of which are lawns or lawns with trees and shrubs or just trees. These three lawn categories alone account for 44 per cent of the total urban area.

Comparing classifications

Most habitat classifications have been made for specific purposes in specific places. However, with the interest in monitoring habitats and understanding changes in both ecosystems and biodiversity, a far greater emphasis has been placed on habitat monitoring and on comparisons between one region and another. This has led to much discussion of appropriate ways of classifying habitats. For urban areas, the problem is that habitats tend to be described by the type of land use of arrangements of buildings, rather than by floristic, physiognomic, hydrological and ecological features. This difficulty is well-illustrated when different classifications are compared (Table 36.2). Many of the columns describe habitats in terms of what occupies the ground, while Shimwell's classification uses vegetation to describe the habitat. There is an underlying notion that a distinction should be made between totally artificial habitats and those where nature has been modified by human action. This idea has been incorporated into some of the more recent international classifications.

Table 36.3 The ten most important cover types in the City of Troy, NY, USA (after Matthews *et al.* 1988)

Cover type	*Number*	*Area ha*	*Area % of city*
Lawns/shrubs and trees	49	698	31
Surfaced	69	318	14
Lawns	54	252	11
Deciduous trees	25	211	9
Rivers	9	138	6
Old field/deciduous trees	5	112	5
Lawns/surfaced	13	110	5
Old field	16	77	3
Lawns/deciduous trees	3	49	2
Disturbed	9	44	2
Total of above	252	2,009	88
Total for city	304	2,266	100

Developing international urban habitat classifications

Visual interpretation of aerial photographs, supported by field checking has been replaced in nearly all practical attempts to classify habitats by the automated processing of multi-spectral earth observation data. Essentially this involves recognizing attributes of every pixel that can be used as surrogates for aspects of the earth's surface. For example, grassland may be distinguished from woodland. Within each category, a further subdivision can be made, for example into deciduous and evergreen woodland. Eventually these can be built up into a hierarchical classification. This is the principle behind most classifications based on remotely sensed data.

Other classifications may rest on the statistical analysis of attributes for a set of sampling units, for example a grid of 10×10 m quadrats. Information on such attributes as soils, number of different plant species, canopy density, the proportion of impermeable surface and proportion of native species can then be used to group spatial units by their similarity on different criteria. Thus for any study area, sets of similar sampling units can be made, and descriptive names, similar to those used in the older less objective habitat classifications, can be given to them.

What ever is attempted by way of classification, efforts to preserve and enhance both habitats and biodiversity in the urban environment have to be based on reliable evidence. Decision making requires sound evidence of why particular habitats are significant. In Europe it is recognized that an essential tool for obtaining policy-relevant information is an accessible Europe-wide geo-referenced inventory of habitat distribution, status and trends, built upon harmonized habitat and landscape classifications. Such work involves selecting appropriate remote sensing methods, including appropriate scales and levels of resolution, and the development and testing of spatially explicit indicators for habitat assessment and monitoring (Mander *et al.* 2005).

Urban aspects of the European Nature Information System (EUNIS)

To help create useful databases for future environmental management and the protection of valued habitats to sustain biodiversity, the European Union has been developing common habitat classifications. The European Nature Information System (EUNIS) habitat classification is a comprehensive pan-European system to facilitate the harmonized description and collection of data across Europe through the use of criteria for habitat identification; it covers all types of habitats from natural to artificial, from terrestrial to freshwater and marine. EUNIS is also cross-comparable with CORINE and with the Natura 2000 habitat system (Davies *et al.* 2004).

EUNIS defines a 'habitat' as: 'a place where plants or animals normally live, characterized primarily by its physical features (topography, plant or animal physiognomy, soil characteristics, climate, water quality, etc.) and secondarily by the species of plants and animals that live there' (Davies *et al.* 2004: 1). Different habitats occur at different scales. While moss and lichen tundra or deep-sea mud habitats may be of vast extent, others, such as cave entrances or springs, spring brooks and geysers, are much smaller. Most, but not all EUNIS habitats, can be said to be 'biotopes': 'areas with particular environmental conditions that are sufficiently uniform to support a characteristic assemblage of organisms' (Davies *et al.* 2004: 1). A few EUNIS habitats such as glaciers and highly artificial non-saline standing waters may have no living organisms other than microbes. These features, although not strictly habitats, are included for completeness. The EUNIS habitat classification covers the whole of the European land and sea area, i.e. the European mainland as far east as the Ural Mountains, including offshore islands (Cyprus; Iceland but not Greenland), and the archipelagos of the European Union Member States (Canary Islands, Madeira and the Azores), Anatolian Turkey and the Caucasus (Davies *et al.* 2004).

The broadest classification in the EUNIS system comprises the level 1 units:

A Marine habitats
B Coastal habitats
C Inland surface waters
D Mires, bogs, fens
E Grasslands and lands dominated by forbs, mosses or lichens
F Heathland, scrub and tundra
G Woodland, forest and other wooded land
H Inland unvegetated or sparsely vegetated habitats
I Regularly or recently cultivated agricultural, horticultural and domestic habitats
J Constructed, industrial and other artificial habitats.

Habitats in any one of these major categories could occur within, or immediately adjacent to, an urban area. Categories I and J are however the habitats most affected by human activity. Category I has two level 2 units: I1 Arable lands and market gardens, and I2 Cultivated areas of parks and gardens. It is important to note that domestic and other gardens, parks and city squares are treated as complexes comprising combinations of units from any other level 1 unit (Davies *et al.* 2004). Thus a park could contain some wooded habitat of category G, some grassland of category E, a pond or lake from category C and some buildings of category J, as well as flower beds in category I. This approach is extremely helpful in classifying urban habitats.

Category J has the following level 2 units:

J1 Buildings of cities, towns and villages
J2 Low density buildings
J3 Extractive industrial sites
J4 Transport networks and other constructed hard-surfaced areas
J5 Highly artificial man-made waters and associated structures
J6 Waste deposits

Each category has further subdivisions (Table 36.4).

As the standard habitat classification for Europe, EUNIS has been used in many vegetation studies, including the analysis of alien species distributions (Chytrý *et al.* 2008, 2009). Mapping based on habitat types worked well both because the highest level of neophyte invasive plants were associated with urban areas and because habitat types were much better predictors of the level of invasion than other environmental and propagule pressure variables.

EUNIS habitat types are used in DAISIE (the European programme on Delivering Alien Invasive Species Inventory for Europe) (DAISIE 2009). For every species listed, the EUNIS habitat types in which it is found are provided (e.g. Tables 36.5 and 36.6).

The practicality of the EUNIS classification has been demonstrated in many aspects of urban ecology from the analysis of changing bird habitats to the study of the habitats of invasive insect disease vectors. Although not the only current classification, EUNIS now works as an accessible database which should make the monitoring of future changes in urban ecology in Europe much easier than it was in the past. It particularly helps researchers and practitioners to recognize the ecological and biodiversity value of informal and adventitious urban greenspace as well as that of more deliberately planned and managed open spaces. In the UK, development plans still tend to accord higher 'green values' to conventional parks and playing fields than to high-biomass sites not listed, or indicated on plans, as 'official' open spaces. The derelict and temporarily vacant

Table 36.4 Subdivisions of the major classes of the J category of the EUNIS habitat classification

J1	*J2*	*J3*	*J4*	*J5*	*J6*
J1.1 Residential buildings of city and town centres	J2.1 Scattered residential buildings	J3.1 Active underground mines	J4.1 Disused road, rail and other constructed hard-surfaced areas	J5.1 Highly artificial saline and brackish standing waters	J6.1 Waste resulting from building construction or demolition
J1.2 Residential buildings of villages and urban peripheries	J2.2 Rural public buildings	J3.2 Active opencast mineral extraction sites, including quarries	J4.2 Road networks	J5.2 Highly artificial saline and brackish running waters	J6.2 Household waste and landfill sites
J1.3 Urban and suburban public build	J2.3 Rural industrial and commercial sites still in active use	J3.3 Recently abandoned above-ground spaces of extractive industrial sites	J4.3 Rail networks	J5.3 Highly artificial non-saline standing waters	J6.3 Non-agricultural organic waste
J1.4 Urban and suburban industrial and commercial sites still in active use	J2.4 Agricultural constructions		J4.4 Airport runways and aprons	J5.4 Highly artificial non-saline running waters	J6.4 Agricultural and horticultural waste
J1.5 Disused constructions of cities, towns and villages	J2.5 Constructed boundaries		J4.5 Hard-surfaced areas of ports	J5.5 Highly artificial non-saline fountains and cascades	J6.5 Industrial waste
J1.6 Urban and suburban construction and demolition sites	J2.6 Disused rural constructions		J4.6 Pavements and recreation areas		
J1.7 High density temporary residential units	J2.7 Rural construction and demolition sites		J4.7 Constructed parts of cemeteries		

Table 36.5 The EUNIS habitats occupied by alien bryophytes in the UK. Note the large percentages in the category J urban habitats (after Essl and Lambdon 2009)

EUNIS habitats	*Bryophyte habitats*	*% of species*
I2	Gardens	36
H	Roadsides	20
J	Walls, buildings	20
C	Freshwater (incl. littoral)	18
J	Ruderal habitats	13
H	Rocks	13
J	Greenhouses	11
D	Mires, bogs, fens	11
G	Broadleaved woodlands	11
I	Arable land	9
G	Conifer woodlands	9
A	Dunes, coastal habitats	4
C	Freshwater	2

sites of former industrial and institutional sites can readily be included when using EUNIS and similar classifications. Further details on practical habitat mapping are provided in the next chapter (Chapter 37).

Conclusions

Habitat analysis involves a range of skills and techniques:

- sampling techniques for describing plant species distributions and community structure and diversity, including the quadrant and transect-based, point-intercept, and plotless methods;
- identification of common and dominant indicator plant species of wetlands and uplands, identification of hydric soils;
- use of soil, topographic, and geologic maps and aerial photographs in deriving a site description and site history; and
- graphic and statistical methods including GIS applications for analysing and presenting the field data.

Habitat analysis has to be undertaken at a scale appropriate to the purpose of the project or enquiry. Broadly, four scales can be used, although finer subdivisions with these four are possible:

- The country or province (state) level;
- The eco-regional level (to examine the potential suitability of a given landscape for the survival of a species;
- The district or city-region level (for planning purposes and action on biodiversity);
- The micro-habitat level (dealing with individual patches or parcels within the urban mosaic) (see Chapter 21).

Analysis requires a habitat classification which can be through using the field and earth observation data on ecological attributes to perform a statistical analysis of associations between different variables, by techniques such as cluster analysis or factor analysis, or by examining to which

Table 36.6 Association of invasive plant species with particular EUNIS urban habitats (compiled from data in DAISIE 2009)

J1 Buildings of cities, towns and villages	*Acacia dealbata*, silver wattle, blue wattle	*Ailanthus altissima* Swingle, tree of heaven	*Ambrosia artemisiifolia*, common ragweed	*Campylopus introflexus*, heath star moss	–	–	–
J2 Low density buildings	*Ambrosia artemisiifolia*, common ragweed	*Campylopus introflexus*, heath star moss	*Oxalis pes-caprae*, Bermuda buttercup	–	–	–	–
J4 Transport networks and other constructed hard-surfaced areas	*Acacia dealbata*, silver wattle, blue wattle	*Ailanthus altissima* Swingle, tree of heaven	*Ambrosia artemisiifolia*, common ragweed	*Cortaderia selloana*, pampas grass	*Impatiens glandulifera*, Himalayan balsam	–	–
J5 Highly artificial man-made waters and associated structures	*Impatiens glandulifera*, Himalayan balsam	*Oxalis pes-caprae*, Bermuda buttercup	*Robinia pseudoacacia*, black locust	–	–	–	–
J6 Waste deposits	*Acacia dealbata*, silver wattle, blue wattle	*Ambrosia artemisiifolia*, common ragweed	*Campylopus introflexus*, heath star moss	*Cortaderia selloana*, pampas grass	*Fallopia japonica*, Japanese knotweed	*Oxalis pes-caprae*, Bermuda buttercup	*Prunus serotina*, black cherry

category of an existing classification the attributes suggest a given observed site should be allocated. For urban areas various descriptive classifications have been widely used, but the European Nature Information System (EUNIS) habitat works well.

Further attention needs to be paid to making classifications and analyses usable by practitioners in open space planning and biodiversity management, because some patches of urban land change their character rapidly and others remain relatively stable. Rapid change is typified by the transition of land from industrial use to derelict land, to land restoration, to new construction; stability by some suburban gardens that, apart from normal maintenance, remain unchanged for decades, or the slow change in old neglected cemeteries that eventually become parts of the urban forest. Whatever the rate of change, monitoring it and keeping records linked to a spatially registered database (in a GIS) will help to ensure that the evidence needed to support recommendations for particular policies for urban nature is available and accessible.

References

Chytrý, M., Maskell, L.C., Pino, J., Pysek, P., Vilà, M., Font, X. and Smart, S.M. (2008) 'Habitat invasions by alien plants: a quantitative comparison between Mediterranean, subcontinental and oceanic regions of Europe', *Journal of Applied Ecology*, 45(2): 448–58.

Chytrý, M., Pysek, P., Wild, J., Pino, J., Lindsay C. Maskell, L.C. and Vilà, M. (2009) 'European map of alien plant invasions based on the quantitative assessment across habitats', *Diversity and Distributions*, 15(1): 98–107.

Coleman, A. (1961) 'The Second Land Use Survey: Progress and Prospect', *The Geographical Journal*, 127(2): 168–80.

Coleman, A and Maggs, K.R.A (1965) *Land Use Survey Handbook*, fourth (Scottish) edition, Margate: Isle of Thanet Geographical Association.

Coleman, A. and Shaw, J. E. (1980) *Land Utilisation Survey: Field Mapping Manual. Second Land Utilisation Survey of Britain*, London: King's College.

DAISIE (ed.) (2009) *Handbook of Alien Species*, Dordrecht: Springer Netherlands.

Davies, C.E., Moss, D. and Hill, M.O. (2004) *EUNIS Habitat Classification Revised 2004*, Copenhagen: European Environment Centre, European Topic Centre on Nature Protection and Biodiversity.

Dickman, C.R. and Doncaster, C.P. (1987) 'The ecology of small mammals in urban habitats: 1 Populations in a patchy environment', *Journal of Applied Ecology*, 56(2): 629–40.

Dorney, R.S. (1979) 'The ecology and management of disturbed urban land', *Landscape Architecture* (May): 268–72.

Douglas, I. (1983) *The Urban Environment*, London: Edward Arnold.

Douglas, I. (2008) 'Environmental change in peri-urban areas and human and ecosystem health', *Geography Compass*, 2(4): 1095–137.

Duvigneaud, P. (1974) 'L'ecosystème urbs', *Mémoires de la Société de Botanique de Belgique*, 4: 5–35.

Essl, F. and Lambdon, P.W. (2009) 'Alien Bryophytes and Lichens of Europe', in DAISIE (ed.), *Handbook of Alien Species in Europe*, Dordrecht: Springer Netherlands, pp. 29–41.

Gilbert, O.L. (1989) *The Ecology of Urban Habitats*, London/New York: Chapman and Hall.

Goode, D. (1986) *Wild in London*, London: Michael Joseph.

Herrold, M., Gardner, M., Hadley, B. and Roberts, D., (2002) 'The spectral dimension in urban land cover mapping from high resolution optical remote sensing data', *Proceedings of the 3rd Symposium on Remote Sensing of Urban Areas, June 2002, Istanbul, Turkey*, online, available at: www.geogr.uni-jena.de/~c5hema/pub/istanb_herold_gardn_hadl_roberts.pdf

Lengyel, S., Kobler, A., Kunar, L., Franstad, E., Henry. P.-Y., Babij, V., Gruber, B., Schmeller, D. and Henle, K. (2008) 'A review and a framework for the integration of biodiversity monitoring at the habitat level', *Biodiversity and Conservation*, 17(14): 3341–56.

Mander, U., Mitchley, J., Xofis, P., Keramitsoglou, I. and Bock, M. (2005) 'Earth observation methods for habitat mapping and spatial indicators for nature conservation in Europe', *Journal for Nature Conservation*, 13(2–3): 69–73.

Marren, P. (2002) *Nature Conservation: A Review of the Conservation of Wildlife in Britain 1950–2001*, London: HarperCollins.

Matthews, M.J., O'Connor, S. and Cole, R.S. (1988) 'Database for the New York State Urban Wildlife Habitat Inventory', *Landscape and Urban Planning*, 15(1–2): 23–37.

Nicol, C. and Blake, R. (2000) 'Classification and use of open space in the context of increasing urban capacity', *Planning Practice and Research*, 15(3): 193–210.

Niemelä, J. (1999) 'Ecology and urban planning', *Biodiversity and Conservation*, 8(1): 119–31.

Ridd, M.K. (1995) 'Exploring a V-I-S (vegetation-impervious surface-soil) model for urban ecosystem analysis through remote sensing: comparative anatomy for cities', *International Journal of Remote Sensing*, 16(12): 2165–85.

Schulte, W., Sukopp, H. and Werner, P. (1993) 'Flächendeckende Biotopkartierung im besiedelten Bereich als Grundlage einer am Naturschutz orientierten Planung', *Natur und Landschaft*, 68(10): 491–526.

Scott, R. (2004) *Wild in Belfast*, Belfast: Blackstaff Press.

Shimwell, D.S. (1983) *A Conspectus of Urban Vegetation Types*, unpublished, Manchester: School of Geography, University of Manchester (see the Appendix in P. Bullock and P.J. Gregory (eds) *Soils in the Urban Environment*, Oxford: Blackwell: 171–2).

Shirai, N., Setojima, M., Hoyano, A. and Yun, D. (1998) 'Urban vegetation cover mapping with vegetation cover', *GIS Development Proceedings ACRS 1998*, online, available at: www.gisdevelopment.net/aars/acrs/1998/ps2/ps2015pf.htm.

Shochat, E., Stefanov, W.L., Whitehouse, M.E.A. and Faeth S.H. (2004) 'Urbanization and spider diversity: influences of human modification of habitat structure and productivity', *Ecological Applications*, 14(1): 268–80.

White, J.G., Antos, M.J., Fitzsimons, J.A. and Palmer, G.C. (2005) 'Non-uniform bird assemblages in urban environments: the influence of streetscape vegetation', *Landscape and Urban Planning*, 71(2–4): 123–35.

Zerbe, S., Muarer, U., Schmitz, S. and Sukopp, H. (2003) 'Biodiversity in Berlin and its potential for nature conservation', *Landscape and Urban Planning*, 62(3): 139–48.

37
Urban habitat type mapping

Peter J. Jarvis

Introduction

It should be axiomatic that decisions in town planning are based on comprehensive information and appraisal of what already exists. It is evident, however, that in many towns and cities decisions that greatly affect the urban fabric, the urban landscape and urban quality of life have been made, and continue to be made, on incomplete, inappropriate, marginally relevant or partial data. Urban planning in general, environmental planning in particular, and nature conservation and amenity planning most of all should demand a comprehensive, frequently updated set of relevant information. Much of this information needs to be given a spatial framework, including distributions and distributional interrelationships. In a word, the information needs to be mapped. And these maps need to be in a format that can be accessed, reviewed, interpreted and used.

In order to use the green environment as a component in the decision-making process in urban planning and, reciprocally, in order to use the planning process as an opportunity and tool in the enhancement of the green environment it is necessary to know what actually exists, what the planning and conservation goals are, and what the options are for achieving these goals. The green environment itself must be seen as including biodiversity, habitat type (biotope), habitat quality, and the structure and pattern of the landscape. Together, these components provide what might be thought of as the 'green capital' of the city.

This green capital ranges from the presence of common and abundant species to the rare and restricted; from relict ancient woodland to small manicured lawn; from large semi-natural vegetation to tiny wasteland plots; and from integrated green networks to isolated green islands. Quantifying and mapping the green capital is essential for identifying, monitoring and assessing the consequences of planning decisions, bearing in mind that urban form and function are both inherently dynamic. Mapping has increasingly become dependent on geographic information systems (GIS) or uses expert systems such as Conservation Management System (www.esdm.co.uk) and can enable decision makers to set out appropriate goals, identify realistic choices, implement suitable environmental impact assessments, and monitor results.

Air photographs, remote sensing and geographic information systems

Satellite remote-sensing data and their analysis have increasingly been used since the 1970s having replaced the air photograph as a means of constructing various kinds of map. Habitat mapping is one of the many uses to which GIS can be put (Maantay and Ziegler 2006). Even though systems such as Système pour l'Observation de la Terre (SPOT) have only moderate spatial resolution they have proven useful in mapping urban areas, for example in Brisbane (Phinn *et al.* 2002), but it remains difficult to do more than simply separate vegetated areas from soil cover and impervious surfaces (Ridd 1995; Herold 2002). The 2 m resolution of images from the KVR-1000, however, makes this Russian Cosmos satellite-based system particularly suitable for studying urban areas. In comparing high-resolution airborne scanner data for Berlin with satellite data for Seoul, Kim *et al.* (2005), for example, conclude that both have advantages in mapping urban habitats. Satellite imagery is also useful for mapping urban habitat types that are widespread and important but inaccessible, for example gardens (Mathieu *et al.* 2007).

Air photographs remain a valuable tool. Their use in the spatial delimitation of 'urban morphology types' might be limited, for example in a study of Manchester, UK, by Gill *et al.* (2008) interpretative analysis of such images could not differentiate between woodland types, and for 'open habitats' could not go beyond distinguishing formal recreation, formal open space, informal open space and allotments. However, 'surface cover analysis' was more usefully able to distinguish between buildings, other impervious sites, tree, shrub, mown grass, rough grass, cultivated land, water and bare soil.

Differences between rural habitats and urban equivalents

One problem with classifying and mapping urban habitats is that most show important differences in content, structure and disposition compared with their non-urban equivalents. For example, relict ancient woodlands in towns tend to have an impoverished ground flora compared with their rural analogues; they are often relatively small in area and are not infrequently isolated, with implications for the dispersal of some plant and less motile animal species; and they have a greater amount of 'edge habitat' (with implications for disturbance) and less 'core habitat' than outside the built-up area. Another problem is that there exists a number of habitats that are essentially only ever found, or are only really important, in towns and cities, for example garden lawns, street trees, playing fields, waste lots and railway marshalling yards. And plant communities often reflect recombinant ecology, where species have 'recombined' into assemblages unique to particular urban locations or environments (Barker 2000; and see Chapter 17 this volume).

At the very least it is important to have descriptive information on what is present, whether land-use, habitat type, landscape characteristics or species distributions. A number of exercises, however, have also usefully attempted to place some level of 'value' on what is there.

Mapping species and species associations

Species distributions and, consequently, spatial representations of biodiversity can only be assessed by field survey which is time-consuming and labour-intensive, and often requires a high level of taxonomic expertise. Results are generally presented as presence-absence maps on grids of 1 km×1 km or 2 km×2 km. Such maps become particularly valuable if there are opportunities to relate species distributions to key environmental variables, using GIS systems, thus giving some degree of explanation. Many species are valuable as biological indicators, identifying particular environmental conditions.

Distributional data collected using the same sampling procedures over a number of years can be an important tool for monitoring environmental changes. Information presented as dot maps may be useful in identifying and explaining population as well as distributional changes (Harris and Rayner 1986), or indeed potential disease threat, as with modelling the possible spread of rabies in urban foxes (Saunders *et al.* 1997).

Single maps, however, often represent data collected over a number of years, and might not be able to distinguish between, on the one hand, a single sighting some years previously of a species no longer present and, on the other hand, multiple sightings of a common, regularly-occurring species. It is nevertheless clear that cartographic representations of distribution data are an important tool in urban conservation and planning, and have been since at least Teagle's snapshot of Britain's Birmingham and Black Country conurbation in *The Endless Village* (Teagle 1978), possibly the earliest cogent example of a survey-based argument for urban nature conservation in a particular region.

More interestingly, presence/absence data for 829 vascular plant species in Plymouth, UK, have been analysed to establish major species assemblages and to examine their spatial distribution across the city (Kent *et al.* 1999). TWINSPAN groups lying close to the first axis of variation showed a correlation between variation in plant species assemblages and both the process of urban development and the historical changes in urban structure. The second axis seemed to be related to a set of remnant semi-natural habitats within Plymouth that can be considered as 'hotspots' for survival.

In another example of the application of mapping to both urban planning and nature conservation, Nakamura and Short (2001) examined the relationships between the distribution patterns of 165 endangered species (99 plant species and 66 animal species) and land-use planning in Chiba City, southeast of Tokyo, Japan. Distribution maps were analysed in terms of green cover and zoning categories, and suggested that regional diversity depended heavily on areas in which traditional landscapes remain relatively intact.

Where comparable data are available for a number of time periods it is possible to analyse floristic or faunistic changes in a variety of ways. Godefroid (2001), for example, has taken advantage of 1 km × 1 km grid floristic data for Brussels, Belgium, from 1943 onwards to demonstrate an increasingly nitrophilous and shade-tolerant flora, and a flora also increasingly dominated by introduced species.

Land-use and land-cover mapping in towns

Land use generally refers to the different uses to which land is put, from a predominantly human perspective. If they have no direct human use, categories tend to be fairly general. Examples are forestry, lowland agriculture and residential land. *Land cover maps* represent the dominant landscape cover in a particular area, and do not adopt a particularly human or natural perspective. Examples are coniferous trees, arable fields, built-up areas. Owen *et al.* (2006: 311) argue that their land classification of the UK West Midlands provides 'an impartial basis for a wide range of environmental and ecological surveys'. *Habitat maps* detail the distribution of particular species (usually plant) assemblages, but usually lack detail in urban areas because assemblages are often small in scale, complex and reflective of often unique ecological outcomes.

During the 1930s the Land Utilisation Survey of Britain created a detailed record of all major land-uses in England, Wales and southern Scotland. Data were surveyed on 1:10,560 (6 inches to a mile) Ordnance Survey maps. Information, which included urban areas, was then published on a set of 169 map sheets using OS 1:63,360 (1 inch to a mile) maps, displaying the information according to a set of colour codes. Discrimination of urban land-uses was inevitably weak, but different kinds of open space were distinguishable, for example grassland, woodland (coniferous, deciduous, mixed), orchards and wetlands.

A similar exercise was undertaken in the 1960s allowing comparison with the survey undertaken 30 years earlier, and therefore providing an opportunity to categorize and quantify change. This timeline of historical data has continued with ecological surveys undertaken by the Institute of Terrestrial Ecology in 1978 and the Countryside Surveys of 1984, 1990 and 1998 using air photo and satellite imagery and ground-truthing.

There has been a respectable post-war history of aerial photographs having been used to acquire information on the land use of towns (Pownall 1950), for example in Leeds (Collins and El-Beik 1971). GIS has been instrumental in the production of environmental atlases, for example in the Randstad region of the Netherlands (Canters *et al.* 1991) and in St Petersburg, Russia (Chertov *et al.* 1996).

Phase 1 and cognate habitat surveys

Phase 1 surveys provide

> a standardised system for classifying and mapping wildlife habitats in all parts of Great Britain, including urban areas.... The aim of Phase 1 survey is to provide, relatively rapidly, a record of the semi-natural vegetation and wildlife habitat over large areas.
>
> *(Nature Conservancy Council 1990: 7)*

The habitat classification is based on vegetation, with topographic and substrate features included where appropriate, and augmented by as complete a species list as possible in and at the time of survey. For sites perceived to be of importance for ecological or other reasons, a follow-up Phase 2 survey describes the plant communities in greater detail, wherever possible in terms of the UK National Vegetation Classification – something that is not always realistic in an urban setting.

Phase 1 surveys have been undertaken in many urban areas in Britain, early examples being such work during the early 1980s preliminary to the production of nature conservation strategies in such locations as London, the West Midlands, Leicester and Nottingham (Graf 1986; Jarvis 1996). Throughout the UK, habitat surveys have also provided information for the many local and regional biodiversity action plans that were produced in the latter part of the 1990s and the first few years of the twenty-first century. Maps resulting from Phase 1 surveys are akin to the spatially-incomplete (selective) mapping described later.

A modification of the Phase One survey was used in London in 1984–5 to compile a map-based inventory of sites of potential value for nature conservation. An initial desk study using air photography resulted in over 1800 'sites' being selected for survey, totalling about 20 per cent of the land area of Greater London. For each site information was collected on the types of habitat present and the dominant species, richness of plant species, presence of rare or unusual species, current land-use and accessibility. Habitats were mapped at a scale of 1:10,000. Information has since been updated through periodic resurveys of individual London boroughs (GLC 1985; Goode 1989, 1999).

This survey provided the basis for the first comprehensive strategic nature conservation plan for Greater London and provided the starting point for selection of Sites of Importance for Nature Conservation. A standardized set of criteria was used for comparing and evaluating sites. Details of these criteria and the way in which they have been applied are given in Goode (1999, 2005).

Mapping tree cover

It is also often useful for planning purposes to survey just one particular kind of habitat (or land use), an obvious one in the context of planning for a green environment being that of woodland. Measurements of tree cover using remote sensing are also valuable in providing data used in urban forest management (Nowak 1993; Nowak *et al.* 1996), and in identifying opportunities for public recreation and other kinds of social benefit (Dwyer *et al.* 1992). Certainly tree-cover data when combined with ground-sampling of species composition, tree height, trunk diameter and tree health enhances the opportunities for planning and the management of this green resource in towns.

Where there are long-term projects involving local authority and/or community-based tree planting schemes it is important to maintain a record of what has been planted, and where – evident in the UK at a regional level, for example schemes undertaken under the aegis of the National Urban Forestry Unit (www.nufu.org.uk) or London's Million Trees Campaign, launched in June 2002 (www.treesforcities.org/page.php?id=76).

Biotope and habitat mapping

Most ecological mapping of use to the urban planner and environmental manager is of habitats or biotopes. Habitats and biotopes are not infrequently used synonymously, though the latter is more accurately limited to describing an area with boundaries within which plants and animals can live (Dahl 1908). Forman (1995) similarly refers to a discrete environmental area characterized by certain conditions and populated by a characteristic biota.

In their assessment of how urban biotopes represent a valuable tool in natural conservation, Starfinger and Sukopp (1994) stress how the green open spaces of towns represent modifications of older biotopes.

Selective mapping

The earliest biotope mapping programmes of the 1970s restricted themselves to the habitats of endangered plant and animal species in natural and semi-natural ecosystems. Mapping was therefore selective. An example was the 'Mapping of habitats worthy of protection in the city', undertaken in Munich in 1978 (Brunner *et al.* 1979; Duhme *et al.* 1983). Mapping identified areas with at least some of the following characteristics: species-rich areas and the occurrence of at least some rare species; areas with high structural diversity; and areas of importance for informal recreation and with opportunities for urban dwellers to have contact with nature.

Biotope mapping in Munich was based on its flora, but in Augsburg (Bichlmeier *et al.* 1980; Müller and Waldert 1981) biotope mapping incorporated information on flora, birds, herpetofauna, beetles, butterflies and dragonflies.

A major problem of early survey work was its cost in terms of labour and money. What was needed was an approach to mapping which was quick and inexpensive while retaining its ability to indicate 'value'. One attempt at this compromise approach was undertaken in Düsseldorf by Wittig and Shreiber (1983), who used a point-scoring system associated with four criteria – period of development (i.e. age and continuity of the biotope), area, rarity, and function as habitat. A major advantage of this methodology is that it requires no botanical or zoological survey, simply the recording of vegetation structure, and therefore information can be recorded by non-specialists taking a relatively short period of time.

Comprehensive (spatially-complete) mapping

Ecological mapping has tended to focus on pre-determined 'significant' habitat (Freeman 1999). Since value (for people, plants and wildlife) is context-dependent, however, recent work has often focused on comprehensive, spatially-complete habitat or biotope mapping. The initial map is purely descriptive and non-evaluative.

Seventeen biotope mapping projects had been undertaken in West German cities by 1980; by the mid-1990s this number had risen to over 160 (Starfinger and Sukopp 1994). The success of the German methodology has led to its application elsewhere, for example in Japan where a study began in 1996 in the urban agglomeration of Tokyo (Müller 1998) and in Brazil (Weber and Bede 1998).

A different approach to spatially-complete mapping using overlays of vegetation and land-use has been adopted in the urban wildlife programme developed in the USA by the Department of Environmental Conservation's Division of Fish and Wildlife in 1974, and exemplified by work in New York State (Matthews *et al.* 1988, see also Chapter 36 this volume).

Spatially-complete habitat mapping has been essayed in New Zealand by Freeman and Buck who aimed

> to produce a map that would accommodate the diverse highly modified habitats characteristic of Dunedin and that would incorporate all types of urban open space ranging from indigenous habitats, such as forest, to exotic habitats such as lawns, and residential gardens.
>
> *(Freeman and Buck 2203: 161)*

The project developed a hierarchical classification that considered both land use and habitat, and the resulting map was the first to record all land uses in any New Zealand city at a large scale (1:3000).

Patches, edges, connectivity and corridors: the language of landscape ecology

Habitat corridors are a component part of the geometry of landscape ecology. They provide links between spatially separate habitat patches – opportunities for movement for many plants and non-flying animals, and facilitating movement within a sympathetic habitat for many birds and flying insects.

In their study of Shanghai, China, Zhang *et al.* (2004) demonstrated that as the city became more and more urbanized so there was a reduction in the average size of the habitat/biotope patches, with particularly sharp decreases in the number of large patches. This in turn led to an increase in patch density and the amount of habitat edge. The whole urban landscape became more complex in its pattern, and there was a pronounced decrease in landscape connectivity. As the landscape became geometrically more complex so, ecologically, it became increasingly fragmented.

Work in Oslo, Norway, again used GIS to demonstrate how urbanization has led to pronounced changes in habitat amount, type and distribution pattern (Pedersen *et al.* 2004). In Stockholm, Sweden, Löfvenhaft *et al.* (2002) used a biotope classification in their attempt to provide urban planners with context-sensitive planning tools which would be sufficiently flexible to allow individual responses to the environmental conditions of each individual area. Four types of planning category were identified, each with a different set of requirements for spatial planning and management.

Value-added urban landscapes

It is all very well identifying and mapping what is present in terms of habitat and its spatial configuration, but three key interrelated questions remain:

- In the planning and management of green space, how possible is it to compromise between the often conflicting demands on an area by plants and wildlife on the one hand, and people (for amenity, recreation and education) on the other?
- Is it possible to define different kinds of 'value' that green open spaces may have and to quantify these?
- Can we use any such quantification in measuring enhanced or reduced 'value' following land-use change and reconfiguration of the geometry of landscape compartments or biotopes?

An attempt to quantify 'value', measure habitat change and integrate ecological importance and amenity/recreation importance in an urban landscape made by Young and Jarvis (2001a; 2003) used a 'Habitat Value Index' (HVI) which evaluated each habitat patch individually then used GIS to view this patch within the context of its immediate site (neighbourhood) and also the regional landscape. Predicted future changes in HVI could be computed for each habitat patch or group of patches. That the information was stored in a GIS allowed further quantitative analysis of habitat fragmentation and the structural heterogeneity of the area (Young and Jarvis 2001b, 2001c). A GIS–based approach to evaluation has also been used by Herbst and Herbst (2006) in examining the importance of wasteland sites as urban wildlife areas in Birmingham, UK, and Leipzig, Germany.

Green public open space

The provision of urban green space, its accessibility and its attractiveness have all been studied in four urban areas in Flanders, northern Belgium – Antwerp, Ghent, Aalst and Kortrijk (Van Herzele and Wiedemann 2003). Green spaces should, if possible, have multiple uses: people use such landscapes – parklands, playing fields, urban forests, etc. – often without regard to their original purpose.

A practical guide to mapping and assessing the resource and implementing local standards for the provision of accessible natural greenspace in towns and cities is provided by the Centre of Urban and Regional Ecology (Handley *et al.* 2003) on behalf of English Nature in an attempt to provide a means of establishing urban areas within or outside English Nature-recommended availability of public open space. The need to consider site quality is recognized, a small high-quality greenspace having greater value than a large one of poorer quality. An example of such mapping for a real area is shown for Sheffield at www.greenstructureplanning.eu/COSTC11/Sheffield/sh-basicfacts.htm.

Conclusions

There remain a number of limitations concerning the acquisition, storage, cartographic representation, analytical modelling, interpretation and use of spatial data appropriate to planning for a green urban environment. Nevertheless a great deal of progress has been made, particularly in the last 30 or so years. The rise of and access to technologies such as satellite imagery, aerial photographs and various geographic information systems means that it is becoming easier not only to gather data but also to analyse it, and model scenarios pertinent to the conservation and enhancement of urban nature. The increasing use of GIS, web sites and computerized databases also means that data can be updated, transferred and shared more readily, and a more informed and professional approach adopted.

References

Barker, G. (ed.) (2000) *Ecological Recombination in Urban Areas: Implications for Nature Conservation*, Peterborough: English Nature.

Bichlmeier, F., Brunner, M. Patsch, J., Mück, H. and Wenisch, E. (1980) 'Biotope mapping in the city of Augsburg', *Garten & Landschaft*, 7(80): 551–9.

Brunner, M., Duhme, F. Mück, H., Patsch, I. and Wensich, E. (1979) 'Kartierung erhaltenswerter Lebensräume in der Stadt', *Das Gartenamt, Sonderdruck*, 28: 1–8.

Canters, K.J., Den Herder, C.P., De Veers, A.A., Veelenturf, P.W.M. and De Waal, R.W. (1991) 'Landscape-ecological mapping of the Netherlands', *Landscape Ecology*, 5(3): 145–62.

Chertov, O.G., Kuznetsov, V.I. and Kuznetsov, V.V. (1996) 'The Environmental Atlas of St Petersburg City area, Russia', *Ambio*, 25(8): 533–5.

Collins, W.G. and El-Beik, A.H.A. (1971) 'The acquisition of urban land use information from aerial photographs of the City of Leeds', *Photogrammetrica*, 27(2): 71–92.

Dahl, J. (1908) 'Grundsätze und Grundbegriffe der biozönotischen Forschung', *Zoologischer Anzeiger*, 33: 349–53.

Duhme, F. and 15 others (1983) *Kartierung schutzwürdiger Lebensräume in München* Munich: TU.

Dwyer, J.F., McPherson, E.G., Schroeder, H.W. and Rowntree, R.A. (1992) 'Assessing the benefits and costs of the urban forest', *Journal of Arboriculture*, 18(5): 227–34.

Forman, R.T.T. (1995) *Land Mosaics: The Ecology of Landscapes and Regions*, Cambridge: Cambridge University Press.

Freeman, C. (1999) 'GIS and the conservation of urban biodiversity', *Urban Policy and Research*, 17(1): 51–60.

Freeman, C. and Buck, O. (2003) 'Development of an ecological mapping methodology for urban areas in New Zealand', *Landscape and Urban Planning*, 63(3): 161–73.

Gill, S.E., Handley, J.F., Ennos, A.R., Pauleit, S., Theuray, N. and Lindley, S.J. (2008) 'Characterising the urban environment of UK cities and towns: a template for landscape planning', *Landscape and Urban Planning*, 87(3): 210–22.

GLC (1985) *Nature Conservation Guidelines for London, Ecology Handbook 3*, London: Greater London Council.

Godefroid, S. (2001) 'Temporal analysis of the Brussels flora as indicator for changing environmental quality', *Landscape and Urban Planning* 52(4): 203–24.

Goode, D.A. (1989) 'Urban nature conservation in Britain', *Journal of Applied Ecology*, 26(3): 859–73.

Goode, D.A. (1999) 'Habitat survey and evaluation for nature conservation in London', *Deinsea* (Natural History Museum of Rotterdam) 5: 27–40.

Goode, D.A. (2005). 'Connecting with nature in a capital city: The London Biodiversity Strategy', in T. Trzyna (ed.), *The Urban Imperative*, Sacramento, CA: California Institute of Public Affairs and IUCN, pp. 75–85.

Graf, A. (1986) 'Stadtbiotopkartierung in England', *Landschaft + Stadt*, 18: 120–7.

Handley, J., Pauleit, S., Slinn, P., Barber, A., Baker, M., Jones, C. and Lindley, S., for English Nature (2003) *Accessible Natural Green Space Standards in Towns and Cities: A Review and Toolkit for their Implementation, Research Report No. 526*, Peterborough: English Nature.

Harris, S. and Rayner, J. (1986) 'Urban fox (*Vulpes vulpes*) population estimates and habitat requirements in several British cities', *Journal of Animal Ecology*, 55(2): 575–91.

Herbst, H. and Herbst, V. (2006) 'The development of an evaluation method using a geographic information system to determine the importance of wasteland sites as urban wildlife areas', *Landscape and Urban Planning*, 77(1–2): 178–95.

Herold, M., Scepan, J., Müller, A. and Günther, S. (2002) 'Object-oriented mapping and analysis of urban land use/cover using IKONOS data', *Proceedings 22nd EARSEL symposium "Geoinformation for European-wide integration*, Rotterdam: Millpress Science.

Jarvis, P.J. (1996) 'Planning for a green environment in Birmingham', in A.J. Gerrard and T.R. Slater (eds), *Managing a Conurbation: Birmingham and its Region*, Studley: Brewin Books, pp. 90–100.

Kent, M., Stevens, R.A. and Zhang, L. (1999) 'Urban plant ecology patterns and processes: a case study of the flora of the City of Plymouth, Devon, UK', *Journal of Biogeography* 26(6): 1281–98.
Kim, H.O., Lakes, T., Kenneweg, H. and Kleinschmit, B. (2005) 'Different approaches for urban habitat type mapping – the case study of Berlin and Seoul', in M. Moeller and E. Wentz (eds), *3rd International Symposium Remote Sensing and Data Fusion over Urban Areas*, Lemmer, Netherlands: GITC.
Löfvenhaft, K., Björn, C. and Ihse, M. (2002) 'Biotope patterns in urban areas: a conceptual model integrating biodiversity issues in spatial planning', *Landscape and Urban Planning*, 58(2): 223–40.
Maantay, J. and Ziegler, J. (2006) *GIS for the Urban Environment*, Redlands, CA: ESRI Press.
Mathieu, R., Freeman, C. and Aryal, J. (2007) 'Mapping private gardens in urban areas using object-oriented techniques and very high-resolution satellite imagery', *Landscape and Urban Planning*, 81(3): 179–92.
Matthews, M.J., O'Connor, S. and Cole, R.S. (1988) 'Database for the New York State Urban Wildlife Habitat Inventory', *Landscape and Urban Planning*, 15: 23–37.
Müller, N. (1998) 'Assessment of habitats for natural conservation in Japanese cities – procedure of a pilot study on biotope mapping in the urban agglomeration of Tokyo', in J. Breuste, H. Feldmann and O. Ohlmann (eds), *Urban Ecology*, Berlin: Springer-Verlag, pp. 631–5.
Müller, N. and Waldert, R. (1981) 'Erfassung erhaltenswerter Lebensräume für Pflanzen und Tiere in der Stadt Augsburg – Stadtbiotopkartierung', *Natur und Landschaft*, 56(11): 419–29.
Nakamura, T. and Short, K. (2001) 'Land-use planning and distribution of threatened wildlife in a city of Japan', *Landscape and Urban Planning*, 53(1): 1–15.
Nature Conservancy Council (1990) *Handbook for Phase 1 Habitat Survey: A Technique for Environmental Audit*, Peterborough: UK Joint Nature Conservancy Council.
Nowak, D.J. (1993) 'Historical vegetation change in Oakland and its implications for urban forest management', *Journal of Arboriculture*, 19(5): 313–19.
Nowak, D.J., Rowntree, R.A., McPherson, E.G., Sisinni, S.M., Kermann, E.R. and Stevens, J.C. (1996) 'Measuring and analyzing urban tree cover', *Landscape and Urban Planning*, 36(1): 49–57.
Owen, S.M., MacKenzie, A.R., Bunce, R.G.H., Stewart, H.E., Donovan, R.G., Stark, G. and Hewitt, C.N. (2006) 'Urban land classification and its uncertainties using principal component and cluster analysis: a case study for the UK West Midlands', *Landscape and Urban Planning* 78(4): 311–21.
Pedersen, Å.Ø., Nyhuus, S., Blindheim, T. and Krog, O.M.W. (2004) 'Implementation of a GIS-based management tool for conservation of biodiversity within the municipality of Oslo, Norway', *Landscape and Urban Planning*, 68(4): 429–38.
Phinn, S., Stanford, M., Scarth, P., Murray, A.T. and Shyy, P.T. (2002) 'Monitoring the composition of urban environments based on the vegetation-impervious surface-soil (VIS) model by subpixel analysis techniques', *International Journal of Remote Sensing*, 23(20): 4131–53.
Pownall, L.L. (1950) 'Photo interpretation of urban land use in Madison, Wisconsin', *Photogrammetric Engineering and Remote Sensing*, 16(3): 414–26.
Ridd, M.K. (1995) 'Exploring a V-I-S (vegetation-impervious surface-soil) model for urban ecosystem analysis through remote sensing: comparative anatomy for cities', *International Journal of Remote Sensing*, 16(12): 2165–85.
Saunders, G., White, P.C.L. and Harris, S. (1997) 'Habitat utilization by urban foxes (*Vulpes vulpes*) and the implications for rabies control', *Mammalia*, 61: 497–510.
Starfinger, H. and Sukopp, H. (1994) 'Assessment of urban biotopes for nature conservation', in E.A. Cook and H.N. Van Lier (eds), *Landscape Planning and Ecological Networks*, Amsterdam: Elsevier, pp. 89–115.
Teagle, W.G. (1978) *The Endless Village* Shrewsbury: Nature Conservancy Council.
Van Hertzele, A. and Wiedemann, T. (2003) 'A monitoring tool for the provision of accessible and attractive urban green spaces', *Landscape and Urban Planning*, 63(2): 109–26.
Weber, M. and Bede, L.C. (1998) 'Comprehensive approach to the urban environmental status in Brazil using the biotope mapping methodology', in J. Breuste, H. Feldmann and O. Ohlmann (eds), *Urban Ecology*, Berlin: Springer-Verlag, pp. 636–40.
Wittig, R. and Schreiber, K.-F. (1983) 'A quick method for assessing the importance of open spaces in towns for urban nature conservation', *Biological Conservation*, 26(1): 57–64.
Young, C.H. and Jarvis, P.J. (2001a) 'A simple method for predicting the consequences of land management in urban habitats', *Environmental Management*, 28(3): 375–87.
Young, C.H. and Jarvis, P.J. (2001b) 'Measuring urban habitat fragmentation: an example from the Black Country, UK', *Landscape Ecology*, 16(7): 643–58.
Young, C.H. and Jarvis, P.J. (2001c) 'Assessing the structural heterogeneity of urban areas: an example from the Black Country (UK)', *Urban Ecosystems*, 5(1): 49–69.

Young, C.H. and Jarvis, P.J. (2003) 'A multicriteria approach to evaluating habitat change in urban areas: an example from the Black Country, West Midlands (UK)', in J. Brandt and H. Vejre (eds), *Multifunctional Landscapes Vol. II: Monitoring, Diversity and Management* Southampton: WITPress, pp. 59–74.

Zhang, L., Wu, J., Zhen, Y. and Shu, J. (2004) 'A GIS-based gradient analysis of urban landscape patterns of Shanghai metropolitan area, China', *Landscape and Urban Planning*, 69(1): 1–16.

38

Invasive species and their response to climate change

Ian Douglas

Invasive plants

Invasive plants are typically non-native plants that aggressively out-compete other vegetation. Some aggressive native plants can act in an invasive manner in disturbed environments. When invasive, non-native species grow outside of their natural range they are not controlled by the natural interactions of predators, parasites, diseases and competition from other plants. Invasive plants, or novel species, have aggressive reproductive qualities such as rapid growth, abundant seed production, widespread seed dispersal and vigorous vegetative spreading. Invasive plants are highly adaptable and are able to tolerate a wide range of habitat conditions.

Novel species of this kind are high frequency immigrants into urban areas, creating some potential for high, but not necessarily appreciated, biodiversity (Rebele 1994). Novel species often cause species composition changes that erode biodiversity, for example through competition and predation.

Urban natural areas are often the most deeply affected because they are more disturbed and the parcels of land are smaller. The urban wilds that contain the most exotic invasive plants are those that were once cleared for housing plots, roads, or industrial use. These kinds of sites were abandoned and managed to lie fallow long enough to regrow with a mix of natives and exotics.

Many invasive plants were introduced to North America, Australia and New Zealand during European settlement. These plants were introduced intentionally for aesthetics and food value; however, others were introduced unintentionally along with crop seeds or in ship ballasts. People travelling to other countries and continents brought back plants and seeds to beautify their gardens and add to botanical collections. Many such plants escaped and spread to become nuisances. Invasive plants continue to be introduced today through the horticultural trade and by spreading from established communities. Invasive plants on private properties find their way into parklands by wind; flooding; seeding by wildlife; tracking in seeds caught on pets, clothing and shoes; and the dumping of fill, garden waste and compost.

Many alien species are found in cities throughout the world (Franeschi 1996; Kent *et al.* 1999; Niemelä 1999). The numbers are slightly less in Mediterranean cities than generally elsewhere in Europe (Chronopoulos and Christodoulakis 1996; Celesti Grapow and Blasi 1998). In Almería City, Spain, alien species accounted for 17 per cent of the 106 plants enumerated (Dana *et al.* 2002). The general reasons for these high proportions of invasive species are not always clear. They are complex, depending on how the plants are introduced and the way in which activities in cities create opportunities for them to spread. One factor is the urban climate, with its greater aridity and warmer heat island than the surrounding countryside (Davison 1977; Kunick 1980; Rebele 1994). However, the disturbance regime in the city is also highly important (Fox and Fox 1986), especially the intensity and frequency of disturbance (Montenegro *et al.* 1991). Severe disturbance, such as that caused by bombing, can create great opportunities for invasive species to prosper.

Good examples of how particular species are initially introduced and then subsequently spread through a combination of deliberate human actions and opportunities created by disturbance are provided by the Rose-bay willow herb (*Epilobium augustifolium*) and the ragworts (*Senecio* spp.) in the UK. The Oxford ragwort (*S. squalidus*) now widespread in London (Burton 1974; Fitter 1945) was brought from the volcanic ash fields of Sicily to the Oxford Botanic Gardens in 1794. Seeds carried from the gardens enabled the ragwort to reach London by 1867 where it flourished on areas damaged by bombing in the 1939–45 world war. Rose-bay willow-herb invades any unused city land, especially where buildings have been demolished. However it is a relatively recent introduction, being first recorded in Cheshire, for example, near Crewe railway station in 1905 (Newton 1971).

Japanese knotweed (*Fallopia japonica*) has become a major problem, but it was awarded a gold medal in 1847 by the Society for Agriculture and Horticulture at Utrecht as the most interesting new ornamental of the year (Bailey and Connolly 2000). The earliest records for its cultivation are for gardens near Leeds (1881) and Altrincham, Greater Manchester (1883). Its earliest recorded escape from a garden was from Melrose Abbey, Scotland. From the beginning of the twentieth century, the spread of Japanese knotweed began to be a problem. In East Cornwall by 1930 it was known as 'Hancock's Curse', perhaps because it was sold by a local nursery, Hancocks of Liskeard, in the 1900s.

In 1998, a survey of 400 km^2 of Swansea City found that *F. japonica* occupied some 99 ha spread across a variety of habitats. It had been noted as a frequent garden escape in Glamorgan as early as 1907 (Bailey and Connolly 2000). Now, it alters the structure and dynamics of stream detritus food webs, and reduces the light available for primary producers on stream beds because of its dense canopy (Lecerf *et al.* 2007). It causes serious problems for river managers, amenity areas and nature reserves, with much money being spent on its control. In 1981, Japanese knotweed and Giant Hogweed were proscribed by the Wildlife and Countryside Act, making it an offence to dispose of them in the wild. Dumping of material from construction sites on to derelict land was one of the major contributors to its spread.

Himalayan Balsam (*Impatiens glandulifera* Royle) was introduced into the British Isles in 1839, was recorded in Middlesex as a naturalized alien in 1855 and thence spread rapidly, being recorded in Manchester in 1859. It spread so rapidly that it was given weed status in 1898, less than 60 years after its first introduction (Perrins *et al.* 1993). It is an annual herb with serious nuisance potential since it is believed that it may be able to out-compete even hardy, well-established, native perennial species in riparian habitats (Beerling and Perrins 1993). Part of its success comes from its popularity as a garden plant, it being frequently transported to new localities. Its explosive seed pods send out about 600 seeds per plant. But most of those seeds fail to germinate. Nevertheless, it can spread through germination alone at a rate of two to three metres per year.

Himalayan balsam has altitudinal and latitudinal limitations that influence its time to flowering and height of growth. However, having been in Europe for 150 years, the plant may be adapting to conditions at the fringes of its present distribution. Measured variations in growth and phenology of Himalayan balsam plants grown from seeds for a 20 degree latitudinal range across Europe may be genetically based and correlated with ecophysiological requirements for flower initiation (Kollmann and Bañuelos 2004). Climate change may also influence its ability to spread to higher latitudes.

Control measures for Himalayan balsam are problematic. Some UK local wildlife trusts organize "balsam bashing" events to help control the plant. However in some circumstances, such efforts may cause more harm than good (Hejda and Pyšek 2006). Destroying riparian stands of Himalayan Balsam can open up the habitat for more aggressive invasive plants such as Japanese knotweed and aid in seed dispersal (by dropped seeds sticking to shoes). Riparian habitat is suboptimal for *I. glandulifera*. Spring or autumn flooding destroys seeds and plants. The best way to control the spread of riparian Himalayan Balsam may be to decrease eutrophication, so allowing the better-adapted local vegetation that gets outgrown by the balsam on watercourses with high nutrient loads to rebound naturally. For areas away from streams, the best option is to manually clear affected areas. In 2010, the UK government announced that it would allow a natural predator to be used to help control Himalayan balsam. The effectiveness and outcome of such action will be watched closely by all concerned with urban nature and the countryside.

Invasive animals and insects

Many insects and animals have spread as a result of transport of foodstuffs from distant areas. Cockroaches such as *Blatta orientalis* and *Periplaneta Americana* introduced long ago are now widespread in houses and buildings in Europe (Owen 1978). The ship rat (*Rattus rattus*) in countries on both sides of the Atlantic Ocean tends to live inside buildings adjacent to docks or, in a few major cities such as central London, in a dense zone of multi-storied, centrally heated, largely non-residential buildings interspersed with restaurants, canteens and other sources of food (Bentley 1959). The feral pigeon, now a characteristic feature of large cities in northwest Europe, originally domesticated by humans from the wild rock dove (*Columba livia*), was allowed to escape from captivity when meat supplies became more abundant through improved methods of preservation and distribution (Lever 1977). Control of pigeons in cities costs the USA an estimated $1.1 billion/yr. This is based on a cost of at least $9 per pigeon per year and an assumption of 1 pigeon per ha in urban areas, or approximately 0.5 pigeons per person. These control costs exclude the environmental damage and disease impacts associated with pigeons, which carry such diseases as parrot fever, ornithosis, histoplasmosis, and encephalitis (Pimentel 2005).

Irresponsible breeding and release of pet raccoons (*Procyon lotor*) has led to their naturalization in Japan. Already naturalized in 42 of Japan's 47 prefectures, raccoons numbers have increased remarkably in urban areas, where they can find food and hide easily (Ikeda *et al.* 2004). Raccoons that escaped in Kamakura city, with its many old wooden houses, Buddhist temples and Shinto shrines invaded the air spaces and built nests under the eaves and floors of some of these old buildings. Raccoons displace other species and are now being controlled under the 2005 Alien Invasive Species Act.

In Europe, raccoons were intentionally released for hunting and its fur in Germany. Escapes from fur farms, zoological gardens and from animal husbandries have occurred in several countries: France, Russia and several other countries of Europe. Now spread throughout the areas from France to Poland and the Netherlands to Hungary, the raccoon is expected to spread towards south-eastern Europe. It has a preference for woody habitats adjacent to fresh water or for urbanized areas (Michler *et al.* 2004).

Many insects have been brought into countries and continents through the transport of food and plants from other continents. Some spread naturally, others move as agricultural products are carried from producers to markets. Among the most invasive insects now in Europe that occupy habitats in urban areas, are some that use a narrow range of plants, such as the Colorado beetle and the horse chestnut leaf-miner (Table 38.1) and others that are more generally found such as the harlequin ladybird.

Table 38.1 The most invasive alien insect species in urban areas in Europe (compiled from data in DAISIE 2009)

Species	*Common name*	*EUNIS urban habitats*	*Issues*
Aedes albopictus	Asian tiger mosquito	J6	Mostly opportunistic container breeder capable of using any type of artificial water container, especially discarded tyres, tin cans and plastic buckets. Vector for at least 22 arboviruses
Aphis gossypii Glover	cotton aphid, melon aphid	I2	Impact especially important on courgette, melon, cucumber, aubergine, strawberry, cotton, mallow and citrus
Arion vulgaris	Spanish slug	I2	Most important slug pest in Europe causing severe damage to horticultural plants in private and public gardens
Cameraria ohridella	horse chestnut leaf-miner	I2, J	A single tree can host up to 10^6 leaf-miners that severely damage horse chestnut trees. Defoliation results in smaller seeds
Frankliniella occidentalis	western flower thrips	I2	Present in glasshouses in N and central Europe but already in the field in S Europe. Continuous and rapid spread since 1980. Flowers and foliage of a great number of economically important crops are affected including urban garden centres
Harmonia axyridis	harlequin ladybird	I2, J1	Introduced intentionally as a biocontrol agent for aphids and unintentionally in horticultural/ornamental material. Pest of orchard crops (apples and pears) and grapes
Leptinotarsa decemlineata	Colorado beetle	I2	Serious pest of potatoes and of other solanaceous plants such as tomato, aubergine, tobacco and peppers. Capable of adapting to different climatic conditions
Linepithema humile	Argentine ant	I2	Associated with disturbed, human-modified habitats in its introduced range. Competes with other arthropod species for resources (e.g. for nectar with bees). Regarded as a nuisance for tourism at some places on the Mediterranean coast
Liriomyza huidobrensis	serpentine leaf miner	I2	Present outdoors in S Europe, but mainly a glasshouse pest in N Europe, affects cut flower and vegetable crops
Spodoptera littoralis	African cotton leaf worm	I2	One of the most destructive agricultural lepidopteran pests within its subtropical and tropical range. In N Africa it damages vegetables, in Egypt cotton and in S Europe plant and flower production in glasshouses or vegetables and fodder crops

Other invasives in cities

Alien bryophyte species in Europe are widely found in habitats that are frequently disturbed by human activity, especially gardens, roadsides, and walls (Essl and Lambdon 2009). Urban bryophyte species are likely to occur in cities throughout the world. Human-aided dispersal, and the novelty and homogeneity of the urban environment across the world are likely to have played a major role in forming this distribution pattern, and many species are likely to be alien in parts of their range. Alien bryophytes and lichens can alter ecosystem functioning occasionally by stabilizing soils, binding leaf litter, altering decay rates, and creating humid microhabitats which affect the composition of microfaunal communities. Relatively little is known about these impacts at the microhabitat scale (Essl and Lambdon 2009).

The three most invasive fungi in Europe all affect trees in cities, particularly Dutch elm disease and cypress cankers (Table 38.2). While ink disease is highly sensitive to frost and is generally thermophilic, with the minimum temperature for growth 5–6°C, optimum 24–28°C and maximum 32–34°C, its range is likely to spread with global warming.

Invasive species in different cities

Berlin has 839 native and 593 alien species of plant (McNeely 2001). Of the species in Berlin 41 per cent are aliens, compared to an average of 40.3 per cent in a survey of Central European cities (Pyšek 1998), 32.3 per cent in Plzen, Czech Republic (Chocholoušková and Pyšek 2003) and 32.3 per cent in Plymouth, UK (Kent *et al.* 1999). In Brussels, 20 per cent of all species were neophytes (post-1880 introductions) compared to 22.9 per cent in Plymouth (Kent *et al.* 1999), 17 per cent in Plzen, Czech Republic (Chocholoušková and Pyšek 2003) and 25.2 per cent in Central European cities generally (Pyšek 1998). There are considerable differences in the most successful neophytes between Berlin and Brussels (Table 38.3), perhaps reflecting environmental contrasts between the two cities.

Table 38.2 The three most invasive alien fungi species in urban areas in Europe (compiled from data in DAISIE, 2009)

Species	*Common name*	*EUNIS urban habitats*	*Issues*
Ophiostoma novo-ulmi	Dutch elm disease	I2	responsible for the second pandemic of Dutch elm disease. Introduced to Great Britain *c.*1960, probably with a shipment of elm logs. Twenty-five million elm trees die in the UK. Few mature elms left in all Europe
Phytophthora cinnamomi	ink disease	I2	A soil-borne pathogen causing disease on many woody hosts. Natural habitats in France, Italy, Spain and Portugal, in other countries occurs mainly in nurseries
Seiridium cardinale	cypress canker	I2	Has caused the loss of millions of cypress trees in S Europe. Cypress trees today still embellish historical sites and gardens. Major issue for tree nurseries

Table 38.3 The most successful neophytes in Berlin and in Brussels and Chicago (with frequency of occurrence) (after Godefroid, 2001 and Gulezian and Nyberg, 2010). Frequency is the percentage of sample units or grid squares in which the species was found.

Berlin	*Brussels*	*Frequency (Brussels) %*	*Chicago*	*Frequency (Chicago) %*
Acer negundo	*Conyza canadensis*	79	*Cirsium vulgare*	100
Ailanthus altissima	*Acer platanoides*	74	*Rhamnus cathartica*	100
Bidens frondosa	*Matricaria discoidea*	73	*Alliaria petiolata*	94
Clematis vitalba	*Fallopia japonica*	71	*Arctium minus*	94
Matricaria discoidea	*Buddleja davidii*	70	*Rosa multiflora*	94
Parietaria pensylvanica	*Mercurialis annua*	68	*Melilotus alba*	81
Prunus serotina	*Robinia pseudoacacia*	67	*Setaria glauca*	75
Robinia pseudoacacia	*Galinsoga ciliata*	66	*Phragmites australis*	69
Rumex thyrsiflorus	*Oxalis fontana*	53	*Dipsacus silvestris*	69
Sisymbrium loeselii	*Solidago gigantea*	51	*Ailanthus altissima*	13
Solidago canadensis	*Lamium galeobdolon subsp. Montanum*	49	–	–

A west-east transect through Chicago, USA, was used to document and analyse patterns of abundance of ecologically problematic invasive plants across a gradient of spatial urban structure. Correlations of the urban variables with the abundance of invasive plants were expected to identify if and how land use planning affects abundances of ecologically problematic species (Gulezian and Nyberg 2010). Most invasive species were found in a wide range of urban land uses. Burdock however was negatively correlated with the proportion of commercial or industrial land. It was also negatively correlated with income, suggesting that perhaps the inhabitants of wealthy districts have the resources to remove the tightly adhering seeds more effectively. Pets and larger urban mammals such as opossums (*Didelphis virginiana*) and raccoons (*Procyon lotor*) are probably responsible for the spread of burdock in cities like Chicago (Gulezian and Nyberg 2010).

The lack of a general relationship between invasive plants and land use in Chicago suggests that the diversity of situations that plants colonize occurs at a finer scale than land use mapping. Within a single garden or derelict factory site, the variety of substrates and canopy density will create many micro-habitats. Scattering of seeds by dispersal agents, such as birds or small mammals, as well as a host of human activity will not overall be related to land use. This information will not help landscape planners to avoid invasive plants, but will suggest all gardeners should be vigilant.

Climate change and invasive species

Cities are experiencing the consequences of both changes in global trade patterns as a consequence of economic and social globalization and the consequences of global environmental changes, driven largely by the burning of fossil fuels. The climatic changes are already having impacts on the flowering of plants in cities (Fitter and Fitter 2002). In Boston, MA, USA, over the last 100 years the trend has been for plants to flower progressively earlier. Now, in an average year, they flower eight days earlier than they did around 1900 (Primack *et al.* 2004). Of 100 plant species examined in Washington DC from 1970 to 1999 (Abu-Asab *et al.* 2001), 89 per cent flowered earlier and 11 per cent later, compared to an overall mean advancement of 2.4 days in 30 years, and extreme shifts of –46 to 110 days. Notably, in one alien species, the introduced shrub *Buddleja davidii* (Buddlejaceae), flowering was delayed during the 1990s by 36 days compared with the 1954–1990 mean. The average first flowering date of 385 British plant species advanced by 4.5 days during the 1990s compared with the previous four decades: 16 per cent of species flowered significantly earlier in the 1990s than previously, with an average advancement of 15 days in a decade (Fitter and Fitter 2002). These phenological changes, along with other effects of climate change such as changes in geographic range, will alter population-level interactions and community dynamics and have far-ranging consequences for evolution and ecosystem dynamics.

Global environmental change has implications for the spread of invasive species and pests in all countries. The paragraphs that follow examine the indicators of what is likely to happen in the British Isles. Similar trends could be expected elsewhere. However, as the Buddleia example above shows, not everything will change in the same way. The patient observations of year-to-year changes in phenology and ecology will always be needed.

Climate change and invasive species in urban areas

With climate change, non-native species from adjacent areas may cross frontiers and become new elements of the biota (Walther *et al.* 2002). Globally, many species are already being affected by the slight climatic changes brought about by global warming. Thermophilous plants have begun to spread from gardens into surrounding open spaces (Dukes and Mooney 1999; Walther 2000). Pest distributions are beginning to change and diseases of plants and animals are spreading, not merely because of alterations to the climate, but through the combined effect of the whole range of human impacts on the environment and ecosystems. Many pathogens of terrestrial and marine taxa are sensitive to temperature, rainfall and humidity, creating synergisms that could affect biodiversity. Climatic warming may increase pathogen development and survival rates, disease transmission and host susceptibility. Although most host-parasite systems are expected to experience more frequent or severe disease impacts with warming, a subset of pathogens might decline with warming, releasing hosts from disease. Since 1990, changes in El Niño-Southern Oscillation events have had a detectable influence on marine and terrestrial pathogens, including coral diseases, oyster pathogens, crop pathogens, Rift Valley fever and human cholera (Harvell *et al.* 2002).

Warmer and drier conditions would be likely to lower the incidence of diseases such as white pine blister rust and scleroderris canker that require cool, wet conditions for infection, and in the case of scleroderris canker, an extended cold period. However, warmer and drier conditions will be likely to increase root rot damage, particularly by *Heterobasidion annosum*, which is restricted in its range by a cooler climate. Similarly, diseases such as oak wilt, which have had a serious impact on oak species in the United States, would become a greater problem with a warmer climate (Boland *et al.* 2004). As the climate changes, new combinations of host-stress-saprogens might give rise to new types of decline diseases, particularly in tree species. This poses particular problems for the urban forest and especially for street trees.

The interaction between pathogens and hosts under changing climatic conditions will affect tree health. Climate change could alter stages and rates of development of the pathogen, modify host resistance, and result in changes in the physiology of host-pathogen interactions. The most likely consequences are shifts in the geographical distribution of host and pathogen and altered crop losses, caused in part by changes in the efficacy of control strategies. Climate change will add another layer of complexity and uncertainty for managers of urban ecosystems.

Understanding the emergence of infectious plant diseases (EIDs) requires knowledge of host-parasite biology. However, it is possible to identify broad trends in plant EID emergence by analysing hypothesized environmental drivers (Anderson *et al.* 2004). The emergence of plant EIDs is driven mainly by anthropogenic environmental change (such as introductions, farming techniques and habitat disturbance). For plant EIDs, these changes are those largely related to trade, land use and severe weather events (predicted to increase in frequency and severity owing to anthropogenic global climate change). Although only a fraction of the parasite community of an invasive host is introduced during host introduction events, pathogen introduction is the major driver of plant EID emergence. Endemic species might be particularly vulnerable to introduced pathogens with which they have not coevolved.

Despite the complexity of climate changes and the biotic responses to them, some broad trends can be suggested. Climate change can lead to disease emergence through gradual changes in climate (e.g. through altering the distribution of invertebrate vectors or increasing water or temperature stresses on plants) and a greater frequency of unusual weather events (e.g. dry weather tends to favour insect vectors and viruses, whereas wet weather favours fungal and bacterial pathogens). Thus, climate change can lead to the emergence of pre-existing pathogens as major disease agents or can provide the climatic conditions required for introduced pathogens to emerge (Anderson *et al.* 2004). The ranges of several important crop insects, weeds and plant diseases have already expanded northward. Global climate change is likely to change the distribution and abundance of arthropod vectors and to increase the frequency of unusual weather events, one of the major drivers of plant EIDs.

Cane toads in Australian urban areas

The spread of cane toads in tropical Australia has long been a problem. Cane toads appear to be most successful in open habitats associated with human disturbances, such as roadsides and suburban developments (Lever 2001). Statistical modelling of future cane toad colonization in Australia supported this observation and indicated that most large coastal Australian cities are susceptible to cane toad colonization and hence make ideal targets for control efforts (Urban *et al.* 2007). Cane toads distributions will be influenced by climate change, but already they are moving into areas at the extreme temperature ranges of their distribution, into areas of higher maximum monthly temperatures and other areas of lower minimum temperatures. This has an important warning about forecasting the impact of climate change on invasive species. Invasion dynamics may include a fourth stage (after arrival, establishment and spread). Once adaptation and niche expansion are sufficient, an expanding invasive population may experience adaptive genetic variation (Urban *et al.* 2007).

Pests and diseases under climate change

In the UK, as elsewhere, climate change will affect pests and diseases, influencing plant and animal life in public and private urban areas of managed and natural vegetation from gardens to nature reserves and derelict land. This will have cost implications associated with control or extra urban greenspace management. Already several insect pests common further south in Europe, such as the lily beetle, berberis sawfly and rosemary beetle, have become established in Britain (Prior 2003). Termites have already been found in Cornwall (Bisgrove and Hadley 2002). A comprehensive review of the effects of climate change on insect pests is given by Cannon (1998). A major report on the effects of climate change on gardens (Bisgrove and Hadley 2002) contains an excellent bibliography on, and a good summary of, likely changes to the major pests liable to be found in UK gardens.

Early work on predicting the likely spread of agricultural insect pests in Europe (Porter *et al.* 1991) showed potential shifts of up to 1220 km north in the distribution of the European corn borer (*Ostrinia nubilalis*). However, the authors of this work also noted that changes in climate would not only result in changes in insect distribution, but would possibly lead to increased over-wintering, changes in population growth rates, increases in the number of generations, extension of the development season, changes in plant-pest synchrony, changes in interspecific interactions and increased invasion by migrant plants.

Climate change impact on British urban gardens and greenspaces

Global warming in England is a reality. Spring 2002, for example, had temperatures in February to April some 2.6°C above the 30-year average (Sparks and Collinson 2003). The precipitation gradient across Britain – the ratio of winter precipitation in Scotland to summer precipitation in south-east England – has increased markedly in recent decades (Porley and Hodgetts 2005). Although this warming may enable tender plants to be grown outside in gardens where hitherto such species would not have been expected to flourish, unusual pests and diseases may increase too, and this climate change also promises extremes of weather conditions, be it flood, drought or storm (Hepper 2003). The changing climate is already creating ideal conditions for the spread of insects such as the red spider mite and new vine weevil species. Fungal diseases thrive with the wet winter conditions, and *Phytophthora* has already decimated some of the country's historic yew hedges.

Response of insects to climate change

a **Carbon dioxide**
A doubling of carbon dioxide levels could increase plant growth by 40 to 50 per cent. More robust plants would then be likely to show greater freedom of flowering and be more resistant to pest attack.

b **Higher temperatures**
Higher temperatures will favour the spread of over-wintering pests and the survival of introduced pests or those moving northwards. Not only will the range and distribution of butterflies shift dramatically (Fox *et al.* 2001), but also they will appear two to three weeks earlier. The same will be true for pest species of insects. On the other hand, warmer winter temperatures are already leading to greater winter activity in bee populations and are resulting in poorer winter survival. Warmer winter and spring temperatures will increase the over-wintering survival of aphid species in the UK and, in some cases, advance the appearance of winged adults by as much as a month (Bisgrove and Hadley 2002).

c **Wetter winters**
Short-term high levels of soil water, coupled with milder temperatures will increase the risk of wet areas around tree roots and thus the occurrence of root fungi such as *Phytophthora* (Bisgrove 2003).

d **Drier summers**
Reduced water supplies will lead to increases in cell sap concentration. Sucking insects and mites will have a more concentrated food supply and may increase more rapidly. In dry conditions, red spider mite numbers increase in greenhouses. In 1995, the dry summer saw severe impacts of red spider mite and aphids on such outdoor crops as lettuce, apple and raspberry (Orson 1999). Water stress in plants leads to higher nutrient concentrations in the sap, which in turn increases the growth rates and fecundity of the aphids feeding on the plants.

Predator number would also respond to an increase in food supply. Raindrops that dislodge aphids or damage their feeding parts form one of the most important natural controls of aphid populations. Whether lower summer rainfall will mean less removal of aphids will depend on how the rainfall is distributed. Heavier, but less frequent storms could do more damage to aphid populations (Bisgrove and Hadley 2002).

e **More frequent extremes**
More drought conditions and increased flooding, especially in winter, will change the character of habitats and thus the relative proportions of different kinds of pest.

f **Changes in wind speed**
Occasional high winds associated with more frequent storms of gale conditions would affect large trees with leafy canopies in exposed positions. Wind variability may lead to the introduction and dispersal of some species from the south.

g **Sea level rise**
Sea level rise will be greatest in southeastern Britain where the land is subsiding. There will be negligible effects in the northwest due to continuing isostatic rebound of that part of the country. In the southeast, a rise of 15 to 85 cm will affect some coastal urban areas, while changing flood protection works will have important impacts on urban and peri-urban vegetation. Changing wetness of some areas will alter insect breeding grounds.

Changes to plants

Changes in vegetation in response to global warming will occur over decades or centuries. Human activity changing land use and agriculture in response to climate change will open opportunities for invasive species and thus lead to changes in the structure of vegetation. However, vegetation does not move as a unit. Different species respond to change at varying rates (Woodward and Beerling 1997). This will lead to new associations of plants and altered ecosystems, including the associations in managed urban forests, parks and gardens.

Extant colonies of the Red Listed European endemic moss *Zygodon gracilis* in the Yorkshire Dales National Park are found at significantly higher altitudes than extinct colonies (Headley and Rumsey 2003). The increase in average winter temperatures by 0.5°C since 1900 may be sufficient to influence this oceanic-alpine species. The moss *Grimmia tergestina* also appears to be responding to climate change. A species with a predominantly Mediterranean distribution, first noted in Britain in 1966, it was found in 2003 at Great Orme's Head in North Wales, alongside another Mediterranean species, *G. orbicularis*. These northward advances may well indicate the influence of climate change (Porley and Hodgetts 2005).

Changes in larger animals

Warmer winters will favour an increase in the number of grey squirrels. Because the grey squirrel is particularly fond of the bark of beech trees, climate change is likely to increase the difficulty of maintaining or regenerating plantings of beech in gardens, parks and woodlands. Fifty years' work on regeneration of beech woodland on Box Hill, Surrey has had little success because bark stripping has destroyed the young trees. Grey squirrels have also caused widespread damage to sycamore and Japanese maples in National Trust and other gardens (Bisgrove and Hadley 2002).

Rats are now surviving winters in Scandinavian cities. Black rats are alleged to have increased in numbers in UK cities since 1995. UK studies suggest that urban rat numbers have increased since the 1970s, possibly due to changes in the way local authorities and water utilities mange their control. Domestic premises with rat infestations are more likely in the older parts of urban areas with poorer environments and where, it is argued, it is also likely that there will be lower levels of complaint due to greater tolerance of rats even when they are known to be present (Battersby *et al.* 2002). Climate change appears to be merely one of a number of factors influencing the numbers and distribution of rats in the UK. Indeed, reports of tawny owls breeding during mild winters and feeding on small mammals, including rats, suggest that close attention has to be paid to changing inter-species relationships as warmer winters alter ecological conditions.

Understanding of particular plant pests

a **Lily beetle**
Coming from southern Europe and first established in Britain in 1940, its spread accelerated in the warm years of the 1990s.

b **Red spider mite**
Widely known as an indoor greenhouse pest, this mite spreads outdoors in hot dry summers. Climate change would make it much more common in gardens.

c **Berberis sawfly**
During the 1990s, possibly assisted by warm weather conditions, *A. berberidis* spread northwards through Belgium and the Netherlands. Severe infestations of the sawfly occurred around Arnhem in the Netherlands in 2000. It was discovered in a private garden in Essex in 2002 where it had been present for two years. A survey of the local area found the pest in three other locations in the north-west London area: Middlesex, Hertfordshire and Buckinghamshire. It leads to defoliation of *Berberis thunbergii*, *B. vulgaris* and *Mahonia* species in May–June and late July–September (Prior 2003).

d **Rosemary beetle**
First reported outdoors at Wisley in 1994, colonies and individuals have been found across south-eastern England between Norfolk and Surrey. As temperatures rise, it will become a major issue for all gardeners.

e **New species of vine weevil**
Probably imported on plants from the Mediterranean areas, two species of weevil appear to be doing well in gardens in south-west London. They cause extensive damage to the edges of leaves of evergreen garden plants.

f **Horse chestnut leaf mining moth**
Since its discovery in Wimbledon in 2002, the moth has spread to Surrey, Hertfordshire and Oxfordshire. The larvae feed on horse chestnut leaves, causing them to wither and produce less conkers. Trees growing in streets may be more badly affected than those in parks and gardens, due to the warmer microclimate in built-up areas.

g **Umber moth caterpillars**
Mottled umber moth caterpillars devoured an area of woodland in South Yorkshire. Probably a combination of a mild spring and lack of predators was to blame for the freak occurrence.

h **Green vegetable bugs** (*Nezara viridula*)
These bugs attack a broad range of crops, from soft fruits to potatoes and beans, damaging fruit, transmitting disease and leaving plants open to attack by other pests. Three colonies of the shield-shaped green vegetable bug, a crop pest usually found in much warmer climates, are living and breeding in London. After two colonies feeding on tomato plants in London were identified at the Natural History Museum, a third colony was discovered near King's Cross Station, London, at a nature reserve. These bugs are a particular problem in the Mediterranean, the Middle East, Australia, North America and Africa, and could be a significant pest if, as the evidence suggests, they become resident in the British Isles.

i **Wisteria scale** (*Eulecanium excrescens*)
In the autumn of 2001, soft-scale insects *Eulecanium excrescens* caused leaf loss and dieback on a 16m tall Wisteria plant in a private garden in Vauxhall, London. *E. excrescens* has since been found in at least eight other locations in Chelsea, Vauxhall, Earls Court and Chiswick. An Asian species, which has also been introduced to the USA, it was unrecorded in the EU prior to its detection in London and the origin of the UK outbreak is unknown. It is a highly polyphagous species and is recorded feeding on most deciduous orchard trees and many deciduous ornamental trees, although Wisteria appears to be the preferred host. The extent of the infestation and the large size of the affected Wisteria, means that eradicating *E. excrescens* from the UK is not considered viable. The arrival of this pest is most probably through some form of introduction on plant material and not related to climate change. Its survival may be.

j **Elaeagnus sucker** (*Cacopsylla fulguralis*)
The elaeagnus sucker, which also originates from Asia, was first reported in Europe from France in 1999, and is now present in the British Isles. This pest has already caused widespread damage to Elaeagnus hedges in Guernsey, and has also spread to Jersey, and parts of England. The nymphs and adults suck out the sap from the underside of leaves, and excrete honeydew, which falls onto the leaves and encourages the growth of sooty moulds. The covering of this black fungus prevents leaves from photosynthesizing, causing the plant to die back. Outbreaks of this pest were found in quick succession in different parts of the UK from Brighton to York, indicating a very rapid invasion

k **Cottony cushion scale** (*Icerya purchasi*)
I. purchasi, which is considered to be of Australian origin, is polyphagous on a wide range of woody plants and has spread throughout the tropics, subtropics and the Mediterranean. There have been eight findings of the scale outdoors in the UK. First detected on an *Acacia dealbata* tree in central London, and the population is still present having successfully overwintered. In August 2002, numerous immature and adult *I. purchasi* were detected on ornamental plants belonging to 23 genera and causing severe damage to Acacia, Choisya, Hebe, Laurus and Pyracantha. *I. purchasi* appears to be naturally spreading northwards perhaps as a consequence of climate change. In 1999, a severe infestation was reported in the Jardin des Plantes, Paris, and the number of interceptions of the species in trade have increased since 1998. However, it is unlikely to be able to establish widely in the UK beyond the sheltered, warm microclimate of the capital and the south-west.

l **Cypress gall mite** (*Trisetacus chamaecyparis*)
In September 2002, the eriophyid mite, *T. chamaecyparis*, was found for the first time in Britain, causing die-back of a leylandii hedge in a private garden in Cheshire. *T. chamaecypari* was first described in Canada and is also present in the USA in coastal Oregon, California and Pennsylvania; this is the first time it has been recorded on leylandii. It is probably an introduction unrelated to climate change.

Forecasting future climatic impacts on urban pests

A good example of the forecasting of the spread of pests under climate change is the modelling of the expansion of oak disease (Bergot *et al.* 2004). For red oak, the zones most favourable for *P. cinnamomi* survival (in 90 per cent of years) may extend a further 100km eastward from the Atlantic coast. The northward extension from the Mediterranean zone is confined to the Rhone valley, between the mountainous zones of the Massif Central and the Alps. The same and even amplified trends are observed for *P. cinnamomi* survival in pedunculate, or common, oaks. As a consequence of higher *P. cinnamomi* survival, for both oak species, the zones at high disease risk may extend about 1,000 km inland from the Atlantic coast. This potential expansion of the disease (canker symptoms) of one to a few hundred kilometers in one century is of the same order of magnitude the projected potential range shifts for plants in the next century (Bergot *et al.* 2004).

Climate scenario mapping has been widely used to predict the impact of climate change on pest distributions. It has been criticized for ignoring dispersal and interactions between species, such as parasitism, predation and competition (Baker *et al.* 2000). Climate mapping studies should be used to determine whether the establishment of breeding colonies is possible. This follows the principle that for every organism, there is a fundamental niche, produced by physical environmental factors, which differs from the actual realized niche that is determined by biotic factors. What has to be remembered in the urban environment is that the physical factors are constantly being changed by biotic (anthropic) actions, which cumulatively create the urban heat island effect, but which also produce a whole series of thermal, moisture and biogeochemical conditions that create a host of fundamental (micro-) niches in which introduced and invasive species can survive.

With our increasing understanding of the introduction of species through both international passenger traffic and international trade flows, from super tankers discharging ballast waters to tiny insects on imported fruit and flowers, a major change in urban biodiversity is taking place. In 1995, the UK Central Agricultural Science Laboratory in York received over 5,000 samples of invertebrates intercepted on plants and plant products imported into England and Wales and 560 taxa belonging to 169 families were recognized. However, this would have been but a small proportion of all the insects coming into the countries (Baker *et al.* 2000). It indicates that natural dispersal might be a minor factor in spreading insects to hew habitats. Climate change is just one factor in a complex issue of multiscale human impacts on global, regional, local and micro-ecology. A truly interdisciplinary approach to predicting the consequences for urban pests in required.

Conclusion

Maintaining urban ecosystems requires an ability to deal with external change, not only those resulting from human behaviour, but also those from natural processes, or from natural responses to ecosystem disturbance at all scales from the local to the global. With some invasive species having become extremely costly problems, and with the risk of new problems emerging as a consequence of global environmental change, urban ecology faces major challenges in both helping society adapt to climate change, and in deciding what sort of managed and wild ecological conditions should be present in urban areas. These decisions will reflect differing tastes and ideals about what an ecological environment is like, but in order for the chosen options to be successful there will always be a need for good, practical and well-communicated scientific information.

References

Abu-Asab M.S., Peterson, P.M., Shetler, S.G. and Orli, S. S. (2001) 'Earlier plant flowering in spring as a response to global warming in the Washington, DC, area', *Biodiversity and Conservation*, 10(4): 597–612.

Anderson, P.K., Cunningham, A.A., Patel, N.G., Morales, F.J., Epstein, P.R. and Daszak, P. (2004) 'Emerging infectious diseases of plants: pathogen pollution, climate change and agrotechnology drivers', *Trends in Ecology and Evolution*, 19(10): 534–44.

Bailey, J.P. and Connolly, A.P. (2000) 'Prize-winners to pariahs – A history of Japanese Knotweed *s.l.* (Polygonaceae) in the British Isles', *Watsonia*, 23: 93–110.

Baker, R.H.A., Sansford, C.E., Jarvis, C.H., Cannon, R.J.C., MacLeod, A. and Waiters, K.F.A. (2000) 'The role of climatic mapping in predicting the potential geographical distribution of non-indigenous pests under current and future climates', *Agriculture, Ecosystems and Environment*, 82(1–3): 57–71.

Battersby, S.A., Parsons, R. and Webster J.P. (2002) 'Urban rat infestations and the risk to public health', *Journal of Environmental Health Research*, 1(2): 4–12.
Beerling, D.J. and Perrins, J.M. (1993) '*Impatiens glandulifera* Royle (*Impatiens roylei* Walp.)', *Journal of Ecology*, 81(2): 367–82.
Bentley, W.E. (1959) 'The distribution and status of Rattus rattus L. in the United Kingdom in 1951 and 1956', *Journal of Animal Ecology*, 28(2): 299–308.
Bergot, M., Cloppet, E., Pérarnaud, V., Déqué, M., Marçais, B. and Desprez-Loustau, M.-L. (2004) 'Simulation of potential range expansion of oak disease caused by *Phytophthora cinnamomi* under climate change', *Global Change Biology*, 10(9): 1539–52.
Bisgrove, R. (2003) 'Climate change impacts on public and private gardens in the UK', *Proceedings: Impacts of Climate Change on Horticulture: Developing a research and education outreach agenda. A symposium held in conjunction with the Centennial Conference of the American Society for Horticultural Science, Saturday, October 4, 2003 Providence, RI.* 27–31, online, available at: www.cleanair-coolplanet.org/information/horticulture/horticulture_proceedings.pdf [accessed 15 May 2010].
Bisgrove, R. and Hadley, P. (2002) *Gardening in the Global Greenhouse: The Impacts of Climate Change on Gardens in the UK*, Technical Report, Oxford: UKCIP.
Boland, G.J., Melzer, M.S., Hopkin, A., Higgins, V. and Nassuth, A. (2004) 'Climate change and plant diseases in Ontario', *Canadian Journal of Plant Pathology*, 26: 335–50.
Burton, J.A. (1974) *The Naturalist in London*, Newton Abbott: David and Charles.
Cannon, R.J.C. (1998) 'The implications of predicted climate change for insect pests in the UK, with emphasis on non-indigenous species', *Global Climate Change*, 4(7): 785–96.
Celesti Grapow, L. and Blasi, C. (1998) 'A comparison of the urban flora of different phytoclimatic regions of Italy', *Global Ecology and Biogeography Letters*, 7(5): 367–78.
Chocholoušková, Z. and Pyšek, P. (2003) 'Changes in composition and structure of urban flora over 120 years: a case study of the city of Plzeň', *Flora – Morphology, Distribution, Functional Ecology of Plants*, 198(5): 366–76.
Chronopoulos, G. and Christodoulakis, D. (1996) 'Contribution to the urban ecology of Greece: the flora of the city of Patras and the surrounding area', *Botanica Helvetica*, 106: 159–76.
DAISIE (2009) *Handbook of Alien Species in Europe*, Dordrecht: Springer Netherlands.
Dana, E.D., Vivas, S. and Mota, J.F. (2002) 'Urban vegetation of Almería City, Spain – a contribution to urban ecology in Spain', *Landscape and Urban Planning*, 59(4): 203–16.
Davison, A.W. (1977) 'The ecology of *Hordeum murinem L.* Part 3. Some effects of adverse climate', *Journal of Ecology*, 65(2): 523–30.
Dukes, J.S. and Mooney, H.A. (1999) 'Does global change increase the success of biological invaders?' *Trends in Ecology and Evolution*, 14(4): 135–9.
Essl, F. and Lambdon, P.W. (2009) 'Alien bryophytes and lichens of Europe', in DAISIE (ed.) *Handbook of Alien Species in Europe*, Dordrecht: Springer Netherlands, pp. 29–41.
Fitter, A.H. and Fitter, R.S.R. (2002) 'Rapid changes in flowering time in British plants', *Science*, 296(5573): 1689–91.
Fitter, R.S.R. (1945) *London's Natural History*, London: Collins.
Fox, M.D. and Fox, B.J. (1986) 'The susceptibility of natural communities to invasion', in R.H. Groves and J.J. Burdon (eds), *Ecology of Biological Invasions: An Australian Perspective*, Canberra: Australian Academy of Science, pp. 57–66.
Fox, R., Warren, M.S., Harding, P.T., McLean, I.F.G., Asher, J., Roy, D., and Brereton, T. (2001) *The State of Britain's Butterflies Butterfly Conservation*, Wareham: CEH and JNCC.
Franceschi, E.A. (1996) 'The ruderal vegetation of Rosario City, Argentina', *Landscape and Urban Planning*, 34(1): 11–18.
Godefroid, S. (2001) 'Temporal analysis of the Brussels flora as indicator for changing environmental quality', *Landscape and Urban Planning*, 52(4): 203–24.
Gulezian, P.Z. and Nyberg, D.W. (2010) 'Distribution of invasive plants in a spatially structured urban landscape', *Landscape and Urban Planning*, 95(4): 161–8.
Harvell, C.D., Mitchell, C.E., Ward, J.R., Altizer, S., Dobson, A.P., Ostfeld, R.S. and Samuel, S.D. (2002) 'Climate warming and disease risks for terrestrial and marine biota', *Science*, 296(5576): 2158–62.
Headley, A. and Rumsey, F. (2003) *The Status and Population of Nowell's Limestone Moss (Zygodon gracilis* Wils.*) in Britain*, Peterborough: unpublished report to English Nature.
Hejda, M. and Pyšek, P. (2006) 'What is the impact of *Impatiens glandulifera* on species diversity of invaded riparian vegetation?' *Biological Conservation*, 132(2): 143–52.
Hepper, F.N. (2003) 'Phenological records of English garden plants in Leeds (Yorkshire) and Richmond

(Surrey) from 1946 to 2002. An analysis relating to global warming', *Biodiversity and Conservation*, 12(12): 2503–20.

Ikeda, T., Asano, M., Matoba, Y. and Abe, G. (2004) 'Present status of invasive alien raccoon and its impact in Japan', *Global Environmental Research*, 8(2): 125–31.

Kent, M., Stevens, R.A. and Zhang, L. (1999) 'Urban plant ecology patterns and processes: a case study of the flora of the City of Plymouth, Devon, UK', *Journal of Biogeography*, 26(6): 1281–98.

Kollmann, J. and Bañuelos, M.J. (2004) 'Latitudinal trends in growth and phenology of the invasive alien plant *Impatiens glandulifera* (Balsaminaceae)', *Diversity and Distributions*, 10(5–6): 377–85.

Kunick, W. (1980) 'Comparison of the flora of some cities of the central European lowlands', in R. Bornkamm, J.A. Lee and M.R.D Seaward (eds), *Urban Ecology: Proceedings of the 2nd European Ecology Symposium, Berlin*, Oxford: Blackwell, pp. 13–22.

Lecerf, A., Patfield, D., Boiché, A., Riipinen, M. P., Chauvet, E. and Dobson, M. (2007) 'Stream ecosystems respond to riparian invasion by Japanese knotweed (*Fallopia japonica*)', *Canadian Journal of Fisheries and Aquatic Sciences*, 64(9): 1273–83.

Lever, C. (1977) *The Naturalized Animals of the British Isles*, London: Hutchinson.

Lever, C. (2001) *The Cane Toad: The History and Ecology of a Successful Colonist*, Otley: Westbury Academic and Scientific Publishing.

McNeely, J. (ed.) (2001) *The Great Reshuffling: Human Dimensions of Invasive Alien Species*, Gland, Switzerland/ Cambridge, UK: IUCN.

Michler, F.U., Hohmann, U. and Stubbe, M. (2004) 'Aktionsräume, Tagesschlafplätze und Sozialsystem des Waschbären (*Procyon lotor* Linné 1758) im urbanen Lebensraum der Großstadt Kassel (Nordhessen)', *Beiträge zur Jagd- und Wildtierforschung*, 29: 257–73.

Montenegro, G., Teillier, S., Arce, P. and Poblete, V. (1991) 'Introduction of plants into the Mediterranean-type climate area of Chile', in R.H. Groves and F. di Castri (eds), *Biogeography of Mediterranean Invasions*, Cambridge: Cambridge University Press.

Newton, A. (1971) *Flora of Cheshire*, Chester: Cheshire Community Council.

Niemelä, J. (1999) 'Ecology and urban planning', *Biodiversity and Conservation*, 8(1): 119–31.

Orson, J.H. (1999) *A Review of the Impact of the Hot, Dry Summer of 1995 on the Main Agricultural and Horticultural Enterprises in England and Wales*, Wolverhampton: ADAS.

Owen, D. (1978) *Towns and Gardens*, London: Hodder & Stoughton.

Perrins, J., Fitter, A. and Williamson, M. (1993) 'Population biology and rates of invasion of three introduced *Impatiens* species in the British Isles', *Journal of Biogeography*, 20(1): 33–44.

Pimentel, D. (2005) 'Environmental consequences and economic costs of alien species', in S. Inderjit (ed.), *Invasive Plants: Ecological and Agricultural Aspects*, Basel: Birkhäuser, pp. 269–76.

Porley, R. and Hodgetts, N. (2005) *Mosses and Liverworts*, London: Collins.

Porter, J.H., Parry, M.L. and Carter, T.R. (1991) 'The potential effects of climatic change on agricultural insect pests', *Agricultural and Forest Meteorology*, 57(1–3): 221–40.

Prior, C. (2003) *The Impacts of Climate Change on Gardens in the UK: Pest and Disease Threats*, Wisley: RHS Advisory Service.

Primack, D. Imbres, C., Primack, R.B., Miller-Rushing, A.J. and Del Tredici, P. (2004) 'Herbarium specimens demonstrate earlier flowering times in response to warming in Boston', *American Journal of Botany*, 91(8): 1260–4.

Pyšek, P. (1998) 'Alien and native species in Central European urban floras: a quantitative comparison', *Journal of Biogeography*, 25(1): 155–63.

Rebele, F. (1994) 'Urban ecology and special features of urban ecosystems', *Global Ecology and Biogeography Letters*, 4(6): 174–87.

Sparks, T. and Collinson, N. (2003) 'Wildlife starts to adapt to a warming climate', *Biologist*, 50(6): 273–6.

Urban, M.C., Phillips, B.L., Skelly, D.K. and Shine, R.L. (2007) 'The cane toad's (*Chaunus* [*Bufo*] *marinus*) increasing ability to invade Australia is revealed by a dynamically updated range model', *Proceedings of the Royal Society B: Biological Sciences*, 274(1616): 1413–19.

Walther, G.-R. (2000) 'Climatic forcing on the dispersal of exotic species', *Phytocoenologia*, 30(3–4): 409–30.

Walther, G.-R., Post, E., Convey, P., Menzel, A., Parmesan, C., Beebee, T.J.C., Fromentin, J.-M., Hoegh-Guldberg, O. and Bairlein, F. (2002) 'Ecological responses to recent climate change', *Nature*, 416, 389–95.

Woodward, F.I. and Beerling, D.J. (1997) 'The dynamics of vegetation change: health warnings for equilibrium "dodo" models', *Global Ecology and Biogeography Letters*, 6(6): 413–18.

39

Urban biogeochemical flux analysis

Nancy B. Grimm, Rebecca L. Hale, Elizabeth M. Cook, and David M. Iwaniec

Biogeochemical fluxes—the movement of materials among components of ecosystems and between different ecosystems—are fundamental ecological processes. Biogeochemical fluxes usually are defined for elements, rather than compounds. Biogeochemical cycles involve changes in form due to chemical, physical and organism-mediated processes, physical or organism-mediated translocation, and utilization in biotic production. Mass is conserved in biogeochemical cycles, and, in theory, any atom released by biota to the environment will eventually enter the biotic component of biogeochemical cycles again. Investigation of the reciprocal controls of elements and biotic processes at a variety of scales has been fertile ground for ecological investigation for decades.

Human activities have accelerated biogeochemical cycles, especially over the past 150 years, and cities in particular are implicated as foci of altered nutrient and non-essential element cycles. People (who are concentrated in cities) alter urban biogeochemistry in two main ways: (1) directly by creating new fluxes of materials or enhancing existing ones; and (2) indirectly by altering environmental conditions and thereby rates of biogeochemical processes. In this chapter we discuss urban biogeochemical fluxes at various scales and present examples of approaches to quantifying biogeochemical fluxes that are unique to urban systems.

A simple framework

In the urban ecosystem, materials exchange between air, water, land, and groundwater compartments or storage pools, and they cycle within these pools (Figure 39.1). An urban ecosystem also is linked biogeochemically with other ecosystems, from which it imports materials and to which it exports natural or human-made products and wastes (input-output arrows in Figure 39.1). Fluxes within and among compartments are strongly influenced by engineering and the built environment, the demographic trends of cities, and the behavior of individual or collective urban inhabitants in designing and maintaining landscapes (Kaye *et al.* 2006). For example, built structures and, in particular, impervious surfaces have profound influences on hydrologic transport of materials during storms (Arnold and Gibbons 1996; Paul and Meyer 2001; Walsh *et al.* 2005). Trends in China toward smaller households increase per capita urban carbon emissions associated with heating the same space for fewer people (Liu *et al.* 2003). A

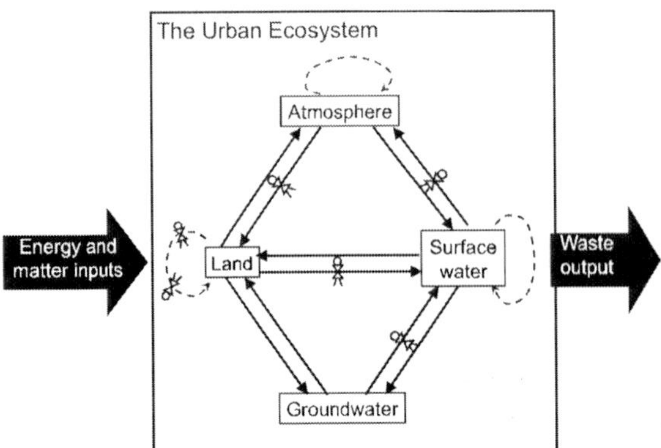

Figure 39.1 Simple conceptual framework for biogeochemical fluxes in urban ecosystems. Major fluxes between the urban ecosystem and other ecosystems are the inputs from external ecosystems, including imported food, fuel, fertilizer, building materials, goods, and animal feed; outputs carried by water or wind vectors or those deliberately exported from the system. Solid arrows denote internal fluxes among four major compartments, and dashed arrows indicate the recycling of nutrients within them. Human activities in the urban ecosystem control, alter, or enhance fluxes among compartments (denoted by converting the classic 'bow tie' control symbol to a stick figure). For example, groundwater pumping hastens the transfer of materials from the subsurface to surface compartment. Modification of the land surface, especially increases in impervious areas, and the construction of drainage infrastructure alter the fluxes of materials between land and surface water, both through changes in mobility of nutrients and changes in hydrology. Adapted from Kaye *et al.* (2006).

desire to have a lush, green lawn prompts people to purchase and apply large amounts of nitrogen (N), phosphorus (P), and potassium (K) to their home lots (Robbins 2007).

This chapter is organized to follow the elemental fluxes (mostly carbon [C], N, and P): (1) between the major compartments (e.g. land – atmosphere – water) within urban ecosystems; (2) within urban terrestrial and aquatic systems; and (3) between urban ecosystems and other systems (e.g. the surrounding native ecosystem, an agricultural production area outside the city, as well as distant systems connected by trade) via imports and outputs. We adopt this simple framework to explore the changes that might be expected with urbanization, but we emphasize that biogeochemical fluxes can be examined at multiple scales. Furthermore, a mass-balance accounting of inputs and outputs enables us to understand whether urban ecosystems are sources or sinks of any given material. Urban areas are both responsible for, and respond to, global-scale changes in biogeochemical cycles (Grimm *et al.* 2008a). As centers of transportation and industry, cities are important sources of carbon dioxide (CO_2) and other greenhouse gases that affect global climate (Molina and Molina 2004; Pataki *et al.* 2006). Air pollution from cities influences nutrient cycling and primary production in adjacent ecosystems and even at remote locations. Therefore, after considering the biogeochemical fluxes among and within major compartments, we discuss the ways in which scientists quantify biogeochemical relationships between urban ecosystems and nearby and remote ecosystems. We return to the question of scale in a brief chapter conclusion.

Biogeochemical fluxes among compartments

Elements move among the atmospheric, terrestrial, and aquatic compartments of the urban ecosystem. Nutrients and other elements can enter urban ecosystems from atmospheric deposition, and, likewise, emissions to the atmosphere are redistributed within the city and comprise an important output from the urban ecosystem. Outputs via this land–atmosphere pathway greatly exceed inputs (e.g. Kaye *et al.* 2004). Hydrological vectors deliver nutrients from the land to aquatic ecosystems in a primarily unidirectional flux; from there, materials may be retained, transformed, exported from the system, or transferred to groundwater. Groundwater–surface water fluxes are mostly human mediated.

Fluxes to and from the atmosphere

Elevated atmospheric concentrations of pollutants result from urban activities, but the fate of these pollutants depends upon chemical reactions in the urban atmosphere and transport and deposition processes. Carbon dioxide (CO_2) is a direct product of fossil-fuel combustion, whereas nitrogen oxides (NO_x) are formed from atmospheric nitrogen gas (N_2) as a by-product of combustion processes. Transportation is the dominant source of these gases, accounting for 80 percent of CO_2 and NO_x emissions in greater Toronto, Ontario (Sahely *et al.* 2003) and of CO_2 emissions in Phoenix, Arizona (Koerner and Klopatek 2002). Per capita emissions increase in many cities with economic growth, as vehicle ownership increases, or with changes in urban form that extend driving distances (Decker *et al.* 2000; Kennedy *et al.* 2007; Grimm *et al.* 2008a). Per capita CO_2 emissions increased 28 percent in Sydney, Australia between 1970 and 1990 (Sahely *et al.* 2003), and more than doubled in Hong Kong between 1971 and 1997 (Warren-Rhodes and Koenig 2001). In contrast, cities can reduce per capita emissions by encouraging public transportation, discouraging sprawl, and promoting fuel-efficient vehicles.

Biogenic gas fluxes from urban soils/land typically differ from those of surrounding ecosystems. For example, Kaye *et al.* (2004) and Hall *et al.* (2008) found that N_2O emissions from lawns in Fort Collins, Colorado and Phoenix, Arizona exceeded those from native grasslands and remnant desert, respectively. CO_2 emissions from urban land in Phoenix were also much greater than those from the surrounding desert (Koerner and Klopatek 2002). Urbanization not only changed the sources and rates of gas emissions, but also the capacity of land to act as a sink for gases. Methane (CH_4) fluxes include both biogenic sources and sinks and also are enhanced by human modifications to the environment (e.g. in landfills and sewers), combustion (from incineration and vehicle exhaust), or from natural-gas distribution systems. Oxidation of CH_4 by natural ecosystems may be an important sink; in a study in the front range of Colorado, Kaye *et al.* (2004) found that lawns took up less methane than native grasslands. Similarly, lawns in Baltimore, Maryland, had minimal uptake of CH_4 compared to surrounding temperate forest soils (Groffman and Pouyat 2009). Biogenic gas fluxes can be highly variable within cities as well: lawns in Phoenix, Arizona have significantly higher N_2O fluxes than xeric, gravel-rock residential landscapes (Hall *et al.* 2008).

Although biogenic fluxes are usually dwarfed by fossil fuel-derived emissions, they still may be substantial on a regional basis. For example, high CO_2 flux from grass lawns in an urban arid ecosystem (Phoenix, Arizona) contributed up to ~10 percent of regional CO_2 emissions (Koerner and Klopatek 2002). Interestingly, human respiration accounted for 1.6 percent of total CO_2 emissions in this same study. Gaseous N emissions from urban lawns can represent up to 30 percent of regional N_2O fluxes, and these systems collectively consumed nearly 5 percent of regional soil CH_4 (Kaye *et al.* 2005).

Atmospheric emissions from the world's cities extend well beyond their boundaries, contributing to the rise in global greenhouse gas concentrations (Grimm *et al.* 2008a; Karl *et al.* 2009). The extent to which urban pollutants are transported depends upon the properties of the materials (e.g. form, deposition velocity) and local and regional atmospheric dynamics. Upper atmospheric circulation redistributes energy and matter across the globe. Urban heat islands that characterize many urban areas may induce photochemical smog formation and create local air-circulation patterns that promote dispersion of pollutants away from their source (Gregg *et al.* 2003). Atmospheric reactive N gases, such as NO_x and ammonia (NH_3), may be transported long distances to affect downwind ecosystems (McDonnell *et al.* 1997; Lovett *et al.* 2000; Fenn *et al.* 2003; Driscoll *et al.* 2003; Ortiz-Zayas *et al.* 2006; Lohse *et al.* 2008). Wet deposition may be further enhanced in adjacent ecosystems due to the complex of higher atmospheric concentrations of pollutants and urban heat island induced increased rainfall rates downwind of cities (Shepherd *et al.* 2002).

Urban atmospheric pollutants that are not transported beyond the urban boundary are deposited to terrestrial and aquatic systems within urban areas. Atmospheric deposition of N is elevated in urban areas. Gaseous, aerosol, and particulate forms of N (dry deposition) may exceed wet deposition, especially in arid regions (Fenn *et al.* 2003). Organic C deposition, mostly as a complex mixture of aerosols including fatty acids and alkanes derived from incomplete combustion and cooking (Schauer *et al.* 1996), also is enhanced in cities. In a recent laboratory study, Kaye *et al.* (in press) found that soil microbes were able to use deposited organic C, but they reported that deposition rates were too low to enhance microbial processes on a regional scale. A correlation of NO_3^- and organic C deposition with dust-derived P and particulate base cations was found in a Phoenix study (Lohse *et al.* 2008), which suggests that urban dust increases local deposition by scrubbing acidic gases and particles from the atmosphere. Lovett *et al.* (2000) reported a similar finding in the New York City region. Dust derives from construction activities and as a by-product of combustion; complex atmospheric interactions and reactions can trap pollutants that are generated within cities and redeposit materials to the land. Enhanced deposition may act as a fertilizer, stimulating plant productivity and soil biogeochemical cycling (Oren *et al.* 2001; D'Antonio and Mack 2006; Shen *et al.* 2008). On the other hand, myriad chemical substances, especially ozone, have negative effects (Chameides *et al.* 1994; Gregg *et al.* 2003), such that the net impact on urban ecosystem processes of fluxes from the atmosphere is not always clear.

Fluxes to aquatic recipient systems

Biogeochemical fluxes from the land to surface waters are well described by a classical approach in ecosystem ecology: the small watershed approach (Likens and Bormann 1995). Although the small watershed approach has been used to assess whole-system dynamics of terrestrial ecosystems such as forests, it also yields the magnitude of outputs to recipient systems such as streams, lakes, estuaries, and coastal marine ecosystems where problems of eutrophication from excess nutrient inputs may be acute (Carpenter *et al.* 1998). The approach is beginning to be applied to urban watersheds to describe the transfer of materials from urban uplands to streams during both base flow and storms. In many cases, urban watersheds are compared to non-urban watersheds. The mass balance equation, input = output + change in storage (ΔS), is quantified with both human-mediated and non-human inputs of nutrients and output in streamflow, and ΔS is estimated by difference. As a result, ΔS includes both material accumulated in plants and soils and any unmeasured outputs, such as plant harvest or gas loss via denitrification or volatilization. Wollheim *et al.* (2005) compared the N budgets of two suburban Boston,

Massachusetts, watersheds that were 16 and 79 percent urbanized. Inputs of N were 650 percent higher, and exports in stream water 45 percent higher, in the urban compared with the forested watershed. Inputs to this suburban Massachusetts system were from atmospheric deposition, leakage from septic systems, and fertilizer additions to lawns (Williams *et al.* 2004). Groffman *et al.*'s (2004) N mass balance for a suburban Baltimore, Maryland, watershed showed that despite high retention (ΔS), rate of N delivery to streams was high because inputs were high. Inputs in the Baltimore watershed were dominated by deliberate human additions (fertilizer) and high atmospheric deposition.

Because of extensive human modification of urban flowpaths, the contributing areas of urban watersheds are not always predictable from topographic boundaries. Road networks, storm drainage infrastructure, designed retention systems, and other aspects of the built environment change urban hydrology and thereby affect (usually enhance) movements of materials to streams and rivers (Walsh *et al.* 2005; Kaye *et al.* 2006). The mosaic of land covers, spatial arrangement of impervious surface, and drainage pathways influence pollutant (N, P, and organic C) and sediment concentrations of receiving waters (Paul and Meyer 2001; Goonetilleke *et al.* 2005; Walsh and Kunapo 2009). Nutrient loading to urban streams also depends on the presence of wastewater treatment plants, septic tanks, combined sewer overflows, and stormwater drainage (Paul and Meyer 2001; Walsh and Kunapo 2009). In a Phoenix study comparing storage on impervious surfaces to the runoff concentrations of dissolved organic carbon (DOC), N, and P, Hope *et al.* (2004) found that DOC and P were transported in proportion to storage, whereas much less N was transported than would be predicted. Also in Phoenix, Lewis and Grimm (2007) found that land use was a significant predictor of N concentrations and loads in runoff from individual storms, but interacted with broader-scale climate variables such as antecedent dry days and event size. The authors highlighted the need to understand what catchment features would lead to reduced loads, given the high degree of variability among the small urban watersheds they studied. Urban planning that takes into account the impacts of such aspects of land configuration would be an important step forward in controlling pollutant loading for urban environments.

Groundwater as a biogeochemical sink

Not all exports from urban watersheds reach large rivers or the sea. Especially in arid and semi-arid regions, groundwater may be recharged at mountain fronts and through ephemeral streambeds. Furthermore, in more humid regions fertilizer leaching through the soil column to shallow groundwater is a significant problem. Leaks from septic tanks and sewer lines may be sources of nutrients to shallow groundwater (Lerner 2002). Thus, surface water, land surface, and belowground infrastructure may represent routes of pollutant introduction to groundwater. The location of an urban ecosystem relative to geomorphic features and underlying bedrock geology may determine the extent of threats to groundwater quality from urban activities.

Extensive modifications of hydrologic systems and built infrastructure in cities again play a role in determining biogeochemical fluxes among compartments. Intended to reduce the magnitude and impact of storms, retention and detention structures are increasingly prevalent components of urban drainage systems. These features certainly retard the movement of water to streams, but also may be important sites of nutrient processing and removal (Zhu *et al.* 2004) or of transport to groundwater. In many cities, sanitary and storm drainage systems are established in the same locations, which inadvertently promotes undesirable connections between the two systems. Leaking sewer systems can account for surprisingly large loadings to streams and shallow groundwaters in cities with aging infrastructure (Lerner 2002).

Biogeochemical cycling within urban ecosystems

Urban ecosystems comprise a heterogeneous matrix of land covers and land uses that influence biogeochemical cycling (Wu and David 2002; Grimm and Redman 2004; Band *et al.* 2005; Cadenasso *et al.* 2007). Small-scale urban studies often quantify patterns and processes within discrete land-use categories (Groffman *et al.* 2005; Jenerette *et al.* 2006; Baker *et al.* 2007), in which land ownership and jurisdiction is known. If the land-use category approach is blended with the watershed approach described earlier, researchers should be able to 'unpack' the components and drivers that significantly affect biogeochemical fluxes between compartments. Urbanization can indirectly affect biogeochemical cycles via increased CO_2 concentrations, the increased temperatures of the urban heat island, altered wind patterns, and altered hydrology. However, characteristics of cities themselves, such as urban form, may drive changes in biogeochemical fluxes through soils and vegetation (Pataki *et al.* 2006). Urban form encompasses the spatial and structural patterns of an urban area, including the extent of impervious areas, vegetation patterns, and management practices (Churkina 2008).

Terrestrial ecosystems: residential, commercial, and other land uses

Turfgrass is the largest irrigated crop in the United States, covering 10–16 million hectares (Milesi *et al.* 2005). Urban ecosystems comprise a heterogeneous matrix of land covers and land uses that influence biogeochemical cycling (Wu and David 2002; Grimm and Redman 2004; Band *et al.* 2005; Cadenasso *et al.* 2007). In arid Phoenix, high organic C concentration in soils reflects previous agricultural land use (Lewis *et al.* 2006; Kaye *et al.* 2008). Similarly, residential land use leaves a legacy on soil properties, such that residential age is positively correlated to soil organic matter (Lewis *et al.* 2006; Golubiewski 2006).

Soil N and P in urban lawns also are significantly influenced by human management factors such as property age, high chemical and water inputs, and increased organic matter (e.g. Kaye *et al.* 2008). Lawns show high rates of N cycling and retention (Raciti *et al.* 2008; Groffman *et al.* 2009), and thus accumulation of soil N over time (Lewis et al. 2006). Turfgrass lawns in Baltimore had larger pools of NO_3^- and NH_4^+, and higher rates of mineralization and nitrification, than surrounding forests (Raciti *et al.* 2008). In Phoenix, net rates of nitrification and mineralization in residential lawns were shown to be much higher than in other land-cover patches (Zhu *et al.* 2006). Bennett *et al.* (2005) found variation in P concentration between land uses in Wisconsin, but a more homogeneous concentration within land uses. In that study, residential lawns had P concentrations that were intermediate between the prairie and agricultural sites, which had high fertilization rates (Bennett *et al.* 2005).

An innovative new concept for evaluating biogeochemical fluxes at the scale of individual households was introduced recently by Baker *et al.* (2007). Their 'household flux calculator' (HFC) evaluates inputs, outputs and accumulation of C, N, and P at the small residential-parcel scale, allowing scaling to citywide fluxes. The advantage of this household-scale tool is that biogeochemical fluxes can be directly linked to human behaviors, such as commuting patterns, food consumption, and energy use. At a slightly larger scale, Codoban and Kennedy (2008) evaluated material inputs and outputs of four Toronto neighborhoods that differed in housing types and predominant forms of transportation. However, biogeochemical fluxes were not explicitly considered in their analysis. Currently, there is no nested dataset on household, neighborhood, municipal, and metropolitan biogeochemical fluxes, but we can imagine that such a series would be an invaluable tool to evaluate human controls on urban biogeochemistry at different levels of human organization (e.g. from household-level landscaping decisions to municipal irrigation and fertilizer-use regulations).

It should be clear from this brief description of selected studies within urban ecosystems that human management has a profound and lasting impact on biogeochemical dynamics of small-scale landscapes. The very creation of urban residential and commercial landscapes, with their assembled communities of non-native vegetation (Walker *et al.* 2009), dictates that biogeochemical dynamics will differ from non-urban ecosystems. Fertilization, irrigation, and organic-matter additions enhance productivity, accelerate biogeochemical cycles, and alter soil-atmosphere exchanges (e.g. Livesley *et al.* 2010). Biogeochemical inputs may be dominated by human-mediated transport of material goods (e.g. food, fertilizer, fuel, building material, etc.) into cities, and retention of these materials may be localized within the built environment. Indirect effects of larger-scale, contextual changes in local climate or atmospheric chemistry can accelerate or alter nutrient cycling independent of human management. Increased temperature associated with the urban heat island may enhance microbial process rates, reduce soil moisture by increasing evaporation rates, or extend the growing season (Imhoff *et al.* 2004; Zhu and Carreiro 2004).

In addition to residential landscapes, urban open space (usually remnant native ecosystem patches or grassy parks) and highly modified commercial/industrial land uses have characteristic biogeochemistry. Remnant native (e.g. forest, desert, etc.) ecosystems within urban areas and other open spaces are often only indirectly affected by humans (for example, via atmospheric deposition) because they are minimally managed. Studies have shown that they do not resemble surrounding native ecosystems in terms of biogeochemical processes (e.g. Hall *et al.* 2008).

Mass balance models for urban watersheds consistently show substantial retention of N, even though exports (loads to streams) are high. Groffman *et al.* (2004) reported 75 percent of inputs to a suburban Baltimore watershed were retained. Comparisons of urban, suburban, and forested watersheds in the same region showed that up to 85 percent of inputs were retained in the developed watersheds compared to >90 percent in the forested watersheds (Kaushal *et al.* 2008). However, during a wet year the retention capacity of the former was only 35 percent, whereas retention in the forested watershed remained >90 percent. A similar comparison in Massachusetts also showed lower retention in a wet year (65–85 percent in the urban compared to 93–97 percent in the forest watershed in wet vs. dry years, respectively; Wollheim *et al.* 2005). The low retention of wet years in both studies could be due to several factors. First, inputs may increase to the point that they exceed the retention capacity of the system. Alternatively, different contributing areas may come into play during wet as compared to dry years. Finally, climatic factors may alter the rates or controls on processes that are responsible for N retention.

Aquatic ecosystems

Biogeochemical cycles in streams and riparian areas are highly altered by urbanization. The drivers of these changes are both direct (elevated nutrient and toxic concentrations) and indirect (changes in stream morphology, hydrology, and riparian or upland ecosystem structure). Indirect catchment activities may be most important in explaining biogeochemical function of streams left in their quasi-natural states (e.g. Groffman *et al.* 2005; Walsh *et al.* 2005). In cases of profound human alteration of hydrologic systems (e.g. Grimm *et al.* 2004; Roach *et al.* 2008; Roach and Grimm 2009), direct human manipulation of water sources and nutrient sources and sinks is the most important driver.

Nutrient concentrations of urban streams are typically elevated; however, these increases have been mitigated to some extent by improvements in wastewater treatment technology, regulation of point sources, and retrofitting of stormwater systems (Paul and Meyer 2001; Grimm *et al.*

2005; Walsh *et al.* 2005). Urbanization has a slightly lesser effect on stream N concentration than does agriculture (USGS 1999). Yet, urban streams do experience high N inputs from wastewater effluent, fertilizer runoff, leaching from sewers and septic systems, groundwater pumping, and atmospheric deposition (Groffman *et al.* 2005; Grimm *et al.* 2005; Makepeace *et al.* 1995; Paul and Meyer 2001; Walsh *et al.* 2005). Elevated P concentration in urban streams comes from sewage effluent (Bedore *et al.* 2008; David and Gentry 2000) and fertilizer runoff (La Valle 1975; Paul and Meyer 2001). Particulate P concentration in particular correlates with urbanization (Paul and Meyer 2001).

Whereas unaltered forested or grassland/ shrubland streams are able to respond to increases in N concentration by increasing N retention processes such as uptake and denitrification (Bernhardt *et al.* 2003), the efficiency of N retention is reduced in N-loaded streams (Mulholland *et al.* 2008). Although rates of both uptake and denitrification increase in urban streams in response to increased nitrate concentration (Groffman *et al.* 2005; Mulholland *et al.* 2008, 2009), retention (as a percentage of input) is lower than in non-urban streams (Grimm *et al.* 2005; Meyer *et al.* 2005). Although little work has been done on P retention in urban streams, the theoretical basis for expecting lower efficiency of biotic P removal is the same as for N. However, aquatic P dynamics often are under abiotic control, especially since urban P loads are primarily in particulate form (Paul and Meyer 2001), and the role of abiotic processes in urban stream P dynamics is poorly known.

Why do urban streams exhibit reduced retention efficiency? Geomorphic simplification (Grimm *et al.* 2005), increased peak flows (Walsh *et al.* 2005), and/or loss of benthic organic matter substrate in urban streams (Meyer *et al.* 2005) all contribute to this impaired function. We caution, however, that generalizing across all urban streams is problematic when some other context, such as land-use history or regional geology, may drive stream response. Indeed a multi-biome comparison of stream N dynamics that included 27 urban streams showed that stream nitrate concentration and discharge were more important than land use (urban, agricultural, or unaltered) in determining N retention capacity and efficiency (Mulholland *et al.* 2008, 2009; Hall *et al.* 2009).

Increased N and P concentrations can increase productivity in streams (Elsdon and Limburg 2008) and alter rates of nutrient cycling (Groffman *et al.* 2005). These two elements are the most common limiting nutrients in aquatic ecosystems. In an urban lake chain in Phoenix metropolitan region, these two elements were alternately limiting to primary production: when N was abundant under baseflow conditions, P was limiting; following floods when P concentration increased but high N baseflow concentration was diluted by floodwaters, N was limiting (Roach and Grimm 2009).

Organic matter is an important substrate for microbial growth and therefore can have strong control on biogeochemical processes (Groffman *et al.* 2005). Coarse particulate organic matter (CPOM) is lower in urban streams than in non-urban streams because scouring floods with elevated peak flows remove the material (Booth and Jackson 1997, Walsh *et al.* 2005, Aldridge *et al.* 2009), and because loss of riparian vegetation reduces organic inputs (Lepori *et al.* 2005). Dissolved organic carbon (DOC) concentration also is altered in urban streams: DOC concentration increases with urbanization (Paul and Meyer 2001; Izbicki *et al.* 2007; Westerhoff and Anning 2000) and imperviousness (Hatt *et al.* 2004; Harbott and Grace 2005) but may decrease if wetlands are eliminated (Williams *et al.* 2005). Urbanization also alters the chemical composition of DOC (Izbicki *et al.* 2007), including human-created organic compounds. Many persistent organic pollutants can be recovered from urban wastewater and even stormwater; research on the sources, transformations, fate, and impacts of these compounds is needed.

Riparian ecosystems are among the most biogeochemically active components of natural landscapes (Dahm *et al.* 1998; McClain *et al.* 2003), but urbanization can remove riparian zones entirely, drastically reduce their extent, or change their hydrodynamics. Urban riparian areas have lower organic-matter content than non-urban areas (Groffman *et al.* 2002) and may be sources of N to streams (Groffman *et al.* 2002, 2003), depending on changes in geomorphology and hydrology. Microbial denitrification is an important process for removing nitrate from streams and riparian zones, but it may be curtailed in urban riparian zones as lowered water tables, stream bank incision, and decreased subsurface flow lead to reduced contact on nitrate-rich waters with denitrifying microbes (Groffman *et al.* 2002, 2003).

Whole-ecosystem analysis of urban biogeochemical fluxes

Understanding biogeochemical cycling within urban ecosystems will contribute to mechanistic knowledge of controls on biogeochemical processes and how they are affected by humans, both deliberately and inadvertently (Figure 39.1). However, placing this understanding within a regional, continental, or global context requires that we develop and use metrics for understanding fluxes at the level of the whole ecosystem, and then compare these metrics among urban ecosystems differing in their characteristics and setting (Grimm *et al.* 2008b). There are three primary approaches in current use: nutrient budgets/mass balance, ecological footprints, and urban metabolism analyses. Interestingly, each approach has its proponents and each derives from a particular intellectual history. The mass-balance approach, especially for C, N, and P, is most closely aligned to ecosystem ecology (Kaye *et al.* 2006), whereas the ecological footprint and urban metabolism concepts (which are discussed together here) derive more from engineering and, in particular, the relatively new field of industrial ecology (Graedel and Allenby 2003).

Nutrient budgets for whole cities

Mass-balance models—nutrient budgets—can be constructed at various scales, including the watershed scale described previously, or the whole-city scale. Mass balances quantify inputs and outputs of elements or materials, and yield information about whether a city is a source (input < output) or sink (input > output) for the material. Inferences about how the material is transformed by urban activities are possible. Mass balance also is used to identify potential sinks and sources of an element within the system (Baker *et al.* 2001, Kaye *et al.* 2006). Whereas in non-urban ecosystems human-mediated fluxes can often be safely ignored or are included in deposition measurements, in urban ecosystems these inputs dominate. Urban mass-balance studies must also consider human drivers such as population size, climate, urban form, and social factors (Kaye *et al.* 2006). Changes in mass balance may be direct (e.g. imports of fertilizer) or indirect (e.g. changes in nutrient cycling due to changes in urban hydrology or climate). The few existing whole-city nutrient budgets uniformly point to strong control of urban biogeochemical fluxes by humans, massive inputs, and high throughput (i.e. both input and output rates are high). Retention (ΔS), or whether a city is a source or sink, varies depending upon the material being considered and the city—its characteristics as well as its context.

Although a complete mass balance for C has yet to be done for any city, it is clear that cities are major sources of CO_2 and large contributors to its global enrichment in the atmosphere. Extremely large carbon imports of fuel, including oil, gas, coal, or biofuel, dominate the C budget. Based upon a large body of research on urban C sequestration (e.g. in trees), a general consensus has emerged that any sink capacity of the urban forest is overwhelmed by CO_2 emissions from combustion (Pataki *et al.* 2006). Boyle and Lavkulich (1997) quantified changes in C storage during urbanization in the Vancouver, BC (Canada) region. They showed that C storage dramatically declined as wetlands and forests rich in organic matter were replaced with urban built structure. This change contrasts sharply with the increased soil C storage found in arid Phoenix metro (Lewis *et al.* 2006; Kaye *et al.* 2008). Ecosystem process models, such as CENTURY and its derivatives, are beginning to be applied to urban systems (Bandaranayake *et al.* 2003; Qian *et al.* 2003; Shen *et al.* 2008). None has yet successfully integrated soil and vegetation fluxes (e.g. photosynthesis and respiration) with human-mediated fluxes and stocks (e.g. imports of fossil fuel, food, building materials; Churkina 2008), but this cross-fertilization of models will be needed to advance understanding of urban biogeochemical dynamics.

People are responsible for the vast majority of nutrient inputs to urban ecosystems. Approximately 90 percent of N and P inputs are human-mediated in Phoenix and the Gävle region of Sweden (Baker *et al.* 2001 and Nilsson 1995, respectively). Urban P fluxes are food-related (Nilsson 1995; Faerge *et al.* 2001), whereas N fluxes are increasingly related to combustion (i.e. NO_x production), which is often dominated by transportation in urban areas (Björklund *et al.* 1999; Baker *et al.* 2001; Warren-Rhodes and Koenig 2001). Few P budgets exist for urban ecosystems, although Faerge *et al.* (2001) constructed a P budget for Bangkok province, Thailand that showed P accumulation in the ecosystem and large exports in surface water. Human decisions affect the capacity for nutrient retention and the amounts and forms of outputs/exports; however, scaling from an understanding of household dynamics to the whole ecosystem is problematic because new agents appear at different scales (i.e. individual decision makers versus institutions). In theory, knowledge of the fluxes at nested, hierarchical scales, the spatial configuration or 'land architecture' (*sensu* Turner 2010) of the region, and an understanding of cross-scale influences should allow scaling to the whole system.

Two detailed N budgets, one for the central Arizona region (the Central Arizona–Phoenix Long-Term Ecological Research [CAP LTER] study area; Baker *et al.* 2001) and one for Hangzhou, China (Gu *et al.* 2009), illustrate both commonalities and idiosyncrasies of urban N budgets. The CAP study encompassed the Phoenix metropolitan area, surrounding agricultural lands, and desert, and separately considered the urban, agricultural, and desert components of the landscape. Inputs to the urban system were dominated by human-controlled fluxes and fixation due to combustion (NO_x production), and were an order of magnitude higher than those to the desert ecosystem. The urban ecosystem was estimated to accumulate 17–21 GgN/y (~17–21 percent of inputs). In rapidly urbanizing Hangzhou, 73 percent of inputs were human controlled, including food, chemical N, combustion-fixation, and fertilizers, and comparable to the Phoenix budget, 17 percent of inputs accumulated in the ecosystem. Outputs from both ecosystems were primarily to the atmosphere (88 percent in CAP [66 percent denitrification and 22 percent NO_x] and 69 percent in Hangzhou). The two ecosystems differed in hydrologic output: export in surface water accounted for only 3 percent of inputs for the CAP ecosystem, but 22 percent in Hangzhou. Surface hydrologic output from Phoenix is tightly controlled via hydrologic modifications that retain water in the urban system (Lauver and Baker 2000). Wastewater N is recycled within the Phoenix ecosystem and only 4 percent of wastewater N exits in surface water (Lauver and Baker 2000). As an even more striking contrast to both Phoenix and Hangzhou's hydrologic outputs, Bangkok, Thailand's urban N budget is characterized by low retention (3 percent of input) and very high export to surface water (97 percent) (Faerge *et al.* 2001).

Temporal changes in Hangzhou's N budget over recent decades may be emblematic of changes in rapidly growing China. Hangzhou's N accumulation rate increased 120 percent and both inputs and outputs increased 130 percent between 1980 and 2004, associated with increased affluence of expanding urban populations and their habits (e.g. fertilizer use; Gu *et al.* 2009). Surface-water export and products export from Hangzhou increased 450 percent and 800 percent, respectively. These increases in throughput are significant in that the extent of change may outstrip the capacity of receiving systems to handle them, at least on a short time scale.

Urban metabolism

In an analogy to human or other organismal metabolism, Wolman (1965) introduced the idea of urban metabolism: an accounting of the material inputs to a city and its outputs as wastes. On the face of it, urban metabolism is very similar to the mass-balance/material budget approach described above. However, the analogy has taken on its own life and cities may be erroneously likened to actual organisms. Furthermore, the vagueness of the concept makes accounting difficult: what are the units to be used in accounting for inputs of very different materials (e.g. water, fiber, nitrogen)? The beautiful illustration of Duvigneaud and Denaeyer-De Smet (1977; Figure 39. 2) exemplifies this challenge well: units of energy, mass of compounds, mass of elements, and mass of substances are included. If metabolism studies do not identify the breakdown of specific elements, they will be difficult to compare among cities. In general, data incompatibility hinders such comparisons (Kennedy *et al.* 2007). However, comparisons of temporal trends with current and past metabolism data are appropriate for a single city (Warren-Rhodes and Koenig 2001). Kennedy *et al.* (2007) evaluated general temporal trends for seven cities, and found increases in material and energy use and wastewater generation. Increases in efficiency or decreases in emissions due to regulation can be identified using this approach. For example, water and energy efficiency improved in Toronto over time (Sahely *et al.* 2003), and air emissions decreased in other cities, especially SO_2 and particulate materials (Kennedy *et al.* 2007).

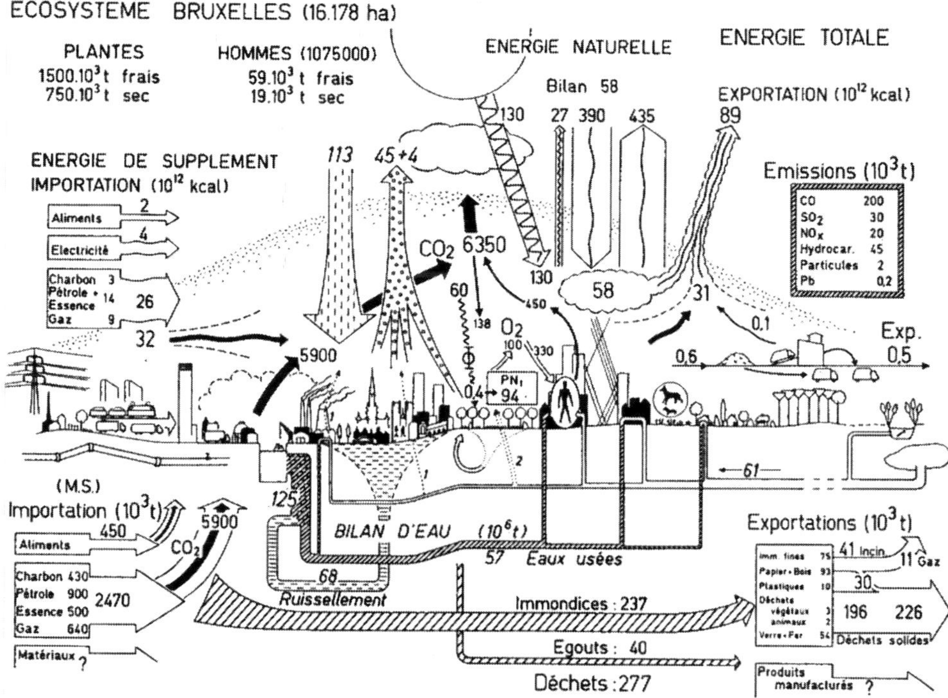

Figure 39.2 Diagram showing the complex and multi-faceted flows into and out of an urban ecosystem that comprise 'urban metabolism.' (*source:* Duvigneaud and Denaeyer-De Smet 1977).

As a contrast to mass-balance or metabolism approaches, the ecological footprint (EF) concept derives a single metric to describe human impact on the environment. The EF is defined as the total land area required to supply all of the material needs of a given population and to absorb all of its wastes (Rees and Wackernagel 1996). The EF can be calculated at any scale, with reference to a specific population (e.g. a family, a school district, a city, a nation), and can provide a powerful illustration of the extent of urban impact. Dietz *et al.* (2007) ranked the footprints of the world's nations and again emphasized the applicability of the IPAT model, ascribing importance to affluence and technology. Urban EFs may be hundreds of times that of the cities themselves; the EF for Phoenix is 180 times the (extensive) land area occupied by the urban population, for example (Grimm *et al.* 2002).

The EF is useful primarily as a heuristic tool rather than as a representation of reality; limitations include the inability of the model to account for possible benefits of urbanization such as dense populations (Kaye *et al.* 2006; Grimm *et al.* 2008a), and accounting inaccuracies (e.g. double counting and failure to account for trade and transportation). Moreover, negative impacts of cities are emphasized, but because it treats population as a homogeneous entity, the method cannot account for changing urban efficiencies from technological or design improvements, or changes in human behavior. Combining the household flux calculator (Baker *et al.* 2007) with the EF concept would be a useful starting point for considering the multi-scaled impact of urban activity.

Grimm *et al.* (2008b) called for a framework to understand urban ecosystem function that would be based upon variation among cities across gradients reflecting their own characteristics (size, shape, demography) as well as the biophysical and societal environments in which they are embedded. Such a framework could organize comparison of metabolism, mass balance, or ecological footprints among cities, as a tool to enhance understanding of direction and magnitude of change across cities.

Conclusions: scale and urban biogeochemistry

In this chapter, we have reviewed our knowledge of the fluxes of nutrient elements, between and within urban compartments and between urban ecosystems and other ecosystems with which they are linked (Figure 39.1). The general picture that emerges is that urban ecosystems do cycle these materials but fluxes are greatly enhanced by material inputs and the activities of people that alter biotic and abiotic 'actors' in the biogeochemical play. The urban ecosystem may be simultaneously a sink for materials as well as a large source of materials, via certain vectors (e.g. wind, water), to adjacent or even distant ecosystems.

Because urban ecosystems are ultra-heterotrophic (i.e. their total ecosystem respiration (all biota, including people, plus fossil-fuel burning or 'industrial' metabolism) exceeds primary production (photosynthesis) by a very large amount; Collins *et al.* 2000), their potential impacts on ecosystems to which they are linked are huge. The question of how far the impacts of cities extend has been approached in a variety of ways, including the EF models discussed above. A different approach is to examine patterns and processes along an urban-rural gradient, asking how biogeochemical processes and fluxes change as one moves away from the center of urban activity (McDonnell and Pickett 1990; McDonnell *et al.* 1997; Zhu and Carreiro 2004). In any case, most approaches reveal that even though urban areas cover <3 percent of the global land surface (MEA 2005), their impacts are far reaching. As centers of human habitation, transportation, and industry, urban areas are a major source of CO_2 and other greenhouse gases globally (Decker *et al.* 2000; Grimm *et al.* 2008a). As centers of consumption, cities import resources from global networks of sources (Kennedy *et al.* 2007).

To link the growing understanding of controls on urban biogeochemical cycles with the global impacts of cities, we need to develop a clearer understanding of the cross-scale interactions between individual human decisions or choices and biogeochemical fluxes, but we also must account for heterogeneity in the drivers of these relationships among the world's cities. Wide variation in the characteristics of cities themselves (e.g. their size, growth rate, wealth, governance structure, urban form), as well as the regions in which they are found (e.g. climate, topography, land-use characteristics, ecosystems, social structure, national wealth, culture, etc.), present major challenges for scaling urban biogeochemistry to the globe. Two situations illustrate the variation, both of which have important consequences for urban impacts on external ecosystems: municipal services and fuel consumption.

Significant biogeochemical differences between cities in developed and developing countries can be traced to differences in services such as sewage and solid waste removal and available resources for technological innovation (e.g. wastewater treatment, recycling; Decker *et al.* 2000; Kennedy *et al.* 2007). Many cities in the developing world are growing at rates that outstrip the capacity of governments to provide these services, especially when informal settlements (shanty towns, favelas, and so forth) spring up around these cities (e.g. McDonald 2008). Population size and growth rate interacts with urban wealth to determine fluxes such as wastewater; both treatment level and volume of wastewater tend to increase with increasing wealth (Decker *et al.* 2000).

Fuel use associated with concentrated urban living is a major source of greenhouse gases. Urban form, combined with wealth and technology, may determine the contribution of vehicular transportation to the global rise in CO_2. These factors vary greatly among cities; in wealthy countries personal automobile use is lower in dense cities (typical of Europe and a few American examples, such as New York) than in those typified by sprawl; in countries experiencing rapid economic growth, automobile use is greatly increasing; in the less-developed world transportation may be a minor contributor. The type of fuel used in different cities also influences regional to global effects. For example, Mexico City is primarily fueled by liquefied petroleum gas (LPG), Buenos Aires and London use natural gas, much of China is dependent upon coal, and some of the fast-growing cities of India are still using wood biofuels. Waste products from these fuels, or potential leaks in delivery systems, will have different consequences for greenhouse gases and atmospheric chemistry.

The challenges ahead in understanding urban biogeochemical fluxes and their role in global biogeochemical cycles are great. To promote the development of an effective theoretical framework and enhance our understanding, we suggest that research continue at the several scales we have discussed in this chapter while also seeking to quantify the influences and controls that cross scales. We suspect many of these will be identified when considering the explicit role that human decisions, carried out at levels of societal organization from the household to an international community of governments, have played and will continue to play in altering biogeochemical fluxes.

References

Aldridge, K.T., Brookes, J.D., and Ganf, G.G. (2009) Rehabilitation of stream ecosystem functions through the reintroduction of coarse particulate organic matter, *Restoration Ecology*, 17(1): 97–106.

Arnold, C.L. and Gibbons, C.J. (1996) Impervious surface coverage: The emergence of a key environmental indicator, *Journal of the American Planning Association*, 62(2): 243–58.

Baker, L.A., Hartzheim, P., Hobbie, S., King, J., and Nelson, K. (2007) Effect of consumption choices on fluxes of carbon, nitrogen and phosphorus through households, *Urban Ecosystems* 10(2): 97–117.

Baker, L.A., Hope, D., Xu, Y., Edmonds, J., and Lauver, L. (2001) Nitrogen balance for the central Arizona-Phoenix (CAP) ecosystem, *Ecosystems*, 4(6): 582–602.

Band, L.E., Cadenasso, M.L., Grimmond, C.S., Grove, J.M., and Pickett, S.T.A. (2005) Heterogeneity in urban ecosystems: Patterns and process, in G.M. Lovett, M.G. Turner, C.G. Jones, and K.C. Weathers (eds.), *Ecosystem Function in Heterogeneous Landscapes*, New York: Springer, pp. 257–78.

Bandaranayake, W., Qian, Y.L., Parton, W.J., Ojima, D., and Follett, R.F. (2003) Estimation of soil organic carbon changes in turfgrass systems using the century model, *Agronomy Journal*, 95(3): 558–63.

Bedore, P.D., David, M.B., and Stucki, J.W. (2008) Mechanisms of phosphorus control in urban streams receiving sewage effluent, *Water, Air, and Soil Pollution,* 191(1–4): 217–29.

Bennett, E.M., Carpenter, S.R., and Clayton, M.K. (2005) Soil phosphorus variability: Scale-dependence in an urbanizing agricultural landscape, *Landscape Ecology*, 20(4): 389–400.

Bernhardt, E.S., Likens, G.E., Buso, D.C., and Driscoll, C.T. (2003) In-stream uptake dampens effects of major forest disturbance on watershed nitrogen export, *Proceedings of the National Academy of Sciences of the United States of America*, 100(18): 10304–8.

Björklund, A., Bjuggren, C., Dalemo, M., and Sonesson, U. (1999) Planning biodegradable waste management in Stockholm, *Journal of Industrial Ecology*, 3(4): 43–58.

Booth, D.B. and Jackson, C.R. (1997) Urbanization of aquatic systems: Degradation thresholds, stormwater detection, and the limits of mitigation, *Journal of the American Water Resources Association*, 33(5): 1077–90.

Boyle, C.A. and Lavkulich, L. (1997) Carbon pool dynamics in the Lower Fraser Basin from 1827 to 1990, *Environmental Management*, 21(3): 443–55.

Cadenasso, M.L., Pickett, S.T.A., and Schwarz, K. (2007) Spatial heterogeneity in urban ecosystems: Reconceptualizing land cover and a framework for classification, *Frontiers in Ecology and the Environment*, 5(2): 80–8.

Carpenter, S.R., Caraco, N.F., Correll, D.L., Howarth, R.W., Sharpley, A.N., and Smith, V.H. (1998) Nonpoint pollution of surface waters with phosphorus and nitrogen, *Ecological Applications*, 8(3): 559–68.

Chameides, W.L., Kasibhatla, P.S., Yienger, J., and Levy, H. (1994) Growth of continental-scale metro-agro-plexes, regional ozone pollution, and world food-production, *Science*, 264(5155): 74–7.

Churkina, G. (2008) Modeling the carbon cycle of urban systems, *Ecological Modeling*, 216(2): 107–13.

Codoban, N. and Kennedy, C.A. (2008) Metabolism of Neighborhoods. Journal of Urban Planning and Development 134(1):21-31

Collins, J.P. Kinzig, A., Grimm, N.B., Fagan, W.F., Hope, D., Wu, J.G., and Borer, E.T. (2000) A new urban ecology, *American Scientist*, 88(5): 416–25.

Dahm, C.N., Grimm, N.B., Marmonier, P., Valett, H.M., and Vervier, P. (1998) Nutrient dynamics at the interface between surface waters and ground waters, *Freshwater Biology*, 40(3): 427–51.

D'Antonio, C.M. and Mack, M.C. (2006) Nutrient limitation in a fire-derived, nitrogen-rich Hawaiian grassland, *Biotropica*, 38(4): 458–67.

David, M.B. and Gentry, L.E. (2000) Anthropogenic inputs of nitrogen and phosphorus and riverine export for Illinois, USA, *Journal of Environmental Quality*, 29(2): 494–508.

Decker, E.H., Elliott, S., Smith, F.A., Blake, D.R., and Rowland, F.S. (2000) Energy and material flow through the urban ecosystem, *Annual Review of Energy and the Environment*, 25: 685–740.

Dietz, T., Rosa, E.A., and York, R. (2007) Driving the human ecological footprint, *Frontiers in Ecology and the Environment,* 5(1): 13–18.

Driscoll, C.T., Whitall, D., and Aber, J.D. (2003) Nitrogen pollution in the northeastern United States: Sources, effects and management options, *BioScience*, 53(4): 357–74.

Duvigneaud, P. and Denayeyer-De Smet, S. (1977) L'écosystème urbs, in P. Duvigneaud and P. Kestemont (eds.), *L'écosystème Urbain Bruxellois*, Bruxelles: Traveaux de la Section Belge du Programme Biologique International, pp. 581–97.

Elsdon, T.S. and Limburg, K.E. (2008) Nutrients and their duration of enrichment influence periphyton cover and biomass in rural and urban streams, *Marine and Freshwater Research*, 59(6): 467–76.

Færge, J., Magid, J., and Penning de Vries, F.W.T. (2001) Urban nutrient balance for Bangkok, *Ecological Modelling*, 139(1): 63–74.

Fenn, M.E., Haeuber, R., Tonnesen, G.S., Baron, J.S., Grossman-Clarke, S., Hope, D., Jaffe, D.A., Copeland, S., Geiser, L., Rueth, H.M., and Sickman, J.O. (2003) Nitrogen emissions, deposition, and monitoring in the western United States, *BioScience*, 53(4): 391–403.

Golubiewski, N.E. (2006) Urbanization increases grassland carbon pools: Effects of landscaping in Colorado's front range, *Ecological Applications*, 16(2): 555–71.

Goonetilleke, A., Thomas, E., Ginn, S., and Gilbert, D. (2005) Understanding the role of land use in urban stormwater quality management, *Journal of Environmental Management,* 74(1): 31–42.

Graedel, T.E. and Allenby, B.R. (2003) *Industrial Ecology*, second edition, Upper Saddle River, NJ: Prentice-Hall.

Gregg, J.W., Jones, C.G., and Dawson, T.E. (2003) Urbanization effects on tree growth in the vicinity of New York City, *Nature*, 424(6945): 183–7.

Grimm, N.B. and Redman, C.L. (2004) Approaches to the study of urban ecosystems: The case of Central Arizona-Phoenix, *Urban Ecosystems*, 7(3): 199–213.

Grimm, N.B., Arrowsmith, R.J., Eisinger, C., Heffernan, J., Lewis, D.B., MacLeod, A., Prashad, L., Roach, W.J., Rychener, T., and Sheibley, R.W. (2004) Effects of urbanization on nutrient biogeochemistry of aridland streams, in R. DeFries, G. Asner, and R. Houghton (eds.), *Ecosystem Interactions with Land Use Change, Geophysical Monograph Series 153*, Washington, D.C.: American Geophysical Union, pp. 129–46.

Grimm, N.B., Baker, L.A., and Hope, D. (2002) An ecosystem approach to understanding cities: familiar foundations and uncharted frontiers, in A.R. Berkowitz, C.H. Nilon, and K.S. Hollweg (eds.), *Understanding Urban Ecosystems: A New Frontier for Science and Education*, New York: Springer, pp. 95–114.

Grimm, N.B., Faeth, S.H., Golubiewski, N.E., Redman, C.L., Wu, J., Bai, X., and Briggs, J.M. (2008a) Global change and the ecology of cities, *Science*, 319(5864): 756–60.

Grimm, N.B., Foster, D., Groffman, P., Grove, J.M., Hopkinson, C.S., Nadelhoffer, K., Pataki, D.E., and Peters, D.P.C. (2008b) The changing landscape: Ecosystem responses to urbanization and pollution across climatic and societal gradients, *Frontiers in Ecology and the Environment*, 6(5): 264–72.

Grimm, N.B., Sheibley, R.W., Crenshaw, C.L., Dahm, C.N., Roach, W.J., and Zeglin, L.H. (2005) N retention and transformation in urban streams, *Journal of the North American Benthological Society*, 24(3): 626–42.

Groffman, P.M., Bain, D.J., Band, L.E., Belt, K.T., Brush, G.S., Grove, J.M., Pouyat, R.V., Yesilonis, I.C., and Zipperer, W.C. (2003) Down by the riverside: Urban riparian ecology, *Frontiers in Ecology and the Environment*, 1(6): 315–21.

Groffman, P.M., Boulware, N.J., Zipperer, W.C., Pouyat, R.V., Band, L.E., Colosimo, M.F. (2002) Soil nitrogen cycle processes in urban riparian zones, *Environmental Science and Technology*, 36(1): 4547–52.

Groffman, P.M., Dorsey, A.M., and Mayer, P.M. (2005) N processing within geomorphic structures in urban streams, *Journal of the North American Benthological Society*, 24(3): 613–25.

Groffman, P.M., Law, N.L., Belt, K.T., Band, L.E., and Fisher, G.T. (2004) Nitrogen fluxes and retention in urban watershed ecosystems, *Ecosystems*, 7(4): 393–403.

Groffman, P.M. and Pouyat, R.V. (2009) Methane uptake in urban forests and lawns, *Environmental Science and Technology*, 43(14): 5229–35.

Groffman, P.M., Williams, C.O., Pouyat, R.V., Band, L.E., and Yesilonis, I.D. (2009) Nitrate leaching and nitrous oxide flux in urban forests and grasslands, *Journal of Environmental Quality*, 38(5): 1848–60.

Gu, B.J., Chang, J., Ge, Y., Ge, H.L., Yuan, C., Peng, C.H., and Jiang, H. (2009) Anthropogenic modification of the nitrogen cycling within the Greater Hangzhou Area system, China, *Ecological Applications*, 19(4): 974–88.

Hall, R.O., Tank, J.L., Sobota, D.J., Mulholland, P.J., O'Brien, J.M., Dodds, W.K., Webster, J.R., Valett, H.M., Poole, G.C., Peterson, B.J., Meyer, J.L., McDowell, W.H., Johnson, S.L., Hamilton, S.K., Grimm, N.B., Gregory, S.V., Dahm, C.N., Cooper, L.W., Ashkenas, L.R., Thomas, S.M., Sheibley, R.W., Potter, J.D., Niederlehner, B.R., Johnson, L.T., Helton, A.M., Crenshaw, C.M., Burgin, A.J., Bernot, M.J., Beaulieu, J.J., and Arango, C.P. (2009) Nitrate removal in stream ecosystems measured by N-15 addition experiments: Total uptake, *Limnology and Oceanography*, 54(3): 653–65.

Hall, S.J., Huber, D., and Grimm, N.B. (2008) Soil N_2O and NO emissions from an arid, urban ecosystem, *Journal of Geophysical Research – Biogeosciences*, 113: G01016, doi:10.1029/2007JG000523.

Harbott, E.L. and Grace, M.R. (2005) Extracellular enzyme response to bioavailability of dissolved organic C in streams of varying catchment urbanization, *Journal of the North American Benthological Society*, 24(3): 588–601.

Hatt, B.E., Fletcher, T.D., Walsh, C.J., and Taylor, S.L. (2004) The influence of urban density and drainage infrastructure on the concentrations and loads of pollutants in small streams, *Environmental Management*, 34(1): 112–24.

Hope, D., Naegeli, M.W., Chan, A., and Grimm, N.B. (2004) Nutrients on asphalt parking surfaces in an arid urban environment, *Water, Air, and Soil Pollution: Focus*, 4(2–3): 371–90.

Imhoff, M.L., Bounoua, L., DeFries, R., Lawrence, W.T., Stutzer, D., Tucker, C.J., and Ricketts, T. (2004) The consequences of urban land transformation on net primary productivity in the United States, *Remote Sensing of Environment*, 89(4): 434–43.

Izbicki, J.A., Pimentel, I.M., Johnson, R., Aiken, G.R., and Leenheer, J. (2007) Concentration, UV-spectroscopic characteristics and fractionation of DOC in stormflow from an urban stream, Southern California, USA, *Environmental Chemistry*, 4(1): 35–48.

Jenerette, G.D., Wu, J.G., Grimm, N.B., and Hope, D. (2006) Points, patches, and regions: Scaling soil

biogeochemical patterns in an urbanized arid ecosystem, *Global Change Biology*, 12(8): 1532–44.
Karl, T.R., Melillo, J.M., and Peterson, T.C. (eds.) (2009) *Global Climate Change Impacts in the United States*, New York: Cambridge University Press.
Kaushal, S.S., Groffman, P.M., Band, L.E., Shields, C.A., Morgan, R.P., Palmer, M.A., Belt, K.T., Swan, C.M., Findlay, S.E.G., and Fisher, G.T. (2008) Interaction between urbanization and climate variability amplifies watershed nitrate export in Maryland, *Environmental Science and Technology*, 42(16): 5872–8.
Kaye, J.P., Burke, I.C., Mosier, A.R., and Guerschman, J.P. (2004) Methane and nitrous oxide fluxes from urban soils to the atmosphere, *Ecological Applications*, 14(4): 975–81.
Kaye, J.P., Eckert, S.E., Gonzales, D.A., Allen, J.O., Hall, S.J., Sponseller, R.A., and Grimm, N.B. (in press) Can atmospheric carbon deposition stimulate microbial respiration in desert soils? *Journal of Environmental Quality*.
Kaye, J.P., Groffman, P.M., Grimm, N.B., Baker, L.A., and Pouyat, R.V. (2006) A distinct urban biogeochemistry? *Trends in Ecology and Evolution*, 21(4): 192–9.
Kaye, J.P., Majumdar, A., Gries, C., Buyantuyev, A., Grimm, N.B., Hope, D., Jenerette, G.D., Zhu, W.X., and Baker, L. (2008) Hierarchical Bayesian scaling of soil properties across urban, agricultural, and desert ecosystems, *Ecological Applications*, 18(1): 132–45.
Kaye, J.P., McCulley, R.L., and Burke, I.C. (2005) Carbon fluxes, nitrogen cycling, and soil microbial communities in adjacent urban, native and agricultural ecosystems, *Global Change Biology*, 11(4): 575–87.
Kennedy, C., Cuddihy, J., and Engel-Yan, J. (2007) The changing metabolism of cities, *Journal of Industrial Ecology*, 11(2): 43–59.
Koerner, B. and Klopatek, J. (2002) Anthropogenic and natural CO_2 emission sources in an arid urban environment, *Environmental Pollution*, 116(Suppl. 1): S45–S51.
Lauver, L. and Baker, L.A. (2000) Mass balance for wastewater nitrogen in the central Arizona-Phoenix ecosystem, *Water Research*, 34(1): 2754–60.
La Valle, P.D. (1975) Domestic sources of stream phosphates in urban streams, *Water Research*, 9(10): 913–15.
Lepori, F., Palm, D., and Malmquist, B. (2005) Effects of stream restoration on ecosystem functioning: detritus retentiveness and decomposition, *Journal of Applied Ecology*, 42(2): 228–38.
Lerner, D.N. (2002) Identifying and quantifying urban recharge: a review, *Hydrogeology Journal*, 10(1): 143–52.
Lewis, D.B. and Grimm, N.B. (2007) Hierarchical regulation of nitrogen export from urban catchments: Interactions of storms and landscapes, *Ecological Applications*, 17(8): 2347–64.
Lewis, D.B., Kaye, J.P., Gries, C., Kinzig, A.P., and Redman, C.L. (2006) Agrarian legacy in soil nutrient pools of urbanizing arid lands, *Global Change Biology*, 12(4): 703–9.
Likens, G.E. and Bormann, F.H. (1995) *Biogeochemistry of a Forested Ecosystem*, second edition, New York: Springer.
Livesley, S., Dougherty, B., Smith, A., Navaud, D., Wylie, L., and Arndt, S. (2010) Soil-atmosphere exchange of carbon dioxide, methane and nitrous oxide in urban garden systems: impact of irrigation, fertiliser and mulch, *Urban Ecosystems*, 13(3): 273–93, online, available at: http://dx.doi.org.ezproxy1.lib.asu.edu/10.1007/s11252-009-0119-6.
Liu, J., Daily, G.C., Ehrlich, P.R., and Luck, G.W. (2003) Effects of household dynamics on resource consumption and biodiversity, *Nature*, 421: 530–3.
Lohse, K.A., Hope, D., Sponseller, R., Allen, J.O., and Grimm, N.B. (2008) Atmospheric deposition of carbon and nutrients across an arid metropolitan area, *Science of the Total Environment*, 402(1): 95–105.
Lovett, G.M., Traynor, M.M., Pouyat, R.V., Carreiro, M.M., Zhu, W.X., and Baxter, J.W. (2000) Atmospheric deposition to oak forests along an urban-rural gradient, *Environmental Science and Technology*, 34(20): 4294–300.
Makepeace, D.K., Smith, D.W., and Stanley, S.J. (1995) Urban stormwater quality: Summary of contaminant data, *Critical Reviews in Environmental Science and Technology*, 25(2): 93–139.
McClain, M.E., Boyer, E.W., Dent, C.L., Gergel, S.E., Grimm, N.B., Groffman, P.M., Hart, S.C., Harvey, J.W., Johnston, C.A., Mayorga, E., McDowell, W.H., and Pinay, G. (2003) Biogeochemical hot spots and hot moments at the interface of terrestrial and aquatic ecosystems, *Ecosystems*, 6(4): 301–312.
McDonald, R.I. (2008) Global urbanization: can ecologists identify a sustainable way forward? *Frontiers in Ecology and the Environment*, 6(2): 99–104.
McDonnell, M.J. and Pickett, S.T.A. (1990) Ecosystem structure and function along urban rural gradients: An unexploited opportunity for ecology, *Ecology*, 71(4): 1232–7.

McDonnell, M.J., Pickett, S.T.A., Groffman, P., Bohlen, R., Pouyat, W.V., Zipperer, W.C., Parmelee, R.W., Carreiro, M.M., and Medley, K. (1997) Ecosystem processes along an urban-to-rural gradient, *Urban Ecosystems*, 1(1): 21–36.

Meyer, J.L., Paul, M.J., and Taulbee, W.K. (2005) Stream ecosystem function in urbanizing landscapes, *Journal of the North American Benthological Society*, 24(3): 602–12.

Milesi, C., Running, S., Elvidge, C., Dietz, J., Tuttle, B., and Nemani, R. (2005) Mapping and modeling the biogeochemical cycling of turf grasses in the United States, *Environmental Management*, 36(3): 426–38.

MEA (Millennium Ecosystem Assessment) (2005) *Ecosystem and Human Well-Being: Current state and trends*, Washington, D.C.: Island Press.

Molina, M.J. and Molina, L.T. (2004) Megacities and atmospheric pollution, *Journal of the Air and Waste Management Association*, 54(6): 644–80.

Mulholland, P.J., Hall, R.O., Sobota, D.J., Dodds, W.K., Findlay, S.E.G., Grimm, N.B., Hamilton, S.K., McDowell, W.H., O'Brien, J.M., Tank, J.L., Ashkenas, L.R., Cooper, L.W., Dahm, C.N., Gregory, S.V., Johnson, S.L., Meyer, J.L., Peterson, B.J., Poole, G.C., Valett, H.M., Webster, J.R., Arango, C., Beaulieu, J.J., Bernot, M.J., Burgin, A.J., Crenshaw, C., Helton, A.M., Johnson, L., Niederlehner, B.R., Potter, J.D., Sheibley, R.W., and Thomas, S.M. (2009) Nitrate removal in stream ecosystems measured by 15N addition experiments: denitrification, *Limnology and Oceanography*, 54(3): 666–80.

Mulholland, P.J., Helton, A.M., Poole, G.C., Hall, R.O., Hamilton, S.K., Peterson, B.J., Tank, J.L., Ashkenas, L.R., Cooper, W.L., Dahm, C.N., Dodds, W.K., Findlay, S., Gregory, S.V., Grimm, N.B., Johnson, S.L., McDowell, W.H., Meyer, J.L., Valett, J.M., Webster, J.R., Arango, C., Beaulieu, J.J., Bernot, M.J., Burgin, A.J., Crenshaw, C., Johnson, L., Niederlehner, B.R., O'Brien, J.M., Potter, J.D., Sheibley, R.W., Sobota, D.J., and Thomas, S.M. (2008) Stream denitrification across biomes and its response to anthropogenic nitrate loading, *Nature*, 452: 202–5.

Nilsson, J. (1995) A phosphorus budget for a Swedish municipality, *Journal of Environmental Management*, 45(3): 243–53.

Oren, R., Ellsworth, D.S., Johnsen, K.H., Phillips, N., Ewers, B.E., Maier, C., Schafer, K.V.R., McCarthy, H., Hendrey, G., McNulty, S.G., and Katul, G.G. (2001) Soil fertility limits carbon sequestration by forest ecosystems in a CO2-enriched atmosphere, *Nature*, 411: 469–72.

Ortiz-Zayas, J.R., Cuevas, E., Mayol-Bracero, O.L., Donoso, L., Trebs, I., Figueroa-Nieves, D., and McDowell, W.H. (2006) Urban influences on the nitrogen cycle in Puerto Rico, *Biogeochemistry*, 79(1–2): 109–33.

Pataki, D.E., Alig, R.J., Fung, A.S., Golubiewski, N.E., Kennedy, C.A., McPherson, E.G., Nowak, D.J., Pouyat, R.V., and Romero Lanko, P. (2006) Urban ecosystems and the North American carbon cycle, *Global Change Biology*, 12(11): 2092–102.

Paul, M.J. and Meyer, J.L. (2001) Streams in the urban landscape, *Annual Review of Ecology and Systematics*, 32: 333–65.

Qian, Y.L., Bandaranayake, W., Parton, W.J., Mecham, B., Harivandi, M.A., and Mosier, A.R. (2003) Long-term effects of clipping and nitrogen management in turfgrass on soil organic carbon and nitrogen dynamics: The Century model simulation, *Journal of Environmental Quality*, 32(5): 1694–700.

Raciti, S.M., Groffman, P.M., and Fahey, T.J. (2008) Nitrogen retention in urban lawns and forests, *Ecological Applications*, 18(7): 1615–26.

Rees, W. and Wackernagel, M. (1996) Urban ecological footprints: Why cities cannot be sustainable – and why they are a key to sustainability, *Environmental Impact Assessment Review*, 16(4–6): 223–48.

Roach, W.J. and Grimm, N.B. (2009) Nutrient variation in an urban lake chain and its consequences for phytoplankton production, *Journal of Environmental Quality*, 38(4): 1429–40.

Roach, W.J., Heffernan, J.B., Grimm, N.B., Arrowsmith, J.R., Eisinger, C., and Rychener, T. (2008) Unintended consequences of urbanization for aquatic ecosystems: a case study from the Arizona desert, *BioScience*, 58(8): 715–27, doi:10.1641/B580808.

Robbins, P. (2007) *Lawn People: How Grasses, Weeds and Chemicals Make Us Who We Are*, Philadelphia, PA: Temple University Press.

Sahely, H.R., Dudding, S., and Kennedy, C.A. (2003) Estimating the urban metabolism of Canadian cities: Greater Toronto Area case study, *Canadian Journal of Civil Engineering*, 30(2): 468–83.

Schauer, J.J., Rogge, W.F., Hildemann, L.M, Mazurek, M.A., Cass, G.R., and Simoneit, B.R.T. (1996) Source apportionment of airborne particulate matter using organic compounds as tracers, *Atmospheric Environment*, 41 (Suppl.1): 241–59.

Shen, W.J., Wu, J.G., Grimm, N.B., and Hope, D. (2008) Effects of urbanization-induced environmental changes on ecosystem functioning in the phoenix metropolitan region, USA, *Ecosystems*, 11(1): 138–55.

Shepherd, J.M., Pierce, H., and Negri A.J. (2002) Rainfall modification by major urban areas: Observations from spaceborne rain radar on the TRMM satellite, *Journal of Applied Meteorology*, 41(7): 689–701.
Turner, B.L. (2010) Sustainability and forest transitions in the Southern Yucatan: The land architecture approach, *Land Use Policy*, 27(2): 170–9.
U.S. Geological Survey (1999) *The Quality of Our Nation's Waters—Nutrients and Pesticides*, U.S. Geological Survey Circular 1225, online, available at: http://pubs.usgs.gov/circ/circ1225/pdf/index.html.
Walker, J.S., Grimm, N.B., Briggs, J.M., Gries, C., and Dugan, L. (2009) Effects of urbanization on plant species diversity in central Arizona, *Frontiers in Ecology and the Environment*, 7(9): 465–70.
Walsh, C.J. and Kunapo, J. (2009) The importance of upland flow paths in determining urban effects on stream ecosystems, *Journal of the North American Bentham Society*, 28(4): 977–90.
Walsh, C.J., Roy, A.H., Feminella, J.W., Cottingham, P.D., Groffman, P.M., and Morgan, R.P. (2005) The urban stream syndrome: current knowledge and the search for a cure, *Journal of the North American Benthological Society*, 24(3): 706–23.
Warren-Rhodes, K. and Koenig, A. (2001) Escalating trends in the urban metabolism of Hong Kong: 1971–1997, *Ambio*, 30(7): 429–38.
Westerhoff, P. and Anning, D. (2000) Concentrations and characteristics of organic carbon in surface water in Arizona: influence of urbanization, *Journal of Hydrology*, 236(3–4): 202–22.
Williams, M., Hopkinson, C., Rastetter, E., and Vallino, J. (2004) N budgets and aquatic uptake in the Ipswich River basin, northeastern Massachusetts, *Water Resources Research*, 40: W11201, doi:10.1029/2004WR003172.
Williams, M., Hopkinson, C., Rastetter, E., Vallino, J., and Claessens, L. (2005) Relationships of land use and stream solute concentrations in the Ipswich River basin, northeastern Massachusetts, *Water, Air, and Soil Pollution*, 161(1–4): 55–74.
Wollheim, W.M., Pellerin, B.A., Vorosmarty, C.J., and Hopkinson, C.S. (2005) N retention in urbanizing headwater catchments, *Ecosystems*, 8(8): 871–84.
Wolman, A. (1965) The metabolism of cities, *Scientific American*, 213(3): 179–90.
Wu, J.G. and David, J.L. (2002) A spatially explicit hierarchical approach to modeling complex ecological systems: theory and applications, *Ecological Modelling*, 153(1–2): 7–26.
Zhu, W.X. and Carreiro, M.A. (2004) Temporal and spatial variations in nitrogen transformations in deciduous forest ecosystems along an urban-rural gradient, *Soil Biology and Biochemistry*, 36(2): 267–78.
Zhu, W.X., Dillard, N.D., and Grimm, N.B. (2004) Urban nitrogen biogeochemistry: status and processes in green retention basins, *Biogeochemistry*, 71(2): 177–96.
Zhu, W.X., Hope, D., Gries, C., and Grimm, N.B. (2006) Soil characteristics and the accumulation of inorganic nitrogen in an arid urban ecosystem, *Ecosystems*, 9(5): 711–24.

40

Urban metabolism analysis

Shu-Li Huang and Chun-Lin Lee

What is urban metabolism?

Post-industrial cities are characterized by their routine use of energy as a driving force to power production, transportation of goods, construction of buildings and infrastructure as well as domestic comfort. The existence and maintenance of a city and its internal structure depend on the flow of goods and services into, out of, and throughout that city. The materials, energy, and food supplies brought into cities; the transformation of these inputs within the cities; and the products and wastes sent out from the cities are often referred to as urban metabolism, a concept first suggested by Wolman (1965).

Metabolism is a concept adopted from biology, which refers to the physiological processes within a living organism that describes the energy flow connected to the conversion of matter for reproduction. Extending this concept to the social sciences, metabolism can be seen as a main feature in the analysis of human interactions with the natural environment. Like human metabolism, the physical and biological processes of a city system transform inflows of energy and materials into useful products, services, and wastes. The complete metabolism of a city consists of many inputs such as food, fuel, clothing, durable goods, electricity, construction materials, and services (Figure 40.1). The linear metabolic system of the modern city is different from nature's circular metabolism, where every output by an organism is also an input that renews and sustains the living environment. The metabolic cycle is not completed until the residues of daily consumption have been removed and disposed of adequately with minimum nuisance and hazard to life.

Kennedy *et al.* (2007) reviewed eight urban metabolism studies conducted since 1965 and identified metabolic processes that threaten the sustainability of cities. The major environmental problems and associated social costs of an urban ecosystem are related to the rapid increase of resource inputs for urban consumption and the disposal of construction waste, both of which are nuisances to urban dwellers. Much more radical changes in the urban metabolism are required to make cities more ecologically viable.

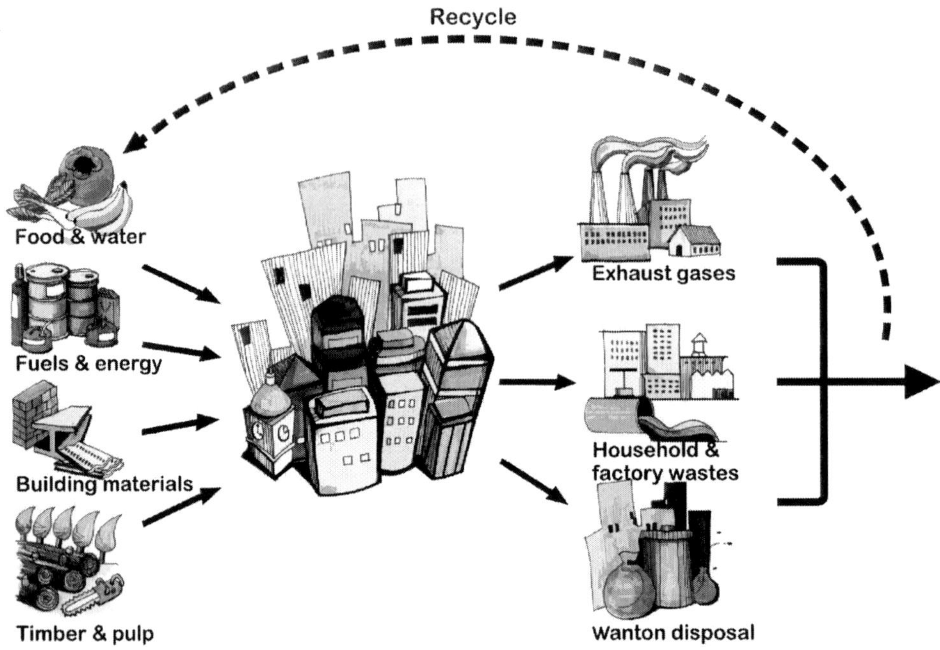

Figure 40.1 The complete metabolism of a city.

Material flow analysis

The ways cities function and the standards of living that they provide, determine the amounts and types of resource use. The analysis of urban metabolism has largely been dominated by quantification of material inflows and outflows. Material flow analysis is an analytic tool which can be used to examine the stocks and flows of material inflow and outflow from a given system. The material flows of urban metabolism can be measured in terms of the rate of mass input from the natural environment to outputs of residues from a city. Material flows analysis can provide a framework for analyzing urbanization processes and the way cities are transforming the earth's ecosystems as a consequence of human activities.

Wolman (1965) pointed out that provision of adequate water supply, effective disposal of sewage control and control of air pollution are the three metabolic problems that have become more acute as a result of urban growth. He used a hypothetical American city with a population of one million to quantify the inflows of water, food, and fuel, and outflows of sewage, solid waste, and air pollutants. Similar to Wolman's approach, Boyden *et al.* (1981) quantified the material inflows and outflows of Hong Kong. Their study examined the urban ecology of Hong Kong and its relationships with Hong Kong's social characteristics.

In order to provide a more convenient tool for summarizing the sustainability of a socioeconomic system, the complex process of material flow analysis has been compiled and aggregated into Material Flow Accounts (MFA) (Eurostat 2001). In the framework of MFA, the 'total material requirement' also includes indirect flows of 'unused extraction'; namely flows that do not enter the economy under consideration but are mobilized to produce goods and services consumed. The unused extraction, for example soil excavation, also plays an important role in assessing society's impact on the environment. Material flow analysis is the most productive application of socioeconomic metabolism analysis. However, the material relationship between humans and their environment cannot simply be viewed from an assessment of input-output processes (Fischer-Kowalski and Haberl 1997). Newman (1999) expanded the concept of material flow analysis of an urban system to include aspects of livability for demonstrating the practical meaning of sustainability. The transformation of materials into economic assets for sustaining urban metabolism and maximizing their usefulness for human societies must be analyzed as well. In addition, this measure does not allow for comparison of the qualitative usefulness of different materials to socioeconomic systems (Huang *et al.* 2006).

Life-cycle assessment

The concept of 'life-cycle' began to receive environmental researchers' attention after the first report on life-cycle analysis was proposed by the Coca-Cola Company in 1969 to analyze the environmental impacts of their products through all stages of their 'life'—manufacturing to final disposal. On the one hand life-cycle analysis was originally adopted to analyze specific products, but now it is used to examine material consumption, electricity use, and waste management for businesses, industries, and cities. Formalized by the Swiss Federal Laboratories for Materials Testing and Research (EMPA), Life-Cycle Assessment (LCA) has become an important tool for exploring and comparing metabolism of businesses and cities (EMPA 1984). LCA can be defined as

> a technique for assessing the environmental aspects and potential impacts associated with a product by compiling an inventory of relevant inputs and outputs of a product system; and evaluating the potential environmental impacts associated with those inputs and outputs; and interpreting the results of the inventory analysis and impact assessment phases in relation to the objectives of the study.
>
> *(ISO 2006: iii)*

As presented in Figure 40.2, LCA has two major components: process analysis and input-output analysis. The former emphasizes tracking material and energy flows during production processes. The approach produces more accurate results for the short term than for the long term. One major difficulty, however, with process analysis is that it usually produces an extremely complex system with substantial data requirements. Input–output analysis focuses on raw material inflows and waste outflows at all stages of production and has an understandable system framework. Both of them can be useful tools for exploring urban metabolism by tracing impacts through whole supply chain of industries.

LCA has matured as an analytical method to the point where it has been used to assess not only the complex life cycle of a business, but also the current metabolism of cities. Mora (2007) explored the relationship between life cycle and sustainability of engineering works; and the impact of urban growth and urban infrastructure on the environment through the consumption of raw materials and energy. Lenzen (2008) used LCA to show that conservation, efficiency and reductions in the overall material metabolism of economic activity can be as effective as purely technologically-driven changes in reducing greenhouse gas emissions.

Energetics of urban metabolism

Most current research on socioeconomic metabolism ignores energy flow because of the difficulty of comparing materials and energy with the same units. Aggregating material flows according to

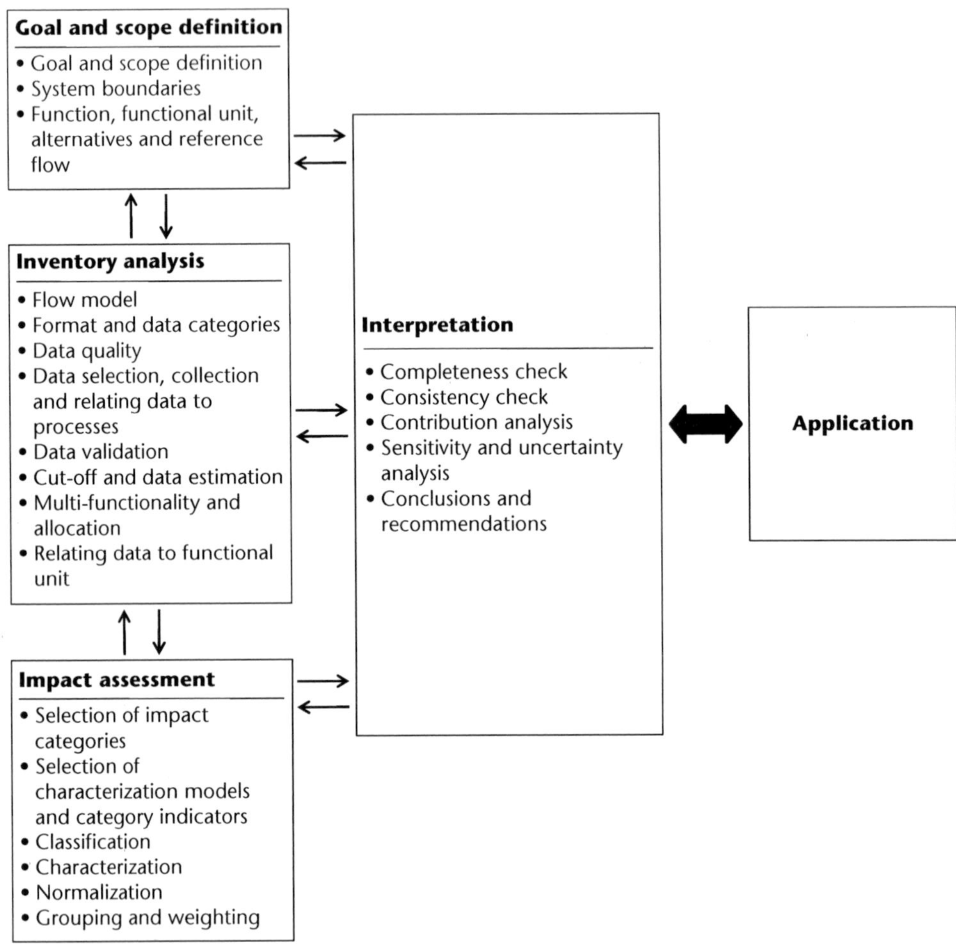

Figure 40.2 LCA framework (authors' interpretation of description in ISO 2006).

their mass neglects the relative contribution associated with the values of materials with different qualitative contents (Huang *et al.* 2006). Energy flows are the common biophysical measure connecting ecosystems and economic systems (Hall *et al.* 1986). Consequently, understanding the role of energy flows through an economic system is necessary to complement and illuminate the study of socioeconomic metabolism as well as to provide a common value basis to its evaluation.

The addition of energy flows in the analysis of socioeconomic metabolism should go beyond the mere accounting of energy flows. It should incorporate energetic principles to assess the relative contribution of various material and energy flows. In order to evaluate the contributory value of different material flows to the ecological economic system, a new accounting system is required that can assess the biophysical value of resources to the economic system. Based on the general system principles and laws of thermodynamics, Odum (1971, 1983) designed a set of energy circuit symbols (Figure 40.3) for describing the interactions of ecosystem components via energetic flows. Odum formulated a unifying theory of system ecology of values (1971, 1988, 1996) and introduced the terms, emergy and transformity. Emergy is defined as *all the available energy that was used in the work of making a product in units of one type of energy*; transformity is *the emergy of one type required to make a unit of energy of another type* (Odum 1996: 288–9). Most often solar energy units are used (solar equivalent joules, sej) to measure these parameters. Solar transformities have been calculated for a wide variety of renewable and nonrenewable energies, physical resources, and commodities (e.g. Brown and Bardi 2001; Brandt-Williams 2002; Campbell *et al.* 2005; Odum 1996, 2000; Odum *et al.* 2000). It has been suggested that an energy system diagram using the energy symbols in Figure 40.3 should be drawn to provide an overview of any study area and its subject of study, and to identify the sources of flows and major processes. An emergy synthesis table can then be developed to quantify the energy content or mass of the identified flows. For the purpose of accounting for the varied qualities of energy content inherent in the material and energy flows of a socioeconomic system, the energy content (e.g. joule) or mass of a flow can be multiplied by its solar transformity to obtain its solar emergy in solar emergy joules (sej) (Figure 40.4). These values can then be substituted for the units of mass in the material flow analysis. Emergy indices, such as ratio of imported material flows to total emergy used or per capita total material use, can be calculated for policy evaluation. Further details on the concept and procedure of emergy synthesis can be found in Odum (1996), and Brown and Ulgiati (2004).

Huang *et al.* (2006) combined material flows and energy flows by incorporating emergy synthesis to provide an overview of the socioeconomic metabolism of Taiwan from 1981 to 2001. The differences between the results obtained from material flow analysis and emergy synthesis suggest that qualitative characteristics of materials and energy flows cannot be neglected. Huang and Chen (2009) have applied Odum's emergy concept to integrate energy and material flows in the study of the socioeconomic metabolism of the Taipei area. Urban sprawl in Taipei was also taken into consideration in order to study its relationships with the changing socioeconomic metabolism of the region.

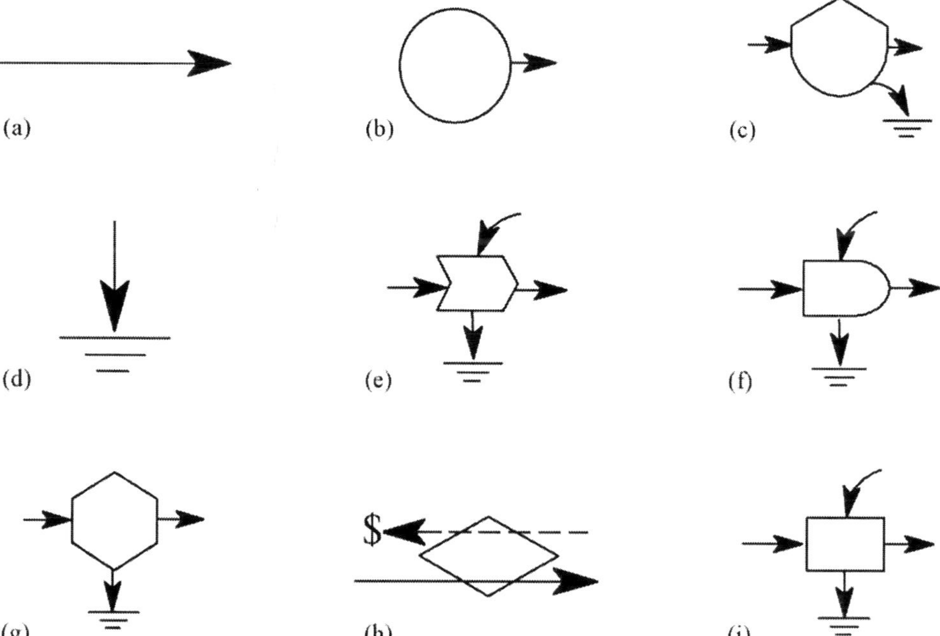

Figure 40.3 Energy circuit symbols: (a) energy circuit; (b) source; (c) tank; (d) heat sink; (e) interaction; (f) producer; (g) consumer; (h) transaction; (i) box (*source:* Huang *et al.* 2006: 173).

Item	Raw Data	Solar Transformity (sej/unit)	Solar Emergy	Em$
⋮				
Rain	$6.56*10^{17}$ J	15,444 J/sej	$101.33*10^{20}$ sej	$2,571.99*10^{6}$ $
⋮				
Limestone	$1.13*10^{13}$ g	$1*10^{9}$ g/sej	$113.31*10^{20}$ sej	$2,867*10^{6}$ $
⋮				

Figure 40.4 Emergy synthesis table (*source:* Huang *et al.* 2006: 173).

Application: ecological footprint and land use/cover change

Past studies on urban metabolism have provided important insights about energy and material consumption, nutrient and hydrology cycling, and the waste management of cities. More recently, concern over the impacts of urban life in a global context has given rise to a focus on the ecological footprints of cities (Wackernagel and Rees 1996). Ecological footprint measures the total amount of land required to provide resources and energy for a given population under a specified living standard. The land required to meet these needs is categorized into energy land, consumed land, farm land, and forest land. Girardet has calculated the resource use in greater London and estimated the city's ecological footprint to be 125 times the area city occupies (Sustainable London Trust 1996).

Urban metabolism not only has a unique pattern of energy and material cycle, it has also been noted that there are close relations between land use/cover change and socioeconomic metabolism. Land use is the most important socioeconomic driver of ecosystem change. Land use/cover change is a dynamic process influenced by complex interactions between socio-economic drivers and biophysical conditions (Lambin *et al.* 1999). Although land cover is modified by societies to address any number of important needs, such modifications can lead to an acceleration in the consumption of energy and result in an increase of indirect material flows, such as erosion. Urban metabolism cannot be analyzed adequately without considering land use/cover change because resource consumption, asset accumulation and waste emissions aspects of urban metabolism involve complex processes of land change (Turner *et al.* 1993).

Haberl *et al.* (2001) adopted Ecological Footprint (EF) to convert the use of selected materials in a country into the area needed to sustain material flows. Krausmann and Haberl (2002) proposed Human Appropriation of Net Primary Production (HANPP) to analyze the relation between land change and metabolism. Additionally, the HANPP has been combined with Material Flow Accounting (MFA) to provide a macroscopic image between urban metabolism land change for a city or region. Huang *et al.* (2006) and Lee *et al.* (2008) further explored socioeconomic metabolism and land use change based on emergy synthesis. The spatial temporal dynamics between resource consumption and land use and land cover change can also be analyzed using GIS and spatial modeling to explore the relationships between land use/cover change and urban metabolism. The importance of each of these approaches cannot be understated. HANPP, EF, and emergy synthesis have all become important methods for exploring urban metabolism and land change, and they will remain so for some time into the future.

References

Boyden, S., Millar, S., Newcombe, K., and O'Neill, B. (1981) *The Ecology of a City and its People: The Case of Hong Kong*, Canberra: Australia National University Press.

Brandt-Williams, S. (2002) *Handbook of Emergy Evaluation Folio 4: Emergy of Florida agriculture*, Gainesville, FL: Center for Environmental Policy, University of Florida.

Brown, M.T. and Bardi, E. (2001) *Handbook of Emergy Evaluation Folio 3: Emergy of Ecosystems*, Gainesville, FL: Center for Environmental Policy, University of Florida.

Brown, M.T. and Ulgiati, S. (2004) 'Emergy and environmental accounting,' in C. Cleveland (ed.), *Encyclopedia of Energy*, Amsterdam: Elsevier, pp. 329–53.

Campbell, D.E., Brandt-Williams, S.L., and Meisch, M.E.A. (2005) *Environmental Accounting Using Emergy: Evaluation of the State of West Virginia*, Report No. EPA/600/R-05/006. Narragansett, RI: U.S. Environmental Protection Agency, Atlantic Ecology Division.

EMPA (1984) *Ecological Report of Packaging Material*, Duebendorf: The Swiss Federal Laboratories for Materials Testing and Research (EMPA).

Eurostat (2001) *Economy-Wide Material Flow Accounts and Derived Indicators: A Methodological Guide*, Luxembourg: Eurostat, European Commission.

Fischer-Kowalski, M. and Haberl, H. (1997) 'Tons, joules, and money: modes of production and their sustainability problems,' *Society and Natural Resources*, 10(1): 61–85.

Haberl, H., Erb, K.-H., and Krausmann, F. (2001) 'How to calculate and interpret ecological footprints for long periods of time: the case of Austria 1926–1995,' *Ecological Economics*, 38(1): 25–45.

Hall, C.A.S., Cleveland, C.J., and Kauffmann, R. (1986) *Energy and Resource Quality: The Ecology of the Economic Process*, New York: John Wiley and Sons.

Huang, S.-L. and Chen, C.-W. (2009) 'Urbanization and socioeconomic metabolism in Taipei,' *Journal of Industrial Ecology*, 13(1): 75–93.

Huang, S.-L., Lee, C.-L., and Chen, C.-W. (2006) 'Socioeconomic metabolism in Taiwan: Emergy synthesis versus material flow analysis,' *Resources, Conservation and Recycling*, 48(2): 166–96.

ISO (2006) *Environmental Management – Life Cycle Assessment – Principles and Framework: International Standard 14040*, Geneva: International Organization for Standardization (ISO).

Kennedy, C.A., Cuddihy, J., and Engel-Yan, J. (2007) 'The changing metabolism of cities,' *Journal of Industrial Ecology*, 11(2): 43–59.

Krausmann, F. and Haberl, H. (2002) 'The process of industrialization from the perspective of energetic metabolism: Socio-economic energy flows in Austria 1830–1995,' *Ecological Economics*, 41(2): 177–201.

Lambin, E.F., Baulies, X., Bockstael, N., Fischer, G., Krug, T., Leemans, R., Moran, E.F., Rindfuss, R.R., Sato, Y., Skole, D., Turner II, B.L., and Vogel, C. (1999) *Land-Use and Land-Cover Change (LUCC): Implementation Strategy*, Stockholm: International Geosphere–Biosphere Programme (IGBP) and International Human Dimensions Programme on Global Environmental Change (IHDP).

Lee, C.-L., Huang, S.-L., and Chan S.-L. (2008) 'Biophysical and system approaches for simulation land-use change,' *Landscape and Urban Planning*, 86(2): 187–203.

Lenzen, M. (2008) 'Sustainable island businesses: a case study of Norfolk Island,' *Journal of Cleaner Production*, 16(18): 2018–35.

Mora, E.P. (2007) 'Life cycle, sustainability and the transcendent quality of building materials,' *Building and Environment*, 42(3): 1329–34.

Newman, P.W.G. (1999) 'Sustainability and cities: Extending the metabolism model,' *Landscape and Urban Planning*, 44(4): 219–26.

Odum, H.T. (1971) *Environment, Power and Society*, New York: John Wiley and Sons.

Odum, H.T. (1983) *Systems Ecology*, New York: John Wiley and Sons.

Odum, H.T. (1988) 'Self-organization, transformity, and information,' *Science*, 242(4882): 1132–9.

Odum, H.T. (1996) *Environmental Accounting: Emergy and Environmental Decision Making*, New York: John Wiley and Sons.

Odum, H.T. (2000) *Handbook of Emergy Evaluation Folio 2: Emergy of Global Processes*, Gainesville, FL: Center for Environmental Policy, University of Florida.

Odum, H.T., Brown, M.B., and Brandt-Williams, S. (2000) *Handbook of Emergy Evaluation Folio 1: Introduction and Global Budget*, Gainesville, FL: Center for Environmental Policy, University of Florida.

Sustainable London Trust (1996) *Creating a Sustainable London*, London: The Trust.

Turner II, B.L., Moss, R.H., and Skole, D.L. (1993) *Relating Land Use and Global Land Cover Change: A Proposal for an IGBP-HDP core Project*, IGBP Report No. 24 and IHDP Report No. 5, Stockholm: International Geosphere-Biosphere Programme and Human Dimensions of Global Change Programme.

Wackernagel, M. and Rees, W. (1996) *Our Ecological Footprint: Reducing Human Impact on the Earth*, Gabriela Island: New Society Publishers.

Wolman, A. (1965) 'The metabolism of cities', *Scientific American*, 213(3): 179–88.

Part 6
Applications and policy implications

Introduction

Ian Douglas

All work on urban ecology has to consider the policy dimensions of managing and planning cities. This final section looks at the practical applications of urban ecology and ways in which ecological revitalization can stimulate socioeconomic revitalization by bringing together through wild flower planting, community forestry and stream restoration projects and by creating a greater sense of social well-being. The final ten chapters look at what planning policies can achieve and what needs to be done in the future, including the new ideas about eco-towns and eco-cities.

Natural England's Urban Greenspace standards provide a set of benchmarks for ensuring access to places of wildlife interest. These standards recommend that people living in towns and cities should have

a an accessible natural greenspace less than 300 metres (five minutes' walk) from home;
b statutory Local Nature Reserves at a minimum level of one hectare per thousand population; and
c at least one accessible 20 hectare site within two kilometres of home; one accessible 100 hectare site within five kilometres of home; and one accessible 500 ha site within ten kilometres of home.

The 31 biggest towns and cities in the Netherlands have agreed on a guideline standard of 75m^2 greenspace per dwelling, but not every local authority will achieve this target. In France, there are no national standards for greenspace, but many cities have followed the example of Rennes in adopting a scheme of differentiated greenspace management that recognizes that the main purpose of any particular space may be for a particular function, from a flower bed to a biodiversity conservation area. The recreational goals of the Rennes scheme include: (a) the possibility of practicing sports activities, recreational activities in open air, relaxation activities in greenspace close to a person's home (e.g. parks, gardens, public gardens, sports fields); and (b) to be in permanent contact with diversified natural spaces (e.g. lawns, meadows and woodlands). John Box shows that aspirational open space standards are commonplace in strategies, plans and frameworks for guiding the spatial planning of towns, cities and regions. Standards and targets for urban greenspace are ideal for the initial stages of planning large-scale development at a regional or sub-regional scale that incorporates green networks and green infrastructure. However, the implementation of such open-space standards may be planned and strategic or be opportunistic and piecemeal, or some combination of the two, depending on circumstances. Less common are studies of the results of the incorporation of such standards into spatial plans or strategies. The lack of mechanisms for sharing information on implementation is widely recognized as a weakness.

The concept of a biosphere reserve was first developed in 1974 by UNESCO's Man and the Biosphere (MAB) Programme as a place that conserves biological diversity, promotes economic development, and maintains cultural values. The concept could be used to integrate the current multiplicity of initiatives and designations. For example, national parks may be covered by different local biodiversity action plans and different planning regulations from those applying to their close-lying urban hinterland – even though they exist in the same bio-geographic zone and often the same river catchment. In turn, the towns and cities adjacent to the national park may have different local authorities with different policies on open space, environmental education, nature conservation and outdoor recreation. Linking all these together in the biosphere reserve (BR) could constructively help to achieve more sustainable nature conservation and economic and recreational values from natural areas in and around cities. Good examples are the Mata Atlantica rainforest biosphere reserve around São Paulo in Brazil and the series of national parks on the hills around Barcelona, Spain. Pete Frost and Glen Hyman discuss the applicability of the concept to urban areas, using the example of Durban, South Africa. They note that in the UK, it has been proposed that BR might come in three forms: Bordering the city; Permeating the city; or ultimately the city as Biosphere Reserve. In the first case the city would lie alongside a highly protected area and would be encompassed by the Transition Area, in the second case protected areas such as wetland or river systems would run into or through the city, and in the final case the city would surround one or more highly protected areas.

Ecosystem services must lie at the centre of any adaptation policy and the impacts of climate change on human well-being are essentially mediated by natural systems. The biophysical features of greenspace in urban areas, through the provision of cooler microclimates and reduction of surface water runoff, offer potential to help adapt cities for climate change. A Greater Manchester, UK, case study showed that adding 10 per cent green cover to areas with little green, such as the town centre and high density residential areas keeps maximum surface temperatures at or below the 1961–1990 baseline temperatures up to, but not including, the 2080s High Emissions scenario. On the other hand, if 10 per cent green is removed maximum surface temperatures by the 2080s High Emissions scenario are 7°C and 8.2°C warmer in high density residential and town centres, respectively, compared to the 1961–1990 current form case. Increasing green cover by 10 per cent in the residential areas would reduce runoff from these areas from a 28 mm precipitation event, expected by the 2080s High, by 4.9 per cent; increasing tree cover by the same amount reduces the runoff by 5.7 per cent. Green areas also mitigate the effects of increases in storm runoff. Trees modify the effects of wind and vegetation plays a major role in coastal defences on salt marshes, mangroves and sand dunes in coping with rising sea levels. Ian Douglas discusses how this information can now be used to frame guidelines to help decision makers to (a) plan new urban areas appropriately and (b) figure out how they might retro-fit the elements needed to improve resilience and/or comfort levels for residents within existing urban areas.

Sustainable urban drainage systems (SUDS) can be designed to take account of the ecological context of the site by recognizing adjacent habitats and their ecological functions, and thus to enable the SUDS to support a wider scale strategic ecological role. By simply using a range of aquatic plant species the wetland SUDS can achieve the three cornerstones of quantity, quality and amenity. However, if the selection of those plant species, together with specifications for wetland topography and soil type, is guided by ecological principles, the opportunity exists to optimize the nature conservation value of the SUDS system. Peter Worrall and Sarah Little note that by their very nature, SUDS will be constrained in the ecological services they can provide because of the character and nature of their source water. However, even from an ecological perspective, 'sub-optimal' habitats have the potential to support the delivery of local and national biodiversity objectives as well as providing a service to urban drainage management.

A comparison of runoff from a site prior to urban development, when it was covered by woodland and after the introduction of impervious covers shows dramatic changes. Prior to development up to 40 per cent of precipitation is intercepted by vegetation and turned back into the atmosphere through evapotranspiration, approximately 50 per cent infiltrates to replenish groundwater and only 10 per cent becomes runoff. After 75 to 100 per cent of a site have been converted into impervious surfaces, evapotranspiration is reduced and infiltration reduced to 15 per cent, increasing runoff from an initial 10 per cent of rainfall to 55 per cent, causing overbank flow and bank erosion in most streams. Joachim Tourbier explains that to maintain the pre-development water balance in urbanizing areas in regard to runoff, infiltration and evapotranspiration has since become a global effort. Increasingly the water balance that existed under a woodland condition is being considered as a standard to obtain. In the USA the state of Maryland is currently considering site design criteria that consider 'woods in good condition' as the 'pre-development' hydrology condition to be restored by incorporating into land development measures that can retain and slowly release equivalent volumes. This can be accomplished by single measures such as basins and rain-gardens or by an interconnected series, or a 'treatment train' of practices. When measures to infiltrate runoff are selected to be components of such a treatment train, preference is being given in Germany to practices that use vegetation and the upper soil horizon to upgrade water quality. In Germany efforts are also being made at the federal UBA (*Umweltbundesamt*) to control stormwater runoff at new construction sites by replicating runoff volumes and peaks found in woodland conditions, including its evapotranspiration. The single most effective measure to accomplish this are vegetated roofs that evapotranspirate precipitation through a soil-substrate and a vegetation cover. The novelty here are extensive roof covers that can be installed on sloping and flat roofs, have a fully saturated weight of only 15 pounds per square foot, and thus are not heavier than the standard gravel layer used on many roofs. Subsequently no structural improvements need to be made to existing roofs. Vegetated roof covers consist of foliage, a growth medium and root zone, and a drain layer. The annual runoff volume reduction is 50 per cent or more. Vegetated roofs are a component of runoff attenuation, lessening a substantial amount of runoff prior to concentration. Other measures are porous bituminous concrete pavement, permeable interlocking concrete paving block, and prefabricated grid pavers that can be seeded with grass. All of the measures mentioned above help to abate implications of climate change and can bring a visual enhancement to cities when they reflect the interests of user groups and become a component of city planning. Most successful case examples have practiced the type of stakeholder participation now being called for by the European Commission.

The benefits from urban wildlife and urban nature conservation are well-known, but are difficult to quantify. The multiple benefits of greenspace straddle the areas of many administrations and professional interests and thus principles of integrated environmental management have to be brought into play. Ecological revitalization can stimulate socioeconomic revitalization by bringing people in underserved (poor) neighbourhoods together through community forestry and stream restoration projects. These projects foster the development of community cohesion, which leads to community interest in improved city services. Increases in services lead to improvements in environmental and socioeconomic conditions and create positive feedback for neighbourhood revitalization, reversing the negative spiral of population loss with consequent environmental and social degradation which leads to further population loss. Such approaches demonstrate how attention to urban nature assists in achieving broader socioeconomic goals. This in turn is good for business. If then both community and business can work together to help maintain the quality of greenspace, either through practical involvement in working parties or through financial sponsorship the problem of greenspace upkeep can be tackled, thus gaining benefits for nature and for business and society. At the moment in many situations, grants to initiate and establish urban greenspace projects are easier to obtain than long-term funding for the maintenance of such areas. Getting local authorities to invest in greenspace management may depend on local residents' inputs and enthusiasm.

There are numerous green spaces designated as 'therapeutic' in Europe and the UK (healing gardens, care farms, ecotherapeutic projects). This perspective veers people into a vision of these spaces as having a specific and possibly 'medicalized', single function. By so doing it reduces or limits the potential multifunctionality inherent in these 'special' green spaces. Ambra Burls argues that much more creative scope exists to broaden the functional perspective, based on synthesizing the intrinsic parallel uses of therapeutic green spaces. Further exploration of these uses, based on real examples, leads to synergies and connections with access and maintenance of urban green spaces by citizens for citizens, aiding the formation of urban 'green grids' and healthy biospheres; public health outcomes through active assessment of health impact on the local community (Health Impact Assessment – HIA); stigma abating, safety and community identity; citizenship, stewardship and local networks; and sustainability education, multicultural integration and interdisciplinary networks. Connectedness and social inclusion, management and prevention of mental and physical health problems are some of the most incisive outcomes of therapeutic green spaces. However it can be seen that much more is achieved than predefined outcomes for the individuals who seek therapy. The general public is likely to benefit from wide-ranging outcomes by utilizing a local greenspace otherwise only purportedly identified as therapeutic. These spaces simultaneously enhance the personal growth of diversely able individuals and improve the quality of both the ecology and green infrastructure of urban sprawls. They also directly contribute to public health, community cohesion and social capital.

Urban greenspaces have to be seen in the wider context of regional conservation networks. Just as river valleys through cities can link the city core to the countryside beyond, so urban nature reserves can be seen as a part of a linked hierarchical system, as separate core areas within a regional urban biosphere reserve, such one that might be developed around a partially industrialized river estuary. For example in the Netherlands, the Randstad (the major urban agglomeration in the west of the Netherlands) is particularly short of recreational greenspace. The aim of government policy is to create an additional 16,000 hectares of public green space by 2013 – but this is less than half the amount needed to satisfy demand. Planned regional cycleways and footpaths giving access to open space are part of this effort. Many initiatives remain piecemeal. Joe Ravetz uses examples from the European RLUREL project to examine peri-urban change and the role of green infrastructure in a series of European cities. He recommends that urban ecology should extend its remit to a 'peri-urban ecology', just as cities and urban structures are extending to a wider hinterland. This raises questions for the practice of landscape ecology itself: on scale and hierarchy: on connectivity and conservation: and on effective management. But it is clear that the agenda has multiple layers, from the physical environment and techniques for management, to the economic and political, to the sociocultural and structural. Ravetz's chapter swings over to the latter end of the spectrum, in pursuit of the question – what or who is the peri-urban landscape for? How should society decide on what to do, whether production, leisure and tourism, new urban development or nature conservation? How do current tangible pressures and discourses about globalization or localization enter this debate?

Biodiversity needs to be mainstreamed into all new urban development. Despite current pressures for housing and the intensification of residential building in existing urban areas, the case for incorporating biodiversity in strategic urban planning is strong. For example, the City of Cape Town, in a metropolitan setting and with a population of 3.1 million, is within the small area that comprises the Cape Floristic Kingdom, one of only six floristic kingdoms on Earth, and the concentration of endemic plant species is without parallel in an urban setting. Working with people living near remaining natural areas, the municipal government has adopted a strategy to protect this resource.

Living in harmony with nature has always had to be managed alongside the need to create shelter and protection from climatic extremes, wild animals and the risks associated with contact with other human beings. Much understanding and experience has been accumulated about the risks from inappropriate design, whether they are geophysical (fogs, subsidence, landsliding, flooding, earthquake or storm surge), biological (pest infestations, disease vectors, wood decay, mould outbreaks) or chemical (air pollution, water pollution, soil contamination). Their appreciation involved understanding the ground the city is built upon, the climate of the area in which the city is being built, the local hydrological regime, the prevailing ecosystem dynamics and the likely future environmental changes, especially the implications of climate change. Such things still play a small role in the professional training of architects, planners, engineers and city managers. They are not well understood by local government councillors who sit on planning committees. Thus all too frequently, the appropriate questions are not asked when development applications are considered and district plans are prepared. Education is part of the key to improving the role of ecology in urban planning. The other key is to get public support for urban ecology. People are enthusiastic about particular local projects, but often cannot become enthusiastic about the overall strategic picture for a whole city or region. Bringing all the civil society groups together in stronger partnerships with local government helps to sustain the awareness and enthusiasm for sound ecological planning and management of cities.

Ambitious new ecocities are now being built or planned in Abu Dhabi and Ras-El Khamina. In China, ecocities are administrative units with 'economically productive and ecologically efficient industries, systematically responsible and socially harmonious culture, and physically beautiful and functionally vitalized landscapes'. 'Ecoscape' planning integrates green space, blue space (water), and red space (built-up areas). These ideas, and the technologies they involve offer new alternatives for future urban development, and perhaps also for retrofitting into existing cities. This broad debate about urban greening is termed 'ecopolis development' and implies new opportunities for urban ecology and urban living. Rusong Wang and his co-authors point out that efficiency, equity and vitality are the three dominating agents in ecopolis development. Its main driving forces are energy, money, power and spirit. Competition, symbiosis and self-reliance are the main mechanisms for maintaining sustainability. The key instrument for ecopolis planning is eco-integration in the metabolism of energy, water and materials, in cultivating eco-industry, ecoscape and eco-culture; in the total coordination of system contexts in time, space, quantity, structure and order; total design of development goals in wealth, health and faith; and total cooperation between decision makers, entrepreneurs, researchers and the general public.

41

Delivering urban greenspace for people and wildlife

John Box

Introduction

Greenspace (including parks and wildspaces) in urban areas are the places where people have the contact with nature that is important for well-being and quality of life (Rohde and Kendle 1997; Douglas 2008; Maller *et al.* 2008; see also Chapters 32 and 33 this volume). Therefore, ensuring adequate opportunities for people to come into contact with nature in their everyday lives should result in direct benefits to their health and happiness and hard evidence for this is becoming available (Fuller *et al.* 2007; Pretty *et al.* 2007; Mitchell and Popham 2008).

High urban land values require a significant commitment by the landowner for land where the primary function is nature conservation. It is unusual for a private landowner to set aside land for wildlife purposes due to the high value of urban land and, therefore, most urban natural greenspaces to which the public have access are found on land owned by a public body such as a local authority or local council or a voluntary organisation such as a wildlife trust. But much can be achieved for wildlife and people through the promotion of multifunctional urban greenspace where multiple land uses are recognised (Barker 1997; Commission for the Built Environment (CABE) 2004). The value of such multifunctional urban greenspaces can therefore be costed in terms of environmental services (e.g. flood regulation, air quality amelioration), thus increasing the notional or theoretical land value of a given urban greenspace.

An excellent example of the results of costing the environmental services provided by greenspace in an urban area has been set out in the pioneering open space and environmental services plans for Durban in South Africa, the Durban Metropolitan Open Space System (or D'MOSS) (1999); eThekwini Environmental Services Management Plan (2001, 2003); eThekwini Municipality and Local Action for Biodiversity (2007)].

Urban greenspace provision is usually seen in terms of quantitative standards (unit area of greenspace per resident or household), or accessibility standards (set areas of greenspace within set distances from every resident). For example, the 31 largest towns and cities in the Netherlands have agreed on a guideline of 75 m^2 green space per dwelling (van Egmond and Vonk 2007). Aarhus, the second largest city in Denmark, set a standard defined in the Green Structure Plan that no dwelling should be more than 500 m from a green area of at least 6,000 m^2 (reported in

Figure 41.1 Landscape structure planting around Telford Central station provides a sense of arrival in a green town as well as screening and noise reduction.

CABE 2004: 25). However, quantity and accessibility are not the whole story because the quality of the resource is also significant in terms of the benefits derived by the public. For example, research for the Scottish government into minimum standards for open space has proposed that open space standards should address a qualitative standard as well as a quantitative standard and an accessibility standard (Ironside Farrar Ltd 2005).

Aspirational open space standards are commonplace in strategies, plans and frameworks for guiding the spatial planning of towns, cities and regions. Standards and targets for urban greenspace are ideal for the initial stages of planning large-scale development at a regional or sub-regional scale which incorporates green networks and green infrastructure (TCPA 2004). Examples include the Green Network of Telford in the West Midlands of England (Box *et al.* 2001), the East London Green Grid Framework (Greater London Authority 2008), and the Green Space Plan 2000 for Tokyo (CABE 2004: 16).

The implementation of such open space standards may be planned and strategic or be opportunistic and piecemeal – or some combination of the two depending on circumstances. What is less commonly undertaken are studies of the results of the incorporation of such standards into spatial plans or strategies. The lack of mechanisms for sharing information on implementation is recognised as a weakness by the research undertaken by CABE Space into the experiences in urban greenspace management of eleven cities across the world (CABE 2004: 90).

This chapter examines the evolution of a set of standards for the provision of natural urban greenspace in towns and cities in England and assesses one of them – the supply of designated nature reserves by local authorities – to see how effective it has been over the period from 1993 to 2006.

Accessible natural greenspace standards – a case study

A set of targets (standards or benchmarks) for accessible natural greenspace in towns and cities has been promoted since 1996 by Natural England (previously English Nature), which is the official nature conservation agency in England (English Nature 1996; Natural England 2010):

- an accessible natural greenspace, of at least two hectares in size, no more than 300 metres (five minutes' walk) from home;
- at least one accessible 20 hectare site within two kilometres of home;
- one accessible 100 hectare site within five kilometres of home;
- one accessible 500 hectare site within ten kilometres of home;
- statutory Local Nature Reserves at a minimum level of one hectare per thousand population.

These targets for ensuring access by people living in towns and cities to places of wildlife interest include a mixture of quantitative and accessibility standards, as well as a qualitative standard in Local Nature Reserves which are a statutory designation where the primary land use must be

Figure 41.2 A wooden dragon constructed by children at Plants Brook LNR in Birmingham shows that all forms of wildlife can be appreciated even in a high quality designated site.

nature conservation which is managed both for its inherent qualities and also for enjoyment by the public and local residents. These targets are derived from the UK tradition of urban planning, open-space hierarchies and recreational standards and also take account of the emerging understanding of the ecology of urban areas and the need to conserve important wildlife habitats and geological features.

The term 'natural greenspace' in the targets includes the full range of richness and diversity of urban greenspaces which can range from small sites awaiting redevelopment and which have been colonised by spontaneous assemblages of plants and animals to much larger areas such as the substantial islands of countryside surrounded by urban development found in most urban areas in Britain. Such urban greenspaces are likely to be multifunctional by providing ecosystem services (such as flood regulation, amelioration of temperature, noise and air quality), recreational areas, landscape quality and places for plants and animals to live.

These targets in relation to people and wildlife in urban areas were novel in the UK when they were first published in 1993 (Box and Harrison 1993) and the intellectual foundation was undertaken by work undertaken through the UK Man and the Biosphere Urban Forum (www.ukmaburbanforum.co.uk). Subsequent research refined the targets (Harrison *et al.* 1995) which were adopted by the statutory nature conservation agency for England (English Nature 1996, 2004; Natural England 2010) and their use disseminated in various publications (Barker 1997; TCPA 2004). Technical and institutional barriers for the implementation of such an urban greenspace model have been identified (Handley *et al.* 2003). A toolkit was produced for local authorities (Handley *et al.* 2003) who are envisaged as being the key agencies for applying the targets at a local level through local planning policies and local development frameworks.

A broadly similar process is being undertaken in Wales where accessible natural greenspace standards established by the Countryside Council for Wales (Centre for Urban and Regional Ecology 2002; Countryside Council for Wales 2006) are being promoted by the Welsh Assembly Government through the environment strategy for Wales (Welsh Assembly Government 2006: 42 and 43) and planning advice for open spaces (Welsh Assembly Government 2009: 9). These accessible natural greenspace standards are not yet given such official recognition in Scotland where there is guidance on greenspace quality (Greenspace Scotland 2008). However, research for the Scottish government into minimum standards for open space (Ironside Farrar Ltd 2005) proposed that open space standards should address a qualitative standard, a quantitative standard and an accessibility standard – these standards are addressed by the accessible natural greenspace standards used in England and Wales.

Setting targets is the easy part. Their implementation by local authorities may be visible at a local level but they are hard to monitor at a regional or national level because of a lack of appropriate mechanisms in the UK. However, one target – the provision of designated nature reserves by local authorities in England – can be measured over time because the data is collected both locally and nationally.

Local Nature Reserve is a statutory designation in the UK whose origin lies with the enjoyment of nature (Wild Life Conservation Special Committee 1947; English Nature 1991). Local Nature Reserves (LNRs) are designated by local authorities and can be chosen to reflect local priorities as opposed to the national priorities reflected in the selection of National Nature Reserves (Barker and Box 1998). Indeed, local authorities can hold the view that LNRs should be established because the natural features of a site are of special interest 'by virtue of the use to which the public puts them for quiet enjoyment and appreciation of nature' (English Nature 1991: 3). The current position in respect of LNRs across the UK is set out by Box *et al.* (2007).

The target for statutory LNRs at a minimum level of one hectare per thousand population is a simple and appealing measure that allows local authorities to establish a nature reserve on a formal statutory basis on land that they own, lease or over which they have a long-term management agreement. Funding to assist this programme was established by English Nature through the *Wildspace!* grants programme for LNRs in 2001, financed largely by a National Lottery award from the New Opportunities Fund (now the Big Lottery Fund) under its Green Spaces and Sustainable Communities programme. By the time the programme ended in October 2006, almost £7 million of *Wildspace!* grants had been spent to encourage more and better LNRs to be established by projects working for people, places and nature (English Nature 2005).

Local Nature Reserves are best seen as nodes in multifunctional green networks. Such a view places them in a landscape context, values them as part of the environmental resources of the county or district and draws attention to their excellence as sites of nature conservation value (Barker 1997). LNRs are usually identified in the local development plans produced by local authorities in the UK which have a statutory basis and demonstrate the existing land uses in the area of a local authority as a means of guiding the location of future developments. The demonstration of a positive landuse for LNRs has important practical benefits by clearly indicating that there is no potential for other land uses, such as built development, on these sites. Such a positive land use allocation helps to move away from the idea, particularly in urban areas, that nature conservation only occurs on land which has no other use or which no one wants.

The original article by Box and Harrison (1993) set out data from a sample of 25 urban local authorities in England whose provision of LNRs in 1993 ranged from 1 ha of LNR for 889 residents (Canterbury) to 1 ha of LNR for 170,500 residents (Camden). This baseline dataset was updated just over ten years later with data on the number and area of LNRs in each of the same sample of 25 urban local authorities as at December 2006 (Table 41.1) (Box 2007).

By 2006, there were significant improvements in the supply of Local Nature Reserves with some local authorities achieving order of magnitude or even greater increases in their provision over a period of little more than a decade (Barnet, Derby, Gloucester, Leicester, Newcastle-upon-Tyne). Of these, Leicester City Council has increased its provision of LNRs by a factor of 67 from 1 ha for 135,300 residents in 1993 to 1 ha for 2,014 residents in 2006. For some local authorities, the population has increased but the total area of LNRs has remained essentially unchanged and the provision per thousand residents has therefore actually decreased (Haringey, Portsmouth, Southampton, Southwark). It is notable that the provision in Leeds has remained static since 1993 at just over 1 ha for 1,100 local residents, but the total area of over 600 ha of LNRs in Leeds in 1993 was far ahead of other local authorities in England at that time and still remains exceptional.

Conclusions

Standards, targets and guidelines can all be used to turn policy into practice. But implementation is the key to real success and legislation and regulation or financial incentives are the most effective drivers. Costing the environmental services provided by multifunction greenspaces in urban areas can be an effective way of countering arguments that built development is required to realise the inevitably high urban land values (e.g. eThekwini Municipality and Local Action for Biodiversity 2007 and associated references; The Trust for Public Land and Philadelphia Parks Alliance 2008). Another driver can be competition – in this case, increasing the supply of Local Nature Reserves in some towns and cities through monitoring of the results of a sample of local authorities over a period of time – as long as the results can be published in places where the results can be readily seen by the target organisations, in this case the planning journals that local authority planners read.

Table 41.1 Provision of Local Nature Reserves in a selection of urban local authorities in England in 1993 and 2006

Local authority	*1993*			*2006*			
	Population[1]	*LNRs (total area, number)*[2]	*Population/ area*	*Population*[3]	*LNRs (total area, number)*[4]	*Population/area*	*Comments*
<1,000 residents per ha LNR (in 2006)							
Gloucester	91,800	4.3 ha (2)	21,349	109,885	169.5 ha (7)	648	Large improvement and achieved target
Canterbury	127,100	143 ha (3)	889	135,278	177.7 ha (10)	761	Improving and achieved target
Wakefield	306,300	313 ha (7)	979	315,172	401.5 ha (10)	785	Improving and achieved target
Norwich	120,700	52.5 ha (5)	2299	121,550	136.2 ha (8)	892	Improving and achieved target
Stoke-on-Trent	244,800	82 ha (1)	2985	240,636	246.4 ha (9)	977	Improving and achieved target
Range 1,000:1 to 5,000:1 (2006)							
Dudley	300,400	181.7 ha (4)	1653	305,155	274.6 ha (7)	1111	Improving and target in sight
Leeds	674,400	605.4 ha (5)	1114	715,402	613.0 ha (8)	1167	Static – but there was a very large area of LNR in 1993
Sandwell	282,000	30.3 ha (2)	9307	282,904	205.8 ha (9)	1375	Large improvement
Coventry	292,500	48 ha (3)	6094	300,848	216.7 ha (14)	1388	Improving
Derby	214,000	9.3 ha (1)	23,011	221,708	143.2 ha (7)	1548	Large improvement
Portsmouth	174,700	119 ha (1)	1468	186,701	119.0 ha (1)	1569	Getting worse

Plymouth	238,800	105 ha (5)	2274	240,720	146.1 ha (7)	1648	Improving
Peterborough	148,800	51.4 ha (2)	2895	156,061	81.2 ha (5)	1922	Improving
Barnet	283,000	4.9 ha (1)	57,755	314,564	158.5 ha (6)	1985	Large improvement
Leicester	270,600	2 ha (1)	135,300	279,921	139.0 ha (7)	2014	Large improvement
Newcastle-upon-Tyne	263,000	8 ha (1)	32,875	259,936	113.0 ha (6)	2300	Large improvement
Liverpool	448,300	21 ha (1)	21,348	439,473	134.1 ha (3)	3277	Large improvement
Hereford	49,800	6.1 ha (2)	8164	50,149	14.4 ha (3)	3483	Improving
Range 5,000:1 to 10,000:1 (2006)							
Haringey	187,300	36.2 ha (3)	5174	216,507	32.6 ha (3)	6641	Getting worse
Southwark	196,500	29.9 ha (1)	6572	244,866	32.4 ha (4)	7558	Getting worse
Birmingham	934,900	39.5 ha (4)	23,668	977,807	102.6 ha (7)	9530	Large improvement
Range 10,000:1 to 50,000:1 (2006)							
Southampton	194,400	14 ha (1)	13,886	217,445	14.0 ha (1)	15,532	Getting worse
Oxford	109,000	2.2 ha (2)	49,545	134,248	6.4 ha (3)	20,976	Improving
Islington	155,200	2.5 ha (1)	62,080	175,797	5.3 ha (3)	33,169	Improving
Greater than 50,000:1 (2006)							
Camden	170,500	1 ha (1)	170,500	198,020	1.85 ha (4)	107,038	Improving

Notes

1 Population data are preliminary 1991 Census figures (*Whitaker's Almanac* 1993).
2 LNR areas and numbers for April 1993 (English Nature data).
3 Population data are 2001 Census figures.
4 LNR areas and numbers for December 2006 (Local Authority data).

Like air and water, wildlife is assumed to be a free resource that we take for granted and which can be adversely affected without direct economic payment. However, the protection and continued enjoyment of natural resources does entail costs to individuals and to society. Legislation, planning guidance and public attitudes are continually driving the burden of these costs away from the victim and the taxpayer and onto the consumer and the shareholder where they rightfully belong. Over the past 40 years the conservation of nature in Europe has focused on the protection of rare habitats and species, rather than on the overall losses of biodiversity due to specific developments. Wildlife legislation has not yet set limits for changes in species or populations in relation to the development of individual sites – either due to the initial land-take or to subsequent disturbance. Increased emphasis on environmentally sustainable development may offer a more appropriate mechanism to achieve net gains in biodiversity for individual projects.

Environmentally sustainable development demands that environmental capital is not diminished from one generation to the next. The next generation will only know what it finds and will not be able to fully comprehend past losses. Therefore, important urban greenspaces that are rich in wildlife need systems which can deliver good site management in order to maintain the quality of the resource in the long term. Large sites are more likely to be able to accept multiple use without damage and can provide a greater variety of opportunities for local people to use and enjoy. But in many urban areas the severe constraints of high land values and existing land uses mean that only small sites are practicable as urban greenspaces. It is increasingly being recognised however that even very small urban greenspaces are valuable not only in terms of their ecological and educational benefits but also in supporting more sustainable communities, for example through their contribution to people's health and well-being.

Some may argue that there is no room for more publicly accessible urban greenspace in crowded urban areas. But why not create these areas? A number of nature conservation strategies in the UK have recognised the concept of areas which are deficient in wildlife habitats to which the public have reasonable access. Indeed one of the main aims of the nature conservation strategy produced 25 years ago for the metropolitan county of the West Midlands (West Midlands County Council 1984) was to ensure that all residents had reasonable access to wildlife habitats. The strategy identified 'Urban Deserts' based on areas where residents were more than 1 km away from accessible wildlife habitats; habitat creation was seen as being very important in these areas which were called 'Wildlife Action Areas'. Such a methodology has been used in other nature conservation strategies for urban areas, for example London (Greater London Authority 2002) and Birmingham (Birmingham City Council 1997).

The challenge is for local authorities and public bodies to turn places that they own, such as mown amenity grassland, into more interesting and stimulating natural greenspace and to incorporate accessible natural greenspace into new developments through spatial planning and through working with those involved in the new developments.

Creative management of biodiversity at the local level is demonstrated on an international scale by the ICLEI (Local Governments for Sustainability) initiative known as Local Action for Biodiversity which is supported by the UNEP Urban Environment Unit and IUCN (www.iclei.org/index.php?id=lab). Tokyo has significantly less greenspace per person (6.1 m^2/person) than London (26.9 m^2/person) and the Green Space Plan 2000 for Tokyo aims to develop 400 ha of green space by 2015 (CABE 2004: 16). Paris has a goal that all citizens can live within 500 m of a greenspace which has resulted in a programme to create new greenspaces within identified areas of deficiency, including creating small greenspaces by buying derelict houses (CABE 2004: 82–3). Some 16,000 ha of public greenspace are proposed to be created by 2013 in the Randstad, the major urban area in the west of the Netherlands that includes Amsterdam,

Figure 41.3 Regularly mown grassland in the Town Park in Telford in 1990.

Figure 41.4 The same area in 2005 showing scrub habitats created by natural regeneration after mowing ceased.

Rotterdam, The Hague and Utrecht (van Egmond and Vonk 2007). Rennes in France has pioneered a programme of differential management regimes for greenspace in the city which is based on ecological awareness and results in the development of wildlife areas within the city, control of water and soil pollution and increasing biodiversity while reducing the overall cost of open space management.

In conclusion, standards for environmental quality and targets for enhancing and protecting biodiversity which include both wildlife and people – such as those for accessible natural greenspace in the UK – can be powerful levers for change and their use to influence behaviour should not be underestimated.

Acknowledgements

The data on Local Nature Reserves in Table 41.1 was first published in 2007 in *Town and Country Planning*, 76(May), 160–162 (Box 2007). Comments and information from colleagues are gratefully acknowledged, in particular Pete Frost (Countryside Council for Wales) and Ian Angus (Scottish Natural Heritage).The views in this article are my own and are not derived from any project in which Atkins is involved.

References

Barker, G. (1997) *A Framework for the Future: Green Networks with Multiple Uses In and Around Towns and Cities*, English Nature Research Report No. 256, Peterborough: English Nature (now Natural England), online, available at: http://naturalengland.etraderstores.com/NaturalEnglandShop/R256 [accessed 30 January 2009].

Barker, G.M.A. and Box, J.D. (1998) 'Statutory local nature reserves in the United Kingdom', *Journal of Environmental Planning and Management*, 41(5): 629–42.

Birmingham City Council (1997) *Nature Conservation Strategy for Birmingham*, Birmingham: Birmingham City Council and Land Care Associates Ltd, online, available at: www.birmingham.gov.uk/cs/Satellite?c=Page&childpagename=Development%2FPageLayout&cid=1223092715237&pagename=BCC%2FCommon%2FWrapper%2FWrapper [accessed 30 January 2009].

Box, J. (2007) 'Increasing the supply of local nature reserves', *Town and Country Planning*, 76(May): 160–2.

Box, J., Berry, S., Angus, I., Cush, P. and Frost, P. (2007) 'Planning local nature reserves', *Town and Country Planning*, 76(November): 392–5.

Box, J., Cossons, V. and McKelvey, J. (2001) 'Sustainability, biodiversity and land use planning', *Town and Country Planning*, 70(7/8): 210–12.

Box, J.D. and Harrison, C. (1993) 'Natural spaces in urban places', *Town and Country Planning*, 62(9): 231–5.

CABE (Commission for Architecture and the Built Environment) (2004) *Is the Grass Greener? Learning from International Innovations in Urban Green Space Management*, London: Commission for Architecture and the Built Environment, online, available at: www.cabe.org.uk/default.aspx?contentitemid=479 [accessed 30 January 2009].

Centre for Urban and Regional Ecology (CURE) (2002) *Developing Standards for Accessible Natural Greenspace in Towns and Cities*, Manchester: CURE [available from Pete Frost at the Countryside Council for Wales – p.frost@ccw.gov.uk].

Countryside Council for Wales (2006) *Providing Accessible Natural Greenspace in Towns and Cities*, Bangor: Countryside Council for Wales [available from Pete Frost at the Countryside Council for Wales – p.frost@ccw.gov.uk].

Douglas, I. (2008) 'Psychological and mental health benefits from nature and urban greenspace', in G. Dawe and A. Millward (eds), *Statins and Greenspaces: Health and the Urban Environment*, Manchester: Urban Forum of the UK Man and the Biosphere Committee, pp. 12–22, online, available at: www.ukmaburbanforum.co.uk/docunents/papers/statinsandGreenspaces.pdf [accessed 30 January 2009].

Durban Metropolitan Open Space System (D'MOSS) (1999) online, available at: www.ceroi.net/reports/durban/response/envman/dmoss.htm [accessed 26 February 2009].

English Nature (1991) *Local Nature Reserves in England*, Peterborough: English Nature (now Natural

England), online, available at: http://naturalengland.etraderstores.com/NaturalEnglandShop/NE301 [accessed 30 January 2009].

English Nature (1996) *A Space for Nature*, Peterborough: English Nature (now Natural England), online, available at: http://naturalengland.etraderstores.com/NaturalEnglandShop/IN46 [accessed 23 October 2010].

English Nature (2004) *Local Nature Reserves: Places for People and Wildlife*, Peterborough: English Nature (now Natural England), online, available at: http://naturalengland.etraderstores.com/NaturalEnglandShop/ST112 [accessed 30 January 2009].

English Nature (2005) *Wildspace! – Waking you up to Wildlife,* Peterborough: English Nature (now Natural England), online, available at: http://naturalengland.etraderstores.com/NaturalEnglandShop/WS1 [accessed 30 January 2009].

eThekwini Environmental Services Management Plan (2001, 2003) Durban, South Africa: Environmental Management Branch, Development Planning and Management Unit, the eThekwini Municipality; illustrated summary available online at: www.cbd.int/doc/presentations/cities/mayors-01/mayors-01-southafrica-03-en.pdf [accessed 26 February 2009].

eThekwini Municipality and Local Action for Biodiversity (2007) *eThekwini Municipality Biodiversity Report 2007*, Durban: eThekwini Municipality and Vlaeberg, South Africa: Local Action for Biodiversity, online, available at: www.iclei.org/index.php?id=9236 [accessed 8 March 2009].

Fuller, R.A., Irvine, K.N., Devine-Wright, P., Warren, P.H. and Gaston, K.J. (2007) 'Psychological benefits of greenspace increase with biodiversity', *Biology Letters*, 3(4): 390–4.

Greater London Authority (2002) *Connecting with London's Nature: The Mayor's Biodiversity Strategy*, London: Greater London Authority, Appendix 1, A1.2.13, 118, online, available at: http://legacy.london.gov.uk/mayor/strategies/biodiversity/docs/strat-full.pdf [accessed 23 October 2010].

Greater London Authority (2008) *East London Green Grid Framework Supplementary Planning Guidance*, London: Greater London Authority, online, available at: http://static.london.gov.uk/mayor/strategies/sds/docs/spg-east-lon-green-grid-08.pdf [accessed 23 October 2010].

Greenspace Scotland (2008) *Greenspace Quality: A Guide to Assessment, Planning and Strategic Development*, Stirling: Greenspace Scotland and Glasgow: Clyde Valley Green Network Partnership, online, available at: www.greenspacescotland.org.uk/qualityguide [accessed 30 January 2009].

Handley, J., Pauleit, S., Slinn, P., Lindley, S., Baker, M., Barber, A. and Jones, C. (2003) *Accessible Natural Green space standards in Towns and Cities: A Review and Toolkit for their Implementation*, English Nature Research Report 526, Peterborough: English Nature (now Natural England), online, available at: http://naturalengland.etraderstores.com/NaturalEnglandShop/R526 [accessed 30 January 2009].

Harrison, C., Burgess, J., Millward, A. and Dawe, G. (1995) *Accessible Natural Greenspace in Towns and Cities: A Review of Appropriate Size and Distance Criteria*, English Nature Research Report 153, Peterborough: English Nature (now Natural England), online, available at: http://naturalengland.etraderstores.com/NaturalEnglandShop/R153 [accessed 30 January 2009].

Ironside Farrar Ltd (2005) *Minimum Standards for Open Space*, Edinburgh: Scottish Executive Social Research, online, available at: www.scotland.gov.uk/Publications/2005/07/18104215/42175 [accessed 30 January 2009].

Maller, C., Townsend, M., Henderson-Wilson, C., Pryor, A., Prosser, L. and Moore, M. (2008) *Healthy Parks, Healthy People: The Health Benefits of Contact with Nature in a Park Context: A Review of Relevant Literature*, second edition, Melbourne: Deakin University and Parks Victoria, online, available at: www.parkweb.vic.gov.au/1process_content.cfm?section=99&page=16 [accessed 23 May 2009].

Mitchell, R. and Popham, F. (2008) 'Effect of exposure to natural environment on health inequalities: an observational population study', *The Lancet*, 372(9650): 1655–60, doi:10.1016/S0140-6736(08)61689-X

Natural England (2010) *'Nature Nearby': Accessible Natural Greenspace Standards*, Peterborough: Natural England, online, available at: http://naturalengland.etraderstores.com/NaturalEnglandShop/NE256 [accessed 23 October 2010].

Pretty, J., Peacock, J., Hine, R., Sellens, M., South, N. and Griffin, M. (2007) 'Green exercise in the UK countryside: effects on health and psychological well-being, and implications for policy and planning', *Journal of Environmental Planning and Management*, 50(2): 211–31.

Rohde, C.L.E. and Kendle, A.D. (1997) *Human Well-being, Natural Landscapes and Wildlife in Urban Areas: A Review*, English Nature Science No. 22, Peterborough: English Nature (now Natural England).

TCPA (2004) *Biodiversity by Design – A Guide for Sustainable Communities*, London: Town and Country Planning Association, online, available at: www.tcpa.org.uk/pages/biodiversity-by-design.html [accessed 30 January 2009].

The Trust for Public Land and Philadelphia Parks Alliance (2008) *How Much Value Does the City of Philadelphia Receive from its Park and Recreation System?*, Washington: The Trust for Public Land and Philadelphia: Phililadelphia Parks Alliance, online, available at: www.philaparks.org/ [accessed 6 March 2009].
van Egmond, P.M. and Vonk, M. (eds) (2007) *Nature Balance 2007: Summary*, Bilthoven: Netherlands Environmental Assessment Agency, online, available at: www.mnp.nl/en/publications/2007/Nature_balance_2007.html [accessed 30 January 2009].
Welsh Assembly Governnment (2006) *Environment Strategy for Wales*, Cardiff: Welsh Assembly Government, pp. 41–43, online, available at: http://wales.gov.uk/topics/environmentcountryside/epq/envstratforwales/?lang=en [accessed 30 January 2009].
Welsh Assembly Governnment (2009) *Planning Policy Wales Technical Advice Note 16: Sport, Recreation and Open Space*. Cardiff: Welsh Assembly Government, online, available at: http://wales.gov.uk/topics/planning/policy/tans/tan16e/?lang =en [accessed 21 February 2009].
West Midlands County Council (1984) *The Nature Conservation Strategy for the County of the West Midlands*, Birmingham: West Midlands County Council.
Wild Life Conservation Special Committee (1947) *The Conservation of Nature in England and Wales*, Command 7122, London: HMSO.

42

Urban areas and the biosphere reserve concept

Pete Frost and Glen Hyman

Introduction

Recent decades have been marked by great advances in our knowledge of how urban systems function within – and indeed form central components of – their ecological contexts. The preceding chapters in this volume demonstrate the breadth and depth of this knowledge – the complexity of which, quite understandably, does not easily translate into scale-appropriate public policy. Yet, as unabated urbanization continues, cities increasingly serve as the front line of our global efforts to protect biological diversity, and more generally, to manage the impacts of global change. In many instances, the approach offered by UNESCO's Biosphere Reserve concept can help urban decision makers better integrate science into their policies, leading to better informed action, more liveable cities and healthier environments. This chapter will detail this potential for applying the Biosphere Reserve concept to urban areas, provide examples of this concept at-work in a variety of urbanized settings, and discuss some challenges and opportunities for the future.

Theory: what the BR concept can add to urban areas

Right from their invention in the late 1970s Biosphere Reserves (BRs) were ahead of their time. Designed to be 'super nature reserves' conserving pristine examples of the biosphere's great ecosystems, they rapidly metamorphosed into 'living laboratories for sustainable development' before the term 'sustainable development'. They went beyond the idea of a nature reserve as an island of biodiversity separated from human life, to that of a natural system where people and nature lived in mutually supportive balance. The UNESCO Man and the Biosphere (MAB) programme that originated the concept also had a project on Urban Systems and in the early 1990s workers began to seriously consider if these two strands could be linked and the Biosphere Reserve concept applied to urban areas.

What are biosphere reserves?

BRs are intended to show how to reconcile conservation of biodiversity with its sustainable use (UNESCO 1996). Each BR in the World Network (WNBR) must include ecological systems representative of a major biogeographic region and significant for biodiversity conservation. They must actively include people (the UK asked for the island of St Kilda to be removed from the WNBR because it no longer held a resident human population – it remains a UNESCO World Heritage Site). BRs are developed around a zonation system (Figure 42.1) that has itself been built into protected area management practice across the globe. The Core Area (or areas) contains high biodiversity value and is subject to limited human intervention (e.g. traditional use by indigenous peoples, non-destructive research, ecotourism). The surrounding or contiguous Buffer Zone(s) contains more human activities such as sustainable agriculture and helps sustain the Core Area, and both of these are surrounded by a Transition Area (also known as a Zone of Cooperation) where sustainable resource management practices are promoted and developed. In turn, the whole BR serves three functions: a Conservation Function for landscapes, ecosystems, species and genetic variation; a Development Function that fosters human development which is socio-culturally and ecologically sustainable; and a Logistic Support Function for demonstration projects, research, environmental education, training and monitoring – linking issues of conservation and sustainable development from the local to the global level (UNESCO 1996).

In practice the BR concept has proved extremely flexible and has been applied on scales ranging from vast expanses of tropical forest to a single rooftop (Kim 2004)! The concept has been applied across ecosystems ranging from tundra/taiga to equatorial deserts but no BR has yet been admitted to the WNBR with an explicitly urban Core Area.[1] Because BRs are not 'protected areas' (Dudley 2008: 4), but a system for integrating protected areas with human use to their mutual benefit, practitioners across the world have suggested that BR principles could be used as a framework for the management of ecosystem goods and services in towns and cities (Douglas and Box 2000).

Why apply the biosphere reserve concept to towns and cities?

There is a plethora of initiatives available to a city interested in improving aspects of its environmental performance,[2] so why should a municipality choose to apply the BR concept above all others? With money and time in short supply, any initiative must: bring obvious benefits to the citizens; multiply the value of any resources invested in it; and do things other initiatives cannot. Any scheme that does not do these things will waste resources, duplicate efforts and lose political support.

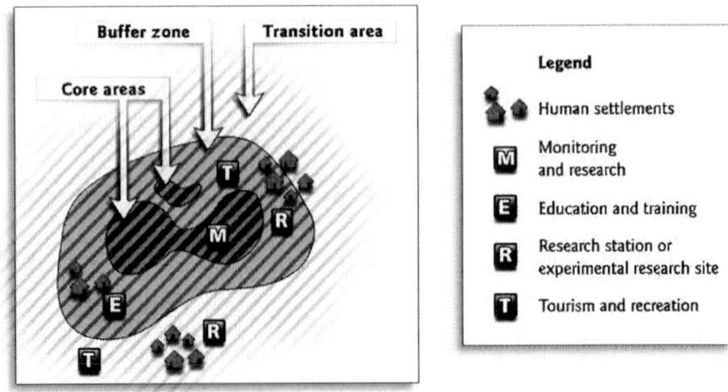

Figure 42.1 Biosphere Reserve Zonation (reproduced by permission of the UNESCO UK National Commission).

Making green space protection politically desirable

Unlike other initiatives, Biosphere Reserves are designed to explore approaches to both conservation and sustainable development (UNESCO 1996). Biodiversity is at the heart of the concept, but so too is the need for human development, with both set in a very clear spatial planning framework: this combination of features makes the BR approach unique, setting it apart from other initiatives. Showing that people as well as nature benefit can make a Biosphere Reserve acceptable where pure 'nature conservation' measures would not be. The groundbreaking Durban Management of Open Spaces System (D'MOSS)[3] gave a spatial approach to safeguarding ecosystem goods and services – including biodiversity. However, there were still tensions between environmental planning and the desperate need for the city to accommodate an influx of people escaping rural poverty in post-apartheid South Africa. Now that sustainable development has been integrated into Durban's city-wide Integrated Development Planning process, it is hoped that such tensions can be resolved more easily. In municipalities with less well developed environmental management practices the BR approach could provide a holistic, city-wide structure for resolving resource conflicts using scientific evidence of the human benefits conferred by green spaces and biodiversity.

Providing a mutually acceptable forum for sustainable development

City governance is never straightforward and often highly fragmented – the New York metropolitan region is believed to have in the region of 2000 jurisdictions spread over three states (Alfsen-Norodom *et al.* 2004) whilst Greater London has 32 boroughs, each responsible for its own unitary development plan (see www.london.gov.uk). Concerted action for sustainable development is difficult in situations where overlapping municipalities compete for authority with each other and resist the diminution of their authority by top-down schemes imposed by national government. In such situations it is easier for authorities to enter into voluntary agreements with each other – especially if such agreements are based on existing frameworks supervised by respected third-parties. Biosphere Reserves provide exactly this 'politically neutral platform' (Dogsé 2004) for collaboration. BRs remain under the sovereign jurisdiction of the member state, but in order to remain a member of the WNBR the 'rules of the club' have to be adhered to. Local authorities lose nothing by signing-up to become part of a BR with their neighbours, but to retain the advantages they have to collaborate with them.

Tying in to a globally respected 'brand'

UNESCO manages the well-known World Heritage List (WHS) to which most nations aspire to nominate their own natural and cultural treasures. UNESCO's Biosphere Reserve designation is even more exclusive with only 564 BRs as of December 2010. For many municipalities the opportunity to be associated with an attractive and well-respected brand like UNESCO will be incentive enough to investigate joining the WNBR. This brand attractiveness may succeed in engaging city authorities on conservation issues, whereas a purely 'environmental' approach may fail due to its perceived concentration on issues which are not of immediate political priority when compared with, for example poverty alleviation or economic development.

Access to a global knowledge-base

The Task Force on Quality Economies in Biosphere Reserves[4] provides policy advice and guidance on obtaining key benefits such as the labelling, branding and marketing of Biosphere

Reserves' goods and services. This task force has generated some of the most popular products of the WNBR – for obvious reasons. MAB has supported the creation of a series of regional and thematic networks to assist information and good practice to spread around the WNBR. Sometimes networks have spun-off from collaboration with non-UNESCO institutions. The URBIS network is one such collaboration of cities interested in investigating how to apply the BR concept in practice, and it is coordinated via the Stockholm Resilience Centre.[5] The network includes researchers in twelve cities across the globe and aims to re-establish the connection between people and their natural environment through innovative forms of knowledge networking and governance of the urban landscape.

How would the BR concept be applied to a city?

The 'pure' BR arrangement of a Core Area surrounded by a Buffer Zone, set within a Transition Area hardly exists in any rural BR today, and would certainly not fit neatly into a city, no matter how much green space existed there. A series of 'models' for the spatial arrangement of a BR containing one or more cities have been proposed (Dogsé 2004; Kim 2004). All acknowledge the reality that cities will have multiple areas of high nature conservation interest, that these may be 'buffered' from human pressure in different ways than a circumscribed area (e.g. by protecting watersheds/catchment zones upstream), and that the Zone of Cooperation would encompass novel entities not normally considered significant in rural areas (e.g. green roofs and green walls). In the UK it has been proposed that BRs might come in three forms: bordering the city; permeating the city; or ultimately the city as Biosphere Reserve. In the first case the city would lie alongside a highly protected area and would be encompassed by the Transition Area, in the second case protected areas such as wetland or river systems would run into or through the city, and in the final case the city would surround one or more highly protected areas (Frost 2001).

The governance of such a system would need to be voluntary, for the reasons discussed already, and highly participative to ensure all stakeholders remained engaged with the process. Canada already uses BRs in areas of potential resource conflict because of UNESCO's stipulation on stakeholder engagement in their management arrangements and Chicago Wilderness gives a glimpse of how a coalition can be built to tackle natural resource conservation in a large conurbation.[6] In reality, municipalities are developing their own response to the BR concept within the flexible guidelines laid out by MAB, as the examples in the following section show.

Practice: how urban areas respond to BR status

Since the early 1990s, an increasing number of urban areas have been implementing the BR concept, in diverse political and ecological landscapes: as integral parts of Brazil's Mata Atlântica BR (785,000 km^2; total population ≈ 100 million people), both the City of Florianopolis and the greater São Paulo region have begun – at quite different scales – to mainstream the BR concept into their municipal decision making. In Australia, however, the state of Victoria has explicitly proscribed such an outcome, leaving only indirect channels for BR status to influence planning for metropolitan Melbourne. In South Africa, the Western Cape Province has already incorporated the BR concept into provincial planning regulations, as NGOs work together with multiple spheres of government, to implement this UNESCO status on Cape Town's urban fringes. We will now deal with these southern hemisphere examples in turn.

Santa Catarina Island (Florianopolis), Brazil

The City of Florianopolis (population 350,000) lies mostly on Santa Catarina Island (523 km^2), in the south-east of Brazil. In 1993, this island was designated as part of the Mata Atlântica BR, and more recently, local government planners have been working to integrate the BR zones into the municipal planning scheme. As a pilot site of the Mata Atlântica 'Urban Forests' project, these efforts have benefited from national level support, and by 2007, the Municipal Master Plan of the Municipality was totally integrated into the BR zonation. Particularly noteworthy are the additional zones which were incorporated into the BR, providing additional conceptual breadth to the concept, ensuring that it fully articulates Florianopolis' urban reality (Figure 42.2).

Rather than viewing the urban environment as a threat from which the natural landscape must be protected by a series of buffers, the zoning scheme developed in Santa Catarina recognizes the cultural importance of its urban areas, while maintaining and enriching their transitions to the neighbouring ecosystems that support them. Accordingly, in this application, traditional core and buffer zones (Figure 42.1) have been reproduced from both natural and urban perspectives, with a shared transition zone, where these different land uses find complementary balance.

At an urban scale, participatory processes jointly implement the Local Agenda 21, the Biosphere Reserve Action Plan and the Municipal Master Plan.

As part of the larger Mata Atlântica BR, this sophisticated urban interface contributes to the continental-scale ecosystem management. But within Florianopolis, where the BR concept is fully articulated by the urban planning scheme, it remains somewhat misleading to speak of a partnership between BR and city, for on this island, they act as one.

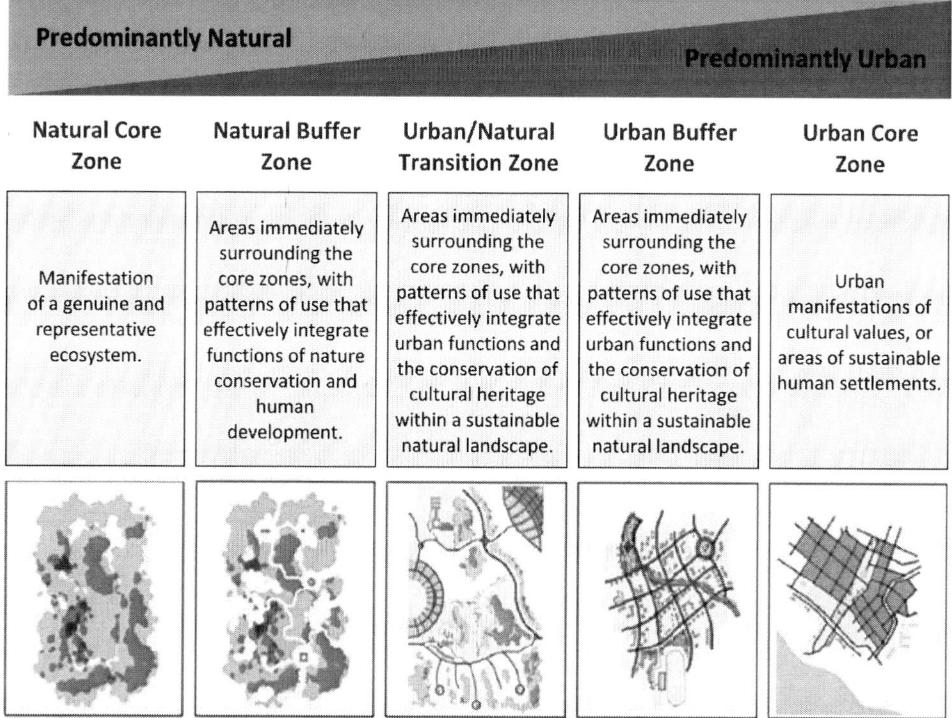

Figure 42.2 Local adaptation of BR zones to manage the urban and natural reality of Florianopolis (adapted from Rubén Pesci, Fundación CEPA 2005).

São Paulo City Green Belt Biosphere Reserve, Brazil

Similarly to Santa Catarina Island in the south, the Mata Atlântica BR boundaries also include most of metropolitan São Paulo, and the forests which surround it (Figure 42.3). However, with a population of some 23 million people (living in 78 distinct municipalities), this conurbation presents a very different scale of management challenges to those in other parts of the BR. While São Paulo plays an important role in the management of the Mata Atlântica BR, in a unique institutional arrangement, the Green Belt which surrounds it (23,317 km^2) bears its own – totally overlapping – BR designation, allowing for an urban-scale expression of the BR concept within a continental scale initiative.

The urban landscape of the São Paulo City Green Belt is incredibly complex, and the conservation tasks most daunting. For example, some 30 per cent of the city's population lives in illegally or informally constructed housing (Ancona 2007: 2), the growth of which remains concentrated at its urban fringe, further degrading remaining remnants of the native Atlantic Forest. Some historical attempts to regulate this urban expansion (in particular near freshwater sources) have had but limited success, as restrictions placed on legal construction lowered land prices, making the 'protected' sites even more attractive for illegal housing (Marcondes 1996, in Torres *et al.* 2005: 6).

Managed in cooperation with local governments, scientists and civil society, the São Paulo City Green Belt BR offers a flexible framework with which to respond to such challenges throughout the metropolitan region – not by prescribing actions for individual municipal

Figure 42.3 An urban forest: the São Paulo metropolis (background) and Cantareira State Park, a BR core zone (foreground) (photograph by Francisco Honda).

governments to take, but by engaging their decision makers, who then incorporate the BR concept into their daily work. One recent example of this approach can be found in the ongoing revision to the BR's zoning and boundaries.

In 2006, well over a decade after joining the World Network, many changes had taken place within the Green Belt. To bring the BR's boundaries up to date, a technical team from the BR Secretariat began its spatial review of the BR, relying on GIS data and stakeholder consultations. At the periphery, the BR's outer limits were adjusted to better reflect watershed boundaries; throughout the BR, newly declared protected areas were incorporated into as Core and Buffer Zones; and on the coast, a significant marine component was added. Using specially developed urban criteria, this process began to formalize a network of urban green spaces (zoning them as Buffer), and granted additional visibility to poorly protected lands near freshwater sources (now also Buffer Zones).

A more general aim of this process was to totally align the BR zones with municipal land-use schemes, functionally down-scaling the concept for ground-level implementation. Although availability of GIS data and limited time precluded the full implementation of these criteria across all 78 municipalities within the BR, the master plan for the City of Santo André (south-west of São Paulo) is already finely integrated into the BRs zones, and other local governments will follow in due course (RBMA 2008: 137–148). Indeed, the very act of participating in the BR rezoning builds collaborative relationships with municipal urban planners, whose decisions directly impact the long-term health of their cities, and thus the greater Green Belt region.

Still, within Green Belt, there is some debate about the relationship of the BR status to São Paulo's densely urban centre. Some argued that the most densely urban areas should be designated as 'special' transition areas. Others preferred to exclude such zones from the BRs boundaries, to concentrate resources on areas of higher conservation values. To achieve consensus on the latest rezoning proposal (in 2008), the question was left open. However, for all practical purposes, the interplay between this megalopolis' urban cores and rural fringes render the distinction moot. No matter where the lines on the map are drawn, São Paulo is a complete region, and good planning requires collaborative decisions taken across sectors and jurisdictions. With full support from Federal, State and local spheres of government, the São Paulo Green Belt BR contributes to this integration.

Mornington Peninsula and Western Port (Melbourne), Australia

In late 2002, the Mornington Peninsula and Western Port (2142 km^2, population 180,000) became Australia's first addition to the WNBR in over two decades. This area includes some of the fastest growing communities on Melbourne's urban fringe, a deep-water harbour scheduled for a major expansion and a large Ramsar-protected marine environment (Figure 42.4). Obviously, such parameters require balanced management, in order to respond effectively to the interrelated (and seemingly conflicting) development and conservation priorities they present. While the BR concept can provide a framework well suited to this challenge, as this case will show us, merely obtaining this status is insufficient to allow its potential to be realized: full involvement of statutory actors remains a prerequisite for success.

Although the first suggestion of seeking UNESCO BR status for this area came from Protected Area managers, the initiative was largely driven by a group of community volunteers, in cooperation with staff from the local government of the Mornington Peninsula. As the proposal developed, these proponents approached the state government ofVictoria, whose approval would be required for the nomination to go forward. This did not come easily.

As nearly 20 years had passed since a BR had been designated in Victoria, there were no standing procedures for how the state should evaluate such a proposal, nor did the bureaucrats

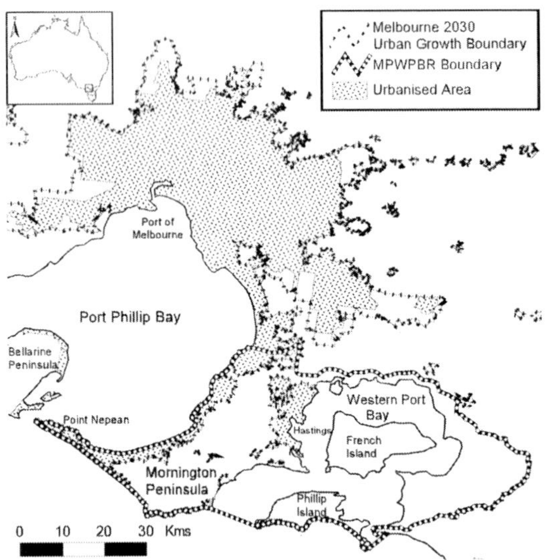

Figure 42.4 Geographic proximity: Melbourne and the MPWPBR (map by Glen Hyman).

and policymakers clearly understand how a BR would affect the activities of existing agencies. As the state government hesitated to make decisions in this policy vacuum, they quickly compiled a State Government Policy on Biosphere Reserve Proposals (State of Victoria 2002), notable provisions of which include:

- that any new biosphere reserve will be community-driven, self-funded, and based on voluntary participation by individuals;
- that designation of a BR shall not restrict other land uses approved by the state; and
- that should a biosphere reserve be considered not to be fulfilling its functions or to lose community support, the state government may ask the commonwealth government to de-list the biosphere reserve.

The policy also explicitly underscores the state's intention that designation of a BR will not 'lead to increased commonwealth control over state matters' (State of Victoria 2002). Partly out of concern that other spheres of government – both higher and lower – would use the BR status to undermine state authority, and without much thought to the benefits that BR principles could offer, the overall aim of this rather blunt policy was to limit the potential influence that the BR designation might bear upon state government policies and actions. In this respect, the policy was a resounding success.

For the first year after joining the World Network, the MPWPBR struggled to develop an institutional arrangement that would not run foul of the state. Although a non-profit foundation was eventually established to implement the BR status, with no access to state government funding and but limited participation from its constituent local governments, the activity of the MPWPBR still today remains estranged from the pressing question of how Melbourne's growth can be harnessed to ensure the region's environmental health.

In a 2007 interview (unpublished), an official involved in drafting the State Policy reflected on this quandary:

> When the MPWPBR came along, we all had to relearn, remind ourselves what [BRs] were meant to be, then decide how harmful they will be to autonomy, to decision making ... There was also a slight feeling that it would become a club to hit government over the head about planning controls. It's a reactive policy, and I feel guilty now ... we could have seized the opportunity.

While the MPWPBR Foundation continues its journey towards implementing the MAB Programme, the rails laid down by state government's initial policy have made for a tremendous detour. As metropolitan Melbourne's development and conservation challenges grow ever more acute, the BR concept – as a tool to help find creative solutions – still has much to offer. But without the active support of state-level decision makers, this potential remains difficult to realize.

Cape West Coast Biosphere Reserve (Cape Town), South Africa

Located within South Africa's Western Cape Province, the Cape West Coast BR (3769 km^2, 1.2 million population) includes an expanding industrial harbour, a Ramsar wetland, and some of Cape Town's fastest growing suburbs. Here (unlike the geographically similar Australian example above) the provincial government has totally embraced the BR concept; at the same time (somewhat in contrast to our cases from Brazil) implementation of BR status is vested outside of government, where a non-profit company has been established to coordinate activity under the BR banner. These factors are particular to South Africa, and these differences highlight the extent to which individual sites adapt the BR concept to their specific needs.

The Western Cape Provincial Government's support for Biosphere Reserves is firmly rooted in the region's planning schemes. Since 2000, they have explicitly advocated a planning approach largely based on the BR concept, even subtitling the Bioregional Planning Framework 'Toward the application of Bioregional planning principles and the implementation of UNESCO's Biosphere Reserve Program in land-use planning within the context of Act 7 1999' (PGWC 2000). As part of a broader set of planning tools, the Western Cape has to some degree upscaled the BR concept, attempting to manage the entire province in a Biosphere Reserve manner. In this sense, their support for officially designated BRs – such as the Cape West Coast – comes quite naturally.

In another sense, the particularities of South Africa render the non-governmental sector the most expedient place to house such initiatives. As the BR concept is already present in the region's planning schemes, creating a government body to implement the status would to some extent duplicate existing efforts. Here, the real value added by a BR designation comes from the neutral platform it can provide – free of party-political constraints – from which to broker scientifically informed planning decisions. Non-governmental standing enables this. As well, while governments are often excluded from direct external financing (CEPF, CI, GEF etc.), a non-profit BR company is in a much better position to apply for such project funding. Through this structure, the Cape West Coast BR contributes to local application of the BR concept.

Conclusions

The Magic Feather – managing in a BR manner

An old children's story tells us of Dumbo – an aeronautic elephant – that believes he must tightly hold onto a 'magic feather' in order to fly (Aberson 1939). As it turns out, his feather is quite ordinary and Dumbo could fly all along: he just needed a talisman to give him confidence. In some respects, Biosphere Reserve status works just the same – whether or not a site has formally obtained recognition from UNESCO, the BR approach has proven sound, and there are no natural barriers to keep stakeholders from implementing the concept. But given the competing pressures ordinarily faced by urban decision makers, a little help can enable them to perform extraordinary work. In this sense, UNESCO BR status (or even the process of seeking it) can expand horizons and enable better management outcomes.

In South Africa, the Western Cape Province has a number of other initiatives - some with BR status, others without; all use the concept. In addition to its six UNESCO-designated BRs (all of which have urban components), several of its other ecologically defined regions are managed as though they were BRs, but without having gone through the formal nomination process. Durban thoroughly investigated the option of applying for BR status, but concluded that its open space management system would work just as well without it. There are many programs and tools available for the management of biodiversity, or to promote sustainability. One of the strengths of the BR concept is that it offers a flexible framework, but it may not be for everyone. City of Cape Town decided just this (Stanvliet *et al.* 2004).

After thorough debate at the 3rd World Congress of Biosphere Reserves in 2008, UNESCO MAB no longer proposes to establish a separate category for 'Urban Biosphere Reserves'. But at the same time, as the phenomenon of urbanization remains real and continues to grow, all BRs must work to manage its effects, and to take advantage of its opportunities. Networks will be established and help given to BR struggling with urban questions (UNESCO 2008).

BR designation should enhance, not replace existing efforts

In Australia, fear that a BR designation would erode state autonomy led government officials to undermine its potential for success. In complex urban environments, the need to tread carefully cannot be overstated. These environments are characterized by the presence of lots of historical adversaries, with shared interests. The BR approach can generate cooperation by helping the former recognize the latter. Yet, poor execution can undermine these efforts.

A new way forward? Biosphere Cities Programme

The WNBR has over 30 years of experience, the first two decades of which were concentrated mainly on protected areas in highly rural settings, and consequently the network is almost exclusively populated by rural BRs. MAB has been persuaded that the admission of purely 'urban' BRs would threaten the integrity of both the network and the BR concept. Hence examples can be found such as one cited above, where downtown São Paulo remains formally outside the boundaries of the BR which envelopes it. It has been suggested that a Biosphere Cities programme[7] will bypass this tension, though like its parent, it will remain a 'magic feather'. No special status is needed for cities to go forth – take Durban or the Western Cape Province. The BR principles are sound, and when applied, give good results. This provides the added value, and as more urban regions adopt the concept – in practice – this serves as a testament to the programme's success.

Should the concept alone be insufficient to induce more cities to manage themselves as BRs then there are certain elements of the BR programme which contribute to its attractiveness and could be repeated in a Biosphere Cities scheme. The BR concept's combination of a spatial planning framework, coupled to an emphasis on sustainability in practice, seem to be unique amongst other schemes. However, the word 'Reserve' is acknowledged to be 'difficult'

(UNESCO 1996) and would be a real barrier to the adoption of any scheme for cities. The WNBR is recognized as a key component of the BR concept, which is not yet used to its fullest potential to spread good practice across the globe. Any new scheme would greatly benefit from affiliation to the WNBR purely to learn from places which have decades of experience of implementing the concept. Any such affiliation could not afford to be too tight in order to avoid MAB's concerns about the integrity of the network. Similarly, a scheme for cities would still benefit from the use of biodiversity conservation as an entry qualification: demonstrating a commitment to it would gain a city entry to the scheme, and if biodiversity degraded, the city would no longer be allowed to remain in the scheme. However the main emphasis of the scheme would be to promote sustainability in practice. Association with the highly visible and well regarded UNESCO 'brand' is what makes BR status so highly attractive to many places. Any new scheme would need a sponsor of equal or higher attractiveness in order to make it in to what is a very crowded arena. Administratively, any scheme should be open to all and subject to a transparent, criteria-led application process like the WNBR. Key to international acceptability of the BR concept is its voluntary nature – UNESCO MAB has no hold over BR, and there is no compulsion to follow any rules. The only sanction is that failure to comply with the criteria laid down in the statutory framework will eventually result in expulsion from the WNBR. Consequently any new urban scheme should follow this example if it too is to be acceptable. Finally, following early reports of indigenous peoples being ejected from BR in the 1970s the concept has now been turned on its head to enshrine stakeholder engagement as the central tenet of BR management. Given that the life of a city is dependent on the lives of its people, their engagement in the management of any Biosphere City programme is a pre-requisite for its success, and the improvement of the city for both people and nature.

Notes

1 For a full list of Biosphere Reserves go to http://portal.unesco.org/science/en/ev.php-URL_ID=6433&URL_DO=DO_TOPIC&URL_SECTION=201.html.
2 For examples see www.iclei.org/lab/ for the Local Action for Biodiversity program of ICLEI Local Governments for Sustainability, and www.unep.org/urban_environment/key_programmes/index.asp for the Cities Alliance and Sustainable Cities programs of UNEP.
3 See www.ceroi.net/reports/durban/response/envman/dmoss.htm.
4 See http://portal.unesco.org/science/en/ev.php-URL_ID=6429&URL_DO=DO_TOPIC&URL_SECTION=201.html.
5 See www.stockholmresilience.org/research/researchthemes/urbansocialecologicalsystems.4.aeea46911a3127427980003731.html.
6 For the Canadian Biosphere Reserves Association, see www.biospherecanada.ca/home.asp; and for Chicago, see www.chiwild.org.
7 Download a summary at www.biosphere-research.ca/work_in_progress.htm.

Further reading

São Paulo Forest Institute (2002) *Application of the Biosphere Reserve Concept to Urban Areas: The Case of São Paulo City Green Belt Biosphere Reserve*, São Paulo: São Paulo Forest Institute.

UNESCO (2003) *Urban Biosphere Reserves in the Context of the Statutory Framework and the Seville Strategy for the World Network of Biosphere Reserves*, Paris: UNESCO (a discussion of how 'urban' Biosphere Reserves might work).

UNESCO (2006) *Urban Biosphere Reserves – A Report of the MAB Urban Group*, SC-06/CONF.202/INF.6 Paris: UNESCO (a full consideration of the criteria for choosing Biosphere Reserves with a large urban component, and discussion of issues around Biosphere Reserves in urban areas).

References

Aberson, H. (1939) *Dumbo the Flying Elephant*, Syracuse, NY: Roll-A-Book.

Alfsen-Norodom, C., Boehme, S.E., Clemants, S., Corry, M., Imbruce, V., Lane, B.D., Miller, R.B., Padoch, C., Panero, M., Peters, C.M., Rosenzweig, C., Solecki, W. and Walsh, D. (2004) 'Managing the Megacity for Global Sustainability: The New York Metropolitan Region as an Urban Biosphere Reserve', *Annals of the New York Academy of Sciences*, 1023: 124–141.

Ancona, A.L. (2007) 'Legalizing Informal Settlements in São Paulo, Brazil', paper presented at the Capacity Development Workshop on Decentralization and Local Governance, 7th Global Forum on Reinventing Government, 26–29 June 2007, Vienna, Austria, online, available at: http://unpan1.un.org/intradoc/groups/public/documents/UN/UNPAN026559.pdf [accessed 25 Feb 2009].

Dogsé, P. (2004) 'Toward Urban Biosphere Reserves', *Annals of the New York Academy of Sciences*, 1023: 10–48.

Douglas, I. and Box, J. (eds) (2000) *The Changing Relationship between Cities and Biosphere Reserves*, Manchester: Urban Forum of the UNESCO UK Man and the Biosphere Committee.

Dudley, N. (ed.) (2008) *Guidelines for Applying Protected Area Management Categories*, Gland, Switzerland: IUCN.

Frost, P.T. (2001) 'Urban Biosphere Reserves – Re-integrating People with the Natural Environment', *Town and Country Planning*, 70 (July/August): 213–216.

Kim, K.-G. (2004) 'The Case of Green Rooftops for Habitat Network in Seoul', *Annals of the New York Academy of Sciences*, 1023: 187–214.

Marcondes, M.J.A. (1996) *Urbanização e meio ambiente: os mananciais da metrópole paulista*, São Paulo: FAU/USP, doctoral thesis.

PGWC (2000) *Bioregional Planning Framework for the Western Cape Province*, Cape Town: Provincial Government of the Western Cape.

RBMA (Mata Atlântica Biosphere Reserve) (2008) *Review and Update of the Limits and the Zoning of the Mata Atlântica Biosphere Reserve in Digital Cartographic Base*, São Paulo: Mata Atlântica Biosphere Reserve.

Stanvliet, R., Jackson, J., Davis, G., De Swardt, C., Mokhoele, J., Thom, Q. and Lande, B.D. (2004) 'The UNESCO Biosphere Reserve Concept as a Tool for Urban Sustainability: The CUBES Cape Town Case Study', *Annals of the New York Academy of Sciences*, 1023: 80–104.

State of Victoria (2002) *Biosphere Reserve Proposals State Government Policy*, online, available at: www.dse.vic.gov.au/DSE/nrenpr.nsf/LinkView/E8C72597ED280289CA256C8B00032C0AAE2B49A37062E4B6CA256C8C001E0929.

Torres H., Alves, H. and de Oliveira, M.A. (2005) *São Paulo Peri-Urban Dynamics: Some Social Causes and Environmental Consequences (First Draft)*, online, available at: www.worldbank.org/urban/symposium2005/papers/torres.pdf [accessed 25 Feb 2009], also available at: http://eau.sagepub.com/content/19/1/207.full.pdf+html.

UNESCO (1996) *Biosphere Reserves: The Seville Strategy and the Statutory Framework of the World Network*, Paris: UNESCO.

UNESCO (2008) *Madrid Action Plan for Biosphere Reserves (2008–2013)*, Paris: UNESCO.

43

Urban ecology and sustainable urban drainage

Peter Worrall and Sarah Little

The ecosystem services available through the application of sustainable urban drainage systems (SUDS) are well understood in terms of the provision of water management functions, both quantity and quality; and ecology and amenity resources (CIRIA 2007). However, the development of these systems has, in many countries, been very limited. For example, the twenty-sixth report of the Royal Commission on Environmental Pollution, 'The Urban Environment' (2007) stressed the need in the United Kingdom for changes in practice, legislation and guidance to encourage the use of such systems for the benefit of urban communities. Only a matter of months later the low level of implementation of SUDS contributed to the scale of impact from unprecedented and devastating flooding in parts of England. The report produced in response to these floods, the Pitt Review (2008), clearly identified that the lack of effective surface water management systems, such as SUDS, exacerbated the level of damage and associated costs of the flooding. Significantly, the Pitt Review highlighted the lack of clear ownership and responsibility for SUDS as a key reason for their limited adoption and that by addressing this one factor the wider use of SUDS would be supported.

The momentum and culture to use SUDS (referred to as Best Management Practices in Scotland and the USA and Water Sensitive Urban Design in Australia) is slowly recognising that urban water management for dealing with flooding and runoff quality does not have to occur at the cost of ecological potential. In Seattle, programmes for urban stormwater management have integrated SUDS that have a key target of benefiting the ecological and nature conservation value of the urban greenspace (Johnson and Staeheli 2006). In the United Kingdom, SUDS development so far has been driven by flood management and water quality with biodiversity playing a peripheral, almost incidental gain to a particular scheme (Worrall 2007). However, SUDS that embrace ecology at an early stage and with a deterministic role in the design are likely to be more robust and provide optimum benefits to the local and wider community (Everard 2002).

SUDS using habitats as integral features include green roofs, grass swales, detention ponds, wetlands, marshes and reedbeds. Each of these features may play a role in regulating and storing storm flow, improving urban runoff quality and providing nature conservation and landscape opportunities. Often, if an ecologist is consulted on a SUDS project it will be to specify the types of plant species to be used. This seemingly simple task is, however, critical to the success and long-term functionality of most habitat-based SUDS. In Australia and the USA concerns have been expressed about the 'failure' of wetland SUDS because of the use of inappropriate plants which are not able to tolerate the 'dynamic nature of urban hydrology' (abstract from Greenway *et al.* 2006).

Therefore, a critical role of an ecological approach in the development of SUDS is to define the species and habitat structures that address the essential prerequisite of such systems, i.e. to be sustainable. To embark on this process there are three basic questions to be asked:

- Which plants would be viable in response to the hydrometric and water quality character of the site?
- Which species and assemblages of plants are ecologically appropriate within the nature conservation context of the site in which the SUDS is to be built?
- How can the selection of plants, assemblages of plants and habitat structures provide strategic wildlife and nature conservation services?

Selecting the plants

Most habitat-based SUDS are scaled on the basis of their capacity to retain or manage the design flood event. This often means, for example, constructing a wetland basin, a large proportion of which only acts as a wetland infrequently. Some parts of the basin may be wet all year round whilst others will have varying degrees of wetness. In order to create a list of appropriate plants that can survive in such a system it is necessary to characterise the hydrology of the basin by generating frequency/depth/duration curves for a series of typical annual return rainfall events (see Figure 43.1).

Such an analysis of the hydrological character of the SUDS would enable the ecologist to determine which plants are the most appropriate and capable of surviving the conditions that would be found in the system. Many SUDS wetland basins may face extended periods of drought

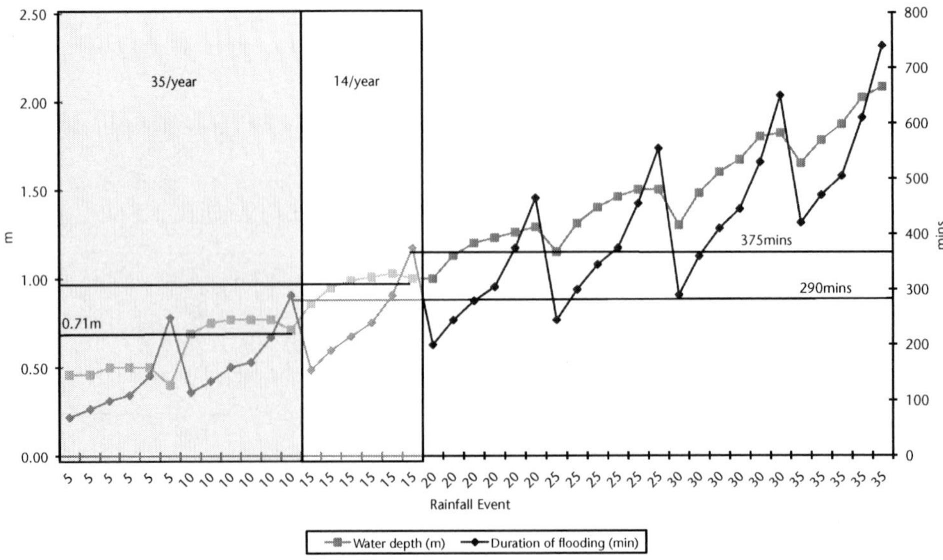

Figure 43.1 Depth frequency duration analysis for SUDS wetlands.

as well as continuous periods of inundation. With these hydrological conditions, consideration would need to be given to plant families such as Polygonaceae or Gramineae (for example *Polygonum amphibium* and *Phalaris arundinacea*), which can tolerate such extreme conditions (Newbold and Mountford 1997). Without undertaking such a hydrological characterisation and equating this to the tolerances of the plants available, the SUDS risks reverting to a species poor, low value and, possibly, high maintenance system.

Once a SUDS 'hydrograph' is understood and the nature of the plant types has been defined, the next step involves identifying groups or assemblages of similarly tolerant species that would collectively function within the hydrological regime of the SUDS. Guidance on this issue tends to be limited but in the UK, *The SUDS Manual* (CIRIA 2007) and *Ecohydrological Guidelines for Lowland Wetland Plant Communities* (Wheeler *et al.* 2004) provide a sound basis for determining appropriate grouping of plant species that are associated with varying hydrometric regimes.

Besides the hydrological character of the SUDS as a determinant of appropriate plant species selection, the pollution tolerance of the plants has also to be taken into account. Urban runoff is often characterised by pulses of high nutrient levels so it is essential that the trophic status of the plants and plant assemblages is considered (Haslam 1990). In addition, metal loadings and other pollutants in urban runoff may be significant, which also should influence the selection of plants. For example, in the borage family (Boraginaceae), *Myosotis scorpioides* will only tolerate chloride within the range of 14–22 $mg l^{-1}$, whereas in the Potamogetonaceae family, *Potamogeton natans* can tolerate chloride at much higher levels, up to 250 $mg l^{-1}$ (Jeffries 1988).

Once a selection of plants has been made on the basis of their appropriateness to the hydrological and pollution characteristics of the SUDS, a further ecological assessment is required in terms of the inter-specific competitiveness, soil associations and successional characteristics of the species (Grime *et al.* 1998). The significance of this lies in understanding the future management requirements and sustainability of the SUDS. Using aggressive species such as *Typha latifolia* in the UK or *Phragmites australis* in the USA needs to be adopted only with a full understanding of how these plants may dominate the system in a relatively short space of time.

Without taking these steps a 'landscaped' urban SUDS may, following construction and commissioning, appear aesthetically pleasing and operationally effective. However, in time, such systems may become limited in their biodiversity with only the most aggressive, pollution tolerant species surviving. In addition, and possibly more importantly, poor plant selection is likely to condemn the SUDS to repeated management problems in order to retain a visually acceptable and biodiverse system.

Selecting plant assemblages and habitat structures

Having determined which assemblages of plants might best fit the physical, chemical and biological characteristics of the SUDS, the next stage in defining the ecology of the SUDS is to develop a rationale for selecting amongst the various options of plants and plant assemblages. The key to this process is to appeal to the local nature conservation character of the surrounding area. This may be achieved through site survey or, more generally, a desk-based study where ecological data is collated for the development site or catchment. Designing the SUDS to conform to and mimic the nature conservation character of the surrounding area supports the long-term sustainability of the new habitat by providing nearby sources of appropriate colonisers. In addition, this approach makes the SUDS more ecologically robust and adaptive, as well as ensuring the capacity for ecological connectivity to other local habitats.

As an alternative to using the nature conservation character of the surrounding area as the guide to the habitat selection and species for the SUDS, appeal could be made to local and national Biodiversity and Habitat Action Plans (in the UK see www.ukbap.org.uk) or policies and aspirations of the local Wildlife Trusts (in the UK see www.wildlifetrusts.org). Not only would this mean that the SUDS would be contextually appropriate from an ecological perspective but the system may take advantage of offering significant enhancements to biodiversity objectives, something the local planning authority and community may be keen to see.

Having the right plant groups is one requirement but another relates to the types of habitat structures and management regimes that may be integrated into the scheme. The structural balance between open water and reedbed in a wetland SUDS can significantly influence the capacity of the SUDS to support diversity in invertebrate fauna and bird life (see Figure 43.2).

Again, providing targeted plant assemblages and a suitable management regime may mean that the SUDS can sustain populations of protected or endangered species, enabling the SUDS to contribute to wider ecological aspirations, policies and strategies.

By selecting the most ecologically appropriate species, assemblages and structures, the ecologist will be directly influencing the nature and extent of management of the SUDS system. It is a myth that habitat-based SUDS require specialist and expensive maintenance. The opposite is in fact the case, with an ecologically well-designed SUDS system providing a low maintenance option compared with orthodox landscape development approaches. A grass swale, for example, under orthodox management may be cut several times a year, whereas a habitat-based SUDS swale using subsoils and flower rich grassland mixes may need only one cut per year.

SUDS and strategic wildlife opportunities – a case study?

The reality of climate change is leading key nature conservation bodies to consider how strategic planning may be used to facilitate the response of wildlife to the changes that

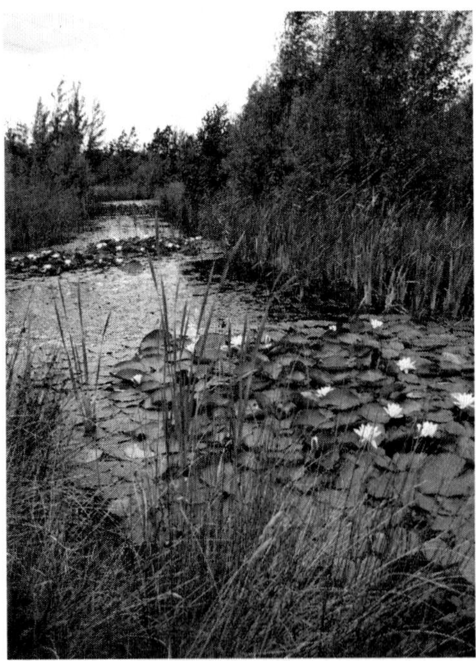

Figure 43.2 Structurally diverse SUDS wetland.

will occur in the distribution of environmental conditions. In the UK this momentum is stimulating the production of vision statements for a range of habitats and species assemblages which could be used to inform the nature of SUDS developments across the British Isles (www.wetlandvision.org. uk). As the impacts of climate change begin to manifest themselves in ecological responses, a key function of our urban greenspaces will be to facilitate the movement of species. This would be achieved through the implementation of green infrastructure networks and habitats acting as 'stepping stones' for species movement. The role SUDS could play in this process within the urban context is potentially significant as long as the design of the SUDS is guided by ecological principles and thinking.

For a SUDS to achieve ecological sustainability, it is proposed that plants and habitats are selected that are viable in hydrological and water quality terms; that they are ecologically appropriate within the nature conservation context of the site; and that the SUDS and its associated species provide a means of achieving strategic wildlife goals. However, even if these goals are achieved it may be argued that the ecological viability of habitat-based SUDS is compromised by the nature of the runoff they receive from the built environment. It is well documented that habitat and species diversity may decline in response to increased nutrient and pollutant loading (Moss 1998). However, many wetland habitats are highly tolerant of and resistant to environmental stresses but it is not well understood how long SUDS habitats can retain viable nature conservation values in the face of continuous long-term pollutant loadings.

This issue has been investigated in the UK at a SUDS constructed to protect a nature reserve at Potteric Carr, owned by the Yorkshire Wildlife Trust. This nature reserve (a Site of Special Scientific Interest) on the urban fringes of the city of Doncaster, comprises a complex of wetland habitats ranging from reedbeds to wet woodlands. At over 200 hectares, the site depends on a pumped water system to sustain the mosaic of wetland features. In recent years, the limited availability of water for managing the habitats has become critical. Urban runoff and sewage works effluent runs through the site but its water quality is not sufficiently good to support the wildlife designated habitats of the Reserve. To address this, in 1998, the Trust constructed a SUDS wetland system that would divert the urban runoff into a series of wetland cells that would treat the water and make it more appropriate for use in supporting the sensitive habitats within the Reserve.

A primary concern of the Reserve managers was the risk posed by the high levels of phosphorus in the urban runoff. This prompted the installation of brick-rubble filters at the inlet to each of the five wetland SUDs cells to adsorb as much phosphorus as possible before entering the wetland cells (see Figure 43.3).

Within the first three years of operation the SUDS removed phosphorus from an average inlet concentration of 2.79 mgl^{-1} to less than 0.5 mgl^{-1} at the outlet (Huckson 2004). With targeted planting of wetland species and the provision of structural diversity within the reedbed cells, the SUDS wetland rapidly evolved into a diverse habitat supporting breeding birds, small mammals, reptiles and amphibians. Two years after commissioning, detailed surveys found a good assemblage of invertebrate species, primarily from rapid colonisers such as beetles, mayflies, diptera and hemiptera (bugs) (Bateson 2000). Many of the species recorded were from pollution sensitive families suggesting that the water quality within the cells was good.

When the survey was repeated in September 2007 it appeared that the taxon richness of the aquatic macro-invertebrates using the SUDS had increased but this masked the fact that the assemblage had shifted towards a domination of pollution tolerant species. Nevertheless, the SUDS in 2007 supported at least three invertebrate species of conservation interest. This shift in the ecology of the aquatic invertebrates may be as a result of the continuous environmental stresses imposed by the quality of the source water to the SUDS. However, a key factor that may

Figure 43.3 Reedbed SUDS at Potteric Carr.

be influencing this apparent shift towards a more pollution tolerant species assemblage is the management of the SUDS. Like so many SUDS, the Potteric Carr system has to some extent suffered from the 'build and abandon' approach. This has meant that the operational assumptions that were inherent in the design have not been fully met, with management requirements being overlooked.

The Potteric Carr SUDS is over ten years old and despite an apparent shift in the invertebrate populations towards more pollution tolerant species, the wetland habitat still provides significant ecological resources, as well as meeting water quality objectives and to a lesser extent flood management services.

By their very nature, SUDS will be constrained in the ecological services they can provide because of the character and nature of their source water. However, even from an ecological perspective, 'sub-optimal' habitats have the potential to support the delivery of local and national biodiversity objectives as well as providing a service to urban drainage management.

Green roofs

Green roofs are one of a number of technically feasible improvements that can be made to urban environments to provide support for native biodiversity, both within the building designs themselves, and also more widely through the provision of greenspaces (see also Chapter 44 this volume). Not only are these features important in terms of improving resource use efficiency, careful design and integration into the design process but will also enhance local and regional biodiversity through the creation of suitable, interconnected, semi-natural habitat patches throughout urbanised areas.

Benefits associated with such areas, ideally considered as strategic elements within a wider green infrastructure network, are wide-ranging, including the provision of: (i) additional habitat for urban wildlife (i.e. Biodiversity Action Plan species known to associate with urban/brownfield habitats); and (ii) multifunctional uses such as stormwater control, building insulation and noise reduction. These natural greenspaces can also serve to improve health and well-being, social cohesion, increase property values and reduce management costs.

Types

In recent years, largely through an increased emphasis on urban greenspaces and the drive towards sustainable building design, a clear definition of the term 'green roof' has arisen, incorporating three broad categories: 'Intensive', 'Semi (or Simple)-Intensive' and 'Extensive'.

First, intensive green roofs, characterised by deep (>100 cm) substrates, high biomass and intensive management regimes, are designed to emulate parks or domestic gardens at roof level. These predominantly ornamental systems present the opportunity for diverse planting schemes, including trees and shrubs, and are often designed with public access in mind through provision of a range of recreational functions such as garden space, patios, rooftop restaurants, and even small golf courses. Depending on the thickness of the soil layer and the structural components of the building, a rooftop may require additional support features to bear the weight of the green roof. In general, and adding to the overall weight and complexity of such systems, intensive roofs require some form of irrigation built into the design.

Second, semi-extensive green roofs are designed as an intermediate stage between intensive and extensive roof systems. Such roof areas are designed to be lower cost, but with an intermediate substrate depth (typically 10–25 cm) and a requirement for regular maintenance. These design constraints aim to emulate more natural urban habitat conditions, but allow for a greater diversity of plants to be grown as compared with extensive systems. In general, such green roof systems incorporate a range of both native and ornamental plant species and are designed to be visually appealing as their overriding function.

Finally, extensive roofs are designed to have largely native species planting, less overall biomass, and shallower substrates (5–15 cm). Such roofs tend to be sown with (local) wildflower mixes or sedum matting or are left to colonise naturally. In general, extensive green roofs are designed to be lightweight and easily retrofitted to existing roof surfaces; to require little or no maintenance; and to be relatively inexpensive to establish. By virtue of their harsh growing conditions, these systems favour low-growing, stress-tolerant ruderal vegetation (such as stonecrops – *Sedum* spp.) and can thus be used to replicate brownfield biodiversity.

Three types of extensive roof system are regularly incorporated into urban developments: sedum roofs (see Figure 43.4) (created from pre-grown sedum mats, cuttings, or plug plants with a substrate depth of 2–10 cm), meadow roofs (visually striking systems created with diverse seed mixes of native wildflowers and grasses with a substrate depth of 7–10 cm), and brown/biodiversity roofs (designed to recreate more natural and often local habitats using the by-products of the development process such as crushed brick and subsoil which are left to colonise naturally over time or seeded with wildflowers).

Other terms in general use include: roof top gardens, eco-roof, vegetated roof, building integrated habitat, and plant based surface systems:

- **Rooftop gardens**: As the name suggests, these are designed to comprise a range of ornamental planters without the need for substrate, with a focus on public/private access to urban greenspace.

Figure 43.4 Sedum green roof at a Sheffield school.

- **Eco-roof**: an overarching term for roof surfaces that aim to serve some ecological function, from vegetated roofs to those with photovoltaic cells installed.
- **Vegetated roofs**: a synonym for green roof describing those roof surfaces where vegetation has been incorporated.
- **Building integrated habitat**: refers to any intentionally established habitat on buildings, including additional features such as bat, bird and insect boxes.
- **Plant-based surface systems**: A largely scientific term and synonym for green roof proposed by Tapia Silva *et al.* (2006).

Biodiversity benefits

Despite a growing recognition of the potential value of green roofs in urban environments, empirical data on their biodiversity benefits are not yet widely available. Nonetheless, many benefits are repeatedly reported within the available literature, including: the provision of new habitats in largely urbanised environments, replacement of brownfield habitats lost to development, creation of green corridors linking existing habitats, facilitation of wildlife movement and dispersal, and provision of refugia for declining and rare species. One of the key functional roles of the green roof system is to replicate brownfield habitats, considered important for a range of species including bees, beetles, birds, flies, lichens, spiders and moths (Grant *et al.* 2003). For example, Jones (2002) recorded 48 invertebrate species on sedum roofs in Canary Wharf (London), including *Helophorus nubilis*, a scarce 'crawling water beetle', *Chlamydatus evanescens*, a nationally rare leaf bug, *Erigone aletris*, a North American spider recently naturalised in the UK, and *Pardosa agrestis*, a nationally scarce wolf spider.

In addition to providing valuable habitat for wildlife in general, green roofs can host a number of species of interest that are rare or scarce in other habitats. Consequently, green roofs could play an important role not only in creating additional wildlife spaces in urban areas but also in the conservation of rare or endangered species. For example, across London, green roofs were designed to mimic the conditions found on the derelict sites favoured by the black redstart. Initially termed 'brown roofs', these roofs were constructed from recycled crushed concrete and brick aggregate and were allowed to be colonised naturally (Gedge 2003).

Other benefits

To the extent that the roof surface can be transformed into useful space, the building becomes economically and functionally more efficient and can have a more benign effect on the surrounding landscape including a wide range of environmental, amenity, health, building fabric, economic and educational benefits as follows (updated from Grant *et al.* 2003).

Environment

- Attenuation of stormwater run-off.
- Run-off attenuation reducing sewer overflows (up to 75 per cent of rainwater can be stored in the short term (Grant *et al.* 2003)).
- Potential for cleaning and recycling grey water.
- Absorption of air pollutants and dust (air purification).
- Amelioration of the 'urban heat island' effect through increased humidity, reflectance and shade (flat gravel covered roofs may be up to 21°C hotter than vegetated roofs (Kaiser 1981)).
- Absorption of urban/industrial noise (a green roof can reduce sound within a building by 8 dB or more when compared with a conventional roof (Grant *et al.* 2003)).
- Absorption of electromagnetic radiation.
- Absorption of greenhouse gases (particularly CO_2) and giving off oxygen.

Amenity

- Additional options for designers.
- Masking of grey and uniform roofing materials.
- Screens equipment.
- Aesthetic appeal.
- Access to greenspace.

Health

- Psychological benefits of contact with nature.
- Improved air quality – helps to reduce lung disease.
- Improved water quality (as part of Sustainable Urban Drainage Systems).

Building fabric

- Protecting the roof from UV and mechanical damage.
- Improvements to thermal insulation. Savings in fuel costs have been estimated at 2 l fuel oil/ m^2/year for a typical green roof in Germany (ZinCo 2000).
- Meeting building standards by improving the energy performance of buildings (i.e. Part L of the Building Regulations, Code for Sustainable Homes and BREEAM).

Economic

- Extended life expectancy of roof membranes (in many cases, a green roof can extend the lifespan of a roof up to three times that of a traditional roof).
- Increased property prices/rents.
- Potential reduction in water/sewer charges.
- Reduced heating and air conditioning costs through decreased energy requirements.

Education

- Green roofs can provide outdoor classrooms in inner city areas.

Ecological modifications

Ecologists have begun looking for alternatives to widely used sedum mats that incorporate microhabitats customised for particular species and/or more closely mimic natural habitats, with varied microtopography, scattered rocks, rubble, dead wood, and more diverse vegetation. Increasing the structural diversity of the vegetation on green roofs will create a patchwork of niches to attract higher faunal species diversity. This can be achieved by varying substrate depth and creating slight undulations, perhaps using shallow hollows to collect rainwater. For instance, sedum can be planted onto thinner substrate depths, whilst the deeper substrate areas will support the meadow mix. In addition to a more diverse mix of plant species, varying substrate depths will provide a heterogeneous matrix of drier and damper areas to benefit a high invertebrate diversity. Similarly, encouraging moss growth by leaving patches of bare substrate or providing wood piles/rockeries will create invertebrate microhabitats and provide potential foraging and nesting areas for birds.

Management

Management and monitoring of green roofs is critical to the realisation of their full potential. In this regard, general management of the green roof will include:

- An annual check on plant growth, with re-planting if areas of greater than 1 m^2 become bare and devoid of growth;
- Regular checks (minimum quarterly) on the rainfall harvesting and diversion to the meadow area of the green roof;
- Annual weeding and debris clearance from the roof.

References

Bateson, D. (2000) *Survey of Freshwater Macro and Micro Species Developing in a Water Filtration System: Potteric Carr Nature Reserve*, York: Yorkshire Wildlife Trust.

CIRIA (2007) *The SUDS Manual*, CIRIA Publication C697, London: CIRIA.

Everard, M. (2002) 'Stemming the tide: flooding and sustainable urban drainage', *Surveyor*, 21 February 2002.

Gedge, D. (2003) 'From rubble to redstarts', *Proceedings of the First Annual Greening Rooftops for Sustainable Communities Conference, Awards and Trade Show, Chicago* (CD-ROM), online, available at: www.greenroofs.org/index.php/component/content/article/10-miscarchive/108.

Grant, G., Engleback, L. and Nicholson, B. (2003) *Green Roofs: Their Potential for Conserving Biodiversity in Urban Areas*, English Nature Research Report 498, Peterborough: English Nature.

Greenway, M., Jenkins, G.A. and Polson, C. (2006) 'Wetland design to maximise macrophyte establishment and aquatic biodiversity', in A. Deletic and T. Fletcher (eds) *Proceedings, 7th International Conference on Urban Drainage Modelling and the 4th International Conference on Water Sensitive Urban Design*, Clayton, Vic.: Monash University.

Grime, J.P. Hodgson, J.G. and Hunt, R. (1998) *Comparative Plant Ecology: A Functional Approach to Common British Species*, London: Unwin Hyman Ltd.

Haslam, S.M. (1990) *River Pollution. An Ecological Perspective*, London: Belhaven Press.

Huckson, L. (2004) *Potteric Carr: Reed Bed Filtration System*, prepared for Yorkshire Wildlife Trust and sponsored by Scott Wilson, Leeds, Ref D104557.

Jeffries, M. (1988) *Water Quality and Wildlife: A Review of Published Data*, London: Nature Conservancy Council.

Johnson, R.L. and Staeheli, P. (2006) *Stormwater Low Impact Development Practices*, World Environmental and Water Resources Congress 2006 and see online at: www.cityofseattle.net/UTIL/groups/public/@spu/@usm/documents/webcontent/spu02_020004.pdf

Jones, R.A. (2002) *Tecticolous Invertebrates: A Preliminary Investigation of the Invertebrate Fauna on Green Roofs in Urban London*. Peterborough: English Nature, London.

Kaiser, H. (1981) 'An attempt at low cost roof planting', *Garten und Landscaft*, January: 30–3.

Moss, B. (1998) *Ecology of Fresh Waters: Man and Medium, Past to Future*, third edition, Oxford: Blackwell Science.

Newbold, C. and Mountford, O. (1997) *Water Level Requirements of Wetland Plants and Animals*, English Nature Freshwater Series, No. 5, Peterborough: English Nature.

Pitt, M. (2008) *The Pitt Review: Implementation and Delivery Guide*, London: Cabinet Office.

Royal Commission on Environmental Pollution (2007) *Twenty Sixth Report: The Urban Environment Cm7009*, Norwich: The Stationery Office.

Tapia Silva, F.O., Wehrmann, A., Henze, H.-J. and Model, N. (2006) 'Ability of plant-based surface technology to improve urban water cycle and mesoclimate', *Urban Forestry and Urban Greening*, 4(3–4): 145–58.

Wheeler, B.D., Gowing, D.J.G., Shaw, S.C., Mountford, J.O. and Money, R.P. (2004) *Ecohydrological Guidelines for Lowland Wetland Plant Communities: Final Report*, Bristol: Environment Agency.

Worrall, P. (2007) 'The role of ecology in SUDS', *Proceedings of the SUDSnet National Conference 2007, Coventry University TechnoCentre. Nov 14th 2007*, Dundee: Sudsnet, pp. 96–101, online, available at: http://sudsnet.abertay.ac.uk/documents/SUDSnetConference2007ProceedingsBOOK.pdf.

ZinCo (2000) *Green Roofs Planning Guide*, sixth edition (manufacturer's brochure), Unterensingen, Germany: ZinCo GmbH.

44

Green roofs, urban vegetation and urban runoff

Joachim T. Tourbier

Climate change is affecting urban areas through the extremes of flash flooding caused by more intensive rainfall and through heat islands, formed in the inner cities during summer months. Consequences are severe. Between 1998 and 2004, flooding catastrophies in Europe caused damage in excess of 25 billion EUR and the loss of 700 lives (Commission of the EC 2006). Severe summer heat has caused fatalities in major cities, such as in Paris. Both problems are made more severe by a particular factor affiliated with urban areas: impervious surfaces. Paved surfaces and impervious rooftops speed up the runoff of stormwater, causing urban drainage problems and flooding and excessive heat in summer. "Extensive" green roofs are an exciting new technology that permits retrofitting of existing urban areas to alleviate these problems. The common denominator is evaporation and evapotranspiration through plants, lessening the volume of runoff that causes flooding and permitting vegetation to live up to its role as the cooling system of the planet earth.

Green architecture and green technologies have been a theme of the new millennium. In antiquity, the hanging gardens of Semiramis, built by Nebuchadnezzar II, were actually an early example of a green roof and were one of the Seven Wonders of the World (Wikipedia 2009a). Any roof in a temperate climate is actually a potential habitat for vegetation, such as the succulent houseleek (*Sedum sempervivum*). To date, green roofs were made possible by a technological breakthrough of the DuPont Company, which invented a woven plastic polymer that not only formed the basis for nylon stockings, but also for geofabrics invented to seal sanitary landfills. The use of geofabrics as a root blocking membrane for vegetation on roofs was quickly realized in Germany. Here a colleague in landscape construction, Professor Liesecke of the University of Hannover (Liesecke *et al.* 1989), conducted research concerning lightweight soils involving blown shale, making an important contribution to the special role Germany soon assumed in developing a green roof industry.

Conventional roofs with a bituminous cover have a high shrink-swell ratio during the temperature extremes of the day, tend to crack and require regular replacement. The German contribution was the development of "extensive" green roofs, which have a thin layer of lightweight soil that can be installed over existing roofs without having to change or upgrade their structural support. This offers a tremendous opportunity to retrofit and green existing structures in cities, thereby realizing multiple benefits of green roofs, including: reduced runoff and flood management; reducing energy demands in heating and cooling houses; improving the longevity and replacement costs of roofs; and benefiting city climate all while also providing for aesthetics and wildlife benefits in downtown areas. The latter includes ground nesting birds and various types of bees that feed on the nectar of blooming plants.

Green roofs provide a new form of urban vegetation, as well as integrated, sustainable management of urban drainage. In this chapter we will explore pre- and post-development conditions of urban areas, showing what causes heat and urban drainage problems. We will further investigate a concept of maintaining pre-development conditions by sustaining the water cycle and energy budget of cities. A key to bringing about the use of green roofs, as a new technology, is to maximize performance aided through design principles of multiple benefits and multiple uses. Green roofs are of value to society and can rely on community action and support, involving institutional aspects of implementation. The wide extent to which this technology is already being used will be demonstrated through examples of installations.

Pre- and post development conditions of urban areas

We know that stormwater runoff, and peak and volume increases, are inextricably linked to increases in impervious coverage: rooftops, sidewalks, parking lots, highways and other roadways. To see how this is so, one need only think of the natural storage capacities of an average undisturbed woodland. When it rains in the forest, up to 40 percent of the precipitation is intercepted by tree surfaces and evapotranspirated, and up to 50 percent of precipitation is contained in depression storage, where it recharges groundwater supplies. A mere 10 percent of precipitation is turned into runoff. Contrast this to a land surface covered by 75 to 100 percent impervious surfaces. Here, just 15 percent is contained in depression storage to infiltrate groundwater, and fully 55 percent is converted to runoff. This is summarized in the diagram (shown in Figure 44.1)

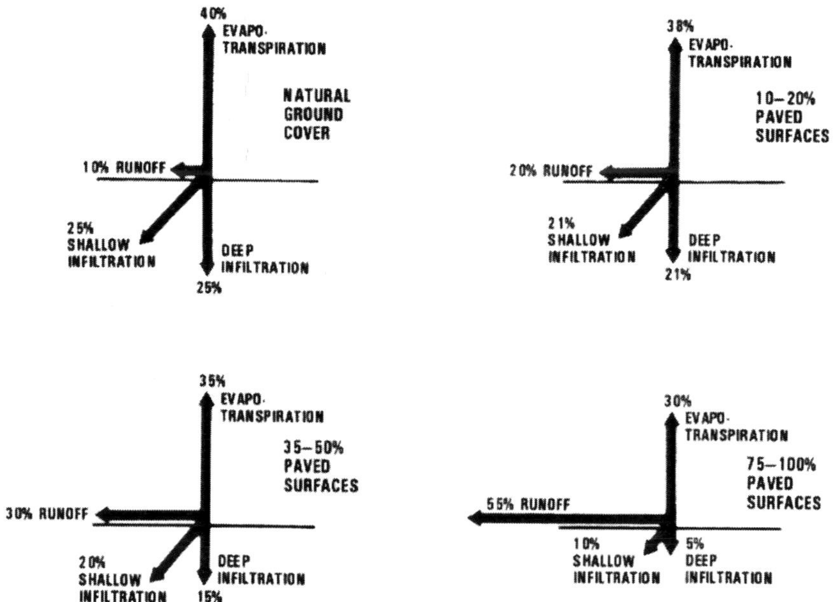

Figure 44.1 Typical hydrograph changes due to increases in impervious surface areas (*source:* EPA 1983, after Tourbier and Westmacott 1981: 3).

that was co-developed by the author and printed in a publication by the US Environmental Protection Agency, presenting the results of the nationwide urban runoff program (EPA 1983).

When such changes are forced onto the natural hydrology of a given area, the effects are numerous and detrimental. More stormwater runoff concentrates in a short time to form high volumes that surge in drainage ways and urban streams. The higher peaks cause streambank erosion, downstream siltation and degradation of the stream's habitat, reducing or even eliminating the diversity and total number of fauna and flora living in and near the water.

Water quality is also negatively impacted by impervious surfaces. As runoff flows over these surfaces, it captures pollutants such as hydrocarbons, pesticides and nutrients, and carries them downstream, where they pollute water resources, fish and wildlife habitat, and community water supplies. Though these "non-point source pollutants" are not as noticeable as larger scale "point" sources of pollution (such as those stemming from industrial sources), they can be every bit as destructive. Now, you may reason that in urban areas we have sewage treatment plants that will prevent such pollution. It turns out that most sewage treatment plants for combined sewers that transport a combination of domestic wastes and street runoff have been sized for "dry weather flow." During storm events untreated sewage is discharged into receiving waters through "combined sewer overflows". Some engineers may point to the benefits of rainwater flushes, making the sewers "self cleansing", and to a dilution factor. Higher water quality standards though now have made combined sewer overflows an issue that is being recognized around the globe. Solutions that have been planned for big cities like London, Washington DC and Toronto are deep cavity storage in bedrock, and later treatment, costing billions for each city to be financed through user fees. This is most lucrative for sewage authority shareholders, but not a sustainable solution. Green infrastructure and non-structural solutions are an option that reduces the volume of untreated combined sewer overflows into the nation's streams.

The core of the problem is the common misconception that rainwater runoff is best collected and transported in underground sewers. In the dense cities of the past this was more appropriate than in the green cities envisioned now. Today, though, many city drainage codes still require that stormwater be collected by curbs and gutters and directed into inlets and combined sewer systems leading to an "end-of-pipe treatment" with massive discharges.

A concept of maintaining pre-development conditions

Decentralized "Blue-Green Technologies" are an alternative, first advocated at the University of Delaware Water Resources Center (Tourbier 1980), that use stormwater to green cities, permitting it to seep into the ground, irrigate plants, be cleansed by the vegetation and evaporate back into the atmosphere, closing the hydrologic cycle. These Blue-Green technologies have gained wide support from public interest groups, being termed Decentralized Stormwater Management, Best Management Practices (BPPs), Low Impact Development (LID), and Sustainable Urban Drainage Systems (SUDS) and more recently green infrastructure such as urban open space including green roofs. In existing cities such practices have the problem of requiring space where space is at a premium. An exception is the greening of roofs, which offers boundless opportunities for retrofitting structures that already exist, attenuating the generation of runoff at its source. Today over 70 percent of the global population lives in cities, with a tendency to rise to 80 percent and more. A high percentage of the "cities of tomorrow" are yet to be built. At a time of global warming there will be higher frequencies of intensive precipitation to increase water runoff and flooding. New standards should be set to incorporate measures into land development to mitigate adverse environmental impacts. What materializes as the potential standard is to maintain the water balance, or pre-development conditions (shown in Figure 44.1) to be equal to the *ex ante* post-development condition at development sites (Tourbier 2003). This means to hold back and slowly release peaks through depression storage, to recharge groundwater and to evaporate and evapotranspirate and to cleanse runoff. To maintain pre-development peaks on development sites has already been a requirement of some US drainage codes since the 1990s. Now increasingly the water balance that existed under a woodland condition is being viewed as a standard. In Germany efforts are also being made at the federal UBA (*Umweltbundesamt*) to control stormwater runoff at new construction sites by replicating runoff volumes and peaks found in woodland conditions, including evapotranspiration (Sieker *et al.* 2008).

In the US the state of Maryland is considering site design criteria that consider "woods in good condition" as the "pre-development" hydrology condition to be restored by incorporating into land development measures that can retain and slowly release equivalent volumes (Maryland 2008). Such standards have implications on design. No single, end of pipe treatment measure can achieve them, but a "treatment train" of practices needs to combine measures in a linear fashion. In existing cities the installation of measures requires retrofitting. The single most effective measure to accomplish this is vegetated roofs that evapotranspirate precipitation through a soil substrate and a vegetation cover. Extensive roof covers can be installed on existing sloping and flat roofs. They also have a fully saturated weight of only 15 pounds per square foot, and thus are not heavier than the standard gravel layer used on roofs.

Maximizing performance

Vegetated Roof Covers (VRC) function as a veneer of living vegetation that controls runoff at its source, while offering additional practical benefits. VRCs, also known as "green roofs," "ecoroofs" and "roof gardens," fall into two broad categories: extensive (Figure 44.2) and intensive systems.

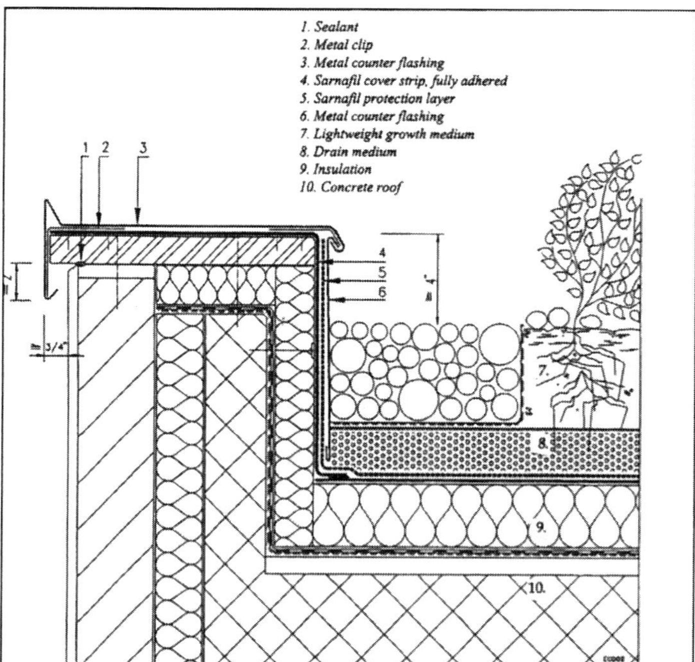

Figure 44.2 Cross-section through an extensive roof cover (*source:* Miller 1998).

Extensive green roofs have a thin, lightweight cover and are expected to be self-sustaining, requiring a minimum of maintenance (Wikipedia 2009b). In most instances, extensive VRCs are not intended to be walked on, except when required for maintenance activities. Intensive VRCs, on the other hand, are landscaped roof environments that provide recreational as well as practical functions. In general, systems in which the principal root zone is deeper than four inches (10.2 cm) are considered intensive VRCs. They tend to be labor-intensive and require fertilization and irrigation.

Conventional roofs generate the highest amount of concentrated runoff from developed sites. By mimicking natural hydrologic processes, VRCs can achieve runoff characteristics that closely approximate open-space conditions. Plants are essential elements in the functioning of such roof covers, intercepting and delaying rainfall runoff by holding precipitation in the plant foliage and absorbing water in the root zone causing an eventual transpiration reduction in total runoff volume of 50 percent or greater.

VRCs consist of three primary layers: 1) foliage, 2) growth medium and principal root zone, and 3) a special drain layer (Figure 44.3) or drainage medium (Figure 44.2). These overlay the waterproofing membrane of the roof deck. Each layer typically incorporates subsystems that are included to provide specific performance features (e.g. water retention, insulation, dimensional stability). The drain layer (Figure 44.3) conveys excess water from the roof and prevents the ponding of precipitation directly on the roof membrane. Roof covers can be designed to detain water within the roof profile. In this chapter, the term "VRC" pertains to fully drained roof designs that do not permit ponding on rooftops.

In order to minimize weight, a drain layer can consist of a synthetic material. During the initial stages of a rainfall, the foliage intercepts nearly all of the precipitation. But as the rain continues, water percolates into the growth medium and the plant root zone. When the volume of rainfall exceeds the growth medium's field capacity, runoff begins for the first time. Field capacity refers to the quantity of water that will not freely drain from the growth medium, measured at the moisture content by volume at a capillary tension of 0.33 bars.

The effectiveness of VRCs in managing runoff peak rates is highly dependent on the characteristics of the design storm. Typically, VRCs are designed to control runoff peak rates for

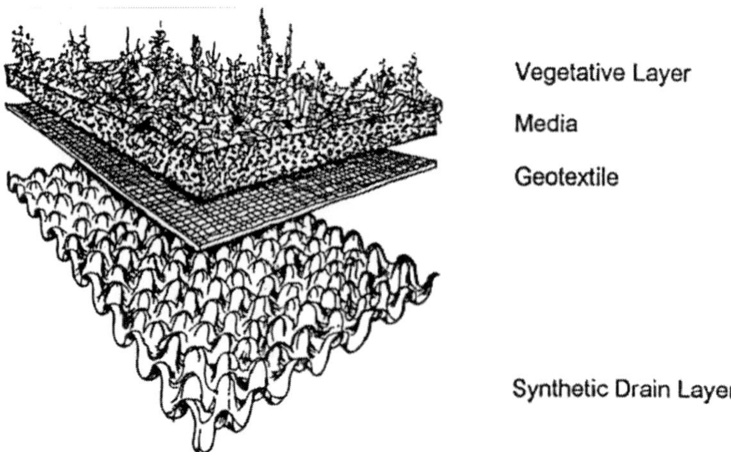

Figure 44.3 An egg-carton like synthetic material may be used as a drain layer (*source:* ZinCo GmbH).

storms with magnitudes equal to the two-year return frequency rainfall event, or smaller. Since well over 95 percent of storms are smaller than the two-year storm, this represents a significant level of protection. The weight of the VRC is a critical consideration, especially where retrofit installations are under consideration. Their saturated weight must be considered in addition to other design loads, including maximum snow loads.

In many instances, building owners will want to determine the potential savings associated with the added insulation value of the VRC. These savings will be greatest for single-story buildings where the roof represents the largest surface for heat transfer. The underlying waterproofing membrane must be of high quality and in good condition. Careful inspection of the membrane, seams, and flashing is required prior to installing the VRC. Once installed, maintenance of the underlying waterproofing membrane is no longer required. The VRC functions as a protective layer for the roof, shielding it from ultra-violet light, extremes of temperature, and mechanical damage.

The growth medium plays three roles in the VRC:

1 Supporting the growth of the foliage layer;
2 Intercepting and retaining precipitation; and
3 Controlling the rate of percolation of rainfall into the underlying drainage layer.

Depending on the physical properties of the growth medium, a wide range of performance characteristics can be achieved. Specifications for the growth medium should include:

1 Thickness;
2 Saturated infiltration capacity (hydraulic conductivity);
3 Matric potential versus moisture content relationship (including the moisture content at field capacity);
4 Porosity; and
5 Saturated unit bulk unit weight.

Finer-grained media tend to support higher field capacities and lower infiltration capacities, characteristics that are advantageous when peak rate attenuation is important. A limited number of plants can thrive in the extreme conditions that typify roof environments. Effective plant species must exhibit the following:

1 A tolerance for mildly acidic conditions and poor soil;
2 A preference for well-drained conditions and intense sunlight;
3 A toleration for dry soil; and
4 An ability to colonize vigorously.

Both annuals and perennials can be used. In temperate climates, many species of stonecrop (*Sedum spektabile, Sedum sempervivum, Sedum lineare, Sedum reflexum, Sedum spurium*) are especially well adapted. Other plants that are commonly used in conjunction with extensive VRCs include ornamental grasses (e.g. *Festuca glauca*), yarrow (*Achillea taygetea*), lavender (*Lavendula angustifolia*), meadow sage (*Salvia officinalis*), allium (*Allium moly*), and thyme (*Thymus lanuginosus*).

Plants can be established using several techniques, including:

1 Hydro-seeding;
2 Broadcast seeding and the direct setting of sedum cuttings; and
3 Planting of nursery propagated plugs.

In addition to providing a powerful tool for controlling and treating runoff, VRCs offer other benefits. They:

1 Reduce heat transfer across the roof, thereby lowering energy costs for heating and cooling. These savings will be greatest for single-story buildings where the roof represents the largest surface for heat transfer;
2 Reduce radiated heat from roofs, thus moderating urban microclimates and urban heat-islands effects;
3 Reduce the rate at which the underlying roof membrane weathers. VRCs shield roofs from the effects of ultra-violet radiation, wind, and temperature variations. As a result, the service life of roofs protected with VRCs is typically doubled, reducing both costs and landfills;
4 Conserve energy;
5 Reduce sound reflection and transmission. A VRC with a 12 cm substrate can reduce sound by 40 dB (GRHC 2009);
6 Create wildlife habitat;
7 Visually enhance roofs, and increase the potential for recreational use of roof space; and
8 Conserve space that would otherwise be required to construct detention ponds and related stormwater management facilities. In particular, commercially valuable space can be maximized in densely developed areas.

Design principles of multiple benefits and multiple uses

The performance of VRCs can be calibrated by design, and so can the performance of other sustainable drainage measures, where detention capacity, release rates, cleansing capacity and other effects can be made to perform to standards. Increasingly projects are subject to an *ex ante* evaluation through an effectiveness control, comparing pre- and post-development conditions. Single performance factors, though, usually are not determinants in the decision-making process by clients (Miller 2009). Rather, a combination of factors, including aesthetics, drives the decision. Green buildings offer environmental, social and economic benefits and multiple uses (Figure 44.4) that can be expressed through design criteria.

The concept of sustainability broadens the way we look at the built urban environment to include ecologic, social and economic considerations. A study on success control for stormwater management (Tourbier 2001) considered:

1 **Ecological criteria** – including the reduction of runoff peaks, groundwater recharge, water quality, fauna, flora, soil, and energy use;
2 **Social criteria** – include public health and safety, public education, and participation of stakeholders and use of local resources;
3 **Efficiency of construction, operation and maintenance** – considers a minimal use of capital and a maximum efficiency of performance, use of self-regulating systems, cost effectiveness, integrated, decentralized systems, resilience to future change; and
4 **Enhancement of the urban environment** – includes consideration of natural- and cultural heritage, creation of natural areas, enhancement through hard edge design, multiple uses of facilities, biologic continuity, conveyance in floodways, buffer strips along swales, stormwater corridors as greenways, and observation appeal through designed accessibility.

It was recommended that alternative systems be rated in a matrix showing whether the criteria were (a) fully, (b) partially or (c) insufficiently achieved.

Figure 44.4 The Post Giro Facility in Munich, Germany (*photo:* Optigrün International AG).

The "Blue-Green Technologies" approach (Tourbier 2003) is providing design criteria for drainage systems, placing them in linear fashion for:

1 lessening the volume of runoff close to its source;
2 improving the quality of runoff by filtering it through vegetation;
3 maintaining groundwater recharge;
4 detaining flood peaks;
5 integrating a combination of management practices with existing water features; and
6 protecting and restoring streams (and wetlands?) as natural stormwater conveyance systems.

Measures that supplement one another and are arranged in sequence in the path of runoff concentration are linked in "treatment trains," in which the size of each measure will be reduced by the capacity and performance of the measure preceding it. Because stormwater management is not simply a matter of peak flow control, such combinations are useful and practical.

The first criterion, lessening the volume of runoff close to its source, is of particular interest for existing urban communities and can be achieved through the greening of roofs. When criteria outlined here are applied to alternative stormwater management systems VRCs outperform others and should be made a component of every sustainable drainage system.

Institutional aspects of implementation

To mitigate flooding, reduce pollution from combined sewer overflows, reduce energy needs, enhance cities and to improve the quality of urban life are points of interest for both the public and to public institutions. In any jurisdiction there are government agencies that have been given statutes that enable them to respond to public concerns through regulations. In addition there are non-government organizations (NGOs), such as environmental groups and professional organizations that often are a driving force for government to act.

Government cannot mandate the installation ofVRCs or any other particular Best Management Practice, but aid their use through drainage guidelines and through financial incentives. Examples for redevelopment in inner cities are the city of Philadelphia, requiring a lessening of runoff by 20 percent (Miller 2008) and the Emscher Lippeverband in the Ruhr valley in Germany where runoff is to be reduced by 15 percent (Kaiser 2005). This can be achieved by intercepting and reducing the volume of runoff before it becomes concentrated by greening roofs and by breaking runoff generating surfaces into smaller, manageable drainage areas, reducing the length of slope as a component of the equation to calculate runoff.

A component of planning that is being given new emphasis by the European Community is stakeholder participation, being particularly called for in the UN/ECE Best Practice Document (UN/ECE 2003), the UN/ECE Guidelines for Sustainable Flood Prevention (UN/ECE 2000) and the Water Framework Directives (WFD 2000). Public participation in the past usually was an after-the-fact activity where the public was informed when a plan had already been formulated. Stakeholder participation now involves the definition of stakeholder groups to not only include level (A) community groups, but also level (B) property managers and level (C) government

Figure 44.5 The University of Warsaw library has become a landmark of the city (*photograph:* ZinCo GmbH).

agencies to function as teams (Tourbier and Ashley 2007) to be involved throughout the course of a project. An early form of such cooperation was the involvement of the Dresden University of Technology (Germany), with the University of Warsaw (Poland), developing options for greening the 6,400 square meter roof of the university library to be built in the university's botanical garden. On invitation by the director of the botanical garden the chair of Landscape Construction and ten students from the TU Dresden developed five options for greening the roof and in 1995 conducted a workshop in Warsaw to involve city planners, university administrators, and interest groups. A design was accepted and today the greened roof of the library (Figure 44.5) has become one of Warsaw's landmarks.

Examples of installations

There are many international examples of implementation. A large expanse of VRC is the Ford Motor Company River Rouge Plant in Dearborn, Michigan (Klinkenborg 2009). Here plants cover 42,000 m^2 of the assembly plant roof. The Green Roof Research Program of Michigan State University monitored the installation (2009). With the help of Roofscapes Inc. the Chicago City Hall building has been retrofitted with 2,450 m^2 of VRC to demonstrate that green roofs help to reduce urban air temperatures. In the United Kingdom Rolls-Royce Motor Cars have one of the largest VRCs in Europe. Motivation here was to blend buildings into the surrounding countryside. In France a new museum, the L'Historial de la Vendée at Les Lucs-sur-Boulogne has a VRC of about 8,000 m^2. In the year 2006 alone about 14.5 million square meters of green roof were installed in Germany (Tourbier 2007). In the United States the Green Building Council has been formed. It offers the "Leadership in Energy and Environmental Design (LEED)" as a green building rating system to encourage and accelerate global development of sustainable green building and development practices (USGBC 2009). There is no question that, at a time of climate change, the greening of buildings has become a new global directive.

References

Commission of the EC (2006) *Proposal for a Directive of the European Parliament and of the Council on the Assessment and Management of Floods*, Brussels: European Commission.

EPA (1983) *Results of the Nationwide Urban Runoff Program, Volume 1-Final Report*, Washington DC: U.S. Environmental Protection Agency, Water Planning Division.

GRHC (Green Roofs for Healthy Cities) (2009) *Green Roofs for Healthy Cities North America*, Toronto, ON: GRHC, online, available at www.greenroofs.org [accessed February 10 2009].

Kaiser, M. (2005) "Near Natural Stormwater Management as a Model in the Sustainable Development of Settlements," in J.T. Tourbier and J. Schanze (eds.), *Urban River Rehabilitation Proceedings – International Conference on Urban River Rehabilitation*. Dresden: Technische Universität Dresden and Leibniz Institute of Ecological and Regional Development.

Klinkenborg, V. (2009) "Up on the Roof: A Lofty Idea is Blossoming in Cities around the World, where Acres of Potential Green Space Lie Overhead," *National Geographic Magazine*, May, online, available at: http://ngm.nationalgeographic.com/2009/05/green-roofs/klinkenborg-text.

Liesecke, H.-J., Krupka, B., Lösken, G., and Brüggemann, H. (1989) *Fundamentals of Roof-Greening (Grundlagen der Dachbegruenung)*, Berlin, Germany: Platzer Verlag.

Maryland (2008) *The Stormwater Management Act of 2007 (Act), Proposed Redevelopment Policy, Draft – July 25, 2008*, Baltimore, MD: Department of the Environment.

Michigan State University (2009) *The Green Roof Research Program at Michigan State University*, Department of Horticulture, online available at: www.hrt.msu.edu/greenroof [accessed February 10, 2009].

Miller, C. (1998) *Vegetated Roof Covers: A New Method for Controlling Runoff in Urbanized Areas*, Pennsylvania Stormwater Management Symposium, October 21–22, 1998, Villanova, PA: Villanova University.

Miller, C. (2008) *Impact of Local Stormwater Management Regulations, Preliminary Draft*, Philadelphia: Roofscapes, Inc.

Miller, C. (2009) Personal Communication, January, 2009.

Sieker, F. (2008) *Ingenieurgesellschaft Prof. Dr. Sieker mbH, Festlegung bundeseinheitlicher Regelungen für das Regenabwasser (FKZ 206 26 301)*, Dessau-Roßlau: Umweltbundesamt.

Tourbier, J.T. (1980) *Stormwater Management Alternatives: Papers concerning "Blue-Green Technology" turning stormwater from a liability into an asset*, Newark, DE: University of Delaware, Water Resources Center.

Tourbier, J.T. (2001) "Indicators of Success for Ecologically Sustainable Management of Stormwater," *Wissenschaftliche Zeitschrift der TU Dresden*, Spring: 104–10.

Tourbier, J.T. (2003) *Blue Green Technologies: Integrated practices to manage stormwater as an asset*, Geraldine R. Dodge Foundation, USA, Madison, NJ: Great Swamp Watershed Association.

Tourbier, J.T. (2007) "Developments in Environmental Landscapes in the EU," paper presented at the Annual Green Roof Seminar – Roofscapes, Philadelphia, PA, March 19, 2007.

Tourbier, J.T. and Ashley, R. (2007) "Integrative Teaching and Learning Modules for the Application of On-Site Urban Flood Resilience Planning," in E. Pasche (ed.), *Special Aspects of Urban Flood Management: Hamburger Wasserbau Schriften*, Hamburg: TU Hamburg Harburg, pp. 85–102.

Tourbier, J.T. and Westmacott, R. (1981) *Water Resources Protection Technology: A Handbook of Measures to Protect Water Resources in Land Development*, Washington, DC: Urban Land Institute.

UN/ECE (2000) *Guidelines for Sustainable Flood Prevention, 2000*, Athens: United Nations and Economic Commission for Europe.

UN/ECE (2003) *Best Practice for Flood Prevention, Protection and Mitigation*, United Nations and Economic Commission for Europe, Best Practice Document, Athens: UN/ECE.

USGBC (US Green Building Council) (2009) *website*, online available at: www.usgbc.org [accessed January 4, 2009].

WFD (2000) *Directive 2000/60/EC of the European Parliament and of the Council of 23 October 2000 Establishing a Framework for Community Action in the Field of Water Policy*, Official Journal of the European Communities, L 327 (December 22, 2000), pp. 1–73.

Wikipedia (2009a) "Babylon," online available at: http://en.wikipedia.org/wiki/Hanging_Gardens_of_Babylon [accessed February 10, 2009].

Wikipedia (2009b) "Green roof," online available at: http://en.wikipedia.org/wiki/Green_Roof [accessed February 10, 2009].

ZinCo GmbH (n.d.) *Extensive Dachbegrünung mit System* Unterensingen: ZinCo-GmbH, online, available at: www.zinco.de/downloads/planungshilfen_pdfs/Extensive_Dachbegruenung.pdf [accessed September 20, 2010].

45

The role of green infrastructure in adapting cities to climate change

Ian Douglas

Adaptation

The words adaptation and mitigation have specific meanings in the context of climate change through their use in the United Nations Framework Convention on Climate Change (UNFCC). Adaptation refers to the 'adjustments that are possible in practices, processes, or structures of systems to projected or actual changes of climate. Adaptation can be spontaneous or planned, and can be carried out in response or in anticipation of change in climate conditions' (Watson *et al.* 1996: 863). Mitigation is defined as 'an anthropogenic intervention to reduce the emissions or enhance the sinks of greenhouse gases' (Watson *et al.* 1996: 869).

Adaptation has sometimes been used as a synonym for 'human adjustment' in the way people respond to the environmental hazards (Burton 2002). This is indeed the sense in which it applies to societal responses to climate change, the conscious altering of behaviour or our immediate environment to cope with changing conditions.

One of the ways in which cities can adapt to climate change is to modify the intensity of the urban heat island by using trees and vegetation to shade the surface and provide evaporative cooling (Voogt 2002). However, the great natural contrasts in the climatic settings of the world's urban areas should always be borne in mind. Although since 1950 the world's cities have tended to look more and more like one another, with similar architecture, similar transportation and even similar clothing styles, the use of vegetation to adapt to climate change has to be locally specific to the regional climate and the intensity of the heat island effects in individual cities. Reliance on heating or cooling to keep building interiors at similar temperatures throughout the affluent parts of cities everywhere is a major source of greenhouse gas emissions. Reduction of these emissions can be a major form of urban adaptation to climate change. The actual strategy for emission reduction will vary from one city to another.

Perceived role of the urban forest and urban greening

Urban trees can be aesthetically pleasing and also help to reduce air pollution and heat island effects, altering local meteorological conditions and providing shelter from wind.

Most cities would benefit from an expansion of their tree cover from a few parks and street trees into an urban forest that would provide a denser canopy and thus energy savings from shade in the summer and protection from wind in winter (White 2002). Trees help to reduce the adverse impacts that urbanization has on the hydrological cycle and also modify the intensity of the urban heat island. Additional tree cover could reduce stormwater run off by 4 to 18 per cent depending on the situation, soils and topography of a city. This reduction would also help reduce stream pollution as the pollutants end up being held in the soils and deposits beneath the vegetation rather than being carried into watercourses (Bass 2002). Where higher urban temperatures accelerate the formation of ozone, urban forestry can help reduce the severity of smog episodes. A reduction of the urban heat island intensity by 2 to 4°C in the Los Angeles area could have the same effects as converting half the local vehicle fleet to zero emissions (Bass 2002).

Even small green spaces, like neighbourhood parks, can have important effects on the urban climate (Heidt and Neef 2008). Inner-city green spaces are particularly helpful in improving air quality through the trapping of pollutants. In small parks, the amount, kind and ratio of trees and shrubs are important. A 'protection plantation' consisting of rows of trees forming high hedges, with shorter bushes between the rows, is more efficient at filtering out air particulates than a forest of the same size consisting only of trees. A small park with both trees and shrubs may bind up to $68\,t\,ha^{-1}\,yr^{-1}$ of dust. Even small numbers of trees in high-density neighbourhoods can help to decrease the amount of dust in the air. Trees in a street cause local minor air turbulence that helps to disperse pollutants and so reduce the risk of inversions and smog. Greenspaces of 50 to 100 m depth improve air quality for up to 300 m from their outer edges (Meyer 1997).

Green roofs also contribute to urban climate modification (see Chapters 43 and 44 this volume) and to adaptation of individual buildings to climate change. During warm weather, green roofs reduce the amount of heat transferred through the roof, thereby lowering the energy demands of the building's cooling system. The heat transfer through a green roof in Singapore over a typical day was less than 10 per cent of that of a reference roof. Research in Japan found reductions in heat flux of the order of 50 per cent per year, and work in Ottawa found a 95 per cent reduction in annual heat gain (Oberndorfer *et al.* 2008).

Yet, trees can introduce new hazards. They can accelerate desiccation of clay soils, leading to shrinkage and subsidence. They can also add to property damage during high winds (Scott *et al.* 1996). The likelihood of higher and more frequent insurance claims for property damage from high winds under the increased storminess associated with climate change is a cost that will require national planning and new integrated financial arrangements (Dlugolecki 1996). In Montreal, Canada, many ice-laden trees fall over or suffer broken branches, blocking roads and damaging property. In Toronto, Canada, trees provide more habitat for raccoons, a pest that is likely to become a host for rabies infestations (White 2002). In areas of dry, hot summers, such as the Mediterranean or south-western USA, urban forests provide tinder for wildfires whose frequency has increased since 1990.

The concept of green infrastructure

The green infrastructure is 'an interconnected network of green space that conserves natural ecosystem values and functions and provides associated benefits to human populations' (Benedict and McMahon 2002: 12). The green infrastructure should operate at all spatial scales from urban centres to the surrounding countryside (URBED 2004).

Many regard greenspace as a crucial component of urban landscapes, not least for countering the urban heat island, reducing flood risk, improving air quality and promoting habitat availability and connectivity (Wilby 2007). Gardens contribute to this greenspace significantly, but their performance as habitat is, like that of other vegetated areas, susceptible to climate change.

Modelling the climate change adaptation benefits of urban greenspace

The Urban Forest Effects model (UFORE) quantifies species composition and diversity, diameter distribution, tree density and health, leaf area, leaf biomass and other structural characteristics (Nowak and Crane 2000). It also calculates hourly urban forest Volatile Organic Compound (VOC) emissions (emissions that contribute to ozone formation), total carbon stored and net carbon sequestered annually by urban trees, and hourly pollution removal by the urban forest and associated percentage improvement in air quality. The model requires four sets of data: field, tree cover, meteorological and pollution. The field data include land use, ground and tree cover, building attributes, individual tree species, stem diameter at breast height, tree height, crown width and dieback percentage. In the USA, digital hourly meteorological data from the National Climatic Data Center and hourly pollution concentration data from the Environment Protection Agency are used in calculating emissions and pollutant removal.

In 1994, trees in New York City removed an estimated 1,821 metric tons of air pollution at an estimated value to society of $9.5 million (Nowak and Crane 2000). Air pollution removal by urban forests in New York was greater than in Atlanta and Baltimore, but pollution removal per m^2 of canopy cover was fairly similar among these cities (New York: 13.7 $g\,m^{-2}yr^{-1}$; Baltimore: 12.2 $g\,m^{-2}yr^{-1}$; Atlanta: 10.6 $g\,m^{-2}yr^{-1}$). In terms of the impact of the urban forest on the carbon cycle, carbon storage in the Brooklyn, NY, estimated at 172,4000 t (9.4 $t\,ha^{-1}$) urban forest is equivalent to the amount of carbon emitted by the borough's population in about five days as assessed from average per capita emission rates (Nowak *et al.* 2002). Thus urban trees have a small impact on mitigating climate change by sequestering carbon dioxide. Rackham (2006) notes that exhorting people to plant trees to sequester carbon is like telling them to drink more to hold down rising sea level. The value of trees is in helping people in cities to adapt through their ability to shade, cool and reduce air pollution.

Modelling work based on data for Greater Manchester, UK (Gill *et al.* 2007), suggests that the use of urban greenspace offers significant potential for moderating the increase in summer temperatures expected with climate change. Adding 10 per cent green in high-density residential areas and town centres kept maximum surface temperatures at or below 1961–1990 baseline levels up to, but not including, the predicted conditions of the 2080s. Greening roofs in areas with a high proportion of buildings, for example in town centres, manufacturing, high-density residential, distribution and storage, and retail areas, also appeared to be an effective strategy to keep surface temperatures below the baseline level for all time periods and emissions scenarios.

On the other hand, the modelling work highlights the dangers of removing green from the conurbation. For example, if green cover in high-density residential areas and town centres is reduced by 10 per cent, surface temperatures will be 7°C or 8.2°C warmer by the 2080s High in each, when compared to the 1961–1990 baseline case; or 3.3°C and 3.9°C when compared to the 2080s High case where green cover stays the same.

Action on urban greenspace for adaptation to climate change

Developers and design teams are already encouraged to incorporate greenspace in their planning and to use shading and green roofs to help reduce the urban heat island effect. The underlying assumption is that urban people will want to spend more time outdoors in natural spaces as temperatures rise (Wilby 2007). In pursuing greenspace policies, managers must consider the net effect of tree growth on atmospheric CO_2. Nearly all the carbon sequestered will eventually be converted to CO_2 when the trees decompose. As a result, the benefits of carbon sequestration will be relatively short-lived if the forest structure is not maintained. Looking after the forest structure is likely to involve the regular use of fossil fuels, which in some circumstances, depending on frequency and extent of maintenance operations, could lead to the emission of more carbon than is sequestered by the trees (Nowak *et al.* 2002). Another issue is water use to sustain urban greenspace. In arid climates, where increasing shade is a great benefit, increased water demand may counter other adaptation strategies aimed at reducing water consumption (Grimm *et al.* 2008). A third consideration is increased fire risk associated with urban and peri-urban vegetation, especially under higher temperatures and greater drought risk. In discussing adaptation to climate change in Cape Town, South Africa, the removal of pine plantations was considered a possible strategy for reducing fire risk, especially where climate change would make them less productive (Mukheiber and Ziervogel 2007). The use of greenspace to adapt to climate change has to be appropriate to local conditions, both biophysical and socio-cultural. Nevertheless, for many cities extension of the urban vegetation cover brings major benefits.

Thus, one possible adaptation strategy to increasing temperatures is to preserve existing areas of greenspace and to enhance it where possible, whether in private gardens, public spaces or streets. For example, in areas where social housing in being redeveloped or where brownfield sites are being prepared for new dwellings, significant new greenspaces should be created.

However, in many existing urban areas where the built form is already established, it is not feasible to create large new greenspaces. Thus, greenspace will have to be added creatively by making the most of all opportunities, for example through the greening of roofs, building façades and railway lines, street tree planting, and converting selected streets into greenways. Priority should be given to areas where the vulnerability of the population is highest. A study in Merseyside found that vegetation, and in particular tree cover, is lower in residential areas with higher levels of socioeconomic deprivation (Pauleit *et al.* 2005). The socio-economic deprivation index used included variables relating to health deprivation. Such populations will therefore be more vulnerable to the impacts of climate change. One caveat to the potential of green cover in moderating surface temperatures is the case of a drought, when grass dries out and loses its evaporative cooling function. Output from the daily weather generator used suggests that with climate change there will be more consecutive dry days and heatwaves of longer duration in summer (BETWIXT 2005; Watts *et al.* 2004a, 2004b).

One possible adaptation strategy for temperate latitudes facing drier summers would be drought-resistant plantings. In the UK this would involve planting vegetation, such as trees, that is less sensitive than grasslands to drought. Trees are common in open spaces in the Mediterranean. Tree species which are less sensitive to drought can be chosen from warmer parts of the temperate zones, such that they will still evapotranspire and provide shade. Site conditions for trees in streets may need improving so that there is sufficient rooting space for the trees. In addition, irrigation measures must be considered to ensure that they have an adequate water supply. This could be through rainwater harvesting, the re-use of greywater, making use of water in rising aquifers under cities where present, and floodwater storage. Unless adequate provision is made there will be conflict as greenspace will require irrigating at the same time as water supplies are low and restrictions may be placed on its use. Ironically, measures which are currently in hand to reduce leakage in the water supply system may reduce available water for street trees which are critically important for human comfort in the public realm (Gill *et al.* 2007).

Conclusion: the need for a holistic approach

Adaptation to climate change is part of the package of ecosystem services provided by the urban green infrastructure. It is worth remembering three propositions put forward by Bennett *et al.* (2009):

1 Relationships among multiple ecosystems services are better identified and assessed by integrated social-ecological approaches than with either social or ecological data alone.
2 Understanding the mechanisms behind simultaneous response of multiple services to a driver and those behind interactions among ecosystem services can help identify ecological leverage points where small management investments can yields substantial benefits.
3 Managing relationships among ecosystem services can strengthen ecosystem resilience, enhance the provision of multiple services, and help avoid catastrophic shifts in ecosystem service provision (Bennett *et al.* 2009: 1398–9).

In everything to do with using urban greenspace and the science of urban ecology to help urban areas adapt to climate change, it has to be remembered that:

> rather than being a technical issue, of the need for more information or better practice ... the interpretation and implementation of climate protection locally is a political issue, where different actors and groups seek to have their understanding of the problem, and its solutions, acted upon.
>
> *(Bulkeley and Betsill 2003: 185)*

References

Bass, B. (2002) 'Greening of cities', in I. Douglas (ed.) *Causes and Consequences of Global Environmental Change: Volume 3 of Encyclopaedia of Global Environmental Change*, Chichester: Wiley, pp. 356–62.

Benedict, M.A. and McMahon, E.T. (2002) 'Green infrastructure: smart conservation for the 21st century', *Renewable Resources Journal*, 20(3): 12–17.

Bennett, E.M., Peterson, G.D. and Gordon, L.J. (2009) 'Understanding relationships among multiple ecosystem services', *Ecology Letters*, 12(12): 1394–404.

BETWIXT (2005) *Built Environment: Weather scenarios for investigation of Impacts and eXTremes. Daily time-series output and figures from the CRU weather generator*, online, available at: www.cru.uea.ac.uk/cru/projects/betwixt/cruwg_daily/ [accessed 4 April 2010].

Bulkeley, H. and Betsill, M.M. (2003) *Cities and Climate Change: Urban Sustainability and Global Environmental Governance*, London: Routledge.

Burton, I. (2002) 'Adaptation strategies', in M.K. Tolba (ed.), *Responding to Global Environmental Change: Volume 4 of Encyclopaedia of Global Environmental Change*, Chichester: Wiley, pp. 80–5.

Dlugolecki, A.F., Clark, K.M., Knecht, F., McCaulay, D., Palutikov, J.P. and Yambi, W. (1996) 'Financial services', in R.T. Watson, M.C. Zinyowera and R.H. Moss (eds), *Climate Change 1995: Impacts, Adaptations and Mitigation of Climate Change: Scientific-Technical Analyses*, Cambridge: Cambridge University Press, pp. 539–60.

Gill, S., Handley, J., Ennos, R. and Pauleit, S (2007) 'Adapting cities for climate change: the role of the green infrastructure', *Built Environment*, 3(1): 115–33.

Grimm, N.B., Faeth, S.H., Golubiewski, N.E., Redman, C.L., Wu, J., Bai, X. and Briggs. J.M. (2008) 'Global change and the ecology of cities', *Science*, 319: 756–60.

Heidt, V. and Neef, M. (2008) 'Benefits of urban green space for improving urban climate', in M.M. Carreiro, Y.-C. Song and J. Wu (eds), *Ecology, Planning, and Management of Urban Forests: International Perspectives*, New York, NY: Springer, pp. 84–96.

Meyer, J. (1997) *Die zukunftsfaehige Stadt. Nachhaltige Entwicklung in Stadt und Land*, Düsseldorf, Germany: Werner Verlag.

Mukheiber, P. and Ziervogel, G. (2007) 'Developing a municipal adaptation Plan (MAP) for climate change: the city of Cape Town', *Environment & Urbanization*, 19(1): 143–58.

Nowak, D.J. and Crane, D.E. (2000) 'The Urban Forest Effects (UFORE) Model: Quantifying urban forest structure and functions', in M. Hansen and T. Burk (eds), *Integrated Tools for Natural Resources Inventories in the 21st Century: Proceedings of the IUFRO Conference, General Technical Report NC-212*, St Paul, MN: US Department of Agriculture, Forest Service, North Central Research Station, pp. 714–20.

Nowak, D.J., Crane, D.E., Stevens, J.C. and Ibarra, M. (2002) *Brooklyn's Urban Forest General Technical Report*

NE-290, Newton Square, PA: US Department of Agriculture, Forest Service, Northeastern Research Station.

Oberndorfer, E., Lundholm, J., Bass, B., Coffman, R.R., Doshi, H., Dunnett, N., Gaffin, S., Köhler, M. Liu, K.K.Y. and Rowe, R. (2008) 'Green roofs as urban ecosystems: ecological structures, functions, and services', *BioScience*, 57(10): 823–32.

Pauleit, S., Ennos, R. and Golding, Y. (2005) 'Modeling the environmental impacts of urban land use and land cover change – a study in Merseyside, UK', *Landscape and Urban Planning*, 71(2–4): 295–310.

Rackham, O. (2006) *Woodlands*, London: Collins.

Scott, M.J., Aguilar, A.G., Douglas, I., Epstein, P.R., Liverman, D., Mailu, G.M., Shove, E., Dlugolecki, A.F., Hanaki, K., Huang, Y.J., Magadza, C.H.D., Olivier, J.G.J., Parikh, J., Peries, T.H.R., Skea, J. and Yoshino, M. (1996) 'Human settlements in a changing climate: impacts and adaptation', in R.T. Watson, M.C. Zinyowera and R.H. Moss (eds), *Climate Change 1995: Impacts, Adaptations and Mitigation of Climate Change: Scientific-Technical Analyses*, Cambridge: Cambridge University Press, pp. 399–426.

URBED (2004) *Biodiversity by Design – A Guide for Sustainable Communities*, London: Town and Country Planning Association.

Voogt, J.A. (2002) 'Urban heat island' in I. Douglas (ed.) *Causes and Consequences of Global Environmental Change: Volume 3 of Encyclopaedia of Global Environmental Change*, Chichester: Wiley, pp. 660–6.

Watson, R.T., Zinyowera, M.C. and Moss, R.H. (eds) (1996) *Climate Change 1995: Impacts, Adaptations and Mitigation of Climate Change: Scientific-Technical Analyses*, Cambridge: Cambridge University Press.

Watts, M., Goodess, C.M. and Jones, P.D. (2004a) *The CRU Daily Weather Generator*, Norwich: Climatic Research Unit, University of East Anglia.

Watts, M., Goodess, C.M. and Jones, P.D. (2004b) *Validation of the CRU Daily Weather Generator*, Norwich: Climatic Research Unit, University of East Anglia.

White, R.R. (2002) 'Sustainable city policies', in M.K. Tolba (ed.) *Responding to Global Environmental Change: Volume 4 of Encyclopaedia of Global Environmental Change*, Chichester: Wiley, pp. 406–11.

Wilby, R.L. (2007) 'A review of climate change impacts on the built environment', *Built Environment*, 33(1): 31–45.

46

Creative use of therapeutic green spaces

Ambra Burls

Introduction

There are numerous green spaces designated as 'therapeutic' in Europe and the UK (healing gardens, care farms, ecotherapeutic projects). This perspective veers people into a vision of these spaces as having a specific and possibly 'medicalised', single function. By so doing this reduces or limits the potential multifunctionality inherent in these 'special' green spaces. Much more creative scope exists to broaden the functional perspective of therapeutic green spaces, based on synthesising their intrinsic parallel uses. Further exploration of these uses leads to synergies and connections with:

- Access to and maintenance of urban green spaces by citizens for citizens, aiding the formation of urban 'green grids' and healthy biospheres
- Sustainability education, multicultural integration and interdisciplinary networks
- Stigma abating, safety and community identity
- Citizenship, stewardship and local networks
- Public health outcomes through active assessment of the health impact on the local community (Health Impact Assessments – HIA)

This chapter will consider the extended functionality of therapeutic urban green spaces, based on some good practice 'showcase' examples.

There is a general assumption that therapeutic green spaces, often known as 'healing gardens' or 'therapeutic landscapes', are spaces designed for specific clinical or rehabilitative functions, for people with disabilities or ill-health. This is the case for some of the evidence-based designed green spaces, which are usually associated with healthcare settings (AHTA 2007). They are often seen as a retreat, a place of respite and usually only accessible to those people who are considered to be eligible to use them by virtue of their role as a patient, visitor or staff member. Where such spaces are not located within healthcare settings, but are nevertheless designated as therapeutic, they unfortunately often become associated with the 'separateness' of the above. Whether a care farm or an urban green space or any other type of 'therapeutic green space', they run the risk of being perceived by the general public and, alas, many specialist disciplines or agencies, as spaces reserved for those who are referred there by virtue of their physical or psychological conditions. Many people also believe that these spaces are awash with medicinal plants or that the plants have to have a specific symbolism and sensory quality held in their scent, texture or aesthetic qualities. All of these factors can be expressly true and are valuable from the point of view of their specific therapeutic value; however there is much broader scope for considering that therapeutic values can co-exist with other and more multifunctional values. These spaces are also spaces where the 'feel good factor' can be achieved by all in the local community. The key in unearthing the broader characteristics and uses of these green spaces is in eschewing the conventional view and embracing the more holistic and broader social facets inherent in them. The East London Green Grid Primer Paper (GLA 2006: 11) illustrates 'multifunctionality', asserting that:

> Green infrastructure projects can deliver multiple objectives: they can frame and shape the growth of sustainable communities, to strengthen their image and identity; they help cities to adapt to climate change by reducing flood risk and overheating; they promote access to open space, nature, culture and sport, improving the offer to visitors and quality of life for all.

Based on living examples of urban green spaces, chiefly designated as therapeutic, this chapter reveals how many more creative outcomes of multifunctionality can derive from the activities carried out within these spaces.

Access and maintenance of urban green spaces

Accessibility in urban design is an element of one of the core themes of the World Health Organization (WHO) Phase V (2009–2013) (WHO 2009a). To make urban green spaces accessible to citizens can enhance cultural interchange, lead to increased safety and encourage active living. Access to green spaces in cities has been found to benefit health and well-being and to enable residents to better manage the stress often experienced in large urban areas (Fuller and Gaston 2009). One therapeutic green space in London (MIND Meanwhile Wildlife Garden in Kensington and Chelsea) has become a 'showcase' of good practice in integrating the maintenance of, and access to, a piece of urban green space, which is public, open and dedicated specifically to the provision of wildlife habitat. This 'social enterprise' works closely with the local borough council to maintain access, thus the participants, as citizens, provide fellow citizens with an area for their enjoyment, socialising, active living and restoration. At the same time there are therapeutic outcomes for the project participants, who in this specific case have mental health problems. Through vocational rehabilitation and targeted training the people involved achieve personal goals such as skills development, mainstream environment related qualifications, reintegration into a work environment, increased self-esteem, and employability. They also progress along a continuum from ill-health to well-being in physical, psychological and social terms (Burls 2005, 2007a, 2007b). The added value in this multifunctional backdrop is in the fact that they essentially provide their community with a 'triple bottom line' service (Elkington 1994) where they, as 'stakeholders', are engaged in obtaining an additional continuum of values of societal success in economic, ecological and social terms. They are vehicles for the interdependent interests of human capital, natural capital and economic capital, based on the conscientious and sustainable framework of the 'social enterprise'. This philosophy does not endanger or exploit any of the stakeholders, but engenders safety, security and cohesiveness within the community.

In terms of urban ecology the clear outcomes from therapeutic green spaces are almost always overlooked by many policymakers across disciplines. Preconceived and fixed ideas about the scope of such spaces are one of the most insidious reasons for the wrongful dismissal of the multifunctional value of therapeutic green spaces. The UK agency Natural England is engaged in providing citizens with access to 'green space' whether it is a local park, a wildlife garden or even an area of woodland close to where they live (RUDI 2009). In examining these objectives alongside the outcomes of the London therapeutic project, there is clear indication that it meets the following Natural England targets:

- provides a greenspace within 300 m of a large number of the borough homes;
- supports an increase in priority species and habitats in an urban area which is otherwise highly populated;
- provides a wild area and an open space to meet the needs of both nature and people;
- helps to cope with the effects of climate change and extreme weather events;
- is designed to ensure it fits into its landscape setting.

Moreover the Greater London Authority's document (GLA) (2006) describing the 'Green Grid' concept aims to provide 'residents and workers with a multi-functional network of strategic open space and in turn improved quality of life'. The Authority foresaw the creation of 'new public spaces, the enhancement of existing open spaces and improvements to the links in between'. The aims of the Green Grid are to 'create a network of interlinked, multi-functional and high quality open spaces'. Given that the London project discussed here works to maintain a small portion of a bigger town park, adjoining the Grand Union Canal, then it also meets the scope of the Green Grid concept. It is part of 'a living network of open spaces, river and other corridors connecting urban areas'. The London project therefore meets some of the main aims of the Green Grid by:

- providing and enhancing an existing public open space;
- providing public access along waterways and green areas;
- providing informal recreational uses and landscape, promoting healthy living;
- providing and enhancing existing wildlife sites;
- providing beautiful, diverse and managed green infrastructure to the highest standards for people and wildlife;
- responding to the dual drivers of climate change and future development.

By far the most important issue in this is that the above provisions are not the fruit of a statutory service, provided by specialised professionals, but a spin-off from people with disabilities, seeking to improve their well-being and fostering recovery from their conditions. The public is influenced either directly or indirectly by their actions and example and, as this happens, they develop a strong wish to be directly involved in changing attitudes and educating the visiting public about the benefits of access to urban green spaces. Citizens who would otherwise be 'socially silent' instead directly encourage other citizens to be involved in using and running healthy urban green spaces.

Sustainability education, multicultural integration and interdisciplinary networks

Within the integrative and participatory processes described above are several important factors in the interface of potentially marginalised groups and their community, particularly the direct outcomes from raising public awareness of the value of urban wildlife. This comes not by the use of traditional, static interpretation boards often found in deserted or uncongenial 'wild' urban green spaces, but, as at Meanwhile Wildlife Garden by having interactive interpretation or 'interpretation people' (Scoffield 2009). The daily presence of green space 'stewards', working for wildlife and their verbal transactions with the public foster dialogue, interest, and a wish by the public to be concerned in safeguarding this local asset. City parks and urban green spaces have positive benefits on the health and well-being of people (See Chapters 32 and 33 this volume), economic revival and reduction in crime and antisocial behaviour, however people are not always au fait with the intrinsic value of wildlife, the buffer effect of mature native trees on local climate, the reduction in stress levels for people who come into contact with wildlife in urban green spaces. People generally know that being in the fresh air is beneficial, but this is often an abstract notion; these 'interpretation people' can be the source of the information which will crystallise that notion and create 'a significant experience of nature' (GLA 2006) for the general public. This is essentially sustainability (or environmental literacy) education and is done by citizens for citizens, in an informal way, in situ and with evidence at hand. They as 'interpretation people' are the living examples of personal involvement with the community and of the benefits of using and caring for wildlife in urban green spaces. They are a functional resource to redress the balance of sustainable living, working with the natural systems that support us (Baines 2006). They are important role models not only as ordinary citizens, but moreover as people who are normally seen as disadvantaged by society at large and here they are demonstrating their own self-driven social inclusion and their eagerness to contribute to their community's quality of life, defying their 'disadvantaged' label.

There are strong cultural connections between the green environment and people; for suburban people marginalised by typical 'minority difficulties', whether ethnic or otherwise, access to wildlife and natural green spaces can be a strong channel for reconnection and integration. The existence of a kinship represented in those 'interpretation people' who themselves have experience of marginalisation, through their disability, acts as a direct outreach strategy. Their stories and information generate a kind of mutual exposure which validates commonalities and which creates an atmosphere of awareness for different cultural contexts and social circumstances of exclusion. Such comparative learning creates an awareness of diversity and facilitates dialogue (Wong 2004). This means that those 'interpretation people' are, in practice, playing a key role. They are a powerful source of reciprocal insights, thus breaking down barriers, encouraging the active participation and social integration of minority or disadvantaged groups in what is essentially their own community, their green spaces and local wildlife. This in turn can reduce or even eliminate their social exclusion.

In our world of complexity, the interconnectedness between individuals, groups and communities can be a powerful source of solutions to many problems. Synthesis thinking (Brown 2007) is a strategy that moves away from typical disciplinary boundaries, towards interdisciplinary skills of 'collective thinking'. Brown (2007: 187) states: 'Anyone seeking to guide social change will find a most powerful ally in a synthesis of the decision-making knowledge cultures and the combined answers to commonsense question'.

Brown's model of 'nested knowledge' outlines how the 'constructions of reality' can derive from the knowledge of individuals. 'Individual knowledge' extends to intersect with other types of knowledge. The latter, often attributed to traditional experts in the fields of science, technology and research, seems to often remain in disciplinary 'silos' (Cunningham 2006). However Brown indicates that 'holistic knowledge' is at 'the core' of 'nested knowledge' and should take account of the often tacit lay knowledge, derived from lived experiences. Therefore true interdisciplinarity should include such lay and tacit knowledge. In similar ways the Expert Patients Programmes (Department of Health UK 2007), now common in many world health services, are based on individuals developing the confidence and motivation to use their own skills, information and experience to take effective control over living with their disabilities and their lifestyles. Thus they become proactive in aiding services to respond appropriately to individuals or groups with specific problems. This generic lay-led approach is a vast resource of problem-solving and coping skills which individuals can share with others. It goes without saying, therefore, that those 'interpretation people' working in therapeutic green spaces such as Meanwhile Wildlife Garden, can also be an invaluable source of knowledge, through their shared lived experiences, for many of those disciplines separately or collectively concerned with the interface of the environment and health. The report *Collaborating across the Sectors* (Cunningham 2006) highlights how disciplinary 'silos' are a great impediment to society, the economy and the environment and that those who are disciplinary 'boundary spanners' are often not fully heard, yet 'they are the key to some of the most promising research'. An example of such 'boundary spanners' is in the participants (both service users and practitioners) working in the London project; they can lead to cross-disciplinary innovation such as that evidenced in recent research on ecotherapy (Burls 2007a, 2007b, 2008). Their tacit knowledge and lived experiences acted as the principal research resource in the development of new higher education curricula in ecotherapy for practitioners, which have now come to fruition in the UK. The multifunctional value of therapeutic green spaces can therefore metaphorically and literally reconnect those interdisciplinary, cultural and public education links which are generally separated by illusory and socially derived distinctions.

Stigma abatement, safety and community identity

'Disadvantaged' people using therapeutic green spaces can be role models, a source of knowledge and lay educators, opening up channels of communication with the general public about the environment and the benefits of embracing nature from the backdrop of their own experience. The kinship of 'disadvantage' can be a powerful driver for ethnic minorities or other marginalised individuals to become involved in their community, enjoy wildlife and take advantage of their local green spaces. These tacit or explicit connections have an added social value in buffering the effect of stigma. People with disability or who are socially disadvantaged, directly involved as 'stewards' in maintaining, caring and providing access to a healthy urban green space for the general public, change the perception of the public towards their disability. Furthermore the public seems to react positively to relating their green space to the 'real' people who work there. Their green space is no longer an impersonal piece of 'council property'; it becomes '*their* green space'. Communal activities then soon develop and community identity starts to take shape in the green space which was previously perhaps only a place to stroll through. The green space becomes of symbolic value in time and space. This in itself is a great outcome, but this kind of social development not only abates stigma: it also reduces or even eliminates those problems of negative perception about green spaces by the public who may otherwise convey fear for personal safety, of vandalism and antisocial behaviour. Kaplan (Chapter 31 this volume) reports on the complexities of attitudes and management philosophies of public green space and how these have changed over time. Therapeutic green spaces may bring a fresh viewpoint on green space use and alleviate some of the concerns and prejudices the public hold about parks and accessible urban nature. These ordinary citizens, who use their green space as a therapeutic tool, are also the very providers of access to it, fostering a certain unspoken cohesiveness. Experience from the London project seems to indicate that even the most ill-mannered of individuals within the community seem to revert to being respectful, most of the time, out of a sense of solidarity, consideration and cohesion.

Citizenship, stewardship and local networks

There is explicit and direct reference to projects wherein 'working with nature' involves users' direct engagement and contribution to the design, management, restoration and maintenance of public green spaces (Burls and Caan 2005; Burls 2005, 2007a, 2007b, 2008). Some studies talk of the 'distinct conservation objectives' for therapeutic conservation projects (Hall 2004: 7) set to 'promote and assist in conservation to jointly benefit wildlife and people undergoing recovery'. It is evident that therapeutic green spaces have values that are sometimes hidden by flawed assumptions; values which would become much more obvious if one were to take a more holistic view. Their value can manifest through symbolic or physical expressions in response to the struggles people face within their social systems. Collective and individual struggles can unite and reconnect. Nature's response to the blunders of our society can have a certain metaphoric resonance with people who themselves may feel a sense of kinship with its struggle. Networks of people can grow from rekindling of collective memory, which locates a community in time and space and therefore gives it meaning (McCreary 2006).

Urban green spaces can be, and are, first and foremost social spaces, where people reconnect through the 'doing of things' in the absence of stereotypes predisposed by their social situation. When these local networks can be established at grassroots level, by community groups or even individuals, they portray a certain spontaneity and meaning which often have the power to reconstruct lost connections of people with their community, with their environment and even with the wider world.

Citizens with disabilities or who are disadvantaged, but motivated to enhance their own mental health and well-being, become stewards of their therapeutic green space and this, in turn, gives them the impetus and motivation to place themselves in a role of facilitators of holistic thinking through very direct and simple actions. The tending of a public urban green space for therapy, well-being and socialisation, brings them a sense of self-esteem and skills development which far outweighs their own personal health outcomes. They encourage others to become engaged and participate, to create local networks, to belong and develop a sense of place. They are the embodiment of citizenship and social capital (Burls and Caan 2004). Whilst providing a 'triple-bottom-line' service for their community, they level out social differences and prejudice or stigma and they are proactive in this, without the need for this onus to be on the traditional statutory bodies. This makes for *sustainable* local networks, which are *sustainably* healthy and which foster a collective approach to *whole systems sustainability*, from the grassroots.

Public health outcomes

Among the research asserting the public health benefits of urban green spaces, the European Public Health Alliance (EPHA) (2008) reports that 'access to green, open spaces is a determinant of public health and protects us from strokes, heart disease and perhaps even reduce stress'. They 'will have implications for the future of urban planning and could encourage councils and local authorities to introduce and protect green spaces in our cities'. This in turn would 'encourage people to be more active' (EPHA 2008). There is further evidence that viewing nature (Kaplan 2001; Kuo and Sullivan 2001; Ulrich 1984); being in the presence of nearby nature (Cooper Marcus and Barnes 1999; Hartig and Cooper Marcus 2006; Ulrich 1999); or active participation and involvement with nature (Reynolds 2002; Yerrell 2004; Frumkin 2001; Pretty *et al.* 2005) are all beneficial to our health in different ways. The public health white paper *Choosing Health* (Department of Health 2004: 79) refers to them as 'schemes that support people in gardening or local environmental improvement while providing opportunities for exercise and developing social networks'. As these environments are instrumental in promoting good health and reducing health inequalities, then it should be evident that green spaces are naturally therapeutic for the general public. However the presence of direct 'stewardship' by people such as the Meanwhile Garden project participants should strengthen this further. Their social enterprise approach should be valued by policymakers and urban planning authorities for contributing to the ecological and public health targets they meet.

These are but a few of the important links that can be drawn between mental, physical and community public health and green space. Professionals in both green space and health arenas have been considering the importance of sharing their experiences and expertise to develop mutually credible initiatives and policies. However there is a felt need for guidance on how to measure the health and equity impacts of green space so that it might begin to receive greater recognition in improving health and reducing health inequalities. Furthermore there are now visible and proficient groups of citizens such as the diversely able users of therapeutic green spaces who can directly participate in carrying out assessments of the benefits of these spaces for their community. As stakeholders they are in a powerful position to extend the evidence-base so that it becomes fully participatory and democratic, based on the 'nested knowledge' concept, which Val Brown (2007) alludes to in her 'synthesis thinking' approach to inter-disciplinarity. This concept brings into play a certain 'equity of knowledge' value which puts individual and lay knowledge in the mix of 'holistic knowledge' and holistic knowledge at the core of a synthesis of 'local community knowledge', 'specialised knowledge' and 'organisational knowledge'. This in turn produces 'collective and networked knowledge' which are 'nested' and which contribute to problem solving and the 'co-creation' of health for people and places. It can be envisaged therefore how those people who derive a direct benefit from therapeutic green spaces should be a source of holistic knowledge. They should in fact be encouraged to participate in a democratic and direct way to developing evidence based knowledge which can enhance and 'co-create' public health.

A straightforward approach to determining how much therapeutic green spaces contribute to the general public's health would be to systematically conduct health impact assessments (HIA) (WHO 2009b).

HIA are based on four values:

- '**Democracy** – allowing people to participate in the development and implementation of policies, programmes or projects that may impact on their lives.'
- '**Equity** – HIA assesses the distribution of impacts from a proposal on the whole population, with a particular reference to how the proposal will affect vulnerable people (in terms of age, gender, ethnic background and socio-economic status).'
- '**Sustainable development** – that both short and long term impacts are considered, along with the obvious and less obvious impacts.'
- '**Ethical use of evidence** – the best available quantitative and qualitative evidence must be identified and used in the assessment. A wide variety of evidence should be collected using the best possible methods.'

It is even more creative to use an HIA if it involves people with therapeutic needs, using and working in such green spaces as Meanwhile Gardens in London. They could be the very stakeholders whose involvement the WHO encourages. The WHO deems the use of HIAs important because they promote cross-sectoral working beyond the health sector and are a participatory approach that values the views of all sectors of the community. HIAs provide clear and transparent information for decision makers to engage with members of the public affected by a particular proposal. The views of the public can sit alongside other evidence from expert opinion and scientific data, with each being presented and valued equally in order to respond constructively to their concerns.

The HIA's aims are summarised as follows:

- It provides the best available evidence to decision makers, going beyond published reviews and research papers, to include the views and opinions of key players who are involved or affected by a proposal.
- It improves health and reduces inequalities by using a wider model of health, working across sectors to provide a systematic approach for assessing how the proposal affects a population; recommendations can specifically target improvement of health, particularly for vulnerable groups.
- It is a positive approach aiming to maximise potential health benefits.
- It is appropriate for policies, programmes and projects, at many different levels. Its flexibility allows projects, programmes and policies to be assessed at either a local, regional, national or international level – making HIA suitable for almost any proposal.
- It offers timeliness by reaching decision makers in order to influence decision-making processes with either comprehensive (longer) or rapid (shorter) HIAs, well before decisions about a proposal are made.
- It links with sustainable development and resource management by integrating assessment of sustainable development perspectives including: health, education, employment, business success, safety and security, culture, leisure and recreation, and the environment. Drawing on the wider determinants of health, and working across different sectors HIA play an important role in the sustainability agenda.
- It is a participatory approach which can serve many potential users, including:

  a Decision-makers
  b Commissioners of the HIA in order to consult widely and gather differing views, build capacity and develop strong partnerships
  c HIA workers, who actually carry out the individual components of the HIA, which may include consultants, local staff from a wide variety of organisations and the community
  d Stakeholders, who want their views to be considered by decision-makers.

Greenspace Scotland (2008) has also recently provided a more specific guide to help people conduct a HIA of greenspace (whether these are green space policies, strategies, plans, frameworks, programmes and projects). The guide contains:

- background information on green space
- outlines of the current green space policy context in Scotland
- reviews of international research evidence on green space and health
- suggestions on questions which will help apply the evidence to specific green space or green space-related proposals
- guidance on how to use evidence in a HIA
- short case studies describing completed HIAs of green space
- sources of data and further information on green space

The recently updated paper *Healthy Parks, Healthy People* (Maller *et al.* 2008) also clearly outlines the value of healthy green spaces in enhancing of the surrounding community.

Maller *et al.* (2008) indicate how parks play a vital role in providing space for exercise and physical activity, which is important for the management and prevention of depression and anxiety. They are also a hub of community activity – facilitating social connectedness and inclusion. Tilt and Matsuko and Sullivan (Chapters 32 and 33 respectively, this volume) further emphasise the many health outcomes which access to urban green spaces can achieve for the general public.

This consolidates what is seen in spaces such as Meanwhile and other urban green spaces, notwithstanding their explicitly stated therapeutic role for specific groups or individuals. Thus such therapeutic green spaces should be clearly and fully considered as intrinsic resources in the wider infrastructure of urban green spaces.

Conclusion

Connectedness and social inclusion, management and prevention of mental and physical health problems are some of the most incisive outcomes of therapeutic green spaces. However it can be seen that much more is achieved than predefined outcomes for the individuals who seek therapy. The general public is likely to benefit from wide-ranging outcomes by utilising a local green space otherwise only purportedly identified as therapeutic.

These spaces simultaneously enhance the personal growth of diversely able individuals and improve the quality of both the ecology and green infrastructure of urban sprawls. They also directly contribute to public health, community cohesion and social capital. By tending a wildlife garden the participants of Meanwhile in London act as leaders of a different kind: they 'lead to serve'. This model of leadership formulated by Greenleaf (2003) is shaped by a combination of personal growth, teamwork, ethical and caring behaviour. The actions and decision making which emerge from this are not solely self-serving as one might imagine under the notion of therapy, but 'other-centred' and lead to very tangible benefits for both the environment and people at large.

References

AHTA (American Horticultural Therapy Association) (2007) *Position Paper on Definitions of Healing Gardens*, E.R. Messer (ed.), King of Prussia, PA: AHTA.

Baines, C. (2006) 'Green Grid: Environment and Economics', in GLA, *East London Green Grid Primer*, London: Greater London Authority.

Brown V.A. (2007) *Leonardo's Vision: A Guide to Collective Thinking and Action*, Sydney, NSW: Sense Publishers.

Burls A. (2005) 'New landscapes for mental health', *Mental Health Review*, 10(1): 26–9.

Burls A. (2007a) 'People and green spaces: promoting public health and mental well-being through ecotherapy', *Journal of Public Mental Health*, 6(3): 24–39.

Burls A. (2007b) 'Meanwhile Wildlife Garden DVD', in *With Nature in Mind*, London: Mind Publications.

Burls A. (2008) 'Seeking nature: a contemporary therapeutic environment', *International Journal for Therapeutic Communities*, 29(3): 228–44.

Burls A. and Caan, A.W. (2004) 'Social exclusion and embracement: a helpful concept?' *Primary Health Care Research and Development*, 5(3): 191–2.

Burls A. and Caan, A.W. (2005) 'Editorial: human health and nature conservation', *British Medical Journal*, 331(7527): 1221–2.

Cooper Marcus, C. and Barnes, M. (eds) (1999) *Healing Gardens: Therapeutic Benefits and Design Recommendations*, New York: John Wiley and Sons.

Cunningham, S. (2006) *Collaborating across the Sectors: Isolating Researchers in Disciplinary Silos Discourages Innovative Research*, posted in Australian Policy Online: Creative Economy, 14 December 2006, online, available at: www.creative.org.au/webboard/results.chtml?filename_num=127548 [accessed 5 December 2008].

Department of Health (2004) *Choosing Health: Making Healthy Choices Easier*, London: The Stationery Office.

Department of Health (2007) *The Expert Patients Programme*, www.dh.gov.uk [accessed 30 April 2009].

Elkington, J. (1994) 'Towards the sustainable corporation: Win-win-win business strategies for sustainable development', *California Management Review*, 36(2): 90–100.

EPHA (European Public Health Alliance) (2008) *Green Spaces Reduce Health Gap*, online, available at: www.epha.org/a/3318 [accessed 30 April 2009].

Frumkin, H. (2001) 'Beyond toxicity: human health and the natural environment', *American Journal of Preventive Medicine*, 20(3): 47–53.

Fuller, R.A. and Gaston, K.J. (2009) 'The scaling of green space coverage in European cities', *Biology Letters*, 5(3): 352–5, doi: 10.1098/rsbl.2009.0010.

GLA (2006) *East London Green Grid Primer*, London: Greater London Authority, Architecture and Urbanism Unit.

Greenleaf, R.K. (2003) *The Servant-Leader Within: A Transformative Path*, New York: Paulist Press.

Greenspace Scotland (2008) *Health Impact Assessment of Greenspace: A Guide* (Health Scotland, Greenspace Scotland, Scottish Natural Heritage and Institute of Occupational Medicine), Edinburgh: Greenspace Scotland.

Hartig T. and Cooper Marcus, C. (2006) 'Healing gardens: places for nature in health care', *The Lancet*, 368: S36–S37, doi:10.1016/S0140-6736(06)69920-0.

Kaplan, R. (2001) 'The nature of the view from home: psychological benefits', *Environmental Behavior*, 33(4): 507–42.

Kuo, F.E. and Sullivan, W.C. (2001) 'Environment and crime in the inner city: does vegetation reduce crime?' *Environment and Behavior*, 33(3): 343–67.

Maller C., Townsend M., St Leger, L., Henderson-Wilson C., Pryor A., Prosser L. and Moore M. (2008) *Healthy Parks, Healthy People: The Health Benefits of Contact with Nature in a Park Context: A Review of Relevant Literature*, second edition, Geelong, Vic.: Deakin University and Parks Victoria.

McCreery S. (2006) 'Landmarks, Corridors and (Green) Grids' in GLA, *East London Green Grid Primer*, London: Greater London Authority Architecture and Urbanism Unit.

Pretty, J., Griffin, M., Peacock, J., Hine, R., Sellens, M. and South, N. (2005) *A Countryside for Health and Wellbeing: The Physical and Mental Health Benefits of Green Exercise*, Sheffield: Countryside Recreation Network, online, available at: www.countrysiderecreation.org.uk/publications/record.php?show=145.

Reynolds, V. (2002) *Well-Being Comes Naturally: An Evaluation of the BTCV Green Gym at Portslade, East Sussex*, Oxford: Oxford Brookes University.

RUDI (Resource for Urban Design Information) (2009) *website*, online, available at: www.rudi.net/node/20080 [accessed 21April 2009].

Scoffield, D. with Burls A. (2009) Unpublished conference presentation at International Conference on Human Ecology from 29 June to 3 July 2009 at the University of Manchester, UK.

Ulrich R.S. (1984) 'View through a window may influence recovery from surgery', *Science*, 224(4647): 420–1.

Ulrich R.S. (1999) 'Effects of gardens on health outcomes: theory and research' in C. Cooper Marcus and M. Barnes (eds), *Healing Gardens: Therapeutic Benefits and Design Recommendations*, New York, NY: John Wiley, pp. 27–86.

WHO (World Health Organization) (2009a) *Phase V (2009–2013) of the WHO European Healthy Cities Network: Goals and Requirements*, World Health Organization Regional Office for Europe.

WHO (World Health Organization) (2009b) *Health Impact Assessment (HIA)*, online, available at: www.who.int/hia/about/why/en/index.html.

Wong J.L. (2004) 'Reflections: nature for people and people for nature', in T. Trzyna (ed.), *The Urban Imperative*, California Institute of Public Affairs and IUCN, online, available at: www.interenvironment.org/pa/jlwong.htm [accessed 30 April 2009].

Yerrell P. (2004) *National Evaluation of BTCV Green Gym*, Oxford: Oxford Brookes University.

47

Peri-urban ecology

Green infrastructure in the twenty-first century metro-scape

Joe Ravetz

Introduction

Large parts of Europe and the developed world are not quite urban or rural, but somewhere between – an emerging 'peri-urban' landscape, in the fringes and hinterlands of cities and city-regions. This reflects a more networked mobile society, but one which also needs new kinds of local character and quality of life. The frequently rapid changes in such areas bring both problems and opportunities in urban ecology – so much that we could propose an extended concept of 'peri-urban ecology'.

In the fringes and hinterlands of most cities there are typical challenges which compound together – climate change impacts and soil erosion, road traffic, ageing population, landscape stress, urban-rural migration, farm restructuring, tourism impacts and pressure for urban development. Such a peri-urban space – a 'metro-scape' – may be the most common type of living and working situation in the twenty-first century. In some parts of the world it will be an arena for affluence and conspicuous consumption. In others it will be a fractured zone of poverty and displacement, a front line between the problems of the city and the countryside.

Underlying this is the changing nature of the city and urban expansion itself. As well as the physical growth of urban form, there is a wider economic and social dynamic of change, across the whole global urban system. This in turn changes the relationship between urban and rural, and between 'urban form' and 'urban ecology'. A new kind of geography begins to emerge, looking beyond the conventional divide of 'urban' and 'rural', to a territorial perspective in which the 'peri-urban' is the central feature.

Such 'peri-urban' territories are generally in a state of rapid flux and structural transition. We can identify at least three types of dynamic change at work:

- Metropol-ization: an 'urban infrastructure' transition which is diffused and networked across wider peri-urban and rural areas;
- 'Cognitive capitalism': a 'cultural-structural' transition – new patterns of globalizing economic and social activities and divisions;
- Peri-urban ecology, or a spatial 'green infrastructure' transition. This reflects a countervailing force of 'localization', and changing social and cultural relationships between a more affluent urbanized population and its landscape surroundings.

This chapter focuses on the third of these, the peri-urban ecology, but keeping in mind the context of the first and second. We aim to outline the agenda for a 'peri-urban ecology', as both a technical question, and a creative zone of social and cultural experimentation. First we provide an outline of the situation across Europe, and an overview of trends now emerging. Then we take a case study approach, drawing upon current EU-funded research on the peri-urban, the FP6 project PLUREL 'Peri-Urban Land-Use Relationships' (EC 036921). Then we explore each of the three transitions above, and their implications for peri-urban ecology. Finally there are some directions for future opportunities in peri-urban policy.

We focus here on the developed world (i.e. the North): the agenda for developing countries (the South) would be a chapter in itself (McGranahan *et al.* 2002). But many of the initiatives described here would be relevant across the whole spectrum of North and South (Roberts *et al.* 2009).

Context

From urban to peri-urban ecology

Something is happening in peri-urban areas. There is the obvious physical expansion of cities, with land-take, infrastructure, pollution and waste on the urban edges and fringes. There is also a wider process of urbanization, modernization, restructuring and globalization of social and economic activities in the larger areas of peri-urban and rural hinterland.

This has major impacts on land-use and landscape ecology: but land-use change is one aspect of broader economic, social and cultural change. Local rural economies or self-contained market towns are shifting rapidly to new roles as satellites and outposts of a much larger and fast moving city-region system.

Land and landscape often become degraded by agricultural change, and fragmented by roads and other infrastructure. Fringe developments such as garden centres tend to increase car traffic and urban sprawl. Incoming commuters tend to swamp local communities, while bus services and village amenities close down.

The governance system in such areas is often diffused between many units and organizations: by its nature it falls between urban policy and rural policy. The result is that the special problems and opportunities of peri-urban areas are often underplayed and under-resourced. At the same time there are many new types of ecological habitats, landscape types, partnership governance, social enterprises and other creative responses. The peri-urban ecology can be seen as a zone of innovation and possible subversion, as well as a zone of fragmentation and competition.

The global urban agenda

Overarching the issues of the peri-urban is the challenge of the global urban system. At present, 3.3 billion people live in urban centres across the globe; by 2030 this number is predicted to reach five billion, with 95 per cent of this growth in developing countries (UN Population Fund 2007). While mega-cities dominate the development agendas, overall growth in urban centres of ten million or more inhabitants is expected to level out; over the next ten years, cities of less than 500,000 will account for half of all urban growth. The United Nations has made detailed *business as usual* or default projections for urbanization at the international level, as in Figure 47.1 (UN Department of Economic and Social Affairs 2004).

The projections show that current levels of urban growth are likely to continue almost unabated. They also show a remarkable turning around of historic trends, from rapid growth to

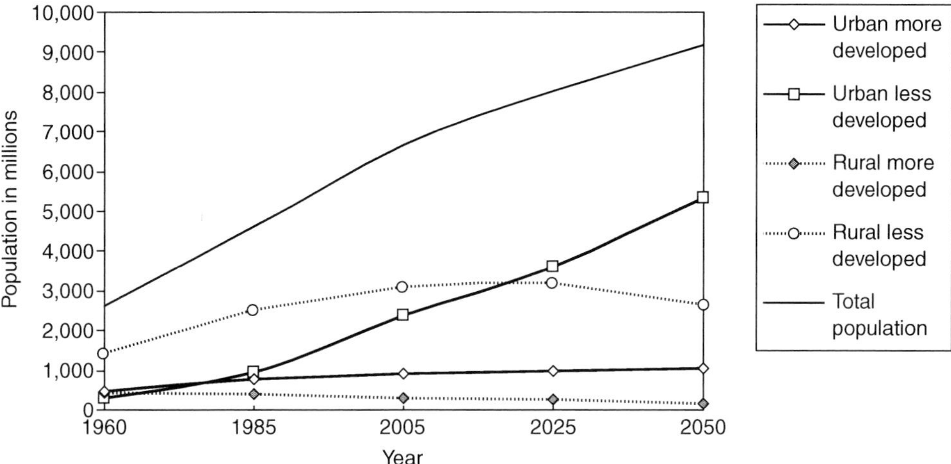

Figure 47.1 World urbanization prospects (*source:* UN Department of Economic and Social Affairs 2004).

reduction in the rural populations, in both medium- and less-developed countries. The urban-rural balance is of course dependent on pressures and policies in both rural and urban areas: development aid, agriculture, water, climate change, health care and so on.

However this headline focus on the largest agglomerations can easily miss the many smaller urban systems in the 500,000–5 million band, where the rate of agglomeration is proceeding more rapidly in many instances. As to what kind of urban environments can be expected, on current trends, the proportion of slums and/or informal settlements could increase to over half of the world's urban population by 2030 (Neuwirth 2005: Davis 2005). The form of urban agglomerations is also changing: the former hard edges of urban built form are shifting to a more fragmented and diffused pattern (Angel *et al.* 2005). So it is feasible and perhaps probable, that the majority of the world's future urban dwellers will be in quasi-temporary shacks, lacking fixed systems such as water, sanitation and electricity, even while mobile phone and other transient networks thrive, and spread out in a peri-urban pattern which is not quite rural or urban (Webster and Lai 2004).

This prospect raises the contrast between cities with the 'green agenda' of the 'North', versus the 'brown agenda' of the 'South' (McGranahan *et al.* 1996, 2004). There are massive differences in material wealth and environmental conditions, but these extremes are part of a complex spectrum. There is actually not a clear division between the South or North, as the cities of many transitional economies, and those of Latin American and Asian countries, are somewhere between. Also, each type of city is interdependent on the other – many of the consumer goods that drive the urban metabolism in the North are produced by low cost labour in cities of the South (Roberts *et al.* 2009).

Urban fringe and the Edge City

In the peri-urban areas, such differences can be multiplied up, as such areas often act as filters and residues for urban systems with various kinds of dysfunctionality. In both North and South there are peri-urban pockets of extreme wealth, side by side with concentrated deprivation and exclusion. And while conventional urban development has generally assumed that urbanization means higher density continuous built form, it is likely that the future will be rather different in many places.

The urban fringe of many metropolitan areas is a tangle of raw materials and residues from the urbanization process – economic innovation with landscape change, and opportunity alongside dereliction. The growth of 'edge cities' is indicative of a new kind of networked *metropol-ization*, sweeping away the previous pattern of cities surrounded by countryside, and producing new and more diffused urban forms (Garreau 1991). For instance, in Tyson's Corner near Washington, DC, a former rural crossroads is now the centre of a major spread of business, retail and housing development covering 100 km^2. In such areas, reducing pollution and re-using vacant land are worthwhile goals, but little is achieved without looking at the wider dynamics of the peri-urban fringe and hinterland (Wood and Ravetz 2000).

One of the primary driving forces is often the city or regional airport – the first determinant of location and value in the globalized economy, just as highways were in the twentieth century, and railways in the nineteenth century. The concept of the *aerotropolis* is an organizing principle for a trend that is generally ad hoc and unplanned at present. But in size, throughput and value added, the airport centred development zone is overtaking the role of the Central Business District, to form the primary *technopole* axis of the region (Kasarda 2004; Castells and Hall 1994). The environmental management of such sites poses special challenges, both locally and globally.

Overall the dynamic of *metropol-ization* is a combination of economic, technological and social change, which is (sometimes) mediated by various forms of spatial planning. The environment then becomes a crucial factor in shaping the result in spatial development; high quality and high value environments are favoured for business parks and leisure parks; low quality and polluted residual environments are preferred for 'bad neighbour' urban infrastructure, such as landfill sites, sewage plants and power stations (Blair 1987).

There are many negative effects of such economic and spatial growth. For farming and land resources, these dynamics can often produce instability and lack of investment; the productivity of conventional farming drops near to urban areas, small producers are overridden by commercial interests, and there is much land vacancy and fragmentation (McGranahan *et al.* 2004). In response to such problems there are moves to reclaim peri-urban landscapes for urban community use (Nicholson-Lord 1987). There is a movement in both South and North to reconnect the urban food market with the potential of the urban fringe to provide food and livelihoods. For instance, in the Greater Manila region, various planning frameworks have aimed to retain a patchwork of intensive agriculture as part of a mixed use and diverse economy across the wider conurbation. However there are still the typical problems of rising land values, labour markets and water resource demands between rural and urban communities (Junde and Zaide 1996).

Definition of the peri-urban

The wording 'peri-urban', of French origin (*peri-urbain*) is used here. This includes the fringe at the edge of the built area, and extends into a broader zone of urban influence. This can be seen in open arable or green areas, as in Warsaw, Koper and Montpellier; or it can merge into a wider urban agglomeration, such as in Haaglanden, Manchester or Leipzig. The functional concept of a 'rural-urban region' (RUR) has been defined by the PLUREL project at the 'NUTS x' level (a combination of NUTS 2 and 3, which accounts for the anomaly in the size of NUTS 2 in some countries – see Haase *et al.* 2011). This can be used to analyse the features of such extended peri-urban territory, based on employment patterns, economic self-containment, journey time from urban centres, or bio-regional and landscape types. For more densely populated areas of Europe and North America, such peri-urban areas can then emerge rapidly to cover much larger agglomerations: it is arguable that large parts of western Europe are peri-urban, titled by some as 'metro-scapes' (Hall and Pain 2006; Bender 2007; Giannini 1994; Greater Helsinki Vision 2007; Kraffczyk 2004)).

Perhaps the most common feature is that the peri-urban is ignored for its own intrinsic values and resources, and seen merely as an opportunity space or dumping ground for urban problems, and as a cheap source of land and resources. It is also a territory of great contrasts, containing both wealthy and deprived communities: both advanced and retro-facing activities: and zones of rapid growth alongside zones of fixed conservation. Further exploration of these structural dilemmas follows in the Dynamics section.

Policy responses to peri-urban issues

A crucial theme in spatial development is that public policy can be a powerful counterpart to market forces; not so much that planners are popular heroes, rather that most private sector development is dependent on public investment in roads, drainage, utilities and so on. One of the foundations of twentieth century spatial planning, the UK Green Belt, is practiced around the world under different guises and policy regimes. However in all peri-urban areas, there is increasing economic pressure for motorways, airports, business parks and other infrastructure, and much Green Belt land is damaged and polluted. There are policy responses such as landscape areas, green wedges and river valleys, but it is clear that green belt policies and functions and boundaries may need to be re-defined (Elson *et al.* 1993; CPRE 2007). Some common problems and opportunities include:

- degraded or derelict land where diversification and/or development would promote enhancement and after-use;
- rural areas where diversification needs leisure or ecological development;
- smaller settlements where development would enhance viability of local services;
- green wedges in the urban area which would benefit from green belt extensions.

At the same time there is a need for long-term stability of protected land boundaries and policies: the slightest hint of uncertainty tends to inflate speculation on 'hope values', and undermine investment in maintenance and agricultural activities. Meanwhile on the ground, large areas of such zones are often degraded and underused, and there is an agenda for working towards a more integrated, diversified and sustainable landscape in the form of an *eco-belt*, within or between cities (Natural England 2010).

Peri-urban relationships

We can look at the peri-urban, not only as a frontier zone between urban and rural, but a system of 'relationships' and interactions. Such relationships can be land-based or human-based: direct or indirect. They can be framed as 'ecosystem services', in direct functional terms such as food, water, tourism: and in socio-cultural terms such as aesthetics and amenity (TEEB 2010). They can also be seen as the driving forces behind settlement types and patterns, landscape types and patterns, and the accumulation of historic layers. We can identify generic types of relationships across the spatial geography types, as in Figure 47.2:

- urban/peri-urban: the relationship of city dwellers to suburbs and hinterland: the core agenda of urban geography and spatial modelling, with flows of commuting and services, and the search for optimal locations for housing and employment;

- peri-urban community: the potential for economic and social development within the peri-urban zone itself. The difference between a commuter 'dormitory' settlement and a vibrant multifunctional market town, can be seen both in land use and in human activity patterns. New kinds of local opportunities are now emerging in post-industrial and restructuring city-regions;
- peri-urban/rural or ecosystem: focusing on the relationships between humans and landscapes. There may be direct interactions between urban and ecosystems. For instance, if a tourist coach drives directly to a national park, this shows how peri-urban communities are often marginalized and vulnerable to forces from outside the territory.

The peri-urban ecology agenda

The physical urban ecology in the 'metro-scape' of the peri-urban hinterland is unique combination of natural ecology, urban form and human activity. It is also apparent as a 'complex system', showing in some cases rapid transitions, phase changes, and emergence of new patterns (Waltner-Toews *et al.* 2009). To understand this means looking at the combination and synergy of many factors, including:

- Environmental factors: geography, climate, water, farming biodiversity and landscape patterns;
- Urban factors: growth and spread, settlement patterns, transport modes;
- Economic factors: growth and change: economics of farming, urban development;
- Social factors: population movement and demographics: social structures: lifestyle and location choices.

In addition to these, it is clear that more intangible and qualitative factors can be at least as important:

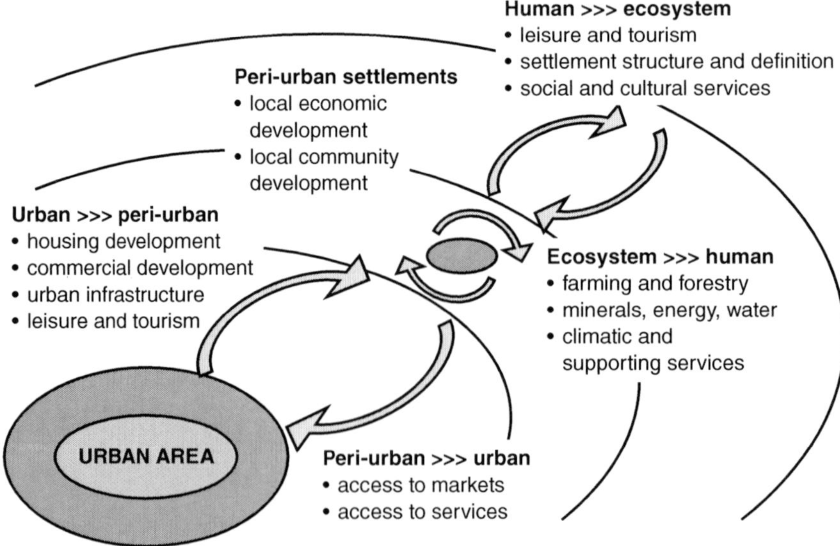

Figure 47.2 Peri-urban land use relationships.

- Social and cultural factors in lifestyles, attitudes, values and ethics
- Governance, structures and processes of decision making;
- Spatial planning, the regime and the actor-networks;
- Discourses of stakeholders, investors and other power brokers;
- Institutional factors: property regime, or the political economy of infrastructure.

This can be summarized in a simplified urban/peri-urban model (Figure 47.3), based on the DP-SIR scheme ('driving forces, pressures, states, impacts and responses') (EEA 1999). External drivers of change, such as climate change, impact on to a typical urban system. This is already growing, both directly in land-take and through associated infrastructure such as energy, water and waste. There are also land-use sectors, such as farming or tourism, which compete for scarce resources. The results downstream can be seen with environmental impacts on soil, air, water and biodiversity: and the policy responses are aimed at one or another of these pressures or driving forces.

A European research programme

The PLUREL ('Peri-urban Land Use Relationships') was a large consortium research project, funded by the EC, coordinated by the University of Copenhagen (EC 036921). It aimed to provide a state of the art analysis of urbanization, landscape ecology and spatial governance for peri-urban areas. It includes advanced urban landuse modelling: detailed policy analysis: and a set of best practice tools and resources (see www.plurel.net for details). The research method is based on a view of peri-urbanization, as not only the outward spread of cities, but a dynamic system of 'land use relationships', as above.

At the centre is a set of in-depth case studies in six EU city-regions: Manchester, UK; Montpellier, France; Haagland, the Netherlands; Leipzig, Germany; Koper, Slovenia; and Warsaw, Poland; together with an external comparison to Hangzhou in China. Each of these is developing

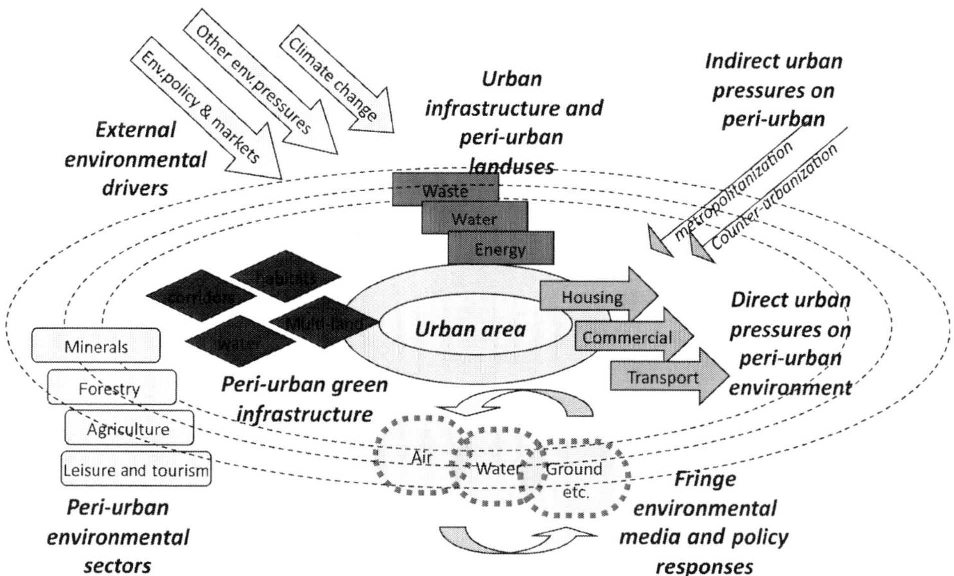

Figure 47.3 Environmental drivers in peri-urban land use.

a multi-level, in-depth perspective on the peri-urban agenda, by understanding the trends in the urban-rural system, and by working closely with stakeholders. The method also follows a scenario perspective on future trends and possibilities over the next 40 years; this explores some topical 'what if' questions, starting at the highest system level:

- Will the new peri-urban metro-scapes become more fragmented, polluted, inaccessible and socially exclusive? OR,
- Will the peri-urban become a more harmonious landscape, with ecological biodiversity, economic activity and many forms of social/cultural quality of life?

There are many more questions and uncertainties, such as on climate change, agriculture, employment patterns and particularly the future of urban-rural linkages. Again the questions are both technical, and socio-cultural-political: should the peri-urban aim to supply its city with food and energy? Are five small towns more sustainable than one large one? And what kind of governance can coordinate and manage investment within the typical fragmentation between competing municipalities?

Peri-urban ecologies around Europe

This set of case study outlines is drawn from the PLUREL project above. Each shows a certain combination of geographical, social and economic factors and dynamics of change. A basic typological sets out two axes:

- High or low rate of urban growth and development
- Northern temperate climate, or southern arid climate.

Within these possibilities, each city-region shows a unique set of responses to the peri-urban ecology agenda, many of them highly creative.

Manchester

The Manchester city-region has a complex geography, with 2.5 million people in the main conurbation and a further 1.5 million people in a wider 'rural-urban region'. The overall density is 1971 persons/km^2 for the inner core, and half that in the outer ring: the overall rate of urbanization (i.e. additional to the existing urban area) is 0.2 per cent per year. This city-region was one of the birthplaces of the industrial revolution, and following the export of most of its heavy industry, has spent the last 50 years on restructuring and reclamation of large areas of DUN land ('derelict, under-used and neglected'). The regional economic performance is below the national average, but there is a basic level of affluence and organization of public services, coupled with a well-established planning system.

This example shows the peri-urban ecology agenda in a Northern post-industrial city-region, with moderate rates of growth following social and economic restructuring (Ravetz and Warhurst 2011).

The result is that the urban fringe and peri-urban areas show many environmental improvements, even while further urban activity and infrastructure puts on additional pressures. There are some positive responses to continuing urban pressure: for instance, the controversial Manchester Airport second runway was planned to cross a river valley of high ecological value: in order to maintain the wildlife corridor, a 200 metre wildlife tunnel was constructed underneath it.

Alongside a complex set of governance and public funding structures is a multitude of local partnerships, networks, forums and other groups, each working on part of a highly connected environmental/economic/social agenda. Organizations such as Pennine Prospects take on a broad agenda in the upland peat bog landscapes of the South Pennines, acting as delivery bodies for the Rural Development Plan for England, which includes support for the 'high level stewardship' landscape conservation under the Common Agriculture Policy.

The Green Infrastructure (GI) agenda is an active policy agenda, with a high quality research programme involving various Community Forest organizations and partnerships, and many wildlife groups and local green groups. However the programme is often underfunded, insecure and short term in outlook, and the 'policy integration' which is needed to bring together many different interests in sustainable landscape management, is hard to achieve. Although the Regional Economic Strategy made a priority of the economic benefits of GI, at times of scarcity such as the 2008–9 economic crisis, the core funding stream is quite vulnerable.

Montpellier

Montpellier is a historic trading post, with a strategic location, good climate, and a rich hinterland. The city shows both a tight historic core and rapid modern expansion. Within the Urban Agglomeration there is a population of 367,000 at a density of 870 persons/km^2. The current rate of urbanization is 0.7 per cent per year (at 250 ha yr^{-1} in the last four years): an affluent urban population is spreading outwards into a rural hinterland that has seen rapid restructuring, and in some cases depopulation.

This case demonstrates the peri-urban ecology in a Mediterranean provincial city-region with high rates of growth and development (Jarrige *et al.* 2011). Behind the affluence and growth agenda there are mounting environmental pressures in the peri-urban areas:

- severe traffic congestion, and pressure from urbanization in natural areas;
- general agricultural crisis, particularly in the wine sector;
- land rent and speculative strategies of land owners;
- barriers to diversification and innovative forms of farming;
- decrease of floodplain forests and increasing vulnerability through urbanization;
- long term quality and availability of water, in an arid climate vulnerable to drought.

This case study has focused on farming, which is undergoing many changes, and in some sectors such as wine, a crisis in the market. Farmers are key stakeholders in the rural-urban system, and their situations and strategies differ according to family assets, farming system and life cycle position. There is an interesting set of choices for farmers, each with implications for the peri-urban area:

- carry on farming and reinforce the estate, buying more farmland where possible;
- farmland partially sold to be developed, with the capital reinvested in peri-urban agricultural production: or capital expatriated towards other economic sectors;
- total liquidation of peri-urban farmland and exit out of farming: or, shifting to farming into a more rural area, or towards Central Europe, or even North Africa;
- waiting strategies, 'land freezing' and fallow management; or land leased in a precarious tenancy for short cycle crops and cereals.
- for non-farming heirs or investors (with outside capital) there is a smaller personal problem, but a growing problem for an underused landscape. For others there is a difficult access to

farmland (high capital cost, growing scarcity of tenant farming, vulnerability of tenancies leading to exclusion of public subsidies).

The policy responses aim to respond to these, by supporting diversification and a range of creative initiatives, and linking them to the multi-level SCOT (Scheme of Territorial Coherence) system of coordinated multi-level planning framework.

- support to farming activities and infrastructures (agricultural hamlet, the local 'wine road' scheme, various festivals and promotions, etc.);
- protection of farmland, via ban on development in agricultural zones;
- management of urban fringe, via the spatial planning limit;
- emerging support regime for innovative forms of farming and rural economy.

One major player for large areas of semi-natural habitat is the Conservatoire du Littoral, a public institution concerned with sustainable land management, similar to the UK National Trust. It has both the financial means and legal power to buy natural open land on the littoral fringe, in order to protect it. Once bought by the Conservatoire, land becomes inalienable and *non aedificandi* (means that it should never be sold or built on), dedicated to environmental preservation or sustainable farming.

Haagland

The Haaglanden city region, the hinterland to Den Haag (The Hague) is on the west coast of the Netherlands, between the conurbations of Rotterdam and Amsterdam. It is one of the most urbanized areas in a small country, and home to nearly one million people at a density of 2500 persons/km^2. A large part of the region is below sea level, and only one third of the area consists of the traditional meadow landscape. The region is home to the largest concentration of greenhouse horticulture in the country, which is more similar to industrial development than agriculture: but these glass houses are now struggling to compete with overseas suppliers in warmer climates. There is no available space for expansion of any land use type, and spatial planning is a major challenge.

This case explores the policy agenda for peri-urban ecology in an affluent high density Northern coastal city-region, with continuous restructuring of settlements and landscapes (Westerink *et al.* 2011). There is a diminishing stock of agricultural land, with effects on landscape and agricultural ecology, reducing capacity for water retention, and continuing pressure for growth in recreational areas.

- There are typical pressures for housing development in attractive locations: problems for low-cost housing in high value areas, and the distortion of values and communities adjacent to Green Belt areas.
- There is progress in conservation policy, against a general trend of pressure from urban development. There are strong Green Belt policies but apart from holding back development, these do not provide encouragement for landscape management or nature conservation.
- There is a shortage of outdoor recreational space, and many agricultural areas are not yet very suitable or accessible for leisure and recreation.
- There are major congestion problems on main routes, and much disruption of local communities and landscape qualities on minor routes. Public transport is generally less viable in peri-urban areas: while any new infrastructure increases site values in peri-urban areas.

In the Netherlands there is a long history of planning in harmony with water – there is little alternative when building below sea level. Green-Blue infrastructure is one of the physical aspects of zoning plans and development masterplans. Some (national) spatial concepts have a very high status, however, such as the Ecological Main Structure (EHS, including Nature 2000), the Green Heart and the Buffer Zones. At the site level there is much good practice in bringing together ecological habitats and corridors, leisure and play areas, climate and flood protection, urban design and ecological amenity. In the Haagland, such Green-Blue infrastructure fits closely up to intensive horticulture in the glass houses. Typical policy responses include:

- Protection of remaining agricultural land use through zoning and subsidies. Further separation of 'glass' and 'grass' types of cultivation;
- Expansion of natural and recreational land, through purchase, reconstruction and zoning protection;
- Enhancement of agricultural ecology through agri-environment schemes;
- Coastal extension and landscape engineering: integration of water retention in urban, greenhouse and agricultural land uses, with water storage and removal strategies.

The Haagland is aware that peri-urban landscape and urban ecology strategies are not only a functional question of farm values or zoning policies: the cultural identity and commitment to heritage is also important. There are various initiatives which build on the long history of fine art in the region, and the strong public appreciation of a sensitive landscape ecology.

On the spatial governance level, there is an ideal concept of integrated master-planning which can be visualized in terms of space and time: 'fast' spaces focus on the airports, business parks, motorways and communications: in contrast, 'slow' spaces involve awareness of ecology, geology and landscape. Various human settlement forms and patterns can then provide an interface between the fast and the slow, in a closely layered structure. This idealized landscape concept then has to be fitted to the reality, where there is often complex and lengthy negotiation between nine independent municipalities which govern the peri-urban ecology.

Koper

Situated on a former rocky island, Koper is the most important city on the Slovenian coast. The coastal zone itself has rapidly developed and urbanized, with a mix of port-related industries, tourism and commercial activities: in the hinterland there is high quality agricultural land and a major nature reserve on the slopes of the Karst. This example is relatively small with a population of 51,000 in the city, surrounded by a large hinterland: however there is a high rate of urbanization at 3 per cent per year (additional to existing urban areas).

This case study shows a more rural peri-urban ecology, in a mediterranean climate and a hilly terrain, responding to diversification under an emerging post-socialist system of governance (Pintar and Perpar 2011).

The main environmental pressures again centre on restructuring and urbanization, with land use change from agricultural to urban, local biodiversity decreasing, and more mono-cultural agricultural production for export. There is continuing urban demand for healthy and green living environments, maintainence of cultural heritage and attractiveness of rural landscapes. The port facility itself has grown by reclaiming former wetland and salt marshes, leaving the historic city marooned by surrounding container parks and distribution warehouses.

Agriculture is also under pressure, because the best quality agricultural land is situated where the interests are also highest for other purposes: mostly for settlement, for the location of industrial and craft zones, for infrastructure (highways) etc. Agriculture is under pressure because of international competition, globalization and the common market for agricultural products in the EU. Farm patterns are also a problem: small farms with very fragmented land, missing land titles and unresolved ownership, less favourable conditions for agriculture in the hinterland, and an ageing farming population.

In response, some of the policies in the peri-urban aim to bring together the demands of rapid modernization with a unique cultural heritage. There is some acceptance of the local development dynamics, as with the *post festum* legalization of constructions without building permission. Many local inhabitants continue to farm (mostly wine, vegetable and olive production, and small cattle breeding in the Karst), investing in supplementary activities on farms (mostly farm tourism), craft and revitalization of ancient traditions. Culture and tradition are being strengthened by the promotion of craftsmanship, tourism, olive oil production, and the production of regional wines, smoked ham, cheese, truffles and other delicacies. Overall this case may be one example of a benign outcome to the peri-urban challenge.

Leipzig

Leipzig is the political and functional centre of the Leipzig-Halle agglomeration, a densely populated region of central eastern Germany. The entire rural-urban-region has about two million inhabitants, with the urban area of around half a million, at a density of 1650 persons/km^2 that continues to decrease. The area has seen massive restructuring: both physical with the demise of coal mining, with economic restructuring, social migration and political change in the post-unification Germany.

This example concerns the peri-urban ecology agenda in a Northern industrial city-region, with high levels of urban shrinkage and 'perforation' – i.e. physical holes in the urban fabric of land and buildings (Bauer *et al.* 2011).

The city-region comprises a compact form, with a polycentric urban area around two larger centres, and two rings of smaller settlements. Typical environmental problems include land consumption, conversion or sprawl, and uncontrolled development encouraged by road infrastructure. The former open-cast coal mining areas are on the way to full remediation but this has required massive public investment. There are critical voices on past funding schemes in support of modernization, such as the home-owners grant, which proved to be conducive to sprawl. Public service providers are concerned about rising infrastructure costs and transport congestion. There are various schemes for integrated spatial planning and investment, which try (not always successfully) to bring together the many interests in a scattered and policycentric agglomeration:

- Regional Development Concepts (REK);
- Integrated Rural Development Concepts (ILE);
- Green Ring initiative;
- New Leipzig Lakeland;
- Water City Leipzig Initiative;
- Lignite Plans (part of regional plan).

A very topical agenda here is the 'greening shrinking city'. Leipzig-Halle suffered rapid depopulation, social change and investment withdrawal, following the reunification of Germany. Under the title of the 'perforated city', many buildings and spaces were abandoned and derelict, and the shape of many urban and peri-urban communities was turned inside-out (as shown vividly in the exhibition and website (www.shrinkingcities.com). Since then, a continuous programme has been meticulously re-engineering not only the physical urban and peri-urban space, but the connections to the economic and social activity around it. Like other German cities this also includes space for summer houses and allotments, in a 'green ring' around the urban area, permitting outdoor leisure and connection between urban dwellers and the forest. Many types of greenspaces and green corridors penetrating through the urban area have also emerged, with a complex patchwork of interim uses, land remediation and habitat re-establishment, and an adaptive management combination of short and long term landscapes.

Warsaw

Warsaw is a regional metropolis with 1.7 million residents, living on an area of 517 km^2. The Warsaw Metropolitan Area (WMA) is located in the Mazowsze Region in the centre of Poland, with over five million population, of which 64 per cent is urbanized. The principal challenge is the rapid pace of modernization, resulting in polarized, chaotic and dysfunctional development of settlements and urban infrastructure. This is then the context and the challenge for 'peri-urban ecology' within a rich agricultural region (Grochowski and Slawinski 2011).

There are three main types of functional settlement-landscape patterns identified in the WMA region:

- type A: suburban zone nuclei: dominated by urban functions; providing job opportunities and services for surrounding rural areas; surrounded by open space; with tangible urban edges and boundaries between built-up areas and open land;
- type B: suburban inner zone: higher density municipality with mixed functions; some open spaces, more rural in terms of landscape, size and functions of settlement units;
- type C: suburban outer zone: lower density with agricultural functions; rural landscape; moderately affected by urban development pressure.

Research in progress is showing a pattern of 'dual-mode' development – on urbanized and on rural areas – that result in rising inequalities in the level of economic development and quality of life. The strongest urban pressure is in the locations with the best environmental qualities (open spaces, forests, valuable land for agricultural production); and particularly affects ecological systems which cross administrative boundaries. Suburbanization, urban sprawl and peri-urban development are not seen as problems from the individual municipality perspective: municipalities only seem to cooperate as a 'community of common problems' than 'community of common interests'.

Area-based strategies now seem to be a promising approach, where sectoral programmes can be tailored for the specific needs of peri-urban areas and other specific territories. Three approach strategies are now being trialled:

- multi-level governance: where long-term (strategic) planning is to be integrated with medium-term and short-term (operational) planning. This aims at building coalitions and consensus among conflicting interests. Success of this strategy depends on proactive involvement of actors like: central government agencies, regional self government, local governments, farmers, land owners, planners, and NGOs;
- harmonization of development: with a spatial concept for integrated and sustainable development across the territory of the WMA. This will aim at coordination of sub-regional development programmes and projects, with a platform for negotiations between different municipalities and counties;

- sectoral programmes: targeted on specific types in the peri-urban area, this will support different strands including rural development, urban infrastructure and green infrastructure. This will also encourage positive spatial planning policies such as the creation of a Green Belt around Warsaw.

The outcome of such current initiatives will emerge in the coming years. One of the benefits of the PLUREL project may be to encourage cross-border exchange and learning, on such challenges as containing urban sprawl and the role of multilevel governance.

Dynamics of the peri-urban metro-ecology

Dynamics of the peri-urban

The cases above, and many others, show that there are powerful dynamics and pressures on the peri-urban ecology and landscape. There are physical processes in the landscape itself; direct urban pressures such as demand for housing and industry; indirect pressures such as those from lifestyle and location choices; and changing socio-cultural-institutional agendas or discourses. Overall there are the dynamics of globalization (globally driven modernization and restructuring): and a countervailing dynamic of localization (newly emerging relationships between individuals, communities, networks, and their socio-cultural landscape surroundings).

The aerial view graphic shows such dynamics, as seen with the example of the peri-urban Manchester city-region (Figure 47.4) (Ravetz and Warhurst 2011).

The later levels in this scheme, we suggest, are hugely important, as they address the very topical questions – what or who is the peri-urban landscape for? What is the policy agenda for land and ecosystems? Large parts of these areas are no longer competitive in agricultural use, and this

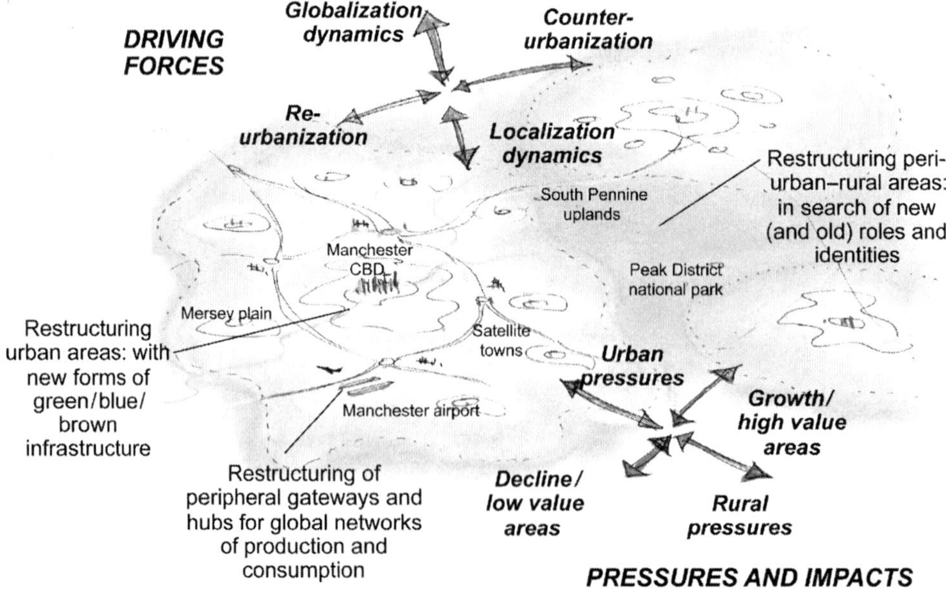

Figure 47.4 Globalization-localization dynamics in the peri-urban: example of Manchester city-region (based on Ravetz in press).

raises wider possibilities – natural conservation; high quality housing; green infrastructure with social enterprise; or more intensive cultivation to provide the city with food or energy. In turn, the local populations may be incoming commuters and mobile knowledge workers, semi-retired or retired; the future role, identity and 'reason for being' for the place and its communities, can be a very open question for an often fragmented but sometimes innovative set of communities. There are economic, ecological, historical, residential and functional agendas, often competing side by side (Gallent 2006).

The most obvious trend and pressure for change is the process of *metropol-ization* of former rural communities – a transition of economic activities, social types and spatial patterns of work and lifestyle (Duany *et al.* 2000). This is driven in many ways by the global networks of *cognitive capitalism* – a transition which is based not only on new economic functions, but on a knowledge based global economic order – with new kinds of social and cultural lifestyles, attitudes and perceptions (Scott 2000, 2006).

There is also a counterpart to the trends of mobility and networked globalization – a new kind of *localization*, in new kinds of attachments to place and landscape. Almost all public surveys find the public desire for green spaces, in small safe communities with good public services. This raises the priority for a peri-urban system of *green infrastructure* of open spaces, corridors, waterways, cycle and horse riding routes, as well as established nature conservation sites (Benedict and McMahon 2001). Such an infrastructure enables newly affluent/mobile communities, alongside lower income local communities, and potentially many other social and ethnic groups, to identify and locate themselves within a common landscape.

The green infrastructure agenda (or in many parts of Europe, 'Green-Blue infrastructure'), has been shown widely to increase economic vitality, investment, health, education, and social well-being (Ecotec 2008). In practice it is often sidelined and squeezed between different parts of the governance system: with the fragmentation of municipal governance, lack of private interest in the public realm, complex patterns of land ownership, and shortage of funds for maintenance.

However there are many successful initiatives that find new ways of combining the roles of public, private, civic and community sectors, such as the community forests and country parks. A new generation of partnership organizations, for example Pennine Prospects in the Manchester city-region, show new possibilities in multilevel, multisectoral networked types of strategy-making. These in turn can help to mobilize local social enterprises; for example, the Todmorden 'Incredible Edible' scheme for local food cultivation (www.incredible-edible-todmorden.co.uk).

Figure 47.4 also shows the tensions between globalization and localization, and their more well-researched spatial dimensions (Champion *et al.* 1998):

- Inward urban pull – urbanization or reurbanization;
- Outward rural pull or urban push – counterurbanization.

Both of these can be seen from analysis of Manchester and other European city-regions. One dynamic is related to social life-cycle effects: the city centre urban renaissance suits younger professionals, while families and older people tend to move outwards to smaller settlements. Such driving forces then produce typical impacts, as shown in the lower part of the figure:

- Impacts of growth and high value development;
- Impacts of decline and low value activity;
- Urbanization and direct urban growth impacts;
- Rural change and restructuring impacts.

Such impacts are often focused on certain places or certain communities, and we can simplify some typical outcomes into two extremes of social and economic life:

- Positive environments contain country parks, horse stables and niche garden centres. 'Winners' drive to work in the sunrise business parks, live in high-value housing, and access the nearby airport and shopping centre.
- Negative environments are next to motorways or landfill sites, with no entry to fenced off derelict land. 'Losers' might live in failing suburbs or regeneration areas, inaccessible to public transport, with falling values, the wrong kind of skills and training, alienated from the leisure-landscapes which surround them.

This is a caricature of a complex situtation, but it does map out the range of problems and opportunities. The point is here to raise the question (which underlies the PLUREL research above) – *'Is it possible to shape a complex city-region and its peri-urban ecology, to help produce better outcomes for both ends of the social/economic spectrum?'*

Cognitive capitalism and spatial ecology

A topical angle on the dynamics of peri-urban ecology comes from a cognitive-cultural-capitalist perspective, as in the writings of Scott (2000, 2006), and drawing on the 'creative classes' thinking of Florida (2004). This focuses on the nature and significance of 'creative cities', which is very relevant to the new types of public identity and engagement needed for green infrastructure.

Up to now the 'creative city' has tended to assume the cultural industries, against a backdrop of waterfront regeneration and artisanal urban quarters. But in reality this links to the wider infrastructure which is demanded implicitly by entrepreneurs and artisans alike – science parks, retail parks, distribution sheds, high quality housing developments and natural areas, all of which are powerful shaping forces in the peri-urban areas. This raises an agenda for the 'creative peri-urban region': this might start with the obvious high value 'islands of privilege', and then search for a new city-region or peri-urban equilibrium where people live, work and pursue leisure within an emerging space of physical, economic and socio-cultural dimensions. This suggests a *self-reinforcing localization dynamic* within the creative-city thesis, and suggests that the creative-city is one where prime activities become focused on a set of *urban cultural cores*, which in physical space might constitute a set of islands in a peri-urban space. Some possibilities for the urban-peri-urban relationship can then be explored:

- *Limited interaction*: an absence of connections between the creative 'inner urban' city and the city-region.
- A *hinterland* model. This suggests that people-movements and firm-movements can link people from the periphery to the cultural, work and business opportunities of the creative city. This is essentially a traditional commuting pattern, reliant in large part on an effective transport system.
- A *polycentric* model. This envisions the development of secondary 'creative satellites', circling the centrally concentrated core. Satellites might be expected to exhibit a degree of autonomy and symbolic differentiation, whilst benefiting from the synergies and spillovers of the core.
- *Hinterland-nodal model*. We might witness both hinterland and nodal dynamics operating simultaneously. A competitive-collaborative mix of relations may well exist in an 'up-scaled' creative city, networking the creative energy across a wider peri-urban area.

Following this through, we can envisage the concept of a 'spatial ecology', as applied to the peri-urban spaces of cities and regions. Such an ecological concept has several levels:

- Spatial inter-dependence of different settlement types, and different landscape types and locations, across a peri-urban city-region (from high value to low value; from sunrise to sunset; from golden gateways to grey sinks, etc.).
- Spatial interdependency of economic/social activities, and their emergence towards concepts of sustainable development. Peri-urban areas are often seeking new kinds of roles and identities but these do not exist in isolation, rather as a component in a larger system or ecology.
- Spatial interdependency of governance processes and citizenship patterns, in the context of an often fragmented peri-urban metro-scape. The situation is well known, of a poor urban district surrounded by rich peri-urban districts who use the services and pay few or no taxes. To move beyond this requires not necessarily a 'top-down' regional authority, but a responsive and responsible means of multisectoral and multilevel coordination.

Figure 47.5 shows the spatial dimension of'places' i.e. spatial concentrations of high or low value, fast or slow change, hubs or gateways, sinks or resources etc.

We can then draw on urban studies, and the more recent study of 'political urban ecology', to explore how relationships which are actualized in the spatial dimension can be polarized or reproduced in socio-economic-cultural hegemony, or in some cases conflict (Hajer 2003; Brand with Thomas 2005; Kaika 2005). In simple terms this revolves around the question of land and territory – ownership, stewardship, rights of access or of exclusion (Shoard 1983). Table 47.1 outlines some of the structural dynamics and conflicts in the peri-urban agenda, and how these can translate into the peri-urban ecology agenda.

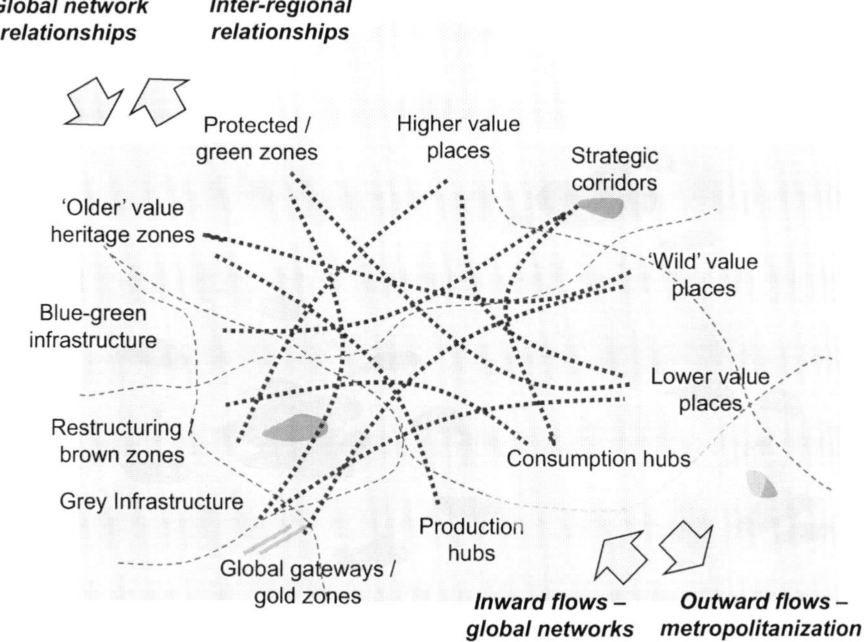

Figure 47.5 'Spatial ecology' of land-use relationships in the peri-urban.

Table 47.1 Structural dynamics and conflicts in the peri-urban: based on Ravetz (2010)

Type of dynamic	*Structural dynamics in the peri-urban*	*Implications for peri-urban ecology, green infrastructure and spatial ecology*
Global pressure with local results	Structural 'metropol-ization' as globalizing/localization tensions	The modernizing city-region, expanding labour market and service catchment, enabled by transport infrastructure, ICT and air travel
Innovation on the economic axis	Peri-urban zones of urban infrastructure, creative enterprise, urban residues etc.	Business parks, science parks, retail parks, utility infrastructure
Innovation and conflict between economic groups	Peri-urban economies as creative destruction of obsolescent economies and communities	Declining town and village centres, shops and services, centralized by cost-efficiency
Competition between economic groups	Peri-urban land use as a frontier for capital accumulation, in the circuit of urban property investment	Property values with/without planning permission, strategic institutional landholdings, land speculation and 'lotting' tactics
Conflict between social groups	Peri-urban land use as class competition for territory and control	Historic aristocratic country estates, public housing on peripheral estates with no services, new high quality gated commuter developments
Conflict between social groups	Peri-urban land use as frontier of dominant urban power and wealth, over rural interests	Landscape given up to road interchanges, energy, water and other infrastructure
Innovation on the social axis	Peri-urban community initiatives as new social movements, socio-cultural enterprise etc.	Urban fringe experiments, community forests, integrated catchment management, peri-urban revitalization, green tourism

Towards a peri-urban policy agenda

Spatial policy or local economic development, of course, will aim to respond to both problems and opportunities – but it can only act with some kind of policy agenda, or definition of 'what is the problem', and 'where lies the solution'? Such questions can easily get political, as every real public policy creates both winners and losers. We can chart out different policy agendas against a framework (Figure 47.6). This shows pressures and demands coming from urban and rural directions, and responses for both a conservation and development agenda (Ravetz 2000: CURE 2003):

- Urban containment/sustainability: with policies such as Green Belt, this uses peri-urban areas as a boundary for the city.
- Urban development and expansion: this focuses on feeding the city with infrastructure such as roads, airports, business parks and retail parks.
- Rural conservation/sustainability: here the countryside is seen as an asset for a mainly urban population, enhancing landscape and ecological qualities.
- Rural development and enterprise: favours small business development which can easily conflict with the conservation agenda.

Such a framework helps us to understand the range of problems and opportunities – and then to see where current policies are working, or with internal conflict, sub-optimal, responsive to future changes and multiple objectives, or simply missing.

This applies in particular to the issues of landscape planning and management, and the balance of urban-rural interests at the strategic or peri-urban scale. The practical agenda focuses on:

- Climate change adaptation and resilience: flood defence and carbon storage;
- Strategic river corridors, catchments, infrastructure corridors;

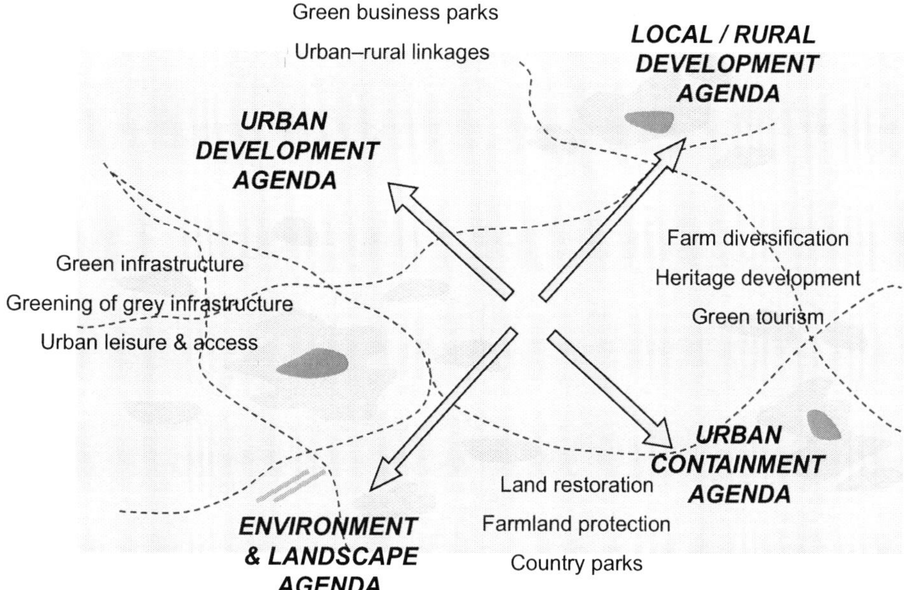

Figure 47.6 Sustainability policy agendas for the peri-urban.

- Sustainable agriculture and forestry practices which enhance landscape ecology;
- Landscape as a generator of quality living environments, and of leisure and tourism opportunities;
- The overall ecological balance of the city within its peri-urban hinterland.

The challenge comes when these aspirations come into conflict with other social or economic needs and demands, or simply with the typical fragmentation and disorganization of the peri-urban areas.

Next steps

Conclusions

Urban ecology clearly needs to extend its remit to a 'peri-urban ecology', just as cities and urban structures are extending to a wider hinterland. This raises questions for the practice of landscape ecology itself: on scale and hierarchy; on connectivity and conservation; and on effective management (see Chapters 21, 22 and 26 this volume).

But it is clear that the agenda has multiple layers, from the physical environment and techniques for management, to the economic and political, to the socio-cultural and structural. In this chapter we have swung over to the latter end of the spectrum, in pursuit of the question – what or who is the peri-urban landscape for? How should society decide on what to do, whether production, leisure and tourism, new urban development or nature conservation? How do current tangible pressures and discourses about globalization or localization enter this debate?

The concept of 'sustainability' then injects a further layer of aspiration, of value seeking, of expectation, and of emerging forms of governance (or not quite emerging yet, as is often the case). If we look more closely at these and similar case studies, we might find the magic word 'sustainability' is often used as a blanket to cover up otherwise open cracks and divisions within society. The peri-urban then seems to act as a kind of chromotograph, separating out in visible circles the residues and unresolved problems of a city-region.

Implications for policy

So to respond to the peri-urban agenda, the implications for strategy and policy development are challenging. These are some promising approaches:

- **Policy integration and partnership**: the peri-urban by its nature requires multilevel horizontal, vertical and sectoral integration between organizations, sectors and levels. This places new demands on policymakers, but the range of opportunities can already be seen, with the growth of new partnership organizations and collaborative planning processes (Healey 1997).
- **Policy enlargement**: as seen with local development and green infrastructure. For example, urban street trees can be a problem for markets and property values, and a problem for 'rational management' public programmes, which are risk averse and cost-driven. Urban trees, and other GI features, rely on a wider scope for community involvement: of values which are more than financial or functional, and on timescales which are more than the short term targets and objectives.
- **Policy mobilization**: this extends the partnership approach towards active investment: including financial, physical and human resources. Mobilizing GI programmes, particularly on a peri-urban or sub-regional level, is a challenge to conventional forms of investment: but

there is clearly added value for many types of stakeholders. How to mobilize this 'extended stakeholder community' is a challenge. It is especially topical at a time of public sector deficits and spending cuts: rising expectations from citizens and communities, and tension or conflict between the pressures of globalization and localization.
- **Policy evaluation and assessment**: each of the above is a challenge for traditional linear models for evaluation and assessment. The new model requires an extended and more holistic view on the policy environment and discourse/regime: alongside a more system-focused approach to physical science in landscape ecology (Waltner Toews *et al.* 2009). This would apply the general approach of 'complex adaptive systems', to link between physical ecosystems and socio-economic-cultural systems.

Overall, from this brief review, an emerging stream of 'peri-urban studies' can be envisaged, just as 'urban studies' was defined over 100 years ago. Its application to 'peri-urban ecology' will be increasingly urgent and topical in the coming century.

References

Angel, S, Sheppard S.C. and Civco, D.L. (2005) *The Dynamics of Global Urban Expansion*, Washington, DC: World Bank, Transport and Urban Development Dept.

Bauer, A, Röhl, D, Haase, D, & Schwarz, N, (2011): Leipzig – between growth and shrinkage: In: Pauleit, S., Bell, S., and Aalbers, C., (eds), *Peri-Urban Futures: Land Use and Sustainability,* Berlin: Springer.

Bender, T, (2007): *The Unfinished City: New York and the Metropolitan Idea:* NY, New York University

Benedict, M.A. and McMahon, E.T. 2001 *Green Infrastructure: Smart Conservation for the 21st Century*, Washington, DC: The Conservation Fund, Sprawl Watch Clearinghouse.

Blair, A.M. (1987) 'Future Landscapes of the Rural-Urban Fringe', in D.G. Lockhart D.G. and B. Ilbery (eds), *The Future of the British Rural Landscape*, Norwich: Geo Abstracts.

Brand, P. with Thomas, M.J. (2005) *Urban Environmentalism: Global Change and the Mediation of Local Conflict*, New York and Oxford: Routledge.

Castells, M. and Hall, P. (1994) *Technopoles of the World: The Making of 21st Century Industrial Complexes*, London: Routledge.

Champion, T., Atkins, D., Coombes, M. and Fotheringham, S. (1998) *Urban Exodus*, a report for CPRE, London: Campaign for the Protection of Rural England.

CPRE (Campaign for the Protection of Rural England) (2007) *2026 – a Vision for our Countryside*, London: CPRE.

CURE (Centre for Urban and Regional Ecology) (2003) *Sustainable Development of the Countryside Around Towns*, Report to the (then) Countryside Agency (now Natural England).

Davis, M. (2005) *Planet of Slums*, London: Verso.

Duany, A., Plater-Zyberk, E. and Speck, J. (2000) *Suburban Nation: The Rise of Sprawl and the Decline of the American Dream*, New York: North Point Press.

EcoTec (2008) *Economic Benefits of Green Infrastructure*, Report to the NWDA, Warrington: Northwest Regional Development Agency.

EEA (European Environment Agency) (1999) *Environmental Indicators: Typology and Overview*, Copenhagen: European Environment Agency.

Elson, M., Walker, S., MacDonald, R. and Edge, J. (1993) *The Effectiveness of Green Belts*, London: DoE, HMSO.

Florida, R. (2004) *The Creative Talent: Cities and the Creative Class*, New York: Routledge.

Gallent, N. (2006) *Planning on the Edge*, Abingdon: Routledge.

Garreau, J. (1991) *Edge City: Life on the New Frontier*, New York: Doubleday.

Giannini E, (1994) *Metroscape:* Melbourne, Royal Melbourne Institute of Technology.

Greater Helsinki Vision (2007): *Metroscape Helsinki 2007-2057: a review of fifty years of successful regional development:* Greater Helsinki Vision (International Ideas Competition): available as of November 2010 on www.greaterhelsinkivision.fi

Grochowski, M., & Slavinski, T. (2011): Warsaw - Spatial growth with limited controls: In: Pauleit, S., Bell, S., and Aalbers, C., (eds), *Peri-Urban Futures: Land Use and Sustainability,* Berlin: Springer.

Haase, D., Piorr, A., Müller, F.,, Bell,. S., Rickebusch, S., van Delden, H., and Loibl, W., (2011): Methods, models and tools – the quantitative part of PLUREL: In: Pauleit, S., Bell, S., and Aalbers, C., (eds), *Peri-Urban Futures: Land Use and Sustainability,* Berlin: Springer.

Hajer, M.A. (2003) *The Politics of Environmental Discourse: Ecological Modernization and the Policy Process*, Oxford: Oxford Scholarship Online.

Hall, P, and Pain, K. (2006) *The Polycentric Metropolis: Learning from Mega-City Regions in Europe*, London: Earthscan.

Healey, P. (1997) *Collaborative Planning: Shaping Places in a Fragmented Society*, London: Macmillan.

Junde, L. and Zaide, P. (1996) Rural–urban transition and urban growth in Shanghai, *Asian Geographer*, 15(1–2): 106–13.

Kaika, M. (2005) *City of Flows: Modernity, Nature, and the City*, New York: Routledge.

Kasarda, J.D. (2004) *Asia's Emerging Airport Cities: Urban Land Asia*, Washington, DC: Urban Land Institute.

Kraffczyk, D, (2004) *The MetroScape: a geography of the contemporary city*, Cambridge MA, Harvard University, Graduate School of Design.

McGranahan, G., Satterthwaite, D. and Tacoli, C. (2004) *Rural–Urban Change, Boundary Problems and Environmental Burdens*, London: International Institute for Environment and Development working paper.

McGranahan, G., Songore, J. and Kjellen, M. (1996) 'Sustainability, poverty and urban environmental transitions', in C. Pugh (ed.), *Sustainability, Environment and Urbanization*, London: Earthscan.

Natural England and CPRE (2009) *Green Belts: a greener future*. Cheltenham, UK: Natural England and CPRE.

Neuwirth, R. (2005) *Shadow Cities: A Billion Squatters, A New Urban World*, London/New York: Routledge.

Nicholson-Lord, D. (1987) *The Greening of the Cities*, London: Routledge.

Pintar, M, & Perpar, A. (2011) Koper – Finding the balance between spatial growth, agricultural land protection and nature conservation: In: Pauleit, S., Bell, S., and Aalbers, C., (eds), *Peri-Urban Futures: Land Use and Sustainability,* Berlin: Springer.

Ravetz, J. (2000) *City-Region 2020: Integrated Planning for a Sustainable Environment*, London: Earthscan (Chinese language version, transl. Jian-Cheng Lin and Tian-Tian Hu Taipei, Chan's Publishing Co., Ltd (Taiwan)).

Ravetz, J. (in press) 'Integrated governance in the peri-urban: experiences from an EU integrated project', *Living Reviews in Landscape Research*.

Ravetz, J. & Warhurst, P. (2011): Towards the new local-global: the emerging peri-urban of the Manchester city-region: In: Pauleit, S., Bell, S., and Aalbers, C., (eds), *Peri-Urban Futures: Land Use and Sustainability,* Berlin: Springer.

Roberts, P., Ravetz, J. and George, C. (2009) *Environment and City: Critical Perspectives on the Urban Environment around the World*, Oxford: Routledge/Taylor and Francis.

Sassen, S. (1994) *Cities in a World Economy*, Thousand Oaks, CA: Pine Forge Press.

Scott, A.J. (2000) *The Cultural Economy of Cities: Essays on the Geography of Image-Producing Industries*. New York: Sage.

Scott, A.J. (2006) 'Creative Cities: Conceptual Issues and Policy Questions', *Journal of Urban Affairs*, 28(1): 1–17.

Shoard, M. (1983) *This Land is Our Land*, London: Paladin.

TEEB (2010) *The Economics of Ecosystems and Biodiversity. Mainstreaming the Economics of Nature: A synthesis of the approach, conclusions and recommendations of TEEB*. New York, UN Environment Programme.

UN Department of Economic and Social Affairs, Population Division (2004) *World Urbanization Prospects: The 2003 Revision*, New York, United Nations.

UN Population Fund (UNFPA) (2007) *State of the World Population 2007; Unleashing the Potential of Urban Growth*, New York, UNFPA.

Waltner-Toews, D., Kay, J.J. and Lister, N.-M. (eds) (2009) *The Ecosystem Approach: Complexity, Uncertainty, and Managing for Sustainability*, New York: Columbia University Press.

Webster, C.J. and Lai, W.C.L. (2004) *Property Rights, Planning and Markets: managing spontaneous cities*, Cheltenham, UK and Northampton, MA: Edward Elgar.

Westerink, J, Aalbers C, Velinova, T (2011) Haaglanden-Squeezed, wet and diverse. In: Pauleit, S., Bell, S., and Aalbers, C., (eds), *Peri-Urban Futures: Land Use and Sustainability,* Berlin: Springer.

Wood, R. and Ravetz, J. (2000) 'Recasting the Urban Fringe', *Landscape Design*, 294(10): 13–16.

48

Biodiversity as a statutory component of urban planning

David Goode

Introduction

The Convention on Biological Diversity was one of several initiatives stemming from the Earth Summit in Rio de Janeiro in 1992, which together formed the basis of an international agreement on sustainable development. Signatories recognised that urgent action was necessary to halt the global loss of species and habitats and that each country had a primary responsibility to conserve and enhance biodiversity within its own jurisdiction. They agreed to develop national strategies, plans and programmes for the conservation and sustainable use of biodiversity. This chapter describes how such programmes have been developed to conserve biodiversity in urban areas, using examples from London and Edinburgh in Britain and from Cape Town in South Africa.

The UK biodiversity action plan

After the Earth Summit in Rio, the UK government set up a national steering group to develop a detailed programme of action. Whilst the group concentrated largely on setting national targets for endangered species and habitats, representatives of local authorities were keen to ensure that the programme would encourage greater public awareness of biodiversity issues and that it would provide a mechanism to implement action on the ground at the local level. The outcome was a far-reaching set of recommendations that radically changed the face of nature conservation in the UK. The steering group report (DoE 1995) recognised that four key components needed to be developed:

- developing costed targets for the most threatened and declining species and habitats
- establishing an effective system for handling biological data at both local and national level
- promoting increased public awareness of the importance of biodiversity, and broadening public involvement
- promoting Local Biodiversity Action Plans as a means of implementing the national plan at a local level.

The resulting UK Action Plan has had a profound effect on the practice of nature conservation by developing clear targets and action plans for conservation of endangered habitats and species.

But it also promoted the development of urban nature conservation, particularly through the production of Local Biodiversity Action Plans. These were intended to ensure delivery of the overall plan at the local level, but they also provided a means of engaging a wider constituency in biodiversity conservation, and catered for local rather than national needs. Most towns and cities have now produced Local Biodiversity Action Plans, reflecting their local priorities. Species and habitats have been selected for priority treatment to meet these needs. It is notable that such plans include a range of habitats and species characteristic of urban areas. Habitat action plans include parks, cemeteries, private gardens, wastelands, railwaysides, docklands and even green roofs. Action plans for species include badger, bat, swift, house martin, black redstart, song thrush, house sparrow and peregrine falcon.

The guidance issued by the UK Local Issues Advisory Group (UK LIAG 1997) sets out the following primary functions of Local Biodiversity Action Plans:

- To ensure that national targets for species and habitats, as specified in the UK Action Plan, are translated into effective local action.
- To identify targets for species and habitats appropriate to the local area, and reflecting the values of people locally.
- To develop effective local partnerships to ensure that programmes for biodiversity conservation are maintained in the long term.
- To raise awareness of the need for biodiversity conservation in the local context.
- To ensure that opportunities for conservation and enhancement of the whole biodiversity resource are fully considered.
- To provide a basis for monitoring progress in biodiversity conservation at both local and national level.

These local biodiversity action plans have been extremely effective in bringing biodiversity conservation firmly into the work of many urban local authorities. Within five years of guidance being issued, most local authorities had embarked on production of a local 'BAP'. With government reorganisation of local planning in 2002 it was announced that Local Biodiversity Action Plans would be included in Community Strategies. All local authorities were therefore required to incorporate biodiversity planning within their community strategy and many urban local authorities found themselves having to address biodiversity issues for the first time. In the case of London, this had already been taken a stage further by placing a requirement on the Mayor of London to include Biodiversity as part of the statutory London Plan (HM Government 1999). This is described more fully in Chapter 8. The general requirement for local authorities was reinforced by a new duty placed on all public bodies, including local authorities, to have regard to biodiversity in their decision-making and in carrying out their functions (HM Government 2006). A similar duty was introduced in Scotland, under the Nature Conservation (Scotland) Act 2004, which requires public authorities to further the conservation of biodiversity. Government guidance for local authorities on implementing the Biodiversity Duty was published by Defra (2007).

One of the key features of Local Biodiversity Action Plans is the development of action plan partnerships, which include a wide spectrum of players all of whom are involved to a greater or lesser extent in delivering the plan. For some of these bodies, like the urban wildlife trusts and the Royal Society for the Protection of Birds (RSPB), conservation is their mainstream activity. They are directly involved in the protection and management of nature reserves, and in developing programmes involving local people. For others like BTCV (formerly the British Trust for Conservation Volunteers), local Groundwork Trusts and regional park authorities, it is a large part of their core activities. But many other bodies are involved for which biodiversity conservation is only a small part of their overall work. These include water companies, housing corporations, transport operators, port authorities, British Waterways and regional health authorities. All these bodies need to be involved in biodiversity conservation as between them they have considerable influence on the extent to which nature can exist and thrive in towns and cities.

Local Biodiversity Partnerships are in most cases led by the relevant local authority. A key to the success of local action plans has been to engage with all these players to ensure that they are directly involved in biodiversity conservation in ways which did not occur before. Individual partners are encouraged to take responsibility for particular action plans. Some towns and cities have gone further by seeking to ensure that partners sign up to a Biodiversity Charter. Many of these existing partnerships provided precisely what was needed for Local Sites Partnerships, as advocated by Defra in its guidance related to local sites for nature conservation (Defra 2006).

Because Local Biodiversity Action Plans operate through partnerships they are particularly appropriate in the urban environment where their approach is consistent with the community-based philosophy of urban nature conservation. But it has to be realised that despite the considerable progress which has been made over the past 20 years, urban conservation still tends to be seen as a minority interest in relation to both the national biodiversity strategy, and mainstream urban development. The problem is that urban biodiversity has three kinds of value:

- the intrinsic value of specialist urban habitats and species
- the value of biodiversity to people living in urban areas
- the value of the ecological services which it provides.

The legislative framework for conservation of biodiversity in the UK was not designed to cater for urban areas and even with the benefit of recent legislation, such as the Natural Environment and Rural Communities (NERC) Act 2006, it can still be argued that none of the values referred to above are catered for effectively.

In the first case, urban habitats tend to be undervalued as they commonly fall outside the categories of habitat recognised nationally as being of nature conservation importance. The physical characteristics and processes of urban areas frequently result in a range of habitats and species of particular urban character that are not generally found in the countryside (see Chapter 35 this volume). Some are virtually restricted to specialised conditions resulting from post-industrial landscapes which are poorly described in UK habitat and vegetation classifications. As a result national policies for biodiversity conservation do not give adequate recognition to these specialist urban habitats. A few outstanding examples have achieved Site of Special Scientific Interest (SSSI) designation, as in the case of invertebrate populations at Canvey Wick in Essex, but these are exceptional cases and there are at present no clear guidelines for evaluating such habitats.

The value of biodiversity for people living in urban areas has become more widely appreciated in recent years, being now an accepted element of nature conservation policy at national level. However, this does not sit easily with a national biodiversity strategy which gives priority for protection to habitats and species of international and national importance. Hardly any of the many important biodiversity sites in towns and cities qualify as SSSI in the national context. Protection of important urban habitats requires other criteria to be recognised, reflecting their social and cultural values. This needs to be given greater recognition in planning guidance, otherwise such sites will receive only limited protection in the planning process. The guidance issued by Defra (2006) on the identification, selection and management of local sites for nature conservation goes a long way to meet this requirement.

Although much progress in urban nature conservation has been made through Local Biodiversity Action Plans, the lack of a strong national policy for urban conservation has resulted in difficulties. It was intended that the national Action Plan should cater for all aspects of biodiversity (by setting national targets for priority habitats and species) but the UK Biodiversity Group failed to do this for urban habitats. Admittedly the need for a broad urban-habitat category was recognised early on, but priority habitats were not defined. This was due largely to the lack of a nationally recognised classification of urban habitats. Several years passed before a study commissioned by English Nature (Tucker *et al.* 2005) recommended a new category of 'Post-industrial sites of High Ecological Quality' as a Priority Habitat. In 2008 it was decided to add this new category to the UK Biodiversity Strategy, under the title 'Open mosaic habitats on previously developed land'. Work has been completed to provide a definition and description of the range of habitats involved. The inclusion of such habitats as a national priority in the UK Biodiversity Strategy now provides a means of safeguarding these specialist urban habitats. The case for their conservation is extremely strong, not least because of the range of critical species, especially invertebrates, which are involved. But because many of these habitats occur within brownfield land there is an urgent need for unambiguous guidance, not only for developers but also for those involved in the identification and protection of important sites.

London Biodiversity Action Plan

The London Biodiversity Action Plan (LBP 1996) has been developed in parallel with, and complementary to, the statutory Biodiversity Strategy produced by the GLA (see Chapter 8). Led initially by the London Ecology Unit, and since 2000 by the Greater London Authority, the Action Plan has developed programmes for the conservation of priority habitats and species that are of particular significance for London. The plan is developed and implemented by the London Biodiversity Partnership, consisting of a wide range of bodies with an interest in biodiversity. These bodies are listed below. In January 2000 the Partnership published an audit of the key habitats and species in London, followed in 2001 by detailed Action Plans for Priority Habitats and Species. Both these documents can be found on the Partnership's website (www.lbp.org.uk). The original plan included 10 action plans for priority habitats and 12 action plans for priority species.

Members of the London Biodiversity Partnership

- Greater London Authority
- London Boroughs Biodiversity Partnership
- BTCV
- Association of London Government
- English Nature
- Community Initiatives Partnership
- London Wildlife Trust
- London Natural History Society
- Royal Parks Agency
- Thames Water Utilities Ltd
- Wildfowl and Wetlands Trust
- Royal Society for the Protection of Birds
- London First
- London Boroughs
- British Waterways
- Environment Agency
- London Underground Ltd
- Thames Estuary Partnership
- Peabody Trust

The London Biodiversity Action Plan

Habitat Action Plans	*Species Action Plans*
Woodland	Bats
Chalk Grassland	Water Vole
Heathland	Grey Heron
Wasteland	Peregrine
Acid Grassland	Sand Martin
Tidal Thames	Black Redstart
Canals	House Sparrow
Churchyards and Cemeteries	Stag Beetle
Private Gardens	Tower Mustard
Parks, Squares and amenity grassland	Mistletoe
	Reptiles
	Black Poplar

The selection of priority habitats for action is based broadly on the following criteria:

- Provides a good opportunity for Londoners to enjoy contact with nature (e.g. private gardens; parks and city squares; churchyards and cemeteries)
- Is an example of UK priority habitat within London (e.g. heathland and acid grassland)
- Supports a diversity of fauna and flora, including uncommon species (e.g. chalk grassland)
- Is under threat in London (e.g. heathland)
- Presents opportunities for habitat restoration and enhancement (e.g. heathland).

Criteria for the selection of priority species include the following:

- Culturally valued and appealing, offering an opportunity to raise greater public awareness of biodiversity (e.g. grey heron, mistletoe)
- Characteristic of London (e.g. black redstart)
- Priority species in UK BAP with a significant population in London (e.g. tower mustard, black poplar, bats, water vole and stag beetle)
- Substantial recent decline (e.g. house sparrow)
- Restricted distribution in London (e.g. reptiles)
- Easy to monitor (e.g. sand martin).

In addition to these specific action plans, the partnership has identified a number of generic actions including site management, habitat and species protection, ecological monitoring, biological recording and communications, for each of which specific programmes have been developed.

The overall programme provides a set of targets for urban nature conservation in the capital aimed at conserving and enhancing the status of critical habitats and species, and encouraging greater public participation in wildlife conservation. Some of the Action Plans may seem very specific, but are actually addressing broad issues of urban development. The plan for black redstarts is a case in point. These nationally protected birds tend to be associated with derelict buildings and urban wasteland. With the pace of urban redevelopment increasing, their habitat is under threat. However, one of the options to cater for this species is to create artificial wasteland habitats on roofs of buildings. Such 'black redstart roofs' or brown roofs are now being developed locally in the Thames Gateway to cater for this species, and with it no doubt a host of other wasteland specialities (Frith and Gedge 2000). The conservation of wasteland habitats could well be facilitated more widely through the use of such roofs, and also through the careful design of new industrial units to accommodate a range of other artificial habitats. Such proposals are advocated by the London Development Agency in its guidance on biodiversity for developers (LDA 2004), and also by the GLA in a detailed report supporting policies for green roofs in the London Plan (GLA 2008). The environmental benefits of green roofs goes well beyond their importance for biodiversity since they have clear benefits for flood alleviation and water conservation (see Chapter 44 this volume), and also for ameliorating the urban microclimate and reducing the heat island effect in cities (Ecoschemes 2003).

As well as the overall London Biodiversity Action Plan, individual London Boroughs have developed their own local action plans which cater for the specific needs of each borough. Examples are those for Tower Hamlets and Westminster, both of which are heavily built-up parts of the capital. In Tower Hamlets (THP 2003) there are few examples of semi-natural habitats and the main emphasis is on habitats which are clearly urban in character, including parks, city squares, cemeteries, gardens, school grounds, the built environment (including green and brown roofs), commercial areas, and brownfield land. Species action plans include bats, black redstart, bird's foot trefoil, brimstone butterfly and a southern European species of jumping spider *Macaroeris nidicolens* for which Mile End Park is the only known locality in the UK. In Westminster (Westminster Biodiversity Partnership 2008), the list of Habitat Action Plans is much the same, but here there is an additional category of veteran trees and decaying wood reflecting the importance of old trees in some of the long-established parks and squares. Emphasis is also placed on improving aquatic habitats of parkland lakes and the Thames foreshore. Species Action Plans include tawny owl, hedgehog, house sparrow, and a nationally rare moth, the buttoned snout moth (*Hypena rostralis*) which occurs along railway and canal corridors and in one of the borough's cemeteries.

Edinburgh Local Biodiversity Action Plan

The city of Edinburgh provides something of a contrast, both in the range of habitats represented and in the presence of several nationally rare species. It is notable that the Action Plan (Edinburgh Biodiversity Partnership 2004) is largely devoted to semi-natural habitats and places little emphasis on urban habitats. The plan drawn up in 2000 attempted to produce action plans for the full range of habitats, but in 2004 they were reduced to eight categories as follows:

- Coastal and marine
- Farmland
- Uplands
- Wetlands
- Rock faces
- Semi-natural grassland
- Urban
- Woodland.

The extraordinary range of habitats within the boundary of the city includes a variety of coastal habitats, notably rocky shores and islands; wetlands, including mesotrophic and oligotrophic lakes and even examples of raised bog; semi-natural grasslands and rock faces on the hills of volcanic rock such as Arthur's Seat and Castle Rock; and extensive woodlands in river valleys and on some of the hills. Many of these habitats are of considerable conservation value and there are numerous notable species associated with them. The original Action Plan identified nearly 100 priority species. Most of these were later subsumed within Habitat Action Plans, but some which have special requirements have retained individual Species Action Plans. An example is the sticky catchfly *Lychnis viscaria* which has a very restricted distribution in the UK. A small number of these plants still grow on Arthur's Seat and the object of the Action Plan is to manage the existing sites to encourage this species, and to reintroduce the plant to other suitable habitats in Edinburgh. Other plants for which Species Action Plans have been produced include juniper, rock whitebeam, maiden pink and adder's tongue fern. Amongst the birds, the swift has received special attention in the Edinburgh Action Plan owing to a recent reduction in its numbers in Scotland. Special efforts are now being made, working with Edinburgh City Council and developers, to provide nest boxes on existing buildings and to incorporate swift bricks in new developments. Other species singled out for special treatment include otter *Lutra lutra* and badger *Meles meles*. After an absence until the early 1990s, otters are gradually recolonising Edinburgh's rivers and the Species Action Plan aims to identify and establish a series of refuge areas with artificial holts where appropriate. Small populations of badgers are present even in heavily built-up areas of the city, especially around Corstorphine Hill and some of the wooded valleys. The Action Plan includes guidance for planners, developers and land managers on their legal responsibilities for badgers in relation to development.

Cape Town Biodiversity Strategy

The City of Cape Town, with a population of over three million and covering 2,500 sq km, lies within the small area that comprises the Cape Floristic Kingdom, one of only six floristic kingdoms on earth, where the concentration of endemic plant species is without parallel in an urban setting. The Cape Floristic Kingdom contains approximately 9,600 plant species, of which 70 per cent are endemic. Within the relatively small area of South Africa where this kingdom occurs, the Fynbos and Renosterveld vegetation groups contain most of the floral diversity and many species are extremely localised in their distribution, with sets of such localised species organised into 'centres of endemism'. The city of Cape Town sits squarely on two such centres of endemism, as a result of which several hundred species are threatened by urban expansion and development. As a global biodiversity hot-spot Cape Town has an international responsibility for biodiversity which is unique for a major city. Of particular importance are the Cape Lowlands which support nearly 1,500 plant species of which 76 are endemic and 131 are Red Data Book species. These lowlands are also home to the majority of the human population including the most disadvantaged people of Cape Town.

Recognising the importance of conserving biodiversity, the city government has committed itself to developing, implementing and actively promoting a citywide Biodiversity Strategy (City of Cape Town 2003). This is one of six priority strategies in the city's Integrated Metropolitan Environmental Policy, adopted in 2001. The Strategy identifies seven objectives that must all be met to adequately conserve the unique biodiversity of Cape Town. These objectives fall under the following headings:

> **Primary biodiversity areas.** Here the strategic objective is the establishment of a network of protected areas that are actively managed for biodiversity. These include 22 protected areas administered by the city. The city is also unusual in that the entire Table Mountain National Park is situated within its boundaries.

Secondary biodiversity areas. These are areas which are not managed primarily for biodiversity, but which serve to connect primary biodiversity areas into a complete and functional network.

Freshwater aquatic systems. Here the objective is to manage wetlands and water courses to provide green corridors as part of the biodiversity network.

Invasive alien species management. The management of invasive alien species is a crucial element of the strategy as this is one of the major causes of extinctions of indigenous species.

Legislation and enforcement. This objective is aimed to ensure that existing and new legislation on biodiversity conservation is enforced and made effective.

Information and monitoring. This objective recognises that up to date information is required for implementation of the Strategy, and that such information is made available to decision makers at all levels.

Education and awareness. The strategy relies heavily on support from the people of Cape Town. This objective is concerned with empowering citizens as well as informing city leaders and officials about their responsibilities regarding the city's biodiversity.

The Cape Town Biodiversity Strategy is described in greater detail by Katzschner *et al.* (2005).

Although the City of Cape Town has demonstrated its strong commitment to biodiversity conservation by the development of this strategy, there are still serious administrative difficulties in its implementation, largely due to a lack of clarity regarding responsibilities for biodiversity. There are also overwhelming social and economic challenges which make progress difficult. The way forward is seen as involving local people in protection and management of biodiversity, and developing an integrated approach in order to avoid, so far as possible, further loss of biodiversity through ill-planned urban development.

Conclusions

These examples from London, Edinburgh and Cape Town illustrate different approaches to biodiversity conservation in major cities. In the UK biodiversity plans involve action to protect the most significant habitats as well as key species which are locally important. The plans are firmly rooted in action by local organisations and communities, very often with strong involvement of businesses and commerce.

The situation in Cape Town is very different. The importance of the Cape flora is recognised internationally. Here there is an outstanding need for biodiversity conservation, yet the city must also deal with major social inequalities and pressure for urban expansion, which will inevitably have consequences for biodiversity. In the face of this the city council has produced a far-sighted biodiversity strategy, which aims to conserve the most important areas, but acknowledges social and economic realities.

To be successful all these examples require adequate legislation and a real commitment on the part of civic leaders to make biodiversity conservation a normal part of city life.

References

City of Cape Town (2003) *Biodiversity Strategy*, online, available at: www.capetown.gov.za/imep/pdf/biodiversity.pdf.

Defra (2006) *Local Sites: Guidance on their Identification, Selection and Management*, London: Department for Environment, Food and Rural Affairs.

Defra (2007) *Guidance for Local Authorities on Implementing the Biodiversity Duty*, London: Department for Environment, Food and Rural Affairs.

DoE (1995) *Biodiversity: The UK Steering Group Report. Vol 1: Meeting the Rio Challenge*, London: HMSO.

Ecoschemes (2003) *Green Roofs: Their Existing Status and Potential for Conserving Biodiversity in Urban Areas*, English Nature Research Reports No. 498, Peterborough: English Nature.

Edinburgh Biodiversity Partnership (2004) *Edinburgh Biodiversity Action Plan 2004–2009:* Edinburgh: City of Edinburgh Council.

Frith, M., and Gedge, D. (2000) 'The black redstart in urban Britain; a conservation conundrum?' *British Wildlife*, 11(6): 381–8.

GLA (2008) *Living Roofs and Walls: Technical Report Supporting London Plan Policy*, London: Greater London Authority.

HM Government (1999) *Greater London Authority Act 1999*, London: The Stationery Office.

HM Government (2006) *Natural Environment and Rural Communities Act 2006*, London: HMG.

Katzschner, T., Oelofse, G., Wiseman, K., Jackson, J. and Ferreira, D. (2005) 'The City of Cape Town's Biodiversity Strategy', in T. Trzyna (ed.), *The Urban Imperative*, Sacramento CA: California Institute of Public Affairs, pp. 91–5.

LDA (2004) *Design for Biodiversity: A Guidance Document for Development in London*, London: London Development Agency.

London Biodiversity Partnership (1996) *Capital Assets: Conserving Biodiversity in London*. London: London Biodiversity Partnership.

London Biodiversity Partnership (2001) *London Biodiversity Action Plan: Volume 2, The Action*, London: London Biodiversity Partnership.

THP (2003) *Draft Local Biodiversity Action Plan for Tower Hamlets*, pre-consultation draft, London: Tower Habitats Partnership, London Borough of Tower Hamlets.

Tucker, G., Ash, H. and Plant, C. (2005) *Review of the Coverage of Urban Habitats and Species within the UK Biodiversity Action Plan*, English Nature Research Report Number 651, Peterborough: English Nature.

UK LIAG (1997) *Guidance for Local Biodiversity Action Plans. Guidance Notes 1–5*, London: UK Local Issues Advisory Group, Local Government Management Board.

Westminster Biodiversity Partnership (2008) *Biodiversity Action Plan*, London: City of Westminster.

49

Making urban ecology a key element in urban development and planning

John Stuart-Murray

…where ecological and aesthetic order are congruent, we have a human ecosystem.

(Lyle 1991: 39)

Introduction

This chapter explores how different theoretical relationships can advance an integration of ecology in the urban environment, through an understanding of different theoretical languages and disciplines. The aim could be achieved in part by recognising of the interdependency of the needs of culture and nature. Comparisons of diverse theories from seemingly opposing fields could then be used as instruments to formulate a common language and a common perception. By examining the vocabularies of urban legibility, landscape urbanism, landscape ecology and biosemiotics, new concepts of cognitive landscape ecology and ecological syntax will be advanced. The terms also imply an engagement with ecology as an active cultural resource, instead of dualist assumptions that human activity is antithetical to ecological diversity.

The city as a silvicultural and/or ecological system

Making and exploring a simple analogy can test the validity of the approach, which seeks to synthesise different fields of knowledge. Urban form and growth can, for example, be compared to ecological elements, structures and processes. The founding of a city and how it responds to its environment can be compared with primary succession. Like primary succession, early urban development can display a degree of randomness and unpredictability, before a more stable and predictable state, with cyclical properties is reached. The comparison can be critically examined. For example, and at a broad level, a city, which in its intrinsic qualities, choice of site and subsequent growth fails to acknowledge the properties and characteristics of its environment, can become unsustainable. A wind-blown seed of *Betula pendula*, which lands on a salt marsh may germinate in a drought year, but it is unlikely to survive in the long term.

In silvicultural terms the city can be broken down into several analogous elements. The high-rise canopy trees that cast shade on life below. The medium-rise shrub layer or understorey and below that the ground layer, which grow beneath the canopy. At the boundary of the forest with other ecosystems lies the edge or ecotone. The ecotone can be compared with the boundary between mass and void, between buildings and the spaces they enclose. For urban designers like Jan Gehl (Gehl 2006) as well as ecologists, this boundary is highly significant, primarily because light levels and interior to exterior connections change rapidly over short distances. An ecotone, which has a diverse and well-structured transition, supports a wide variety of animal species. One, which has an abrupt boundary between canopy trees and open grassland, for example, supports a much lower diversity. Similarly, as Gehl has advanced, the long, divisive and solid facades of many modern financial buildings in business districts, which provide no visual connection between interior and exterior spaces, do not support a diversity of street life. They do not attract those who choose, but do not have to use a city space from necessity – one of Gehl's key criteria for evaluating the success of urban spaces.

The establishment of cities has already been referred to in a manner that likens the process to primary succession. In a post-industrial context, the cyclical process of secondary succession is arguably a more useful concept for the further extension of our analogy. As buildings outlive their functional possibilities, they are demolished and clearings form in the urban canopy, which allow the germination and growth of new development. This is how the city and the ecosystem can sustain themselves. These are what Geddes would have termed neotechnic cities as distinct to a paleotechnic cities (Geddes 1949), where linear and non-regenerative processes dominate.

Urban legibility

Despite its age, *The Image of the City* remains a key text for both landscape architects and urban designers. In it, Lynch (1960) argues from a basis of field observation and experiment that a key to the legibility of urban areas and their consequent navigability is how well people can form a cognitive image of their urban habitat. He goes on to define a number of key elements in the cognitive maps of cities, which he demonstrates are common to both experts and non-experts. These elements comprise edges, paths, landmarks, nodes and districts. All elements can be arranged in hierarchies of scale and importance. Edges can exist between different types of development as well as between urban form and natural or open areas such as rivers or open spaces. Nodes occur when paths meet and intersect. Lynch examined three contrasting cities in the United States; Boston, Jersey City and Los Angeles, and concluded from the sketch maps and cognitive diagrams made by expert and non-expert participants that connectivity between elements was a significant factor in how much an urban area was imageable and therefore legible. An examination of the maps associated with Jersey City, the place that performed least well in the field exercises shows that the links between different cognitive elements were relatively fragmented at the time of the study.

Landscape ecology

A text of equal significance to *The Image of the City* is *Landscape Ecology* by Richard Forman (1986). Landscape ecology as a discipline emerged from the theory of island biogeography (MacArthur and Wilson 1967). The idea of the island, its dimensions, distance from other islands and their effect on species migration and sustainability, has subsequently been reapplied to terrestrial networks. As a result it has become a predictive tool for conservationists dealing with habitat fragmentation, which sometimes can lead to extinctions. However its utility for designers rests on a vocabulary which can be translated into visual form and in so doing links ecological systems and process with spatial structures and patterns. The vocabulary comprises the patch, the corridor and the matrix. Forman has promoted the utility of landscape ecological vocabulary as a tool for landscape planning as spatial patterns strongly influence movements and flow. Stuart-Murray (2007) and Steinitz (2008) have pointed out the similarity between the visual and cognitive language of Lynch and the ecological and spatial lexicon of Forman. For both theorists, the notion of connectivity underpins visual and ecological health of the city or ecosystem.

The notion of the patch is particularly relevant in an ecologically fragmented world. The nature of a patch, particularly its edge, can determine the degree of connectivity in a fragmented system. Straight edges tend to encourage more movement along, rather than across them, than do curvilinear edges. This is strikingly similar to Gehl's observations referred to earlier, about human behaviour along different kinds of urban interfaces, which separate buildings from the street.

Sustainability and contemporary landscape architectural practice

In the recent history of the profession, a divide has developed between those who espouse and avow sustainable and ecological values and those who are driven primarily by visual considerations (Stuart-Murray 2007; Thompson 2000). The ecological approach to landscape pioneered by Alan Ruff (1979) at the University of Manchester and best exemplified by the development of green infrastructure planting, which defined new housing areas at Warrington New Town in Lancashire (Ruff and Tregay 1982) led to a promotion of ecological complexity and diversity as a design objective.

Perhaps the most vociferous opponent of this approach has been the established US designer Martha Schwartz who championed an overtly visual and arts led direction in landscape architecture in the 1990s.

> To understand the relationship of one space with another, one must first establish a sense of orientation in order to recognise new juxtapositions or changes. Simple geometries are thus best suited in landscape as mental maps. Given the nature of our built environment, the use of geometry in the landscape is more humane than disorientation caused by incessant lumps, bumps and squiggles of a stylised naturalism. Geometry allows us to recognise and place ourselves in space and is more formally sympathetic to architecture. Lastly, it deals with our manufactured environments more honestly; geometry itself is a rational construct and thereby avoids the issue of trying to mask our man-made environment with a thin veneer of naturalism.
>
> *(Martha Schwartz in Treib 1993: 263)*

Clearly Schwartz would not advocate any ecological approach to landscape architecture that maximises biodiversity by creating complex networks of patches and corridors and well-structured ecotones. In her specific reference to the Lynchian vocabulary of orientation, mental maps and the placing of people in spaces, she implies that imageability and resultant legibility are compromised by the ecological approach.

Others have eschewed visuality as a prime mover in the design process. In the first edition of *Landscape and Sustainability*, Maggie Roe and John Benson state their preference openly: 'the view offered here [is] of landscape and sustainability, in contrast to landscape as appealing mainly to the eye or aspiring cerebrally to be fine art' (Benson and Roe 2000: 3).

Thompson attempts a rapprochement by advancing the need to develop a new kind of ecologically informed aesthetic and cites the concept of messy ecosystems, orderly frames

(Nasseur 1995) as a way out of the impasse. She argues that since nature is a cultural construct the application of ecology to landscape can sometimes cause a design problem. For nature, in conforming to Western preferences, has often been represented in the picturesque style of serpentine lakes set in smooth and infinite plains of grass punctuated by clumps of trees. Despite being a representation of nature, such a landscape usually has a low species diversity and a simple ecological structure. On the other hand, the kind of complex ecological design favoured by Ruff, has often met with public resistance to the extent that in one of his iconic schemes at the Bijlmermeer housing estate in Holland, the planting has had to be removed in its entirety. Historically then, what may look good is not necessarily ecologically good and the converse can also be true. To counter such perceptions, Nasseur argues that ecological patterns should be translated into a cultural language, which indicates that the design is intended. To do this, ecological functions are set within frames, such as mown grass or a clipped hedge planted around a woodland.

Landscape ecological theory can identify some difficulties with the idea of orderly frames, however. If the patch or node is framed, how permeable is the frame to human and non-human movement? Might not the idea of a tidy frame lead to movement along rather than across it, which landscape ecology identifies as bad practice? How would a network of patches communicate to each other? Would not their tidy frames impede connectivity? How does the frame relate to the ecotone? In terms of an active human engagement with ecology Nasseur's model is problematic as it portrays ecology as a walled off, and therefore alien other world. Since ecology would not be experienced within the frame, a congruence of language and understanding would not be possible.

Landscape urbanism

Landscape urbanism is the term given to an emerging complex of concepts and may offer a more persuasive rapprochement between visual thinkers and environmentalists. Its central prediction is that the cities of the future will be organised spatially according to dynamic natural processes rather than architecturally generated form, where urban mass shapes external volumes. Modernism it argues, has failed to contain the dynamic processes of urban growth and failed to respond to the wider natural systems, which run through both city and landscape (Corner 2006). One of the main exponents of landscape urbanism is Charles Waldheim (2006) who has argued that the concept has come from a growing aesthetic interest, architecture as a discipline is taking in landscape, and how the languages of ecology and biology are informing critical discussions about the growth of cities. Such new critical language conflates the historical division between the town and the country by using similar concepts to analyse and understand both dynamic contexts. In this sense the highway is as ecological as the city. In this way landscape is no longer seen as a stylised and naturalistic compensation for urbanity, but as a necessary part of city infrastructure. Indeed Corner (2006) refers to Olmsted's emerald necklace of parks in Boston, which deal with aspects of the city's hydrology; and Stuttgart's open space network, which air conditions the city's microclimate as examples of best practice.

Landscape urbanism's aesthetically driven approach to urban planning contrasts with the bottom-up approach of McHarg (1969). Here a sieve mapping of the biophysical realities of landscape (which was the equivalent of a manual GIS process and therefore noteworthy for the period), formed the bedrock of his landscape planning proposals. Of course, McHarg can be criticised for omitting the social factors from his analyses. There is also little reference to visual qualities, and regional plans generated by the process appear to be fixed and focused only on rural areas. Again this is contrary to the philosophy of landscape urbanism where ecology is seen as part of and not separate from urban structures. Contrast this fixed approach with landscape architect James Corner talking about the design process (Corner 2004: 2) he employed at the Freshkills landfill site in New Jersey.

1 Accept the often messy and complex circumstances of the given site ... and develop techniques ... for both representing and working with the seemingly unmanageable or inchoate complexities of the given.
2 Address issues of large-scale spatial organisation and relational structuring of parts, a structuring that remains open and dynamic.
3 Deal with time open-endedly, often viewing a project more in terms of cultivation, staging and setting up certain conditions rather than obsessing on fixity, finish and completeness.

Corner suggests that new representations may use ideas from film-making, choreography and music. In this context Raoul Bunschoten has developed a new working technique whose starting point is equivalent to a field ecologist throwing a quadrat onto a sward. Instead of quadrats, beans are thrown randomly onto a map. At each point or beansite, the landscape is analysed according to the processes of erasure, origination, transformation and migration. The same dynamic terms are also used to generate design proposals. An awareness of dynamism is therefore implicit in the process of analysis and the exploration of design proposals. Projects therefore, are open-ended and merely represent a stage in a complex of processes over which the designer has little ultimate control. Bunschoten developed the method in Kyoto, but it could also be used in rural areas as the concepts employed apply to both urban and ecological processes (Bunschoten *et al.* 2001).

Summary, conclusions and directions for future research

This chapter has argued that city form and process are analogous to ecological form and process. It has compared some vocabularies of urban design and landscape ecology. A historic schism between visually led and ecologically led landscape practitioners has been chronicled and analysed. Some solutions have been identified and evaluated. However, the emerging field of landscape urbanism seems to offer the most fertile territory for urban ecology and design to progress. This is because it analyses and understands ecological and cultural phenomena in the same dynamic and open-ended language. It sees no necessary conceptual distinction between cities and the wider landscape processes in which they are enmeshed. It is critical of landscape as a naturalistic compensation for urban life. Instead landscape forms another layer of urban infrastructure. Methods employed by landscape urbanists parallel the scientific approach of ecologists.

However as Barnett has pointed out (2007: 2).

> what do we mean by adaptive and open-ended design? You may kick the thing off, in the manner of artists who work with organic materials, commencing the process but letting the intrinsic nature of the materials and the interaction with context actually shape production; but design is not like art. The artist retains a high degree of control not only over choice of materials, location of making, the production itself ... but also over representations of the work and its reception, as well as the intellectual framework within which the work is considered and understood. The landscape designer, however, is part of an 'interdisciplinary team' with which framing, production and reception decisions are shared. And what is it all worth anyway, if you can't stand back and say, 'I did that?' when there is no I, the doing was a response to forces way beyond anyone's control and, more problematically, there is no that.

So how do you identify what 'that' might be? Coming from the fields of language and psychology and arguing from a metaperspective, Bateson (1980) has argued that there is a need to dissolve the Cartesian dualities of mind and body, of man and nature. Coming from scale of microbiology Witzany (2000) has used the concept of biosemiotics to classify the systems of signs and stimuli, which are common to all life forms from the lower to the higher orders. Just as landscape urbanism as a term has been promoted as a conceptual framework before its potential scope has been realised, it may be that similar new terms could act as catalysts or seeds for further study. Terms like cognitive landscape ecology, legible ecology and ecological syntax might be of use. With these in mind, it may be possible using the established techniques of cognitive ethology to map and model (perhaps using GIS technology), biosemiotic systems, which are common to man, plant and animal.

References

Barnett, R. (2007) 'Charles Waldheim, The Landscape Urbanism Reader', *Junctures: The Journal for Thematic Dialogue*, 8: 118–21.
Bateson, G. (1980) *Mind and Nature: A Necessary Unity*, Cresskill NJ: Hampton Press.
Benson, J.F. and Roe, M.H. (2000) *Landscape and Sustainability*, London: E. and F. Spon.
Bunschoten, C.R., Hoshino, T. and Binet, H. (2001) *Urban Flotsam: Stirring the City*, Rotterdam: Uitgeverij 010.
Corner, J. (2004) 'Not Unlike Life Itself', *Harvard Design Magazine*, 21: 1–3.
Corner, J. (2006) *Terra Fluxus in The Landscape Urbanism Reader*, New York: Princeton Architectural Press.
Forman, R.T.T. (1986) *Landscape Ecology*, New York: Wiley.
Geddes, P. (1949) *Cities in Evolution*, London: Williams and Norgate.
Gehl, J. (2006) *Life Between Buildings*, sixth edition, Copenhagen: Danish Architectural Press.
Lyle, J.T. (1991) 'Can Floating Seeds make Deep Forms?' *Landscape Journal*, 10(1): 37–47.
Lynch, K. (1960) *The Image of the City*, Harvard: MIT Press.
MacArthur, R.H. and Wilson, E.O. (1967) *The Theory of Island Biography*, Princeton, NJ: Princeton University Press.
McHarg, I. (1969) *Design with Nature*, Garden City, NY: The Natural History Press.
Nassseur, J.I. (1995) 'Messy ecosystems, orderly frames', *Landscape Journal*, 14(2): 161–70.
Ruff, A. (1979) *Holland and the Ecological Landscapes*, Stockport: Deanwater Press.
Ruff, A.R. and Tregay, R. (1982) *An Ecological Approach to Urban Landscape Design*, Occcasional Paper Number 8, Manchester: Dept of Town and Country Planning, University of Manchester.
Steinitz, C. (2008) 'Landscape Planning: A Brief History of Influential Ideas', *Journal of Landscape Architecture*, 5: 68–74.
Stuart-Murray, J. (2007) 'The Practice of Sustainable Landscape Architecture', in J.F. Benson and M. Roe (eds) *Landscape and Sustainability*, London: Routledge, pp. 222–36.
Thompson, I.H. (2000) *Ecology, Community and Delight*, London: E & F Spon.
Treib, M. (1993) *Modern Landscape Architecture: A Critical Review*, Cambridge, MA: MIT Press.
Waldheim, C. (2006) 'Landscape as Urbanism', in C. Waldheim (ed.), *The Landscape Urbanism Reader*, New York: Princeton Architectural Press, pp. 35–53.
Witzany, G. (2000) *Life: The Communicative Structure: A New Philosophy of Biology*. Norderstedt: Libri Books on Demand.

50

Towards Ecopolis

New technologies, new philosophies and new developments

Rusong Wang, Paul Downton and Ian Douglas

Ecopolis

With half the world's people now living in cities and 25 megacities or urban agglomerations of over ten million people, the need for applying urban ecological principles and practice has never been greater. Cities are centres of consumption that have been encouraged by globalisation to source raw materials, food and manufactured goods from producers around the world, regardless of environmental cost. The problems of such cities cannot be solved by trying to relocate people, by continuing to spread existing agglomerations further and further into the surrounding countryside, or by creating new growth poles. New forms of city living have to be developed that are more sustainable and allow cities to function more like natural ecosystems. People require a technologically sophisticated lifestyle, with electrification, computers, modern communications and multiple transport options. This is possible if a new form of city living, Ecopolis, is adopted by greening urban areas through the integration of architecture, planning and ecology, essential to the development of truly viable ecological cities.

One of the groundbreaking views of cities as ecosystems was probably first presented in the study of Hong Kong's ecology in the 1970s (Boyden *et al.* 1981) but was largely overlooked by 'western' planners. Today it is not surprising that China is taking the initiative in trying to develop ecological cities as a result of a strong commitment to the development of urban ecology as a science and planning tool since the 1980s (Ma and Wang 1984; Wang 1991, 1994). This chapter examines the ideas on urban ecology and Ecopolis developed and applied in China, compares them with developments elsewhere and looks at future opportunities for applying urban ecology to develop truly green and sustainable cities.

The concept of 'Ecopolis' differs from more traditional concepts of urban sustainability that tend to focus on resource conservation (Downton 2009). For example, in relation to the biosphere, urban sustainability aims to have a 'mostly harmless' relationship with the biosphere, while Ecopolis specifically aims to produce urban areas that are harmoniously integrated into biosphere processes to optimise their functioning for human purposes.

The Ecopolis concept sees city evolution as an adaptive process of the interaction between man and nature. Its natural subsystem, economic subsystem and social subsystem are engaged in a co-evolutionary process, having both positive and negative effects on ecosystem service and human well-being. Because adaptive Ecopolis development (AED) involves both people and institutions affected by the management strategy, it can also be described as social learning (Bandura 1977).

AED recalls that decisions made long ago by architects and urban planners, when streets were first laid out and buildings first constructed, continue, for decades and even centuries, to have a great influence on the amount of waste produced. These decisions include the orientation of the house and the choice of heating-system, which affect emissions directly, or the insulation level which is important for the fuel-consumption and thus for emissions. Sooner or later buildings will be demolished, creating 'demolition waste' that can be recycled when there is an appropriate new use. Thinking in terms of Ecopolis when designing a new building can reduce future demolition-waste. Architects and planners can avoid and reduce many types of environmental damage through appropriate design. They can also greatly influence the green infrastructure and thus the urban ecology of the future.

Adaptive Ecopolis development includes the adaptation to or learning from physical environmental change, technological innovation, economic fluctuation, institutional fragmentation, demographical mobility, behavioural pattern and data uncertainty. On the other hand, the local natural ecosystem under the urbanisation stress has also responded to human society by changing its physical and biological structure, function and process in response to human disturbance. To adapt to different conditions of varying levels of urban socioeconomic conditions, from less developed, fast transition to highly developed, Ecopolis development can vary in its focus on some or all of the following evolutionary goals in practice:

1 *Ecological sanitation* to provide citizens with a clean and healthy environment by encouraging ecologically oriented, affordable and people-friendly eco-engineering for treatment and recycling of human wastes, sewage and garbage, reducing air pollution and noise etc. Eco-sanitation is a kind of man-nature metabolism system dominated by technological and social behaviour, sustained by natural life support systems, vitalised by ecological process. It interacts with the human settlement system, the waste management system, the hygiene and health care system, and the agricultural system. Sanitation is an eco-complex between humans being and their working and living environment (including the sources of food, water, energy and other materials; the sinks for wastes such as odours, faeces, flies, pathogens and fertilisers; flows that could be enhanced by physical, chemical and biological purification; and pools for buffering and maintaining, such as the kitchen, bathroom and toilet within a house, and its social networks (including culture, organisation and technology).
2 *Ecological security* to provide citizens with ecologically safe, basic living conditions: a clean and safe water supply, food, service, housing and protection against disasters. Eco-security in Ecopolis development includes water security (the quality and quantity of drinking water, production water, ecosystem service water), food security (the sufficiency, accessibility and contamination of plant and animal food, vegetables and fruits), settlement security (atmosphere, aquatic and terrestrial pollution, inner environment contamination), disaster reduction (minimisation of epidemics and geological, hydrological and industrial disasters), and life security (physiological and psychological health care and insurance, social violence and traffic accident reduction). This issue is especially urgent for the poor in peri-urban areas in developing countries.
3 *Ecological industry* (EI) with emphasis on the industrial transition from traditional products and profit-oriented industry to function-oriented and closed-loop industry through the coupling of production, consumption, transportation, reduction and regulation. This transition is especially urgent for cities which have achieved a certain level of industrialisation and now need regeneration, upgrading and reform, but which lack the motivation to initiate

such processes. EI encourages an ecological industrial transition according to following strategies (Wang *et al.* 2006):

- *Food web-based horizontal/parallel coupling:* connecting the different production processes and gaining positive benefits by avoiding negative environmental impacts through the sharing of unused resources.
- *Life cycle-oriented vertical/serial coupling:* combining primary, secondary and tertiary industries, consumption and recycling sectors into one EI complex along the life cycle to allow production and materials use to be more systematically responsible.
- *Ecosystem-based regional coupling*: integrating local environment, local community, dominant enterprises and other diversified sectors into one agro-industrial-service ecosystem in order to internalise environmental costs and to assimilate pollutants and minimise their formation within the system itself;
- *Flexible and adaptive structure:* multiple production functions, diversified products and easy-to-change engineering processes rather than rigid, unified and limited ones to adapt to the external change;
- *Functional service rather than products output oriented production:* switching the production focus from products to service with *three kinds of final output:* goods (hardware), services (software) and life style changes (mindware); switching the single goal of production from profits to three dimensions of human well-being: wealth (economic gains, ecological diversity and human resources), health (of companies, customers and ecosystems) and faith (values, awareness, enjoyment, realisation);
- *Capacity building:* Enhancing EI training and capacity building; EI research and development; EI dissemination and cooperation; EI incubation and consultation; EI operation and management; adaptive and comprehensive decision making, sensitive information feedback, and effective networking of knowledge, experience and expertise;
- *Employment enhancement:* increasing rather than decreasing working opportunities by creating more jobs in research and development, service and training within the industrial ecosystem, though processing jobs might be reduced by greater automation;
- *Respecting human dignity:* EI becomes a process of learning and innovation, a socially interactive and self-fulfilling engagement, rather than slavery to a machine and oriented mainly to earn a basic wage. Human ecological gains in morals, beliefs and universal understanding are much more important than sheer material life conditions.

4 *Ecological landscape (ecoscape)* to restore the fragmented and agglomerated ecosystem, humans have to invoke ecological integrity through ecological landscape planning. An ecoscape here is a multidimensional landscape of a social-economic-natural complex ecosystem, combining geographical patterns, hydrological process, biological vitality, anthropological dynamics and aesthetic contexts. Ecoscape integrity put an emphasis on alleviation of the heat island, hydrological deterioration, greenhouse effects and landscape patterns and processes. Ecoscape planning means a kind of integrative learning process to reach a vision of how the ecoscape is coupling, functioning and vitalising, and a kind of integrative design process for physical, ecological and aesthetic innovation. A sustainable ecoscape should be driven by balanced positive and negative forces to promote and sustain its development according to following feng shui (wind and water) principles formulated in ancient China (Wang and Hu 1999):

- ***Totality***: geographical continuity, hydrological circulation, ecological integrity and cultural consistence;
- ***Harmony***: between structure and function, internal and external environment, implicit and explicit layout, nature and humans, objective being and subjective value, material and spiritual goals;
- ***Mobility***: constant wind and water flowing, vertical and horizontal flow, meandering streams, undulating and far stretching, and five basic movements;
- ***Vitality***: luxuriant, flourishing and productive fauna, flora and soil and aquatic biome;
- ***Purity***: clean and limpid water, clean and transparent atmosphere, quiet and secluded surrounding, never overloading its carrying capacity;
- ***Safety***: backed by hill, enclosure, explicit, spacious, openness, easy to disperse and defense, disaster resistance;
- ***Diversity*** and heterogeneity of landscape: ecosystem, species, society and culture;
- ***Sustainability***: negative and positive interlocking feedback, self-reliance, self-maintenance, sufficiency and efficiency, appropriate exploitation and development.

5 *Eco-culture* grounded on the ecological principles of totality, harmony, recycling and self-reliance, eco-culture is an ecologically sound body of learned behaviours common to a given human society that acts rather like a template (i.e. it has predictable form and content), shaping behaviour and consciousness within a human society from generation to generation in the area of systems of meaning (philosophy, science, education); institutions (ways of organising society, legislation, and policymaking, from kinship groups to states and multinational corporations), the distinctive techniques of a group and their characteristic products (architecture, literature, arts and mode of production and consumption); and paradigms or norms (religion, values, morality, way of thinking and feeling). Only when the lifestyle is harmonious with nature in metabolism process, structural pattern and functional development, and human activities are enhancing rather than depleting the life supporting system, is a sustainable development expected to be realised (Wang *et al.* 1996).

To realise these goals, we need an adaptive process to local natural and human ecological conditions, we need to *reshape* our production modes, consumption behaviour, development goals and life meaning, to *reform* fragmented legislative, governance, decision making, planning and management institutions, and to *renovate*, so as to eliminate, the reductionism based on chain-linked technology. Ecopolis is not a utopian concept. It is an accessible goal, an adaptive process, an applicable strategy and technology, a philosophy about survival and development for all people, and a way to plan and manage sustainable human settlements comprehensively. Situated on the sharp contradiction between reductionism and a holistic approach, objective and subjective method, quantitative and qualitative data, and vertical and horizontal interconnection, a **methodological transition towards ecological integration** is underway, *from planning physical being to ecological becoming, from numerical quantification to relationship qualification, from mathematical optimising to ecological learning,* and from operationally *intelligent computers to ecologically intelligent people.*

Technologies

Ecopolis relies on being able to apply a wide range of technologies, each of which is well documented and widely available, but many of which are not well understood by planners and decision makers. The combining together of these technologies in urban design and building construction does not come readily to most established planners, architects and decision makers.

One person who has brought the technologies together is Peter Head in his 2008 Brunel Lecture, which provides a comprehensive review of the technologies, infrastructure systems, planning approaches, policies and delivery mechanisms needed across the world to move all human development onto a sustainable pathway which attempts to stop and reverse the destruction of the eco-system on which we depend for life (Head 2008). The opportunity for low- and middle-income countries is to combine this new economic thinking with the use of technology, without the need for the wasteful steps the high income world has gone through.

The opportunities include the use of recycled water, both for irrigating gardens, allotments and lawns within the city and to provide efficient irrigation of surrounding farmland. Rainfall and runoff water can be collected and stored in cities for use as grey water for secondary uses. Such measures reduce the demand for potable water and the associated energy needed for treatment. They can also help mitigate climate change impacts of more intense storm rainfall on urban flooding.

Other opportunities are found in the energy field. Second generation biofuels from agricultural wastes (such as rice husks, stalks of grain crops and CO_2-absorbing algae) are much more sustainable and can provide a major opportunity to lower emissions and the ecological footprint of cities. Building design can use readily available local materials to reduce energy consumption. The technology exists to make high quality, commercially competitive, fire retardant building walls from wheat and rice straw and for using sheep wool as building insulation. (Improving the thermal efficiency of the existing housing stock in the UK would reduce up to 40 per cent of the country's carbon emissions).

Suitable legislative frameworks can promote local renewable energy production. Germany has made great strides in renewable energy thanks to its feed-in tariff for electricity supplied to the national grid by domestic solar- and wind-power producers. In China, in places with access to the power grid, the energy feed-in legislation now makes it attractive to install local combined heat and power (CHP), photovoltaic electricity (PV), and energy from waste and large-scale wind power in large residential developments. These houses can be designed to reduce demand and will be able to be built at attractive prices once economies of scale become realistic. Chinese appliance standards will cut the nation's residential demand for electricity by 10 per cent and avoid having to build 36 large (1000 MW) coal-fired power plants.

Energy consumption for distributing goods in urban areas can be greatly reduced by using depots around the city perimeter supplied by major intercity rail and road links. Goods from these depots can be delivered to retailers and consumers by zero-emission vehicles with routes planned to minimise travel distances and congestion. Such systems can achieve energy savings of up to 79 per cent and can increase delivery reliability (Head 2008).

Philosophies

Tjallingii (1996) saw 'Ecopolis' building on the works of Anne W. Spirn (1984) and Michael Hough (1984). These views of urban ecology presented convincing arguments that ecology was alive and well in cities, with numerous case studies cited mostly from North America and Europe. In our proposition Ecopolis takes a different tack: in one respect it lays a rigorous theoretical foundation for ecologically sound urban development, then it assumes the challenge of applying the theory in real communities with significant stakeholder involvement. Ecopolis is not primarily concerned with 'ecology in the city', but with the 'ecology of the city'.

Ecopolis presents a powerful and clear strategy framework consisting of three main themes:

1 the Responsible City – the city must not pass on its problems to higher levels or to future generations;

2 the Living City – to integrate the local ecological potential fundamentally with the identity of the city;
3 the Participating City – emphasising involvement of people in the management of their environment.

These main themes are worked out in guiding models for chains, areas and organisation, respectively. This abstract system-level thinking is highly developed, clear and well illustrated with diagrams. It contains a simple but elegant 'ecodevice model' based on the input and output flows characteristic of all ecosystems, coupled with resistance and retention as the four basic regulating mechanisms of ecosystems (Wang 1994).

Ecopolis development should be guided by a set of principles (Downton 2009):

1 Restore degraded land
Use urban development to restore the health and vitality of the land; rehabilitate and maximise the ecological health and potential of land as a consequence of the development of human settlement:

- Clean-up contaminated land
- Heal degraded rural areas
- Re-establish native vegetation
- Encourage farming practices which sustain ecological health
- Introduce green corridors of native vegetation in rural and urban areas

2 Fit the bioregion
Create human settlements which work with the natural cycles of the region.
Conform to the parameters of the bioregion, fit the landscape with the patterns of development which follow the inherent form and limitations of the land, understood in socio-biophysical terms.

- Maintain the natural cycles of water and nutrients in the landscape
- Create buildings and urban form that fit the landscape and respond to the climate
- Conserve water and recycle effluent
- Use locally produced building materials as much as possible
- Respond to the culture of the region – 're-habitation'
- Introduce green corridors of native vegetation in rural and urban areas

3 Balance development
Balance development with the 'carrying capacity' of the land.
Balance the intensity of development against the ecological carrying capacity of the land whilst protecting all viable existing ecological features. Develop and enhance links between urban and rural areas of an integrated city-region approach.

- Reduce the impact of the city on the land beyond its boundaries (the 'ecological footprint')
- Encourage the diversity of land-use: residential, commercial, recreational, educational, etc.
- Develop urban food-producing gardens
- Recognise the place of all living organisms in the environment – urban design for non-human species

4 Create compact cities
Reverse sprawl and stop ad-hoc development from consuming the landscape.

Develop human habitation at relatively high density within inviolable green belts of natural or restored ecologically viable landscape with the overall development density constrained by ecological limits.

- Have clearly identifiable (but not 'hard') boundaries for urban areas
- Provide for most daily needs within the city
- Create 'walkable' cities and promote non-motorised forms of transport
- Develop integrated transport networks which minimise car use
- Access by proximity
- Three-dimensional built form

5 Optimise energy performance
Generate and use energy efficiently.
Operate at low levels of energy consumption, using renewable energy resources, local energy production and techniques of resource reuse. All ecological development should seek to be energy self-sufficient. The primary energy base for development should come from renewable sources.

- Minimise energy consumption
- Use renewable energy of solar and wind power
- Generate power locally
- Reduce fossil fuel consumption
- No nuclear power
- Design buildings with solar access and natural ventilation
- Use effective insulation and 'thermal mass' in buildings
- Climate responsive design

6 Contribute to the economy
Create work opportunities and promote economic activity.
Support and develop ecologically and socially responsible economic activity. Materials and component manufacture should be derived from, or be located in the local bioregion to the maximum practicable extent. Finance for ecological development from ethical sources, exclude financial support derived from exploitative activity. Capital input to ecological development should be local and financial structures should ensure that ownership and control ultimately rests with the users and inhabitants of the development.

- Develop ecologically responsible industries
- Develop exportable 'green technologies' and services
- Create appropriate information technologies
- Provide incentives for innovation and enterprise linked to ecologically responsible performance

7 Provide health and safety
Create healthy and safe environments for all people.
Employ appropriate materials and spatial organisation to create safe and healthy places for people to live, work and play in the context of an ecologically resilient environment.

- Reduce pollution and promote environmental quality
- Ensure a safe water supply, recycle effluent, maintain clean air
- Provide food security – urban agriculture
- Provide habitat for animals and birds

8 Encourage community
Cities are for everyone.
Create cities with strong citizen involvement – community participation, not just consultation. The community should govern itself. Community needs must drive ecological development. Ecological development must meet community requirements including the community of life that is the eco-system.

- Create development as a community driven process
- Ensure community involvement in public administration and management
- Provide community facilities

9 Promote social justice and equity
Equal rights and access to services, facilities and information.
Employ economic and management structures which embody principles of social justice and equity. Ensure equal rights and access to essential services, facilities and information. Alleviate poverty and create work opportunities.

- Involve all levels of the community in development processes
- Provide affordable housing
- Public use of public space
- Direct democracy

10 Enrich history and culture
Respecting the past whilst looking to the future.
Maximise the value of previous worthwhile human endeavour in terms of both heritage and manufactured artefacts.

- Restore and maintain cherished local monuments and landmarks
- Identify and celebrate the spirit of place
- Celebrate and encourage cultural diversity
- Respect indigenous peoples' inhabitation of the land
- Diverse cultural and social groups provide the basis for socially vital cities
- Support and promote cultural diversity, incorporating ecological awareness into all aspects of the making and maintenance of human settlement. Art and craft should be integral to both the construction and the operation of ecological development from the individual site to the city and its region
- The whole process of creating ecological development and its subsequent operation requires education and skill development
- Develop culture by involving all aspects of the arts including music, electronic media and technology
- Develop culture by integrating the arts and sciences with both daily life and special events and occasions
- Promote ecological awareness as part of cultural development
- Support community art and craft events, fairs, fetes and functions and develop festivities and events that relate to the locality.

New developments

Conjugate ecological planning is a form of social-economic-natural complex ecosystem planning that aims to achieve a harmonious relationship between environmental and economic development, between social and natural service, between physical and ecological infrastructure, between local and regional development, between historical and future contexts, between tangible and intangible, positive and negative ecological impacts (Wang and Paulussen 2004). It includes:

1 coordinating urban planning with regional/watershed/hinterland planning to promote sustainable exploitation, use and maintenance of the five natural elements of water, fire, wood, soil and mineral;
2 built-up area (red space) planning with open-space planning (green, blue and brown space) to develop a comprehensive ecosystem service business from urban agriculture, forestry, gardening, wetland and to waste regeneration;
3 demographic planning with quality of life planning, combining cultural heritage conservation with ecological texture, social arteries and veins, and human ecological integrity;
4 red line control to avoid eco-sensitive areas' degradation with feng shui network development to cultivate and promote ecosystem services;
5 two dimensional land use planning with three dimensional ecoscape planning including underground and above ground physical space and multi-ecological carrying capacity (water/air/heat/green) planning;
6 water use for human consumption and production with water use for natural ecosystem maintenance;
7 intensive energy exploitation and utilisation planning with extensive energy dissipation and renewal energy use planning, and pollution control and treatment planning, with eco-service conservation and development to reduce heat island effects, pollution effects, greenhouse effects and citizen disturbance effects;
8 supply-oriented urban metabolism planning and traffic planning with artery planning for waste recycling and regeneration and eco-sanitation planning;
9 traditional vertical and branching network management institution planning with horizontal linkages, comprehensive decision making, system supervision, information feedback and capacity building planning.

The ultimate goal of the conjugate ecological planning is comprehensive wealth, health and faith. Wealth measures the structural state of the monetary assets: natural assets (mineral, water, forestry, soil, air and biodiversity), human resource (manpower and intellect) and social resources (such as institutions and the arts); health measures the functional state: human health, ecosystem health, and risks and opportunities on human beings and their life support system; faith measures the behavioural mode: values, material attitudes (lifestyle, consumption customs, recycling tradition and eco-ethics) and spiritual relations (perceptions, concepts and beliefs towards the totality or supernatural forces) (Wang *et al.*1996). The temporal, spatial, quantitative, structural and functional contexts are the main contents for systematically responsible planning.

Ecopolis development has four stages: concept initiation and comprehensive planning; ecoscape planning and legislation; eco-engineering design and development; ecosystem monitoring and management. Ecopolis development needs motivation from five directions: administrative approval and authority; scientific supervision; industrial sponsorship; citizen participation; and media support and encouragement.

China's vision

China is experiencing rapid urbanisation and industrial transition. The pace, depth, and magnitude of these changes, while bringing benefits to local people, have exerted severe ecological stresses on both local human living conditions and regional life support ecosystem. Urban sustainability can only be assured with a human ecological understanding of the complex interactions among environmental, economic, political and social/cultural factors and with careful planning and management grounded in ecological principles.

Unlike biological communities, human society is a kind of artificial ecosystem dominated by human behaviour, sustained by natural life support system, and vitalised by ecological process. Shijun Ma called it a 'Social-Economic-Natural Complex Ecosystem' (Ma and Wang 1984). Its structure is expressed as an eco-complex between a human being and their working and living settlement (including geographical, biological and artificial environs), their regional environment (including sources for material and energy, sinks for products and wastes, pools for buffering and maintaining) and their social networks (including culture, institution and technology) and economic networks (the primary, secondary and tertiary industries and infrastructural services). Its natural subsystem consists of the Chinese traditional five elements: metal (minerals), wood (living organisms), water (source and sink), fire (energy), soil (nutrients and land). Its function includes production, consumption, supply, assimilation and recycling.

In recent years, a campaign of Ecopolis development has developed in some Chinese cities and towns. Ecopolis is a kind of administrative unit having economically productive and ecologically efficient industry, systematically responsible and socially harmonious culture, and physically beautiful and functionally vivid landscape. It aims to improve its structural coupling, metabolism process and functional sustainability through cultivating an ecologically vivid landscape (ecoscape), totally functioning production (eco-industry) and systematically responsible culture (eco-culture).

The Ecopolis concept that is widely discussed in China is part of the planning of the Dongtan ecocity planning on Chongming island in Shanghai municipality, of the Sino-Singapore Tianjin ecocity planning, and several other cases of ambitious ecocity planning that show the public a vision of a sustainable city. In Huaibei, Anhui Province, and Beijing cities, conjugate ecological planning is trying to develop a harmonious balance between economic benefits and ecological service for ecosystem development.

Based on the ancient Chinese human ecological principles such as the yin and yang (negative and positive forces play upon each other and formulate all ecological relationships), wuxing (five fundamental elements and movements within any ecosystem promoted and restrained with each other), zhong yong (things should not go to their extremes but keep equal distance from them or take a moderate way) and feng shui theory (wind-water theory expressing the geographical and ecological relationship between human settlements and their natural environment) (Wang and Qi 1991). Conjugate ecological planning has been used in planning the Beijing capital Ecopolis. The main goal is to promote eco-sustainability at four levels of natural ecology, economic ecology, human ecology and systems ecology from five kinds of contexts: time, space, quantity, configuration and order.

During the past 25 years, 390 national demonstration Ecopolises have been appraised and named by the Ministry of Environment including prefecture and county level cities such as Yangzhou, Shaoxin, Panjing, Yancheng, Hangzhou, Xuzhou, Guangzhou, Changsha, Haining, Anji, Changsu, Zhangjiagang, Kunshan, Longgang district of Shenzhen, Rizhao and Dujiangyan; in all 32 cities have passed 'environmental model city' appraisal. Finally, 13 provinces initiated eco-province development (Hainan, Jilin, Heilongjiang, Fujian, Zhejiang, Shandong, Anhui, Jiangsu, Hebei, Guangxi, Sichuan, Tianjin and Liaoning). The Ministry of Science and Technology has also appraised and named 108 experimental cities/counties towards sustainable development covering 29 provinces of China. Great progress had been made in these case studies while some lessons and challenges also emerged such as institutional barriers, behavioural bottlenecks and technical malnutrition (Wang *et al.* 2004a, 2004b).

Yangzhou City

Yangzhou City occupies 6638 km^2 in the middle of Jiangsu province, at the junction of the Grand Canal and Yangtze River. The 2500-year old city has 4.47 million inhabitants and was known as the first state in the lower reaches of the Yangtze River. Due to the dramatic changes in urbanisation and industrialisation that have taken place in the Yangtze Delta, especially along the south bank of Yangtze River in the Suzhou-Wuxi-Changzhou corridor, Yangzhou city decided to catch up with pace of development but also to avoid severe environmental pollution and ecological deterioration, by finding an alternative route by implementing China's Agenda 21 and setting its own agenda, an outline for Ecopolis planning, in 2000 and subsequently putting it into action.

The three goals of the Yangzhou Ecopolis Development are:

1 To promote economic transformation from a traditional economy into a resource-type, knowledge-type and network-type sustainable economy with high efficiency. To make the Yangzhou economy flourish, with ecological industries as trendsetters;
2 To promote the development of the regional eco-environment to a vibrant, clean, beautiful, vigorous and sustainable urban ecosystem. To create a good ecological basis for social and economic development;
3 To promote a change from local people's traditional productive activities, lifestyles and values into an environmentally aware, resource-use efficient, harmonious ecosystem and social culture. To create a new generation of urban developers and builders who share a culture, aspiration and desire for a high quality eco-society.

The city decided to reach the following short-term goals within five years: create 250,000 new employment opportunities, increase the average years in education of citizens by 10 per cent, incubate ten large eco-industry groups and 100 enterprises with eco-culture, reach a GDP annual growth rate of 8–10 per cent, recover 1000 ha of abandoned mine areas and wetlands; set up one natural conservation area, activate and purify 200 km of river, and increase urban greenspace in city and towns by 10 per cent.

There are three stages in its implementation:

- First period (2000–2005): structural adjustment; infrastructure construction; construct basic facilities; prior project initiation; first fruits of pilot projects. Integrative power, including social power, economical power and environmental power, reaches highest level in Jiangsu. Primary indices of eco-development reach eco-development pilot zone standard stipulated by the National Environmental Protection Agency.
- Second period (2006–2010): Key pilot district construction (cities and towns, ecological villages, factories, farms and landscape areas) and key eco-projects are achieved and the experiences are extended in the whole region. Key industries (eco-tourism, eco-agriculture, eco-construction, eco-communication, and eco-food) increase greatly. Some key products (eco-food, tourism products) enter the international market. The whole city, an essential Ecopolis, becomes one of the most advanced cities with comprehensive social economic and environmental power in China.
- Third period (2011–2020): urbanisation of rural areas, modernisation of the city and an ecologically aware and active society are achieved. The comprehensive social economic and environmental power of Yangzhou reach an advanced level compared to the rest of the world.

Compared to other countries, China's ecocity development reflects top-down encouragement more than bottom-up persuasion. The advantage of the Chinese way is that if the decision makers are smart enough, the Ecopolis plan will be strongly implemented, otherwise it will be just an oral promise or a utopian ideal.

Ecopolis elsewhere

A group of European cities, including Freiburg, Zurich, Stockholm and Helsinki, has seen it as an asset to become 'greener' or 'sustainable' cities (Kenworthy 2006). Through compact planning, they devote space to urban agriculture, urban forests and community gardens, as well as excellent public transport systems and high levels of walking and cycling. Environmental technologies such as renewable energy and localised management of water help to 'green' these cities. Whether they can fit all the criteria of an Ecopolis is perhaps questionable, but they clearly demonstrate how cities can adapt and change to meet more of the Ecopolis criteria. At a discussion of ideas of eco-villages and eco-cities in India, it was suggested that India's emerging eco-city would possibly be Agra, the city that is home to the Taj Mahal (Malhotra 2010). The prospect is driven by an initiative taken by a community there and a society that regularly discusses issues concerning the environment and brings out solutions to the problems prevailing in the city.

While Ecopolis criteria can be applied to make any existing settlement more sustainable, planned new settlements have the opportunity to 'get it right' (or wrong) from the beginning. If an eco-town, or eco-city, is designed to achieve self-sufficiency in the longer term, then it needs to be designed as such from the start, so that the centre is large enough, for example. It also needs to be surrounded by land suitable for future growth without overriding environmental constraints. The Ecopolis study of Breda, Netherlands, by Tjallingii (1992, 1994) aimed to find answers for the environmental problems of urban design on an international, a national and a local level. Financed by the Dutch National Planning Institute, it led to a report titled: 'Ecologically Sound Urban Development' and a thesis at the Delft University of Technology (Tjallingii 1996).

After examining Freiburg as an example of an existing city that has applied Ecopolis principles, both a small intra-urban Ecopolis development, and a major new city will be discussed.

Freiburg, Germany

Eco-housing, car-free streets and socially conscious neighbours have made the German city of Freiburg a shining example of sustainability. In 2010, the Shanghai EXPO organisers chose the Freiburg district Vauban for the 'Urban Best Practices Area' giving Freiburg Green City the opportunity to present its urban culture to an international audience. Freiburg's efforts to promote sustainability initially focused on excluding private vehicular traffic from the city centre and maintaining a viable public transit system. In the mid-1970s, Freiburg began to improve the quality of life in the city centre by closing the major north-south traffic route to vehicles. This ban was soon extended to most of the city centre, with a fully pedestrianised city centre (save for trams and buses), emerging by 1986 (Ryan and Throgmorton 2003). New multi-storey car parks were banned, while large bicycle parking areas were installed at the entrances to the pedestrianised area. These bike parks were connected to an extensive network of bicycle paths (Lennard and Lennard 1995). These changes contributed to a doubling in the number of trips by public transport in Freiburg between 1984 and 1995, but the growth was mainly due to a cheap travel pass which gave unlimited use at zero marginal financial cost, interpersonal transferability and wide regional validity (Fitzroy and Smith 1998). The expansion in public transport trip demand did not produce any long-term worsening of the operating deficit of the municipal transport company.

In addition to an integrated transport policy, Freiburg has endeavoured to make its own urban redevelopments sustainable. In a former waste treatment site, a new housing development at Rieselfeld kept 30 per cent of the housing for low-income households offering about 800 subsidised rental apartments, dispersed throughout the first section of the development, along with about 600 owner-occupied units. The development uses an estimated 52 per cent less energy than normal dwellings, by using shared building walls, by building zero energy buildings, by improving the power supply (Rieselfeld is connected to a nearby district heating facility), by urging residents to buy more efficient appliances, and by encouraging the use of sustainable transport (Ryan and Throgmorton 2003).

Christie Walk, Adelaide City, South Australia

Many examples of small Ecopolis-type developments exist in the world's cities, some meeting many of the criteria, others just a few. One of the most outstanding comes from South Australia. Christie Walk, in Adelaide city's south-west quarter, was designed to test and demonstrate the processes, plans and principles contained in the ecological city vision of Urban Ecology Australia (UEA). It is named for the long-time social and environmental activist, Scott Christie, partner of Joan Carlin, part of the community of people who have initiated and sustained this unique development. This community has contributed to the design, development and construction of Christie Walk in a commitment to the human values of urban living and ecological responsibility. It is part of the Whitmore Square EcoCity Project – a conceptual strategic framework adopted by UEA for mapping the south-west quarter of the city as a future piece of eco-city. The design brief was based on energy efficiency, the use of renewables and a high overall ecological performance allied to user participation in the design and development process. It was intended to set the parameters for a project able to demonstrate both the physical and organisational aspects of community and ecological development.

The Christie Walk Project won the Silver Prize in the 2006 Ryutaro Hashimoto APFED Awards for Good Practice (an international environment and development award), and was a 2005 finalist in the international BSHF World Habitat Awards. It is a self-directed social experiment undertaken by people who freely chose to be part of an innovative, non-government initiative.

The project consists of 27 dwellings of various types and configurations with healthy non-toxic construction. Gardens run through the entire site in an environment designed to create spaces for both spontaneous and organised meetings and social interaction. It has solar water heating and photovoltaic power generation and captures storm water for irrigation and toilet flushing. The project was designed in a consciously determined relationship to its broader regional contexts. It demonstrates that the concepts, principles and techniques required to create human settlements that fit within the ecological systems of the biosphere whilst sustaining their biogeochemical functionality do exist. It helps to show that the creation of ecocities will depend on cultural change to transform the deep cultural inertia in local government. The project depended on a creative community with shared ideas and preparedness to translate those ideas into activity. Whether the broader community can be more completely involved with a relatively high level of consciousness of its evolutionary role can only be tested in time.

Masdar, Abu Dhabi

One major eco-city project, designed to incorporate many Ecopolis principles, is currently under construction in Abu Dhabi, the capital of the United Arab Emirates (UAE). Masdar City will feature all of the modern conveniences, services and benefits of living in one of the great cities of the world, but in a carbon-neutral environment. It is being built around pedestrians, where open public squares intersect with narrow shaded walkways and connect to homes, schools, restaurants, theatres and shops. The architecture of the city is inspired by the traditional medinas, souks and wind towers of the Arab world. It is moderately dense but only up to five storeys high, planning to house 45,000 people in just 6 km^2. Electricity will come from photovoltaic panels, cooling from concentrated solar power, water from a solar-powered desalination plant, and landscaping irrigated with city grey water. A solar power plant was one of the first installations on the site to begin to be built.

The Masdar Institute of Science and Technology campus will leverage the fullest use of innovation in energy-efficiency, sustainable practices, resource recycling, biodiversity, transportation and green building standards. Every building will be designed and constructed to provide a model for sustainable living and working. The first stage of the project is scheduled to be finished by 2013, with the date of 2020 set for Masdar City to have a critical mass of residents and businesses.

The entire city is built on a 7 m high concrete podium because this creates the space needed for the network of personal rapid transit (PRT) vehicles which get people around. Above is a pedestrianised world. The idea is people park their cars on the edge of the city and walk to the PRT station. The vehicles travel on a network of roads under podium level, magnets set into the ground at regular intervals and a single wire at roof level guide the vehicles. The initial arrival station and the one at the Masdar Institute were nearly finished in early 2010. The vehicles dock in dedicated glass boxes so that passengers can walk straight into the building and up a flight of stairs (using the lifts is discouraged) into the complex above.[1] The progress of this and other applications of the Ecopolis concept will be watched with great interest, and probably will face considerable scepticism. However, the pioneers who attempt that which to many ecologists and environmentalists appears basically logical and necessary, deserve admiration and praise for their attempts. Future generations will need the conjugate planning, the integrated technologies, the green infrastructure, the low-cost, high frequency public transport, the low-energy buildings and the changes in human behaviour that are already happening in pioneering communities. Urban ecology and Ecopolis principles should permeate education, particularly in the built environment and real estate professions.

Conclusion

The city is both the product and the creator of culture. Our lives, wherever we live, are shaped by ideas and events that occur in cities. Yet despite global urbanism, each city has local context, is an expression of its region and depends upon, and interacts with, its immediate surroundings. Cities are rooted in the ecosystems upon which they depend. Some today exploit distant and past ecological conditions to survive, particularly in extremely arid or cold environments. Others retain a close connection with the land and sea adjacent to them. Urban ecology demonstrates how important and efficient those local links can be. The Ecopolis concept specifies how urban development should enhance, and draw great benefit from the local surroundings.

Efficiency, equity and vitality are the three dominating agents in Ecopolis development. Its main driving forces are energy, money, power and spirit. Competition, symbiosis, regeneration and self-reliance are the main mechanisms for maintaining sustainability. The key instrument for Ecopolis planning is eco-integration in the metabolism of energy, water and materials, in cultivating eco-industry, ecoscape and eco-culture; in the total coordination of system contexts in time, space, quantity, structure and order; total design of development goals in wealth, health and faith; and total cooperation between decision makers, entrepreneurs, researchers and the general public.

To achieve Ecopolis requires many more experiments such as the Freiburg adaptation, the Christie Walk Project, or the Masdar Eco-city. It also requires changes in human behaviour. Many steps that move towards Ecopolis goals are being made through top-down legislation, such as European Directives on water, energy and recycling. Financial incentives, such as feed-in tariffs for home-generated electricty or charges for waste sent to landfill, can also promote change. Other changes are at the grass roots. Many have been assembled by UN-Habitat in their best practices. Many others are distributed by leading environmental NGOs. However, the wealth of knowledge needs to get to people in their formative years, needs to be a core element of education and training. When nature in the city is widely accepted and when pedestrians regain the city streets, progress is made.

Notes

1 See: www.building.co.uk/story.asp?storycode=3156540#ixzz0hLEDbZdB and www.building.co.uk/story.asp?storycode=3155838#ixzz0hLDVtZm6 [accessed 5 March 2010].

References

Bandura, A. (1977) *Social Learning Theory*, Englewood Cliffs, NJ: Prentice Hall.

Boyden, S., Millar, S., Newcombe, K. and O'Neill, B. (1981) *The Ecology of a City and its People: The Case of Hong Kong*, Canberra: Australian National University Press.

Downton, P.F. (2009) *Ecopolis: Architecture and Cities for a Changing Climate*, Canberra: CSIRO Publishing and Berlin: Springer.

Fitzroy, F. and Smith, I. (1998) 'Public transport demand in Freiburg: why did patronage double in a decade?' *Transport Policy*, 5(3): 163–73.

Head, P. (2008) *Entering the Ecological Age: The Engineer's Role, Brunel Lecture 2008*, London: Institution of Civil Engineers.

Hough, M. (1984) *City Form and Natural Process*, London: Routledge.

Kenworthy, J.R. (2006) 'The eco-city: ten key transport and planning dimensions for sustainable city development', *Environment and Urbanization*, 18(1): 67–85.

Lennard, S.H.C. and Lennard, H.L. (1995) *Freiburg-im-Breisgau, Germany: Livable Cities Observed: A Source Book of Images and Ideas*, Carmel, CA: Gondolier Press.

Ma, S.J. and Wang, R.S. (1984) 'Socio-economic-natural complex ecosystem', *Acta Ecologica China*, 4(1): 1–9.

Malhotra, R. (2010) 'Agra: the next eco-city?' *Current Science*, 98(9): 1166–7.

Ryan, S. and Throgmorton, J.A. (2003) 'Sustainable transportation and land development on the periphery: a case study of Freiburg, Germany and Chula Vista, California', *Transportation Research Part D*, 8(1): 37–52.

Spirn, A.W. (1984) *The Granite Garden: Urban Nature and Human Design*, New York: Basic Books.

Tjallingii, S. (1992) *Ecologisch verantwoorde stedelijke ontwikkeling*, IBN-DLO Rapport nr 706, Wageningen: Institute for Forestry and Nature Research (IBN).

Tjallingii, S. (1994), *Ecopolis: Strategies for Ecologically Sound Urban Development*, Leiden: Bakhuys.

Tjallingii, S. (1996) *Ecological Conditions: Strategies and Structures in Environmental Planning*, IBN Scientific Contributions 2, Wageningen: Institute for Forestry and Nature Research (IBN), doctoral thesis, Delft University of Technology.

Wang R.S. (1991) 'Towards Ecopolis: urban ecology and its development strategy', *Journal of City and Regional Planning*, 18(1): 1–17.

Wang R.S. (1994) 'Planning the ecological order – a human ecological approach to urban sustainable development', in R.S. Wang and Y.L. Lu (eds), *Urban Ecological Development: Research and Application*, Beijing: China Environmental Science Press, pp. 1–20.

Wang, R.S. and Hu, D. (1999) 'Totality, Mobility and Vitality: Fengshui Principles and their Application to the Blue Network Development of Yangtze Delta', in: *Land and Water: Integrated Planning for A Sustainable Future, Final Report 1998 ISOCARP Congress, Azores, Portugal*, The Hague: ISOCARP: 36–47.

Wang, R.S. and Paulussen, J. (2004) *Beijing Conjugate Ecological Planning: Final Report*, Beijing: Chinese Academy of Sciences.

Wang, R.S. and Qi, Y. (1991) 'Human ecology in China: its past, present and prospects', in S. Suzuki (ed.), *Human Ecology Coming of Age: An International Overview*, Brussels: Free University Brussels Press, pp. 183–200.

Wang, R.S., Hu D. and Wang X. (2004b) *Urban Eco-Services*, Beijing: Meteorological Press.

Wang R.S., Lin S.K. and Ouyang Z.Y. (2004a) *The Theory and Practice of Hainan Eco-Province Development*, Beijing: Chemical Engineering Press.

Wang, R.S., Zhao J.Z. and Ouyang Z.Y. (1996) *Wealth, Health and Faith: Sustainability Studies in China*, Beijing: China's Science and Technology Press.

Wang, R.S., Zhou T., Chen L. and Liu J. 2006. *Fundamentals of Industrial Ecology*, Beijing: Xinhua Press.

Conclusion

Ian Douglas

Urban ecology presents major challenges in the way in which innovative ideas, new organisms, climatic extremes, potential problem substances, political decisions and economic events all impinge on the stability and evolution of plant and animal communities within cities. Not only is concern for urban nature irregularly distributed among the human population, but conditions for plant growth and animal survival vary greatly over scales of centimetres to metres, reflecting contracts in land cover, soil conditions and applications of horticultural chemicals. As the chapters in Parts 2, 3 and 5 of this book show, we have a good understanding of the basic science, the habitats and the ways of analysing these aspects of the urban environment. Part 4 shows that we know a great deal about the ecosystem services and benefits to society that flow from urban nature. Yet, we would find it difficult to forecast what the nature of urban greenspace would be in 100 years' time. The great urban sprawl is essentially a post-1950 problem, as was the decline of urban parks, but in 1950 would we have foreseen the development of concern for green infrastructure and greenspace networks?

Ecosystems are constantly changing. Ecology is a science of emergent phenomena (Alberti *et al.* 2003). Cities are key examples of emergent phenomena, in which each component contributes to but does not control the form and behaviour of the whole. Human aspirations, planning legislation, real estate prices, landform, transport infrastructure and personal mobility all contribute to patterns of urban expansion, atmospheric emissions, urban heat islands and aquatic ecosystem change. Building the ability to bring these human factors into the analysis of urban ecosystem dynamics remains a continuing task for urban ecology. Alberti *et al.* (2003) see the greatest challenge for ecology in the coming decades as to integrate, fully and productively, the complexity and global scale of human activity into ecological research.

The discussions of the applications of urban ecology in Part 6 of this book have shown the benefits of an interdisciplinary approach. They illustrate how partnership between urban ecologists and professionals and civil society organizations, public participation and working together can use ecological understanding to create healthy, liveable, more sustainable cities. Such partnership is needed to help cities to adapt to climate change and all the challenges of the future.

Already some progress has been made, both in Europe and the USA. Work in the USA Long-Term Urban Ecosystem Ecological Research Program has shown that marketing urban nature has gone beyond the mere description of 'leafy suburbs' to using vegetation cover aspirations as a marketing tool. Many have found socio-economic status to be an important predictor of vegetation in urban residential areas. However when additional household characteristics, associated with lifestyle behaviour, are used, they provide better results, at least for vegetation cover (Grove *et al.* 2006). These lifestyle factors suggest that when the manufacturers of lawn-care chemicals market their products to various consumer groups by associating 'community, family and environmental health with intensive turf-grass aesthetics' and fostering household demand for 'authentic experiences of community, family and connection to the non-human biological world through meaningful work' they are understanding the motivations of human responses to the urban environment.

In Europe, the outstanding achievements of Herbert Sukopp and his colleagues in Berlin show how the scientific development and applications of urban ecology go hand in hand. In Berlin this group saw parks and other urban greenspace not merely in terms of their aesthetic qualities or recreational functions but also as biotopes with an ecologically valuable, flora and fauna. The team not only supported the creation of greenspaces in the city, but also suggested how these spaces should be designed and managed to optimize their biodiversity (Lachmund 2007). Even if the projects for urban wildlife biotope protection that Berlin ecologists lobbied for were not fully realized, new concepts of urban ecology entered urban land-use decision making and aroused the imagination of urban residents. Through ecologically based forms of architecture, garden design and ecosystem protection, Sukopp and his colleagues left an enduring imprint on the physical landscape of the city (Lachmund 2007).

Many contributors to this volume have in their own ways contributed to changing the ideas of politicians, planners, developers and the general public on urban greenspace and the importance of urban ecology. Now is the time to strengthen that impact. A firm place for urban ecology has to be established in the education and training of all planners and municipal engineers, to ensure that when urban development proposals come forward, the question of their urban ecosystem impacts is automatically in the minds of the officials involved. I hope that this book will contribute to that continuing education and to the way all of us see and enjoy our urban environment.

References

Alberti, M., Marzluff, J.M., Schulenberger, E., Bradley, G., Ryan, C. and ZumBrunnen, C. (2003) 'Integrating humans into ecology: opportunities and challenges for studying urban ecosystems', *BioScience*, 53(12): 1169–79.

Grove, J.M., Troy, A.R., O'Neil-Dunne, J.P.M., Burch Jr., W.R., Cadenasso, M.L. and Pickett, S.T.A. (2006) 'Characterization of households and its implications for the vegetation of urban ecosystems', *Ecosystems*, 9(4): 578–97.

Lachmund, J. (2007) 'Ecology in a walled city: researching urban wildlife in post-war Berlin', *Endeavour*, 31(2): 78–82.

Index

abiotic factors 39–40
abiotic traits, woodlands 327–9
Abu Dhabi, ecological principles 648–9
accessibility of nature 88–9, 228–30, 395–6, 590–1
acculturation 42
acidic soils 169–70
adaptation 583; insects 270, 291; processes of 45–6; understanding 42; *see also* climate change
Adams, Lowell 85
Adelaide, ecological principles 648
aeration, soil 68–9
aerial photography 293, 471, 479, 481
aerosol effects, precipitation 140–1
aesthetic value of nature 377–84, 385–92, 396, 397–8
African savanna 253–4, 257, 258
Agenda 21 553, 646
aggression, reduction of 416–18
air pollution 115
air quality, effect of 428–9
airport-centred development zones 602
alkaline soils 170
allotments 266
amenity benefits, green roofs 569
American Community Gardening Association 391
Amsterdamse Bos Park 49
ancient woodland 323–4, 326, 327–8, 332, 479
animal urbanism 64–7
animals: co-existence with 70–1; food sources 354–5; gardens 451–3; habitat diversity 266–8; habitat use 353; public attitudes to 356–8; reclaimed grasslands 304–6, 308–9; response to climate change 225–6, 497–8; rights of 68–70; rock outcrop origins 255–7; shelter and microclimate 355; social behaviour and population dynamics 355–6; spatial distribution patterns 353–4; use of passages and culverts 282–3; walls and paved surfaces 242, 247–8; and Western domination of nature 67–8; wetlands 345–6; *see also* feral animals; invasive species
anthropocentric disbenefits of trees 432–3
anthropocosmos model 44–5
anthropogenic factors 39, 495; woodland sites 332
anthropogenic heat flux 109
ANZAC Drive, Christchurch 211
aquatic ecosystems, biogeochemical cycling 509–11
aquatic recipient systems, fluxes to 506–7
architectural practice 632–3
architecture, ecological impacts of 44–5
atmosphere: effect of trees 426–7; fluxes to and from 505–6
attention-deficit hyperactivity disorder (ADHD) 414
Attention-Restoration Theory 417–19, 420
Australia, climate change 495

Babylon, gardens of 378
bacteria, soil 178–9
bank vegetation, management of 298
Barker, George M.A. 78, 86, 90, 276, 479, 537, 540–1
barrier function, corridors 278–9, 285
Beijing 20, 77, 136, 271, 425–8, 645
beneficial/benign plant species 213–14
berberis sawfly 498
Berlin 4–5, 21, 76–8, 84–5, 91, 170, 190, 283, 324, 367, 461, 479, 492–3, 653
bioassay biological testing 318
biodiversity: assessing 56; benefits of green roofs 568–9; effects of urbanization 10–11; factors influencing 453–4; gardening for 454–5; gardens 268–72, 450–5; halting losses to 230–1; open spaces 293–7; rock outcrops 243–5
biodiversity actions plans, UK 621–8
biofuels 640
biogenic volatile organic compounds (BVOCs) 428
biogeochemical flux analysis: among compartments 505–7; biochemical cycling in urban ecosystems 508–11; framework 503–4; scale and urban biogeochemistry 514–15; whole-ecosystem analysis 511–14
biological characteristics: street trees 425–6; urban soils 175–9
biological health models 43–4

biomass structure, woodlands 329–30
biophilia 257
Biosphere Cities Programme 558–9
biosphere reserves (BRs): application to towns and cities 550–2; theory 549; understanding of 549–50; urban areas response to 552–7
biotic factors 39–40
biotic homogenization 11
biotic traits, woodlands 329–32
biotope mapping 85, 482–3
birds: grasslands 308–9; parks 297; stonework habitats 247, 249; wetlands 345–6
Birmingham, wildlife conservation 86, 277, 282, 286
black redstarts 625–6
Blue-Green technologies 574, 579, 609
Bold Moss Colliery, St. Helens 223
botanical exploration 74–6
boundary layer urban heat islands 128–30
Box, John 537–46
Boyden, S. 21, 27, 30, 40–1, 45–6, 77, 522, 636
British Trust for Conservation Volunteers (BTCV) 622–3
Brown, Lancelot "Capability" 378–9
bryophytes 492
Buddleia 494
built environment as habitat 257–8, 466
built-up areas 21
bulk density, soil 166–8
burdock 493
business parks, habitat diversity 267
butterflies 225, 295, 297, 304

California, wildlife corridors 282–3
Camley Street Natural Park, London 55, 466
canals 268
canary-winged parakeets 365–6
cane toads 495
canopy layer urban heat islands: controls on heat island magnitude and dynamics 127–8; genesis 125–6; observation 125; spatial form 126–7; temporal dynamics 127
Cape Town 38, 84, 535, 552, 557–8, 586, 621; biodiversity strategy 627–8
Cape West Biosphere Reserve, Cape Town 557
carabids 177
carbon cycle 179–80
carbon levels 496
carbon sequestration, trees 429, 585–6
cats 308, 357–8, 366–7
Celia Hammond Animal Trust 367
Centre of Urban and Regional Ecology, UK 484
chemical characteristics, urban soils 166–70
chemicals, use of 292
Chicago School of Sociology 27, 76
Chicago Wilderness 59, 552
China: ecological principles 644–7; urban ecology 77, 79
Chinese Academy of Science 79
Choosing Health white paper, UK 594–5
Christchurch, New Zealand 198, 200, 203, 205, 208–12, 269
Christie Walk, Adelaide City 648
circadian rhythms 41
cities: applying biosphere reserve concept to 550–2; climate of 103–16; ecology in 28; ecology of 29; as economic systems 18–20; human dimension in 29; as integrated biosocial systems 22–3; nutrient budgets 511–13; as silvicultural and/or ecological system 630–1; as systems 17–18; as urban ecosystems 20–2
civil disorder and neighbourhood maintenance 402
climate change: and creative conservation 225–6; impact of 89–91; and invasive species 493–500; and rainfall 150–1; response of invasive animals and insects 490–3; response of invasive plants 488–90; role of green infrastructure 583–7
climate mapping 499–500
climate scales 120–2
climate: air pollution 115; anthropogenic heat flux 109; effects of trees 426–7; effects of urbanization 10–11; global, regional and local effects 103–7; net all-wave radiation 108–9; net heat advection flux 113; net heat storage flux 112–13; surface energy and water balance 107–8; turbulent latent heat flux 111–12; turbulent sensible heat flux 109–11; urban heat islands (UHI) 120–30; wind fields 113–15
cloud condensation nuclei (CCN) 136, 140–1
Coalition for a Livable Future 58, 59
Coca-Cola Company 523
cockroaches 490
cognate habitat surveys 481
cognitive capitalism 614–16
colliery spoil 223
colonization: reclaimed land 303–5; walls and paved surfaces 239–40
colonizers, public attitudes to 261
Columbia Region Association of Governments (CRAG) 52
commercial landscapes, biogeochemical dynamics 508–9
commercial systems, cities as 19
Common Agricultural Policy, EU 607
common plants 226
community benefits, street trees 430–1
community building 415–16, 643
community gardens, health benefits 412, 414, 416, 418
community health 408–20
community identity 593
compact cities 641–2
Competitive-Stress tolerant-Ruderal (CSR) theory 292
complex plant associations 207–8
complex systems approach 34

conceptual site model (CSM) 311, 314, 316, 319
conduit function, corridors 278, 281, 285
conferences 84–5
conjugate ecological planning 643–4
connectivity 278–83; mapping 483
conservation: accessible nature 228–30; common wild plants 226; environmental justice 227–8; historical perspective 84–91; liberty in the system 230–1; making a start 221–2; start points not end points 226–7; taking the opportunity 222–5; time for change 225–6; tumbling effects 225
Conservatoire du Littoral, France 6–8
construction, geomorphic impacts during 161–2
contaminated land: contaminant source zones 314; Ecological Risk Assessment (ERA) 315–19; historic influence 313–14; interaction with wider environment 314–15; remediation 320–1; statutory meaning 313
convective process 132–43
Convention on Biological Diversity (CBD) 213, 621
convergence ecology 212–13
convergence effect 137–40
Coquitlam corridor, British Columbia 284
cottony cushion scale 499
Countryside Commission, UK 380
Countryside Council for Wales 540
coupled atmosphere-land surface (CALS) models 137, 139–40
coupled human and natural system (CHANS) 18, 22, 23
course particulate organic matter (CPOM) 510
creative cities 614
creative conservation 221–31
cultivation, understanding 41–2
cultural adaptation 42
cultural benefits, street trees 430–1
cultural factors 39–40
culture, enriching 643
culverts 156, 161, 282–3
cypress gall mite 499

DAISIE programme 472
de-icing salts 169–70, 173–4, 303
Decentralized Stormwater Management, Best Management Practices (BPPs) 574
definitions 7–12, 385–6, 602–3
Defra, UK 226, 320, 432, 623, 624
deliberative recombinants 203, 204–5, 213
Department of Environmental Conservation, US 483
deposition, landforms of 162–3
derelict land 77; as corridors 281; habitat diversity 266–7; restoring 641
Descartes, René 68
desert landscaping 269
detention ponds 153, 161, 283
developed countries: edge cities 601; soil pollution 171–3
development: balancing 641; and ecology 630–5; geomorphic constraints on 159–63
diseases: animals 356, 357, 363; humans 397, 430; and nutrition 46; plants 431–2, 492, 494–6, 498–500
dispersal routes, corridors as 283–4
dissolved organic carbon (DOC) 510
DNA mapping 227
dogs 308, 367
"doughnut city" phenomenon 19
Doxiadis, Constantinos 44–5
drainage: soil 68–9; *see also* sustainable drainage
Dresden University of Technology 581
drought-resistant plants 586
DSPIR scheme 605
dualism 67–8
DuPont Company 572
Durban 532, 537, 551, 558; Management of Open Spaces System (D'MOSS) 551

early urban nature planning 50–3
Earth Summit, Rio de Janeiro (1992) 621
earthworms 176–7
East Bay Regional Park District (EBRPD), San Francisco 55–6
East London Green Grid Primer Paper 590, 591
eco-culture 639
ecocentric disbenefits of trees 431–2
ecological considerations, combining with people considerations 391–2
ecological economics 93–5
ecological factors, grassland 302–6
ecological footprint (EF) 21–2, 31, 514, 526
ecological impacts of architecture and urban ecosystems 44–5
ecological industry (EI) 637–8
ecological interpretations of health 43–4
ecological justice 69–70
ecological landscape (ecoscape) 638–9
ecological modifications, green roofs 570
Ecological Parks Trust, UK 78
Ecological Risk Assessment (ERA) 315–19, 321
ecological sanitation 637
Ecological Screening Levels (Eco-SSLs) 316
ecological security 637
ecological services, soils 164–5
ecological surveys 317
ecologically acceptable exotic plant species, criteria for 213–14
ecologically sound management, open spaces 297–9
Ecology of Indonesia Series 79
ecology: in cities 28; of cities 29; in urban development/planning 630–5
economic activity, promotion of 642
economic benefits, green roofs 570
economic perspective, street trees 435–6
economic risk factors 43–4
economic systems, cities as 18–20
economic valuation 93–5

Ecopolis concept 289–90, 636–8; adaptive Ecopolis development (AED) 637; China 644–7; new developments 643–4; philosophies 640–3; technologies 639–70; worldwide 647–9
ecosystem engineering 255–7, 261–2
ecosystem functions, effects of urbanization 10–11
ecosystem modelling 21, 86
ecosystem services: green infrastructure 89–91
ecosystems 26–7; biogeochemical cycling within 508–11; cities as 20–2, 630–1; ecological impacts 44–5
edge effect, woodland sites 333–4
edges: as boundaries 631; mapping 483
Edinburgh local biodiversity action plan 626–7
education, sustainability 591–3
educational resources 261–2
ekistics 44–5
El Paso water company (EPWC) 155
elaegnus sucker 499
Eliot, T.S. 383
elm trees 429, 435
emerging economies: edge cities 601; soil pollution 171–3
emergy 4, 63, 524–6
endangered species 330, 356, 452–3, 480, 482, 625–6, 627
Endless Village (Teagle) 78, 86, 480
energetics of urban metabolism 523–5
energy flows analysis 22
energy performance, optimising 642
energy use 9, 41
English landscape movement 378–9
English Nature 88–9, 90, 226, 230, 231, 381, 484, 541, 624
Environment Agency (EA) 311, 315, 316, 318, 319
environmental benefits: green roofs 569; street trees 426–31
environmental capital 544
environmental interaction, contaminated land 314–15
environmental justice 58, 227–8
environmental movement 381–2
Environmental Protection Act (1990), UK 313
environmental risk factors 43–4
environmental valuation 94–5
equity issues 58
Ernle Reserve, Christchurch 208–9
ethnic differences, park use 396
Europe: cultural landscapes 211; peri-urban ecologies 606–12
European cities 49
European Environmental Agency (EEA) 93
European Federation of Green Roofs 224
European Habitats Directive (1992) 313
European Nature Information System (EUNIS) 461, 471–4, 476
European Public Health Alliance (EPHA) 594
European Statistical Office (EUROSTAT) 29
European Sustainable Communities report, EU 49
European Symposium of Urban Ecology 85
European Union 79; Expert Group 49; *Green Paper on the Urban Environment* (1990), EU 49; Peri-Urban research programme 605–12
European Water Framework Directives (WFD) 311, 580
evapotranspiration 151–2
evo-deviation, monitoring 45–6
evolutionary adaptation 42
exotic species *see* introduced species
Expert Patients Programmes, UK 592–3
exploitation 66
extraction, landforms of 162
exurban areas 9

false colour IKONOS images 466
farming 607–10
feral animals: cats 357–8, 366–7; dogs 367; parakeets 364–6; pigeons 361–4; pigs 367; terrapins 367–8
fertilisers, use of grasslands 308
filter function, corridors 278, 285
fire suppression 189, 212
Flanders, suburban parks 293–7
flood channels 161–2
flooding 561, 572
Florianopolis, Brazil 552, 553–4
floristic composition, woodlands 329–30
floristic provenance, woodlands 331–2
food production 46, 391
food sources, animals 354–5, 362, 364
footwear, as dispersal mechanism 190–1
foxes 356–7
fragmentation: animal habitats 354; woodland sites 333–4
Freiburg, Ecopolis development 647–8
Freshkills landfill site, New Jersey 634
From doughnut city to café society report 19
fungi 178–9, 305, 492

garden biodiversity, UK 268–9
garden city movement 48
gardening 390–1; for biodiversity 454–5
gardens: age of 454; biodiversity 268–9, 450–5; green space coverage 454; habitat and resource occurrence 454; impact of climate change 496–500; insects in 270; intercontinental movement of tastes 269–70; origins and fashions 268; size of 453; trees 265–6; *see also* community gardens
gaseous pollutants, effects of trees 428
geographic information systems (GIS) 152, 293, 466, 478, 479, 481
geometric traits, woodlands 332–4
Gerard, John 75–6
global ecological balance (GEB) 45
global urban agenda 600–1
global weather, impact of cities 103–7

globalization of rare habitats 257–8
globalization-localization dynamics 612–14
golf courses, habitat diversity 267
Goode, David 54–5
gradient approach 11–12
Grass Roof Company 224
grassland on reclaimed land 301–9
Green Belts, UK 48, 555, 603, 608
'Green Grid' concept 591
green infrastructure: climate change role of 583–7; concept of 584–5; mapping 484; naturalizing role of 56–7; in twenty-first century cities 599–619
Green Infrastructure Network 56
green infrastructure services 89–91
Green Networks 4
green roofs 224, 261; biodiversity benefits 568–70; design principles 578–9; examples of installations 581; institutional aspects of implementation 580–1; role in climate modification 584; types 567–8
green vegetable bugs 498
Greenspace Scotland 596
greenspace: access and maintenance 590–1; action on adaptation to climate change 585–6; and biodiversity 293–7, 454; developing sustainable management 290–3; ecologically sound management measures 297–9; impact of climate change 496–500; integrative concepts 289–90; mapping 484; modelling climate change adaptation benefits 585; politically desirable protection 551; standards for provision of 537–46
greenways *see* wildlife corridors
Greenwich Peninsula Ecology Park, London 101
Grey to Green, Going for Clean Rivers program, Portland 57
groundwater 148–9, 151, 153–5; as biogeochemical sink 507–8
Groundwork Trusts, UK 222, 223, 622–3
Grove, J.M. 11, 23,31, 210, 455, 653
guilds 285
gulls 209

Haagland (Netherlands), peri-urban ecology 608–9
Habitat Action Plans, UK 626, 627
habitat analysis: classification of habitats 465–70; developing international habitat classifications 471–6
habitat diversity, suburbia 266–70
habitat functions, corridors 278–9, 281–2, 285
habitat heterogeneity, woodlands 334
habitat modification 66
habitat structures, SUDS 563–4
habitat templates 253–4
habitat type mapping 478–85
habitat units as templates for biodiversity 293–5
habitat use, animals 353
habitats: biological evolution 252–62; density and biodiversity 454; differences between rural and urban 479
Hackensack Meadowlands, New Jersey 346
Hampstead Heath 75–6
hay strewing 223
Health Impact Assessments (HIAs) 589, 595–6
heavy metal soil contamination 170–3, 343–4
hedges, plant species 205
herbaceous borders: management of 299; plant species 205–7
herbals 74–6
herbicides, use on grasslands 308
herbivory 193
Himalayan Balsam 489–90
historic influence, contaminated land 313–14
historic perspective: effects of precipitation and convection 133–7; nature planning 84–91
history of urban nature studies 74–9
home, exposure to nature 410, 411, 413, 415, 416–17, 418
Hong Kong 4, 21, 30, 39, 45–6, 77, 86, 149, 162, 168, 170, 172, 241, 505, 522, 636; human ecology research 45–6
horse chestnut leaf mining moth 498
horticultural stock 190
hospitals, exposure to nature 411, 413, 415, 418
household flux calculator (HFC) 508
housing estates, landscaping 224, 227–8, 229
Houston Atlas of Biodiversity 59
Houston Wilderness 59
Hudson, W.H. 75
Human Appropriation of Net Primary Production (HANPP) 526
human considerations, and ecological considerations 391–2
human dimension, cities 29
human ecology: applying core principles of 39–41; core principles 39; ecological impacts of architecture/urban ecosystems 44–5; health and quality of urban life 42–4; interdisciplinary and transdisciplinary contributions 46; monitoring evo-deviation and techno-addiction 45–6; understanding adaptation 42; understanding cultivation 41–2
human evolution, landscape context of 253
human health 42–4; benefits of green roofs 569; benefits of nature 227–8, 383, 430; biological and ecological interpretations 43–4; and diet 46; outcomes of therapeutic landscapes 594–7; and physical activity 394–403; provision of 642; psychological and community health 408–20
human health issues, trees 432–5
human relationship with nature 63–71; scientific analysis of 380–1
humans as generalists 258–9
hydrological effects, trees 427
hydrology: changes to rivers 155–6; groundwater in 153–5; modified natural hydrological system 150–3; urban hydrological cycle 148–l; wetlands 342–4; woodlands 328–9

Image of the City (Lynch) 631
impervious surfaces, water runoff 572–4
industrial ecology: brief history of 27; common frontiers of 32–4; current scope of interests 28–9; urban metabolism as linking and/or common approach 29–32
industrial ecosystems 26–7
infectious plant diseases (EIDs) 495
innate adaptation 42
insects: effects on plants 193; in gardens 270; response to climate change 496–7; *see also* invasive species
Institute of Ecology, Berlin 85
Institute of Terrestrial Ecology 481
integrated biosocial systems, cities as 22–3
integrative approach, open spaces 289–90
interdisciplinary contributions, human ecology 46
interdisciplinary networks 591–3
International Biological Program (IBP) 77–8
Intertwine Alliance 58–9
intrinsic value of nature 377–84, 385–92
introduced species: adaptation to local conditions 291–2; adaptive variation 226–7; ecologically acceptable 213–14; gardens 452; and recombinant ecology 198–216; response to climate change 488–90; urban corridors as dispersal routes 283–4; wetlands 344; woodland sites 331–2
invasive species 214; animals and insects 357, 490–3; and climate change 493–500; in different cities 492–3; plants 488–90; public attitudes to 356
invertebrates, soils 177–8

Japanese knotweed 489
Joint Nature Conservancy Committee 226

Kensington Gardens 75
Koper (Slovenia), peri-urban ecology 609–10

Lake Eerie Allegheny Partnership for Diversity (LEAP) 59–60
land clearance 188–9
land cover types, suburbia 265–6
land cover: change 526; mapping 480–1
land use: change 526; comparison 12; mapping 465–76, 480–1
land use planning, role in nature preservation 54
Land Utilisation Survey of Britain 480–1
landfill 223–4
landforms of extraction/deposition 162–3
Landlife 221–2, 225, 227–8, 231
landscape architectural practice 632–3
landscape context of human evolution 253
landscape ecology 57–60
Landscape Ecology (Forman) 631–2
landscape features as habitats 466
landscape perspectives 214–15
landscape planning: assessing biodiversity 56; European models 49; Portland, Oregon 50; learning from others 54–6; regional growth management and landscape ecology 57–60; regional landscape view 50–4; regional planning 49–50; role of green infrastructure 56–7
landscape structure 189; and species availability 191
landscape urbanism 633–4
landscaped parks: biodiversity 293–7; developing sustainable management 290–3; ecologically sound management measures 297–9; integrative concepts 289–90
lawns: grass species 207; in landscaped parks 298–9; preoccupation with 390
Leadership in Environmental Design (LEED) buildings 50, 581
Leipzig, peri-urban ecology 610–11
lichens 305, 492
life-cycle assessment (LCA) 523
lightning, mechanistic linkages among urbanization, rainfall and 137–41
lily beetle 498
lime trees 425, 429, 430
Limits to growth report, Club of Rome 27
Liverpool, wildlife corridors 276–7
living cities 641–3
Living Urban Parks (LUPs) 289–90
Local Biodiversity Action Plans, UK 269, 621–7
Local Governments for Sustainability (ICLEI) initiative 544
Local Issues Advisory Group (LIAG), UK 622
Local Nature Reserves (LNRs), UK 5, 88, 539–43
local weather, impact of cities 103–7
London 4, 21, 38, 54–5, 74–9, 80, 86,101, 148, 152, 164, 169, 200, 222, 224–5, 229, 237, 240–1, 244, 270, 274, 321, 324, 352, 354, 361, 363, 365, 367, 379–80, 425–6, 429–30, 432–5, 453, 462, 481–2, 489–90, 498–500, 526, 538, 544, 551, 568–9, 574, 590–1, 593, 596, 622; biodiversity action plan 624–6; nature conservation 87–9; parks 75–6; values of urban nature 377–84
London Biodiversity Action Plan 622, 625–6
London Biodiversity Partnership 224, 624
London Development Agency 626
London Ecology Committee 88
London Ecology Unit 54–5, 79, 87, 624
London MIND Meanwhile Wildlife Garden 590, 591, 592–3, 595, 597
London Plane trees 169, 425, 429
London Wildlife Trust 55, 87
London's Natural History (Fitter) 77, 84
Long Term Ecological Research (LTER) projects, US 12, 18, 512, 652–3
Low Impact Development (LID) 574

macrobiota 176–7

management: continuity of 292; green roofs 570; landscaped parks and open spaces 290–3, 297–9; reclaimed land 307–8; walls and paved surfaces 245; woodlands 334
Manchester: peri-urban ecology 606–7; wildlife corridors 276–7, 279–80
Manchester University School of Landscape 222
Manhatta Project 48, 60
maple trees 425
Masdar Institute of Science and Technology 649
Masdar, Abu Dhabi 648–9
Mata Atlântica Biosphere Reserve 552, 553–4
Mata Atlântica 'Urban Forests' project 553
Material Flow Accounts (MFAs) 522, 526
Mayor's Biodiversity Strategy, London 87–90
McHarg, Ian 23, 84, 165, 633
megapolises 10
mental inspiration, trees as 430–1
mental models/images 387–8
mesobiota 177–8
metropolitan areas 9
METROpolitan Meteorological EXperiment (METROMEX) 133–4
Michigan State University 581
microbiota 178–9
microclimate 120–30; and animals 355; walls and paved surfaces 244; woodlands 328–9
microhabitat, walls and paved surfaces 245–7
Million Trees Campaign, London 482
modified pH, urban soil 168–70
monk parakeets 365
Montpellier, peri-urban ecology 607–8
moral considerability, boundaries of 68–70
Mornington Peninsula and Western Port BR, Melbourne 555–7
mosses 305
mowing 292, 299, 307–8
multicultural integration 591–3
multifunctional greenspace 537–46, 590

National Climatic Data Center, US 585
National Environment and Rural Communities (NERC) Act (2006), UK 623
National House-Building Council, UK 432
National Institute for Urban Wildlife, US 85
National Nature Reserves, UK 88, 540
National Parks and Access to the Countryside Act (1949), UK 5
National Urban Forestry Unit, UK 482
National Vegetation Classification, UK 226–7, 230, 231, 481
National Wetlands Inventory, US 339
National Wildflower Centre, UK 229, 230
native species: adaptation to local conditions 291–2; gardens 452; honorary species 214; and recombinant ecology 198–216; response to climate change 488–90; wetlands 344; woodland sites 330, 331–2
natural colonisation, reclaimed land 303–5
Natural England 226, 228, 539, 591
Natural Environment Research Council, UK 221
natural greenspace, Areas of Deficiency 88–9
natural history 74–9
natural history societies 76
Natural Resources Group (NRG), New York 60
Natural Resources Lands Inventory (NRLI) 57–8
nature: aesthetic and intrinsic value 377–84, 385–92; accessibility of 88–9, 228–30, 590–1; defining 385–6; human relationship with 63–71; observing 386–7; origins of Western domination of 67–8; and physical human health 394–403; and psychological/community health 408–20; purposes and functions 386–8; rights of 69–70; and spirituality 381–4; "thereness" 387–8
Nature Conservancy Council, UK 78, 86, 222, 380
nature conservation: green infrastructure and ecosystem services 89–91; in London 87–9; role of land use planning 54; through strategic planning 86
Nature Conservation (Scotland) Act (2004) 622
Nature Conservation Strategies (NCS), UK 5
Nature in Neighbourhoods program, Portland 54
nature preferences: and gardening 390–1; preference matrix 388–9; wild and tame 389–90
nature spirituality, origins of 382–4
Nazi environmentalism 67
neighbourhood maintenance 402
neophytes 492–3
net all-wave radiation 108–9
net heat advection flux 113
net heat storage flux 112–13
New Atlas of British and Irish Flora (Defra) 226
new developments, Ecopolis 643–4
new ecological landscapes 221–31
New York 4, 21, 48, 60, 65, 74–5, 77, 84–5, 134, 141, 149, 154, 163, 165, 167–8, 174, 176, 177, 179, 180, 192–4, 200, 237, 340, 342, 346–7, 364, 365–6, 391, 506, 515, 551, 585; urban land use inventory 470
New Zealand, recombinant ecology 198–216
Nexis® UK "all English" news database 434
Nicholson, Max 77
No Child Left Inside movement, US 55, 59
"no-analog" model 259
noise reduction 430
non-native species *see* introduced species
non-urban areas, criteria for separating from urban areas 9–10
Normalized Difference Vegetation Index (NDVI) 398
novel plant/animal associations 198–216
nutrient budgets 511–13
nutrients, wetlands 342–4

oak trees 499–500
"oasis effect" 11
Olmsted, Frederick Law 48, 52, 75, 395, 633
Olmsted, John Charles 50, 75

open spaces *see* greenspace
open systems 27, 41
Ordnance Survey Maps 480–1
Oregon Gap Analysis Program 56
Oregon Natural Heritage Database Project 58
Organization of Economic Cooperation and Development (OECD) 29, 430
Oxford Botanic Gardens 489
oxygen production, trees 429

parakeets, feral 364–6
Paris 4, 21, 49, 74, 76–7, 79, 106, 138, 142, 148, 153, 164, 237, 324, 364
parks: features and users 396–7; health benefits 395, 412, 413, 416, 418; history of 75–6; proximity of 395–6; trees in 204, 265–6, 297–8, 426; *see also* landscaped parks
participating cities 641–3
particulate interception, trees 429
pathogens 494–5
pathological plant species 214
paved surfaces: colour 243–4; colonization 239–40; decay 243; evaporation 152; management 245; microclimate 244; microhabitat 245–7; pollution 244–5; runoff 244, 572–4; stonework communities 240–2; structure 242–3; suburbia 265–6
Pennine Prospects 612
peri-urban areas 21; context 600–6; dynamics of 612–18; Europe 606–12; policy implications 618–19
Peri-Urban Land Use Relationships (PLUREL) project 600, 602, 605–12
peri-urban, definition of 602–3
personal rapid transit (PRT) vehicles 649
pest status, feral pigeons 363–4
pests *see* invasive species
philosophies, Ecopolis 640–3
physical activity: future research 402; and parks 395–7; safety issues 400–12; and vegetation 397–400
physical characterisation, street trees 425–6
physical stressors, plants 192–3, 212
Pickerings Pasture, Widnes 223–4
Pickett, S.T.A. 10–11, 20, 22–3, 34, 80, 165, 287–91, 200, 215, 264, 267, 514
pigeons: grasslands 309; history and distribution 361–2; natural history 362–3; pest status and control 363–4; response to climate change 490
pigs 367
Pitt Review (2008) 561
Planning Policy Statement (PPS), UK 313
planting, reclaimed land 303–5, 306
plants: adaptation to local conditions 291–2; adaptive variation 226–7; availability 190–1; differential performance 191–3; for sustainable drainage 562–4; gardens 451–3; habitat diversity 266–8; provenance 214; reclaimed land 302–6; response to climate change 225–6, 497; rights of 63–71; rock outcrop origins 255–7; traits 193; transpiration 152; walls and paved surfaces 239–49; wetlands 344–5; *see also* invasive species
Plato 67–8
political perspective, street trees 436–7
pollutant interception, trees 428–9
pollutants: effect on plant performance 192; urban soil 170–5; walls and paved surfaces 244–5; water 155; wetlands 342–4; wildlife effects 66
pollution effects, precipitation 140–1
pollution, air 115
polycyclic aromatic hydrocarbons (PAHs) 174–5
ponds: parklands 283, 295, 298; water detention 153, 161, 283
population dynamics, animals 355–6
Portland, Oregon 49, 50–9, 282
Post-industrial sites of High Ecological Quality as a Priority Habitat, UK 624
Potteric Carr Nature Reserve, Yorkshire 565–6
precipitation 150–1; historical perspective of urban effects 133–7; mechanistic linkages among urbanization, rainfall and lightning 137–42; reconciling mechanisms and methods 142; societal implications and recommendations 142–3
production systems, cities as 19–20
prospect and refuge theory 401
protected sites, categories of 88
Psycho-Evolutionary Theory 417–19, 420
psychological health 408–20
psychological perspective on nature 385–92
public attitudes, urban wildlife 356–8
public policy, peri-urban issues 603, 617–19, 619
Puget Sound regional integrated simulation model (PRISM) 22

quality of life 42–4
quarries 162

racoons 356, 357, 490, 584
ragworts 489
railway corridors 274, 281
rainfall, mechanistic linkages among urbanization, lightning and 137–41
rapidly growing cities, suburban mosaic 270–1
rats 490, 497–8
reclaimed land: animal use and interactions 308–9; ecological factors 302–6; management of 298–9, 307–8; woodland use 328
recombinant ecology: case studies 204–9; convergence and stable states 212–13; criteria for ecologically acceptable exotic plant species 213–14; definitions 198–9; framework 199–211; landscape perspectives 214–15; managing the matrix 212; practical pathways to recombinant futures 211–15; socio-political context 209–12
reconciliation ecology 204, 212, 259–60; resources for 261–2
recreation management 292–3
recreational use, grassland 302

recruitment dynamics, woodlands 330–1
recycling 640
red spider mite 498
regeneration dynamics, woodlands 330–1
regional growth management 57–60
regional landscape view 50–4
regional planning 49–54
regional weather, impact of cities 103–7
relational ethics 70–1
remediation, contaminated sites 320–1
remnants 200, 466
remote sensing 293, 466, 471, 479
renewable energy production 640
research schools 85
research, wetlands 346–8
residential landscapes, biogeochemical dynamics 508–9
resource consumption 20, 31
responsible cities 640–3
restoration ecology 203–4, 207–10
ring-necked parakeets 364–5
river corridors 274, 282, 283–4
rivers: changes to 155–6; effects of construction works 161–2; hydrological cycle 153–5
rock outcrop habitats 254–5
Romanticism 380
Rose-bay willow herb 489
rosemary beetle 498
Royal Commission on Environmental Pollution, UK 79, 91, 561
Royal Horticultural Society Plant Finder 269
Royal Society for the Protection of Birds (RSPB), UK 622
rural areas 9
Rural Development Plan for England 6–7
rural habitats, differences from urban habitats 479
rural landscapes, integrating with urban landscapes 578

safety issues: parks 400–2, 593; trees 432–5
Santa Catarina Island (Florianopolis), Brazil 553
São Paulo City Green Belt Biosphere Reserve, Brazil 554–5
Scheme of Territorial Coherence (SCOT) system 6–8
schools, exposure to nature 411–12, 413, 416, 418
seasonal effects, climate change 496–7
Second Land Utilisation Survey of Britain 466
seed dispersal 190–1, 330–1
Sheffield City Wildlife Project 224
shrubberies; management of 297–8; plant species 205
shrubs, reclaimed land 306
Silent Spring (Carson) 77
Singapore, green corridors 271
sink function, corridors 278, 285
site disturbance 66, 188–9, 212, 292; wetlands 344–5
sites of cultural importance 257
Sites of Importance for Nature Conservation (SINCs), UK 86, 87, 89, 481
Sites of Metropolitan Importance for Nature Conservation, UK 88
Sites of Special Scientific Interest (SSSIs) 88, 313, 623–4
slugs 304
snails 242, 304
social attributes of sites 189
social behaviour, animals 355–6
social justice 58, 643
social risk factors 43–4
societal implications, changes in precipitation 142–3
socio-political context, recombinant plant associations 209–11
soil carbon 179–80
soil screening values (SSVs) 316–17
soil urban heat islands 124–5
soils: aeration and drainage 168–9; biological characteristics 175–9; changing perspectives on 165–6; compaction of 166–8, 303; grassland 302–3; inversion technique 225; modified pH 169–70; and plant performance 191–2; pollutants in 170–5; surface crusting and water repellency 168; sustainable urban drainage 564–70; temperature 169; trampling 245; woodlands 327–8; *see also* contaminated land; reclaimed land
source function, corridors 278–9, 285
sowing, reclaimed land 303–5
sown grasslands 3–6
sparrows 309, 356
spatial ecology 614–16
spatial patterns: animal distribution 353–4; woodland sites 332–3
Species Action Plans, UK 626, 627
species availability 190–1
species diversity, open spaces 295–7
species mapping 479–80
species richness, gardens 451–2
spiders 178, 242
spirituality and nature 381–4
spontaneous development 291–2
spontaneous recombinants 201–2
squirrels 193, 497
stable states 212–13
starlings 309, 355, 356
stigma abatement 593
Stockholm Resilience Centre 552
stonework communities 240–2
stormwater drains as wildlife corridors 282–3
strategic planning, conserving nature through 86, 89
street trees 204–5; battle for 434–5; benefits 426–31; disbenefits 431–3; economic/quantitative perspective 435–6; felling and reaction to 439–45; physical and biological characterisation 425–6; political/sustainability perspective 436–7; numbers 434; species 204
structural dynamics/conflicts, peri-urban agenda 615–16

subsidence 162
suburban mosaic: garden diversity 268–70; habitat diversity 266–8; land cover types 265–6; rapidly growing low-latitude cities 270–1
"suburban savanna" hypothesis 258
succession: differential species performance 191–3; primary and secondary 630–1; reclaimed land 303–5; site availability 188–91; in urban landscape 193–4; woodland sites 332
Sukopp, Herbert 15, 27–8, 76, 78, 80, 85, 164, 169, 175, 324, 482–3.
surface energy balance 107–8
surface fluxes 137–40
surface urban heat islands 122–4
sustainability: and climate change 89–91; and contemporary landscape architectural practice 632–3; education for 591–3; strategies for 32, 34; street trees 436–7
sustainable development, forum for 551
sustainable management: open spaces 290–3; parks 297–9
sustainable urban drainage systems (SUDS) 150–1, 320, 533; green roofs 566–70; selection of plant assemblages and habitat structures 563–4; selection of plants 562–3; strategic wildlife opportunities 564–6
Swiss Federal Laboratories for Materials Testing and Research (EMPA) 523
synanthropic influence, woodlands 331–2
System for Observing Play and Recreation in Communities (SOPARC) 396
Système pour l'Observation de la Terre (SPOT) 479
systems, cities as 17–18

Taipei, Taiwan 22, 31, 135, 143, 525
tameness 389–90
Task Force on Quality Economies in Biosphere Reserves 551–2
techno-addiction, monitoring 45–6
technological risk factors 43–4
technologies: advent and acceptance of 45–6; Ecopolis 639–40
Telford New Town, UK 276
temperature: and climate change 496; soil 168; walls and paved surfaces 243–4; *see also* global warming
terrapins, feral 367–8
terrestrial ecosystems, biogeochemical cycling 508–9
territorial behaviour 355
therapeutic landscapes 589–97
thermo-regulation 41
thunderstorm bifurcation 141
topography, woodland sites 327
trampling 189, 245, 293, 328
tranquillity 380–1
transdisciplinary contributions, human ecology 46
transport corridors 224, 283–4
transport policies 647, 649
tree cover: climate change role 583–4; mapping 482
tree felling, media reaction to 439–45
Tree Preservation Orders (TPOs) 437
trees: carbon pools 179; drainage effects 168–9; gardens 265–6; parks 204, 297–8, 426; reclaimed land 306; role in climate change 583–4, 585–6; safety issues 584; transpiration 151–2; woodland sites 329; *see also* street trees
tropical countries, gardens 270–1
Trust for Public Land, US 387–8
Trust for Urban Ecology, UK 78
turbulent latent heat flux 111–12
turbulent sensible heat flux 109–11
turfed grassland 306

UK: accessible natural greenspace standards 539–46; biodiversity action plan 621–4; creative conservation 221–31; garden biodiversity 268–9; impact of climate change 496–500; natural history 74–9; nature conservation planning 86–90; planning policies 79; therapeutic green spaces 589–97
UK Man and the Biosphere (MAB) Urban Forum 78, 540
umber moth caterpillars 498
UN: Environmental Program (UNEP) 29, 544; Framework Convention on Climate Change (UNFCC) 583; projections for urbanization 600–l; UNESCO Man and the Biosphere (MAB) Programmes 4, 21, 27, 39, 45, 85–6, 549–59
underpasses as wildlife corridors 282–3
University of Chicago 44
University of Copenhagen 605
University of Delaware Water Resources Center 574
University of Kent 231
University of Warsaw 581
University of Wolverhampton 223
urban: approaches to encapsulate 11–12; definition of 8; importance of defining 10
Urban Cliff Hypothesis: comparisons with other frameworks 258–60; outline of 253–8; practical applications 260–2
Urban Ecology (journal) 78
Urban Ecology Australia (UEA) 648
urban ecology: brief history of 27; common frontiers of 32–4; conferences and research schools 84–5; current scope of interests 28–9; establishment as separate discipline 78–80; goals and future research agendas 12; urban metabolism as linking and/or common approach 29–32
Urban Forest Effects (UFORE) 585
urban geomorphology 159–62
urban heat islands (UHI) 10, 28, 100–1, 109, 132, 148, 150, 169, 189, 191–2, 353–4, 426–7, 432, 500, 506, 508, 569, 578, 583–5; boundary layer UHI 128–30; canopy layer UHI 125–7;

urban heat islands *continued*
destabilization 137–40; effect on plants 192; effect of trees on 426–7; scales of urban climates 120–2; soil UHI 124–5; surface UHI 122–4
urban landscapes, integrating with rural landscapes 57–8
urban legibility 631
urban metabolism analysis 21–2, 513–14; applications 526; energetics of 523–6; as linking and/or common approach 29–32; understanding of 521–2
urban nature, growth of study of 74–9
urban patches 21, 268, 281; mapping 483; woodlands as 323–35
urban planning: and biodiversity 621–8; and ecology 630–5
urban river surveys (URS) 156
Urban Wildlife Group, West Midlands 86
urbanization, effects of 10–11
URBIS network 552
US: nature provision 48–60; wetlands 338–42, 345; wildlife initiatives 84–5
US Environmental Protection Agency (EPA) 174, 316, 319, 574, 585
US Fish and Wildlife Service 56
US National Environmental Policy Act (1969) 77

vacant land, habitat diversity 266–7
value pluralism 94–5
value-added landscapes 484
Vegetated Roof Covers (VRCs) 575–81
vegetation: complexity and biodiversity 454; health benefits 414; patterns and safety issues 400–2; and physical activity 397–400; and urban run-off 572–81; water and bank 298
vehicles as dispersal mechanisms 190
Victoria Park, London 379
Victorian era 75–6
vine weevil 498
volatile organic compounds (VOCs) 166, 314–15, 428, 585

walls: colour 243–4; colonization 239–40; decay 243; management 245; microclimate 244; microhabitat 245–7; pollution 244–5; runoff 244; stonework communities 240–2; structure 242–3
Warsaw, peri-urban ecology 611–12
waste disposal practices 313–14
waste materials, use of 222–5
water abstraction 154–5
water balance 107–8, 149; maintaining pre-development conditions 574–5; pre- and post-development conditions 573–4
water contamination 314–15
water infiltration 152–3
water management 292–3, 561–70
water quality: and impervious surfaces 574; wetlands 342–4
water repellency, soil 68
water runoff 150–1; maintaining pre-development conditions 574–5; pre- and post-development conditions 573–4; and sustainable urban drainage 561–70; use of green roofs 575–81; walls and paved surfaces 244
water vegetation, management of 298
water, surface detention of 153
West Berlin, biotrope mapping 85
West Midlands, wildlife corridors 280, 281–2, 284, 286
Western domination of nature 67–9
wetland basins, construction of 562–6
wetlands: creative conservation 225; hydrology and water quality 342–4; management of 346; plant and animal communities 344–6; research directions 346–8; in urban context 338–42
Whitmore Square EcoCity Project, Adelaide City 648
whole-ecosystem analysis, biogeochemical fluxes 511–14
wilderness movement 378
Wildfowl and Wetlands Trust 225
wildland areas 9
Wildlife and Countryside Act, UK 489
wildlife corridors 280–2; concept of 277–80; as dispersal routes for introduced species 283–4; habitat diversity 268; mapping 483; need for 225–6; practical applications and tools 284–5; Singapore 271; underpasses and stormwater drains as 282–3; woodland sites 332–3
wildlife gardening 455
wildlife, impact of cats 357–8
wildlife opportunities, SUDS 564–6
wildness 389–90
William Curtis Ecological Park, London 78, 229
wisteria scale 499
wind fields 113–15
wind speed: effect of climate change 497; effect of trees 427
women, use of parks 396, 397, 401
woodlands: diagnostic abiotic traits 327–9; diagnostic biotic traits 329–32; diagnostic geometric traits 332–5; management of 297–8; role in climate change 583–4; water storage capacity 573–4; in wetlands 345; *see also* trees
workplace, exposure to nature 410, 411, 418
World Health Organization (WHO) 590, 595
World Network Biosphere Reserve (WNBR) 549–59
World War II 77, 362

XiXi National Wetland Park, Hangzhou, China 346

Yangzhou Ecopolis Development 646–7
Yorkshire Wildlife Trust 565
Yoyogi Park, Tokyo 327

zoning policies 609
zoos 67

Printed in the USA/Agawam, MA
December 28, 2011
563196.206